Modulation of Oxidative Stress in Heart Disease

Sajal Chakraborti • Naranjan S. Dhalla
Madhu Dikshit • Nirmal K. Ganguly
Editors

Modulation of Oxidative Stress in Heart Disease

Springer

Editors
Sajal Chakraborti
Department of Biochemistry and Biophysics
University of Kalyani
Kalyani, West Bengal, India

Naranjan S. Dhalla
Institute of Cardiovascular Science
St. Boniface Hospital Research Centre
Winnipeg, MB, Canada

Madhu Dikshit
Translational Health Science and Technology Institute
Faridabad, Haryana, India

Nirmal K. Ganguly
Translational Health Science and Technology Institute
Faridabad, Haryana, India

ISBN 978-981-13-8945-0 ISBN 978-981-13-8946-7 (eBook)
https://doi.org/10.1007/978-981-13-8946-7

This Springer imprint is published by the registered company Springer Nature Singapore Pte Ltd.
The registered company address is: 152 Beach Road, #21-01/04 Gateway East, Singapore 189721, Singapore

Prof. Sibaji Raha

Prof. Dipak Kumar Kar

This book is dedicated to Prof. Sibaji Raha (Coordinator of the participation of India at the International Facility for Antiproton and Ion Research, Germany, and Former Director, Bose Institute, Kolkata, India) and Prof. Dipak Kumar Kar (Vice Chancellor, Sidho Kanho Birsha University, West Bengal, and Former Pro-vice Chancellor, Calcutta University, Kolkata, India) for their exceptional contributions and outstanding leadership in scientific administration and management in India. We wish them good health and success in their long fruitful activities.

Preface

> *"Day by day thou art making me worthy of the*
> *simple great gifts that thou gaveast to me unasked-*
> *This sky and the lights, this body and the life and the mind-*
> *saving me from perils of overmuch desire ......*
> *day by day thou art making me worthy of thy full acceptance*
> *......saving me from perils of weak uncertain desire"*
>
> Rabindranath Tagore (Gitanjali: Songs of Offering)

This book describes multidisciplinary approach and demonstrates biochemical mechanisms associated with dysregulation of redox signaling that leads to manifestation of heart diseases. It bridges the gap between fundamental and translational research on the modulatory role of oxidants in different types of heart diseases. It also discusses the spatial and temporal aspects of oxidative stress in cardiovascular system, which are immensely important for development of better strategies for treating heart diseases.

This book contains 28 chapters, which are divided into three subsections. Dr. Craig McLachlan, Dr. Ruhul Abid, Dr. Hasan Sayyad, Dr. Nitish Mahapatra, Dr. Sagartirtha Sarkar, and Dr. Angsuman Bagchi narrate different aspects of ROS-mediated heart diseases in general; while Dr. Sachin Kumar summarizes modulation of ROS by nitric oxide in neutrophils. Dr. Parimala Narne, Dr. Vijay Kutala, and Dr. Sudhiranjan Gupta in their chapters enumerate novel insights on the impact of epigenetic factors and miRNA in oxidative stress-induced heart diseases. In the subsection on the pathophysiology of oxidative stress, Dr. Bodh Jugdutt, Dr. Nevena Jeremic, Dr. Suvro Chatterjee, and Dr. Shyamal Goswami provide notable information on oxidative stress-induced heart failure and cardiac remodeling in oxidant-induced heart diseases; while Dr. Vinu Wilson discusses about the role of oxidative stress in pulmonary hypertension. Dr. Maria Baez eloquently states oxidative stress-induced biomarkers, while Dr. Antonio Bernad describes the role of oxidative stress in cardiac progenitor cell fate determinations. To address therapeutic interventions and pertaining issues in oxidative stress-induced heart diseases, Dr. Gemma Figtree, Dr. Parames Sil, Dr. Mark Ziolo, Dr. Emmanual Douzinas, Dr. Yiannakoulou, Dr. Srinivas Gopala, Dr. Surekha Rani, and Dr. Biaus Samanta made extraordinary effort. They focused on different therapeutic implications in modulating oxidative stress-induced heart diseases. Each chapter of this book is profoundly useful for the researchers to identify targets for drug development and to address different types of heart diseases. In fact, these are the flowers of the spring. So, let "hundreds of flowers blossom and hundreds schools of thought contend."

The goal of this book is to provide some glimpses of the role of oxidative stress in heart diseases along with the current understanding of their prevention and therapeutics. We have tried to keep this book concise, informative, and readable. Putting together all the articles, we believe that the book will be helpful to the postgraduate students and biomedical researchers.

Our sincere gratitude goes to all contributors for their considerable energy, time, and effort to accomplish a complete chapter, which generates no quid pro quo benefit. We are thankful to Mr. G. Senthil Kumar, Dr. Madhurima Kahali, and Mr. Daniel Ignatius Jagadisan (Springer Nature) for their cooperation and support during the preparation of this book.

Kalyani, West Bengal, India Sajal Chakraborti

Contents

Part I Regulation of Oxidative Stress

1 **Modulation of Oxidative Stress in Heart Disease by Uncoupling Proteins** 3
Zakaria A. Almsherqi, Bernita Yeo Hui Li, Yuling Zhou, and Craig S. McLachlan

2 **Oxidant-Dependent and Oxidant-Independent Proangiogenic and Vasomotor Signaling in Coronary Vascular Endothelium** 23
Sarah Aldosari, Maan Awad, May Z. Gao, Isabella G. McCormack, Frank W. Sellke, and Md. Ruhul Abid

3 **Modulation of miRNA in Oxidative Stress-Induced Cardiac Remodeling** 63
Sudhiranjan Gupta

4 **Myocardial Oxidative Stress and Metabolic Diseases** 81
Hassan I. H. El-Sayyad

5 **Impact of Genetic and Epigenetic Factors on the Oxidative Stress in Cardiovascular Disease** 107
Shiva Krishna Katkam, Bobbala Indumathi, Shaik Mohammad Naushad, and Vijay Kumar Kutala

6 **Role of Monoamine Oxidases in Heart Diseases** 129
Vinayak Gupta, Vikas Arige, and Nitish R. Mahapatra

7 **Key Cellular Effectors in ROS-Mediated Cardiac Diseases** 151
Ratul Datta Chaudhuri, Santanu Rana, Kaberi Datta, and Sagartirtha Sarkar

8 **Reactive Oxygen Species Generation in Neutrophils: Modulation by Nitric Oxide** 197
Sachin Kumar and Madhu Dikshit

9 Heat Shock Proteins and Their Associated Oxidative Stress-Induced Heart Disease 215
Sangeeta Mitra, Rakhi Dasgupta, and Angshuman Bagchi

10 Modulation of Oxidative Stress in Cardiovascular Diseases 237
Jay C. Jha, Madhura Bose, and Karin Jandeleit-Dahm

Part II Pathophysiology of Oxidative Stress

11 Oxidative Stress and Heart Failure 257
Bodh I. Jugdutt and Bernadine A. Jugdutt

12 Oxidative Stress in Metabolic Syndrome: Experimental Model of Biomarkers 313
María del Carmen Baez, Mariana Tarán, Mónica Moya, and María de la Paz Scribano Parada

13 Oxidative Stress as a Critical Determinant of Adult Cardiac Progenitor Cell-Fate Decisions 339
Diego Herrero, Susana Cañón, Guillermo Albericio, Susana Aguilar, Rosa María Carmona, Adrián Holguín, and Antonio Bernad

14 Role of Oxidative Stress in Hyperhomocysteinemia-Induced Heart Diseases 365
Nevena Jeremic, Jovana Bradic, Anica Petkovic, and Gregory Weber

15 Nitrosative Stress and Cardiogenesis: Cardiac Remodelling Perturbs Embryonic Metabolome 377
Pavitra Kumar, Lakshmikirupa Sundaresan, and Suvro Chatterjee

16 Oxidative Stress in Pulmonary Artery Hypertension 393
Vinu Wilson and Subir Kumar Maulik

17 Free Radicals and Reactive Oxygen Species in Cardiovascular Pathophysiology: An Overview 403
Shyamal K. Goswami

18 Epigenetic Basis of Oxidative Stress in Diabetic Coronary Atherosclerosis: A Shift in Focus from Genetic Prerogative 419
Parimala Narne

19 Role of Oxidative Stress, Mitochondrial Dysfunction, and Autophagy in Cardiovascular Disease: Its Pathogenesis and Amelioration by Different Small Natural Molecules 457
Sharmistha Chatterjee, Uday Hossain, and Parames C. Sil

20 Parkin Protein: The Missing Link Between Cardiovascular and Parkinson's Disease 489
Angshuman Bagchi

Part III Prevention and Therapeutics

21 New Technologies in Drug Development Provide New Hope in Targeting of Dysregulated Redox Signalling in Cardiovascular Disease 505
Soloman Saleh, Kristen Bubb, and Gemma A. Figtree

22 An Exercise Mimetic That Targets Nitroso-Redox Balance as a Therapeutic for Heart Disease 533
Vikram Shettigar and Mark T. Ziolo

23 Myocardial Injury Secondary to Intestinal Ischemia/Reperfusion or Microbiota Disturbance: Preventive and Therapeutic Concerns 555
Emmanuel E. Douzinas and Aikaterini Apeiranthitis

24 Multi-target Approach for Oxidative Stress Modulation by Aspirin, Salicylates and Other NSAIDs: Clinical Implications in Atherosclerosis 565
Eugenia Yiannakopoulou

25 The Evolving Concept of Mitochondrial Dynamics in Heart: Interventional Opportunities 579
Ashok Sivasailam, Mahalaxmi Ganjoo, Varghese T. Panicker, Vivek V. Pillai, and Srinivas Gopala

26 Pharmacogenetic Implications of Statin Therapy on Oxidative Stress in Coronary Artery Disease 629
Nivas Shyamala and Surekha Rani Hanumanth

27 Basic Mechanisms of Ischemia/Reperfusion Injury Leading to Cellular and Tissue Damage: Therapeutic Implications 645
Emmanuel E. Douzinas and Aikaterini Apeiranthitis

28 Targeting Mitochondria for Therapy of Cardiovascular Disease 671
Biaus Samanta, Satabdi Banerjee, Suman K. Nandy, and Sajal Chakraborti

Correction to: Modulation of miRNA in Oxidative Stress-Induced Cardiac Remodeling C1

About the Editors

Dr. Sajal Chakraborti is a Professor of Biochemistry at the University of Kalyani, West Bengal, India. He received his PhD at the Calcutta University, Kolkata, India, and DSc at the University of Kalyani, Kalyani, West Bengal, India. He did his postdoctoral research at the Johns Hopkins University, Baltimore, University of Utah, Salt Lake City, and New York Medical College, New York, as a Fulbright Fellow. He also visited the University of Florida, Gainesville, Florida, and did research on beta-adrenergic receptor modulation in heart diseases. His research mainly deals with the role of vasoactive agents in regulating vascular tone under oxidant and calcium-signaling phenomena. He has been engaged in teaching and research in biochemistry for the past 40 years.

Dr. Naranjan S. Dhalla is a Professor at the University of Manitoba, Winnipeg, Canada. His expertise includes the subcellular and molecular basis of heart function in health and disease. He has been engaged in research and education in cardiology for over 50 years.

Dr. Madhu Dikshit is currently associated with the Translational Health Science and Technology Institute, Faridabad, Haryana, India. She has contributed to the general area of biochemical pharmacology with special emphasis on redox biology. Dr. Dikshit has been engaged in research for over 40 years.

Dr. Nirmal K. Ganguly, a former Director General of the Indian Council of Medical Research (ICMR), is currently associated with Translational Health Science and Technology Institute, Faridabad, Haryana, India. One important aspect of his research is the understanding of nutritional aspects on heart diseases.

Part I

Regulation of Oxidative Stress

Modulation of Oxidative Stress in Heart Disease by Uncoupling Proteins

1

Zakaria A. Almsherqi, Bernita Yeo Hui Li, Yuling Zhou, and Craig S. McLachlan

Abstract

According to "free-radical theory" of disease, Reactive Oxygen Species (ROS) play a key role in the pathogenesis of several diseases including cardiovascular disease. When the balance between production of free radicals and antioxidant capacity of the cardiac cells is altered due to pathophysiological conditions, oxidative stress is induced. Oxidative stress has been linked to the development of ischemic heart disease, atherosclerosis, congestive heart failure, ischemic-reperfusion injury, and vascular endothelial dysfunction. In this context, antioxidant supplementation would have a positive effect on cardiovascular diseases. However, several clinical trials over the past decades employed different strategies of antioxidant therapies which have failed to achieve favorable results in ameliorating or preventing cardiovascular diseases. Much less attention has been paid to the modulation of ROS production, despite the fact that prevention, rather than cure, would appear to be a logic approach to attenuate the oxidative damage. This chapter intends to highlight the mechanisms of oxidative stress modulation – by Natural or induced mitochondrial uncoupling respiration – in regulating ROS production and its significance in cardiovascular pathophysiological conditions.

Zakaria A. Almsherqi and Bernita Yeo Hui Li have been equally contributed to this chapter

Z. A. Almsherqi (✉)
Rural Clinical School, Faculty of Medicine, University of New South Wales, Sydney, NSW, Australia

Yong Loo Lin School of Medicine, National University of Singapore, Singapore, Singapore
e-mail: z.almsherqi@unsw.edu.au

B. Y. H. Li
School of Medicine, Flinders University, Adelaide, NT, Australia

Y. Zhou · C. S. McLachlan
Rural Clinical School, Faculty of Medicine, University of New South Wales, Sydney, NSW, Australia

S. Chakraborti et al. (eds.), *Modulation of Oxidative Stress in Heart Disease*,
https://doi.org/10.1007/978-981-13-8946-7_1

Keywords
Oxidative Stress · Uncoupling Proteins · Cardiovascular Diseases · Myocardial Infarction · Myocarditis · SREBP · PPARs · AMPK · Myocardial Hypertrophy · Heart Failure · Reperfusion

1.1 Introduction

1.1.1 Mitochondrial Oxidative and Uncoupling Metabolism

Myocardial function depends on the energy that it is able to synthesize and transfer in the form of energy-rich phosphate bonds to fuel excitation-contraction coupling. More than 90% of cardiac cell energy is produced in the mitochondria from oxidative phosphorylation activity [1]. Oxidative phosphorylation is the process by which energy from fuel oxidation is converted to the high-energy phosphate bonds of adenosine triphosphate (ATP). During this process, energy from the oxidation of the tricarboxylic acid cycle is conserved in the form of reduced electron-accepting coenzymes, NADH and $FADH_2$. They are passed through the electron transport chain and the electrons released. These electrons are in turn donated to oxygen, which is reduced to water. The energy released from the reduction of oxygen is used for the phosphorylation of adenosine diphosphate (ADP) to ATP, catalyzed by the enzyme ATP synthase (Fig. 1.1). Thereafter, the hydrolysis of ATP releases energy that can be used for cardiac cell contraction and other essential cellular functions.

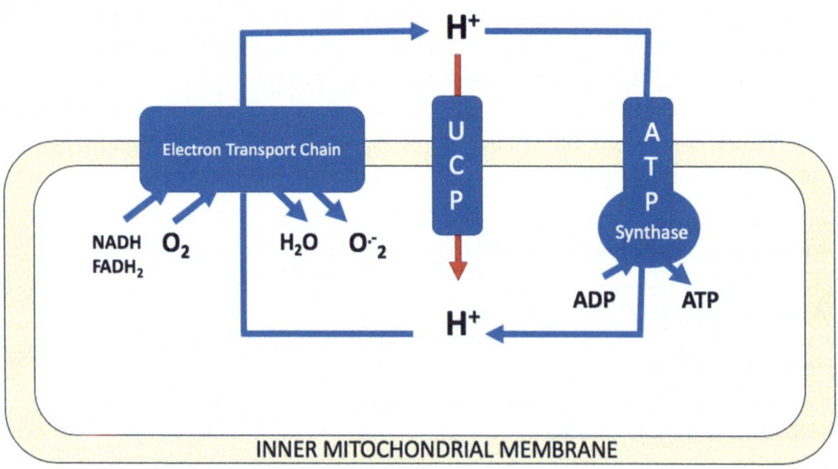

Fig. 1.1 The chemiosmotic proton cycle across the inner mitochondrial membrane. The coupling cycle, consisting of substrate oxidation (NADH, $FADH_2$) and the enzymes of ATP production (ATP synthase), results in coupled oxidative phosphorylation and superoxide ($O^{-\cdot}_2$) generation as a by-product. The uncoupling cycle, consisting of substrate oxidation and proton conductance pathway through the uncoupling proteins (UCPs), results in uncoupled respiration and low levels of superoxide generated

The chemiosmotic hypothesis suggests that the energy for ATP synthesis is provided by the electrochemical gradient across the inner mitochondrial membrane. This electrochemical gradient is maintained by constituents of the electron transport chain (ETC), which acts to pump protons from the mitochondrial matrix to the intermembrane space of the mitochondria as they accept and donate electrons in a prescribed manner. These processes of electron transport and oxidative metabolism in cardiac cells are accompanied by the reduction of oxygen to superoxide and other ROS which are considered as a by-product of mitochondrial respiration.

An appreciation of the ETC and its role in oxidative phosphorylation is essential in the understanding of the significance of uncoupling proteins (UCPs). As one might expect, the rate of ATP synthesis driven by the transmembrane electrochemical gradient is coupled to the rate of the ETC. As the energy demands of the cardiac cell increases and ATP is being utilized, the levels of ADP increases. Proton influx through ATP synthase causes an increase in ATP production which subsequently activates the ETC to restore the electrochemical gradient across the mitochondria inner membrane. In the uncoupling process, as the name suggests, the electrochemical gradient is restored independently of the activity of ATP synthase (Fig. 1.1). The action of uncoupling proteins is based on two main factors: the higher concentration of protons in the intermembrane space than the mitochondrial matrix and the lipid-soluble nature of UCPs. Being lipid-soluble, they are able to traverse the inner mitochondrial membrane, pick up protons, and transport them to the mitochondrial matrix [2]. In this way, the electrochemical gradient across the mitochondrial inner membrane is dissipated, and ATP synthesis disrupted. As a result, ATP is not produced and the energy is released as heat (Fig. 1.1).

A considerable level of basal proton leak, also known as global proton leak, occurs across the inner mitochondrial membrane all the time [3]. While some of this leak is attributed to the action of uncoupling proteins, the permeability of the mitochondrial membrane due to the proteins embedded in its lipid bilayer also contributes to the membrane's leakiness. Approximately 20% of the body's resting metabolic rate is used to maintain the electrochemical gradient that is dissipated by this basal proton leak [4]. This significant mitochondrial proton leak should serve an important function or functions in view of the high energetic cost utilized to maintain it.

1.1.2 Redox Status of Cardiac Cells: The "Uncoupling" Link

ROS are generated as by-products of cellular metabolism and mainly as a part of mitochondrial respiration. Free radicals have a single, unpaired electron, rendering them highly unstable and thus reactive. In cells, they readily react with and oxidize cellular components such as lipids (unsaturated fatty acids), proteins, and DNA. While some ROS are physiologic products of oxidases in peroxisomes, most are produced as part of the mitochondrial oxidative phosphorylation process. Due to their highly reactive nature, these free radicals formed can decay spontaneously. Therefore, under normal physiologic conditions, there are several protective mechanisms in place to neutralize these free radicals and prevent cell injury. Regulation of the mitochondria ROS generation is another cellular mechanism to control ROS levels. When the oxidative load caused by increased ROS production exceeds the antioxidant capacity of these mechanisms, ROS accumulate, resulting in oxidative stress.

At this point, it is important to note that apart from ROS toxicity, ROS also has a physiologic role in cellular signaling. While the mechanisms by which ROS exert their effects are still unclear, it is known that ROS-mediated cellular signaling is achieved by changes in the intracellular redox state as well as the oxidative modification of proteins. Under normal conditions, the enzyme superoxide dismutase (SOD) and glutathione (GSH) peroxidase catalyze the breakdown of free radicals (i.e., $H_2O_2 + 2GSH \rightarrow 2H_2O + GSSG$) to reduce intracellular oxidative stress. The cytosol is usually in a "reduced" state with respect to the extracellular environment. Therefore, the ratio of GSSG (oxidized glutathione) to GSH (reduced glutathione) is an indicator of the oxidative state and hence the oxidative capacity of the cell. By altering this ratio, GSH has been found to play a role in redox signaling. On another hand, agents that increase the mitochondrial respiratory rate, such as ADP (activate ATP synthase) or uncouplers (activate UCPs), are well known to reduce ROS production. An experimental study has shown that UCP3 expression in muscle cells resulted in a decrease in mitochondrial ROS production and may hence be an important factor in ROS regulation [5].

ROS are of interest in clinical research as they have been implicated in many diseases including cardiac pathology [6]. ROS react with a wide variety of compounds such as DNA, proteins, carbohydrates, and lipids in the cardiac cells, causing cellular damage. ROS can also cause a conformational change in proteins via the modification of certain amino acid residues in the functional domain of proteins [7]. Moreover, UCPs seem to play a role in the cellular protective system in tandem with GSH, SOD, and other enzymatic antioxidants.

1.1.3 Types of Uncoupling Proteins and Their Roles in ROS Modulation

UCPs refer to a class of proteins located on the inner mitochondrial membrane. As mentioned previously, UCPs allow protons to cross the inner mitochondrial membrane without the concomitant activation of ATP synthase. There are several types of UCPs observed in different tissues. UCP1, also known as thermogenin, is found in brown adipose tissue and has a role in heat production [8]. UCP2 is rather ubiquitous and found in most cells, while UCP3 is found mainly in skeletal and cardiac muscles. UCP4 and UCP5 have been found in the brain. Table 1.1 summarizes the distribution of the UCPs.

Of clinical significance are UCP2 and UCP3 located in the heart. There is increasing evidence to show that these UCP proteins, by mitochondrial uncoupling, protect the heart by reducing ROS generated by the mitochondria. As a result,

Table 1.1 Distribution of uncoupling proteins

Uncoupling Protein	Distribution
UCP1	Brown adipose tissue
UCP2	Most cells
UCP3	Skeletal muscles
UCP4	Brain
Ucp5	Brain

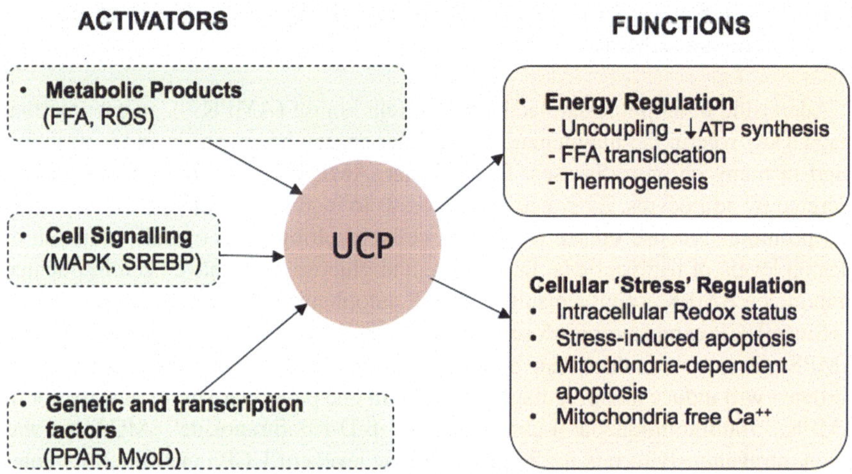

Fig. 1.2 Schematic diagram summarizes the effect of ROS, UCP2, and UCP3 in the major cardiac pathologies. Activation and/or overexpression of UCPs is associated with a lower mitochondrial generation of ROS, ATP production, and consequently potential myocardial contractile dysfunction. Whether these are an adaptive or maladaptive response to stress conditions needs further investigations. Refer to the text for details

cardiomyocyte could be protected from stress-induced apoptosis [9]. UCP3, in particular, has been associated with cellular fatty acid metabolism, with its distribution being most pronounced in muscles with high-fat oxidative capacity (such as cardiac cells) [10]. In physiological or pathological conditions where plasma fatty acid levels increase, UCP3 is upregulated. On the contrary, a decrease in plasma fatty acid levels causes a downregulation of UCP3. These seem to support the hypothesis that UCP3 plays an important role in exporting fatty acids that cannot be oxidized from the mitochondrial matrix, thereby inhibiting the accumulation of fatty acids inside the matrix. In this way, UCP3 provides protection from lipid-induced mitochondrial damage and mitochondria-dependent apoptosis [10].

UCP expression is altered in response to external stressors by a host of transcription factors, in particular, peroxisome proliferator-activated receptor alpha (PPAR-α) and peroxisome proliferator-activated receptor gamma coactivator 1 alpha (PGC1-α). Both transcription factors play an essential role in the response to external environmental stress, such as fasting and physical stress (Fig 1.2).

1.2 UCP Regulation in Cardiac Cell

Activation and/or expression of UCPs is closely monitored by intra- and extracellular factors. Regulation may be achieved by enzymes such as AMP-activated protein kinase (AMPK), proteins such as sterol-responsive element-binding protein (SREBP), and nuclear transcription factors such as PPARα.

1.2.1 UCP Expression and Activation of AMP-Activated Protein Kinase (AMPK)

5′-Adenosine monophosphate-activated protein kinase (AMPK) is an enzyme that plays a key role in the maintenance of cellular energy homeostasis and is dysregulated in many chronic diseases. In particular, AMPK is activated by adiponectin secreted by adipocytes. Activated AMPK leads to increased oxidation of fatty acids and promotes glucose uptake by muscle cells, resulting in an overall reduction of plasma levels of triglycerides, fatty acids, and glucose. ROS have been implicated in inducing AMPK activity, resulting in cell autophagy and apoptosis [11].

Since UCP expression is closely linked to the level of oxidative stress in a cell, AMPK activation – which can be induced in response to an increased oxidative burden – will induce an upregulation of UCP. In one particular study, an activator of AMPK, 5-aminoimidazole-4-carboxamide-1-β-D-ribofuranoside (AICAR), was given intravenously to rats for 28 days, and the levels of UCP3 measured. As compared to the untreated control group, AICAR-treated rats were observed to have an increased UCP3 expression at the mRNA level as well as an increased UCP3 protein content [12].

AMPK-induced upregulation of UCPs is of clinical relevance, especially since many disease states are a result of an increased oxidative burden. Oxidants activate AMPK, which in turn increases the expression of UCP2. This can be seen as a compensatory mechanism which counteracts the increased intracellular oxidative stress via the production of UCP2. In AMPK knockout mice, UCP2 was expressed only weakly in endothelial cells and associated with an endothelial dysfunction. This suggests that AMPK activation is, to a large extent, critical to the transcription of the UCP2 gene [13].

1.2.2 UCP Expression, Sterol-Responsive Element-Binding Protein (SREBP), and Cyclic AMP Response Element-Binding Protein

HMG-CoA reductase is the enzyme that catalyzes the rate-determining step of cholesterol synthesis. Its own production is regulated by a family of sterol regulatory element-binding proteins (SREBPs). The SREBPs are transcription factors that bind to the sterol regulatory element (SRE) upstream of the HMG CoA reductase gene. This binding increases the rate of transcription of HMG CoA reductase when intracellular cholesterol levels are low. When intracellular cholesterol levels are elevated, SREBP binds to SREBP cleavage-activating protein (SCAP) in the membrane of the endoplasmic reticulum. On the other hand, when cholesterol levels fall, SREBP is translocated to the Golgi apparatus where it undergoes proteolytic cleavage, releasing the N-terminal transcription factor domain. This then travels to the nucleus and binds to the SRE. In this way, the level of HMG CoA reductase and hence the level of cholesterol are regulated.

An increase in SREBP expression has been observed to correlate with UCP expression levels. When SREBP-1c was overexpressed, fatty acid synthase, PPARγ, and UCP2 were all upregulated, with a marked decrease in glucose-induced insulin secretion [14]. On the other hand, it has been shown that insulin inhibits cardiac UCP3 expression through activation of the lipogenic factor SREBP-1. Sustained downregulation of cardiac UCP3 by hyperinsulinemia may partly explain the poor prognosis of type 2 diabetic patients after myocardial infarction [5].

Another transcription factor worth mentioning is the cAMP response element-binding protein (CREB). Upon stimulation by adrenaline, cAMP-dependent protein kinase phosphorylates and hence activates CREB. CREB then binds to cAMP response element (CRE) in the promoter regions in genes, thereby activating the synthesis of enzymes involved in gluconeogenesis, for instance. It has been found that although multiple CREs have been associated with many genes, the effect is pronounced in the UCP gene. Some of these CREs are critical for enhancing function, as well as those located near the TATA box promoter perform a regulatory function and control UCP gene expression [15]. In this way, CREB's action is twofold – it causes the synthesis of gluconeogenic enzymes as well as an upregulation of UCP.

1.2.3 UCP Expression and Peroxisome Proliferator-Activated Receptors (PPARs)

PPARs are a subset of nuclear receptors. As their name suggests, these receptors, when activated, are able to cause a proliferation of peroxisomes. A variety of agonists are known to induce peroxisomal proliferation such as nonsteroidal anti-inflammatory agents, hypolipidemic agents, and environmental toxins [16, 17].

Three major isoforms of PPARs are characterized – α, β/δ, and γ. All isoforms of PPARs are expressed in cardiovascular system such as endothelial cells and vascular smooth muscle cells; however, their roles in cardiac function and the outcomes of respective agonists vary significantly. Fatty acids are the endogenous ligands for the α isoform of PPAR ($PPAR_{\alpha}$). When circulating level of fatty acids in the blood increases, $PPAR_{\alpha}$ is activated, resulting in the upregulation of genes involved in fatty acid metabolism. Experiments performed on mice cardiomyocytes have shown that the expression of UCP3 is regulated by $PPAR_{\alpha}$ [18]. Indeed, cardiomyocytes exposed to $PPAR_{\alpha}$ exhibited an increased rate of fatty acid oxidation [19]. A similar increase in UCP3 was also seen in rats treated with a PPARα agonist. In PPARα-deficient rats, a 20-fold decrease in UCP3 expression was observed when compared to their wild-type counterparts. Taken together, this highlights the role of PPARα in regulating cardiac UCP3 expression [18]. PPARβ/δ have also been found to play a role in the regulation of cardiac lipid metabolism and have been implicated to be involved in the pathophysiology of acquired cardiac diseases such as coronary artery disease and rheumatic heart disease [20].

While the expression of UCP3 is PPARα-dependent, UCP2 expression has been found to be PPARα-independent. In the human heart, a positive correlation between levels of free fatty acids (FFA) and levels of UCP2 and UCP3 has been previously observed [21].

It is well known that the heart has virtually no glycogen reserves. Fatty acids are the heart's main source of fuel, although ketone bodies as well as lactate can serve as fuel for the heart muscle. In fact, the heart muscle consumes acetoacetate in preference to glucose [22]. Furthermore, FFA are the natural ligands for PPARα. Therefore, conditions that cause an increase in FFA levels (e.g., diabetes) would be expected to increase levels of UCPs via the PPARα-mediated pathway. However, in rats with streptozotocin-induced diabetes, the level of UCP3 increases without an accompanying increase in cardiac UCP2 [23]. Indeed, post streptozotocin treatment (that induces diabetes), cardiac UCP2 levels remained unchanged in both wild-type and PPARα-deficient rats, while cardiac UCP3 levels increased by 50% only in wild-type rats [23]. Therefore, UCP3 is more responsive to PPARα-mediated upregulation as compared to UCP2.

1.2.4 TNFα and UCP2 Expression and Vascular Damage

Cell-signaling molecules released from the immune system are collectively termed cytokines. They are released in response to a specific stimulus and travel to target cells where they bind to receptors and elicit a response. One of the main cytokines involved in the inflammatory response of the immune system is tumor necrosis factor-alpha (TNFα). It is secreted mainly by macrophages, as well as monocytes, neutrophils, endothelial cells, smooth muscle cells, activated lymphocytes, adipocytes, and astrocytes. TNFα plays a crucial role in the regulation of the cytokine cascade in various inflammatory diseases, and its dysregulation has been implicated in the pathogenesis of diseases such as atherosclerosis, Crohn's disease, sepsis, and diabetes, among many others [24].

Interestingly, a study which involved the administration of a single intravenous dose of TNFα into rats caused a corresponding increase in the expression of UCP2 and UCP3 [25]. The direct relationship between TNFα levels and UCP2 and UCP3 expression is highly suggestive of the role these uncoupling proteins have in the inflammatory response. Indeed, UCP2 and UCP3 may have a role to play in contributing to energetic inefficiency in cardiac cells when the cytokine TNFα is overproduced. Other studies have also shown that TNFα causes an increase in UCP2 mRNA levels in cultured cells, suggesting that UCP2 is a cytokine-inducible gene [26]. This may explain the unsatisfactory results of using TNF inhibitors in cardiac trials [27].

1.3 ROS and UCP Role in Cardiovascular Pathology

Cardiovascular disease (CVD) is common in the general population, and it was estimated to result in more than 17 million deaths worldwide on an annual basis. Identifying factors that play important roles in a person's chances of developing heart disease (risk factors) is therefore essential for preventing and treating CVD. "Classical" cardiovascular risk factors can be classified in different ways, controllable (e.g., lack of physical exercise, high-fat diet) or uncontrollable (age and sex) and major (e.g., high blood pressure, obesity) or minor risk factors (e.g., ECG abnormalities). However, recent advances in our understanding of the molecular mechanisms involved in CVD may change our view on CVD risk factors. Physiological or pathological conditions that are associated with high levels of ROS, cytokines, and UCP regulators could be regarded as new risk factors of CVD. Table 1.2 summarizes some of these biomolecules associated with high risk of CVD.

1.3.1 Atherosclerosis

Atherosclerosis is a type of vascular pathology characterized by endothelial dysfunction and reduced vascular wall compliance and narrowing. Vascular pathology is predominately characterized by the buildup of cholesterol deposits on the inner walls of blood vessels. These raised lesions are known as atherosclerotic plaques or atheromas and protrude into the vessel lumen, causing a decrease in lumen diameter and thereby reducing or obstructing the blood flow. The media underlying the

Table 1.2 Biomolecules associated with high risk of CVD

Oxidized LDL
Oxidized LDL autoantibodies
LDL immune complex
HDL lipoprotein
Lipoprotein A
Lysophosphatidylcholine
Plasminogen activator inhibitor 1
Tissue plasminogen activator inhibitor 1
Fibrinogen
C-reactive protein
Asymmetric dimethylarginine
Nitrites
Serum amyloid A
Homocysteine
Glycosylation
Cytokines

atherosclerotic plaque is weakened, and the plaque itself can rupture, leading to thrombus/embolic formation and arterial occlusion [28].

The pathogenesis of atherosclerosis begins with injury to the endothelial cells of the arterial wall. Chronic endothelial cell injury leads to the thickening of the intima of the vessel wall with an accumulation of lipids such as cholesterol. Macrophages and lymphocytes are recruited to the site of damage and ROS are generated [28]. Following ROS being generated, the ROS will then oxidize low-density lipoprotein (LDL) followed by cholesterol deposition in the plaque, which in turn further stimulates macrophages and causes the endothelial cells to release chemokines, cytokines, and growth factors, all resulting in monocyte recruitment to the lesion [28]. Oxidized LDL produced by ROS is cytotoxic to smooth muscle cells and endothelial cells of the blood vessel wall, contributing to endothelial cell dysfunction. In this way, oxidative stress plays a key role in the progression of atherosclerosis. Since UCP2 has been found to be an important regulator of intracellular ROS production, it has been implicated in the pathophysiology of the development of atherosclerosis. The UCP2-mediated decrease in ROS generation in endothelial cells is a key mechanism by which the progression of atherosclerosis may be interrupted.

Pro-atherogenic factors (e.g., oxidized LDL) associated with an increased oxidative burden can cause endothelial cell apoptosis and initiate the process of atherogenesis. In particular, the toxicity of the oxidized LDL, lysophosphatidylcholine (LPC), on endothelial cells has been established [29]. LPC stimulates ROS generation and promotes inflammation, resulting in injury to endothelial cells. In cultured human aortic endothelial cells, it was observed that an increase in LPC levels caused a concomitant increase in UCP2 expression [29]. This resulted in a suppression of ROS generation and the inhibition of caspase activation, effectively preventing LPC-induced endothelial cell apoptosis. Conversely, when the endogenous expression of UCP2 was suppressed, LPC-induced caspase activation and apoptosis were augmented [30]. Therefore, increasing UCP2 levels in vascular cells may be beneficial in delaying the progression of atherogenesis.

Furthermore, studies have shown that the absence of UCP2 in blood cells promoted the development of atherosclerotic plaques which were collagen-poor and macrophage-rich [31]. Although once thought of as due to plaque quantity, the vulnerability of a plaque to rupture has now been found to be more closely related to its content than its absolute size [32]. Plaques with a high level of vascular smooth muscle cells incorporated into them are less prone to rupture than those with a high level of inflammatory cells such as macrophages and lipids within. This can be attributed to the collagen-synthesizing ability of vascular smooth muscle cells which contribute to the structural integrity of the plaque. On the other hand, matrix metalloproteinases released by inflammatory cells have the opposite effect of degrading collagen and extracellular matrix, causing the destabilization of the plaque [32]. Hence, the significance of UCP in the formation of a collagen-rich plaque can be appreciated as it delays the progression of atherosclerosis.

1.3.2 Ischemic Heart Disease and Myocardial Infarction

Ischemic heart disease (IHD) is characterized by an imbalance between coronary blood supply and cardiac oxygen demand. IHD can progress to cause myocardial infarction (MI). MI is most commonly caused by acute coronary artery thrombosis on the background of a disruption of an atherosclerotic plaque. The disruption of an atherosclerotic plaque via rupture or fissuring results in the formation of a thrombus which can occlude a coronary artery. The occlusive thrombus impedes the coronary flow to the myocardium. A functional consequence that occurs within a minute or so from the onset of ischemia is the loss of myocyte contractility. Changes at the microscopic level include mitochondrial and cell swelling, glycogen depletion, and myofibrillar relaxation [33]. This results in cardiac dysfunction (reduced cardiac wall motion) and decreased cardiac output. As part of the body's compensatory mechanism to maintain cardiac output during an acute infarction, the nonischemic myocardial wall responds by an initial hypercontractility state [34]. However, this might be associated with an "inevitable" increase in mitochondrial ROS generation.

In a study that induced an anterior MI in dogs by ligation of the left anterior descending coronary artery, it was shown that the expression of UCP3 was upregulated in the mitochondria isolated from the posterior, nonischemic myocardium. UCP3 expression levels were also found to be inversely related to ROS levels. This suggests that UCP3 might have a protective effect against ROS production from an ischemic event [35]. However, as the mitochondria become increasingly bioenergetically inefficient, there is a decrease in respiratory coupling and ATP production. Whether this decrease in energy supply affects the contractile function of the nonischemic myocardium warrants further investigation. In support of this observation, upregulation of UCP2 and UCP3 have been found to be associated with increased levels of plasma FFA in patients undergoing coronary arteries bypass surgery [21]. The increased levels of cardiac UCPs are associated with a subsequent energy deficiency characteristic of a failing heart [21]. This association between increased levels of cardiac UCP and a reduction in cardiac efficiency due to the decline in the rate of ATP synthesis has been further observed experimentally in chronically infarcted rat heart [36].

1.3.3 Reperfusion and Its Associated Effects on Mitochondria ROS and Uncoupling

The early cellular changes due to MI are potentially reversible and do not necessarily lead to cell death if prompt reperfusion intervention is carried out. It is only with severe ischemia lasting for more than 20–40 min will irreversible myocyte injury and death ensue. Therefore, the critical window between the onset of ischemia and its progression to cell death is of clinical relevance. In ideal situations where early clinical detection of acute MI is made and subsequent coronary vascular intervention carried out, ischemic myocardium can be salvaged and cell death averted.

Interventional strategies, such as thrombolysis or angioplasty or a coronary artery bypass graft, have the aim of restoring blood and hence oxygen flow to the ischemic myocardium and the ischemic zone close to the infarcted area. However, as straightforward as it sounds, such reperfusion does come at a cost and might, unfortunately, do more harm than good. Myocardial ischemia-reperfusion injury (IRI) occurs when reperfusion of cardiac tissue after an episode of ischemia actually causes damage to the tissues rather than the restoration of normal function [37]. This is because the inflammation associated with ischemia may be exacerbated with reperfusion, mediated in part by the influx of leukocytes and plasma proteins. Activated leukocytes produce, among other substances, reactive oxygen and nitrogen species which worsen tissue injury. Oxidases present in endothelial and parenchymal cells also contribute to the production of ROS as damaged mitochondria cause the incomplete reduction of oxygen. The cell's natural antioxidant defense mechanism is also rendered dysfunctional due to the ischemic insult, favoring the accumulation of free radicals. Moreover, IRI has also been observed with cardiac manipulations resembling beating coronary bypass surgery of the coronary arteries at the posterior wall of the heart with a significant prompt increase in cardiac ROS generation [38].

Ischemic preconditioning (IPC) is a phenomenon first identified by Murry and associate in 1986 [39]. It was observed that repetitive, short, and nonlethal episodes of ischemia actually protected the myocardium from subsequent ischemic insults. Since then, many studies have been done on the usefulness of induced IPC in the clinical setting. Indeed, IPC has been reported to cause a reduction in infarct size as well as the preservation of endothelial cell function.

A reduction in oxygen supply to cardiomyocytes during an ischemic event activates the mitochondrial pathway of apoptosis. Cytochrome c is released and the caspase cascade is activated, leading to nuclear fragmentation and ultimately cell death. This series of events is believed to be caused by an elevation of mitochondrial ROS. In preconditioned cardiomyocytes, the resistance to ischemia is conferred through the action of uncoupling proteins which act to decrease the production of ROS by uncoupling respiration. By reducing the release of cytochrome c from the mitochondria, IPC helps in decreasing IRI-induced apoptosis [40]. Despite the scarce of experimental work on UCP role in IRI, it can be concluded that UCP-mediated proton leak and concomitant mitochondrial uncoupling can protect against myocardial IRI by reducing ROS generation [41] (Fig. 1.3).

1.3.4 Heart Failure

Heart failure (HF), the common endpoint of many cardiac diseases, remains the leading cause of morbidity and death worldwide [42]. HF is associated with an increased production of ROS. Specifically, it has been found that situations such as pressure overload can substantially increase ROS generation in the heart [43]. Under normal conditions, the heart has an antioxidant defense system to combat the rise in ROS. However, when the increase in ROS exceeds the antioxidant capacity of the heart, heart injury ensues. Cardiac UCP2 and UCP3, through their action as uncouplers, allow proton leak across the inner mitochondrial membrane. This

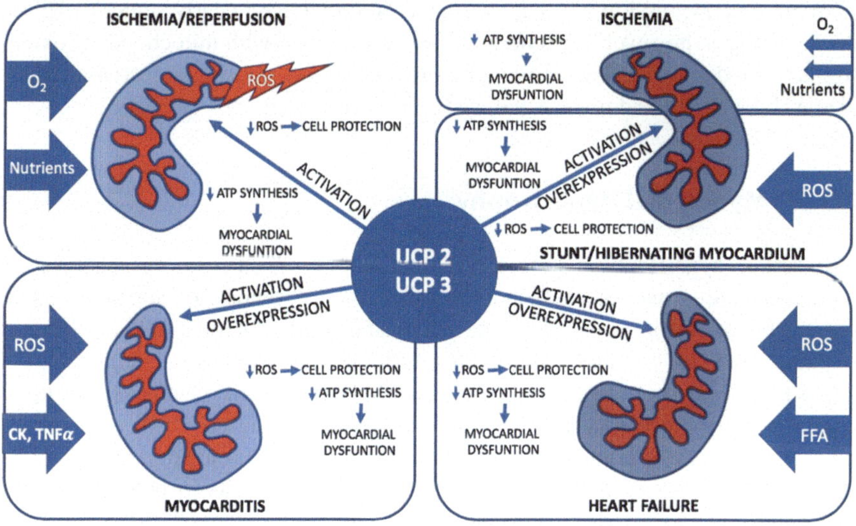

Fig. 1.3 Summary of the main activators of UCPs and their proposed regulatory functions in the cell

results in a reduction in ROS produced by the mitochondria and subsequent cardiomyocyte apoptosis [9].

The importance of UCPs in regulating mitochondrial ROS production and hence cardiac function is demonstrated by the action of doxorubicin, a chemotherapy drug also known as Adriamycin. Doxorubicin is a highly effective chemotherapeutic agent, but its use has been limited by its dose-dependent cardiotoxicity [44]. This is because doxorubicin decreases cardiac UCP2 and UCP3 expression, causing a subsequent increase in oxidant stress in the setting of a failing heart [45]. Again, this points to the close relationship between UCPs and ROS-induced apoptosis in heart failure.

Furthermore, the increased levels of FFA frequently observed in the early stage of HF are directly associated with UCPs. Cardiac UCP2 and UCP3, in particular, are upregulated when FFA levels rise and have been observed to have a greater affinity for unsaturated FAs such as linoleic, arachidonic, and oleic acids as compared to saturated FAs [46]. Several factors associated with an increased metabolic or oxidative stress in cardiac cells and increased FFAs can induce an overexpression of UCP2 and UCP3 in a failing heart. These factors include fasting (with an increase in FA mobilization from adipose tissue), diabetes, and thyroid hormone treatment. UCP3 overexpression may play an important role in exporting fatty acids that cannot be oxidized from the mitochondrial matrix, thereby inhibiting the accumulation of fatty acids inside the matrix and prevents lipid-induced mitochondrial damage and mitochondria-dependent apoptosis [10]. Several experimental studies have established a role of UCP2 and UCP3 in downregulation of programmed cell death and to slow down the progression to heart failure [47, 48].

While overexpression of UCP3 seems to be protective in early stages of HF, in the later stages, however, FA oxidation decreases, along with mitochondrial oxidative activity due to UCP activation and overexpression, causing a decrease in cardiac ATP levels [9] and myocardial dysfunction (Fig. 1.3).

1.3.5 Myocardial Hypertrophy

Myocardial hypertrophy is often a compensatory mechanism in response to increased cardiac workload. Conditions such as the stenosis of aortic valves or chronic hypertension can increase the afterload against which the left ventricle has to pump. This causes an increase in left ventricular wall stress which stimulates the deposition of extracellular matrix and myocardial hypertrophy. Compensatory myocardial hypertrophy can take on two main forms, depending on whether it is developed in response to chronic pressure or volume overload. Chronic pressure overload such as in the case of hypertension or aortic stenosis results in new sarcomeres being added in parallel to existing ones, causing the ventricular wall to thicken without an accompanying increase in chamber size. This is known as concentric hypertrophy. Chronic volume overload such as in the case of mitral or aortic valve incompetence leading to regurgitation results in new sarcomeres being added in series to existing ones, causing an increase in the size of the ventricular chamber in proportion to the increase in wall thickness. This is known as eccentric hypertrophy. Other than pathological cardiac hypertrophy, physiological hypertrophy can also result from exercise without an impairment of cardiac function.

The expression of UCP2 was found to be upregulated in pathological hypertrophy [49] but downregulated in physiological hypertrophy, possibly to promote efficient energy production in the latter condition. In particular, there is evidence that UCP2 levels are increased in response to pressure overload in an attempt to avert apoptosis of cardiomyocytes [50]. Furthermore, a knockout of the UCP2 gene or its inhibition by genipin produced the same effect (in mice) – an attenuation of cardiac hypertrophy caused by pressure overload in part by increasing mitochondrial ATP production and decreasing myocyte apoptosis [51]. The expression of UPC2 has been found to be normalized by the use of two common pharmacological treatments for heart failure, namely, beta-blockers [49] and angiotensin-converting enzyme inhibitors (ACE inhibitors) [52]. Therefore, it has been proposed that blocking UCP2 expression in chronic pathological myocardial hypertrophy may improve cardiac performance and overall myocardial energy production efficiency.

1.3.6 Myocarditis

ROS are produced at high levels in myocarditis. ROS are produced by mitochondrial respiratory chain reactions and enzymes including NADPH oxidases, cyclooxygenase, and xanthine oxidase in the inflammatory cells such as macrophages.

Other systems involved in inflammation and stress response, such as NF-κB, and other cytokine factors also induce oxidative stress in myocarditis.

An elevated ROS production generated in severe inflammation with subsequent ROS-mediated damage beyond what the cellular antioxidant defense system can handle causes heart injury and further exacerbates inflammation. Long-lasting oxidative stress could be one of the pathological mechanisms of cardiac alteration and transformation leading to inflammatory cardiomyopathy and cardiac remodeling. This is where the role of UCPs comes to the fore. They have been seen to regulate ROS production in both cardiac and the inflammatory cells as well. Experimental evidence shows macrophages from UCP2-deficient mice exhibiting a marked increase in ROS generation as compared to their wild-type counterparts [36]. Furthermore, a high level of TNFα caused a corresponding increase in the expression of UCP2 and UCP3 in the rat muscle [25]. This could be regarded as an adaptive response in the setting of myocarditis, which remains open to debate. When ROS levels increased, cardiac function was impaired to a greater extent in UCP3-deficient mice as compared to their wild-type counterparts. The delay in the progression of cardiac dysfunction in the wild-type mice can be attributed to the fivefold increase in UCP3 levels observed [36]. However, this decrease in ROS production via mitochondrial uncoupling comes at a cost of decreased efficiency in ATP synthesis and energy reserve. The observed myocardial dysfunction during myocarditis and HF could be attributed (at least partly) to mitochondrial reduced efficiency. Therefore, it cannot be conclusively determined if upregulation of UCP3 is an adaptive or maladaptive response in myocarditis (Fig. 1.3).

1.4 Modulation of Oxidative Stress as a Potential Therapeutic Target

Overexpression of UCPs have beneficial effects on cardiac energetics regulation, mitochondrial ROS production, calcium handling, and cardiomyocyte apoptosis [53]. Increased UCP expression could also have a positive effect on the cell function in some pathological conditions (e.g., endothelial dysfunction); however, it may have an opposite effects on other CVDs. The associated decrease in ATP synthesis with overexpression of UCP may have deleterious consequences on cardiac function and may worsen the clinical outcomes. Thus, modulation of oxidative stress through UCP regulation in CVD as a therapeutic option should be considered carefully.

Downregulation of UCP in heart failure and subsequent inability of the cell to combat the oxidative burden caused by the failing heart have led to potential therapeutic role of UCP to compact the pathogenesis of the disease. As suggested above, the pathogenesis of heart failure has been characterized by an increase in ROS production and ROS-mediated damage. UCPs are known to prevent ROS accumulation and hence decrease the oxidative burden by limiting ROS production. Furthermore, UCPs may be involved in the detoxification of exogenously produced ROS. In the setting of heart failure, there is an increase in the concentration of circulating FFA. This is positively correlated with an increase in cardiac mitochondrial

UCPs. The metabolic effects of UCPs are cardioprotective in nature and are in fact an adaptive response to the rise in lipid concentration in the mitochondria [54]. Therefore, upregulating UCP expression in heart failure could be of therapeutic benefit.

One of the key regulators of UCP expression in the heart is $PPAR_{\alpha}$. Hence, to upregulate UCP expression, agonists of $PPAR_{\alpha}$ could theoretically be used. Currently, there is an ongoing study in the use of such agonists in the treatment of diabetes which can be extended to the study of their use in heart failure.

AMPK has been implicated to have a protective role (antiapoptotic) as it enhances the survival of cardiomyocytes in response to ischemia and reperfusion. Loss of AMPK activity results in an inability to increase glucose uptake and glycolysis by the cardiomyocytes [55]. Taken together, as a result of AMPK role in mediating the survival of cardiomyocytes during ischemia, is it also a potential target for augmenting the expression of UCP in heart failure.

Another translational approach would be the development of drugs that would induce UCP expression and hence slow the progression of atherosclerosis and endothelial dysfunction. The effects of sitagliptin, a fibrate vegetable extract, have been shown to increase the expression of UCP2 with an associated improvement in mitochondrial biogenesis and function [56]. As such, development of a therapeutic agent that modulates oxidative stress could exert a significant vascular protection.

1.5 Conclusion

Oxidative stress has a crucial role in the initiation and progress of cardiac diseases. Cardiac cells are equipped with powerful antioxidant defense systems and have the ability to mitigate damage by ROS. Clinically, most of the attention has been focused on the potential protective role of antioxidants in CVD. Much less attention has been paid to the mechanisms that regulate ROS production. UCPs seem to be a good candidate to regulate the endogenous ROS; however, activation of UCPs in highly energy-demanding cells such as cardiomyocytes may have a negative effect on the cardiac cell function. Whether these are an adaptive or maladaptive response to stress conditions needs further mechanistic understanding. UCPs play a double-edged sword with favorable and unfavorable consequences. On one hand, activation of UCPs would reduce the mitochondrial ROS generation and, thus, protects the cell from the oxidative damage and improves cell function. On the other hand, activated UCPs would reduce the mitochondrial efficiency to produce ATP and consequently limit the energy supply needed to sustain efficient contraction.

Acknowledgments The authors would like to thank Miss Teresa Cheng for her artwork presented in Figs. 1.1 and 1.2.

References

1. Mootha VK, Arai AE, Balaban RS (1997) Maximum oxidative phosphorylation capacity of the mammalian heart. Am J Phys Heart Circ Phys 272(2):H769–H775
2. Brand MD, Esteves TC (2005) Physiological functions of the mitochondrial uncoupling proteins UCP2 and UCP3. Cell Metab 2(2):85–93
3. NICHOLLS DG (1974) The influence of respiration and ATP hydrolysis on the proton-electrochemical gradient across the inner membrane of rat-liver mitochondria as determined by ion distribution. Eur J Biochem 50(1):305–315
4. Rolfe DF et al (1999) Contribution of mitochondrial proton leak to respiration rate in working skeletal muscle and liver and to SMR. Am J Phys Cell Phys 276(3):C692–C699
5. MacLellan JD et al (2005) Physiological increases in uncoupling protein 3 augment fatty acid oxidation and decrease reactive oxygen species production without uncoupling respiration in muscle cells. Diabetes 54(8):2343–2350
6. Miller D, MacFarlane N (1995) Intracellular effects of free radicals and reactive oxygen species in cardiac muscle. J Hum Hypertens 9(6):465–473
7. Thannickal VJ, Fanburg BL (2000) Reactive oxygen species in cell signaling. Am J Phys Lung Cell Mol Phys 279(6):L1005–L1028
8. Cannon B, Hedin A, Nedergaard J (1982) Exclusive occurrence of thermogenin antigen in brown adipose tissue. FEBS Lett 150(1):129–132
9. Akhmedov AT, Rybin V, Marín-García J (2015) Mitochondrial oxidative metabolism and uncoupling proteins in the failing heart. Heart Fail Rev 20(2):227–249
10. Schrauwen P (2004) The role of uncoupling protein 3 in fatty acid metabolism: protection against lipotoxicity? Proc Nutr Soc 63(2):287–292
11. Cardaci S, Filomeni G, Ciriolo MR (2012) Redox implications of AMPK-mediated signal transduction beyond energetic clues. J Cell Sci 125(9):2115–2125
12. Putman CT et al (2003) AMPK activation increases uncoupling protein-3 expression and mitochondrial enzyme activities in rat muscle without fibre type transitions. J Physiol 551(1):169–178
13. Xie Z et al (2008) Up-regulation of mitochondrial uncoupling protein-2 by the AMP-activated protein kinase in endothelial cells attenuates oxidative stress in diabetes. Diabetes 57:3222
14. DIRAISON F et al (2004) Over-expression of sterol-regulatory-element-binding protein-1c (SREBP1c) in rat pancreatic islets induces lipogenesis and decreases glucose-stimulated insulin release: modulation by 5-aminoimidazole-4-carboxamide ribonucleoside (AICAR). Biochem J 378(3):769–778
15. Kozak UC et al (1994) An upstream enhancer regulating brown-fat-specific expression of the mitochondrial uncoupling protein gene. Mol Cell Biol 14(1):59–67
16. Reddy JK, Krishnakantha T (1975) Hepatic peroxisome proliferation: induction by two novel compounds structurally unrelated to clofibrate. Science 190(4216):787–789
17. Lehmann JM et al (1997) Peroxisome proliferator-activated receptors α and γ are activated by indomethacin and other non-steroidal anti-inflammatory drugs. J Biol Chem 272(6):3406–3410
18. YOUNG ME et al (2001) Uncoupling protein 3 transcription is regulated by peroxisome proliferator-activated receptor α in the adult rodent heart. FASEB J 15(3):833–845
19. Gilde AJ et al (2003) Peroxisome proliferator-activated receptor (PPAR) α and PPARβ/δ, but not PPARγ, modulate the expression of genes involved in cardiac lipid metabolism. Circ Res 92(5):518–524
20. Barger PM, Kelly DP (2000) PPAR signaling in the control of cardiac energy metabolism. Trends Cardiovasc Med 10(6):238–245
21. Murray AJ et al (2004) Uncoupling proteins in human heart. Lancet 364(9447):1786–1788
22. Williamson J, Krebs H (1961) Acetoacetate as fuel of respiration in the perfused rat heart. Biochem J 80(3):540
23. Murray AJ et al (2005) Plasma free fatty acids and peroxisome proliferator–activated receptor α in the control of myocardial uncoupling protein levels. Diabetes 54(12):3496–3502

24. Parameswaran N, Patial S (2010) Tumor necrosis factor-α signaling in macrophages. Crit Rev Eukaryot Gene Expr 20(2):87
25. Busquets Sl et al (1998) In the rat, tumor necrosis factor α administration results in an increase in both UCP2 and UCP3 mRNAs in skeletal muscle: a possible mechanism for cytokine-induced thermogenesis? FEBS Lett 440(3):348–350
26. Lee FJ et al (1999) Tumor necrosis factor increases mitochondrial oxidant production and induces expression of uncoupling protein-2 in the regenerating rat liver. Hepatology 29(3):677–687
27. Kwon HJ et al (2003) Case reports of heart failure after therapy with a tumor necrosis factor antagonist. Ann Intern Med 138(10):807–811
28. Moore KJ, Tabas I (2011) Macrophages in the pathogenesis of atherosclerosis. Cell 145(3):341–355
29. Chang M-C et al (2017) Lysophosphatidylcholine induces cytotoxicity/apoptosis and IL-8 production of human endothelial cells: Related mechanisms. Oncotarget 8(63):106177
30. Lee K-U et al (2005) Effects of recombinant adenovirus-mediated uncoupling protein 2 overexpression on endothelial function and apoptosis. Circ Res 96(11):1200–1207
31. Alves-Guerra JBMM et al. Protective Role of Uncoupling Protein 2 in Atherosclerosis
32. Braganza D, Bennett M (2001) New insights into atherosclerotic plaque rupture. Postgrad Med J 77(904):94–98
33. Jennings RB, Ganote CE (1974) Structural changes in myocardium during acute ischemia. Circ Res 35(3_supplement):III-156–III-172
34. Rechavia E et al (1995) Hyperdynamic performance of remote myocardium in acute infarction: Correlation between regional contractile function and myocardial perfusion. Europ Heart J 16(12):1845–1850
35. Almsherqi ZA et al (2006) Reduced cardiac output is associated with decreased mitochondrial efficiency in the non-ischemic ventricular wall of the acute myocardial-infarcted dog. Cell Res 16(3):297
36. Murray AJ et al (2008) Increased mitochondrial uncoupling proteins, respiratory uncoupling and decreased efficiency in the chronically infarcted rat heart. J Mol Cell Cardiol 44(4):694–700
37. Gottlieb RA et al (1994) Reperfusion injury induces apoptosis in rabbit cardiomyocytes. J Clin Invest 94(4):1621–1628
38. Almsherqi ZA et al (2006) Displacement of the beating heart induces an immediate and sustained increase in myocardial reactive oxygen species. Circ J 70(9):1226–1228
39. Murry CE, Jennings RB, Reimer KA (1986) Preconditioning with ischemia: a delay of lethal cell injury in ischemic myocardium. Circulation 74(5):1124–1136
40. Liu Y et al (2009) Both ischemic preconditioning and ghrelin administration protect hippocampus from ischemia/reperfusion and upregulate uncoupling protein-2. BMC Physiol 9(1):17
41. Bienengraeber M, Ozcan C, Terzic A (2003) Stable transfection of UCP1 confers resistance to hypoxia/reoxygenation in a heart-derived cell line. J Mol Cell Cardiol 35(7):861–865
42. Organisation WH (2017) The top 10 causes of death
43. Misaka T et al (2018) FKBP8 protects the heart from hemodynamic stress by preventing the accumulation of misfolded proteins and endoplasmic reticulum-associated apoptosis in mice. J Mol Cell Cardiol 114:93–104
44. Cappetta D et al (2017) Effects of ranolazine in a model of doxorubicin-induced left ventricle diastolic dysfunction. Br J Pharmacol 174(21):3696–3712
45. Bugger H et al (2011) Uncoupling protein downregulation in doxorubicin-induced heart failure improves mitochondrial coupling but increases reactive oxygen species generation. Cancer Chemother Pharmacol 67(6):1381–1388
46. Jiménez-Jiménez J et al (2006) Fatty acid activation of the uncoupling proteins requires the presence of the central matrix loop from UCP1. Biochimica et Biophysica Acta (BBA)-Bioenergetics 1757(9–10):1292–1296
47. Teshima Y et al (2003) Uncoupling protein-2 overexpression inhibits mitochondrial death pathway in cardiomyocytes. Circ Res 93(3):192–200

48. Perrino C et al (2013) Genetic deletion of uncoupling protein 3 exaggerates apoptotic cell death in the ischemic heart leading to heart failure. J Am Heart Assoc 2(3):e000086
49. Gaussin V et al (2003) Common Genomic Response in Different Mouse Models of β-Adrenergic–Induced Cardiomyopathy. Circulation 108(23):2926–2933
50. Hang T et al (2007) Apoptosis and expression of uncoupling protein-2 in pressure overload-induced left ventricular hypertrophy. Acta Cardiol 62(5):461–465
51. Ji X-B et al (2015) Inhibition of uncoupling protein 2 attenuates cardiac hypertrophy induced by transverse aortic constriction in mice. Cell Physiol Biochem 36(5):1688–1698
52. Murakami K et al (2002) Perindopril effect on uncoupling protein and energy metabolism in failing rat hearts. Hypertension 40(3):251–255
53. Sack MN (2006) Mitochondrial depolarization and the role of uncoupling proteins in ischemia tolerance. Cardiovasc Res 72(2):210–219
54. Hesselink MK, Schrauwen P (2005) Uncoupling proteins in the failing human heart: friend or foe? Lancet 365(9457):385–386
55. Russell RR et al (2004) AMP-activated protein kinase mediates ischemic glucose uptake and prevents postischemic cardiac dysfunction, apoptosis, and injury. J Clin Invest 114(4):495–503
56. Pierelli G et al (2017) Uncoupling protein 2: a key player and a potential therapeutic target in vascular diseases. Oxidative Med Cell Longev 2017:1

Oxidant-Dependent and Oxidant-Independent Proangiogenic and Vasomotor Signaling in Coronary Vascular Endothelium

2

Sarah Aldosari, Maan Awad, May Z. Gao, Isabella G. McCormack, Frank W. Sellke, and Md. Ruhul Abid

Abstract

Depending on their levels, source of generation, and subcellular locations, reactive oxygen species (ROS) are known to have paradoxical effects on coronary vascular endothelium. At low concentrations, ROS contribute to physiological signaling pathways that regulate vascular endothelial cell (EC) growth and survival. At higher concentrations, or with prolonged exposure, ROS can exacerbate endothelial cell injury and trigger apoptosis. In this chapter, oxidant-dependent and oxidant-independent angiogenic and vasomotor signaling pathways will be discussed in-depth, including the structures of oxidant-producing enzymes, their agonists, and their related signaling pathways in EC. Vascular endothelial growth factor (VEGF), a major growth factor involved in the maintenance of EC health, vasomotor tone, and angiogenesis, will also be discussed. VEGF utilizes both reactive oxygen species (ROS)-dependent and ROS-independent arms of EC signaling.

In this chapter, NADPH oxidase (NOX)-induced oxidant-dependent angiogenesis will be discussed in-depth, including the structures of all NADPH oxidase isoforms, agonists, and transcription factors that are involved in proangiogenic signaling pathways. We will also discuss vascular endothelial growth factor (VEGF) signaling pathways that are affected by the upregulation of ROS generation.

Previously, increased levels of ROS were believed to be purely associated with pathological conditions as seen in cardiovascular diseases (CVD). Indeed, ROS are produced in higher levels at sites of inflammation and injury by the

S. Aldosari · M. Awad · M. Z. Gao · I. G. McCormack · F. W. Sellke · M. R. Abid (✉)
Cardiovascular Research Center, Cardiothoracic Surgery Division, Rhode Island Hospital, Brown University Warren Alpert Medical School, Providence, RI, USA
e-mail: ruhul_abid@brown.edu

S. Chakraborti et al. (eds.), *Modulation of Oxidative Stress in Heart Disease*,
https://doi.org/10.1007/978-981-13-8946-7_2

mitochondria and enzymes, such as NADPH oxidases. Recent findings, as to be discussed in this chapter, have contradicted this notion that ROS are purely a part of pathophysiological pathways. Studies have shown that experimentally reducing global ROS levels does not improve vascular function and recovery as expected. Reducing ROS levels instead results in inhibition of endothelial nitric oxide synthase (eNOS) activation and decreased nitric oxide (NO) synthesis in endothelial cells. Rather than improving vascular function, a global decrease in ROS hinders endothelial function, reduces coronary vasodilation, and inhibits angiogenic signaling. Several recent reports suggest that homeostatic and even above physiological levels of subcellular ROS may contribute to optimal endothelial cell and vascular functions. These studies suggested that the beneficial versus detrimental effects of higher levels of ROS are time-, location- and concentration-dependent.

This chapter will shed light on the overwhelming interconnectedness of NOX, growth factors, and vasoactive factors as well as larger-scale oxidant-dependent and oxidant-independent pathways to elucidate the complexity of signaling in coronary vascular endothelium.

Keywords

Oxidative stress · Reactive oxygen species · Vascular endothelium · Cardiovascular diseases · NADPH oxidase · Angiogenesis factor · Endothelial nitric oxide synthase · Endothelium-dependent relaxing factors · Vascular endothelium-dependent relaxation · VEGF

2.1 Introduction

In this chapter, oxidant-dependent and oxidant-independent angiogenic and vasomotor signaling pathways will be discussed in-depth, including the structures of oxidant-producing enzymes, their agonists, and their related transcription factors in endothelial cell (EC). Vascular endothelial growth factor (VEGF) which is a major growth factor involved in the maintenance of EC health, vasomotor tone, and angiogenesis, will also be discussed. VEGF utilizes both reactive oxygen species (ROS)-dependent and ROS-independent arms of EC signaling.

ROS are produced at higher levels at sites of inflammation and injury by enzymes, such as NADPH oxidase. At low concentrations, ROS contribute to physiological signaling pathways that regulate cell growth and cell survival. At higher concentrations, or when exposure is prolonged, ROS can exacerbate endothelial cell injury and trigger apoptosis. In this chapter, NADPH oxidase-induced oxidant-dependent angiogenesis will be discussed in-depth, including the structures of all NADPH oxidase isoforms, agonists, and transcription factors that are involved in proangiogenic signaling pathways. We will also discuss vascular endothelial growth factor (VEGF) signaling pathways that are affected by the upregulation of ROS generation.

Previously, increased levels of ROS were believed to be purely associated with pathological conditions as seen in cardiovascular diseases (CVD). Indeed, ROS are produced in higher levels at sites of inflammation and injury by the mitochondria and enzymes. Recent findings, as to be discussed in this chapter, have contradicted this notion that ROS are purely a part of pathophysiological pathways. Studies have shown that experimentally reducing global ROS levels does not improve vascular function and recovery as expected. Reducing ROS levels instead results in inhibition of endothelial nitric oxide synthase (eNOS) activation and decreased nitric oxide (NO) synthesis in endothelial cells. Rather than improving vascular function, a global decrease in ROS hinders endothelial function, reduces coronary vasodilation, and inhibits angiogenic signaling. Several recent reports suggest that homeostatic and even above physiological levels of subcellular ROS may contribute to optimal endothelial cell and vascular functions. These studies suggested that the beneficial versus detrimental effects of higher levels of ROS are time-, location-, and concentration-dependent.

Notably, VEGF is a major proangiogenic growth factor in endothelial cells required for normal vasculature in homeostasis; yet it can also contribute to tumor growth when uncontrolled. VEGF-mediated signaling pathways seem to be influenced by the levels of ROS. This chapter discusses the VEGF-mediated oxidant-independent and oxidant-dependent pathways that lead to augmented angiogenesis and explores the roles that oxidized low-density lipoproteins play in these signaling cascades. Participating in both oxidant-dependent and oxidant-independent pathways, VEGF responds to oxidative stress by activating the PI3K-Akt-eNOS pathway and NO synthesis, thus inducing vasodilation. The VEGF downstream signaling pathway can also be activated independently of ROS by activating the PLC γ-ERK1/2 axis.

The endothelium itself also plays a significant role in regulating vascular tone in response to different stimuli, such as oxygen consumption and shear stress. For instance, the endothelium can induce a vasodilatory response through the release of vasoactive agents, such as NO, prostacyclin, acetylcholine, and endothelium-derived hyperpolarizing factor. While some of these factors and signaling molecules depend on ROS, we will also discuss the preferential oxidant-independent pathways of vasoactive agents like arachidonic acid metabolites, endothelium-derived hyperpolarizing factors, and acetylcholine. The various mechanisms that govern the interactions between different endothelium-derived vasoactive molecules will also be discussed.

2.2 Regulation of Coronary Vascular Function and Angiogenesis

Maintaining a vascular homeostasis is essential for cardiovascular health and to provide adequate blood supply to the tissues. It is known that the cardiovascular system is considered as a recourse allocation in the presence of trigger factors, such as exercise, hypoxia, and others [1]. Thereby, the body has developed a way for recourse allocation during hypoxic or ischemic crisis, allotted as vasodilation and angiogenesis. Vasodilation or vasorelaxation of the arterioles increases blood flow by reducing resistance. Vascular endothelium plays a critical role in the process by synthesizing NO that acts on the adjacent smooth muscle layer in the arterioles and produces relaxation resulting in vasodilation. Angiogenesis is the physiological process of forming new blood vessels from preexisting vessels [2]. Several regulators play a pivotal role to control angiogenesis. The importance of these regulators, including oxidant-dependent and non-oxidant regulators, in angiogenesis is becoming increasingly apparent. However, pathophysiological responses have shown clinical implications related with variation in the concentration, time exposure, or source of the stimuli. In this chapter, pathways regulating angiogenesis and vascular response have been classified as oxidant-dependent and oxidant-independent pathways.

The process of angiogenesis involves multiple steps that start with the increase in vascular permeability, resulting in the deposition of plasma proteins, such as fibrinogen, fibronectin, and plasminogen in the extracellular matrix [3]. These proteins eventually get activated and form a fibrin-fibronectin gel [3, 4]. The fibrin-fibronectin gel facilitates and acts as a scaffold that supports the migration of endothelial cells and fibroblast within the extracellular matrix (ECM) [3, 5]. The last step in angiogenesis involves remodeling and maturation of the newly formed blood vessel and deposition of new extracellular matrix which contains components that are normally not found in adult ECM [3]. These ECM components include tenascin, larger amounts of hyaluronan and chondroitin sulfate proteoglycans, abnormally glycosylated decorin, and plasma protein-rich interstitial fluid.

2.3 NADPH Oxidase-Derived Oxidant-Dependent Vascular Signaling

2.3.1 Structure and Subcellular Localization of NADPH Oxidases

NADPH oxidase (NOX) is an intracellular membrane-bound enzyme complex that has the capacity to transfer electrons from NADPH to an oxygen molecule, producing the superoxide anion, O_2^- [6–8]. Unlike other reactive oxygen species (ROS)-producing enzymes, ROS production is the sole function of NOX enzyme [6]. Thus, NOX enzyme is considered the major non-mitochondrial source of ROS in various tissues in the body, including coronary endothelium. In the vasculature, there are different isoforms of NADPH oxidases that have been identified, namely, NOX1, NOX2 ($gp91^{phox}$), NOX4, and NOX5 [6, 8–10]. NOX enzymes exhibit distinct isoform-dependent localization, subcellular regulatory subunits, and involvement in physiological function [10]. All four NOX isoforms are found in the endothelium; NOX1, NOX4, and NOX5 are found in vascular smooth muscle cells; and NOX2 and NOX4 are found in adventitial fibroblasts [9, 10]. Vascular NADPH oxidase is composed of membrane-bound and cytosolic regulatory subunits (Fig. 2.1) [10]. Three isoforms, NOX1, NOX2, and NOX4, are associated with the protein $p22^{phox}$. NOX1 requires the cytosolic subunits NOX organizer 1 (NOXO1), NOX activator 1

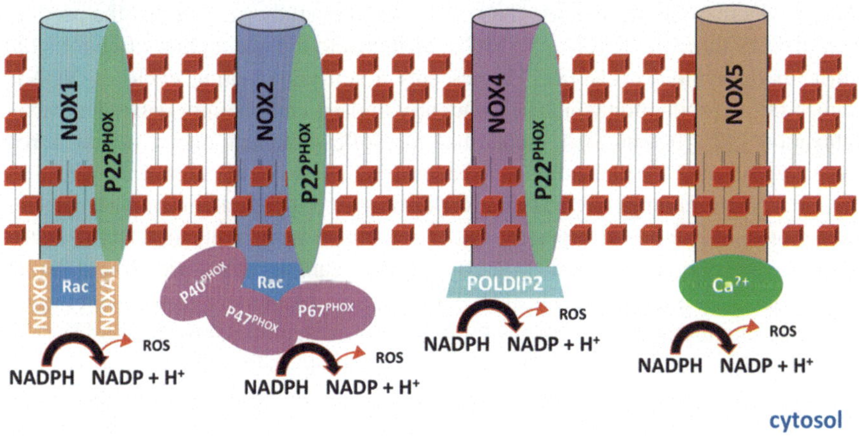

Fig. 2.1 *The structure of vascular NADPH oxidases*. NOX1, NOX2, NOX4, and NOX5 are localized in the membrane and associated with p22^{phox}, except for NOX5. NOX1 activity requires NOXO1, NOXA1, and Rac. NOX2 (gp91^{phox}) activity requires p40^{phox}, p47^{phox}, p67^{phox}, and Rac. NOX4 does not require a cytosolic factor for its activity but can be activated by POLDIP2. NOX5 activity is dependent on Ca^{2+}. All NOX isoforms produce reactive oxygen species by converting NADPH to NADP + H^+, or NADH to NAD + H^+

(NOXA1), and the small G-protein Rac1 for its activity. NOX2 activity is dependent on the cytosolic subunits p40^{phox}, p47^{phox}, p67^{phox}, and Rac1. NOX4, however, is constitutively active and does not require cytosolic factors for its activity but can be activated by polymerase-δ-interacting protein 2 (POLDIP2). The NOX5 isoform is the only NOX isoform not associated with p22^{phox}. NOX5 is a calcium-dependent enzyme which can be regulated by calmodulin [10, 11]. The distinct locations and functions of the NADPH oxidase isoforms have been shown to exert different physiological and pathological effects in vascular homeostasis, involving signal transduction, cell proliferation, and apoptosis. NADPH oxidases are considered to be major sources of ROS in ECs. Since ECs do not depend on their energy production (ATP) on mitochondrial oxidative phosphorylation and instead depend on the glycolytic pathway, the mitochondria are believed to be a minor source of ROS.

2.3.2 NOX1-Induced Oxidant-Dependent Vascular Response

There is compelling evidence that ROS produced in vascular smooth muscle cells (VSMC) influence the function of endothelial cells and the development of several cardiovascular diseases [12–19]. Of interest, NOX1 enzyme is predominantly expressed in VSMC and has been associated with VSMC migration and proliferation, as well as pathological hypertension and neointimal formation [19–22]. Upregulation of NOX1 has been implicated in augmenting and maintaining the angiotensin II vasomotor response [14, 15]. One study showed that in mice treated with angiotensin II, there is impairment of endothelium-dependent vasorelaxation due to a decrease in nitric oxide (NO) [14]. Another study showed that the production of ROS through NOX1 in VSMC contributes to the uncoupling of endothelial nitric oxide synthase (eNOS). This creates a self-perpetuating cycle of NOX1-induced ROS production which subsequently impairs endothelium-dependent vasorelaxation [12]. It was also found that in NOX1-deficient mice, medial aortic hypertrophy and extracellular matrix accumulation were attenuated [9, 11]. Together, these findings suggest that NOX1 plays a critical role in the pathogenesis of angiotensin II-induced hypertension. Other studies have implicated NOX1 in the pathogenesis of atherosclerosis [18, 20–22]. For example, it has been found that in NOX1-deficient mice, the levels of ROS production, neointimal growth, and migration were attenuated in the setting of injury-induced neointimal formation [18, 22]. Thus, selective inhibition of NOX1 may contribute to halting the formation of an atherosclerotic plaque. Further studies should be conducted to examine the role of EC-specific NOX1-derived ROS in human coronary vessels since NOX1 is also expressed in endothelial cells and may play a crucial role in endothelial signaling pathways and phenotypes.

2.3.3 NOX2-Induced Oxidant-Dependent Vascular Response

The NOX2 enzyme has been shown to induce positive effects on the coronary endothelium [8]. Reduced ROS levels have been shown to inhibit coronary vasorelaxation by inhibiting activation of PI3K-Akt-eNOS signaling and did not improve cardiovascular disease outcomes [23, 24]. Conversely, higher levels of ROS have been shown to improve endothelial function in a temporal-dependent manner [25, 26]. Thus, it is imperative to study both the concentration-dependent and time-dependent effects of ROS on pathophysiological states.

In vitro studies have found that endothelial cell-specific NOX2 is involved in the activation of the phosphoinositide 3-kinase (PI3K)-Akt-eNOS-mediated cell growth and survival pathway and in increased vasodilation by producing NO (Fig. 2.2) [7, 27]. However, an in vivo study using EC-specific transgenic animal model has shown that endothelial NOX2 stimulates NO production via activation of the AMPK-eNOS axis through Ca^{2+}-/calmodulin-dependent protein kinase kinase β (CaMKKβ). Activation of this particular axis was associated with improved

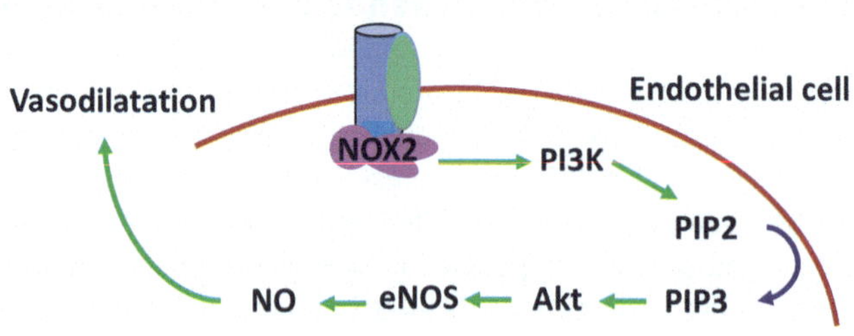

Fig. 2.2 *Vasodilation meditated via endothelial NOX2 enzyme*. NOX2-derived ROS activates PI3K, leading to the activation of Akt family, which in turn stimulates eNOS to produce NO

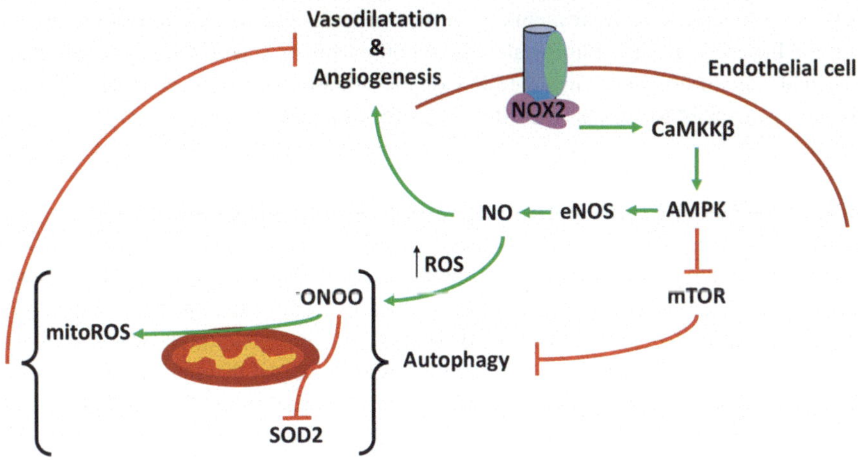

Fig. 2.3 *Vasodilation and angiogenesis mediated via endothelial NOX2-derived ROS using a transgenic animal model.* NOX2 activates CaMKKβ which activates AMPK. AMPK-mediated activation of eNOS leads to NO production. AMPK-mediated inhibition of mTOR leads to a protective autophagy response. After long-term ROS exposure, increased ⁻ONOO formation occurs by ROS and NO, which in turn inhibits vasodilation and angiogenesis

coronary vasodilation and increased endothelial cell proliferation and migration (Fig. 2.3) [27].

Autophagy is an intracellular process which regulates the degradation of proteins, metabolites, and organelles [28]. This process starts by engulfing the cellular components in a double-membrane autophagosome. Subsequently, the autophagosome merges with a lysosome where the components are degraded by the acidic environment [28]. A major part of autophagocytosed materials are recycled to synthesize cellular proteins, enzymes, nucleic acids, etc. The autophagy process exerts a protective effect during cell damage. Increased ROS levels have been shown to stimulate AMPK-mediated inhibition of mTOR (Fig. 2.3) [27], which results in increased autophagy to help recycle the damaged organelles.

Supraphysiological ROS levels exert distinct beneficial and detrimental effects on the coronary endothelium depending on the duration of exposure. After short-term ROS exposure, EC-specific NOX2 activates the CaMKKβ-AMPK-eNOS-NO pathway to produce beneficial effects, such as vasodilation [26]. However, a sustained increase in ROS levels seems to exert deleterious effects on endothelial cells [26]. This sustained increase in ROS results in the formation of peroxynitrite and a decrease in NO bioavailability via increased NO quenching by ROS, rather than direct inhibition of eNOS enzymatic activity. Sustained ROS exposure also results in inactivation of mitochondrial antioxidant MnSOD and thus increase in mitochondrial ROS levels with a decrease in mitochondrial membrane potential. All of these effects have resulted in decreased vasodilation and cell proliferation (Fig. 2.3) [26].

The above-mentioned findings advance our understanding of concentration-dependent and time-dependent roles of NOX2-derived ROS and the communication between subcellular compartments. It further suggests the critical impact on understating disease states and future subcellular compartment-specific treatment modalities that can offer better management of cardiovascular diseases. Hence, future work is warranted to address the effects of subcellular ROS.

2.3.4 NOX4-Induced Oxidant-Dependent Vascular Response

The NOX4 enzyme produces ROS, predominantly hydrogen peroxide (H_2O_2) molecules [8, 19, 29–33]. Compared to the other NOX isoforms, NOX4 is expressed most abundantly in endothelial cells [8]. This enzyme has potential beneficial effects on vascular functions [28, 31–35]. There is a notion that NOX4 may act as a sensor for nuclear redox reactions due to its subcellular location in the human endothelial cell [34]. Functionally, NOX4-derived H_2O_2 activates transforming growth factor β1 (TGF β1) which induces vascular angiogenesis and increases hemoglobin content [35]. NOX4-derived H_2O_2 acts as an endothelial vasodilator and enhances blood flow. One intriguing mechanism by which ROS act as vasodilator is endothelium hyperpolarization [36–38]. This may happen due to Ca^{2+} release from the endothelial endoplasmic reticulum which potentiates the opening of Ca^{2+}-activated K^+ channels. NOX4 also participates in cell survival by inhibiting the activation of apoptotic caspases [35]. One plausible mechanism for this observation is through activation of heme oxygenase-1 (HO-1). HO-1 is an enzyme that confers a vascular protective role through several mechanisms [35]. In NOX4 knockout mice, HO-1 levels were reduced which was accompanied by an increase in apoptosis and endothelial E-selecting expression [35]. To maintain expression of HO-1, NOX4 protects the oxidation of Kelch-like ECH-associated protein (Keap), which in turn prevents degradation of Nrf-1, a transcriptional factor for HO-1. These findings suggest that therapeutic strategies to regulate ROS levels to treat cardiovascular diseases require careful consideration of the source of ROS, for example, NOX4 is unique in its propensity to produce H_2O_2 rather than O_2^-.

2.3.5 NOX5-Induced Oxidant-Dependent Vascular Response

NOX5 enzyme has been found in the endoplasmic reticulum of human microvascular endothelial cells (HMEC-1) and in the vascular wall [39]. It is unique in its structure because it contains an additional N-terminal region that binds calcium, allowing the activation of the enzyme through an increase in intracellular calcium levels [39, 40]. It has different variants, including NOX5α, NOX5β, NOX5γ, and NOX5δ. Some of these variants, including NOX5β, NOX5δ, and a variant lacking the calcium-binding domains (NOX5S), are expressed in the vasculature of HMEC-1 and contribute to endothelial ROS production, cell proliferation, and angiogenesis [11, 39].

In response to thrombin, NOX5 variants were implicated in the production of ROS and the formation of capillary-like structures, which are indications of an NOX5 angiogenic response [24–41]. NOX5β and NOX5S were especially found to be responsible for ROS basal level elevation. This shows that NOX5 may play an equally pivotal role as NOX2 in endothelial responses to thrombin. Although NOX2 requires $p22^{phox}$ for ROS production, NOX5 apparently does not require $p22^{phox}$ [42, 43, 47].

In a patient with atherosclerosis, the level of calcium-dependent NADPH oxidase, NOX5, was markedly elevated compared to non-atherosclerotic subjects [11]. Mechanistically, NOX5 seems to share similarities with NOX4, as NOX5 predominantly releases H_2O_2 in the human vasculature [11]. As mentioned previously, H_2O_2 plays a critical dual role in cell signaling and contributes to atherosclerotic plaque development. Increasing the levels of Ca^{2+} in endothelial cells promoted an increase in NO production; however, one plausible mechanism involved in the development of atherosclerosis is the consumption of NO by ROS [11]. Furthermore, increasing Ca^{2+} in vascular smooth muscle cells triggered the contractile apparatus and a loss of NO due to diffusion into other cells [11]. These events may alter the vascular response to vasoactive hormones. Thus, calcium channel antagonists are commonly employed to treat cardiovascular diseases by reducing intracellular calcium and exert beneficial effects by inhibiting NOX5 enzyme, subsequently preventing oxidant injury.

2.3.6 Mitochondria-Derived Oxidant-Dependent Vascular Response

Mitochondria are endomembrane organelles found in all eukaryotic cells [44, 45]. They orchestrate energy production through respiration via a process known as oxidative phosphorylation [44, 45]. Yet the role of mitochondria goes beyond energy production and includes ROS formation, activation of cellular death, and calcium regulation [46]. They have distinct form and functions depending on the cell type. In an endothelial cell, mitochondria comprise less than 6% of cell volume, which implies that endothelial cells do not rely on mitochondria-derived energy, but rather on anaerobic glycolysis [28, 47]. In fact, mitochondria seem to serve primarily as critical signaling organelles in the vascular endothelium rather than energy powerhouses [46]. For example, mitochondria have been reported to be anchored to the cytoskeleton of coronary endothelial cells and to play roles in angiogenesis in response to shear stress [48]. As intermediary signals, mitochondria are a vital mediator for downstream regulation of angiogenesis-related gene expression, as well as apoptosis [28, 46–48]. Thus, preserving the mitochondrial quality control is essential for optimal function of EC. This is accomplished by biogenesis, dynamics, and mitophagy of the mitochondria [46].

Increased mitochondrial mass is required to carry out different functions. A key mediator which coordinates mitochondrial replication and expression is proliferator-activated receptor-γ coactivator-1α (PGC-1α) (Fig. 2.4) [28]. PGC-1α serves a dual

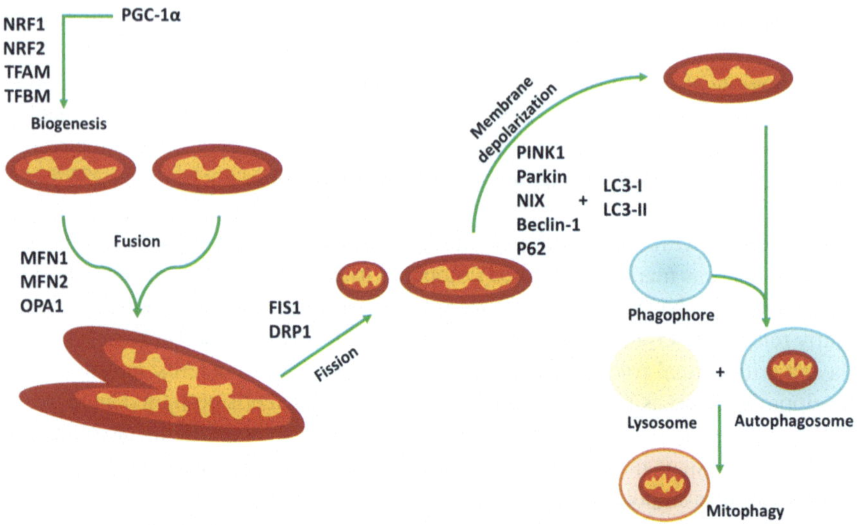

Fig. 2.4 *Illustration of the mitochondrial life cycle and the involvement of mitochondrial dynamic mitophagy mechanisms to maintain quality*. Biogenesis and gene expression are regulated by PGC-1α, which activates NRF1, NRF12, TFAM, and TFBM. The mitochondria undergo cyclic and balanced fusion and fission processes to preserve their function. Fusion is regulated by MFN1, MFN2, and OPA1 to form elongated mitochondrial networks. Similarly, fission is regulated by FIS1 and DRP1 to form smaller organelles. In mitophagy, the process is triggered by mitochondrial membrane depolarization, which causes the accumulation of PINK1, which recruits parkin. P62 is also vital for the formation of the autophagosome and is subsequently degraded during active autophagy. NIX may trigger the uncoupling of Beclin-1 from BLC-2 and BCL-X_L. Beclin-1 and LC3-I are conjugated onto phosphatidylethanolamine to form LC3-II, causing the assembly of autophagosome after the enclosure and isolation of mitochondria in the phagophore. Finally, the autophagosome is incorporated into a lysosome, which initiates the mitophagy process

function in protecting EC against excessive oxidative stress. PGC-1α triggers mitochondrial biogenesis by activating two critical factors, nuclear respiratory factors 1 and 2. It also triggers the expression of mitochondrial transcription factors A and B (TFAM and TFBM) to regulate the expression of mitochondrial DNA [46]. PGC-1α also regulates the expression of VEGF-1 to induce angiogenesis in vascular endothelium [49]. Moreover, it regulates the expression of other genes related to lipid and glucose metabolism, as well as apoptosis [50–52].

Mitochondria undergo cyclic and balanced fusion and fission processes to preserve their integrity (Fig. 2.4) [53]. Fusion of the outer membrane is carried out by the transmembrane GTPases mitofusin-1 and mitofusin-2 (MFN1, MFN2), whereas fusion of the inner membrane is mediated by optic atrophy protein-1 (OPA1). This process is imperative for the distribution of protein, metabolites, and mitochondrial DNA within the mitochondria. Moreover, it is critical for the maintenance of electrical and biochemical connectivity [28]. Fission is controlled by Fission 1 (FIS1) which recruits dynamin-related protein-1 (DRP1) to initiate the process. Fission is essential for cell division and elimination of senescent mitochondria [28]. During apoptosis, fission occurs concomitantly with the release of cytochrome c and outer membrane permeabilization [53, 54]. Emerging evidence has demonstrated that mitochondrial dynamics are critical in patients with cardiovascular risk factors. For example, polymorphism in the OPA1 and MFN2 genes have been implicated in hypertension [55, 56].

Mitochondrial-specific autophagy is referred to as mitophagy. Normal components of the damaged mitochondria can be reincorporated, while dysfunctional progenies are sent for elimination [57]. This process is imperative to yield new daughter mitochondria after damage, as otherwise the mitochondria would undergo apoptosis. An important trigger for mitophagy is membrane depolarization, which under physiological conditions leads to the accumulation of tensin homolog-induced putative kinase protein-1 (PINK1), recruitment of the E3 ubiquitin ligase parkin, and derepression of Beclin-1 (Fig. 2.4) [58]. Mitochondrial surface proteins are ubiquitylated, resulting in the binding and degradation of p62, leading to creation of the autophagosome. The autophagosome requires several proteins for its maturation, including microtubule-associated protein 1 light chain 3 (LC3-I), a ubiquitin-like protein which conjugates with Beclin-1 onto phosphatidylethanolamine to form LC3-II. The action of NIX, which is linked to the mitochondrial membrane and LC3, is also required to target mitochondrial autophagy [58]. During erythroblast differentiation, NIX is activated triggering mitophagy through uncoupling of Beclin-1 from BCL-2 and BCL-X_L [59]. Compelling evidence shows that impaired mitophagy and autophagy are associated with the progression of several vascular diseases, such as atherosclerosis and hypertension [60]. These processes seem to form a protective response in the endothelium by improving vascular functions via breakdown and recycling of damaged cellular components. In Beclin-1/LC3-II-induced autophagy, clearance of oxidized LDL in an endothelial cell was accomplished, reinforcing the protective role of autophagy [61].

Mitochondrial ROS are tightly regulated to carry out physiological functions in EC. Subcellular communications between endogenous cytosolic ROS and mitochondrial ROS have been identified and termed as ROS-induced ROS release [62]. One of the major enzymes that is linked to mitochondrial ROS production in the endothelial cell is NADPH oxidase, specifically NOX4 [63]. Another potential source of mitochondrial ROS is p66Sch, a growth factor adaptor protein that causes the oxidation of cytochrome c, generating H_2O_2 [64–66]. The effects of mitochondrial ROS have been shown to be concentration-dependent and time-dependent [26]. Low levels of mitochondrial ROS are associated with normal vascular responses, such as shear-stress-induced vasodilation, autophagy, and hypoxia signaling [48]. However, the accumulation of mitochondrial ROS has been linked to vascular diseases, for instance, oxidized LDL-mediated dyslipidemia and NOX-derived ROS angiotensin II-induced hypertension [67, 68]. These vascular diseases can be improved by the production of endogenous antioxidants, which halt supraphysiological ROS levels. For example, the mitochondrial antioxidant thioredoxin-2 attenuates the levels of ROS and suppresses angiotensin II-induced hypertension [68]. Excessive mitochondrial ROS levels also have an influence on the development of atherosclerotic plaque and endothelial dysfunction seen in diabetic patients

[69, 70]. In diabetes, high levels of blood glucose induce endothelial dysfunction which alters mitochondrial dynamics [46]. Concomitantly, mitochondrial fragmentation is increased with an increase in organelle fission. Furthermore, the increase in the ROS levels has been shown to blunt cell growth and induce apoptosis. Interestingly, high levels of blood glucose alter mitochondrial membrane potential by activating metalloproteinase 9. The ROS increase also attenuates expression of PGC-1α, thus downregulating mitochondrial biogenesis and gene expression [46].

Following a short-term exposure to increased cytosolic ROS (by NOX) in coronary endothelial cells induces mitochondrial ROS production, and the increased mito-ROS was checked or balanced by increase expression of mitochondrial antioxidant, such as superoxide dismutase 2 (SOD2), resulting in the improvement of vascular functions [26]. On the contrary, long-term exposure to NOX-derived cytosolic ROS has been shown to produce detrimental effects, such as impairment of endothelium vasodilation [26]. During long-term exposure, uncontrolled levels of ROS allow the conversion of NO to the deleterious molecule peroxynitrite, which impairs the function of mitochondrial ROS scavengers.

The mitochondria are organelles that can sense the mechanical forces of blood flow. A key determinant and an oxidant-dependent and oxidant-independent mediator for the development of coronary atherosclerosis are shear stress [71]. Local hemodynamic abnormalities contribute to endothelial dysfunction through mitochondrial ROS. Three forms of blood flow that can influence the function of the endothelial cell include (i) laminar flow, when there is a constant flow velocity; (ii) pulsatile flow, which is unidirectional flow with variation in the magnitude; (iii) and oscillatory flow, bidirectional flow with variation in the magnitude. Laminar and pulsatile flow seem to induce vascular vasodilation by triggering the activation of eNOS and thus production of NO [72–75]. Also, they increase the expression of mitochondrial ROS scavengers, thereby halting the damaging effects of oxidative stress [76]. Mechanistically, laminar flow regulates mitochondrial fission via cytosolic Drp1 recruitment [77]. This, in turn, increases mitochondrial membrane potential and ROS formation, leading to the activation of the antioxidant enzyme PRX3. Furthermore, laminar and pulsatile flow modify autophagosome formation via activation of the AMPK and JNK pathway [78]. Conversely, oscillatory blood flow increases the expression of NADPH enzymes, thereby increasing the levels of ROS [79].

2.4 Oxidant-Dependent and Oxidant-Independent VEGF-Induced Vascular Response

VEGF is the most potent proangiogenic growth factor in EC. It is crucial for normal vascular homeostasis and also contributes to the progression of various diseases by promoting vascular growth [80]. VEGF signaling pathways seem to be influenced by the levels of ROS. Cross talk between NOX2 and NOX4 has been shown to promote development of the angiogenic phenotype in endothelial cells [81]. Both enzymes participate in generating O_2^- and H_2O_2 which contribute to VEGF signaling and angiogenesis [61]. VEGF in turn stimulates NOX2 and NOX4 to induce angiogenesis and other critical endothelial cell functions, such as nitric oxide synthesis and hemostasis [62]. Thus, a reciprocal relationship between ROS and VEGF is crucial in VEGF angiogenic signaling pathway [82].

A positive feedback loop coordinated by NOX-derived and mitochondrial-derived ROS induces an angiogenic response via VEGF [62]. The binding of VEGF on VEGF receptor-2 (VEGFR2) stimulates the autophosphorylation of tyrosine kinase receptor and thus the activation of downstream signaling pathways in the endothelial cell. Once VEGFR2 becomes activated, the ROS-induced axis gets activated (Fig. 2.5). The axis involves, initially, the activation of NOX4 to produce H_2O_2. NOX4-derived H_2O_2 activates NOX2 to produce O_2^- that is rapidly converted to H_2O_2 by the enzyme SOD2. NOX4/NOX2 axis promotes mitochondrial ROS production via phosphorylation of pSer36-p66Sch by either protein kinase C or ERK/JNK in EC, which then gets translocated to the mitochondria. In the mitochondria, pSer36-p66Sch catalyzes transfer of electrons from cytochrome c to oxygen, thereby forming superoxide and then H_2O_2 within mitochondria. ROS then enhance phosphorylation of VEGFR2, thus amplifying ROS signaling, as well as inducing proangiogenic endothelial cell migration and proliferation [62]. Cell migration requires Ca^{2+}, achieved by the glutathiolation of sarcoplasmic/endoplasmic reticulum Ca^{2+}-ATPase (SERCA) found on the endoplasmic reticulum [83].

ROS can modulate protein functions, for example, by causing the oxidation of cysteine thiols [81]. ROS, especially H_2O_2, oxidize cysteine to form disulfide bonds via a sulfonic acid intermediate (Fig. 2.6) [84]. The oxidation of cysteine thiols has been shown to be able to activate or inhibit targeted pathways. For instance, intracellular NADPH-derived ROS have been involved in linking and activating post-VEGF signaling proteins, specifically c-Src, rendering c-Src-PI3K-Akt signaling

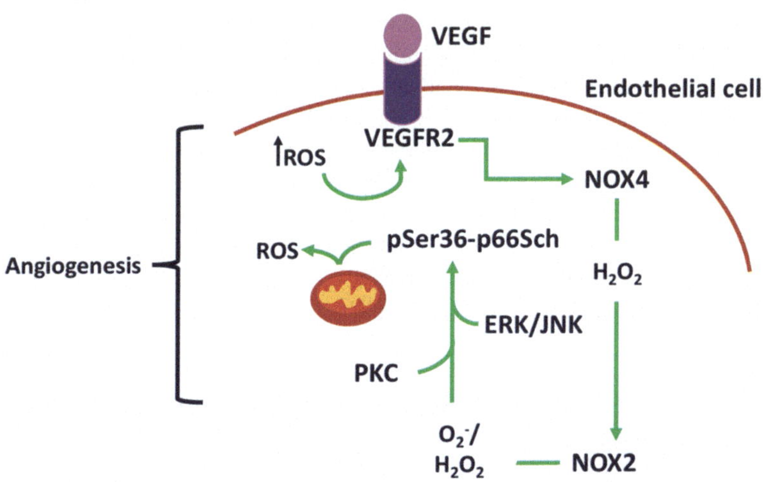

Fig. 2.5 *ROS-dependent VEGF-induced endothelial cell angiogenesis*. The activation of VEGFR2 leads to activation of the ROS release axis. In turn, increase in NOX2-, NOX4-, and mitochondrial-derived ROS leads to increased VEGFR2 activation, the major factor for endothelial angiogenesis

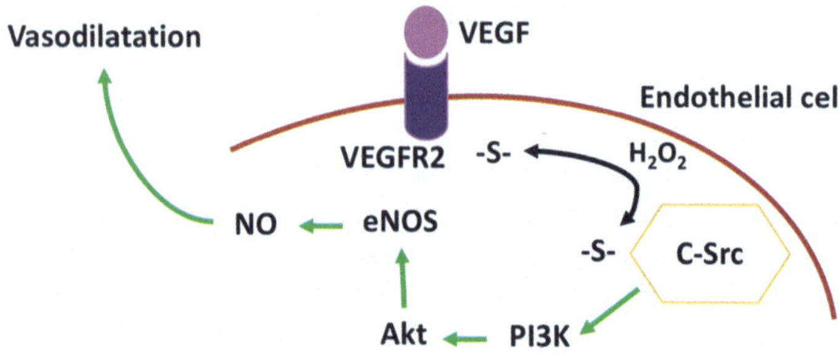

Fig. 2.6 *ROS-dependent VEGF-induced vasodilation*. Increased levels of H_2O_2 induce cysteine oxidation, which activates c-Src. This pathway activates the PI3K-Akt pathway, producing NO via eNOS activation

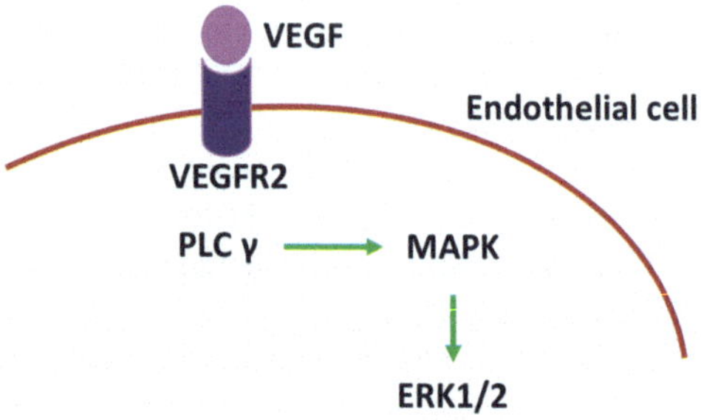

Fig. 2.7 *VEGF downstream signaling pathway can be activated independent of ROS*. This is achieved by activating the PLC γ-ERK1/2 axis, which does not appear to require an oxidation step

ROS-dependent in endothelial cells (Fig. 2.6) [81, 85]. Mechanistically, VEGF induces the downstream PLC γ-ERK1/2 signaling pathway involved in artery-vein specification independent of NADPH-derived ROS [81]. One plausible explanation is that PLC γ activation does not require an oxidative intermediate (Fig. 2.7). In a study using si-$p47^{Phox}$, the level of thrombomdulin activity and decay-accelerating factor (DAF) was measured [85]. They found that VEGF-mediated induction of thrombomdulin activity was abrogated, suggesting that VEGF-mediated thrombomdulin activity is sensitive to NADPH oxidase in endothelial cell. Consistent with the study, VEGF has been shown to protect endothelial cell against complement-mediated lysis independent of NADPH as DAF was not affected by si-$p47^{Phox}$. These findings suggest different NADPH oxidase-dependent and NADPH oxidase-independent signaling pathways for VEGF. In this way, using NADPH medication for therapeutic approach will certainly show selective effects on VEGF signaling pathways.

2.5 Oxidant-Dependent Lipoprotein-Induced Inflammatory Response

Under homeostatic conditions, the endothelium expresses little to no proinflammatory factors, while endothelial dysfunction is linked with vascular lesion, vasoconstriction, and atherosclerosis [86]. Reactive oxygen species (ROS) are produced at sites of inflammation and injury [86]. As such, endothelial cells experiencing oxidative stress show heightened vascular endothelial permeability [86, 87]. As a consequence, oxidative stress promotes leukocyte adhesion in endothelial cells and leads to changes in both signaling pathways and levels of proinflammatory transcription factors [87].

In inflammatory diseases involving endothelial dysfunction, such as atherosclerosis, oxidized phospholipids, which are derived from lipoproteins and oxidatively stressed cell membranes, start to build up. Monocyte chemoattractant protein-1 (MCP-1) is a proinflammatory chemokine that promotes atherogenesis [87]. Its activity increases oxidation of low-density lipoprotein (LDL) in blood vessel walls (Fig. 2.8). A study by Aiello et. al has shown that there was a threefold increase in lipid oxidation in mice with overexpressed MCP-1 transgene as compared to mice with normal levels of expression of MCP-1 mRNA [88]. As such, MCP-1 expression on macrophages progresses atherosclerosis via increasing leukocytes and oxidation of lipids. In the tissue, macrophages initiate a proinflammatory response through production of ROS and growth factors [87]. Summarily, there is evidence that oxidized lipoproteins are not only present in affected mice but also in the atherosclerotic lesions of humans. The oxidized lipoproteins contribute to altering gene expression by increasing VEGF which stimulates collateral arterial growth [82, 87, 89]. The specific factors that elicit lipoprotein oxidation are yet to be identified.

With regard to endothelial dysfunction in diabetes, oxidative stress-induced overexpression of growth factors is associated with neovascularization [89]. Generally, diabetes is characterized by high levels of oxidative stress and accumulation of oxidized LDLs [89]. However, the increase in mitochondrial ROS levels induced by hyperglycemia may drive inflammatory pathways and lead to persistent changes in proinflammatory gene expression [89].

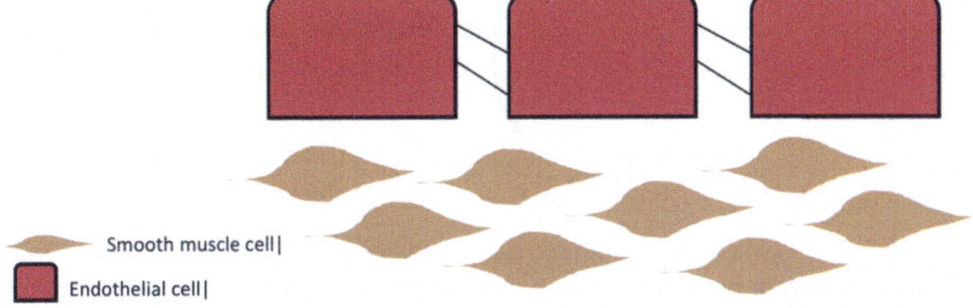

Fig. 2.8 *MCP-1 expression contributes to increased oxidation of low-density lipoproteins.* Rapid accumulation of oxidized LDL (ox-LDL) can lead to atherosclerosis and proinflammatory gene expression

2.6 Oxidant-Dependent NF-kB-Induced Vascular Response

The nuclear factor NF-kB has long been known for its proinflammatory and redox regulation in the vascular endothelial cell [90, 91]. The phosphorylation and activation of NF-kB causes the binding of p50 and p65 subunits into a heterodimer, which is then translocated to the nucleus [92]. Stereoselective activation of NF-kB by one of the arachidonic acid metabolites induces propagation of a proangiogenic signal in the coronary endothelium (see section "Arachidonic Acid") [92]. The cytoplasmic localization of NF-kB is imperative for prompt signal transduction. One study has demonstrated that the activation of this pathway involves protein kinase C [92]. Furthermore, it was found that the levels of ROS were elevated, suggesting the involvement of ROS intermediates in this pathway (Fig. 2.9).

Different pathways involving NF-kB have been implicated in vascular endothelial dysfunction [93, 94]. However, scarcity of evidence exists in humans that implicates detrimental role of NF-kB in chronically impaired endothelial function with aging or obesity [93]. In an obese and middle-aged group, ROS levels were found to be elevated, which may contribute to endothelial dysfunction [93]. One plausible mechanism is the activation of NADPH oxidase via NF-kB, because the levels of ROS were attenuated after the administration of salsalate, an NF-kB inhibitor. In type 2 diabetes, downregulation of NF-kB yields a significant improvement in coronary vascular function [94]. This is achieved by decreasing the activity of mechanisms dependent on PARP-1, Sp-1, and COX2. It is well-known that NF-kB interacts with PARP-1 to form a complex, which then translocates to the nucleus and modulates gene expression. PARP-1 activity is reduced when the NF-kB pathway is inhibited, causing improvement in the vascular function. Furthermore, the p65 NF-kB subunit interacts with Sp-1, which negatively regulates eNOS promoter

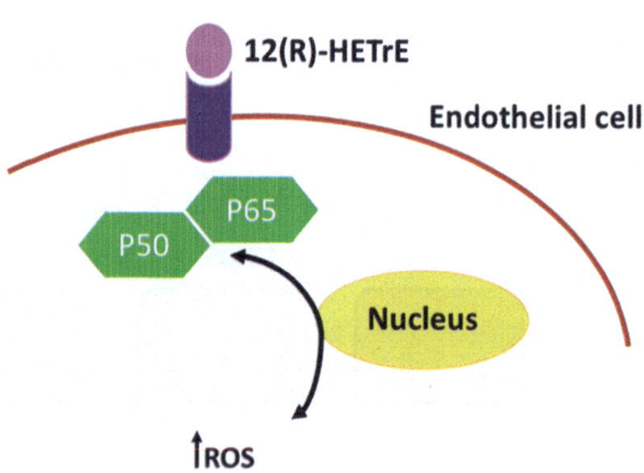

Fig. 2.9 *The role of NF-kB in the angiogenic phenotype of coronary endothelium*. The binding of the arachidonic acid metabolite 12(R)-hydroxyeicosatrienoic acid (12(R)-HETrE) causes the formation and activation of the p50-p65 heterodimer, the major form of NF-kB. This heterodimer then gets translocated to the nucleus, which causes an elevation in ROS levels. This pathway has been implicated to propagate the angiogenic response in the coronary microvascular endothelial cell

activity. Also, since NF-kB is considered a proinflammatory mediator, its inhibition resulted in improvement of vascular function via the inhibition of a downstream signaling molecule, COX2 [94].

2.7 Endothelium-Derived Vasoactive Molecules and Signaling Pathways

The endothelium plays a major role in controlling the blood flow through blood vessels. Regulation of the blood flow by the endothelium is achieved due to EC's sensitivity to various stimuli [94]. These stimulating factors can be mechanical in nature, such as hemodynamic changes. In addition, the endothelium can be activated by various molecules that are produced in different conditions. These molecules include acetylcholine; catecholamines; bradykinin; arachidonic acid derivatives, such as prostacyclin; and many others [93]. Subsequently, the endothelium translates the binding of those molecules into a wide variety of responses through the activation of multiple signaling pathways. Some of these signaling pathways were found to be influenced by the ROS, such as the activation of eNOS and NO production. H_2O_2 was also found to regulate vascular blood flow and affect the contractility of smooth muscle cells through various mechanisms [95].

There are other endothelial signaling pathways that have been recognized to work independently in the absence of ROS. These include prostacyclin signaling pathway and its vasodilatory effect and the production of endothelial-derived hyperpolarizing factors (EDHF), such as epoxyeicosatrienoic acids (EETs) and H_2S [96]. Moreover, the nervous system can affect the vasculature through the activation of muscarinic and adrenergic receptors that are expressed on endothelial cell.

2.7.1 eNOS-Derived NO-Dependent Signaling Pathways

Nitric oxide (NO) is a free radical that diffuses freely across cellular membranes. NO mediates several vascular effects including vasodilation, inhibition of platelet activation, inhibition of leukocyte adhesion to the endothelium, and others [95, 97]. The soluble fraction of guanylyl cyclase (sGC) constitutes the intracellular NO receptor that mediates signal transduction via the generation and elevation of intracellular cyclic guanosine monophosphate (cGMP) levels [95, 98]. cGMP activates cGMP-dependent protein kinase, also called protein kinase G (PKG), which in turn phosphorylates several proteins, such as the vasodilator-stimulated phosphoprotein (VASP) and phospholamban. Phospholamban is known to increase the reuptake of calcium by the endoplasmic reticulum in smooth muscle cells, facilitating muscle relaxation [95, 98, 99]. In addition, the cGMP/PKG signaling pathway results in the activation and opening of calcium-activated K^+ channels (K_{Ca}) and K_{ATP} channels on vascular smooth muscle cells [100, 101]. This indicates that part of NO-induced vasodilation is mediated through hyperpolarization (Fig. 2.10).

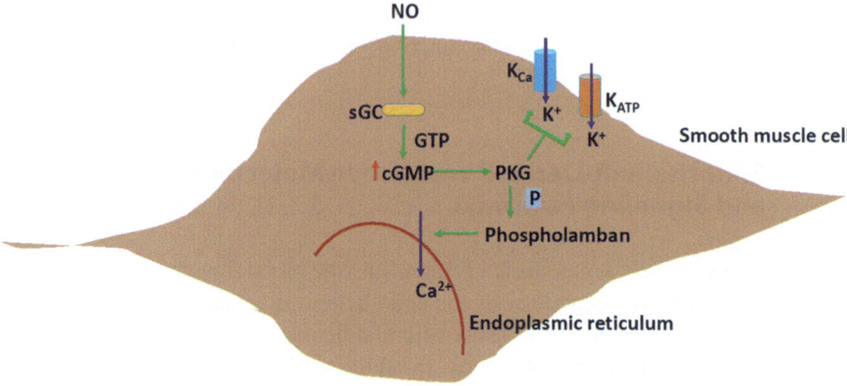

Fig. 2.10 *NO-mediated smooth muscle relaxation*. Endothelial cells produce the potent vasodilator molecule NO, which then diffuses to the underlying smooth muscle cells and activates the soluble fraction of GC enzyme. sGC catalyzes the conversion of GTP into cGMP, which in turn activates PKG. PKG phosphorylates several proteins, such as Phospholamban, which increases the uptake of Ca^{2+} ions by the endoplasmic reticulum that is necessary for smooth muscle contraction. In addition, PKG activates the opening of both K_{ca} and K_{ATP} channels, therefore inducing smooth muscle relaxation through hyperpolarization

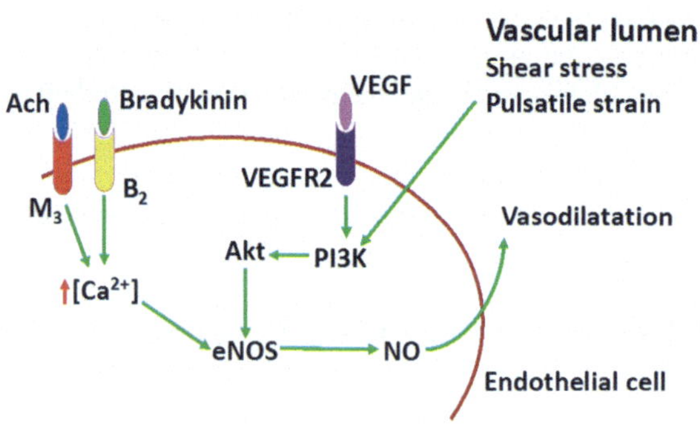

Fig. 2.11 *Mechanism of eNOS activation*. eNOS activation occurs in response to various stimuli, such as increased shear stress, pulsatile strain, and VEGF. Activation of eNOS involves increased calcium levels within the cytosol, which occurs in response to ACh or bradykinin activation. The main physiologic mechanism for NO production involves the phosphorylation of eNOS by Akt. PI3-kinase phosphorylates and activates Akt, which then activates eNOS by making it more sensitive to lower levels of calcium

NO is produced by an enzyme known as NO synthase (NOS). There are three major NOS isoforms, including neuronal NOS (nNOS), inducible NOS (iNOS), and endothelial NOS (eNOS) [95, 102, 103]. The NOS enzyme complex catalyzes the oxidation of nitrogen from L-arginine, forming NO and L-citrulline [95, 102, 104]. Activation of NO synthesis by eNOS is achieved by three main mechanisms. The first involves increased calcium levels within the cytosol, which occurs in response to ACh or bradykinin stimulation [95, 100, 103, 105] (Fig. 2.11). The main physiologic mechanism for NO production involves the phosphorylation of eNOS by protein kinase B (also called Akt). In this pathway, PI3-kinase phosphorylates and activates Akt, which in turn activates eNOS by making it more sensitive to lower levels of calcium [95, 102, 103, 106, 107]. The activation of eNOS by this mechanism occurs in response to various stimuli, such as increased shear stress, pulsatile strain, and VEGF [95, 100, 105–107] (Fig. 2.11). The sphingolipid pathway involved in cellular proliferation and migration also activates eNOS through the interaction between sphingosine-1-phosphate and Akt [95, 102, 103].

Activation of eNOS happens in the caveolae, which are invaginations of the plasma membrane that are rich in cholesterol and sphingolipids but poor in phospholipids [96, 102, 103]. eNOS is found anchored within these caveolae to a scaffolding protein called caveolin-1, which inhibits its activity [95, 108]. Stimuli, such as elevated calcium concentration and Akt-mediated phosphorylation of eNOS, attenuate the interaction between eNOS and caveolin-1, leading to the liberation of the eNOS enzyme from its inhibiting conformation. In fact, it was found that mice lacking the caveolin-1 gene demonstrate increased eNOS activity [95, 109].

There is a clear association between endothelial dysfunction and lack of eNOS activity. Mice that are eNOS-deficient have been found to be more prone to develop several cardiovascular diseases, such as hypertension and atherosclerosis [95, 110, 111]. Studies have shown that the increased coronary blood flow and shear stress in healthy exercising dogs induce a NO-mediated vasodilatory response that is most prominent in the larger coronary blood vessels [112, 113]. The administration of a NO synthase inhibitor blunted the flow-induced vasodilation of those larger vessels, but this effect was counterbalanced by the compensatory vasodilation of the smaller arterioles that decreased the resistance and maintained the blood flow [112]. Similarly, blocking NO production in the smaller coronary arterioles resulted in the activation of other endothelium-mediated vasodilatory pathways. These vasodilatory mechanisms include those that are activated by ACh and bradykinin, as well as the release of endothelium-derived vasodilating agents, such as prostacyclin and EDHF [112, 114–116].

2.7.2 Hydrogen Peroxide-Dependent Vasodilation

A study showed that mice that are deficient in eNOS were found to have an inhibited ACh-dependent vasorelaxation response to catalase [95, 117]. This suggests a role for hydrogen peroxide in inducing vasodilation. H_2O_2 induces smooth muscle relaxation by opening the K_{Ca} channels and hyperpolarizing the cell. In addition, H_2O_2 can mediate vasodilation by creating a disulfide bond in PKG, activating the enzyme in a way that is independent of guanylyl cyclase [95, 118]. Moreover, studies show that H_2O_2 appears to be involved in the autoregulation of blood flow within the small coronary blood vessels in response to ACh and decreased perfusion pressure [95, 119]. Furthermore, it is worth mentioning that H_2O_2 activates PKC, which in turn activates phospholipase A_2 to generate arachidonic acid which can eventually lead to the formation of EETs [95].

2.7.3 Oxidant-Independent Arachidonic Acid Metabolite-Induced Vascular Responses

The cytochrome P450-derived arachidonic acid metabolites, including different epoxyeicosatrienoic acids (EETs) and their corresponding dehydroxy-EETs, have been identified as EDHFs. This was evident in studies that showed that the inhibition of cytochrome P450 attenuated hyperpolarization-dependent vasodilation [95, 120]. Therefore, cytochrome P450, specifically the 2C9 isoform, is considered an EDHF synthase. In addition, cytochrome P450-2C9 generates ROS within coronary endothelial cells, which in turn activates the transcription factor NF-κB, resulting in increased expression of leukocyte adhesion molecules [95, 120]. In response to agonist stimulation from molecules, such as bradykinin, EETs are produced and activate gap junctions between the endothelial and smooth muscle cells, which facilitate their diffusion [95, 121]. Subsequently, EETs, just like other EDHFs, induce smooth muscle relaxation via opening K_{Ca} channels and decreasing intracellular calcium concentration [95, 120].

2.7.3.1 Prostacyclin

Prostaglandins were the first endothelium-derived relaxing factors to be discovered. They were originally identified as anticoagulation factors that are also capable of relaxing the smooth muscle layer in the vasculature [122, 123]. Prostaglandin production starts with the generation of arachidonic acid from membrane-bound phospholipids via activity of phospholipases [122]. Arachidonic acid is then metabolized by different enzymes, yielding various vasoactive molecules. For instance, lipoxygenases act on arachidonic acid, forming lipoxides which are mainly constrictor molecules. Epoxygenases, on the other hand, generate EDHF, which are vasodilatory molecules [122]. Lastly, cyclooxygenases (COX) metabolize arachidonic acid into prostaglandin H_2. This is then acted upon by various synthase enzymes to form prostacyclin, prostaglandin E and F, and thromboxane [122]. There are two major COX isoforms: COX-1, which is produced constitutively, and COX-2, which is produced under specific conditions, such as inflammation. In inflammation, platelets release thromboxane molecules which act on the endothelium and induce COX-2 expression. Consequently, prostacyclin production is increased and accounts for the vasodilatory effect in inflammation [122, 124] (Fig. 2.14). Prostacyclin is also capable of suppressing the release of constrictor molecules from the nerve endings within the vascular wall [122, 125].

Like NO, prostacyclin is a lipid-soluble molecule that diffuses out of the endothelial cell shortly after being synthesized and acts as an anticoagulant and vasodilator. Prostacyclin signal transduction is mediated by activation of the adenylyl cyclase/cAMP system, which removes calcium from the cytoplasm and causes smooth muscle relaxation [100, 122, 126] (Figs. 2.12 and 2.13). In addition, prostacyclin can also induce vasodilation through smooth muscle hyperpolarization [114, 122, 125, 127]. Several studies have investigated the role of hyperpolarization induced by prostacyclin on smooth muscle relaxation [122, 128–131] (Fig. 2.13). A study was conducted in which glibenclamide, an ATP-sensitive K^+ channel blocker,

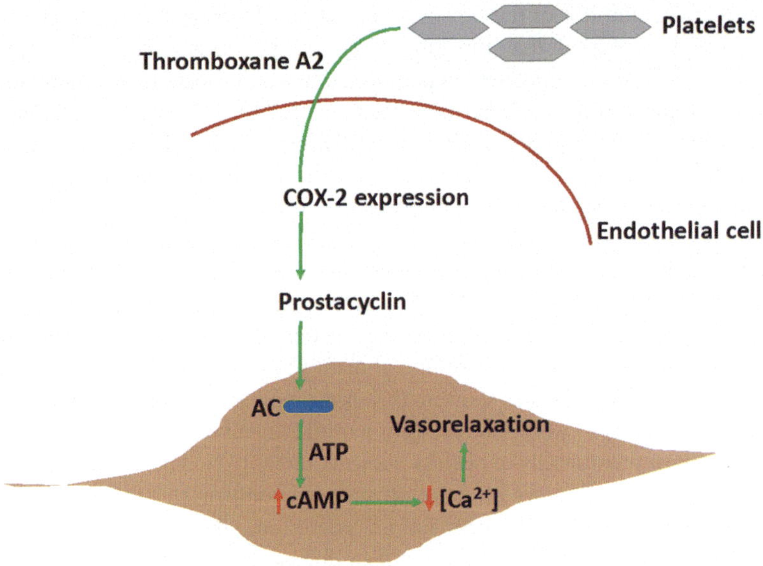

Fig. 2.12 *Platelet-induced vasodilation in inflammation*. Platelet activation during inflammation results in the release of Thromboxane A2, which increases COX-2 expression by endothelial cells. COX-2 overexpression results in the formation of Prostacyclin, which in turn diffuses to the underlying smooth muscle cells and activates AC enzyme. AC catalyzes the conversion of ATP into cAMP, which removes Ca ions from the smooth muscle cell cytoplasm, therefore inducing muscle relaxation

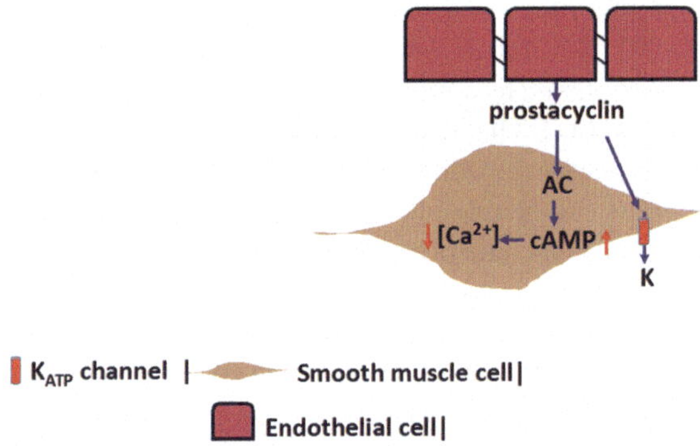

Fig. 2.13 *Mechanisms by which prostacyclin induces smooth muscle relaxation*. Prostacyclin signal transduction is mediated through activation of the adenylyl cyclase/cAMP system, which removes calcium from the cytoplasm and causes smooth muscle contraction. Prostacyclin can also induce vasodilation through smooth muscle hyperpolarization that occurs in response to opening K_{ATP} channels

was used as an inhibitor of the prostacyclin-mediated hyperpolarization and relaxation. The study found that blocking K_{ATP} channels attenuated indomethacin-sensitive, endothelium-dependent hyperpolarization and substantially decreased the relaxation of the coronary arteries of guinea pigs [122, 131]. These results indicate that hyperpolarization plays a role in mediating prostacyclin-induced vascular relaxation, in conjunction with other mechanisms, such as reducing the cytoplasmic calcium concentration and the sensitivity of the contractile apparatus to calcium.

NO deficiency due to pharmacological blockade or genetic knockout of NOS was found to be associated with marked elevation of resting blood pressure, reaching up to a 50 mmHg increase [132, 133]. However, blockade of COX enzyme and subsequent prostacyclin production by indomethacin did not affect resting blood pressure in humans [134]. Furthermore, clinical trials that were done with patients with coronary artery disease showed that prostanoids contribute significantly to vasodilation [100, 134, 135]. Interestingly, a third clinical trial failed to demonstrate a critical role of prostanoids in healthy subjects [100, 137]. These findings support the idea that the vasodilatory effect of prostanoids compensates for NO deficiency in disease conditions, while, under normal conditions, their effect is minimal. In fact, studies have shown that inhibition of cyclooxygenase and prostanoid production with indomethacin had minimal effect on the coronary blood flow in healthy dogs under resting conditions as well as during exercise [100, 137]. Upon coadministration of the eNOS inhibitor L-NAME and indomethacin, the period of reactive hyperemia was shortened, suggesting a possible interaction between NO and prostanoids [100, 138].

As stated earlier, prostacyclin plays an important role in inducing vasodilation and preventing the adhesion of platelets and erythrocytes to the vascular wall, making its role of great importance in diseases that involve thrombus formation, such as atherosclerosis [122]. Several diseases, including atherosclerosis, hypertension, diabetes mellitus, and others, have been found to have a greater tendency for thrombus formation [7, 139–141]. The treatment of these diseases may involve drugs that block prostanoid synthesis including prostacyclin. Suppressing prostacyclin production by such drugs can increase the propensity of thrombus formation. In fact, many treatment regimens include prostacyclin analogues to compensate for prostacyclin underproduction. Alternative treatment approaches should be more specific and target only the harmful prostaglandins and spare prostacyclin. Such drugs could target specific receptors or enzymes that are involved in the signaling or production of undesired molecules [122].

2.7.3.2 Endothelium-Derived Hyperpolarization Factors

There exist variable degrees of vasodilation even after NO and prostacyclin inhibition, suggesting the existence of other endothelial-derived molecules that exert a vasodilatory effect [95, 142]. The fact that this vasodilation response is accompanied by hyperpolarization of the endothelial and smooth muscle cells and is inhibited by blockade of calcium-activated potassium channels (K_{Ca}) or high extracellular K^+ concentration indicates that these molecules mediate their effect through

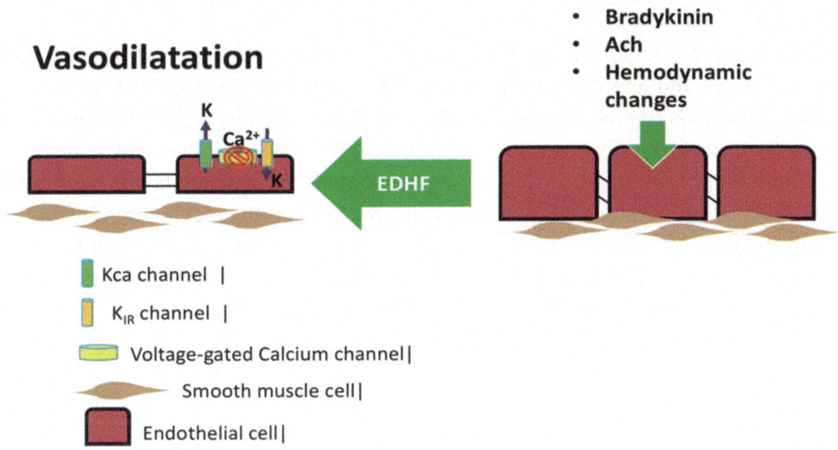

Fig. 2.14 *EDHF-mediated smooth muscle relaxation*. EDHF release induces vasodilation through endothelial cell and smooth muscle cell hyperpolarization. The exposure of the endothelial cell to several vasodilator agonists, such as bradykinin, ACh, or shear stress, leads to the formation and release of EDHF, which then moves through the gap junction to the underlying smooth muscle cells. This results in the opening of Ca-activated K^+ channels and ATP-sensitive K^+ channels, leading to hyperpolarization and smooth muscle relaxation

hyperpolarization [95, 142]. These molecules are called endothelial-derived hyperpolarizing factors (EDHF). Their vasodilatory action is mediated through K^+ efflux due to the opening of K_{Ca} and rectifying K^+ channels (K_{IR}), which prevents the activation of voltage-gated calcium channels, therefore decreasing calcium concentration within the smooth muscle cells and thus reducing muscle contraction [95, 142] (Fig. 2.14). Although the nature of these molecules is not fully understood, several EDHFs have been identified, including arachidonic acid metabolites, hydrogen peroxide, and K^+ ions acting within gap junctions between endothelial and smooth muscle cells [95, 119–121, 143].

EDHF dysfunction has been documented in different diseases affecting the vasculature, such as diabetes mellitus and hypertension [95, 142]. In these conditions, EDHF lose their ability to hyperpolarize and relax the smooth muscle cells, leading to impairment of their vasodilatory influence.

2.7.3.2.1 K^+ Ions and Gap Junctions

Certain molecules, such as Ach, induce endothelial K_{Ca} channel opening leading to the efflux of K^+ ions. These ions activate the Na-K ATPase pump in smooth muscle cells and K_{IR} channels in endothelial cells, resulting in vascular relaxation that occurs through inhibitory mechanisms [95, 121, 142]. As stated earlier, the gap junctions that exist between endothelial and smooth muscle cells contribute to the endothelium-mediated vasodilation via EDHF, allowing electric coupling between the two cell types, as well as the transmission of different chemical mediators [95, 143].

2.7.3.2.2 H_2S

Recent evidence demonstrated that H_2S acts as endothelial-dependent relaxation factor (EDRF) and as EDHF [95]. As an EDRF, H_2S modifies cysteine residues on the K_{ATP} channels of smooth muscle cells in a process called "sulfhydration," leading to their dissociation from their inhibiting ATP molecules and enhancing their binding with phosphatidylinositol 4,5-biphosphate (PIP2), which favors opening of these channels. Opening of K_{ATP} channels results in K^+ efflux and hyperpolarization, which induces vasodilation [95, 144, 145]. Inhibition of H_2S production has been found to decrease ACh-induced endothelial-dependent relaxation, suggesting that H_2S plays a role in mediating this vasodilatory effect [95, 146]. H_2S acts as an EDHF via activating the K^+-dependent calcium channels (SK_{Ca} and IK_{Ca}) on endothelial cells, causing endothelial hyperpolarization and smooth muscle relaxation via the electric coupling transmitted by gap junctions [95, 147]. Moreover, H_2S induces eNOS expression, which increases NO production [95, 148].

2.7.3.2.3 Endothelins

Endothelins are vasoactive peptides produced by vascular endothelial cells. There are three endothelin isopeptides, ET-1, ET-2, and ET-3, of which ET-1 is the most abundant and most active within the coronary vasculature [100, 149, 150]. It is produced by the endothelium in its inactive form known as Big ET-1 or Pro ET-1, which is then acted upon by endothelin converting enzyme (ECE) located on the plasma membrane of endothelial cells, transforming into the active form [100, 149, 152]. ET release is stimulated by many factors acting on the vascular endothelium including bradykinin, high- and low-density lipoproteins, angiotensin II, ischemia, shear stress, and several growth factors [150, 151].

There are two major types of ET-1 receptors, ET_A and ET_B. Both receptors are expressed by vascular smooth muscle cells, while ET_B is the only type expressed in the endothelium [100, 150, 151–153]. Acting on the vascular smooth muscle cells, ET induces vasoconstriction through ET_A and ET_B receptor activation [99, 150] (Fig. 2.9). However, the activation of ET_B receptor on endothelial cells triggers the release of several vasodilators, most importantly NO and prostacyclin, which attenuate the vasoconstrictor influence of ET and even inhibit its production in a negative feedback loop [99, 152, 154] (Fig. 2.9). The vasoconstrictive effect of ET is attenuated during exercise due to the activity of NO and prostacyclin [100, 155]. A study showed that upon administration of Tezosentan, a nonselective ET receptor antagonist, along with inhibition of NOS or COX, the vasodilatory effect of Tezosentan was maintained during exercise [100, 155]. This finding demonstrates that the NO and prostacyclin vasodilatory systems work in coordination with each other to demolish the vasoconstrictor effect of ET during exercise. This also explains why under certain conditions in which NO production is impaired, such as atherosclerosis and endothelial dysfunction, ET-induced vasoconstriction is maximized and contributes to the pathogenesis of these diseases.

The intramyocardial blood vessels have been found to be very sensitive to ET vasoconstrictor influence, and many cardiovascular diseases, such as myocardial infarction, coronary artery disease, heart failure, and others, are associated with increased plasma levels of ET [8]. In order to investigate the effectiveness of blocking ET-induced vasoconstriction in patients with stable angina, a study used Bosentan, a nonselective ET receptor antagonist, as a blocking agent [100, 156]. It was found that Bosentan attenuated the vasoconstrictive influence of ET and increased the vascular diameter. However, this stimulated a simultaneous 10% drop in aortic blood pressure, which caused the coronary blood flow to remain unchanged despite the vasodilatory effect of Bosentan [100, 156].

2.8 Vascular Effects of Oxidant-Independent Nervous System-Derived Molecules

One of the ways by which the nervous system regulates the vascular tone is through the activation of muscarinic and adrenergic receptors that are located on endothelial cells. The parasympathetic system mediates its vascular effects through ACh, while the sympathetic system acts through the release of catecholamines. In this section, we explore some of effects and signaling pathways that are activated by the two systems.

2.8.1 Acetylcholine

The coronary resistance vessels are highly innervated by parasympathetic nerve endings [99]. The effects of vagal stimulation, mediated by acetylcholine (ACh), on coronary vascular tone and blood flow vary among different species. In dogs, the parasympathetic stimulation of the coronary blood vessels results in a vasodilatory response mediated by the release of NO [100, 157, 158]. This effect was found to be blocked by administration of the anticholinergic atropine, indicating that this vasodilatory response was cholinergic in origin [100, 157, 158]. During exercise, the administration of atropine does not affect heart rate or blood flow, indicating that the parasympathetic innervation to the heart during exercise is negligible [100].

In contrast, parasympathetic innervation of the smooth muscle layer of the coronary blood vessels of pigs masks the endothelial vasodilatory effect of ACh, which results in a net vasoconstriction effect [100, 159, 160]. A study has shown that administration of atropine results in a vasodilatory response at rest, but this response was attenuated by increasing levels of exercise since the vagal influence is minimal during exercise [100, 161, 162].

When blood circulates, shear stress is generated on blood vessel walls. In response, the endothelium relaxes the smooth muscle and allows for an increase in arterial diameter via flow-mediated dilatation, in which vasodilators like ACh are released [163]. ACh activates calcium release from internal stores in endothelial cells and stimulates NO production [163]. The endothelium expresses choline acetyltransferase (ChAT), which is responsible for the synthesis of ACh; vesicular acetylcholine transporter (VAChT), which loads ACh into vesicles for secretion; and acetylcholinesterase (AChE), which catalyzes hydrolysis of ACh [163]. Endothelial

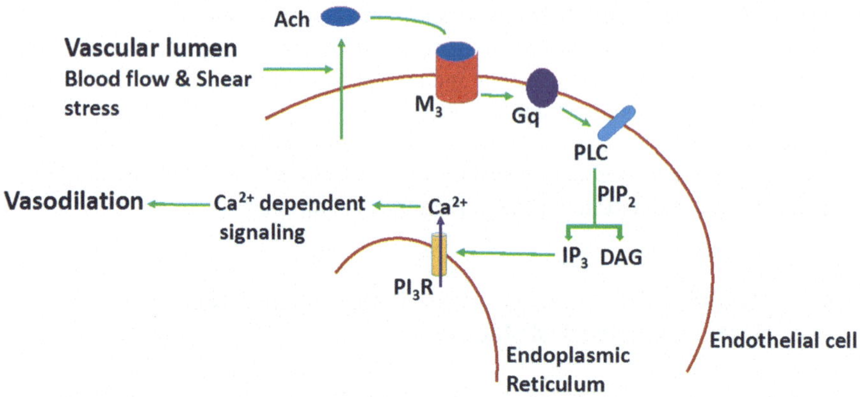

Fig. 2.15 *Mechanical stimulation of endothelial ACh release induces a Ca^{2+}-dependent vasodilatory effect.* Blood flow stimulates endothelial ACh release, which in turn activates M3 cholinergic receptors on endothelial cells. M3 receptor activation results in the formation of IP3 and DAG from PIP2 by the action of PLC. IP3 induces Ca^{2+} release from the endoplasmic reticulum, which results in the activation of Ca^{2+}-dependent signaling pathways that promote vasodilation

organic cation transporters release ACh upon mechanical activation from shear stress [163]. The non-vesicular release of ACh requires production of acetyl Co-A in the mitochondrial matrix, so that the acetyl group can be transferred to choline to produce ACh [163]. Flow activates a predominant M_3 subtype muscarinic receptor/phospholipase C (PLC)/inositol 1,4,5-trisphosphate receptor (IP_3R) signaling cascade [149, 163]. ACh serves as the agonist, activating the muscarinic receptor. PLC then cleaves phosphatidylinositol 4,5-bisphosphate (PIP_2) into diacylglycerol (DAG), and IP_3. IP_3 binds IP_3R, a membrane glycoprotein complex that serves as a calcium ion channel (Fig. 2.15).

Results from other studies show that local ACh release may regulate endothelial cell migration and contribute to angiogenesis [149, 163, 164]. HUVECs exhibit a nonneuronal cholinergic autocrine system [164]. Cholinergic stimulation of muscarinic receptors was shown to regulate angiogenesis and increase tube length as well as complexity [149, 164]. A study by Cooke et al. indicates that there is an endogenous pathway, activated by endogenous acetylcholine or exogenous nicotine, for angiogenesis in response to oxidative stress [164]. Regarding endothelial nicotinic acetylcholine receptors (nAChRs), the selective α7-nAChR antagonist α-bungarotoxin was demonstrated to inhibit endothelial tube formation [164]. In addition, α7-nAChR is upregulated by hypoxia in vitro in endothelial cells experiencing ischemia [165]. Notably, there seems to be interconnection among pathways to angiogenesis modulated by growth factors and pathways mediated by endothelial nAChRs [149, 164]. VEGF-induced endothelial cell migration actually involves nAChR activation, and, in cases where nAChR antagonists are introduced, growth factor-induced endothelial cell migration is also reduced [164].

2.8.2 α_2 Adrenergic Receptors

α_1 Adrenergic receptors are considered the primary adrenergic receptor subtype in mediating the vasoconstrictor effect of catecholamines, though α_2 receptors also appear to be involved [100, 112, 165]. However, the coronary endothelium has been found to harbor α_2 adrenoceptors that induce the release of prostacyclin and NO, resulting in vasodilation [96, 100, 112]. The vasoconstrictor influence of the α_2 receptors on normal coronary blood vessels at rest and even during exercise appears to generally be negligible, rather than being masked by its endothelium-mediated vasodilatory effect [112].

Ishibashi et al. have proposed that myocardial hypoperfusion, due to arterial stenosis, for instance, renders the coronary vessels more susceptible to vasoconstriction during exercise [112]. Their study showed that administration of the NO synthase inhibitor N^{ω}-nitro-L-arginine (LNNA) to exercising dogs with coronary artery stenosis enhanced the previously negligible vasoconstrictor effect mediated by α_2 receptors [112]. The researchers arrived at this conclusion despite the fact that co-inhibition of the α_2 receptors did not return the coronary blood flow back to normal. This finding can potentially be explained by multiple mechanisms, one being that the sympathetic stimulation during exercise might have resulted in the activation of α_1 receptors located on the vascular smooth muscle cells. Since many patients with cardiovascular diseases, such as atherosclerosis and hypertension, have impaired endothelial NO production, based on these findings, these patients are more vulnerable to suffer from myocardial hypoperfusion during exercise due to the enhanced α_2 receptor-mediated vasoconstrictor effect [112].

2.9 Implications in Pathophysiology and Future Research

Endothelial dysfunction is a marker for cardiovascular disease. Endothelial cells induce vasorelaxation by releasing endothelium-derived relaxing factors (EDRFs), such as NO [166]. Upregulating or inhibiting the endothelial cells' ability to release NO leads to endothelial dysfunction and consequently develops into atherosclerosis and coronary disease [166]. In the same vein, the production of the endothelium-derived contractile factors – and subsequent increase in reactive oxygen species – contributes to aging, diabetes, and cardiovascular disease [166].

NO notably prevents abnormal vasoconstriction by controlling the release of endothelin-1 [166]. Without NO fulfilling a protective role, an inflammatory response is induced, leading to atherosclerosis. Endothelial cells upregulate the production of NO by eNOS, an endothelium-specific enzyme isoform that synthesizes nitric oxide [166]. Indeed, experiments have shown that, when the endothelium is removed or dysfunctional, relaxation of coronary arteries no longer occurs, and, instead, vasospasms are observed [166]. NO is able to inhibit EDHF activity by inhibiting EDHF synthase, cytochrome P450. Therefore, conditions that involve endothelial dysfunction and NO depletion, such as atherosclerosis and other cardiovascular diseases, increase the EDHF-dependent vasodilator mechanisms [167]. As such, the importance and contribution of these molecules in regulating blood flow become more prominent. Depending on the arterial size, different endothelial-derived mediators regulate the vascular tone [167]. For instance, in large conductance arteries, such as the aorta and the epicardial arteries, NO is the major contributor to vasodilation. In small arterioles, EDHF predominates along with some other mechanisms, such as NO and adenosine [167].

VEGF inhibitors can lead to chronic hypertension as a side effect, perhaps due to upregulation of endothelin-1 production (as opposed to regulating NO release), and can reduce response to acetylcholine. [166] Hypoxia also increases endothelin-1 synthesis gene expression and synthesis [166]. As a powerful vasoconstrictor, endothelin-1 greatly impacts homeostatic water and sodium levels in the kidneys and increases vascular ROS levels [166]. Thus, endothelin-1 overproduction is linked to endothelial dysfunction and pulmonary hypertension [166, 168]. Aside from increasing oxidative stress, endothelin-1 has also been shown to interfere with lipid metabolism and the NO pathway [166]. Further research on endothelin antagonists could help treat hypertension, high arterial blood pressure, and complications related to chronic kidney disease [166]. There is still uncertainty over short-term versus long-term effects of endothelin-1 antagonism, though long-term antagonism has shown improvement in endothelial function in patients with atherosclerosis [166, 169].

Recently, vasoconstricting prostanoids like COX-1 and COX-2 have been shown to exacerbate endothelium-dependent contractions by activating TP receptors in smooth muscle cells [166, 170]. Future research on antagonists of TP receptors could be done to develop therapy for cardiovascular disorders. A 2015 study found chronic fluoxetine treatment to induce endothelial dysfunction and increased blood pressure through generation of COX-derived prostanoids [168]. These findings warrant further research into the risk fluoxetine treatment poses for aggravating vascular injury.

Finally, miRNA, noncoding RNA that regulate gene expression, are a relatively recent scientific discovery. Studying how miRNA affect the activity or concentrations of the aforementioned EDRFs would illuminate current understanding of eNOS expression. Specifically, miR-155 appears to decrease expression of eNOS [166].

2.10 Conclusion

The findings presented in this chapter suggest the effects of ROS on the vasculature appear to be dependent on the subcellular location, concentration, and duration of ROS exposure. Different isoforms of NADPH oxidase (NOX) enzymes appear to be major contributors of ROS in EC. The four isoforms of NOX are all found in the endothelium at specific subcellular location, and specific activity of each isoforms contributes to physiological and pathological changes in the vasculature. NOX1 plays a critical role in pathogenesis of Ang- II-induced hypertension and possibly in atherosclerosis, and its selective inhibition may mitigate development of atherosclerotic plaques. Endothelial NOX2 activates either the PI3K-Akt-eNOS axis or AMPK-eNOS axis, leading to an increase in NO and subsequent vasodilation, endothelial cell proliferation, and migration. NOX4 essentially generates hydrogen peroxide and is expressed abundantly in the endothelium. NOX4-derived hydrogen peroxide has shown beneficial effects, including inducing vascular angiogenesis. NOX5 is localized in the endoplasmic reticulum and is activated upon increase in intracellular calcium levels. The NOX5 variants also contribute to endothelial ROS production, cell proliferation, and angiogenesis.

The mitochondria are well-known for energy production through oxidative phosphorylation, but they are also integral to ROS formation, signaling cellular death, calcium regulation, and ACh mechanotransduction. Short-term exposure of elevated cytosolic ROS increases ROS production by the mitochondria, counterbalanced by SOD2 activity. Long-term exposure to elevated cytosolic ROS, however, proves to be more detrimental and can lead to impairment in endothelial vasodilation as well as cause conversion of NO to peroxynitrite.

VEGF is critical in inducing angiogenesis. While VEGF-induced vasodilation can take place in a ROS-dependent manner via PI3K-Akt-eNOS pathway, the VEGF signaling pathway can also activate PLC γ-ERK1/2 in an ROS-independent manner. VEGF interacts with proinflammatory factors in a positive feedback loop. Inflammatory stimuli activate TLRs which in turn promote HIF-1α expression, resulting in activation of the VEGF promoter.

Various vasodilating agents respond to shear stress and other stimuli by altering vascular tone and often promoting smooth muscle relaxation. VEGF-induced signaling involves some of these factors, notably ACh. By contrast, Ang-II exaggerates vasocontraction via NO reduction. As was the case with NOX1, Ang-II is involved in the pathogenesis of hypertension. Factors like NO, prostacyclin, endothelin, and EDHF share similar stimuli leading to their production and are interrelated in inducing vasorelaxation.

The overwhelming interconnectedness of NOX, growth factors, and vasoactive factors as well as larger-scale oxidant-dependent and oxidant-independent pathways demonstrate the complexity of signaling in coronary vascular endothelium.

Acknowledgment This work was supported in part by the National Heart, Lung, and Blood Institute (NHLBI) and National Institute of General Medical Sciences (NIGMS) grant R01HL133624-01A1 (MRA), HL46716 and R01 128831 (FWS), and American Heart Association Grant-in-Aid 14GRNT20460291 (MRA).

References

1. Rodríguez I, González M (2014) Physiological mechanisms of vascular response induced by shear stress and effect of exercise in systemic and placental circulation. Front Pharmacol 5:209. https://doi.org/10.3389/fphar.2014.00209
2. Mukhopadhyay D, Zeng H, Bhattacharya R (2004) Complexity in the vascular permeability factor/vascular endothelial growth factor (VPF/VEGF)-receptors signaling. Mol Cell Biochem 264(1/2):51–61. https://doi.org/10.1023/b:mcbi.0000044374.85095.df
3. Folkman J (2006) Angiogenesis. Annu Rev Med 57(1):1–18. https://doi.org/10.1146/annurev.med.57.121304.131306

4. Benezra M, Vlodavsky I, Ishaimichaeli R, Neufeld G, Barshavit R (1993) Thrombin-induced release of active basic fibroblast growth factor-heparan sulfate complexes from subendothelial extracellular matrix. Blood 81(12):3324–3331
5. Risau W (1997) Mechanisms of angiogenesis. Nature 386:671. https://doi.org/10.1038/386671a0
6. Bedard K, Krause K (2007) The NOX family of ROS-generating NADPH oxidases: physiology and pathophysiology. Physiol Rev 87(1):245–313. https://doi.org/10.1152/physrev.00044.2005
7. Feng J, Damrauer SM, Lee M, Sellke FW, Ferran C, Abid MR (2010) Endothelium-dependent coronary vasodilatation requires NADPH oxidase-derived ROS. Arterioscler Thromb Vasc Biol 30(9):1703–1710. https://doi.org/10.1161/ATVBAHA.110.209726
8. Aldosari S, Awad M, Harrington E, Sellke F, Abid M (2018) Subcellular reactive oxygen species (ROS) in cardiovascular pathophysiology. Antioxidants 7(2):14
9. Frey RS, Ushio-Fukai M, Malik AB (2009) NADPH oxidase-dependent signaling in endothelial cells: role in physiology and pathophysiology. Antioxid Redox Signal 11(4):791–810. https://doi.org/10.1089/ars.2008.2220
10. Cholan PM, Cartland S, Kavurma M (2017) NADPH oxidases, angiogenesis, and peripheral artery disease. Antioxidants 6(3):56. https://doi.org/10.3390/antiox6030056
11. Guzik TJ, Chen W, Gongora MC, Guzik B, Lob HE, Mangalat D, Hoch N, Dikalov S, Rudzinski P, Kapelak B, Sadowski J, Harrison DG (2008) Calcium dependent Nox5 NADPH oxidase contributes to vascular oxidative stress in human coronary artery disease. J Am Coll Cardiol 52(22):1803–1809. https://doi.org/10.1016/j.jacc.2008.07.063
12. Dikalova AE, Góngora MC, Harrison DG, Lambeth JD, Dikalov S, Griendling KK (2010) Upregulation of Nox1 in vascular smooth muscle leads to impaired endothelium-dependent relaxation via eNOS uncoupling. Am J Physiol Heart Circ Physiol 299(3). https://doi.org/10.1152/ajpheart.00242.2010
13. Bengtsson SH, Gulluyan LM, Dusting GJ, Drummond GR (2003) Novel isoforms of NADPH oxidase in vascular physiology and pathophysiology. Clin Exp Pharmacol Physiol 30(11):849–854. https://doi.org/10.1046/j.1440-1681.2003.03929.x
14. Gavazzi G, Deffert C, Trocme C, Schäppi M, Herrmann FR, Krause K (2007) NOX1 deficiency protects from aortic dissection in response to angiotensin II. Hypertension 50(1):189–196. https://doi.org/10.1161/hypertensionaha.107.089706
15. Mollnau H, Wendt M, Szöcs K, Lassègue B, Schulz E, Oelze M, Li H, Bodenschatz M, August M, Kleschyov AL, Tsilimingas N, Walter U, Förstermann U, Meinertz T, Griendling K, Münzel T (2002) Effects of angiotensin II infusion on the expression and function of NAD(P)H oxidase and components of nitric oxide/cGMP signaling. Circ Res 90(4). https://doi.org/10.1161/01.RES.0000012569.55432.02
16. Gavazzi G, Banfi B, Deffert C, Fiette L, Schappi M, Herrmann F, Krause KH (2006) Decreased blood pressure in NOX1-deficient mice. FEBS Lett 580(2):497–504. https://doi.org/10.1016/j.febslet.2005.12.049
17. Sadoshima J, Izumo S (1993) Molecular characterization of angiotensin II--induced hypertrophy of cardiac myocytes and hyperplasia of cardiac fibroblasts. Critical role of the AT1 receptor subtype. Circ Res 73(3):413–423. https://doi.org/10.1161/01.RES.73.3.413
18. Lee MY, Martin AS, Mehta PK, Dikalova AE, Garrido AM, Datla SR, Lyons E, Krause KH, Banfi B, Lambeth JD, Lassègue B, Griendling KK (2009) Mechanisms of vascular smooth muscle NADPH oxidase 1 (Nox1) contribution to injury-induced neointimal formation. Arterioscler Thromb Vasc Biol 29(4):480–487. https://doi.org/10.1161/ATVBAHA.108.181925
19. Awad MA, Aldosari SR, Abid MR (2018) Genetic alterations in oxidant and anti-oxidant enzymes in the vascular system. Front Cardiovasc Med 5. https://doi.org/10.3389/fcvm.2018.00107
20. Szöcs K, Lassègue B, Sorescu D, Hilenski LL, Valppu L, Couse TL, Wilcox JN, Quinn MT, Lambeth JD, Griendling KK (2002) Upregulation of nox-based NAD(P)H oxidases in restenosis after carotid injury. Arterioscler Thromb Vasc Biol 22(1):21–27. https://doi.org/10.1161/hq0102.102189

21. Schröder K, Helmcke I, Palfi K, Krause KH, Busse R, Brandes RP (2007) Nox1 mediates basic fibroblast growth factor-induced migration of vascular smooth muscle cells. Arterioscler Thromb Vasc Biol 27(8):1736–1743. https://doi.org/10.1161/atvbaha.107.142117
22. Sheehan AL, Carrell S, Johnson B, Stanic B, Banfi B, Miller FJ (2011) Role for Nox1 NADPH oxidase in atherosclerosis. Atherosclerosis 216(2):321–326. https://doi.org/10.1016/j.atherosclerosis.2011.02.028
23. Sleight P (2000) The HOPE Study (Heart Outcomes Prevention Evaluation). J Renin-Angiotensin-Aldosterone Syst 1(1):18–20. https://doi.org/10.3317/jraas.2000.002
24. Abid MR, Sellke FW (2015) Antioxidant therapy: is it your gateway to improved cardiovascular health? Pharm Anal Acta 6(1):323. https://doi.org/10.4172/2153-2435.1000323
25. Abid M, Kachra Z, Spokes KC, Aird WC (2000) NADPH oxidase activity is required for endothelial cell proliferation and migration. FEBS Lett 486(3):252–256. https://doi.org/10.1016/s0014-5793(00)02305-x
26. Shafique E, Torina A, Reichert K, Colantuono B, Nur N, Zeeshan K, Ravichandran V, Liu Y, Feng J, Benjamin L, Irani K, Harrington OE, Sellke WF, Abid MR (2017) Mitochondrial redox plays a critical role in the paradoxical effects of NAPDH oxidase-derived ROS on coronary endothelium. Cardiovasc Res 113(2):234–246. https://doi.org/10.1093/cvr/cvw249
27. Shafique E, Choy WC, Liu Y, Feng J, Cordeiro B, Lyra A, Arafah M, Yassin-Kassab A, Zanetti A, Clements R, Bianchi C, Benjamin EL, Sellke WF, Abid MR (2013) Oxidative stress improves coronary endothelial function through activation of the pro-survival kinase AMPK. Aging 5(7):515–530. https://doi.org/10.18632/aging.100569
28. Kluge MA, Fetterman JL, Vita JA (2013) Mitochondria and endothelial function. Circ Res 112(8):1171–1188. https://doi.org/10.1161/CIRCRESAHA.111.300233
29. Kim Y-M, Kim S-J, Tatsunami R, Yamamura H, Fukai T, Ushio-Fukai M (2017) ROS-induced ROS release orchestrated by Nox4, Nox2, and mitochondria in VEGF signaling and angiogenesis. Am J Physiol Cell Physiol 312(6):C749–C764. https://doi.org/10.1152/ajpcell.00346.2016
30. Bendall JK, Rinze R, Adlam D, Tatham AL, Bono JD, Channon KM (2007) Endothelial Nox2 overexpression potentiates vascular oxidative stress and hemodynamic response to angiotensin II. Circ Res 100(7):1016–1025. https://doi.org/10.1161/01.res.0000263381.83835.7b
31. Chen L, Hou X, Xiao J, Kuroda J, Ago T, Sadoshima J, Cohen R, Tong X (2014) Both hydrogen peroxide and transforming growth factor beta 1 contribute to endothelial Nox4 mediated angiogenesis in endothelial Nox4 transgenic mouse lines. Biochim Biophys Acta (BBA) – Mol Basis Dis 1842(12):2489–2499. https://doi.org/10.1016/j.bbadis.2014.10.007
32. Datla SR, Peshavariya H, Dusting GJ, Mahadev K, Goldstein BJ, Jiang F (2007) Important role of Nox4 type NADPH oxidase in angiogenic responses in human microvascular endothelial cells in vitro. Arterioscler Thromb Vasc Biol 27(11):2319–2324. https://doi.org/10.1161/atvbaha.107.149450
33. Ray R, Murdoch CE, Wang M, Santos CX, Zhang M, Alom-Ruiz SS, Anilkumar N, Ouattara A, Cave A, Walker S, Grieve JD, Charles LR, Eaton P, Brewer CA, Shah AM (2011) Endothelial Nox4 NADPH oxidase enhances vasodilatation and reduces blood pressure in vivo. Arterioscler Thromb Vasc Biol 31(6):1368–1376. https://doi.org/10.1161/atvbaha.110.219238
34. Kuroda J, Nakagawa K, Yamasaki T, Nakamura K, Takeya R, Kuribayashi F, Imajoh-Ohmi S, Igarashi K, Shibata Y, Sueishi K, Sumimoto H (2005) The superoxide-producing NAD(P)H oxidase Nox4 in the nucleus of human vascular endothelial cells. Genes Cells 10(12):1139–1151. https://doi.org/10.1111/j.1365-2443.2005.00907.x
35. Craige SM, Kai C, Pei Y, Chunying L, Xiaoyun H, Christine C, Shibata R, Sato K, Walsh K, Keaney JF (2011) NADPH oxidase 4 promotes endothelial angiogenesis through eNOS activation. Circulation 124(6):731–740. https://doi.org/10.1161/CIRCULATIONAHA.111.030775
36. Shimokawa H, Morikawa K (2005) Hydrogen peroxide is an endothelium-derived hyperpolarizing factor in animals and humans. J Mol Cell Cardiol 39(5):725–732. https://doi.org/10.1016/j.yjmcc.2005.07.007

37. Matoba T, Shimokawa H, Nakashima M, Hirakawa Y, Mukai Y, Hirano K, Kanaide H, Takeshita A (2000) Hydrogen peroxide is an endothelium-derived hyperpolarizing factor in mice. J Clin Investig 106(12):1521–1530. https://doi.org/10.1172/jci10506
38. Miura H, Bosnjak JJ, Ning G, Saito T, Miura M, Gutterman DD (2003) Role for hydrogen peroxide in flow-induced dilation of human coronary arterioles. Circ Res 92(2). https://doi.org/10.1161/01.res.0000054200.44505.ab
39. Belaiba R, Djordjevic T, Petry A, Diemer K, Bonello S, Banfi B, Hess J, Pogrebniak A, Bickel C, Gorlach A (2007) NOX5 variants are functionally active in endothelial cells. Free Radic Biol Med 42(4):446–459. https://doi.org/10.1016/j.freeradbiomed.2006.10.054
40. Herkert O, Diebold I, Brandes RP, Hess J, Busse R, Görlach A (2002) NADPH oxidase mediates tissue factor–dependent surface procoagulant activity by Thrombin in human vascular smooth muscle cells. Circulation 105(17):2030–2036. https://doi.org/10.1161/01.cir.0000014611.28864.1e
41. Görlach A, Diebold I, Schini-Kerth VB, Berchner-Pfannschmidt U, Roth U, Brandes RP, Kietzmann T, Busse R (2001) Thrombin activates the hypoxia-inducible factor-1 signaling pathway in vascular smooth muscle cells. Circ Res 89(1):47–54. https://doi.org/10.1161/hh1301.092678
42. Cheng G, Cao Z, Xu X, Meir EG, Lambeth J (2001) Homologs of gp91 phox: cloning and tissue expression of Nox3, Nox4, and Nox5. Gene 269(1–2):131–140. https://doi.org/10.1016/s0378-1119(01)00449-8
43. Djordjevic T, Pogrebniak A, Belaiba RS, Bonello S, Wotzlaw C, Acker H, Hess J, Görlach A (2005) The expression of the NADPH oxidase subunit p22phox is regulated by a redox-sensitive pathway in endothelial cells. Free Radic Biol Med 38(5):616–630. https://doi.org/10.1016/j.freeradbiomed.2004.09.036
44. Friedman JR, Nunnari J (2014) Mitochondrial form and function. Nature 505(7483):335–343. https://doi.org/10.1038/nature12985
45. Osellame LD, Blacker TS, Duchen MR (2012) Cellular and molecular mechanisms of mitochondrial function. Best Pract Res Clin Endocrinol Metab 26(6):711–723. https://doi.org/10.1016/j.beem.2012.05.003
46. Tang X, Luo YX, Chen HZ, Liu DP (2014) Mitochondria, endothelial cell function, and vascular diseases. Front Physiol 5:175. https://doi.org/10.3389/fphys.2014.00175
47. Dromparis P, Michelakis ED (2013) Mitochondria in vascular health and disease. Annu Rev Physiol 75(1):95–126. https://doi.org/10.1146/annurev-physiol-030212-183804
48. Caja S, Enríquez JA (2017) Mitochondria in endothelial cells: sensors and integrators of environmental cues. Redox Biol 12:821–827. https://doi.org/10.1016/j.redox.2017.04.021
49. Widlansky ME, Gutterman DD (2011) Regulation of endothelial function by mitochondrial reactive oxygen species. Antioxid Redox Signal 15(6):1517–1530. https://doi.org/10.1089/ars.2010.3642
50. Leone TC, Kelly DP (2011) Transcriptional control of cardiac fuel metabolism and mitochondrial function. Cold Spring Harb Symp Quant Biol 76:175–182. https://doi.org/10.1101/sqb.2011.76.011965
51. Patten IS, Arany Z (2012) PGC-1 coactivators in the cardiovascular system. Trends Endocrinol Metab 23(2):90–97. https://doi.org/10.1016/j.tem.2011.09.007
52. Li J, Zhang Y, Liu Y, Shen T, Zhang H, Xing Y, Zhu D (2015) PGC-1α plays a major role in the anti-apoptotic effect of 15-HETE in pulmonary artery endothelial cells. Respir Physiol Neurobiol 205:84–91. https://doi.org/10.1016/j.resp.2014.10.015
53. Suen D, Norris KL, Youle RJ (2008) Mitochondrial dynamics and apoptosis. Genes Dev 22(12):1577–1590. https://doi.org/10.1101/gad.1658508
54. Landes T, Martinou J (2011) Mitochondrial outer membrane permeabilization during apoptosis: the role of mitochondrial fission. Biochimica Et Biophysica Acta (BBA) – Mol Cell Res 1813(4):540–545. https://doi.org/10.1016/j.bbamcr.2011.01.021
55. Wang Z, Liu Y, Liu J, Liu K, Wen J, Wen S, Wu Z (2011) HSG/Mfn2 gene polymorphism and essential hypertension: a case-control association study in Chinese. J Atheroscler Thromb 18(1):24–31. https://doi.org/10.5551/jat.5611

56. Jin H, Sõber S, Hong K, Org E, Kim B, Laan M, Oh B, Jeong S (2011) Age-dependent association of the polymorphisms in the mitochondria-shaping gene, OPA1, with blood pressure and hypertension in Korean population. Am J Hypertens 24(10):1127–1135. https://doi.org/10.1038/ajh.2011.131
57. Twig G, Hyde B, Shirihai OS (2008) Mitochondrial fusion, fission and autophagy as a quality control axis: the bioenergetic view. Biochimica Et Biophysica Acta (BBA) – Bioenergetics 1777(9):1092–1097. https://doi.org/10.1016/j.bbabio.2008.05.001
58. Green DR, Galluzzi L, Kroemer G (2011) Mitochondria and the autophagy-inflammation-cell death axis in organismal aging. Science 333(6046):1109–1112. https://doi.org/10.1126/science.1201940
59. Youle RJ, Narendra DP (2011) Mechanisms of mitophagy. Nat Rev Mol Cell Biol 12(1):9–14. https://doi.org/10.1038/nrm3028
60. Choi MA, Ryter SW, Levine B (2013) Autophagy in human health and disease. N Engl J Med 368(19):1845–1846. https://doi.org/10.1056/nejmc1303158
61. Zhang Y, Cao Y, Zhang X, Liu H, Tong T, Xiao G, Yang YP, Liu C (2010) The autophagy-lysosome pathway: a novel mechanism involved in the processing of oxidized LDL in human vascular endothelial cells. Biochem Biophys Res Commun 394(2):377–382. https://doi.org/10.1016/j.bbrc.2010.03.026
62. Craige SM, Chen K, Pei Y, Li C, Huang X, Chen C, Shibata R, Sato K, Walsh K, Keaney JF (2011) NADPH oxidase 4 promotes endothelial angiogenesis through endothelial nitric oxide synthase activation. Circulation 124(6):731–740. https://doi.org/10.1161/circulationaha.111.030775
63. Chen F, Haigh S, Barman S, Fulton DJ (2012) From form to function: the role of Nox4 in the cardiovascular system. Front Physiol 3. https://doi.org/10.3389/fphys.2012.00412
64. Paneni F, Cosentino F (2012) P66 Shc as the engine of vascular aging. Curr Vasc Pharmacol 10(6):697–699. https://doi.org/10.2174/157016112803520747
65. Camici GG, Schiavoni M, Francia P, Bachschmid M, Martin-Padura I, Hersberger M, Tanner FC, Pelicci P, Volpe M, Anversa P, Luscher TF, Cosentino F (2007) Genetic deletion of p66Shc adaptor protein prevents hyperglycemia-induced endothelial dysfunction and oxidative stress. Proc Natl Acad Sci 104(12):5217–5222. https://doi.org/10.1073/pnas.0609656104
66. Giorgio M, Migliaccio E, Orsini F, Paolucci D, Moroni M, Contursi C, Pelliccia G, Luzi L, Minucci S, Marcaccio M, Pinton P, Rizzuto R, Bernardi P, Paolucci F, Pelicci PG (2005) Electron transfer between cytochrome c and p66Shc generates reactive oxygen species that trigger mitochondrial apoptosis. Cell 122(2):221–233. https://doi.org/10.1016/j.cell.2005.05.011
67. Hoch NE, Guzik TJ, Chen W, Deans T, Maalouf SA, Gratze P, Weyand C, Harrison DG (2009) Regulation of T-cell function by endogenously produced angiotensin II. Am J Phys Regul Integr Comp Phys 296(2). https://doi.org/10.1152/ajpregu.90521.2008
68. Widder JD, Fraccarollo D, Galuppo P, Hansen JM, Jones DP, Ertl G, Bauersachs J (2009) Attenuation of angiotensin II–induced vascular dysfunction and hypertension by overexpression of thioredoxin 2. Hypertension 54(2):338–344. https://doi.org/10.1161/hypertensionaha.108.127928
69. Yu T, Robotham JL, Yoon Y (2006) Increased production of reactive oxygen species in hyperglycemic conditions requires dynamic change of mitochondrial morphology. Proc Natl Acad Sci 103(8):2653–2658. https://doi.org/10.1073/pnas.0511154103
70. Basta G, Lazzerini G, Turco SD, Ratto GM, Schmidt AM, Caterina RD (2005) At least 2 distinct pathways generating reactive oxygen species mediate vascular cell Adhesion molecule-1 induction by advanced glycation end products. Arterioscler Thromb Vasc Biol 25(7):1401–1407. https://doi.org/10.1161/01.atv.0000167522.48370.5e
71. Chatzizisis YS, Coskun AU, Jonas M, Edelman ER, Feldman CL, Stone PH (2007) Role of endothelial shear stress in the natural history of coronary atherosclerosis and vascular remodeling. J Am Coll Cardiol 49(25):2379–2393. https://doi.org/10.1016/j.jacc.2007.02.059

72. Huddleson JP, Ahmad N, Srinivasan S, Lingrel JB (2005) Induction of KLF2 by fluid shear stress requires a novel Promoter element activated by a phosphatidylinositol 3-kinase-dependent chromatin-remodeling pathway. J Biol Chem 280(24):23371–23379. https://doi.org/10.1074/jbc.m413839200
73. Jin Z, Ueba H, Tanimoto T, Lungu AO, Frame MD, Berk BC (2003) Ligand-independent activation of vascular endothelial growth factor receptor 2 by fluid shear stress regulates activation of endothelial nitric oxide synthase. Circ Res 93(4):354–363. https://doi.org/10.1161/01.res.0000089257.94002.96
74. Kuchan MJ, Frangos JA (1994) Role of calcium and calmodulin in flow-induced nitric oxide production in endothelial cells. Am J Phys Cell Phys 266(3). https://doi.org/10.1152/ajpcell.1994.266.3.c628
75. Silacci P, Formentin K, Bouzourène K, Daniel F, Brunner H, Hayoz D (2000) Unidirectional and oscillatory shear stress differentially modulate NOS III gene expression. Nitric Oxide 4(1):47–56. https://doi.org/10.1006/niox.2000.0271
76. Ai L, Rouhanizadeh M, Wu JC, Takabe W, Yu H, Alavi M, Li R, Chu Y, Miller J, Heistad DD, Hsiai TK (2008) Shear stress influences spatial variations in vascular Mn-SOD expression: implication for LDL nitration. Am J Phys Cell Phys 294(6). https://doi.org/10.1152/ajpcell.00518.2007
77. Bretón-Romero R, Acín-Perez R, Rodríguez-Pascual F, Martínez-Molledo M, Brandes RP, Rial E, Enriquez JA, Lamas S (2014) Laminar shear stress regulates mitochondrial dynamics, bioenergetics responses and PRX3 activation in endothelial cells. Biochimica Et Biophysica Acta (BBA) – Mol Cell Res 1843(11):2403–2413. https://doi.org/10.1016/j.bbamcr.2014.07.003
78. Scheitlin CG, Nair DM, Crestanello JA, Zweier JL, Alevriadou BR (2014) Fluid mechanical forces and endothelial mitochondria: a bioengineering perspective. Cell Mol Bioeng 7(4):483–496. https://doi.org/10.1007/s12195-014-0357-4
79. Sorescu GP, Song H, Tressel SL, Hwang J, Dikalov S, Smith DA, Boyd NL, Platt MO, Lassegue B, Griendling KK, Jo H (2004) Bone morphogenic protein 4 produced in endothelial cells by oscillatory shear stress induces monocyte adhesion by stimulating reactive oxygen species production from a Nox1-based NADPH oxidase. Circ Res 95(8):773–779. https://doi.org/10.1161/01.res.0000145728.22878.45
80. Shibuya M (2008) Vascular endothelial growth factor-dependent and -independent regulation of angiogenesis. BMB Rep 41(4):278–286. https://doi.org/10.5483/bmbrep.2008.41.4.278
81. Lee M, Choy WC, Abid MR (2011) Direct sensing of endothelial oxidants by vascular endothelial growth factor receptor-2 and c-Src. PLoS ONE 6(12). https://doi.org/10.1371/journal.pone.0028454
82. Kim Y-W, Byzova TV (2014) Oxidative stress in angiogenesis and vascular disease. Blood 123(5):625–631. https://doi.org/10.1182/blood-2013-09-512749
83. Evangelista AM, Thompson MD, Bolotina VM, Tong X, Cohen RA (2012) Nox4- and Nox2-dependent oxidant production is required for VEGF-induced SERCA cysteine-674 S-glutathiolation and endothelial cell migration. Free Radic Biol Med 53(12):2327–2334. https://doi.org/10.1016/j.freeradbiomed.2012.10.546
84. Meng T, Fukada T, Tonks NK (2002) Reversible oxidation and inactivation of protein tyrosine phosphatases in vivo. Mol Cell 9(2):387–399. https://doi.org/10.1016/s1097-2765(02)00445-8
85. Abid MR, Spokes KC, Shih S, Aird WC (2007) NADPH oxidase activity selectively modulates vascular Endothelial growth factor signaling pathways. J Biol Chem 282(48):35373–35385. https://doi.org/10.1074/jbc.m702175200
86. Hadi HA, Carr CS, Al Suwaidi J (2005) Endothelial dysfunction: cardiovascular risk factors, therapy, and outcome. Vasc Health Risk Manag 1(3):183–198
87. Aiello RJ, Bourassa PK, Lindsey S, Weng W, Natoli E, Rollins BJ, Milos PM (1999) Monocyte chemoattractant protein-1 accelerates atherosclerosis in apolipoprotein E-deficient mice. Arterioscler Thromb Vasc Biol 19(6):1518–1525. https://doi.org/10.1161/01.ATV.19.6.1518

88. Potz BA, Parulkar AB, Abid RM, Sodha NR, Sellke FW (2017) Novel molecular targets for coronary angiogenesis and ischemic heart disease. Coron Artery Dis 28(7):605–613. https://doi.org/10.1097/mca.0000000000000516
89. Folkman J, Shing Y (1992) Angiogenesis. J Biol Chem 267(16):10931–10934
90. Pierce WJ, Read AM, Ding H, Luscinskas WF, Collins T (1996) Salicylates inhibit IkB-a phosphorylation, endothelial-leukocyte adhesion molecule expression, and neutrophil transmigration. J Immunol 156(10):3961–3969
91. Stoltz RA, Abraham NG, Laniado-Schwartzman M (1996) The role of NF-kappaB in the angiogenic response of coronary microvessel endothelial cells. Proc Natl Acad Sci 93((7):2832–2837. https://doi.org/10.1073/pnas.93.7.2832
92. Pierce GL, Lesniewski LA, Lawson BR, Beske SD, Seals DR (2009) Nuclear factor- B activation contributes to vascular endothelial dysfunction via oxidative stress in overweight/obese middle-aged and older humans. Circulation 119(9):1284–1292. https://doi.org/10.1161/circulationaha.108.804294
93. Kassan M, Choi S, Galan M, Bishop A, Umezawa K, Trebak M, Belmadan S, Matrougui K (2013) Enhanced NF- B activity impairs vascular function through PARP-1-, SP-1-, and COX-2-dependent mechanisms in type 2 diabetes. Diabetes 62(6):2078–2087. https://doi.org/10.2337/db12-1374
94. Deer RR, Heaps CL (2013) Exercise training enhances multiple mechanisms of relaxation in coronary arteries from ischemic hearts. AJP Heart Circ Physiol 305(9):H1321–H1331. https://doi.org/10.1152/ajpheart.00531.2013
95. Harrison DG (1997) Cellular and molecular mechanisms of endothelial cell dysfunction. J Clin Invest 100(9):2153–2157. https://doi.org/10.1172/JCI119751
96. Laurindo FR, Liberman M, Fernandes DC, Leite PF (2018) Endothelium-dependent vasodilation: nitric oxide and other mediators. Endothelium Cardiovasc Dis:97–113. https://doi.org/10.1016/b978-0-12-812348-5.00008-8
97. Lucas KA, Pitari GM, Kazerounian S, Ruiz-Stewart I, Park J, Schulz S, Chepenik PK, Waldman SA (2000) Guanylyl cyclases and signaling by cyclic GMP. Pharmacol Rev 52(3):375–414
98. Koesling D, Friebe A (1999) Soluble guanylyl cyclase: structure and regulation. Rev Physiol Biochem Pharmacol 135:41–65. https://doi.org/10.1007/bfb0033669
99. Quayle JM, Nelson MT, Standen NB (1997) ATP-sensitive and inwardly rectifying potassium channels in smooth muscle. Physiol Rev 77(4):1165–1232. https://doi.org/10.1152/physrev.1997.77.4.1165
100. Su JB, Houël R, Héloire F, Barbe F, Beverelli F, Sambin L et al (2000) Stimulation of bradykinin B1receptors induces vasodilation in conductance and resistance coronary vessels in conscious dogs: comparison with B2receptor stimulation. Circulation 101(15):1848–1853. https://doi.org/10.1161/01.CIR.101.15.1848
101. Michel T, Feron O (1997) Nitric oxide synthases: which, where, how, and why? J Clin Investig 100(9):2146–2152. https://doi.org/10.1172/jci119750
102. Stuehr D, Pou S, Rosen GM (2001) Oxygen reduction by nitric-oxide synthases. J Biol Chem. https://doi.org/10.1074/jbc.R100011200
103. Stuehr DJ, Tejero J, Haque MM (2009) Structural and mechanistic aspects of flavoproteins: electron transfer through the nitric oxide synthase flavoprotein domain. FEBS J. https://doi.org/10.1111/j.1742-4658.2009.07120.x
104. Furchgott RF, Zawadzki JV (1980) The obligatory role of endothelial cells in the relaxation of arterial smooth muscle by acetylcholine. Nature 288(5789):373–376. https://doi.org/10.1038/288373a0
105. Fulton D, Gratton JP, McCabe TJ, Fontana J, Fujio Y, Walsh K et al (1999) Regulation of endothelium-derived nitric oxide production by the protein kinase Akt. Nature 399(6736):597–601. https://doi.org/10.1038/21218
106. Dimmeler S, Fleming I, Fisslthaler B, Hermann C, Busse R, Zeiher AM (1999) Activation of nitric oxide synthase in endothelial cells by Akt-dependent phosphorylation. Nature 399(June):601–605. https://doi.org/10.1038/21224

107. Moncada S, Higgs EA (2006) Nitric oxide and the vascular endothelium. The vascular endothelium I handbook of experimental pharmacology, pp 213–254. https://doi.org/10.1007/3-540-32967-6_7
108. Drab M, Verkade P, Elger M, Kasper M, Lohn M, Lauterbach B et al (2001) Loss of caveolae, vascular dysfunction, and pulmonary defects in caveolin-1 gene-disrupted mice. Science 293(5539):2449–2452. https://doi.org/10.1126/science.1062688
109. Liu VWT, Huang PL (2008) Cardiovascular roles of nitric oxide: a review of insights from nitric oxide synthase gene disrupted mice. Cardiovasc Res. https://doi.org/10.1016/j.cardiores.2007.06.024
110. Huang PL (2000) Mouse models of nitric oxide synthase deficiency. J Am Soc Nephrol 11(Suppl 1):S120–S123
111. Holtz J, Föuml, Rstermann U, Pohl U, Giesler M, Bassenge E (1984) Flow-dependent, endothelium-mediated dilation of epicardial coronary arteries in conscious dogs: effects of cyclooxygenase inhibition. J Cardiovasc Pharmacol 6(6):1161–1169. https://doi.org/10.1097/00005344-198406060-00025
112. Dai XZ, Sublett E, Lindstrom P, Schwartz JS, Homans DC, Bache RJ (1989) Coronary flow during exercise after selective alpha 1- and alpha 2-adrenergic blockade. Am J Phys Heart Circ Phys 256(4). https://doi.org/10.1152/ajpheart.1989.256.4.h1148
113. Komaru T, Lamping KG, Eastham CL, Harrison DG, Marcus ML, Dellsperger KC (1991) Effect of an arginine analogue on acetylcholine-induced coronary microvascular dilatation in dogs. Am J Phys Heart Circ Phys 261(6). https://doi.org/10.1152/ajpheart.1991.261.6.h2001
114. Lamontagne D, König A, Bassenge E, Busse R (1992) Prostacyclin and nitric oxide contribute to the vasodilator action of acetylcholine and bradykinin in the intact rabbit coronary bed. J Cardiovasc Pharmacol 20(4):652–657. https://doi.org/10.1097/00005344-199210000-00020
115. Feletou M, Vanhoutte PM (1988) Endothelium-dependent hyperpolarization of canine coronary smooth muscle. Br J Pharmacol 93(3):515–524. https://doi.org/10.1111/j.1476-5381.1988.tb10306.x
116. Parkington HC, Coleman HA, Tare M (2004) Prostacyclin and endothelium-dependent hyperpolarization. Pharmacol Res. https://doi.org/10.1016/j.phrs.2003.11.012
117. Burgoyne JR, Madhani M, Cuello F, Charles RL, Brennan JP, Schröder E et al (2007) Cysteine redox sensor in PKGIα enables oxidant-induced activation. Science 317(5843):1393–1397. https://doi.org/10.1126/science.1144318
118. Mustafa AK, Gadalla MM, Sen N, Kim S, Mu W, Gazi SK et al (2009) HS signals through protein S-Sulfhydration. Sci Signal 2(96). https://doi.org/10.1126/scisignal.2000464
119. Edwards G, Dora KA, Gardener MJ, Garland CJ, Weston AH (1998) K+ is an endothelium-derived hyperpolarizing factor in rat arteries. Nature 396(6708):269–272. https://doi.org/10.1038/24388
120. Yada T, Shimokawa H, Hiramatsu O, Kajita T, Shigeto F, Goto M et al (2003) Hydrogen peroxide, an endogenous endothelium-derived hyperpolarizing factor, plays an important role in coronary autoregulation in vivo. Circulation 107(7):1040–1045. https://doi.org/10.1161/01.CIR.0000050145.25589.65
121. Matoba T, Shimokawa H, Nakashima M, Hirakawa Y, Mukai Y, Hirano K et al (2000) Hydrogen peroxide is an endothelium-derived hyperpolarizing factor in mice. J Clin Investig 106(12):1521–1530. https://doi.org/10.1172/JCI10506
122. Moncada S, Gryglewski R, Bunting S, Vane JR (1976) An enzyme isolated from arteries transforms prostaglandin endoperoxides to an unstable substance that inhibits platelet aggregation. Nature 263(5579):663–665. https://doi.org/10.1038/263663a0
123. Caughey GE, Cleland LG, Penglis PS, Gamble JR, James MJ (2001) Roles of cyclooxygenase (COX)-1 and COX-2 in prostanoid production by human endothelial cells: selective up-regulation of prostacyclin synthesis by COX-2. J Immunol 167(5):2831–2838. https://doi.org/10.4049/jimmunol.167.5.2831
124. Thapaliya S, Matsuyama H, Takewaki T (2000) Bradykinin causes endothelium-independent hyperpolarisation and neuromodulation by prostanoid synthesis in hamster mesenteric artery. Eur J Pharmacol 408(3):313–321. https://doi.org/10.1016/S0014-2999(00)00776-7

125. Kukovetz WR, Holzmann S, Wurm A, Pöch G (1979) Prostacyclin increases cAMP in coronary arteries. J Cyclic Nucleotide Res 5(6):469–476. Retrieved from http://www.ncbi.nlm.nih.gov/pubmed/231612
126. Schubert R, Serebryakov VN, Mewes H, Hopp HH (1997) Iloprost dilates rat small arteries: role of K(ATP)- and K(Ca)-channel activation by cAMP-dependent protein kinase. Am J Phys Heart Circ Phys 272(3):H1147–H1156. https://doi.org/10.1152/ajpheart.1997.272.3.H1147
127. Siegel G, Stock G, Schnalke F, Litza B (1987). Electrical and mechanical effects of prostacyclin in the canine carotid artery Prostacyclin Stable Analogue Iloprost:143–149. https://doi.org/10.1007/978-3-642-71499-3_17
128. Parkington HC, Tonta MA, Coleman HA, Tare M (1995) Role of membrane potential in endothelium-dependent relaxation of guinea-pig coronary arterial smooth muscle. J Physiol 484(2):469–480. https://doi.org/10.1113/jphysiol.1995.sp020679
129. Wilson VJ, Zarzecki P, Schor RH, Isu N, Rose PK, Sato H et al (1999) Inhibition of endothelium-dependent hyperpolarization by endothelial prostanoids in guinea-pig coronary artery. Br J Pharmacol 126(1):1–10. https://doi.org/10.1038/sj.bjp.0702254
130. Tare M, Parkington HC, Coleman HA (2000) EDHF, NO and a prostanoid: Hyperpolarization-dependent and -independent relaxation in guinea-pig arteries. Br J Pharmacol 130(3):605–618. https://doi.org/10.1038/sj.bjp.0703332
131. Vallance P, Collier J, Moncada S (1989) Effects of endothelium-derived nitric oxide on peripheral arteriolar tone in man. Lancet 334(8670):997–1000. https://doi.org/10.1016/S0140-6736(89)91013-1
132. Huang PL, Huang Z, Mashimo H, Bloch KD, Moskowitz MA, Bevan JA, Fishman MC (1995) Hypertension in mice lacking the gene for endothelial nitric oxide synthase. Nature. https://doi.org/10.1038/377239a0
133. Duffy SJ, Castle SF, Harper RW, Meredith IT (1999) Contribution of vasodilator prostanoids and nitric oxide to resting flow, metabolic vasodilation, and flow-mediated dilation in human coronary circulation. Circulation 100(19):1951–1957. https://doi.org/10.1161/01.CIR.100.19.1951
134. Friedman PL, Brown EJ, Gunther S, Alexander RW, Barry WH, Mudge GH, Grossman W (1981) Coronary vasoconstrictor effect of indomethacin in patients with coronary-artery disease. N Engl J Med 305(20):1171–1175. https://doi.org/10.1056/nejm198111123052002
135. Edlund A, Sollevi A, Wennmalm Å (1989) The role of adenosine and prostacyclin in coronary flow regulation in healthy man. Acta Physiol Scand 135(1):39–46. https://doi.org/10.1111/j.1748-1716.1989.tb08548.x
136. Dai XZ, Bache RJ (1984) Effect of indomethacin on coronary blood flow during graded treadmill exercise in the dog. Am J Physiol 247(3 Pt 2):H452–H458. Retrieved from http://www.ncbi.nlm.nih.gov/pubmed/6433725
137. Puybasset L, Béa ML, Ghaleh B, Giudicelli JF, Berdeaux A (1996) Coronary and systemic hemodynamic effects of sustained inhibition of nitric oxide synthesis in conscious dogs: Evidence for cross talk between nitric oxide and cyclooxygenase in coronary vessels. Circ Res 79(2):343–357. https://doi.org/10.1161/01.RES.79.2.343
138. Andrews TC, Ballantyne CM, Hsia JA, Kramer JH (2001) Achieving and maintaining National Cholesterol Education Program low-density lipoprotein cholesterol goals with five statins. Am J Med 111(3):185–191. https://doi.org/10.1016/S0002-9343(01)00799-9
139. Pravenec M, Kuneš J, Zicha J, Křen V, Klír P (1992) Platelet aggregation in spontaneous hypertension. J Hypertens 10(12):1453–1456. https://doi.org/10.1097/00004872-199210120-00003
140. Nomura S, Suzuki M, Katsura K, Xie GL, Miyazaki Y, Miyake T et al (1995) Platelet-derived microparticles may influence the development of atherosclerosis in diabetes mellitus. Atherosclerosis 116(2):235–240. https://doi.org/10.1016/0021-9150(95)05551-7
141. Busse R, Edwards G, Félétou M, Fleming I, Vanhoutte PM, Weston AH (2002) EDHF: bringing the concepts together. Trends Pharmacol Sci. https://doi.org/10.1016/S0165-6147(02)02050-3
142. Fleming I (2000) Cytochrome P450 2C is an EDHF synthase in coronary arteries. Trends Cardiovasc Med. https://doi.org/10.1016/S1050-1738(00)00065-7

143. Popp R, Brandes RP, Ott G, Busse R, Fleming I (2002) Dynamic modulation of interendothelial gap junctional communication by 11,12-epoxyeicosatrienoic acid. Circ Res 90(7):800–806. https://doi.org/10.1161/01.RES.0000015328.20581.D6
144. Mustafa AK, Sikka G, Gazi SK, Steppan J, Jung SM, Bhunia AK et al (2011) Hydrogen sulfide as endothelium-derived hyperpolarizing factor sulfhydrates potassium channels. Circ Res 109(11):1259–1268. https://doi.org/10.1161/CIRCRESAHA.111.240242
145. Yang G, Wu L, Jiang B, Yang W, Qi J, Cao K et al (2008) H2S as a physiologic vasorelaxant: Hypertension in mice with deletion of cystathionine γ-lyase. Science 322(5901):587–590. https://doi.org/10.1126/science.1162667
146. Wang R, Szabo C, Ichinose F, Ahmed A, Whiteman M, Papapetropoulos A (2015) The role of H2S bioavailability in endothelial dysfunction. Trends Pharmacol Sci 36(9):568–578. https://doi.org/10.1016/j.tips.2015.05.007
147. Kram L, Grambow E, Mueller-Graf F, Sorg H, Vollmar B (2013) The anti-thrombotic effect of hydrogen sulfide is partly mediated by an upregulation of nitric oxide synthases. Thromb Res 132(2). https://doi.org/10.1016/j.thromres.2013.07.010
148. Hasdai D, Kornowski R, Battler A (1994) Endothelin and myocardial ischemia. Cardiovasc Drugs Ther 8(4):589–599. https://doi.org/10.1007/bf00877413
149. Cooke JP (2007) Angiogenesis and the role of the endothelial nicotinic acetylcholine receptor. Life Sci 80(24–25):2347–2351. https://doi.org/10.1016/j.lfs.2007.01.061
150. Rubanyi GM, Polokoff MA (1994) Endothelins: molecular biology, biochemistry, pharmacology, physiology, and pathophysiology. Pharmacol Rev 46(3):325–415. Retrieved from http://www.ncbi.nlm.nih.gov/pubmed/7831383
151. Yanagisawa M, Kurihara H, Kimura S, Tomobe Y, Kobayashi M, Mitsui Y et al (1988) A novel potent vasoconstrictor peptide produced by vascular endothelial cells. Nature 332(6163):411–415. https://doi.org/10.1038/332411a0
152. Sandoval Y-HG, Atef ME, Levesque L-O, Li Y, Anand-Srivastava MB (2014) Endothelin-1 signaling in vascular physiology and pathophysiology. Curr Vasc Pharmacol 12(2):202–214. https://doi.org/10.2174/1570161112666140226122054
153. Folta A, Joshua IG, Clinton Webb R (1989) Dilator actions of endothelin in coronary resistance vessels and the abdominal aorta of the guinea pig. Life Sci 45(26):2627–2635. https://doi.org/10.1016/0024-3205(89)90248-8
154. Wenzel RR, Fleisch M, Shaw S, Noll G, Kaufmann U, Schmitt R et al (1998) Hemodynamic and coronary effects of the endothelin antagonist bosentan in patients with coronary artery disease. Circulation 98(21):2235–2240. https://doi.org/10.1161/01.CIR.98.21.2235
155. Merkus D, Duncker DJ, Chilian WM (2002) Metabolic regulation of coronary vascular tone: role of endothelin-1. Am J Physiol Heart Circ Physiol 283(5):H1915–H1921. https://doi.org/10.1152/ajpheart.00223.2002
156. Osanai T, Fujita N, Fujiwara N, Nakano T, Okumura K (2000) Cross talk of shear-induced production of prostacyclin and nitric oxide in endothelial cells. Am J Phys Heart Circ Phys 278(1):233–238
157. Shen W, Ochoa M, Xu X, Wang J, Hintze TH (1994) Role of EDRF/NO in parasympathetic coronary vasodilation following carotid chemoreflex activation in conscious dogs. Am J Physiol Heart Circ Physiol 267(2). https://doi.org/10.1152/ajpheart.1994.267.2.h605
158. Cowan CL, Mckenzie JE (1990) Cholinergic regulation of resting coronary blood flow in domestic swine. Am J Physiol Heart Circ Physiol 259(1). https://doi.org/10.1152/ajpheart.1990.259.1.h109
159. Furusho N, Araki H, Sakaino N, Nishi K, Miyauchi Y (1988) Effects of perivascular nerve stimulation on the flow rate in isolated epicardial coronary arteries of pigs. Eur J Pharmacol 154(1):79–84. https://doi.org/10.1016/0014-2999(88)90366-4
160. Duncker DJ, Stubenitsky R, Verdouw PD (1998) Autonomic control of vasomotion in the porcine coronary circulation during treadmill exercise: evidence for feed-forward beta-adrenergic control. Circ Res 82(12):1312–1322
161. Duncker DJ, Haitsma DB, Liem DA, Verdouw PD, Merkus D (2005) Exercise unmasks autonomic dysfunction in swine with a recent myocardial infarction. Cardiovasc Res 65(4):889–896. https://doi.org/10.1016/j.cardiores.2004.12.010

162. Wilson C, Lee MD, McCarron JG (2016) Acetylcholine released by endothelial cells facilitates flow-mediated dilatation. J Physiol 594(24):7267–7307. https://doi.org/10.1113/JP272927
163. Dhein S, Wermke A, Von Salisch S, Schlegel F, Stepan H, Dohmen PM et al (2015) Autocrine control of angiogenesis by endogenous acetylcholine in an in vitro model using human endothelial cells: evidence for an autocrine cholinergic system in endothelial cells. J Cardiovasc Pharmacol 65(5):508–515. https://doi.org/10.1097/FJC.0000000000000221
164. Timmermans PB, Wong PC, Chiu AT, Herblin WF, Benfield P, Carini DJ, ... Smith RD (1993) Angiotensin II receptors and angiotensin II receptor antagonists. Pharmacol Rev 45(2):205–251. Retrieved from http://www.ncbi.nlm.nih.gov/pubmed/8372104
165. Cocks TM, Angus JA (1983) Endothelium-dependent relaxation of coronary arteries by noradrenaline and serotonin. Nature 305(5935):627–630. https://doi.org/10.1038/305627a0
166. Vanhoutte PM, Shimokawa H, Feletou M, Tang EHC (2017) Endothelial dysfunction and vascular disease – a 30th anniversary update. Acta Physiol. Blackwell Publishing Ltd. https://doi.org/10.1111/apha.12646
167. Broten TP, Miyashiro JK, Moncada S, Feigl EO (1992) Role of endothelium-derived relaxing factor in parasympathetic coronary vasodilation. Am J Physiol Heart Circ Physiol 262(5). https://doi.org/10.1152/ajpheart.1992.262.5.h1579
168. Simplicio JA, Resstel LB, Tirapelli DPC, D'Orléans-Juste P, Tirapelli CR (2015) Contribution of oxidative stress and prostanoids in endothelial dysfunction induced by chronic fluoxetine treatment. Vasc Pharmacol 73:124–137. https://doi.org/10.1016/j.vph.2015.06.015
169. Yoon MH, Reriani M, Mario G, Rihal C, Gulati R, Lennon R et al (2013) Long-term endothelin receptor antagonism attenuates coronary plaque progression in patients with early atherosclerosis. Int J Cardiol 168(2):1316–1321. https://doi.org/10.1016/j.ijcard.2012.12.001
170. Félétou M, Huang Y, Vanhoutte PM (2011) Endothelium-mediated control of vascular tone: COX-1 and COX-2 products. Br J Pharmacol. https://doi.org/10.1111/j.1476-5381.2011.01276.x

Modulation of miRNA in Oxidative Stress-Induced Cardiac Remodeling

3

Sudhiranjan Gupta

Abstract

Cardiac remodeling is the pathological ramification of myocardium resulting into various cardiac diseases. Oxidative stress represents a self-perpetuating mechanism by producing excess reactive oxygen species (ROS) and plays a critical role in cardiac remodeling. A redox state that maintains the homeostatic balance in the cell is critical in cardiac remodeling. A misbalance of redox state triggers cellular damage and promoting adverse signaling pathways leading to apoptosis. MicroRNAs (miRNAs) are short, 19–21 nucleotides, endogenous noncoding RNAs modulate gene regulation, elicits a vital role in cardiac remodeling including cardiac hypertrophy, fibrosis, myocardial injury and arrythmia via multiple mechanisms. Recent studies indicated that miRNAs are influencing the generation of ROS and modulate antioxidant defense mechanism by regulating antioxidative enzymes and are termed as "redoximiRs." Here, I review the current progress and the mechanisms by which "redoximiRs" regulate cardiac remodeling.

Keywords

Cardiac remodeling · miRNA · Oxidative stress

The original version of this chapter was revised: Figure legends have been updated. The correction to this chapter is available at https://doi.org/10.1007/978-981-13-8946-7_29

S. Gupta (✉)
Central Texas Veterans Health Care System, Temple, TX, USA

VISN 17 Center of Excellence on Returning War Veterans, Doris Miller VA Medical Center, Waco, TX, USA

Department of Biology, Baylor University, Waco, TX, USA
e-mail: Sudhiranjan_Gupta@baylor.edu; Sudhiranjan.Gupta@va.gov

S. Chakraborti et al. (eds.), *Modulation of Oxidative Stress in Heart Disease*,
https://doi.org/10.1007/978-981-13-8946-7_3

3.1 Introduction

Cardiac remodeling is defined as a structural alteration of the heart in response to hemodynamic load. The resulted phenomena exhibited as left ventricular hypertrophy (LVH), interstitial fibrosis and contractile dysfunction leading to heart failure. Evidences suggest that the redox-sensitive signaling pathways contribute a critical role in cardiac dysfunction [1]. The reactive oxygen species (ROS) and reactive nitrogen species (RNS) serve as signaling messengers to mediate the cardiac remodeling process including cardiac hypertrophy, fibrosis, arrhythmia, and heart failure [2, 3].

ROS, a highly reactive and unstable chemical, is generated as intermediates in reduction-oxidation (redox) reactions [4]. Generation of ROS fundamentally damages the cellular membranes, DNA, and enzymes which are involved in basic cellular function and homeostasis [5]. Thus, activation of key mediators/molecules by ROS in the cardiac pathology along with the reduction of its counterpart, the antioxidant molecules, is important in cardiac remodeling process. Additionally, ROS do modulate the extracellular matrix (ECM) function by promoting deposition of increased ECM in the interstitial and perivascular region of the heart and trigger fibrosis [6]. The significance of oxidative stress in cardiac remodeling is crucial at physiological and pathological standpoints and is not completely elucidated.

MicroRNAs (miRNAs) are a class of noncoding endogenous RNA molecules, ~19–22 nucleotides in length, and influence posttranscriptional gene regulation. The miRNAs are discovered in *Caenorhabditis elegans* and are regulatory agents for posttranscriptional level of mRNA [7]. Recent studies illustrated the commanding role of miRNA in modulating transcriptional and translational phases during cardiac remodeling. However, the aberrant expression of miRNA in response to oxidative stress has started emerging. Therefore, this review will discuss the possible connections between ROS-mediated miRNAs and cardiac remodeling, to understand the interactions of these molecular entities during cardiac remodeling. This review further attempts to offer a new approach to understanding cardiac pathogenesis and focuses on the possible interplay between ROS and miRNA.

3.2 Oxidative Stress

3.2.1 ROS, Antioxidant Enzymes, and Redox Balance

Oxidative stress has been shown to play an important role in diverse cardiac pathologies and heart failure [1, 8]. The ROS constitutes several free radicals, such as superoxide anion ($O_2^{\bullet-}$), hydroxyl radical ($^{\bullet}OH$), and non-radicals, and is capable of generating hydrogen peroxide (H_2O_2). The $O_2^{\bullet-}$ contributes primarily to the generation of other ROS, such as H_2O_2 and $^{\bullet}OH$. The $^{\bullet}OH$ is also generated by addition of iron in H_2O_2 called Fenton reaction and from interaction between $O_2^{\bullet-}$ and H_2O_2 called Haber-Weiss reaction. Another strong oxidant is peroxynitrite ($^{\bullet}ONOO^-$). The reaction of nitrous oxide (NO) with $O_2^{\bullet-}$ results in the formation of

$^{\bullet}ONOO^-$ which is detrimental to the cells [9]. It is of note that NO is required for normal cardiac physiology including coronary vasodilation, regulation of contractile function, and inhibition of platelet and neutrophil adhesion and activation, providing a protective role cardiac injury [10, 11].

Increased ROS can disrupt cellular function by DNA damage leading to irreversible cell damage which displayed a wide range of pathological cardiac diseases [12]. Specifically, ROS can significantly compromise cardiac contractile function, modifying excitation-contraction coupling, activating several signaling cascades associated with cardiac diseases, and triggering activation of transcription factors responsible for diverse gene regulation. The ROS can further activate cardiac fibroblast to proliferate and trigger extracellular matrix remodeling. The three major cell types, the myocytes, fibroblasts, and endothelial cells, are the source of ROS generation in the heart. Although their exact contribution to the generation of ROS is unknown, evidence suggests that mitochondrial electron transport chain and non-phagocytic oxidases are likely be the predominant sources in cardiac myocytes [13].

As ROS is harmful to the cells, cells are developing a defense mechanism by synthesizing a set of antioxidant enzymes (e.g., dismutase, catalase, and peroxidase) and small molecules (glutathione, vitamins, etc.) to detoxify the ROS [14]. Among these enzymes, superoxide dismutases (SODs) are one of the main antioxidant enzymes that catalyze the conversion of $O_2^{\bullet-}$ to H_2O_2 and protect cellular damages caused by oxidative stress. In mammalian tissue, three isoforms of SODs, manganese-containing SOD (Mn-SOD), copper-containing SOD (Cu-SOD), and zinc-containing SOD (Zn-SOD), are present [15]. Among them, Mn-SOD is predominantly present in cardiac mitochondria and responsible for ~70% of the overall SOD activity in the heart by controlling O_2^- generation in mitochondria (in myocardium) [16]. Another enzyme, NADPH oxidase (NOX), a multi-subunit enzyme that catalyzes superoxide production by the reduction of oxygen using NADPH or NADH as the electron donor, showed a critical role on cardiac remodeling [17–19]. Therefore, a balance between ROS production and antioxidant systems is critical in cellular homeostasis referred as "redox state." So, the antioxidant enzymes are the body's defense system. The fine-tuning between ROS generation and endogenous antioxidants is essential for redox homeostasis in regular physiological processes.

3.2.2 Mechanism of ROS-Induced Cardiac Remodeling

3.2.2.1 Cardiac Hypertrophy

Cardiac hypertrophy represents an increase in cardiac mass in response to hemodynamic overload. Under this setting, cardiomyocyte undergoes increase in size to compensate the initial thrust, but sustained pressure overload leads to cardiac hypertrophy [20]. The process activates fetal gene program, triggering inflammation, promoting cell death, and eventually impairing the cardiac function. An imbalance between the oxidation and reduction by ROS and antioxidant systems induces oxidative stress at cellular level promoting cardiac diseases like cardiac hypertrophy [21]. The underlying mechanisms of ROS-mediated hypertrophic response are incompletely understood; however, activation of redox-sensitive signaling molecules is documented [22–25]. Three important downstream effectors from MAPK signaling (ERK, p38MAPK, and JNK) are redox-sensitive molecular targets in cardiac hypertrophy [23, 26]. Furthermore, H_2O_2-treated cardiomyocytes in cell culture model releases ROS and suggested to promote apoptosis by triggering mitochondrial permeability pore transition [27, 28]. Recently, NOX have been shown to be a primary source of ROS [29]. NOX is well studied [30] and demonstrated a critical role in cardiac hypertrophy. NOX activity is increased in pressure overload-induced cardiac hypertrophy along with MAPK activation [31]. Genetically modified mice lacking the Nox2 subunit of NADPH oxidase was shown protective in angiotensin II-induced cardiac hypertrophy model [32]. Furthermore, $Nox2^{-/-}$ mice were protected against pressure overload-induced myocardial dysfunction [33]. On the contrary, Nox2-deficient mice subjected to aortic banding developed the same degree of hypertrophy as wild-type mice [34, 35]. Nonetheless, it is imperative that Nox4 oxidase is an important mediator of pressure overload-induced hypertrophy [34]. Interestingly, cardiac-specific knockout of Nox4 elicited reduced hypertrophy and fibrosis in pressure overload model and improved cardiac function apparently by increasing angiogenic activity [36]. However, Nox4-null mice showed opposite phenotype of developing hypertrophy, dilatation and contractile dysfunction [37]. Therefore, it is reasonable to accept that Nox4 plays a protective role in cardiac remodeling in contrast to Nox2.

Another antioxidant enzyme, Mn-SOD (SOD2), has been shown to be protective in cardiac hypertrophy. SOD2 is the primary antioxidant enzyme neutralizing $^{\bullet}O_2^-$ in mitochondria [38]. Overexpression of SOD2 in myocytes showed "super" cardiac performance by enhancing mitochondrial function and metabolic vasodilation [38, 39]. On the other hand, deficiency of Mn-SOD2 showed neonatal lethality and dilated cardiomyopathy indicated that Mn-SOD is essential for normal cellular and biological function of tissues by maintaining the integrity of mitochondrial enzymes [40].

3.2.2.2 Cardiac Fibrosis

Cardiac fibrosis which is a severe pathologic manifestation resulting from diverse stressful situation like pressure overload, ischemic insult, or metabolic stress [41, 42]. The underlying molecular and morphological harmony of cardiac fibrosis is disruption of cardiac architecture by excessive deposition of extracellular matrix (ECM) proteins which include collagens, matrix metalloproteinases (MMP), etc. in the heart leading to cardiac stiffness and impair function [43, 44]. Activation of cardiac fibroblast due to diverse stimuli further contributes in cardiac fibrosis, another manifestation of adverse cardiac remodeling. Although many potential mechanisms have postulated that promote fibrosis over the past decades, the precise molecular mechanisms are not fully understood. Growing body of evidences indicate the involvement of oxidative stress in the development of fibrosis in the heart. Three major intracellular sources of ROS generation are considered in the heart.

They are mitochondria electron-transport chain (ETC), membrane-bound NOXs, and endoplasmic reticulum (ER) [45]. ROS generated by NOXs and their interactions with NO are suggested to activate redox signaling cascade in the pro-fibrotic process leading to heart failure [46, 47]. In doxorubicin-induced cardiac toxicity mouse model, it is reported that Nox2-derived ROS contributes to myocyte death, inflammatory response, cellular infiltration, interstitial fibrosis, and contractile dysfunction [48]. In the same model, contractile dysfunction was attenuated in $Nox2^{-/-}$ mice compared to the WT mice [48]. In angiotensin II-induced cardiac fibrosis model, $Nox2^{-/-}$ mice displayed abrogation of fibrosis indicating a critical role of Nox2 in fibrosis [32, 34]. Furthermore, in vitro studies demonstrated that Nox4 is associated with fibroblast activation and transformation into myofibroblasts in TGF-β1-stimulated human cardiac fibroblast [49]. Using rat neonatal cardiac fibroblast, we have shown that H_2O_2 treatment increases intracellular ROS production and alteration of SODs and catalase levels [50]. Increased ROS further disrupted the mitochondrial membrane potential and subsequently increase the Bax/Bcl_2 ratio favoring cell death [50]. However, the relevance of these results remains to be established in vivo.

3.2.2.3 Cardiac Arrhythmia

Cardiac arrhythmia is defined as abnormal rate or rhythm due to unquiet electrophysiological setting in the myocardium. This include tachycardia (heart rate is too fast) and bradycardia (heart rate is too slow). Atrial fibrillation (AF) which is the most common type of arrhythmia contributed significant morbidity and mortality [51, 52]. In 2010, it is reported that AF affect 2.7–6.1 million people in the United States and 14–16% have died of ischemic stroke [53]. It is reported that AF is linked with increased levels of ROS such as superoxide [54]. ROS promotes structural and electrophysiological remodeling such as ER and mitochondrial dysfunction, abnormal Ca^{2+} handling leading to abnormalities in action potential (AP) conduction, or repolarization culminating into AF [55–57]. In animal studies, it has been reported that NOX played a crucial role for enhancing ROS activity in AF. In a porcine pacing-induced AF model, Dudley et al. demonstrated that AF increased superoxide along with Rac1, NOX, and xanthine oxidase activities [58]. Interestingly, Rac1 overexpression transgenic mice showed AF which was reversed by statin treatment [59]. Furthermore, Reilly S et al. reported that NOX activation may be obligated for early development of AF, whereas mitochondrion and uncoupled eNOS are essential in permanent AF [60]. Analysis of human LA myocardium of AF patients showed significant upregulation of Rac1 GTPase and NOX activity [59, 61]. Furthermore, the role of Nox2 and Nox4 is shown in tachypacing HL-1 atrial myocytes [62, 63]. Tachypacing induced ROS production and TGFβ expression and increased calpain activation for myofibril degradation. In summary, data showed that NOX-derived ROS are involved in the pathogenesis of AF, albeit the role of their specific isoforms remains unclear.

3.2.3 MicroRNA (miRNA) and Cardiac Remodeling

3.2.3.1 miRNA Biogenesis

RNA polymerase II is responsible for the transcription of long primary miRNA transcripts (pri-miRNA) in the nucleus [63]. The pri-miRNAs are then processed by the ribonuclease III, Drosha, to produce a precursor miRNA (pre-miRNA) that is approximately 100 nt long and has a hairpin-like structure. Another critical component of Drosha is RNA-binding protein DiGeorge critical region-8 (DGCR8) which specifically recognizes and facilitates the cleaving process of pri-miRNAs in the nucleus [64]. The pre-miRNA is subsequently translocating to the cytosol by exportin 5. Another ribonuclease III, Dicer, then diced the pre-miRNA further to generate double-stranded RNA [65]. Dicer further then facilitates loading the miRNAs into the RNA-induced silencing complex (RISC) by recruiting Argonaute 2 (Ago2) [66] (Fig. 3.1). The guide miRNA strand binds a complementary or partially complementary sequence in the 3′-untranslated region (UTR) of its target mRNA, followed by translocation to a processing body (P-body) for mRNA degradation where miRNAs guide the RISC to target genes by binding to imperfect complementary sites within the 3′UTRs [67].

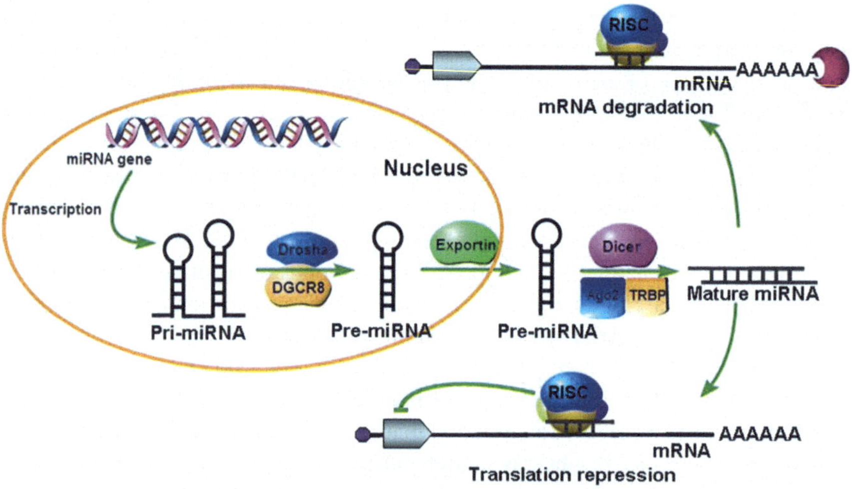

Fig. 3.1 miRNA biogenesis. The miRNAs are transcribed by RNA polymerase II as primary transcript of miRNA (pri-miRNA). The pri-miRNA is the cleaved by RNase III enzyme, Drosha, along with several cofactors including DGCR8 and produces the stem-loop precursor miRNA (pre-miRNA). The pre-miRNA is then exported out of the nucleus by Exportin-5 to the cytoplasm. In the cytoplasm, the pre-miRNA is diced-up by Dicer resulting in miRNA duplex, ~22 nucleotides long. The mature miRNA is incorporated into the RNA-induced silencing complex (RISC) which contains Argonaute (Ago) and is guided to the 3′-UTR of target mRNAs. The gene silencing is achieved by either mRNA degradation or translational repression. (Adopted from Modulation of miRNAs in Pulmonary Hypertension. Gupta S, Li L Int J Hypertens. 2015; 2015:169069)

Studies have shown that miRNAs are dysregulated in cardiac remodeling [68–71]. Furthermore, genetic studies (transgenic and knock-out mouse model) of miRNAs have established their function to cardiac pathologies, but miRNA-mediated ROS modulation is emerging. This review will discuss this important aspect.

3.2.3.2 Redox miRNAs and Redox Homeostasis

It is established that miRNAs are playing a key role in diverse cardiac remodeling process; however, it remains to be elucidated how miRNAs can orchestrate cellular redox homeostasis, a significant contributor in several cardiac pathologies. Recently, it has been reported that miRNAs are able to modulate the redox signaling by direct interaction with nuclear factor-erythroid 2 related factor 2 (Nrf2), Kelch-like ECH-associated protein 1 (Keap1) and CNC homolog 1 (Bach1); the critical transcriptional regulators of ROS [72, 73]. A new subset of miRNAs which regulate redox pathways or regulated by the cellular redox state have been coined as "redox-miRs" [74]. Homeostasis is defined as a self-regulating dynamic process in biological systems which provide a stability by modulating conditions that are favorable for survival. This review highlights the potential role of miRNAs in cellular redox homeostasis.

3.2.3.3 Redox Regulation by miRNAs in Cardiac Remodeling: "MyomiRs"

Redox regulation in various cardiac pathologies and cellular signaling has been well documented [25, 75–77]; however, the roles of miRNA(s) in regulating the ROS pathways are emerging. This review will summarize ROS-mediated miRNA modulation in cardiac remodeling.

Apparently, the first study that link the ROS with miRNA was the Dicer knockdown strategy in human microvascular endothelial cells. The outcome has drastically reduced the ROS production in response to various stimulus like TNFα or vascular endothelial growth factor [78]. Furthermore, reduced Dicer expression was observed under oxidative stress, but liberation of superoxide cannot be ignored for defense response [79]. Also, the Dicer level may be restored once the stress is removed [79] indicating a transient cellular response but the association is remarkable.

Doxycycline (Dox)-induced oxidative stress-mediated myocardial damage significantly increased miR-140-5p along with ROS leading to develop cardiotoxicity [80]. As a result of upregulation of miR-140-5p, the target genes, the Nrf2 and Sirt2, were reduced in Dox-induced rodent models (both rat and mouse). The Dox-induced oxidative damage was exacerbated when treated with miR-140-5p mimic; but the injury was mitigated with miR-140-5p inhibitor treatment [80]. Another study showed that overexpression of miRNA-132 attenuated TAC-induced cardiac hypertrophy and increased antioxidant SOD and Bcl_2, elicits a miRNA-ROS association [81]. Another miRNA, the miR-1, is abundantly expressed in the heart muscle [81] and is reduced in cardiac hypertrophy [82]. It is of note that miR-1 controlled

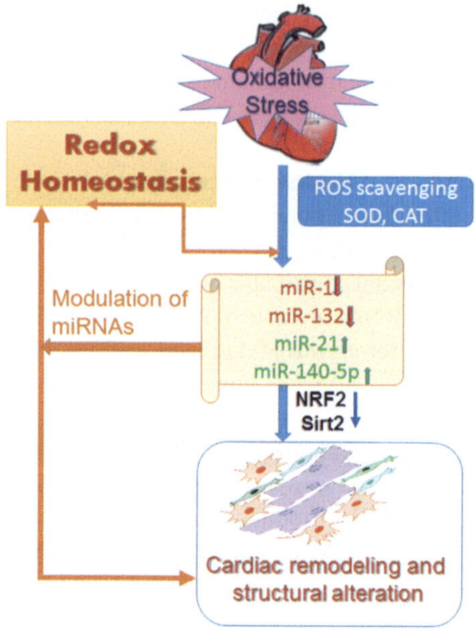

Fig. 3.2 Schematic presentation of ROS-mediated miRNA modulation in cardiac remodeling. ROS can modulate the miRNAs through posttranscriptional regulation of NRF2 and Sirt2 mRNAs. Green arrow indicates upregulatory pattern and red arrow indicated downregulatory pattern

cardiomyocyte growth by negatively regulating calmodulin, Mef2a, and Gata4 and all are critical for cardiac hypertrophy [82]. Furthermore, a recent study indicated ROS-dependent regulation of miR-1 targeting myocardin in cardiac hypertrophy [83]. This study is particularly interesting as it showed miR-1 mimic delivery attenuated TAC-induced cardiac hypertrophy [83, 84]. Together, the data underscore the significance of miR-1 in ROS-dependent cardiac remodeling. Another miRNA, the miR-21, is shown to participate a role in regulating ROS-induced by H_2O_2 [28]. Intracellular ROS was determined in miR-21 mimetic and inhibitor transfected cardiomyocytes and stained with DCFH-DA and DHE, respectively, indicators for H_2O_2 and O_2^- activities. Representatives of confocal microscopy images are shown in Fig. 3.2a. Intracellular ROS level was further measured by fluorimetry. Immunofluorescence images showed a 3.88 ± 0.15-fold ($p = 0.0003$) enhanced ROS activity in H_2O_2-treated cells, compared to unstimulated cells. Transient transfection with miR-21 inhibitor enhanced the ROS level to 4.38 ± 0.14-fold ($p = 0.019$), and miR-21 mimetic suppressed the ROS level by 3.09 ± 0.21-fold ($p = 0.045$), compared to H_2O_2 treatment as shown in Fig. 3.2b. Together, the data suggested that miR-21 has a potential role in regulating the ROS activity under oxidative stress. A schematic presentation of ROS-induced cardiac remodeling is shown in Fig. 3.3.

It is known that an alteration in the intracellular calcium (Ca^{2+}) is a critical regulator in signaling mechanism and responsible for cardiac dysfunction including hypertrophy, apoptosis, or arrhythmia. It has been demonstrated that miR-145 regulates ROS-induced Ca^{2+} overload in cardiomyocytes [85]. Another model, cardiac

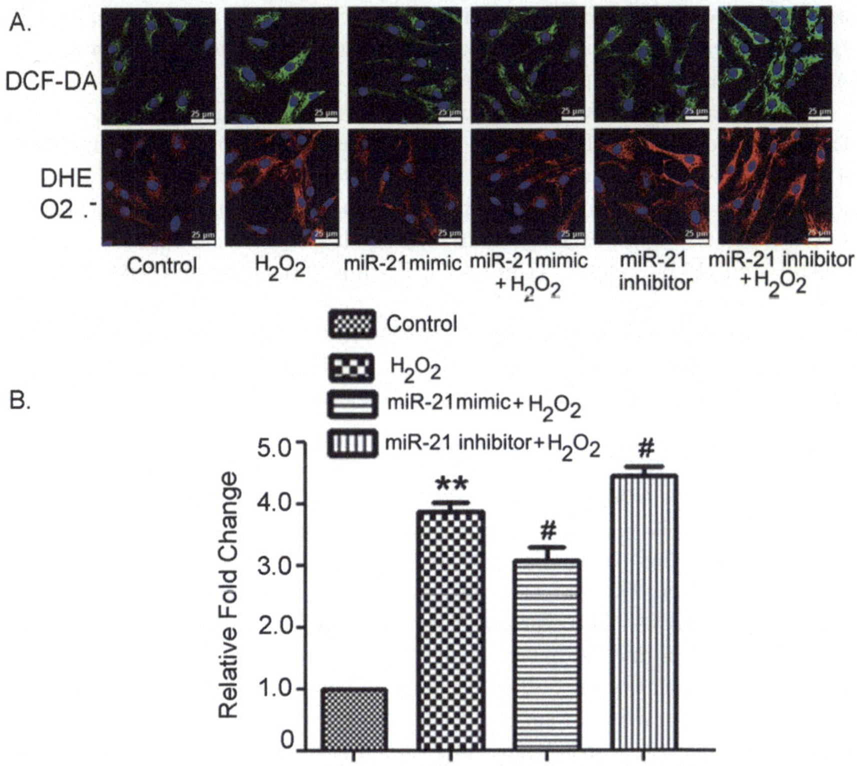

Fig. 3.3 Overexpression of miR-21 attenuates H_2O_2-induced ROS level in neonatal cardiomyocytes. The ROS level was measured in transfected neonatal cardiomyocytes with miR-21 mimetic and inhibitor followed by H_2O_2 treatment for 24 h by confocal microscopy and fluorimetry. (**a**) Representative confocal microscopy images of cardiomyocytes stained with DCFH-DA and DHE, respectively, showing the activity of H_2O_2 and O_2^-. (**b**) Effect of miR-21 mimetic and inhibitor on generation of ROS in cardiomyocytes treated with H_2O_2 by fluorimetry. The data presented are mean ± SE. $^{**}P < 0.01$ vs. control, $^{\#}P < 0.05$ vs. H_2O_2 treatment ($n = 3$). (Adopted from Wei C, Li L, Kim IK, Sun P, Gupta S. Free Radic Res. 2014, Reference 28)

ischemia and reperfusion (I/R), caused cardiomyocyte apoptosis/necrosis, liberated ROS, and caused myocardial injury. As reperfusion increases oxidative stress and tissue damage, the set of miRNAs that are deregulated are often termed as "hypoxamiRs." Several evidences have accumulated a wealth of data linking miRNAs, such as let-7 family, miR-1, miR-133a/b, miR-19a/b, miR-150, miR-195, miR-199a, miR-221, miR-23a/b, and miR-320, to ischemic diseases [for recent comprehensive reviews, see Refs [86–91]]. Hypoxia-inducible factor 1 alpha (HIF1α) is a master transcriptional regulator to hypoxia [92]. Remarkably, miRNAs are critical modulator of HIF1α. For instance, miR-199a is downregulated in cardiac myocytes upon hypoxia, and this reduction is required for rapid upregulation of its target, HIF1α [91]. Another study indicated that therapeutic inhibition of miR-34 showed restoration of ischemia-induced cardiac remodeling and improves recovery [93].

Together, it is speculated that "hypoxamiRs" are critical in the regulation of the HIF system via negative and positive feedback loops and orchestrated a fine-tuning platform in response to the hypoxic response.

As discussed earlier that miRNAs participate in many cardiac diseases, a subset of miRNA is cardiac muscle-specific and have been called myomiRs. The first study on myomiRs was conducted by Sempere et al. [94] listing miR-1, miR-133a, and miR-206 as they are highly enriched both in human and mouse heart and skeletal muscle. Later, few isoforms and miR-208 were added, and currently the list included miR-1-1, miR133a-2, miR1–2/miR133a-1, miR-206, and miR-208 [95–97]. Interestingly, a low level of myomiRs including miR-367, miR-302b/c/d, and miR-499 is reportedly expressed in cardiac tissues [98]. Together, evidence suggested that a complex network of myomiRs posttranscriptional-regulated gene expression may orchestrated overall cardiac growth and function.

3.2.3.4 Redox Regulation by miRNAs in Cardiac Fibrosis: "FibromiRs"

Cardiac fibrosis is a debilitating process and one of the major contributors to the failing heart. Cardiac fibrosis is defined as an excessive deposition of ECM proteins in the interstitium and in perivascular region of the heart resulting the stiffness and impairing the cardiac function [99, 100]. The miRNAs are the regulators of fibrotic processes, and a subset of miRNAs that contribute to this event are called "fibromiRs" [101]. The first study of role of miRNA in cardiac fibrosis was demonstrated by Eva von Rooij et al. and established that miR-208 is a key mediator of cardiac fibrosis [102]. Since then, a large body of evidence suggested the role of several miRNAs in cardiac fibrosis and their potential in therapeutic intervention [103–108]. Previous studies have shown that miR-29 family members are downregulated in a variety of organ fibrosis including cardiac, renal, liver, etc., and considered as a "FibromiR" [102, 104, 109]. There are several reports in ROS-mediated lung, liver, and kidney fibrosis [110–115], but there are little evidence that showed a direct association between miRNA-ROS and cardiac fibrosis; however, the concept is emerging.

A recent report indicated that subclinical lipopolysaccharide (LPS) induced cardiac fibrosis by upregulating collagens, MMP2, MMP9, TIMP1, TIMP2, and periostin but not CTGF [116]. Mechanistically, authors showed a downregulation of miR-29c and upregulation of NOX2 suggesting a link between ROS-miRNA and cardiac fibrosis [116]. Another antioxidant defense molecule, the Nrf2, a major transcriptional regulator, is closely linked with ROS-mediated cardiac remodeling [25, 72] and was regulated by miR-140-5p [80]. Furthermore, a recent study showed a set of miRNAs (miR-27a, miR-28-3p, and miR-34a) which were upregulated and preferentially incorporated into a vesicle (exosome) followed by secretion into the extracellular milieu and eventually reduced the Nrf2 level in myocardial infarction model [117] (Fig. 3.4). The finding is interesting and provides a new mechanism of MI-induced miRNAs that contribute to oxidative stress by inhibiting Nrf2. Data indicated potential new therapy to target miRs and NOX2/Nrf2 warrants further investigation.

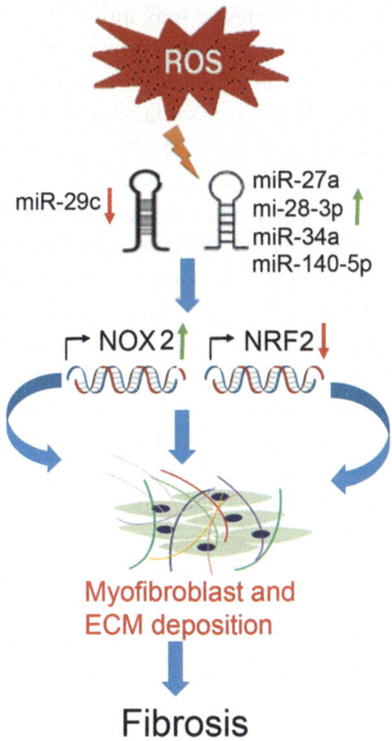

Fig. 3.4 Interconnection of ROS with miRNA expression in the event of fibrosis. ROS can modulate the miRNAs through posttranscriptional regulation of NOX2 and NRF2 mRNA. Green arrow indicates upregulatory pattern, and red arrow indicated downregulatory pattern

3.2.3.5 Redox Regulation by miRNA in Arrhythmia

Accumulating evidence showed involvement of miRNAs in arrhythmogenesis [118]. Seahyoung Lee et al. have tabulated a list of miRNAs by literature review and revealed that ROS-mediated miRNAs are associated with cardiac arrhythmia [119]. The set of miRNAs that showed a possible association with ROS and arrhythmia are miR-1, -19a, -21, -26, -30, -133, -145, and -499. The study connected to ROS-miRNA-arrhythmia is limited, but the possibility that ROS-mediated miRNA modulation contributed either directly or indirectly to arrhythmia still exists. It is of note that the possibility of single nucleotide polymorphism (SNP) in the miRNA coding region for the development of arrhythmia cannot be ignored. There are two reports that showed a correlation between an SNP and atrial fibrillation in Chinese population [120, 121]. Therefore, alteration of miRNA in arrhythmia in response to ROS may indicate a possibility for arrhythmia and warrants further investigation with appropriate experimental conditions or models to understand the underlying mechanism.

3.3 Conclusion

Oxidative stress is continuously showing a significant contributor to cardiac diseases. As a result, its modulation is likely to be attractive at therapeutic standpoint. Accordingly, antioxidant therapies are in the front line for treatment for many decades. Along the line, miRNAs are now floated on the surface for cardiac disease therapeutics. It might be interesting to target miRNA(s) that control ROS and as a result we can achieve two goals in one effort; one is ROS and the other is the specific gene that is associated within it. The beauty of targeting miRNA is that a single miRNA can regulate the expression of several genes and the gene network. Development of specific inhibitor and mimic of miRNA that can be locally or systemically delivered to the specific organ would be a great tool for designing therapeutic molecule. However, a proper and strategic design is required for therapeutic intervention. Furthermore, "off-target" effects should be considered because of multiple gene target and the organ/tissue specificity which is challenging. Although redox-miR-axis is emerging, a significant progress is made in determining the magnitude of several aspects of cardiac remodeling. It seems to take a long run to achieve clinical testing of these agents in cardiac diseases, but a pathway appears to be propitious. The findings showed that miRNAs regulating ROS have uncovered a potential use for treatment of cardiac remodeling via redox-based mechanism of "myomiRs" or "fibromiRs" or together we can say "myofibromiRs."

Moreover, we should acknowledge a new class of noncoding RNA, long noncoding RNA (Lnc RNA), or small nucleolar RNA (snoRNA) which are implicated in a number of diseases including cardiac under oxidative stress condition with a wide range of functions. However, lncRNAs consist of a large number of transcripts may have some difficulties in targeting for specific drug intervention but cannot be ignored.

Acknowledgments The work was partly supported by funds from American Heart Association: Grant-in-Aid (17GRNT33670897) to S Gupta. The material is the result of work supported with resources and the use of facilities at the Central Texas Veterans Affairs Health Care System, Temple, Texas; VISN 17 Center of Excellence on Returning War Veterans, Waco, Texas, USA.

References

1. Takimoto E, Kass DA (2007) Role of oxidative stress in cardiac hypertrophy and remodeling. Hypertension 49:241–248
2. Thomas DD, Ridnour LA, Isenberg JS, Flores-Santana W, Switzer CH, Donzelli S, Hussain P, Vecoli C, Paolocci N, Ambs S, Colton CA, Harris CC, Roberts DD, Wink DA (2008) The chemical biology of nitric oxide: implications in cellular signaling. Free Radic Biol Med 45(1):18–31
3. Afanas'ev I (2011) ROS and RNS signaling in heart disorders: could antioxidant treatment be successful? Oxidative Med Cell Longev 2011:293769
4. Fridovich I (1997) Superoxide anion radical (O2-.), superoxide dismutases, and related matters. J Biol Chem 272:18515–18517
5. Suematsu N, Tsutsui H, Wen J, Kang D, Ikeuchi M, Ide T, Hayashidani S, Shiomi T, Kubota T, Hamasaki N, Takeshita A (2003) α-Oxidative stress mediates tumor necrosis factor-induced mitochondrial DNA damage and dysfunction in cardiac myocytes. Circulation 107:1418–1423
6. Siwik DA, Colucci WS (2004) Regulation of matrix metalloproteinases by cytokines and reactive oxygen/nitrogen species in the myocardium. Heart Fail Rev 9:43–51
7. Lee RC, Feinbaum RL, Ambros V (2003) The *C. elegans* heterochronic gene lin-4 encodes small RNAs with antisense complementarity to lin-14. Cell 75(5):843–854

8. Hill MF, Singal PK (1996) Antioxidant and oxidative stress changes during heart failure subsequent to myocardial infarction in rats. Am J Pathol 148:291–300
9. Pacher P, Beckman JS, Liaudet L (2007) Nitric oxide and peroxynitrite in health and disease. Physiol Rev 87:315–424
10. Tsutsui H, Kinugawa S, Matsushima S (2011) Oxidative stress and heart failure. Am J Physiol Heart Circ Physiol 301(6):H2181–H2190
11. Seddon M, Looi YH, Shah AM (2007) Oxidative stress and redox signaling in cardiac hypertrophy and heart failure. Heart 93(8):903–907
12. Cave A, Grieve D, Johar S, Zhang M, Shah AM (2005) NADPH oxidase-derived reactive oxygen species in cardiac pathophysiology. Philos Trans R Soc Lond Ser B Biol Sci 360(1464):2327–2334
13. Ide T, Tsutsui H, Kinugawa S, Utsumi H, Kang D, Hattori N, Uchida K, Arimura K, Egashira K, Takeshita A (1999) Mitochondrial electron transport complex I is a potential source of oxygen free radicals in the failing myocardium. Circ Res 85:357–363
14. Dröge W (2002) Free radicals in the physiological control of cell function. Physiol Rev 82:47–95
15. Gongora MC, Qin Z, Laude K, Kim HW, McCann L, Folz JR, Dikalov S, Fukai T, Harrison DG (2006) Role of extracellular superoxide dismutase in hypertension. Hypertension 48:473–481
16. Assem M, Teyssier JR, Benderitter M, Terrand J, Laubriet A, Javouhey A, David M, Rochette L (1997) Pattern of superoxide dismutase enzymatic activity and RNA changes in rat heart ventricles after myocardial infarction. Am J Pathol 151:549–555
17. Guo Z, Xia Z, Jiang J, McNeill JH (2007) Downregulation of NADPH oxidase, antioxidant enzymes, and inflammatory markers in the heart of streptozotocin-induced diabetic rats by N-acetyl-L-cysteine. Am J Physiol Heart Circ Physiol 292(4):H1728–H1736
18. Nabeebaccus A, Zhang M, Shah AM (2011) NADPH oxidases and cardiac remodeling. Heart Fail Rev 16(1):5–12
19. Zhang M, Perino A, Ghigo A, Hirsch E, Shah AM (2013) NADPH oxidases in heart failure: poachers or gamekeepers? Antioxid Redox Signal 18(9):1024–1041
20. Swynghedauw B (1999) Molecular mechanisms of myocardial remodeling. Physiol Rev 79:215–262
21. Nakamura K, Fushimi K, Kouchi H et al (1998) Inhibitory effects of antioxidants on neonatal rat cardiac myocyte hypertrophy induced by tumor necrosis factor-alpha and angiotensin II. Circulation 98:794–799
22. Ushio-Fukai M, Alexander RW, Akers M et al (1999) Reactive oxygen species mediate the activation of Akt/protein kinase B by angiotensin II in vascular smooth muscle cells. J Biol Chem 274(32):22699–22704
23. Laderoute KR, Webster KA (1997) Hypoxia/reoxygenation stimulates Jun kinase activity through redox signaling in cardiac myocytes. Circ Res 80(3):336–344
24. Sag CM, Santos CX, Shah AM (2014) Redox regulation of cardiac hypertrophy. J Mol Cell Cardiol 73:103–111
25. Barančík M, Grešová L, Barteková M, Dovinová I (2016) Nrf2 as a key player of redox regulation in cardiovascular diseases. Physiol Res 65(Suppl 1):S1–S10
26. Clerk A, Michael A, Sugden PH (1998) Stimulation of multiple mitogen-activated protein kinase sub-families by oxidative stress and phosphorylation of the small heat shock protein, HSP25/27, in neonatal ventricular myocytes. Biochem J 333:581–589
27. Kang PM, Haunstetter A, Aoki H et al (2010) Morphological and molecular characterization of adult cardiomyocyte apoptosis during hypoxia and reoxygenation. Circ Res 87:118–125
28. Wei C, Li L, Kim IK, Sun P, Gupta S (2014) NF-κB mediated miR-21 regulation in cardiomyocytes apoptosis under oxidative stress. Free Radic Res 48(3):282–291
29. Nakagami H, Takemoto M, Liao JK (2003) NADPH oxidase-derived superoxide anion mediates angiotensin II-induced cardiac hypertrophy. J Mol Cell Cardiol 35(7):851–859
30. Santillo M, Colantuoni A, Mondola P, Guida B, Damiano S (2015) NOX signaling in molecular cardiovascular mechanisms involved in the blood pressure homeostasis. Front Physiol 6:194

31. Li JM, Gall NP, Grieve DJ, Chen M, Shah AM (2002) Activation of NADPH oxidase during progression of cardiac hypertrophy to failure. Hypertension 40(4):477–484
32. Bendall JK, Cave AC, Heymes C, Gall N, Shah AM (2002) Pivotal role of a gp91phox-containing NADPH oxidase in angiotensin II-induced cardiac hypertrophy in mice. Circulation 105:293–296
33. Grieve DJ, Byrne JA, Siva A, Layland J, Johar S, Cave AC, Shah AM (2006) Involvement of the nicotinamide adenosine dinucleotide phosphate oxidase isoform Nox2 in cardiac contractile dysfunction occurring in response to pressure overload. J Am Coll Cardiol 47:817–826
34. Byrne JA, Grieve DJ, Bendall JK, Li JM, Gove C, Lambeth JD, Cave AC, Shah AM (2003) Contrasting roles of NADPH oxidase isoforms in pressure-overload versus angiotensin II-induced cardiac hypertrophy. Circ Res 93:802–805
35. Maytin M, Siwik DA, Ito M, Xiao L, Sawyer DB, Liao R, Colucci WS (2004) Pressure overload-induced myocardial hypertrophy in mice does not require gp91phox. Circulation 109:1168–1171
36. Kang PT, Chen CL, Ohanyan V, Luther DJ, Meszaros JG, Chilian WM, Chen YR (2015) Overexpressing superoxide dismutase 2 induces a supernormal cardiac function by enhancing redox-dependent mitochondrial function and metabolic dilation. J Mol Cell Cardiol 88:14–28
37. Van Deel ED, Lu Z, Xu X, Zhu G, Hu X, Oury TD, Bache RJ, Duncker DJ, Chen Y (2008) Extracellular superoxide dismutase protects the heart against oxidative stress and hypertrophy after myocardial infarction. Free Radic Biol Med 44(7):1305–1313
38. Kuroda J, Ago T, Matsushima S, Zhai P, Schneider MD, Sadoshima J (2010) NADPH oxidase 4 (Nox4) is a major source of oxidative stress in the failing heart. Proc Natl Acad Sci U S A 107(35):15565–15570
39. Zhang M, Brewer AC, Schröder K, Santos CX, Grieve DJ, Wang M, Anilkumar N, Yu B, Dong X, Walker SJ, Brandes RP, Shah AM (2010) NADPH oxidase-4 mediates protection against chronic load-induced stress in mouse hearts by enhancing angiogenesis. Proc Natl Acad Sci U S A 107(42):18121–18126
40. Li Y, Huang TT, Carlson EJ, Melov S, Ursell PC, Olson JL, Noble LJ, Yoshimura MP, Berger C, Chan PH, Wallace DC, Epstein CJ (1995) Dilated cardiomyopathy and neonatal lethality in mutant mice lacking manganese superoxide dismutase. Nat Genet 11(4):376–381
41. Creemers EE, Pinto Y (2011) Molecular mechanisms that control interstitial fibrosis in the pressure-overloaded heart. Cardiovasc Res 89(2):265–272
42. Segura AM, Frazier OH, Buja LM (2014) Fibrosis and heart failure. Heart Fail Rev 19:173–185
43. Takeda N, Manabe I, Uchino Y, Eguchi K, Matsumoto S et al (2010) Cardiac fibroblasts are essential for the adaptive response of the murine heart to pressure overload. J Clin Invest 120:254–265
44. Leask A (2007) TGFβ, cardiac fibroblasts, and the fibrotic response. Cardiovasc Res 74:207–212
45. Dickinson BC, Chang CJ (2011) Chemistry and biology of reactive oxygen species in signaling or stress responses. Nat Chem Biol 7:504–511
46. Nediani C, Raimondi L, Borchi E, Cerbai E (2011) Nitric oxide/reactive oxygen species generation and nitroso/redox imbalance in heart failure: from molecular mechanisms to therapeutic implications. Antioxid Redox Signal 14:289–331
47. Zhang Y, Tocchetti CG, Krieg T, Moens AL (2012) Oxidative and nitrosative stress in the maintenance of myocardial function. Free Radic Biol Med 53:1531–1540
48. Zhao Y, McLaughlin D, Robinson E, Harvey AP, Hookham MB, Shah AM, McDermott BJ, Grieve D (2010) Nox2 NADPH oxidase promotes pathologic cardiac remodeling associated with doxorubicin chemotherapy. Cancer Res 70:9287–9297
49. Cucoranu I, Clempus R, Dikalova A, Phelan PJ, Ariyan S, Dikalov S, Sorescu D (2005) NAD(P)H oxidase 4 mediates transforming growth factor-beta1-induced differentiation of cardiac fibroblasts into myofibroblasts. Circ Res 97:900–907

50. Kumar S, Gupta S (2011) Thymosin beta 4 prevents oxidative stress by targeting antioxidant and anti-apoptotic genes in cardiac fibroblasts. PLoS One 6(10):e26912
51. Alpert JS (2000) Atrial fibrillation: a growth industry in the 21st century. Eur Heart J 21(15):1207–1208
52. Tse G, Yan BP, Chan YW, Tian XY, Huang Y (2016) Reactive oxygen species, endoplasmic reticulum stress and mitochondrial dysfunction: the link with cardiac arrhythmogenesis. Front Physiol 7:313
53. Roger VL, Go AS, Lloyd-Jones DM, Benjamin EJ, Berry JD, Borden WB, Bravata DM, Dai S, Ford ES, Fox CS, Fullerton HJ, Gillespie C, Hailpern SM, Heit JA, Howard VJ, Kissela BM, Kittner SJ, Lackland DT, Lichtman JH, Lisabeth LD, Muc DM, Marcus GM, Marelli A, Matchar DB, Moy CS, Mozaffarian D, Mussolino ME, Nichol G, Paynter NP, Soliman EZ, Sorlie PD, Sotoodehnia N, Turan TN, Virani SS, Wong ND, Woo D, Turner MB (2012) Heart disease and stroke statistics–2012 update: a report from the American Heart Association. American Heart Association Statistics Committee and Stroke Statistics Subcommittee. Circulation 125(1):e2–e220
54. Chang JP, Chen MC, Liu WH, Yang CH, Chen CJ, Chen YL, Pan KL, Tsai TH, Chang HW (2011) Atrial myocardial nox2 containing NADPH oxidase activity contribution to oxidative stress in mitral regurgitation: potential mechanism for atrial remodeling. Cardiovasc Pathol 20(2):99–106
55. Aggarwal NT, Makielski JC (2013) Redox control of cardiac excitability. Antioxid Redox Signal 18(4):4324–4368
56. Tse G, Yeo JM (2015) Conduction abnormalities and ventricular arrhythmogenesis: the roles of sodium channels and gap junctions. IJC Heart Vasc 9:75–82
57. Tse G, Yan BP, Chan YWF, Tian XY, Huang Y (2016) Reactive oxygen species, endoplasmic reticulum stress and mitochondrial dysfunction: the link with cardiac arrhythmogenesis. Front Physiol 7:313
58. Dudley SC Jr, Hoch NE, McCann LA, Honeycutt C, Diamandopoulos L, Fukai T, Harrison DG, Dikalov SI, Langberg J (2005) Atrial fibrillation increases production of superoxide by the left atrium and left atrial appendage: role of the NADPH and xanthine oxidases. Circulation 112(9):1266–1273
59. Adam O, Frost G, Custodis F, Sussman MA, Schäfers HJ, Böhm M, Laufs U (2007) Role of Rac1 GTPase activation in atrial fibrillation. J Am Coll Cardiol 50(4):359–367
60. Reilly SN, Jayaram R, Nahar K, Antoniades C, Verheule S, Channon KM, Alp NJ, Schotten U, Casadei B (2011) Atrial sources of reactive oxygen species vary with the duration and substrate of atrial fibrillation: implications for the antiarrhythmic effect of statins. Circulation 124(10):1107–1117
61. Kim YM, Guzik TJ, Zhang YH, Zhang MH, Kattach H, Ratnatunga C et al (2005) A myocardial Nox2 containing NAD(P)H oxidase contributes to oxidative stress in human atrial fibrillation. Circ Res 97:629–636
62. Yeh YH, Kuo CT, Chan TH, Chang GJ, Qi XY, Tsai F, Nattel S, Chen WJ (2011) Transforming growth factor-β and oxidative stress mediate tachycardia-induced cellular remodeling in cultured atrial-derived myocytes. Cardiovasc Res 91(1):62–70
63. Ha M, Kim VN (2014) Regulation of microRNA biogenesis. Nat Rev Mol Cell Biol 15(8):509–524
64. Hutvágner G, McLachlan J, Pasquinelli AE, Bálint E, Tuschl T, Zamore PD (2001) A cellular function for the RNA-interference enzyme dicer in the maturation of the let-7 small temporal RNA. Science 293(5531):834–838
65. Han J, Lee Y, Yeom KH, Kim YK, Jin H, Kim VN (2004) The Drosha–DGCR8 complex in primary microRNA processing. Genes Dev 18:3016–3027
66. Bartel DP (2009) MicroRNAs: target recognition and regulatory functions. Cell 136(2):215–233
67. Liu J, Valencia-Sanchez MA, Hannon GJ, Parker R (2005) MicroRNA-dependent localization of targeted mRNAs to mammalian P-bodies. Nat Cell Biol 7(7):719–723

68. Chen C, Ponnusamy M, Liu C, Gao J, Wang K, Li P (2017) MicroRNA as a therapeutic target in cardiac remodeling. Biomed Res Int 2017:1278436
69. Kumarswamy R, Thum T (2013) Non-coding RNAs in cardiac remodeling and heart failure. Circ Res 113(6):676–689
70. De Rosa S, Curcio A, Indolfi C (2014) Emerging role of microRNAs in cardiovascular diseases. Circ J 78(3):567–575
71. Divakaran V, Mann DL (2008) The emerging role of microRNAs in cardiac remodeling and heart failure. Circ Res 103(10):1072–1083
72. Hayes JD, Dinkova-Kostova AT (2014) The Nrf2 regulatory network provides an interface between redox and intermediary metabolism. Trends Biochem Sci 39:199–218
73. Levonen AL, Hill BG, Kansanen E, Zhang J, Darley-Usmar VM (2014) Redox regulation of antioxidants, autophagy, and the response to stress: implications for electrophile therapeutics. Free Radic Biol Med 71:196–207
74. Cheng X, Ku CH, Siow RC (2013) Regulation of the Nrf2 antioxidant pathway by microRNAs: new players in micromanaging redox homeostasis. Free Radic Biol Med 64:4–11
75. Lassegue B, Griendling KK (2010) NADPH oxidases: functions and pathologies in the vasculature. Arterioscler Thromb Vasc Biol 30:653–661
76. Cadenas S (2018) ROS and redox signaling in myocardial ischemia-reperfusion injury and cardioprotection. Free Radic Biol Med 117:76–89
77. Kanaan GN, Harper ME (2017) Cellular redox dysfunction in the development of cardiovascular diseases. Biochim Biophys Acta 1861(11 Pt A):2822–2829
78. Shilo S, Roy S, Khanna S, Sen CK (2008) Evidence for the involvement of miRNA in redox regulated angiogenic response of human microvascular endothelial cells. Arterioscler Thromb Vasc Biol 28:471–477
79. Mori MA, Raghavan P, Thomou T, Boucher J, Robida-Stubbs S, Macotela Y, Russell SJ, Kirkland JL, Blackwell TK, Kahn CR (2012) Role of microRNA processing in adipose tissue in stress defense and longevity. Cell Metab 16:336–347
80. Zhao L, Qi Y, Xu L, Tao X, Han X, Yin L, Peng J (2018) MicroRNA-140-5p aggravates doxorubicin-induced cardiotoxicity by promoting myocardial oxidative stress via targeting Nrf2 and Sirt2. Redox Biol 15:284–296
81. Liu X, Tong Z, Chen K, Hu X, Jin H, Hou M (2018) The role of miRNA-132 against apoptosis and oxidative stress in heart failure. Biomed Res Int 2018:3452748
82. Lagos-Quintana M, Rauhut R, Yalcin A, Meyer J, Lendeckel W, Tuschl T (2002) Identification of tissue-specific microRNAs from mouse. Curr Biol 12:735–739
83. Sayed D, Hong C, Chen IY, Lypowy J, Abdellatif M (2007) MicroRNAs play an essential role in the development of cardiac hypertrophy. Circ Res 100:416–442
84. Lee S, Lim S, Ham O, Lee SY, Lee CY, Park JH, Lee J, Seo HH, Yun I, Han SM, Cha MJ, Choi E, Hwang KC (2015) ROS-mediated bidirectional regulation of miRNA results in distinct pathologic heart conditions. Biochem Biophys Res Commun 465(3):349–355
85. Cha MJ, Jang JK, Ham O, Song BW, Lee SY, Lee CY, Park JH, Lee J, Seo HH, Choi E, Jeon WM, Hwang HJ, Shin HT, Choi E, Hwang KC (2013) MicroRNA-145 suppresses ROS-induced $Ca2+$ overload of cardiomyocytes by targeting CaMKIIδ. Biochem Biophys Res Commun 435(4):720–726
86. Fasanaro P, Greco S, Ivan M, Capogrossi MC, Martelli F (2010) microRNA: emerging therapeutic targets in acute ischemic diseases. Pharmacol Ther 125:92–104
87. Fiedler J, Thum T (2013) MicroRNAs in myocardial infarction. Arterioscler Thromb Vasc Biol 33(2):201–205
88. Greco S, Gaetano C, Martelli F (2014) HypoxamiR regulation and function in ischemic cardiovascular diseases. Antioxid Redox Signal 21(8):1202–1219
89. Bertero T, Robbe-Sermesant K, Le Brigand K, Ponzio G, Pottier N, Rezzonico R, Mazure NM, Barbry P, Mari B (2014) MicroRNA target identification: lessons from hypoxamiRs. Antioxid Redox Signal 21(8):1249–1268
90. Azzouzi HE, Leptidis S, Doevendans PA, De Windt LJ (2015) HypoxamiRs: regulators of cardiac hypoxia and energy metabolism. Trends Endocrinol Metab 26(9):502–508

91. Rane S, He M, Sayed D, Vashistha H, Malhotra A, Sadoshima J, Vatner DE, Vatner SF, Abdellatif M (2009) Downregulation of miR-199a derepresses hypoxia-inducible factor-1alpha and Sirtuin 1 and recapitulates hypoxia preconditioning in cardiac myocytes. Circ Res 104:879–886
92. Majmundar AJ, Wong WJ, Simon MC (2010) Hypoxia-inducible factors and the response to hypoxic stress. Mol Cell 40(2):294–309
93. Bernardo BC, Gao XM, Winbanks CE, Boey EJH, Tham YK, Kiriazis H et al (2012) Therapeutic inhibition of the miR-34 family attenuates pathological cardiac remodeling and improves heart function. Proc Natl Acad Sci U S A 109:17615–17620
94. Sempere LF, Freemantle S, Pitha-Rowe I, Moss E, Dmitrovsky E, Ambros V (2004) Expression profiling of mammalian microRNAs uncovers a subset of brain-expressed microRNAs with possible roles in murine and human neuronal differentiation. Genome Biol 5:R13
95. McCarthy JJ (2008) MicroRNA-206: the skeletal muscle-specific myomiR. Biochim Biophys Acta 1779(11):682–691
96. Malizia AP, Wang DZ (2011) MicroRNAs in cardiomyocyte development. Wiley Interdiscip Rev Syst Biol Med 3(2):183–190
97. Horak M, Novak J, Bienertova-Vasku J (2016) Muscle-specific microRNAs in skeletal muscle development. Dev Biol 410(1):1–13
98. Lee EJ, Baek M, Nuovo GJ, Chen C, Schmittgen TD (2008) Systematic evaluation of microRNA processing patterns in tissues, cell lines and tumors. RNA 14:1–8
99. Wynn TA, Ramalingam TR (2012) Mechanisms of fibrosis: therapeutic translation for fibrotic disease. Nat Med 18:1028–1040
100. Davis J, Molkentin JD (2014) Myofibroblasts: trust your heart and let fate decide. J Mol Cell Cardiol 70:9–18
101. Pottier N, Cauffiez C, Perrais M, Barbry P, Mari B (2014) FibromiRs: translating molecular discoveries into new anti-fibrotic drugs. Trends Pharmacol Sci 35:119–126
102. van Rooij E, Sutherland LB, Qi X, Richardson JA, Hill J, Olson EN (2007) Control of stress-dependent cardiac growth and gene expression by a microRNA. Science 316:575–579
103. Wang J, Liew OW, Richards AM, Chen YT (2016) Overview of MicroRNAs in cardiac hypertrophy, fibrosis, and apoptosis. Int J Mol Sci 17(5):pii: E749
104. Creemers EE, van Rooij E (2016) Function and therapeutic potential of noncoding RNAs in cardiac fibrosis. Circ Res 118(1):108–118
105. Jiang X, Tsitsiou E, Herrick SE, Lindsay MA (2010) MicroRNAs and the regulation of fibrosis. FEBS J 277(9):2015–2021
106. He Y, Huang C, Lin X, Li J (2013) MicroRNA-29 family, a crucial therapeutic target for fibrosis diseases. Biochimie 95(7):1355–1359
107. Wei C, Kim IK, Kumar S, Jayasinghe S, Hong N, Castoldi G, Catalucci D, Jones WK, Gupta S (2013) NF-κB mediated miR-26a regulation in cardiac fibrosis. J Cell Physiol 228(7):1433–1442
108. Li L, Bounds KR, Chatterjee P, Gupta S (2017) MicroRNA-130a, a potential antifibrotic target in cardiac fibrosis. J Am Heart Assoc 6(11):pii: e006763
109. Deng Z, He Y, Yang X, Shi H, Shi A, Lu L, He L (2017) MicroRNA-29: a crucial player in fibrotic disease. Mol Diagn Ther 21(3):285–294
110. Liu G, Friggeri A, Yang Y, Milosevic J, Ding Q, Thannickal VJ, Kaminski N, Abraham E (2010) miR-21 mediates fibrogenic activation of pulmonary fibroblasts and lung fibrosis. J Exp Med 207:1589–1597
111. Yang JJ, Tao H, Hu W, Liu LP, Shi KH, Deng ZY, Li J (2014) MicroRNA-200a controls Nrf2 activation by target Keap1 in hepatic stellate cell proliferation and fibrosis. Cell Signal 26(11):2381–2389
112. Fierro-Fernández M, Busnadiego Ó, Sandoval P, Espinosa-Díez C, Blanco-Ruiz E, Rodríguez M, Pian H, Ramos R, López-Cabrera M, García-Bermejo ML, Lamas S (2015) miR-9-5p suppresses pro-fibrogenic transformation of fibroblasts and prevents organ fibrosis by targeting NOX4 and TGFBR2. EMBO Rep 16(10):1358–1377

113. Civantos E, Bosch E, Ramirez E, Zhenyukh O, Egido J, Lorenzo O, Mas S (2017) Sitagliptin ameliorates oxidative stress in experimental diabetic nephropathy by diminishing the miR-200a/Keap-1/Nrf2 antioxidant pathway. Diab Metab Syndr Obes 10:207–222
114. Fierro-Fernández M, Miguel V, Lamas S (2016) Role of redoximiRs in fibrogenesis. Redox Biol 7:58–67
115. Latella G (2018) Redox imbalance in intestinal fibrosis: beware of the TGFβ-1, ROS, and Nrf2 connection. Dig Dis Sci 63(2):312–320
116. Lew WY, Bayna E, DalleMolle E, Contu R, Condorelli G, Tang T (2014) Myocardial fibrosis induced by exposure to subclinical lipopolysaccharide is associated with decreased miR-29c and enhanced NOX2 expression in mice. PLoS One 9(9):e107556
117. Tian C, Gao L, Zimmerman MC, Zucker IH (2018) Myocardial infarction-induced microRNA-enriched exosomes contribute to cardiacNrf2 dysregulation in chronic heart failure. Am J Physiol Heart Circ Physiol 314(5):H928–H939
118. Kim GH (2013) MicroRNA regulation of cardiac conduction and arrhythmias. Transl Res 161:381–392
119. Lee S, Choi E, Cha MJ, Hwang KC (2014) Looking into a conceptual framework of ROS-miRNA-atrial fibrillation. Int J Mol Sci 15(12):21754–21776
120. Li C, Wang F, Yang Y, Fu F, Xu C, Shi L, Li S, Xia Y, Wu G, Cheng X et al (2011) Significant association of SNP rs2106261 in the *ZFHX3* gene with atrial fibrillation in a Chinese Han GeneID population. Hum Genet 129:239–246
121. Su Y, Li J, Chen F, Geng H, Pan M (2014) A polymorphism rs11614913 in pre-microRNAs may be associated with atrial fibrillation in Han Chinese population. J Am Coll Cardiol 64:GW25–e0274

Myocardial Oxidative Stress and Metabolic Diseases

4

Hassan I. H. El-Sayyad

Abstract

Aging, diabetes, obesity, atherosclerosis, hypertension, and dyslipidemia shared similar manifestations of cardiomyopathy. This disease is characterized by pathological, cytological, and molecular alterations of both cardiomyocytes and endothelial cells. The disease altered the cytoplasmic organelle structure and function such as mitochondria, endoplasmic reticulum, Golgi apparatus, and lysosomes. These involved dramatic changes of protein synthesis in endoplasmic reticulum, calcium storage in mitochondria, autophagy in lysosomes, and lipid metabolism in Golgi complex. Enhanced lipid peroxidation, oxidative stress, and release of free oxygen species are the main contributing factors of cell damage and cell death. This review summarized the concept of oxidative stress in cardiomyocytes and role of each cytoplasmic organelle in its development during progress of the disease.

Keywords

Oxidative stress · Myocardium · Aging · Diabetes · Obesity · Cytoplasmic organelles

4.1 Introduction

Cardiac muscle is highly coordinating and pumping blood into the blood vessels carrying oxygen, nutrients, and biochemical components which are essential for the living and function of body cells. Mitochondria and ATP are dispersed in between cardiac muscle fibers and are responsible for aerobic metabolism. Branching of

H. I. H. El-Sayyad (✉)
Department of Zoology, Faculty of Science, Mansoura University, Mansoura, Egypt
e-mail: elsayyad@mans.edu.eg

S. Chakraborti et al. (eds.), *Modulation of Oxidative Stress in Heart Disease*,
https://doi.org/10.1007/978-981-13-8946-7_4

cardiac muscle cells allows it to contract in a wave-like structure. The cardiac contractions are controlled by pacemaker cells which receive signals from the autonomic nervous system and respond to the various hormones. Oxidative stress involves liberation of free radicals, such as $O_2{*}(-)$, H_2O_2, and $*OH$, which are generated extra- or intracellularly and cause toxicity to cardiomyocytes. During myocardial oxidative stress, the generation of ROS occurred due to altered nitric oxide synthase (NOS), NADPH oxidase, xanthine oxidase, and lipoxygenase/cyclooxygenase and the auto-oxidation of catecholamines [1]. Mitochondria play a role on generation of free radicals through an enzymatic source of free radical generation which leads to a catastrophic cycle of mitochondrial DNA damage. The oxygen radical species developed myocyte hypertrophy, apoptosis, and interstitial fibrosis via activating matrix metalloproteinases [2]. Aging [3], obesity, hypercholesterolemia and hyperlipidemia [4], hypertension [5], and diabetes [6] are contributed to myocardial oxidative stress. Also, the role of cytoplasmic organelles in oxidative stress such as mitochondria, rough endoplasmic reticulum, lysosomes, and Golgi apparatus is illustrated.

4.2 Metabolic Diseases and Oxidative Stress

According to World Health Organization (WHO), about 346×10^6 peoples were affected by diabetes worldwide and would be two-fold increase by 2030 [7]. The heart is the active organ, which needs the highest amount of adenosine triphosphate (ATP), for performing its function. Myocardial contraction required ATP and mitochondria for energy utilization, such as ionic pumps and the force-generating sarcomeres [8]. Sarcomeres are highly ordered arrays of molecular motors developed through a finely regulated mechanotransductive mechanism [9]. The poor diabetic control is associated with hyperglycemia and alterations of insulin and fatty acids, which altered cardiomyocyte function [10]. Impaired glucose uptake is correlated with disturbances of the insulin signaling pathway. Impaired cardiac activity and insulin-activated phosphorylation of Akt are more detected [11].

Intra- and extracellular high glucose level led to the development of advanced glycation end products (AGEs) derived from the collagenous protein materials leading to ventricular stiffness, the marker of diabetic cardiomyopathy [12]. Hyperglycemia is mediated by posttranslational modifications such as O-GlcNAcylation of Ca^{2+} proteins and kinases which have been shown to play a major role in cardiac impairment in diabetes [13]. Diabetes cardiomyocytes is associated with alterations of Ca^{2+} and consequently the formations of SERCA2 leading to withdraw Ca^{2+} from the cytosol during the diastolic contraction. Also, it affected myofilament Ca^{2+} sensitivity and increase ryanodine receptor-mediated Ca^{2+} release from the sarcoplasmic reticulum leading to arrhythmia susceptibility [14]. Diabetic cardiomyopathy developed from abnormal changes in the myocardial substrate and energy metabolism. It is affected by the reduction of calcium ion levels impairing the relaxation-contraction process and developed diastolic and systolic dysfunction. Sarcoplasmic reticulum (SR) represents the main part of cardiac excitation-contraction supplied with Ca^{2+} via the SERCA2a (sarcoplasmic/endoplasmic reticulum calcium-ATPase 2a). The decreased activity of SERCA2a and increase of Ca^{2+} release facilitated to the development of diabetic cardiomyopathy [15].

Diabetes have been associated with vasodilatation [16] as expressed by the increase of thrombomodulin, von Willebrand factor (vWF), selectins, and type IV collagen, the markers of endothelial damage [17] which predicted the vascular pathological changes. Also, the overexpression of protein kinase C (PKC) and elevated levels of angiotensin II (ANG-II), AGEs, plasminogen activity inhibitor-1 (PAI-1), are observed in the damaged endothelial cells [18]. Also, diabetes-induced myocardial infarction is associated with overexpression of cleaved caspase-3 expression, the marker of apoptosis. Also, it increased endoplasmic reticulum stress through the elevated level of the serum creatine kinase and lactate dehydrogenase activities [19]. Alterations in the cell metabolism, oxidative stress, formation of extracellular matrix proteins, endoplasmic reticulum stress, AGE, and steatosis (accumulation of lipid droplets) are involved in cell death diagnosed by the diabetes [15]. Also, there is a reduction of cardiac contractile function, impaired autophagic process, and increased ER stress and phosphorylation of signaling molecules Akt and mTOR [20].

Obesity is characterized by accumulated body fat, resulted from the high food intake, missing of physical exercise, and genetic factors. Obese individuals exhibited increased level of hormones, nonesterified fatty acids, glycerol, and cytokines. Obesity impaired the function of B cells and consequently developed diabetes [21], hypertension, atherosclerosis, and cardiovascular disease [22, 23]. Increase lipid intake enhanced the endoplasmic reticulum stress, abnormal autophagic process, de novo ceramide synthesis, oxidative stress, inflammation, and changes in gene expression [24]. These factors implicated in the impairing of the systolic and diastolic dysfunction progress the development of both coronary heart disease and heart failure.

Although the glucotoxicity exhibited the cardiotoxicity, lipotoxicity developed the disease via another way. Fatty acids (FFAs) are the main sources of cell energy. They enter the mitochondria by the enzyme carnitine palmitoyltransferase 1 (CPT1) and undergo further β-oxidation to generate acetyl-CoA, the main substrate for the Krebs cycle. It is also used in ATP synthesis. Excess FFA is abnormally involved in ER stress, mitochondrial dysfunction, calcium dysregulation and cell death, and consequently liberation of reactive oxygen species (ROS) [25]. The cellular FFA uptake is controlled by CD36. Intracellularly, saturated FFAs enhanced the production of cytosolic and mitochondrial reactive oxygen species (ROS). Palmitate overload induced oxidative stress through triggering abnormal endoplasmic reticulum (ER), Ca^{2+} release and consequently reduction of the ER Ca^{2+} stores [26]. Experimental animal model like Sprague-Dawley rats fed on high-fat, low-carbohydrate diet (HFLCD) for 7 weeks resulted in obesity associated with depleted myocardial glycogen and also levels of both plasma adiponectin and insulin. Hyperlipidemia reduced the myocardial antioxidant gene transcript and activities of catalase and superoxide dismutase which coincides with increased xanthine oxidase activity and the gene of NADPH oxidase-4 transcript. Also, the decrease of myocardial mitochondrial DNA led to the marked reduction cardiac mitochondrial factors such as nuclear respiratory factor-1 and transcription factor A-mitochondrial [27]. Also, in obese pigs, cardiomyopathy is manifested by a marked increase of heart weight parallel with interstitial and perivascular fibrosis, increased lipid accumulation, oxidative stress, and decrease of the antioxidant enzymes. HFD overexpressed Grp94, CHOP, caspase 12, p62, and LC3II and increased the ratio of LC3II to LC3I in the left ventricle (LV) of the 5-month-old Lee-Sung (MetS) pigs [28]. Palmitate, stearate, and oleate are the most abundant FFAs, accounting for 70–80% of total plasma FFAs [29]. Palmitate overload developed type 2 diabetes, liberating the ROS generation and Ca^{2+}-mediated cardiac pathogenic changes. Palmitate cytotoxicity facilitated the release of calcium ions from the endoplasmic reticulum (ER) and increased its concentration in the cytosolic mitochondria which may accelerate ROS generation and increase permeability transition (PT) pore opening [26].

Diabetic cardiomyopathy is closely associated with hypertension and dyslipidemia which induce dysfunctional remodeling, myocardial fibrosis, and diastolic dysfunction and systolic impairment. Also, it is known that the impaired insulin metabolic signaling increased the oxidative stress and reduced nitric oxide bioavailability, AGE products, cardiomyocyte extracellular matrix stiffness, mitochondrial dysfunction, renin-angiotensin-aldosterone system, inflammation, cardiac autonomic neuropathy, endoplasmic reticulum stress, microvascular dysfunction, and a myriad of cardiac metabolic abnormalities [30]. Also, obesity led to insulin resistance, and type 2 diabetes developed cardiomyopathy. It is characterized by structural and functional alterations and interstitial fibrosis without coronary artery disease or hypertension. In the old stage, the diastolic and systolic functions are impaired [31]. The diseases suppressed the insulin receptor signaling through hyperactivation of c-Jun N-terminal kinase (JNK) and subsequent serine phosphorylation of insulin receptor substrate-1 (IRS-1). Mice deficient in X-box-binding protein-1 (XBP-1), a transcription factor, promotes the ER stress response and develops insulin resistance [32]. There is a close association between the obesity and endoplasmic stress expressed by tumor necrosis factor (TNF)-alpha and insulin receptor signaling such as c-Jun NH2-terminal kinase (JNK) and inhibitor of nuclear factor-kappaB kinase-mediated transcriptional which inhibits insulin function [33, 34]. The endoplasmic reticulum protein TXNDC5 induces cardiac fibrosis by increase folding of extracellular matrix protein and cardiac fibrosis via redox-sensitive c-Jun N-terminal kinase signaling [35].

Although other factors are involved, aging is closely associated to type 2 diabetes or insulin resistance [36]. Aging is associated with the increase of myocardial remodeling and altering cardiac reserve leading to myocardial damage [37], increasing left ventricular wall thickness, chamber size, and prolonged diastole. These cooperated in the impairing of the myocardial contractility [38].

Aged vascular smooth muscle cells (VSMCs) are associated with the sharp rise of interleukin-6 leading to overexpression of nuclear factor κ-light chain-promoted B-cell activation and production of mitochondrial O2 [39]. Nair and Ren [40] mentioned that aging attenuated the growth hormone (insulin-like growth factors) signaling, lack of DNA replication, and repair of histone acetylation which predicted loss of ventricular function. Aging impaired autophagic process, leading to the accumulation of damaged proteins and organelles. Age-related cardiovascular dysfunction is characterized by liberated reactive oxygen species (ROS) and superoxide by low-grade inflammation such as left ventricular hypertrophy, fibrosis, diastolic dysfunction, endothelial damage, decreased vascular elasticity, and increased vascular stiffness [41]. Collagen types I, II, III, IV, V, and VI, elastin, fibronectin, laminin, and fibrinogen represent the cardiac fibroblasts activated ECM proteins [42]. Although these materials provide structural support to the myocardium, its ECM accumulation mediated the diastolic dysfunction and increase myocardial stiffness [43]. There is a dynamic balance between ECM synthesis and degradation by matrix metalloproteinases (MMPs) which impaired during aging process [44]. The increase of transforming growth factor-β (TGF-β) and connective tissue growth factor (CTGF) levels represents the main element in ECM synthesis in aging mice that express myocardial fibrosis and loss of cardiac diastolic function [45]. Oxidative stress is also associated with myocardial damage resulted from impacting proteins, lipids, and RNA and DNA of cells as well as increase of oxidative stress leading to mitochondrial damage and release of cytochrome c, causing cell death [41].

4.3 Endothelium and Oxidative Stress

The endothelium is a metabolically active thin single cell layer lines the internal lumina of the vasculature and separate the blood circulation from the vascular smooth muscle cells (VSMCs) and maintained the vasodilatation and vasoconstriction. The endothelium performs its function via the expression of the vasodilators and vasoconstrictors which promote vascular homeostasis, cellular adhesion, vessel wall inflammation, and angiogenesis [46]. Dysfunctions of the endothelium led to impair the vasoconstriction and increase inflammation via the depletion of endothelial nitric oxide synthase, endoplasmic reticulum stress, overproduction of vascular endothelial growth factor inflammatory pathways, and oxidative stress. All of these are involved in the development of diabetes, coronary artery disease, peripheral arterial disease, stroke, and microvascular complications [47]. The microvascular complications in diabetes and hypercholesterolemia may lead to retinopathy [48], while macrovascular complications are illustrated in myocardial [49] and peripheral vascular disease [48].

Endothelial cells and macrophages are tightly contact and contributed to the modulation of vascular function. In adult, the endothelial cells possess signals for the progressive differentiation and characterization of macrophages such as expression of Tie2 and CD206/Mrc1. These macrophages accelerate angiogenesis and effectively engage in tight associations with endothelial cells in vivo [50].

The endothelial (ECs), myeloid, and cardiomyocyte cells have the great ability to switch from generating ATP through oxidative phosphorylation to glycolysis and to change from one energy source to another [51]. Hyperlipidemia is closely linked to hyperglycemia and insulin resistance leading to the development of cardiovascular disease [52]. Atherothrombosis is developed from the downregulation of endothelial cells which facilitated attraction, binding, and aggregation of monocytes. These are activated cell adhesion molecules on the endothelial cells together with the chemokines, platelet-activating factor, CCL2 and CCL5 [53]. Cardiomyopathy, related to impairing of coronary endothelial cell (EC), myocardial necrosis, and fibrosis are the main causes of mortality in diabetic patients [54].

Dysfunction of the dynamic balance between the pro- and anti-inflammatory regulators of macrophages and activation of inflammatory macrophages facilitated the development of obesity, metabolic syndrome, and diabetes [55]. Mitochondrial oxidative stress (mitoOS) amplified macrophages by promoting NF-κB-mediated entry of monocytes and other inflammatory processes during the progress of human atherosclerosis [56]. Also, it is involved in the activation of inflammatory macrophage leading to the adipocyte dysfunction. These also illustrated the therapeutic implications for obesity, metabolic syndrome, and diabetes [55]. Activated macrophages are the main source of reactive oxygen species, reactive nitrogen species, and peroxynitrite which are generated through the *respiratory burst.* Also, the liberated pro-inflammatory cytokines, such as tumor necrosis factor-α, initiated NF-κB and activated protein-1 translocation enhancing the liberation of free radicals in macrophages. These epigenetic factors accelerated the development of metabolic diseases [57].

The dynamic balance between the endothelium-dependent of either contracting (EDCFs) or relaxing factors (EDRFs) and endothelium-dependent hyperpolarizing factors is impaired during aging (EDHF) (endothelin-1 (ET-1)). It is a result from the depletion of NO, the endothelium vasorelaxant, and production of cyclooxygenase and the vasoconstrictor molecules [58]. Depletion of L-arginine, the substrate for NO production, or activity of eNOS, the enzyme that synthesizes NO led to decreased production of NO in endothelial cells. It led to of the downregulation of tetrahydrobiopterin (BH4), the cofactor of eNOS. Many authors reported its depletion during aging [59].

4.4 Cytoplasmic Organelles and Oxidative Stress

Reactive oxygen (ROS) and/or nitrogen species (RNS) are observed in many biological processes as well as during the development of hypertension, ischemia/reperfusion injury, diabetes mellitus, atherosclerosis, stroke, cancer, and neurodegenerative disorders. Their overexpression predicted the early marker of cardiomyocyte cell death. Mitofusin-2 (Mfn-2) protein manages the mitochondria and endoplasmic reticulum (ER). Hyperglycemia decreased the cardiac function and activated ER interactions and mitochondrial apoptotic pathways via increase ER and mitochondrial stress factors, apoptotic proteins, cytochrome *c*, and mitochondrial permeability transition pore (mPTP) opening [60]. Type 2 diabetes-associated cardiomyopathy exhibited abnormal twofold higher level of adiponectin, the marker of lipotoxicity [61].

Oxidative stress plays a role in ischemia/reperfusion injury, atherosclerosis, and aging. Type 2 diabetes and other metabolic diseases are associated with ROS, protein aggregation, and glycosylation defects. The existence of more than 20 protein disulfide isomerase family members such as endoplasmic reticulum oxidoreductin 1 (ERO1) and NOX4-knockout mice clearly mentioned the development of stress in

ER and GA. Oxygen (O_2) itself, NADPH oxidase (NOX) formed ROS, and pH changes appear to be of importance and indicate the balance of intercompartmental communication [62]. There is a close relationship between the overexpression of free radicals and oxidative stress-associated cardiomyopathies. Mitochondria represent the main target for the primary generation of superoxide radicals. However, the Golgi apparatus (GA) cooperated in modifying, sorting, and packaging macromolecules for cell metabolism. GA play a great role in Ca^{2+}/Mn^{2+} homeostasis, cell apoptosis, sphingolipid metabolism, signal transduction, and antioxidation [23]. Experimentally induced diabetes in both metallothionein overexpression transgenic and wild-type (WT) mice exhibited cardiac ER stress expressed by chaperones and cell death by CCAAT/enhancer-binding protein (C/EBP) homologous protein (CHOP) and cleaved caspase-3 and caspase-12 after 2 and 5 months [63].

Intracellular hyperglycemia increased the production of free radicals leading to the overexpression of nuclear poly(ADP-ribose) polymerase and inhibits GAPDH, the predictor of diseased signaling pathways. ROS and poly(ADP-ribose) polymerase are also decreased the expression of sirtuin, PGC-1α, and AMP-activated protein kinase activity. These mediated the loss of mitochondrial biogenesis, increased ROS production, and altered normal synchronization of glucose and lipid metabolism. Also, there is a progressive increase of nuclear transport of proatherogenic transcription factors, transcription of neutrophil enzyme initiating NETosis and peptidylarginine deiminase 4, and activates the NOD-like receptor family, pyrin domain-containing 3 inflammasome. Insulin resistance led to the increase of ROS within the cardiomyocyte through increase fatty acid flux and oxidation. There are a marked activation of cardiomyopathy markers such as nuclear receptor PPARα and nuclear translocation of forkhead box O1. ROS impair the balance between mitochondrial fusion and fission through increasing of fission process, decreasing metabolic capacity, loss of mitochondrial electron transport chain, and ATP synthesis [64]. Sirtuin 1 is an NAD^+-dependent histone deacetylase that regulates endoplasmic reticulum stress and cardiomyocyte apoptosis through impairment of cardiac contractility. Twelve-month-old $Sirt1^{-/-}$ mice exhibited overexpression of nitric oxide synthase and increased endoplasmic reticulum stress and apoptosis in the myocardium leading to impairment of cardiac contraction [65].

4.5 Endoplasmic Reticulum and Oxidative Stress

The endoplasmic reticulum (ER) is a highly branched tubular organelle regularly oriented parallel with each other and interconnected with the nuclear envelope. It is important in cellular homeostasis, signal transduction, folding protein via formation of disulfide bonds, and transmission to Golgi apparatus [66] as well as provides a lot of Ca^{2+} responsible for protein folding [67, 68]. It is the most dynamic organelle that determines cell survival or death. It contains protein chaperones and enzymes such as Grp78 (BiP), Grp94, protein disulfide isomerase (PDI), calnexin, and calreticulin, which are involved in protein folding. Folded proteins are transferred to the Golgi organelle, while incompletely folded ones are transported in the ER to complete the folding process or delivered to the cytosol to undergo degradation [69, 70]. Two kinds of endoplasmic reticulum are known. Rough ER play a great role in protein synthesis, folding, modification, and transport. The other type is smooth ER which involved in the biosynthesis of lipids and steroids, metabolism of carbohydrates, and regulation of calcium intracellular homeostasis [69].

The lumen of the ER and its cytosol possesses the increased ratio of GSSG/GSH with a characteristic oxidizing environment which facilitates disulfide bond formation and prevents accumulation of unfolded protein [71]. Glutathione is formed in the cytosol in a reduced form by NADPH-dependent reaction and glutathione reductase [72]. Glutathione works as a thiol-disulfide redox buffer, and the average ration of GSH/GSSG predicted the cellular redox state. The level of reduced glutathione to oxidized form is >50:1 in the cytoplasm and 1:1 to 3:1 in the ER lumen (Hwang and Sinskey 1992). It determined the functions of PKR-like endoplasmic reticulum kinase (PERK) and transcription factor 4 (ATF4), which reduce ROS via transcriptional regulation [73].

There is a close-related function between ER and mitochondria especially during the oxidative stress and inflammation. Both organelles shared the production of the cellular reactive oxygen species (ROS) [74]. In the mitochondria, the ROS is developed from the by-product of oxidative phosphorylation-associated ATP production, whereas in ER, ROS is formed during disulfide bond formation of folded protein [75]. This led to the depletion of the anti-oxidative defense of cells [76]. Also, the calcium ions predicted the correlation between mitochondria and ER. During oxidative stress, calcium ion levels are altered. These are accompanied by disruption of the membrane potential, altered pH, decreased the production of ATP, and opened the permeability transition pores, causing cytochrome c leakage into the cytoplasm. Also, the endoplasmic reticulum (ER) is the site of calcium storage and disulfide bond formation of protein folding. The endoplasmic reticulum stress increased the production of reactive oxygen species via altering the activities of protein disulfide isomerases, endoplasmic reticulum oxidoreductin-1, and reduced glutathione, and mitochondrial electron transport chain proteins are altered [77].

Unfolded or misfolded protein contents of endoplasmic reticulum activate the endoplasmic stress as a result of hyperglycemia, oxidative stress, ischemia, disturbance of calcium homeostasis, and overproduction of abnormal proteins, leading to increase of unfolded protein response (UPR) and cardiac cell apoptosis [78, 79]. Protein kinase R-like endoplasmic reticulum kinase (PERK), inositol-requiring kinase 1 (IRE1), and activating transcription factor 6 (ATF6) are the main proximal transducers of ER stress detected at the molecular level. These inhibit the protein translation, enhance protein-folding capacity, and augment ER-associated degradation to refold denatured proteins and restore cellular homeostasis [80, 81].

Calcium storage and protein folding are detected mainly in the ER. Abnormal changes of both oxidative stress and intra-ER Ca^{2+} activated the ER stress leading to liberation of free radicals. Increased level of calcium ion can transfer into mitochondria through IP3R-enriched MAM to activate the mitochondrial function. Decreased ER calcium channels and enhanced calcium release from the ER into the cytoplasm increased the more production of reactive oxygen species from mitochondria [82].

Also, the ER stress-associated oxidative stress signaling is carried out via PERK-mediated activation of ATF4 and nuclear factor erythroid 2-related factor 2 (NRF2)—the transcription factor responsible for the antioxidant cell response [83]. ER stress altered the ER protein-folding metabolism which led to abnormal accumulation of misfolded proteins in the lumen of ER by exposing to hydrophobic amino acid domain and consequently activated apoptotic pathways [70, 84, 85]. ER-associated degradation (ERAD) machinery or autophagic process plays a great role in degraded the accumulation of misfolded or immature proteins [86]. The ER stress is composed of four main steps: (1) increased protein synthesis carried out to prevent protein aggregation/accumulation, (2) transcriptional induction of ER chaperone genes to accelerate protein folding, (3) transcriptional induction of ERAD genes to increase ERAD ability/capacity, and (4) induction of cell death to remove affected cells. Accumulation of misfolded proteins exerted proteotoxicity. Senescence is a complex cell phenotype induced by telomere attrition, DNA damage, and activation of some oncogenes leading to oxidative stress. It is characterized by a cell hypertrophy, complete cell cycle arrest, and the formation of a secretome enriched in pro-inflammatory [87].

Endoplasmic reticulum stress is essential in coronary artery disease and cardiac hypertrophy. The ER transmembrane illustrates the accumulation of unfolded proteins and increases the transcriptional and translational pathways via specific sensors. Oxidative stress is balanced by antioxidant defense systems managed by the unfolded protein response (UPR). Nuclear factor-E2-related factor (Nrf2) is a regulator of cellular resistance to oxidants and interrelated with the UPR sensor called pancreatic endoplasmic reticulum kinase. The interventions against ER stress and Nrf2 activation reduce the myocardial infarct size and cardiac hypertrophy in animals exposed to I/R injury and pressure overload, respectively [88].

Aging in senescence-accelerated prone mice exhibited activation along with reduced expression of 14-3-3η protein, the downstream mitogen-activated protein kinase-mediated ER stress, apoptosis, and DNA damage in the cardiomyocytes of SAMP8 mice [89]. In an experimental obese mother mice which fed on a high-fat diet (60% fat) for 15 weeks, the abnormal myocardium of E12.5 embryos is developed. It is characterized by ventricular septal defects and persistent truncus arteriosus parallel with increasing oxidative stress markers, such as superoxide and lipid peroxidation, and endoplasmic reticulum stress markers. Additionally, the levels of phosphorylate protein kinase, RNA-like endoplasmic reticulum kinase, phosphorylated IRE1α, phosphorylated eIF2α, C/EBP homologous protein and binding immunoglobulin protein, endoplasmic reticulum chaperone gene expression, XBP1 messenger RNA splicing, and cleaved caspase-3 and caspase-8 are increased in heart embryos [90]. Ischemic heart disease is a stress condition characterized by pathological alterations and triggers cardiac cell death. Recent evidence revealed that ER stress is involved in the development and progression of various heart diseases, such as cardiac hypertrophy, ischemic heart diseases, and heart failure. The endothelium of atherosclerosis upregulated genes associated with endoplasmic reticulum stress. Except PERK, ER transmembrane signal transducers IRE1alpha and ATF6 alpha were activated and coincide with expression of protein-folding enzymes and chaperones, the markers of ER stress [91].

Also, the ER stress activated the release of Bip from the sensor proteins, and UPR after dimerization and autophosphorylation of PERK and IRE1α, and consequently managed intramembrane proteolysis of ATF6. PERK activation initiated factor 2α (eIF2α) phosphorylation and decreased both global protein synthesis and formation of the eIF2-containing initiation complex. The increase translation of activating transcription factor 4 (ATF4) manages pro-apoptotic factor C/EBPα homologous protein (GADD153 or CHOP) mediating the direct transcriptional induction and translocation to the ER membrane of a pro-apoptotic BH3 protein of the BCL-2 [92, 93].

Unfolded protein (UPR) activates ER protein kinase (PERK), transcription factor 6 (ATF6), and the inositol-requiring enzyme 1 (IRE-1) which dissociate the luminal chaperone BiP/GRP78 from the luminal proteins. These were associated

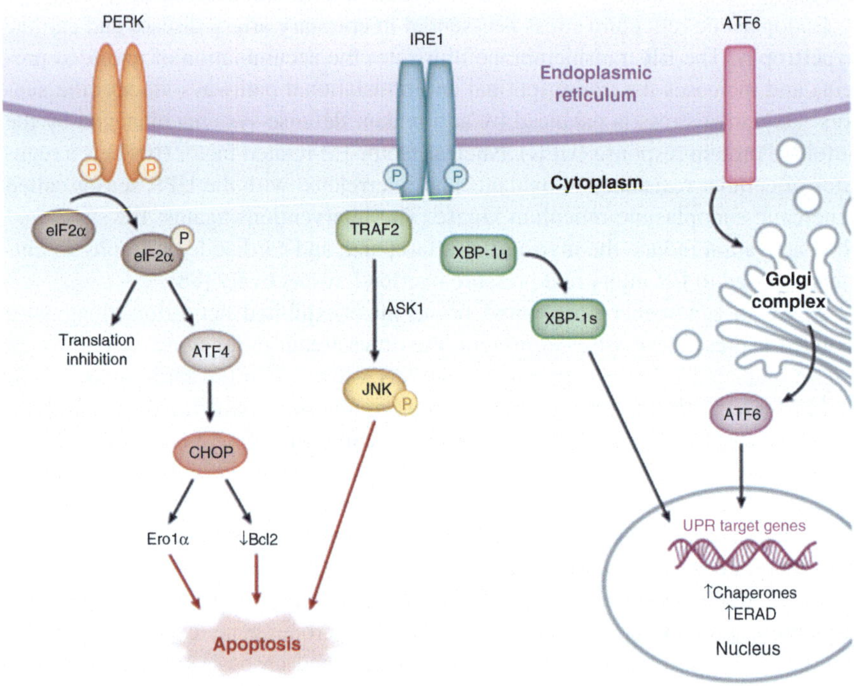

Fig. 4.1 Diagram illustrated endoplasmic reticulum (ER) stress. Accumulated of unfolded proteins within the lumen of ER such asIRE1, PERK, and ATF6 and released signals that manage cell function. PERK, protein kinase RNA-like ER kinase; *IRE1* inositol-requiring protein 1, *ATF6* activating transcription factor 6, *eIF2α* eukaryotic translation initiation factor 2 alpha, *CHOP* C/EBPα homologous protein, *Bcl-2* B-cell lymphoma 2, *Ero1α* ER oxidase 1 alpha, *TRAF2* TNF receptor-associated factor 2, *JNK* c-Jun N-terminal kinase, *ASK1* apoptosis signal-regulating kinase 1, *XBP1* X-box-binding protein 1, *ERAD* endoplasmic reticulum-associated degradation, *UPR* unfolded protein response (Ozcan and Tabas [95])

with depletion of protein synthesis and increase the transcription of genes encoding protein-folding enzymes, ER chaperones, and components of the ER-associated degradation (ERAD) [94] (Fig. 4.1).

4.6 Mitochondria

Aerobic metabolism of the cardiomyocyte required mitochondria and almost more than 90% of the intracellular ATP which are formed after combustion of fatty acids, glucose, and lactate. The cardiomyocyte metabolism required increased level of degraded adenosine triphosphate (ATP). Fatty acids supply approximately 60–90% of the energy to the heart and are acquired about 10–40% from the oxidation of pyruvate (through glycolysis and lactate). Also, glucose is important in production of ATP, which is transported from the bloodstream to internal structure of the mitochondria in the form of two main components: pyruvate and nicotinic adenine dinucleotide (NADH). These two metabolites are further transported into the central part, in the presence of oxygen and produce ATP (adenosine triphosphate) [96].

Mitochondria are abundant in between the myofibrils and just below the sarcolemma of the cardiomyocytes. Metabolism of fatty acid and glucose carried mainly in the mitochondria. Mitochondria are the main energy supplemented organelles, in the form of ATP, and important for cardiac function after combustion of fatty acids, glucose and lactate. Calcium (Ca^{2+}) is released from the sarcoplasmic reticulum (SR) and activates the excitation-contraction (E-C). Abnormal Ca^{2+} accumulation interfered with impaired mitochondrial function, decreased ATP formation, and increased liberation of reactive oxygen species (ROS). The diastolic SR Ca^{2+} production activated mitochondrial Ca^{2+} production and development of murine model of myocardial infarction. Two kinds of Ca^{2+} released channels are detected on cardiac SR, namely, type 2 ryanodine receptors (RyR2s) and type 2 inositol 1,4,5-trisphosphate receptors (IP3R2s). In murine models, mutation of type 2 ryanodine receptors (RyR2s) cause or inhibit leakage of SR Ca^{2+} leak which results in mitochondrial Ca^{2+} overload and morphological abnormality [97].

Mitochondria are in contact with the Ca($^{2+}$) production and uptake of channels of [Ca($^{2+}$)-ATPase (SERCA)] in the sarcoplasmic reticulum (SR). Their reactive-free radicals manage Ca($^{2+}$) cycling in the cardiomyocytes. The authors proposed that free radicals induced overproduction of Ca($^{2+}$) via promoting SR Ca($^{2+}$) release, which increases other Ca($^{2+}$) channels and dysregulates Ca($^{2+}$), leading to abnormal action potential (AP) which altered the mitochondrial and SR unction. Morphological abnormalities of AP are affected by increase of free radicals and AP firing and SR-mitochondria distance [98].

Cardiomyopathy increased the consumption of fatty acids (FAs) via FATP1 cell surface transporters [99] and consequently abnormal lipid accumulation [100], mitochondrial dysfunction [101], and increase oxidative stress [102].

Aging developed dysfunctional cardiac mitochondria is resulted from production of less ATP. Abnormal reduction of mitochondrial function exhibited the reduction of mitochondrial elements, abnormal mitochondrial morphology, opening of the mitochondrial permeability transition pore, missing activity of the electron transport chain, and increased ROS production [103].

Aging is involved in damaging of mitochondria DNA, impairing bioenergetics efficiency, increasing apoptosis, and developing the inflammation processes [104]. In situ electrocardiograms of aged human, ECGs, showed significant prolongations in RR and QT intervals in the aged rats. Light and electron microscopic investigations of the affected myocardium illustrated marked increases in muscle fiber radius and deposition of collagen fibers as well as apparent flattened and partial local splitting in elastic lamellas in the aorta, parallel with disruption of mitochondria and lysosomes within the myofilaments in cardiomyocytes [105]. Abnormal lipid accumulation in cardiomyocytes of obese and diabetic patients altered the peroxisome proliferator-activated receptor-γ(PPARg) and increased the calcium transient amplitude and sarcoplasmic reticulum (SR) calcium stores in cardiomyocytes. These are associated with the arrhythmia and sudden cardiac death [106]. Mitochondria are important elements in the obesity through lipid β-oxidation, ATP production, oxidative stress, and inflammation. MicroRNAs (miRs) are involved in adipocyte differentiation, insulin action, and fat metabolism. Also, miRs are important regulators of mitochondrial function by either modulating of both mitochondrial proteins and metabolic process in the context of obesity [107].

Diabetic cardiomyopathies decreased the metabolism of cardiac mitochondria impairing the contractility and reducing the cardiac efficiency. These may result from the increase in cardiac fatty acid oxidation and glycolysis [108]. Diabetes-associated hyperglycemia and hyperlipidemia led to mitochondrial dysfunction via inducing mitochondrial fission and generation of reactive oxygen species (ROS) [109]. Also, it is associated with altering mitochondrial biogenesis and respiratory function such as mitochondrial abnormalities and reduced gene expression of mitochondrial oxidative phosphorylation (OXPHOS) [110] and altering peroxisome-proliferator-activated receptor (PPAR) gamma and coactivator-1α (PGC-1α [111].

Diabetes-related cardiomyopathy in Sprague-Dawley (SD) rats for 4, 8, and 12 weeks increased fast sugar and HbA1c levels coincides with decreased left ventricular pressure (LVP), heart rate (HR), and altered inflammatory markers of plasma IL-1 and IL-4. Also the levels of cardiac4-HNE and Bax mRNA were increased; meanwhile ALDH2 activity and Bcl-2 mRNA levels and Bcl-2/Bax mRNA ratios were decreased. The activity of ALDH2 was decreased correlated with increased inflammation, oxidative stress, and the occurrence of cardiomyocyte cell death (Wang et al., 2013). High-fat diet increased the plasma and myocardial lipids (TG, LDL-C, TL, and PL), MDA, and CK-MB in rabbits especially in male more than in females which coincide with disrupting myocardial junctional complexes and mitochondrial size [112].

The diabetic myocardium is characterized by alteration in cardiomyocyte signaling and increase in circulating glucose, insulin, and fatty acids [10]. Extracellular hyperglycemia enhanced the development of advanced glycation end products (AGEs) on collagenous protein, contributing to the ventricular stiffness, as well as of the AGEs-associated intracellular proteins leading to diabetic cardiomyopathy [12]. Hyperglycemia-related posttranslational modifications such as

O-GlcNAcylation of key Ca^{2+} proteins and signaling kinases have been shown to induce diabetic cardiomyopathy [13]. Ca^{2+} disturbance prolonged the SERCA2-mediated Ca^{2+} removal from the cytosol during diastole contraction, altering myofilament Ca^{2+} sensitivity, and Ca^{2+} leak from the sarcoplasmic reticulum leading to cardiac arrhythmia [14]. Decreased cardiac insulin stimulated the phosphorylation of Akt is evident in diabetic rodent models [11].

Aging-related cardiomyopathy increased the somatic mtDNA mutations [113], disorganized nuclear and mitochondrial DNA integrity [114] causing epigenetic alterations, mitochondrial dysfunction, and genomic instability [115], and increased oxidative damage-associated cardiac dysfunction. Two kinds of mitochondria are detected in the myocardium such as subsarcolemmal mitochondria and interfibrillar mitochondria which are implicated in age-related cardiac mitochondrial disease [116]. The mitochondria exhibited five kinds of protein markers managing mitochondrial fission and fusion including mitofusins 1 and 2 (MFN1, MFN2), optic atrophy 1 (OPA1), dynamin-related protein 1 (DRP1), and fission 1 (FIS1). Mitochondrial fusion is activated by two Mfn isomers, Mfn1 and Mfn2 and OPA1; meanwhile FIS1 and DRP1 are needed for mitochondrial fission [117, 118]. MFN1 and MFN2 present on the outer membrane with the N-terminal GTPase [119]. OPA1 is present in an intermembrane space protein [120] and promotes mitochondrial fusion [121]. In addition, Mfn2 is also found in the endoplasmic reticulum (ER) membrane where it facilitated ER-mitochondria junctions [122]. DRP1 present in a cytosolic pool, but a fraction localizes to puncta on mitochondria [123]. Cytosol FIS1 molecule possesses a single transmembrane domain at the C-terminal end [123, 124].

A transgenic mouse model of myocardial lipotoxicity exhibited abnormal overproduction of ACSL1 (long-chain acyl-CoA synthetase 1) in the cardiomyocytes associated with myocardial fatty acid uptake and involved in atrophy of the mitochondria. This mitochondria dysfunction activates palmitoyl-carnitine oxidation and generation of the free radicles. Palmitate exposure to ventricular cardiomyocytes of neonatal rat activated the mitochondrial respiration, mitochondrial polarization, and ATP synthesis. However, long-term exposure to palmitate (>8 h) enhances ROS generation and accelerates loss of the mitochondrial reticulum and increased mitochondrial fission. This activated ubiquitination of AKAP121 (A-kinase anchor protein 121) and decreased phosphorylation of DRP1 (dynamin-related protein 1) at Ser637 and altered proteolytic processing of OPA1 (optic atrophy 1) [125].

Myocardial ischemia-reperfusion injury (I/R) exerted the mitochondrial dysfunction assessed by the increase of O_2^- and decrease state 3 respiration as a result of cysteine oxidation on mitochondrial complex I [126]. Mitochondrial oxidative stress (mitoOS) activated the macrophages by managing NF-κB-mediated entry of monocytes and other inflammatory processes during the human atherosclerosis [56].

Diabetes and hypertension activated inner membrane anion channels of mitochondria and depolarization, decreasing signaling pathways and altering the cardiac action potential leading to increase release of mitochondria ROS (AP) [127]. Intracellular hyperglycemia increased the ROS production which inhibited GAPDH and increase of nuclear poly(ADP-ribose) polymerase predicting the early glycolytic pathogenic signaling pathways. Also, ROS and poly(ADP-ribose) polymerase reduce sirtuin, PGC-1α, and AMP-activated protein kinase activity impaired the mitochondrial biogenesis. This enhances overexpression of the nuclear receptor PPARα and nuclear translocation of forkhead box O1, which developed cardiac disease. Also, ROS led to increase the mortality among diabetes through increasing of ryanodine receptor phosphorylation and depletion of sarco/endoplasmic reticulum Ca(++)-ATPase transcription. This is contributed to the increased risk of fatal cardiac arrhythmias associated with diabetic cardiac autonomic neuropathy [64].

Also, mitochondrial Ca^{2+} uptake is needed for energy supply to maintain the anti-oxidative capacity in a reduced form to prevent liberation of ROS. Mitochondria uptake Ca^{2+} through the mitochondrial Ca^{2+} channel, which is present in a multiprotein complex. During cardiomyopathy-related oxidative stress, ROS is liberated from the mitochondria which may activate another ROS liberation from neighboring ones through cellular network of redox signaling. Although low values of ROS may serve biological function, higher ones exerted cardiomyopathy through redox-sensitive kinases and cell death [128, 129].

Hyperglycemia increased oxidative stress via increase of the vascular defects. The mitochondrial NAD(P)H oxidase is involved in liberation of ROS in the vasculature. Increased production of NAD(P)H oxidase in diabetes may decrease the intracellular levels of NADPH, the main precursor for endothelial NO synthase (eNOS) and antioxidant systems. Increase liberation of ROS leads to development of uncoupling eNOS, mitochondrial dysfunction, and impaired antioxidant defenses due to impairment of intracellular NADPH (Gao and Mann, 2009). In vitro studies of neonatal rat cardiomyocytes subjected to high levels of glucose (25 mmol/L) for 24 h markedly elevated CM-H(2)DCFDA fluorescence, which is inhibited by 1,2-bis (o-aminophenoxy) ethane- N,N,N′,N′-tetraacetic acid tetraacetoxymethyl ester (BAPTA-AM), a $(Ca(^{2+}))(i)$ chelator. These findings mentioned that high glucose level upregulated phosphorylated CaMKII expression and $(Ca(^{2+}))(i)$ due to NCX activation and consequently increase liberation of ROS [130].

Both mitochondria and lysosomes exhibit great role in mediating oxidative stress [131] through mitochondrial oxidative phosphorylation, the generator of superoxide radicals during electron transport [132]. Lysosomes contain high concentrations of redox-active iron that can catalyze the homolytic splitting of hydrogen peroxide, which produces the reactive hydroxyl radical by the Fenton reaction [133].

4.7 Golgi Apparatus

Golgi apparatus (GA) is important in Ca^{2+} homeostasis, especially during Ca^{2+} stress (Southall et al. 2006). Golgi apparatus contains three important elements which promote both Ca^{2+} regulation and transportation [134]. These are the sarcoplasmic/endoplasmic reticulum Ca^{2+} ATPase (SERCA), the inositol 1,4,5-trisphosphate receptors (IP3R), and the plasma membrane-Ca^{2+} ATPase

(Pmr1p) [135]. The activity of both IP3R and SERCA can be promoted by ROS/RNS or the cellular redox state [136]. It is known that caspases promoted degradation of the Golgi proteins, such as golgin-160, GRASP65, p115, syntaxin 5, GM130, and giantin, and may cause GA fragmentation and apoptosis (Walker et al. 2004). Golgi protein golgin-160 is proteolyzed by caspase-2, and the other ones are cleaved by caspase-3 or caspase-7. Increased production of caspase-resistant golgin-160, GRASP65, and p115 initiated the kinetics of GA fragmentation [137]. Thus caspases may trigger oxidative stress-related signals leading to fragmentation and apoptosis of GA [138]. It is known that mitochondria are responsible for the generation of superoxide radicals. However, the GA participates in modifying, packaging, and sorting macromolecules for either cell secretion or internal cell function. It is involved in the oxidative stress, via damaging proteins, lipids, and DNA, and consequently alters Ca($^{2+}$)/Mn($^{2+}$) homeostasis, cell apoptosis, sphingolipid metabolism, and signal transduction [23].

Cardiomyocyte cell death associated Golgi oxidative stress causing secretion of eIF5A resulted from tyrosine sulfation and resulted in development of myocardial ischemia/reperfusion (but not ischemia alone) [139]. Mitochondria and endoplasmic reticulum (ER) are affected GA. Oxidative stress of GA possessed signal transduction pathway through the PKR-like ER kinase/activating transcription factor 4 pathway (ATF4) which is the regulator of amino acid metabolism. ATF4 is regulated by the gene involved in the biosynthetic enzyme for cysteine and cystathionine γ-lyase (CSE) and maintains the redox homeostasis [140]. Iron promotes function of many ATPases such as the Na^+, K^+-ATPase, and the Ca^{2+} ATPase. Also, ATP-binding cassette (ABC) families are transmembrane proteins localized in all the plasma membrane of intracellular Golgi apparatus, lysosomes, peroxisomes, and endoplasmic cardiomyocytes against oxidative stress [141].

Iron overload decreases the Ca^{2+} ATPase activity and impairs the sarco/endoplasmic reticulum, increasing Ca^{2+} and the GA Ca^{2+} levels, which manage nitration, enzyme oxidation, and fragmentation [142]. Iron regulatory protein IRPs were found in the cytosol of endoplasmic reticulum and GA and can be affected by cell stress or iron status [143]. Hyperglycemia induces ligand-independent phosphorylation of vascular endothelial growth factor receptor 2 (VEGFR2), the main component in GA and consequently impaired VEGFR2 at the cell surface mediated by Src family kinases [144]. Calcium signaling pathways interact with ROS causing intracytoplasmic influx of Ca^{2+} from the extracellular source as well as from the ER or sarcoplasmic reticulum (SR) via the plasma membrane and the ER/SR channels, respectively. Low level of ROS acts as signaling involved in different cellular processes such as cell growth and death. Dysfunction of these systems induced pathogenesis of various diseases [145]. Increased cytosolic Ca^{2+} altered Ca^{2+}–sensitive enzymes leading to enhancement of mitochondria-derived ROS/RNS generation. This led to disruption of the respiratory chain [146] (Fig. 4.2).

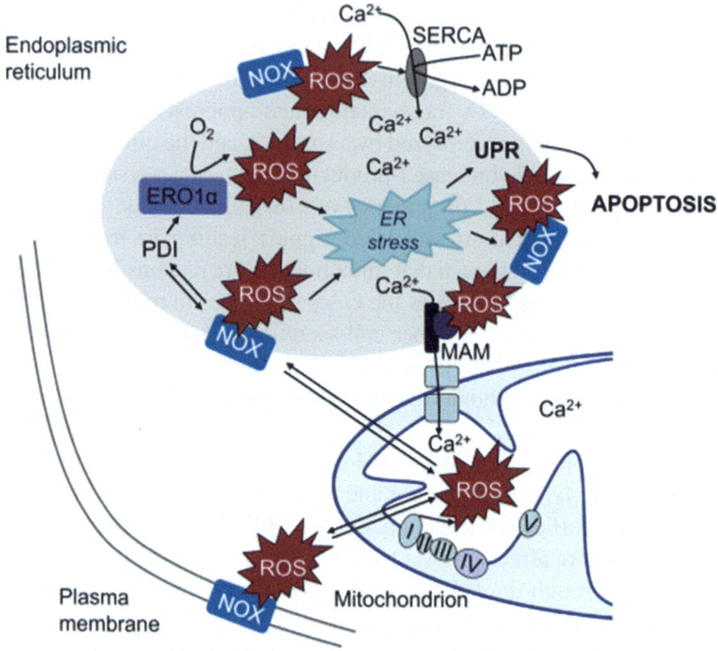

Fig. 4.2 Diagram illustrated ER stress-associated calcium cross stalk and ROS. Reactive oxygen species (ROS) are liberated between PDI and ERO1α. PDI is involved in ROS-generating NADPH oxidases (NOX). NOX-derived ROS modulates SERCA activity by overexpression of calcium ion in ER and activate the unfolded protein response (UPR). Mitochondrial ROS can affect NOX increasing ROS and calcium load in the ER leading to apoptosis. Abbreviations: *SERCA* sarco/endoplasmic reticulum Ca^{2+}ATPase, *MAM* mitochondria-associated ER membranes, *NOX* NADPH oxidase, *PDI* protein disulfide isomerase, *ERO1α* endoplasmic reticulum oxidoreductin-1 [147]

4.8 Lysosomes

Lysosomes are acidic cytoplasmic organelles containing almost 60 types of hydrolases which hydrolyze extracellular materials by endocytosis and intracellularly by autophagy. The hydrolyzed materials are transmitted by lysosomal channels of either specific exporters or trafficking vesicular membrane that maintain membrane potential, ion homeostasis, membrane trafficking, and nutrient sensing. Lysosomal storage diseases are related to impairment of autophagy process (LSDs) [148]. The lysosome is an active organelle exhibiting low ratio of copper and iron released during hydrolysis of metalloproteins. Its acidity and increase content of thiols keep iron in a reduced (ferrous) state, which can react with endogenous or exogenous hydrogen peroxide. During abnormal autophagic process, the development of lipofuscin pigment is detected and predicts the age-related diseases. Increased oxidative stress enhanced permeability of the lysosomal membrane and disrupted relocation of the cytosolic contents of iron and hydrolytic enzymes, leading to apoptosis or necrosis [133].

Autophagy is also involved the breakdown of SQSTM1/p62 protein and reduced the function of senescent cells. This was confirmed by rapamycin, MTOR-dependent autophagy activators and PP242 [149]. The autophagic process is activated by the development of a phagosystem from the mitochondria, endoplasmic reticulum, and lysosomes, which phagocytosed organelles to from the autophagosome. It is promoted by phosphatidylinositol-3-kinase (PI-3 K) and beclin-1 [150]. Also the autophagy is important during myocardial stress. Mitochondrial fission and missing of mitochondrial membrane potential are contributed for autophagic degradation of mitochondria. Maintained hypertension altered the structure and function of cardiomyocyte mitochondria leading to degradation via autophagy [151].

Aging of cardiac myocytes undergoes alterations via the accumulation of waste products such as defective mitochondria, aberrant cytosolic proteins, and intralysosomal lipofuscins. These illustrated the defective activity of autophagy and impaired activities of calpains and proteasomes. The hypertrophied mitochondria are characterized by impairing both ATP synthesis and inner membrane potential and liberation of reactive oxygen species. This autophagic turnover of enlarged and damaged mitochondria is carried out leading to increase oxidative stress and cardiomyocyte cell death. Also cardiomyopathies may result from mitochondrial DNA mutations, as a result of abnormal accumulation of non-eliminated mitochondria by autophagy. The hydrolysis of iron-saturated ferritin increased lysosomal oxidative stress and promoted myocardial damage in hemochromatosis [152, 153]. Diabetes, obesity, and dyslipidemia exhibited abnormalities of autophagic process altering cardiomyocyte homeostasis leading to myocardial disease [154–156]. Cardiomyopathy-related obesity possessed lipotoxicity, inflammation, oxidative stress, apoptosis, and sympathetic overactivation [157]. 7beta-hydroxycholesterol (7betaOH), a cholesterol oxidation metabolite formed during atherosclerotic lesions, was found to increase cell death through lysosomal and mitochondrial damage and production of free radicals [158].

Cardiac aging is manifested by mitochondrial dysfunction, aggregation of misfolded proteins, hypertrophy, and fibrosis. Also, it may be attributed to cardiomyocyte-associated damaging of both mitochondria and lysosomes. Mitochondrial alterations exhibited structural deformation and hypertrophy, while lysosomes possessed autophagic turnover of mitochondria and accumulate lipofuscin pigment [152, 159]. Also, inhibition of receptor of advanced glycation end products increases vascular cell damage and development of atherosclerosis. The cardiac antioxidant capacity was upregulated as detected by overproduction of superoxide dismutase and sirtuin mRNA expressions. Abnormal mitochondrial structure and function, cathepsin L activity, and mitochondrial fission protein Drp1 and Fis1 were increased in $RAGE^{-/-}$ mice sustained autophagy-lysosomal flux [160].

Mice model lacking the lysosomal cysteine protease cathepsin L (CTSL) developed a cardiomyopathy (DCM). It is manifested by swollen heart chamber, fibrosis, and lacking contractility [161]. Aging-related impairing of cell function is characterized by DNA damage, aggregation of impaired organelles, increase liberation of free reactive oxygen species, and accumulation of oxidized proteins and lipids. Autophagy is an important quality control pathway and is necessary to maintain cardiac homeostasis and to adapt to stress. Autophagy is an important biological process, and impairing of its function has been contributed to aging models [162]. Mitochondrial dysfunction and cell death have been predicted by the apoptotic BH3 protein Bnip3. It is also a potent inducer of autophagy. Oxidative stress and increased intracellular Ca ($^{2+}$) level have been reported to induce autophagy, compared to Bnip3-induced autophagy independent of antioxidant treatment or Ca ($^{2+}$) [163].

4.9 Conclusion and Future Direction

This review outlines the sources of free radicals and progress of oxidative stress in cardiomyocytes during aging and metabolic diseases. The disease disrupts the microconstituents of mitochondria, endoplasmic reticulum, Golgi complex, and lysosomes changing their structure and function and shared in the overproduction of free oxygen species. There different opinion of applying natural product for scavenging the liberated free radicals and improving the structure and function of the cytoplasmic organelles consequently improved the cardiomyocytes.

References

1. Misra MK, Sarwat M, Bhakuni P et al (2009) Oxidative stress and ischemic myocardial syndromes. Med Sci Monit 15(10):RA209–RA219
2. Tsutsui H, Kinugawa S, Matsushima S (2009) Mitochondrial oxidative stress and dysfunction in myocardial remodelling. Cardiovasc Res 81(3):449–456
3. Obas V, Vasan RS (2018) The aging heart. Clin Sci (Lond) 132(13):1367–1382
4. Mayyas F, Alzoubi KH, Al-Taleb Z (2017) Impact of high fat/high salt diet on myocardial oxidative stress. Clin Exp Hypertens 39(2):126–132
5. Brito R, Castillo G, González J et al (2015) Oxidative stress in hypertension: mechanisms and therapeutic opportunities. Exp Clin Endocrinol Diabetes 123(6):325–335
6. Zhao MX, Zhou B, Ling L et al (2017) Salusin-β contributes to oxidative stress and inflammation in diabetic cardiomyopathy. Cell Death Dis 8(3):e2690
7. World Health Organization (WHO) (2011) and C. D. Bode, Media Centre-Fact Sheets, Diabetes. http://www.who.int/mediacentre/factsheets/fs312/en/index
8. Pasqualini FS, Nesmith AP, Horton RE et al (2016) Mechanotransduction and metabolism in Cardiomyocyte microdomains. Biomed Res Int 2016:4081638
9. Kontrogianni-Konstantopoulos A, Ackermann MA, Bowman AL et al (2009) Muscle giants: molecular scaffolds in sarcomerogenesis. Physiol Rev 89(4):1217–1267
10. Varma U, Koutsifeli P, Benson VL et al (2018). Molecular mechanisms of cardiac pathology in diabetes – experimental insights Biochim Biophys Acta 1864(5 Pt B):1949–1959
11. Shum M, Bellmann K, St-Pierre P et al (2016) Pharmacological inhibition of S6K1 increases glucose metabolism and Akt signalling in vitro and in diet-induced obese mice. Diabetologia 59:592–603
12. Mellor KM, Brimble MA, Delbridge LM (2015) Glucose as an agent of post-translational modification in diabetes—new cardiac epigenetic insights. Life Sci 129:48–53
13. McLarty JL, Marsh SA, Chatham JC (2013) Post-translational protein modification by O-linked N-acetyl-glucosamine: its role in mediating the adverse effects of diabetes on the heart. Life Sci 92:621–627

14. Pereira L, Ruiz-Hurtado G, Rueda A et al (2010) Mitochondrial fission and autophagy in the normal and diseased heart. Curr Hypertens Rep 12(6):418–425
15. Mishra PK, Ying W, Nandi SS et al (2017) Diabetic cardiomyopathy: an immunometabolic perspective. Front Endocrinol (Lausanne) 8:72. https://doi.org/10.3389/fendo.2017.00072
16. Basha B, Samuel SM, Triggle CR et al (2012). Endothelial dysfunction in diabetes mellitus: possible involvement of endoplasmic reticulum stress?. Exp Diab Res Article ID 481840, 14 pages. https://doi.org/10.1155/2012/481840
17. Calles-Escandon J, Cipolla M (2001) Diabetes and endothelial dysfunction: a clinical perspective. End Rev 22(1):36–52
18. Bohm F, Pernow J (2007) The importance of endothelin-1 for vascular dysfunction in cardiovascular disease. Cardiovasc Res 76(1):8–18
19. Yu L, Li S, Tang X et al (2017a) Diallyl trisulfide ameliorates myocardial ischemia-reperfusion injury by reducing oxidative stress and endoplasmic reticulum stress-mediated apoptosis in type 1 diabetic rats: role of SIRT1 activation. Apoptosis 22(7):942–954
20. Pei Z, Deng Q, Babcock SA et al (2018) Inhibition of advanced glycation endproduct (AGE) rescues against streptozotocin-induced diabetic cardiomyopathy: role of autophagy and ER stress. Toxicol Lett 284:10–20
21. Al-Goblan AS, Al-Alfi MA, Khan MZ (2014) Mechanism linking diabetes mellitus and obesity. Diabetes Metab Syndr Obes 7:587–591
22. Hall JE, do Carmo JM, da Silva AA et al (2015) Obesity –induced hypertension : interaction of neurohumoral and renal mechanisms. Circ Res 116(6):991–1006
23. Jiang Z, Hu Z, Zeng L et al (2011) The role of the Golgi apparatus in oxidative stress: is this organelle less significant than mitochondria? Free Radic Biol Med 50(8):907–917
24. Sletten AC, Peterson LR, Schaffer JE (2018) Manifestations and mechanisms of myocardial lipotoxicity in obesity. J Intern Med doi. https://doi.org/10.1111/joim.12728
25. Choi CS, Savage DB, Kulkarni A et al (2007) Suppression of diacylglycerol acyltransferase-2 (DGAT2), but not DGAT1, with antisense oligonucleotides reverses diet-induced hepatic steatosis and insulin resistance. J Biol Chem 282:22678–22688
26. Ly LD, Xu S, Choi SK et al (2017) Oxidative stress and calcium dysregulation by palmitate in type 2 diabetes. Exp Mol Med 49(2):e291
27. Liu J, Lloyd SG (2013) High-fat, low-carbohydrate diet alters myocardial oxidative stress and impairs recovery of cardiac function after ischemia and reperfusion in obese rats. Nutr Res 33(4):311–321
28. Li SJ, Liu CH, Chu HP et al (2017) The high-fat diet induces myocardial fibrosis in the metabolically healthy obese minipigs-The role of ER stress and oxidative stress. Clin Nutr 36(3):760–767
29. Hagenfeldt L, Wahren J, Pernow B et al (1972) Uptake of individual free fatty acids by skeletal muscle and liver in man. J Clin Invest 51:2324–2330
30. Jia G, Hill MA, Sowers JR (2018) Diabetic cardiomyopathy: an update of mechanisms contributing to this clinical entity. Circ Res 22(4):624–638
31. Nishida K, Otsu K (2017) Inflammation and metabolic cardiomyopathy. Cardiovasc Res 113(4):389–398
32. Ozcan U, Cao Q, Yilmaz E et al (2004) Endoplasmic reticulum stress links obesity, insulin action, and type 2 diabetes. Science 306(5695):457–461
33. Nakamura T, Furuhashi M, Li P et al (2010) Double-stranded RNA-dependent protein kinase links pathogen sensing with stress and metabolic homeostasis. Cell 140(3):338–348
34. Hotamisligil GS (2005) (2005) role of endoplasmic reticulum stress and c-Jun NH2-terminal kinase pathways in inflammation and origin of obesity and diabetes. Diabetes 54(Suppl 2):S73–S78
35. Shih YC, Chen CL, Zhang Y et al (2018) Endoplasmic reticulum protein TXNDC5 augments myocardial fibrosis by facilitating extracellular matrix protein folding and redox-sensitive cardiac fibroblast activation. Circ Res 122(8):1052–1068
36. Veronica G, Esther RRM (2011) Metabolic syndrome: early development and aging. J Diabetes Metab S2:002. https://doi.org/10.4172/2155-6156.S2-002

37. Strait JB, Lakatta EG (2012) Aging-associated cardiovascular changes and their relationship to heart failure. Heart Fail Clin 8:143–164
38. Yang X, Sreejayan N, Ren J (2005) Views from within and beyond: narratives of cardiac contractile dysfunction under senescence. Endocrine 26:127–137
39. Csiszar A, Sosnowska D, Wang M et al (2012) Age-associated proinflammatory secretory phenotype in vascular smooth muscle cells from the non-human primate Macaca mulatta: reversal by resveratrol treatment. J Gerontol A Biol Sci Med Sci 67(8):811–820
40. Nair S, Ren J (2012) Autophagy and cardiovascular aging: lesson learned from rapamycin. Cell Cycle 11(11):2092–2099
41. Wu J, Xia S, Kalionis B et al (2014) The role of oxidative stress and inflammation in cardiovascular aging. Biomed Res Int 2014:615312
42. DeQuach JA, Mezzano V, Miglani A et al (2010) Simple and high yielding method for preparing tissue specific extracellular matrix coatings for cell culture. PLoS One 5:e13039
43. Ouzounian M, Lee DS, Liu PP (2008) Diastolic heart failure: mechanisms and controversies. Nat Clin Pract Cardiovasc Med 5:375–386
44. Bujak M, Ren G, Kweon HJ et al (2007) Essential role of Smad3 in infarct healing and in the pathogenesis of cardiac remodeling. Circulation 116(19):2127–2138
45. Reed AL, Tanaka A, Sorescu D et al (2011) Diastolic dysfunction is associated with cardiac fibrosis in the senescence-accelerated mouse. Am J Physiol Heart Circ Physiol 301:H824–H831
46. Triggle CR, Ding H (2011) The endothelium in compliance and resistance vessels. Front Biosci 3:730–744
47. Suganya N, Bhakkiyalakshmi E, Sarada DV et al (2016) Reversibility of endothelial dysfunction in diabetes: role of polyphenols. Br J Nutr 116(2):223–246
48. El-Sayyad HI, Al-Haggar MS, El-Ghawet HA et al (2012) Cardiomyopathy and angiogenesis defects of Wistar rat fetuses of diabetic and hypercholesterolemic mothers. Nutrition 28(7–8):e33–e43
49. El-Sayyad HIH, El-Sherbiny M, Sobh MA et al (2011) Protective effects of Morus alba leaves extract on ocular functions of pups from diabetic and hypercholesterolemic mother rats. Int J Biol Sci 7:715–728
50. He H, Xu J, Warren CM et al (2012) Endothelial cells provide an instructive niche for the differentiation and functional polarization of M2-like macrophages. Blood 120(15):3152–3162
51. Galvan-Pena S, O'Neill LA (2014) Metabolic reprograming in macrophage polarization. Front Immunol 5:420
52. Veilleux A, Grenier E, Marceau P et al (2014) Intestinal lipid handling: evidence and implication of insulin signaling abnormalities in human obese subjects. Arterioscler Thromb Vasc Biol 34:644–653
53. Martin J, Collot-Teixeira S, McGregor L et al (2007) The dialogue between endothelial cells and monocytes/macrophages in vascular syndromes. Curr Pharm Des 13(17):1751–1759
54. Voulgari C, Papadogiannis D, Tentolouris N (2010) Diabetic cardiomyopathy: from the pathophysiology of the cardiac myocytes to current diagnosis and management strategies. Vasc Health Risk Manag 6:883–903
55. Thomas D, Apovian C (2017) Macrophage functions in lean and obese adipose tissue. Metabolism 72:120–143
56. Wang Y, Wang GZ, Rabinovitch PS et al (2014) Macrophage mitochondrial oxidative stress promotes atherosclerosis and NF-κB-mediated inflammation in macrophages. Circ Res 114(3):421–433
57. Castaneda OA, Lee SC, Ho CT et al (2017) Macrophage in oxidative stress and models to evaluate the antioxidant functions of dietary natural compounds. J Food Drug Anal 25(1):111–118
58. Herrera MD, Mingorance C, Rodriguez-Rodriguez R et al (2010) Endothelial dysfunction and aging: an update. Ageing Res Rev 9:142–152

59. Pierce GL, Larocca TJ (2008) Reduced vascular tetrahydrobiopterin (BH4) and endothelial function with ageing: is it time for a chronic BH4 supplementation trial in middle-aged and older adults? J Physiol 586(11):2673–2674
60. Yang F, Yu X, Li T et al (2017) Exogenous H2S regulates endoplasmic reticulum-mitochondria cross-talk to inhibit apoptotic pathways in STZ-induced type I diabetes. Am J Physiol Endocrinol Metab 312(3):E190–E203
61. Schrieks IC, Nozza A, Stähli BE et al (2018) Adiponectin, free fatty acids, and cardiovascular outcomes in patients with type 2 diabetes and acute coronary syndrome. Diabetes Care pii:dc180158. https://doi.org/10.2337/dc18-0158
62. Mennerich D, Kellokumpu S, Kietzmann T (2018) Hypoxia and reactive oxygen species as modulators of endoplasmic reticulum and Golgi homeostasis. Antioxid Redox Signal doi. https://doi.org/10.1089/ars.2018.7523
63. Xu J, Wang G, Wang Y et al (2009) Diabetes and angiotensin II- induced cardiac endoplasmic reticulum stress and cell death: metallothionein protection. J Cell Mol Med 8A:1499–1512
64. Shah MS, Brownlee M (2016) Molecular and cellular mechanisms of cardiovascular disorders in diabetes. Circ Res 118(11):1808–1829
65. Hsu YJ, Hsu SC, Hsu CP et al (2017) Sirtuin 1 protects the aging heart from contractile dysfunction mediated through the inhibition of endoplasmic reticulum stress-mediated apoptosis in cardiac-specific Sirtuin 1 knockout mouse model. Int J Cardiol 228:543–552
66. Shen X, Zhang K , Kaufman RJ (2004) The unfolded protein response – a stress signaling pathway of the endoplasmic reticulum. J Chem Neuro 28(1):279–292
67. Schroder M, Kaufman RJ (2005) ER stress and the unfolded protein response. Mutat Res Fundam Mol Mech Mutagen 569(1–2):29–63
68. Robinson PJ, Pringle MA, Woolhead CA et al (2017) Folding of single domain protein entering the endoplasmic reticulum precedes disulfide formation. J Biol Chem 292(17):6978–6986
69. Ron D, Walter P (2007) Signal integration in the endoplasmic reticulum unfolded protein response. Nat Rev Mol Cell Biol 8:519–529
70. Smith MH, Ploegh HL, Weissman JS (2011) Road to ruin: targeting proteins for degradation in the endoplasmic reticulum. Science 334:1086–1090
71. Schonthal AH (2012) Endoplasmic reticulum stress: its role in disease and novel prospects for therapy. Scientifica (Cairo) 2012:857516
72. Van der Vlies D, Makkinje M, Jansens A et al (2003) Oxidation of ER resident proteins upon oxidative stress: effects of altering cellular redox/antioxidant status and implications for protein maturation. Antioxid. Redox Signal 5:381–387
73. Moroder L, Besse D, Musiol HJ et al (1996) Oxidative folding of cystine-rich peptides vs. regioselective cysteine pairing strategies. Biopolymers 40:207–234
74. Cullinan SB, Diehl JA (2004) PERK-dependent activation of Nrf2 contributes to redox homeostasis and cell survival following endoplasmic reticulum stress. J Biol Chem 279:20108–20117
75. Malhotra JD, Kaufman RJ (2011) ER stress and its functional link to mitochondria: role in cell survival and death. Cold Spring Harb Perspect Biol 3:a004424
76. Zeeshan HM, Lee GH, Kim HR et al (2016) Endoplasmic reticulum stress and associated ROS. Int J Mol Sci 17:327
77. Cao SS, Kaufman RJ (2014) Endoplasmic reticulum stress and oxidative stress in cell fate decision and human disease. Antioxid. Redox Signal 21:396–413
78. Bhandary B, Marahatta A, Kim HR et al (2012) An involvement of oxidative stress in endoplasmic reticulum stress and its associated diseases. Int J Mol Sci 14(1):434–456
79. Yang Q, Gao H, Dong R et al (2016a) Sequential changes of endoplasmic reticulum stress and apoptosis in myocardial fibrosis of diabetes mellitus-induced rats. Mol Med Rep 13(6):5037–5044
80. Yang R, Jia Q, Liu XF et al (2016b) Effect of hydrogen sulfide on oxidative stress and endoplasmic reticulum stress indiabetic cardiomyopathy. Zhongguo Ying Yong Sheng Li Xue Za Zhi 32(1):8–12

81. Liu Z, Lv Y, Zhao N et al (2015) Protein kinase R-like ER kinase and its role in endoplasmic reticulum stress-decided cell fate. Cell Death Dis 6:e1822
82. Wang X, Xu L, Gillette TG et al (2018) The unfolded protein response in ischemic heart disease. J Mol Cell Cardiol 117:19–25
83. Rieusset J (2018) The role of endoplasmic reticulum-mitochondria contact sites in the control of glucose homeostasis: an update. Cell Death Dis 9:388
84. Cullinan SB, Diehl JA (2006) Coordination of ER and oxidative stress signaling: the PERK/Nrf2 signaling pathway. Int. J. Biochem. Cell Biol 38:317–332
85. Fazi B, Melino S, De Rubeis S et al (2009) Acetylation of RTN-1C regulates the induction of ER stress by the inhibition of HDAC activity in neuroectodermal tumors. Oncogene 28:3814–3824
86. Stankov K, Stanimirov B, Mikov M (2013) Cellular responses to endoplasmic reticulum stress. Biologia Serbica 35(1–2):15–23
87. Araki K, Nagata K (2012) Protein folding and quality control in the ER. Cold Spring Harb Perspect Biol 4:a015438
88. Yoshida H (2007) ER stress and diseases. FEBS J 274:630–658
89. Cominacini L, Mozzini C, Garbin U et al (2015) Endoplasmic reticulum stress and Nrf2 signaling in cardiovascular diseases. Free Radic Biol Med 88(Pt B):233–242
90. Sreedhar R, Giridharan VV, Arumugam S et al (2010) Role of MAPK-mediated endoplasmic reticulum stress signaling in the heart during aging in senescence-accelerated prone mice. Biofactors 42(4):368–375
91. Wu Y, Reece EA, Zhong J et al (2016) Type 2 diabetes mellitus induces congenital heart defects in murine embryos by increasing oxidative stress, endoplasmic reticulum stress, and apoptosis. Am J Obstet Gynecol 215(3):366.e1–366.e10
92. Civelek M, Manduchi E, Riley RJ et al (2009) Chronic endoplasmic reticulum stress activates unfolded protein response in arterial endothelium in regions of susceptibility to atherosclerosis. Circ Res 105(5):453–461
93. Puthalakath H, O'Reilly LA, Gunn P et al (2007) ER stress triggers apoptosis by activating BH3-only protein Bim. Cell 129:1337–1349
94. Ghemrawi R, Battaglia-Hsu SF, Arnold C (2018) Endoplasmic reticulum stress in metabolic disorders. Cells 7(6):pii:E63. https://doi.org/10.3390/cells7060063
95. Foufelle F, Ferré P (2007) Unfolded protein response: its role in physiology and physiopathology. Med Sci (Paris) 23(3):291–296
96. Ozcan L, Tabas I (2012) Role of endoplasmic reticulum stress in metabolic disease and other disorders. Annu Rev Med 63: 317–328
97. Stanley WC (2001) Changes in cardiac metabolism: a critical step from stable angina to ischaemic cardiomyopathy. Eur Heart J Suppl 3(O):O2–O7
98. Santulli G, Xie W, Reiken SR et al (2015) Mitochondrial calcium overload is a key determinant in heart failure. Proc Natl Acad Sci U S A 112(36):11389–11394
99. Li Q, Su D, O'Rourke B et al (2015) Mitochondria-derived ROS bursts disturb Ca^{2+} cycling and induce abnormal automaticity in Guinea pig cardiomyocytes: a theoretical study. Am J Physiol Heart Circ Physiol 308(6):H623–H636
100. García-Rúa V, Otero MF, Lear PV et al (2012) Increased expression of fatty-acid and calcium metabolism genes in failing human heart. PLoS One 7:e37505
101. Leichman JG, Aguilar D, King TM et al (2006) Association of plasma free fatty acids and left ventricular diastolic function in patients with clinically severe obesity. Am J Clin Nutr 84:336–341
102. Neubauer S (2007) The failing heart – an engine out of fuel. N Engl J Med 356:1140–1151
103. Boudina S, Abel ED (2007) Diabetic cardiomyopathy revisited. Circulation 115:3213–3223
104. Boengler K, Kosiol M, Mayr M et al (2017) Mitochondria and ageing: role in heart, skeletal muscle and adipose tissue. J Cachexia Sarcopenia Muscle 8(3):349–369
105. Martín-Fernández B, Gredilla R (2016) Mitochondria and oxidative stress in heart aging. Age (Dordr) 38(4):225–238

106. Olgar Y, Degirmenci S, Durak A et al (2018) Aging related functional and structural changes in the heart and aorta: MitoTEMPO improves aged-cardiovascular performance. Exp Gerontol 110:172–181
107. Joseph SK, Nakao SK, Sukumvanich S (2006) Reactivity of free thiol groups in type-I inositol trisphosphate receptors. Biochem J 393:575–582
108. Murri M, El Azzouzi H (2018) MicroRNAs as regulators of mitochondrial dysfunction and obesity. Am J Physiol Heart Circ Physiol 13. https://doi.org/10.1152/ajpheart.00691.2017
109. Fillmore N, Mori J, Lopaschuk GD (2014) Mitochondrial fatty acid oxidation alterations in heart failure, ischaemic heart disease and diabetic cardiomyopathy. Br J Pharmacol 171(8):2080–2090
110. Duncan JG (2011) Mitochondrial dysfunction in diabetic cardiomyopathy. Biochim Biophys Acta 1813(2011):1351–1359
111. Mootha VK, Lindgren CM, Eriksson KF et al (2003) PGC-1alpha-responsive genes involved in oxidative phosphorylation are coordinately downregulated in human diabetes, Nat. Genet 34(2003):267–273
112. Finck BN, Kelly DP (2006) PGC-1 coactivators: inducible regulators of energy metabolism in health and disease. J Clin Invest 116(3):615–622
113. Judge S, Leeuwenburgh C (2007) Cardiac mitochondrial bioenergetics, oxidative stress, and aging. Am J Physiol Cell Physiol 292(6):C1983–C1992
114. Sibouakaz D, Othmani-Mecif K, Fernane A et al (2018) Biochemical and ultrastructural cardiac changes induced by high-fat diet in female and male Prepubertal rabbits. Anal Cell Pathol (Amst). 2018:6430696. https://doi.org/10.1155/2018/6430696
115. Kujoth GC, Hiona A, Pugh TD et al (2005) Mitochondrial DNA mutations, oxidative stress, and apoptosis in mammalian aging. Science 309(5733):481–484
116. Gredilla R, Bohr VA, Stevnsner T (2010) Mitochondrial DNA repair and association with aging – an update. Exp Gerontol 45(7–8):478–488
117. Lopez-Otin C, Blasco MA, Partridge L et al (2013) The hallmarks of aging. Cell 153:1194–1217
118. Koshiba T, Detmer SA, Kaiser JT et al (2004) Structural basis of mitochondrial tethering by mitofusin complexes. Science 305:858–862
119. Chan DC (2006) Mitochondrial fusion and fission in mammals. Annu Rev Cell Dev Biol 22:79–99
120. Rojo M, Legros F, Chateau D et al (2002) Membrane topology and mitochondrial targeting of mitofusins, ubiquitous mammalian homologs of the transmembrane GTPase Fzo. J Cell Sci 115:1663–1674
121. Olichon A, Emorine LJ, Descoins E et al (2002) The human dynamin-related protein OPA1 is anchored to the mitochondrial inner membrane facing the inter-membrane space. FEBS Lett 523(1–3):171–176
122. Cipolat S, Martins de Brito O, Dal Zilio B et al (2004) OPA1 requires mitofusin 1 to promote mitochondrial fusion. Proc Natl Acad Sci U S A 101:15927–15932
123. de Brito OM, Scorrano L (2008) Mitofusin 2 tethers endoplasmic reticulum to mitochondria. Nature 456:605–610
124. Smirnova E, Griparic L, Shurland DL et al (2001) Dynamin-related protein Drp1 is required for mitochondrial division in mammalian cells. Mol Biol Cell 12:2245–2256
125. Mozdy AD, McCaffery JM, Shaw JM (2000) Dnm1p GTPase-mediated mitochondrial fission is a multi-step process requiring the novel integral membrane component Fis1p. J Cell Biol 151:367–380
126. Tsushima K, Bugger H, Wende AR et al (2018) Mitochondrial reactive oxygen species in Lipotoxic hearts induce post-translational modifications of AKAP121, DRP1, and OPA1 that promote mitochondrial fission. Circ Res 122(1):58–73
127. Kang PT, Chen CL, Lin P et al (2018) Mitochondrial complex I in the post-ischemic heart reperfusion-mediated oxidative injury and protein cysteine sulfonation. J Mol Cell Cardiol pii:S0022-2828(18)30680–1

128. Tse G, Yan BP, Chan YW et al (2016) Reactive oxygen species, endoplasmic reticulum stress and mitochondrial dysfunction: the link with cardiac arrhythmogenesis. Front Physiol 7:313
129. Bertero E, Maack C (2018) Calcium signaling and reactive oxygen species in mitochondria. Circ Res 122(10):1460–1478
130. Hassanpour SH, Dehghani MA, Karami SZ (2018) Study of respiratory chain dysfunction in heart disease. J Cardiovasc Thorac Res 10(1):1–13
131. Nishio S, Teshima Y, Takahashi N et al (2012) Activation of CaMKII as a key regulator of reactive oxygen species production in diabetic rat heart. J Mol Cell Cardiol 52(5):1103–1011
132. Repnik U, Turk B (2010) Lysosomal–mitochondrial cross-talk during cell death. Mitochondrion 10:662–669
133. Andreyev AY, Kushnareva YE, Starkov AA (2005) Mitochondrial metabolism of reactive oxygen species. Biochem Mosc 70:200–214
134. Kurz T, Eaton JW, Brunk UT (2010) Redox activity within the lysosomal compartment: implications for aging and apoptosis. Antioxid Redox Signal 13(4):511–523
135. Michelangeli F, Ogunbayo OA, Wootton LL (2005) A plethora of interacting organellar Ca^{2+} stores. Curr Opin Cell Biol 17:135–140
136. Missiaen L, Raeymaekers L, Dode L et al (2004) SPCA1 pumps and Hailey– Hailey disease. Biochem Biophys Res Commun 322:1204–1213
137. Joseph LC, Subramanyam P, Radlicz C et al (2016) Mitochondrial oxidative stress during cardiac lipid overload causes intracellular calcium leak and arrhythmia. Heart Rhythm 13(8):1699–1706
138. Lane JD, Lucocq J, Pryde J et al (2002) Caspase-mediated cleavage of the stacking protein GRASP65 is required for Golgi fragmentation during apoptosis. J Cell Biol 156:495–509
139. Hampton MB, Fadeel B, Orrenius S (1998) Redox regulation of the caspases during apoptosis. Ann N Y Acad Sci 854:328–335
140. Seko Y, Fujimura T, Yao T et al (2015) Secreted tyrosine sulfated-eIF5A mediates oxidative stress-induced apoptosis. Sci Rep 5:13737. https://doi.org/10.1038/srep13737
141. Sbodio JI, Snyder SH, Paul BD (2018) Golgi stress response programs cysteine metabolism to confer cytoprotectionin Huntington's disease. Proc Natl Acad Sci U S A 115(4):780–785
142. Burke MA, Ardehali H (2007) Mitochondrial ATP-binding cassette proteins. Transl Res 150(2):73–80
143. Sousa L, Pessoa MTC, Costa TGF et al (2018) Iron overload impact on P-ATPases. Ann Hematol 97(3):377–385
144. Patton SM, Piñero DJ, Surguladze N et al (2005) Subcellular localization of iron regulatory proteins to Golgi and ER membranes. J Cell Sci 118(Pt 19):4365–4373
145. Warren CM, Ziyad S, Briot A et al (2014) A ligand-independent VEGFR2 signaling pathway limits angiogenic responses in diabetes. Sci Signal 7(307):ra1
146. Wang H, Joseph JA (2000) Mechanism of hydrogen peroxide-induced calcium deregulation in PC12 cells. Free Radic Biol Med 28:1222–1231
147. Brookes PS, Yoon Y, Robotham JL et al (2004) Calcium, ATP, and ROS: a mitochondrial love–hate triangle. Am J Physiol Cell Physiol 287:C817–C833
148. Görlach A, Bertram K, Hudecova S et al (2015) Calcium and ROS: a mutual interplay. Redox Biol 6:260–271
149. Xu H, Ren D (2015) Lysosomal physiology. Annu Rev Physiol 77:57–80
150. Tai H, Wang Z, Gong H, Han X, Zhou J, Wang X et al (2017) Autophagy impairment with lysosomal and mitochondrial dysfunction is an important characteristic of oxidative stress-induced senescence. Autophagy 13(1):99–113
151. Nakatogawa H, Suzuki K, Kamada Y et al (2009) Dynamics and diversity in autophagy mechanisms: lessons from yeast. Nat Rev Mol Cell Biol 10:458–467
152. Iglewski M, Hill JA, Lavandero S et al (2010) Mitochondrial fission and autophagy in the normal and diseased heart. Curr Hypertens Rep 12(6):418–425
153. Terman A, Brunk UT (2005) The aging myocardium: roles of mitochondrial damage and lysosomal degradation. Heart Lung Circ 14(2):107–114

154. Terman A, Kurz T, Gustafsson B et al (2008) The involvement of lysosomes in myocardial aging and disease. Curr Cardiol Rev 4(2):107–115
155. Sciarretta S, Volpe M, Sadoshima J (2012) Is reactivation of autophagy a possible therapeutic solution for obesity and metabolic syndrome? Autophagy 8:1252–1254
156. Sciarretta S, Boppana VS, Umapathi M et al (2015) Boosting autophagy in the diabetic heart: a translational perspective. Cardiovasc Diagn Ther 5(5):394–402
157. Wang F, Jia J, Rodrigues B (2017) Autophagy, metabolic disease, and pathogenesis of heart dysfunction. Can J Cardiol 33(7):850–859
158. Ren SY, Xu X (2015) Role of autophagy in metabolic syndrome-associated heart disease. Biochim Biophys Acta 1852(2):225–231
159. Laskar A, Miah S, Andersson RG et al (2010) Prevention of 7β-hydroxycholesterol-induced cell death by mangafodipir is mediated through lysosomal and mitochondrial pathways. Eur J Pharmacol 640(1–3):124–128
160. Shirakabe A, Ikeda Y, Sciarretta S et al (2016) Aging and autophagy in the heart. Circ Res 118(10):1563–1576
161. Yu Y, Wang L, Delguste F et al (2017b) Advanced glycation end products receptor RAGE controls myocardial dysfunction and oxidative stress in high-fat fed mice by sustaining mitochondrial dynamics and autophagy-lysosome pathway. Free Radic Biol Med 112:397–410
162. Petermann I, Mayer C, Stypmann J et al (2006) Lysosomal, cytoskeletal, and metabolic alterations in cardiomyopathy of cathepsin L knockout mice. FASEB J 20(8):1266–1268
163. Leon LJ, Gustafsson ÅB (2016) Staying young at heart: autophagy and adaptation to cardiac aging. J Mol Cell Cardiol 95:78–85

Impact of Genetic and Epigenetic Factors on the Oxidative Stress in Cardiovascular Disease

5

Shiva Krishna Katkam, Bobbala Indumathi, Shaik Mohammad Naushad, and Vijay Kumar Kutala

Abstract

Cardiovascular disease (CVD) is one of the most common causes of deaths globally. Oxidative stress is documented to be one of the potential risk factors for CVD. This review focuses on genetic and epigenetic factors contributing toward CVD risk. Certain genetic variants in oxidants, i.e., NADPH oxidase p22Phox rs4673, eNOS (-786 T>C, 894 G>T, and 27bp VNTR), MPO (-463 and -129 GA-genotypes), XO 69901 A>C, COX2 (rs5277 and rs20417), and ALOX15 (rs2619112 and rs7217186), were reported to increase CVD risk. Similarly, genetic variants in antioxidants, i.e., SOD1 (rs9974610, rs10432782, rs1041740), SOD2 (V16A, C24T), GPX1 Pro198Leu, NQO1 C609T, PON1 Q192R, and TXNIP (rs7212 and rs7211), were also shown to exhibit positive association with CVD risk. Apart from oxidants and antioxidants, folate and xenobiotic metabolic pathways were also investigated due to their direct influence on synthesis, methylation, and repair of DNA. Among the functional variants of folate pathway, GCPII H475Y, MTHFR C677T, and MTRR A66G were reported to increase CVD risk, thus influencing *S*-adenosylmethionine/*S*-adenosyl homocysteine ratio. Two genetic variants, i.e., cSHMT C1420T and TYMS 5'-UTR 28bp tandem repeat, were shown to confer protection by inducing the futile folate cycle and by increasing the flux of folate toward remethylation of homocysteine, respectively. Among the xenobiotic variants, CYP1A1 CAC and TAC haplotypes and GSTT1 and GSTM1 null variants contribute toward CVD risk by inducing quinone and semiquinone synthesis and by preventing conjugation of glutathione, respectively. These two process trigger mutagenicity. CYP1A1 TAC

S. K. Katkam · B. Indumathi · V. K. Kutala (✉)
Department of Clinical Pharmacology & Therapeutics, Nizam's Institute of Medical Sciences Hyderabad, Hyderabad, Telangana, India

S. M. Naushad
Biochemical Genetics & Pharmacogenomics Lab, Sandor Life Sciences Pvt Ltd, Hyderabad, Hyderabad, Telangana, India

S. Chakraborti et al. (eds.), *Modulation of Oxidative Stress in Heart Disease*,
https://doi.org/10.1007/978-981-13-8946-7_5

haplotype and CYP4F2 rs2108622 confer protection against CVD. Hypermethylation of ER-alpha, ER-beta, p15(INK4b), FOXP3, and DDAH2 was associated with CVD risk, which is attributed to altered ER-signaling, increased oxidative stress, and impaired synthesis of Treg cells. Factor VII hypermethylation confers protection against CVD. In a nut shell, interplay of genetic and epigenetic factors in oxidant–antioxidant system modulates susceptibility to CVD.

Keywords
Cardiovascular diseases · Oxidants · Folate pathway · Xenobiotic pathway · Epigenetics · Methylation

5.1 Introduction

According to the heart disease and stroke statistics 2018 update, the cardiovascular disease is the primary cause of death and accounts for nearly 836,546 deaths in the United States. On an average, one American dies for every 38 s, accounting to 2300 deaths each day due to cardiovascular diseases (CVD) [1]. The CVD is also a major health burden in India and is the leading cause of mortality accounting age-standardized CVD death rate of 272 per 100,000 populations which is higher than the global average of 235 per 100,000 population [2]. The CVD is a multifactorial disorder that mainly consists of diseases such as coronary artery disease (CAD), heart failure, stroke, and hypertension. It is influenced by several factors including obesity, diabetes, smoking, lifestyle, family history, age, genetic predisposition, and oxidative stress [3].

The reactive oxygen species (ROS) are generated from a diverse group of biological sources including mitochondria electron transport [4, 5], NADPH oxidase [6], nitric oxide synthase, myeloperoxidase, xanthine oxidase, and cyclooxygenase and lipoxygenase [7, 8]. The ROS at physiological levels are important signaling molecules regulating many processes in the cardiovascular system to maintain cardiovascular homeostasis [9]. Increased ROS levels have been linked to initiation, progression, and clinical consequences of cardiovascular diseases including atherosclerosis, arrhythmia, myocardial infarction, cardiac hypertrophy, cardiomyopathy, heart failure, hypoxia–reoxygenation, systemic and pulmonary hypertension, and ischemia–reperfusion injury [10]. The molecular mechanism of free radical-induced cardiovascular risk is mediated through the interaction of ROS with biomolecules such as lipids, proteins, and DNA, thus contributing to cellular damage, necrosis, and apoptosis [11, 12]. However, antioxidant defense mechanisms comprising of enzymatic and nonenzymatic defense systems counteract the free radical-mediated oxidative damage. The enzymatic defense system comprises of superoxide dismutase (SOD), catalase, glutathione peroxidase, thioredoxin, and peroxiredoxin, while the nonenzymatic antioxidant defenses include vitamin E, vitamin C, and glutathione, which in turn confer protection against CVDs [3, 8, 13, 14].

In recent years, several genetic risk factors have been identified which play a crucial role in the initiation and progression of CVD, as these genes code for various enzymes involved in redox reactions. Genetic variations may alter the baseline expression and structural and functional changes of the enzyme and thereby contributing to the imbalance between the prooxidant and antioxidants ultimately leading to increased oxidative stress.

5.2 Oxidants

5.2.1 NADPH Oxidase

Different isoforms of NADPH oxidases, i.e., NOX1, NOX2, NOX4, and NOX5, are present in the vasculature. NOX1 and NOX2 are associated with angiotensin II-induced vascular response, which results in uncoupling of eNOS with reduced levels of BH4 [13, 14]. Angiotensin II induces vasoconstriction, increases blood pressure, enhances vascular smooth muscle cell proliferation, and participates in the process of ROS generation, thus resulting in vascular damage [15]. NOX4 isoform is expressed in endothelial and vascular smooth muscle cells. It generates hydrogen peroxide without the need of activating cytosolic factors [16]. Hydrogen peroxide results in enhanced activation of eNOS, VEGFR2, p38MAPK, and AMPK-α pathways which are crucial for angiogenesis [17]. p47phox is an important cytosolic subunit of NADPH oxidase enzyme whose knockout results in reduced ROS production, PI3K-Akt-eNOS, mediated NO production and VEFG-induced vasodilatation [18]. The p22phox rs4673 polymorphism was shown to be associated with 1.53-fold risk for CAD by impairing the function of p22phox protein [19].

5.2.2 Nitric Oxide Synthase (eNOS)

Nitric oxide (NO) is one of the major effectors molecules that play a crucial role in cardiovascular physiology by inducing endothelium-dependent vasorelaxation and inhibit platelet aggregation, vascular smooth muscle cell proliferation, and leukocyte adhesion [20]. Nitric oxide synthase (NOS) exists in three different isoforms: endothelial (eNOS), inducible (iNOS), and neuronal (nNOS). These isoforms require transfer of electron from oxygen to catalyze the conversion of L-arginine to L-citrulline to produce NO. Among these isoforms, nNOS [21] and eNOS [22] are protective against atherogenesis, while increased iNOS expression produces excess amount of NO and is involved in aggravating atherosclerotic plaque formation [23].

Three functional polymorphisms have been well studied in the eNOS gene: a single nucleotide polymorphism -786T >C (rs2070744) at 5′ flanking region of the eNOS gene that reduces the eNOS gene promoter activity by approximately 50 percent [24, 25], a missense mutation 894G >T (Glu298Asp, rs1799983) located in exon 7 leading to amino acid substitution at position 298 associated with the eNOS activity consequently influencing the NO levels, and a 27-bp VNTR (4b/a) polymorphism located in intron 4 which is a source for small interference RNA (siRNA) spliced during eNOS pre-mRNA processing able to suppress eNOS expression. A recent meta-analysis showed the significant association with all the three SNPs with CAD in European, Middle Eastern, Asian, Asian–Indian, and African ancestries (Glu298Asp, OR = 1.28–1.52, $p = 0.0001$, T786-C, OR = 1.34–1.42, $p = 0.0001$, and 4b/a, OR = 1.19–1.41, $p = 0.002$) [26]. The subgroup analysis has revealed that Glu298Asp and 4b/a have the strongest association among the Middle Easterners, whereas T786-C showed the highest risk for CAD among subjects of Asian ancestry [26].

5.2.3 Myeloperoxidase

Myeloperoxidase (MPO) is expressed in neutrophils and monocytes, catalyzes the formation of ROS, and is involved in inflammation, thus contributing to tissue damage. However, under physiological conditions, these oxidation products derived from MPO play an important role in host defense. Two common SNPs in the promoter of MPO gene, i.e., -463G/A and -129G/A, were reported to affect the binding of the transcriptional factor specificity protein1 (SP1) and thus influencing the expression of MPO [27]. The GA-genotypes at -463 and -129 positions were shown to increase the CAD risk by 1.53- and 1.94-folds, respectively [28]. A meta-analysis comprising of 3491 cases and 7293 controls has confirmed that MPO -463 G>A polymorphism as a potential risk factor for CAD; however, MPO -129 G>A did not emerge as a risk factor in the pooled analysis [29].

5.2.4 Xanthine Oxidase

Xanthine oxidase (XO) catalyzes the oxidation of hypoxanthine to xanthine and further oxidizes xanthine to uric acid. These biochemical reactions require oxygen, forming the superoxide molecules, which are converted to hydrogen peroxide. XO is associated with endothelial dysfunction and is elevated during ischemia–reperfusion injury [30]. Allopurinol, a potent inhibitor of XO, was reported to reduce the incidence of myocardial infarction to the extent of fivefolds in a meta-analysis [31]. The XO SNPs, i.e., rs11904439 and rs148756340, were reported to increase the incidence of hypertension by 1.31- and 1.69-folds [32]. Additionally, SNPs in XO, i.e., 47686C>T, XO 69901A>C, and 67873A>C, were also reported to increase hypertension in Japanese population. [33] Among the three SNPs in XO, 69901A>C was associated with carotid artery atherosclerosis ($p = 0.03$) [33] (Table 5.1).

5.2.5 Cyclooxygenase

Cyclooxygenase (COX) coverts the arachidonic acid into hydroperoxy-endoperoxide, PGG2, which is reduced to form hydroxyl-endoperoxide PGH2, the precursor for eicosanoid synthesis [34]. The eicosanoids play a crucial role in

Table 5.1 Functional genetic variants associated with oxidative stress system and predictions of their impact using SIFT and PolyPhen tool

Gene	rs number	Amino acid change	ROS system	SIFT	PolyPhen
SOD2	rs4880	Val16Ala	Antioxidant system	Tolerated	Benign
MTHFR	rs1801133	Ala222Val	Antioxidant system	Deleterious	Probably damaging
MTRR	rs1801394	Iso22Met	Antioxidant system	Deleterious	Probably damaging
PON1	rs854560	Leu55Met	Antioxidant system	Deleterious	Benign
PON1	rs662	Gln192Arg	Antioxidant system	Tolerated	Benign
p22phox	rs4673	Tyr72His	ROS production system	Tolerated	Benign
eNOS	rs1799983	Glu298Asp	Antioxidant system	Tolerated	Benign
GPx1	rs1050450	Pro198Leu	Antioxidant system	Deleterious	Benign
GCPII	rs61886492	His475Tyr	Antioxidant system	Tolerated	Benign
SHMT1	rs1979277	Leu474Phe	Antioxidant system	Deleterious	Benign
CYP1A1	rs1048943	Iso462Val	Antioxidant system	Tolerated	Benign
CYP1A1	rs1799814	Thr461Asn	Antioxidant system	Tolerated	Benign

atherosclerotic process. COX2 expression is increased in atherosclerotic lesions, specifically in macrophages and foam cells [35]. COX-2 rs5277 C-allele carriers were reported to have increased risk for major adverse cardiac and cerebrovascular events, more specifically for CAD [36]. The G-765C COX2 polymorphism (rs20417) is associated with less frequent occurrence of multivessel CAD [37].

5.2.6 Lipoxygenase (LOX)

Lipoxygenase (LOX) pathway metabolizes arachidonic acid to form pro-inflammatory leukotrienes, which induce vasoconstriction and increase risk for atherosclerosis [38]. In animal models, the knockout of the arachidonate 5-lipoxygenase (ALOX5) was reported to induce resistance against the development of atherosclerosis [39]. Higher levels of ALOX5 were reported in advanced plaques [40]. The Sp1 addition/deletion polymorphism in the promoter region of ALOX5 was reported to increase CAD risk by 4.47-folds by influencing LDL and HDL levels [41]. ALOX15 rs2619112 GA and rs7217186 CT variants are associated with 2.27- and 3.41-fold increased risk for CAD [42]. The 5-lipoxygenase-activating protein

(ALOX5AP) HapB and HapC haplotypes are reported to increase the CAD risk by 1.67- and 2.41-folds, respectively [43]. The epistatic interactions among the ALOX5, ALOX5AP, and MPO were found to act synergistically in contributing to ischemic stroke [44].

5.3 Genetic Variants in Antioxidant Enzymes

5.3.1 Superoxide Dismutase

Superoxide dismutase (SOD) is one of the most crucial antioxidant enzymes that scavenge ROS by converting superoxide to hydrogen peroxide. There are three isoforms of SOD: (i) cytoplasmic CuZn SOD (SOD1), (ii) mitochondrial MnSOD (SOD2), and (iii) extracellular SOD (SOD3). Plasma levels of SOD1 and SOD2 were reported to be elevated in patients with CAD, while SOD3 levels showed no significant association with CAD [45]. Three variants in SOD1, i.e., rs9974610, rs10432782, and rs1041740, are reported to increase the risk for cardiovascular disease [46]. The SOD2 V16A polymorphism was reported to increase the risk for cardiovascular disease independent of gender, smoking, blood pressure, cholesterol, and glycemic index [47]. The SOD2 C24T showed a significant association with the CAD risk in the presence of TT-genotype, while CC and TC genotypes were found to have protective role [48]. The SOD3 R231G polymorphism was shown to influence CAD risk with RG and GG genotypes as risk factors that contribute to severity of CAD and risk of myocardial infarction by lowering α-tocopherol levels [49].

5.3.2 Glutathione Peroxidase and Catalase

Glutathione peroxidase (GPx) isoenzyme and catalase are the part of the second line of antioxidant defense which converts H_2O_2 and ROOH into water and alcohol. Erythrocyte GPX-1 activity <23.9 U/gHb and GPX1 Pro198Leu polymorphism contribute to 4.72- and 2.14-fold risk for CAD [50]. An inverse association was observed between the GPx1 activity and risk for myocardial infarction and severity of CAD [50]. The Pro198Leu (rs1050450 C/T) variation of GPx-1 influences the enzyme activity due to the structural conformation of the active site region of the enzyme [51]. The C198T GPx-1 and C609T NQO1variants contribute toward CAD risk in type II diabetes mellitus [52]. The catalase gene C/T promoter polymorphism at position -262 influences the expression of mRNA and protein [53]. Ultimately, these functional variants may affect the overall enzyme activity and antioxidant capacity leading to oxidative stress and CVD.

5.3.3 Paraoxonase 1

Paraoxonase 1 (PON1), an antioxidant glycoprotein, is majorly synthesized in the liver and secreted into the bloodstream, where it is associated with the high-density lipoproteins (HDL) [54]. The major function of PON1 includes the prevention of LDL and HDL from oxidative processes, inactivates the toxic products resulting from the oxidation of LDL (ox-LDL), prevents the accumulation of ox-LDL, stimulates cholesterol efflux from macrophages through HDL, and suppresses the differentiation of monocytes into macrophages, thereby preventing the formation of atherosclerotic plaques [55–59]. Two missense mutations, L55M (rs854560) and Q192R (rs662), were well studied in association with the heart disease which affects the levels of PON1 and catalytic efficiency, respectively. The rs854560 "T-allele" encodes methionine leading to elevated levels of paraoxonase, whereas "A-allele" encodes leucine with low paraoxonase activity. Previous studies have revealed that AA (55LL) genotype is associated with the increased risk of insulin resistance, blood pressure, increased carotid artery intima-media thickness, increased lipoprotein-associated phospholipase A2 activity (HDL-Lp-PLA(2), and therefore cardiovascular risk [60–63]. PON1 192RR genotype exhibits higher enzyme activity which decreases in the following order QQ > QR > RR, with almost very low paraoxonase activity, and is associated with the RR genotype leading to coronary atherosclerosis [64].

5.3.4 Genetic Variants Associated with the Folate Pathway

The folate pathway or one-carbon metabolic pathway regulates the synthesis, repair, and methylation of DNA and also negates the oxidative stress by acting in synergy with phase II enzymes of xenobiotic metabolic pathway. The supplementation with 5-methyltetrahydrofolate was found to improve the flow-mediated dilatation, a marker of endothelial function [65]. Homocysteine, the by-product of this metabolic pathway was reported to undergo autoxidation to generate free radicals [66]. Hyperhomocysteinemia is a well documented risk factor for CAD [67, 68].

Glutamate carboxypeptidase II (GCPII) enzyme is required for the conversion of folylpolyglutamate to folylmonoglutamate for the intestinal absorption of folate. The C1561T polymorphism (rs61886492, H475Y) of GCPII in which histidine 475 is substituted with tyrosine leading to reduced GCPII activity resulted in low blood folate and higher homocysteine levels (Hcy) [69]. GCPII C1561T was shown to increase in the risk for CAD by 2.71-folds and is associated with the increased oxidative stress [70].

Serine hydroxymethyltransferase1 (SHMT1) is a key enzyme involved in the folate metabolism that catalyzes one-carbon transfer from serine to tetrahydrofolate to form 5,10-methylenetetrahydrofolate. The SHMT1 C1420T polymorphism (Leu474Phe) was reported to reduce CAD risk to 50% by lowering oxidative stress [70].

Methylenetetrahydrofolate reductase (MTHFR) and thymidylate synthase (TYMS) are the two crucial rate-limiting enzymes that dictate whether folate (5,10-methylene tetrahydrofolate) flux should be directed toward synthesis of *S*-adenosylmethionine (SAM) or toward thymidylate synthesis. Both these steps are essential for DNA methylation and synthesis, respectively. MTHFR converts 5,10-methylene tetrahydrofolate to 5-methyltetrahydrofolate, which in turn is required for remethylation of homocysteine. The MTHFR 677 C>T polymorphism induces thermolability in MTHFR thus contributing to the dissociation of active dimer into inactive monomers with subsequent loss in FAD-binding capacity. The presence of MTHFR 677 C>T polymorphism is associated with the homocysteine elevation and thereby increases the CAD risk by 1.61-folds [71]. The TYMS 5′-UTR 28bp tandem repeat, which affects the transcription of TYMS, was shown to reduce CAD risk by 34% [71] and was shown to lower the oxidative stress [70].

The remethylation of homocysteine is catalyzed by methionine synthase (MTR) using 5-methyl tetrahydrofolate as the substrate and methylcobalamin as the cofactor. The reductive methylation of cobalamin is carried out by methionine synthase reductase (MTRR). MTRR A66G polymorphism was reported to increase CAD risk 1.92-folds [71] by increasing oxidative stress [70].

5.3.5 Thioredoxin-Interacting Protein and Nrf2

Thioredoxin-interacting protein (TXNIP) is a binding protein of thioredoxin (TRX), which acts as an oxidative stress modulator by inhibiting antioxidant capacity of TRX [72] and by interacting with transcription factors such as Nrf2 [73]. Two SNPs in TXNIP, i.e., rs7212 and rs7211, were reported to increase the CAD risk by 1.26- and 1.23-folds, respectively [74]. The smoking and alcohol intake were shown to interact with TXNIP rs7212 and increase the risk for CAD by 3.7-folds [74]. The SNP in TXNIP rs7212 was reported to influence the TXNIP mRNA expression, plasma TXNIP, and malondialdehyde levels [74]. The downregulation of Nrf2/ARE was reported in CAD cases with increased oxidation of phospholipid 1-palmitoyl-2-arachidonyl-sn-glycero-3-phosphoryl- choline [75].

5.3.6 Genetic Variants of Xenobiotic Metabolism

Xenobiotic metabolic pathway is the crucial pathway for detoxification of certain endogenous and several exogenous polycyclic aromatic hydrocarbons (PAH). This detoxification process is executed in two phases: (i) phase I involving the activation of the xenobiotic agent into an electrophile or a nucleophile and (ii) phase II involving detoxification of electrophile or a nucleophile through conjugation with glutathione or through quenching the formation of quinones from catechol compounds by *O*-methylation. Hyperinducibility of phase I and impaired activity of phase II enzymes cause the oxidative lesions in DNA due to the formation of PAH–DNA adducts. Several genes regulating this pathway were explored for their association with CVD risk. CYP1A1 CAC and TGC haplotypes were reported to increase CAD risk by 1.72- and 2.05-folds, respectively, while TAC haplotype was reported to reduce the risk for CAD by 44% [76]. Two SNPS in CYP2D6, i.e., C2850T and G1846A, are associated with 2.07 and 1.70-folds increased risk for CAD [77]. A meta-analysis revealed the inverse association of CYP4F2 rs2108622 with CAD risk [78]. Among the phase II genetic variants, the GSTT1 null variant showed independent risk for CAD (OR: 2.53, 95% CI: 1.55-4.12). These variants were shown to have higher levels of 8-oxo-dG; this could be due to the defective detoxification and increased ROS production [76]. The GSTM1 null genotype was reported to increase the CAD risk by 1.35-fold in a meta-analysis comprising of 10595 cases and 13782 controls [79]. The GSTT1/GSTM1 null haplotype was reported to be a potential risk factor for myocardial infarction [80].

5.3.7 Epigenetic Factors Contributing to Coronary Artery Disease (CAD) Risk

Cardiovascular diseases are not only the consequences of genetic variations, but also due to epigenetic aberrations and alterations. The epigenetic mechanisms are able to alter the gene expression (can enhance or silence) without altering the DNA sequence. The capability of cells to transmit their tissue and stage-specific gene expression patterns to daughter cells without mutation of the DNA sequence, thereby making these changes reversible. These processes are crucial in normal development and differentiation of distinct cell lineages in the adult organism [81]. They can be modified by exogenous influences, and as such, they can contribute to alterations of phenotype or pathophenotype. The basic epigenetic regulatory mechanisms includes methylation of CpG islands in the DNA (carried out by DNA methyltransferases), posttranslation modification of histone proteins (PTMs) (carried out by various enzymes, namely, histone acetyl transferases, histone deacetylases, histone methyl transferases, and histone demethylases), and small noncoding RNA-based mechanisms, i.e., microRNAs. Many studies have shown the link between the involvement of epigenetic factors and the cardiovascular diseases such as myocardial infarction, cardiac hypertrophy, atherosclerosis, and heart failure [82, 83] (Fig. 5.1).

5.3.8 Altered DNA Methylation Affecting the Gene Expression

The DNA methylation of cytosine at promoter sites is to downregulate the expression by directly blocking the binding of specific transcription factors of genes by modifying the accessibility of the transcriptional machinery to DNA. The process of

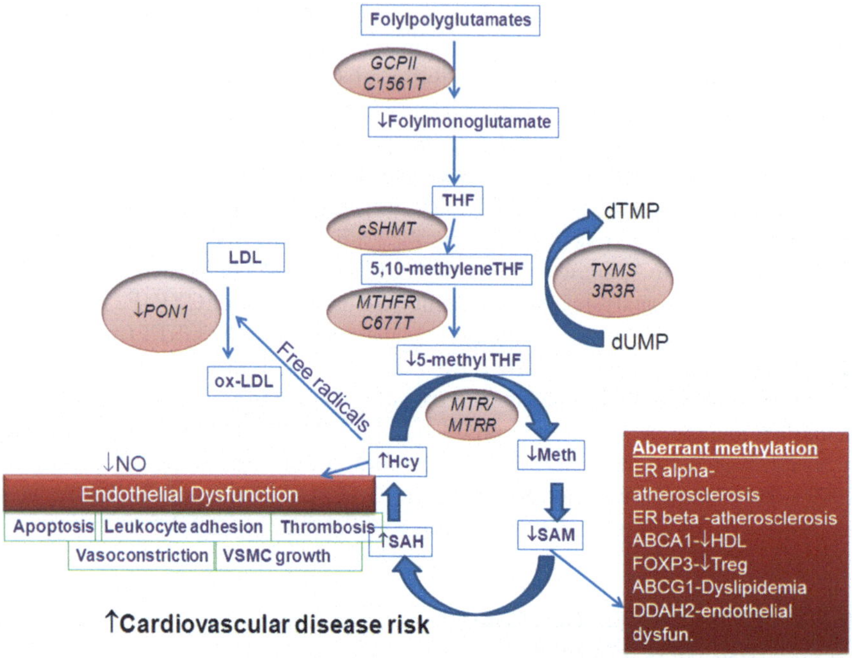

Fig. 5.1 Schematic representation of different candidate genes and epigenetic mechanisms contributing to oxidative stress and cardiovascular disease risk

DNA methylation is catalyzed by the enzyme known as DNA methyltransferase (DNMTs: DNMT1, DNMT3a, and 3b), which utilizes *S*-adenosyl methionine (SAM) as methyl donor. Differential methylation has been observed in the candidate genes that regulate the biological processes underlying cardiovascular diseases like diabetes, hypertension, atherosclerosis, and inflammation [84–87]. The lower levels of methyl donor *S*-adenosyl methionine (SAM), 5-methyltetrahydrofolate, and higher plasma homocysteine levels were reported in CAD patients [88]. Homocysteine is known to be an independent risk factor for CAD [89]. Elevated blood homocysteine levels were correlated with reduced DNA methylation in peripheral blood lymphocytes isolated from patients with vascular disease [90]. The endothelial dysfunction and altered DNA methylation patterns were also demonstrated in animal models with elevated homocystene levels [91, 92]. Higher global DNA methylation was observed in CAD patients whose serum homocysteine levels were found to be >12.5 μM [93]. Hyperhomocysteinemia was shown to induce expression of p66shc via hypomethylation of its promoter thus resulting in increased oxidative stress and reduced bioavailability of nitric oxide [94]. The aberrations in folate metabolism can induce altered gene expression of extracellular superoxide dismutase (EC-SOD), glutathione-*S*-transferase (GST)P1, and BCL2/Adenovirus E1B 19 KDa protein-interacting protein 3 (BNIP3), thus contributing to the increased oxidative stress and increased susceptibility to CAD [95].

The heart cells are able to express constitutive nitric oxide synthases, eNOS and nNOS [96]. The constitutively expressed eNOS is hypomethylated in the endothelium but heavily methylated in smooth muscle cell lines, whereas iNOS expression is increased in atherosclerotic plaque neointima due to inflammatory conditions but downregulated by methylation in most of the tissues [97]. Estrogen receptors (ERs) are present in the coronary arterial wall on both endothelial and smooth muscle cells and may play an important role in protection against atherosclerosis [98]. Estrogens have shown the protective role against oxidative stress mediated by ER-α. In men, deficiencies in ER-α may lead to accelerated atherosclerosis [99]. The hypermethylation of ER-α and ER-β was observed in coronary atherosclerotic plaques in comparison to normal aorta [100, 101]. The promoter methylation of ER-α has been demonstrated to increase with age and reach nearly a complete methylation level in the elderly [100].

The epigenetic modulator, 3-deazaadenosine, was shown to prevent smooth muscle cell proliferation and neointima formation by interfering with Ras methylation [102]. The factor VII promoter methylation confers protection against CAD as it lowers factor VIIa levels [103]. The methylation at ATP-binding cassette A1 (ABCA1) locus was shown to lower HDL levels and increase the risk for CAD [104]. The methylation at p15 (INK4b) locus was shown to increase CAD risk by influencing the expression of antisense noncoding RNA in the INK4 locus (ANRIL) [105].

Low-density lipoprotein L5, the most negatively charged subfraction of low-density lipoprotein, is capable of inducing apoptosis and was shown to inhibit fibroblast growth factor-2 (FGF-2) by inducing hypermethylation of its promoter [106]. Aspirin was found to attenuate the adverse effects of L5 by lowering the FGF2 expression [107]. The oxLDL response element in FGF2 promoter that is responsible for methylation-induced repression of FGF2 [108]. LDL stimulates the binding of the DNA methyl-CpG-binding protein-2 and histone methyltransferase enhancer of Zeste homolog 2 whereas decreases the binding of the KLF2 transcriptional activator, i.e., myocyte-enhancing factor-2, to the KLF2 promoter in endothelial cells [109]. The downregulation of KLF2 by LDL leads to a dysfunctional, hypercoagulable endothelium. Regulatory T (Treg) cells have been shown to play a protective role in experimental models of atherosclerosis. Hypermethylation of the transcription factor forkhead box P3 (FOXP3) was shown to decrease Treg cells and increase the risk for acute coronary syndrome [110]. Hyperlipidemia was shown to be associated with methylation of ATP-binding cassette, subfamily G, member 1 (ABCG1), lipase, hepatic (LIPC), and phospholipid transfer protein (PLTP) [111]. Hypomethylation of the ADP receptor P2Y12 was reported to be associated with clopidogrel resistance in CAD patients [112]. Hypermethylation of dimethylarginine dimethylaminohydrolase 2 (DDAH2) was shown to impair the function of endothelial progenitor cells, thus playing an important role in the pathophysiology of CAD [113]. Smoking and air pollution have shown to influence methylation and cardiovascular risk. The smoking-related methylation pattern in the coagulation factor II (thrombin) receptor-like 3 (F2RL3) gene was reported to influence the prognosis of CAD [114]. Further, differentially methylated regions (DMRs) were observed in genes, i.e., TCN2 promoter, CBS 5'UTR, AMT, and PON1 involved in folate pathway [115]. Exposure to air pollution, an established risk factor for ischemic heart disease and stroke, is associated with reduced blood methylation of LINE-1, increased CDKN2B, and MAGEA1 methylation [116].The smoke induced lower LINE-1 DNA methylation and also alterations in methylation of specific genes in buccal mucosa samples obtained from children [117], indicating that DNA methylation may be one of the mechanisms linking exposure to the pollutants and the development of CVD.

In CAD cases, differentially methylated regions were observed in the intronic region of complement component 1, q subcomponent-like 4 (C1QL4) genes and upstream region of the coiled-coil domain containing 47 (CCDC47) and transforming growth factor, beta receptor III (TGFBR3) genes [118]. A recent systemic review on DNA methylation and CAD identified that the candidate genes such as ABCG1 and FOXP3 are hypomethylation, whereas ESR-α gene was hypermethylated in CHD [119]. Additionally, the EWAS identified 84 genes showing differential methylation patterns in relevance to obesity, inflammation, and lipid and carbohydrate metabolism influencing the risk of CHD [119]. Candidate gene approaches are much needed in order to understand the differential methylation patterns at specific loci contributing to cardiovascular risk as they are influenced by external stimuli.

5.3.9 Histone Modifications and Effect on Gene Expression

Histones have protruding N-terminal tails which can undergo posttranslational modifications (so-called histone modifications/marks) [120]. There are various types of modifications including acetylation (ac), methylation (me), phosphorylation (P), ubiquitination (ubi), and sumoylation (SUMO). Methylations at lysine 9, 27, and 36 on histone H3 generally lead to reduced gene expression, while methylations of lysine 4 and 79 on histone H3 and lysine 20 on histone H4 cause the increased expression of genes. The dynamics of these marks are mediated through histone methyl transferases that place methyl groups and histone demethylases removes methylation. Acetylation of lysines on histones H3 and H4 is associated with increased transcription of genes. Acetyl groups are placed by histone acetyl transferases and removed by histone deacetylases [121]. Histone modifications have been shown to influence the transcription and gene expression [122]. Posttranslational modifications of histones also include their binding to specific proteins and mediate processes such as gene expression, apoptosis, and DNA damage repair [123]. Elevated levels of ROS arising from alterations in cellular metabolism and inflammatory responses constitute a key risk state for DNA damage. DNA repair requires dynamic changes in surrounding chromatin, including changes in nucleosome positioning and histone modifications [124, 125]. Epigenetic alterations have been shown to be induced by the ROS and H_2O_2, where DNMT1 becomes more tightly bound to chromatin after H_2O_2 treatment and thereby alter the methylation status of CpG regions [126]. In a study, eNOS upregulation was associated with an increased H3 and H4 histone acetylation in the eNOS promoter in neonatal rodent-persistent PH of the newborn model [127].

Histone acetyltransferase (HAT) activities have shown to possess a positive role in cardiac hypertrophy (CH) as demonstrated by the HAT activity of the transcriptional co-activators CREB-binding protein (CBP) and p300. The overexpression of CBP or p300 in cardiomyocytes resulted in hypertrophy, whereas that was absent in overexpressive mutant CBP and p300 lacking HAT activity [128]. Cardiac hypertrophy (CH) has also been linked with histone methylation, in particular with H3K9 [129–131] and H3K4 methylations [132, 133] in animal models. Histone modifications are also involved in heart failure; in a genome-wide histone methylation of heart tissues, Kaneda and colleagues reported that tri-methylated histone H3H4 and H3K9 were altered in heart failure [134].

Histone acetylases and deacetylases (HDACs) localize to the sites of DNA damage induced by oxidative free radicals to facilitate repair by increasing the access of repair proteins to the break site [131]. A family of methyl-CpG-binding proteins has recently been recognized that specifically bind to methylated CpGs, thereby contributing to transcriptional repression by recruiting histone-modifying proteins which include the MBD protein family (MBD1,MBD2, MBD4, and MeCP2), Kaiso and Kaiso-like proteins, and SRA domain proteins (e.g., SUVH9 and SUVH2) [124]. After ischemia/reperfusion in caveolin knockout mice, there is an increase in histone methylation and is associated with an increase in the HDAC activity as well as an elevated level of HMT G9a protein [135]. Decreased expression of sirtuin-1 was observed in caveolin knockout mice and a reduction in the translocation of Foxo-3a to the nucleus [135]. Further supporting the cardioprotective role caveolin, the caveolin knockout mice had decreased ventricular function and increased apoptosis of cardiomyocyte cells in the setting of ischemia and reperfusion [135].

5.3.10 RNA-Mediated Gene Silencing

MicroRNAs (miRNA) are highly conserved, small, noncoding RNAs (20–40 nucleotides long), which inhibit translation or decrease the mRNA stability by binding to specific sites usually in the 3′-untranslated region (3′UTR) of target regions of the genome. RNA-induced silencing complex, or RISC, is a multiprotein complex that incorporates one strand of miRNA. In neonatal cardiomyocytes, the overexpression of miR-23a, miR-23b, miR-24, miR-195, or miR-214 induced cardiac hypertrophy, whereas overexpression of miR-133 inhibited the phenotype [136, 137]. Evidence that redox signaling in cells is subject to regulation by miRNA was shown through Dicer knockdown in human microvascular endothelial cells [138]. This was demonstrated by lower inducible production of ROS when activated with phorbol ester, tumor necrosis factor-α, or vascular endothelial growth factor. The miRNA deficiency caused by Dicer knockdown specifically downregulated both p47phox expression and ROS production. Thus, p47phox of the NADPH oxidase complex has been identified as a target of miRNAs [138]. miRNA-29b can reduce the expression of DNMT enzymes and thereby affect the global methylation status [139]. Wang et al. demonstrated that miRNA-152 can knockdown DNMT1 in human aortic SMCs, leading to hypermethylation of the ER-α gene [140]. HDAC expression can be regulated by miRNAs, such as miRNA-449a [141]. miRNA-33 predominantly targets the gene encoding the ATP-binding cassette transporter ABCA1, which is involved in cellular cholesterol mobilization. The HIF-responsive miRNA-210 was shown to be ubiquitously expressed in the hypoxic cell and tissue types [142]. miR 155/miR22 targets eNOS and STAT5A in endothelial cells and is involved in neovascularization [143, 144] and miR-10a/miR-181b targets HOXA1,M βTRC, AP3K7, and KPNA1 in endothelial cells and influences inflammation and endothelial dysfunction [145, 146].

Though the studies have shown the possible role of posttranslation modification of histone proteins and mRNA-based expression in the regulation of candidate genes in cardiovascular disease, however, the redox contribution of several of these miRNAs and histone protein modifications are unclear; further studies are warranted to elucidate the precise role of these epigenetic factors in cardiovascular disease.

References

1. Benjamin EJ, Virani SS, Callaway CW, Chamberlain AM, Chang AR, Cheng S, Chiuve SE, Cushman M, Delling FN, Deo R, de Ferranti SD, Ferguson JF, Fornage M, Gillespie C, Isasi CR, Jiménez MC, Jordan LC, Judd SE, Lackland D, Lichtman JH, Lisabeth L, Liu S, Longenecker CT, Lutsey PL, Mackey JS, Matchar DB, Matsushita K, Mussolino ME, Nasir K, O'Flaherty M, Palaniappan LP, Pandey A, Pandey DK, Reeves MJ, Ritchey MD, Rodriguez CJ, Roth GA, Rosamond WD, UKA S, Satou GM, Shah SH, Spartano NL, Tirschwell DL, Tsao CW, Voeks JH, Willey JZ, Wilkins JT, Wu JH, Alger HM, Wong SS, Muntner P, American Heart Association Council on Epidemiology and Prevention Statistics Committee and Stroke Statistics Subcommittee (2018) Heart disease and stroke statistics-2018 update: a report from the American Heart Association. Circulation 137(12):e67–e492
2. Prabhakaran D, Jeemon P, Roy A (2016) Cardiovascular diseases in India: current epidemiology and future directions. Circulation 133(16):1605–1620
3. Li H, Horke S, Förstermann U (2014) Vascular oxidative stress, nitric oxide and atherosclerosis. Atherosclerosis 237(1):208–219
4. Adam-Vizi V, Chinopoulos C (2006) Bioenergetics and the formation of mitochondrial reactive oxygen species. Trends Pharmacol Sci 27(12):639–645
5. Rigoulet M, Yoboue ED, Devin A (2011) Mitochondrial ROS generation and its regulation: mechanisms involved in H(2)O(2) signaling. Antioxid Redox Signal 14(3):459–468
6. Cave A, Grieve D, Johar S, Zhang M, Shah AM (2005) NADPH oxidase-derived reactive oxygen species in cardiac pathophysiology. Philos Trans R Soc Lond Ser B Biol Sci 360(1464):2327–2334
7. Tardif JC (2003) Oxidative stress and coronary heart disease. Cardiology Rounds, 7. www.cardiologygrounds.org
8. Paravicini TM, Touyz RM (2008) NADPH oxidases, reactive oxygen species, and hypertension: clinical implications and therapeutic possibilities. Diabetes Care 31(Suppl 2):S170–S180
9. Zhang M, Shah AM (2014) ROS signalling between endothelial cells and cardiac cells. Cardiovasc Res 102(2):249–257
10. Brown DI, Griendling KK (2015) Regulation of signal transduction by reactive oxygen species in the cardiovascular system. Circ Res 116(3):531–549
11. Hockenbery DM, Oltvai ZN, Yin XM, Milliman CL, Korsmeyer SJ (1993) Bcl-2 functions in an antioxidant pathway to prevent apoptosis. Cell 75(2):241–251

12. Buttke TM, Sandstrom PA (1994) Oxidative stress as a mediator of apoptosis. Immunol Today 15(1):7–10
13. Sheehan AL, Carrell S, Johnson B, Stanic B, Banfi B, Miller FJ (2011) Role for Nox1 NADPH oxidase in atherosclerosis. Atherosclerosis 216:321–326
14. Judkins CP, Diep H, Broughton BR, Mast AE, Hooker EU, Miller AA et al (2010) Direct evidence of a role for Nox2 in superoxide production, reduced nitric oxide bioavailability, and early atherosclerotic plaque formation in ApoE−/− mice. Am J Physiol Heart Circ Physiol 298:H24–H32
15. Taubman MB (2003) Angiotensin II: a vasoactive hormone with ever-increasing biological roles. Circ Res 92:9–11
16. Ray R, Murdoch CE, Wang M, Santos CX, Zhang M, Alom-Ruiz S et al (2011) Endothelial Nox4 NADPH oxidase enhances vasodilatation and reduces blood pressure in vivo. Arterioscler Thromb Vasc Biol 31:1368–1376
17. Craige SM, Chen K, Pei Y, Li C, Huang X, Chen C et al (2011) NADPH oxidase 4 promotes endothelial angiogenesis through endothelial nitric oxide synthase activation. Circulation 124:731–740
18. Barry-Lane PA, Patterson C, Merwe MV, Hu Z, Holland SM, Yeh ET et al (2001) P47phox is required for atherosclerotic lesion progression in Apo−/− mice. J Clin Invest 108:1513–1522
19. Mazaheri M, Karimian M, Behjati M, Raygan F, Hosseinzadeh CA (2017) Association analysis of rs1049255 and rs4673 transitions in p22phox gene with coronary artery disease: a case-control study and a computational analysis. Ir J Med Sci 186(4):921–928
20. Madamanchi NR, Runge MS (2013) Redox signaling in cardiovascular health and disease. Free Radic Biol Med 61:473–501
21. Kuhlencordt PJ, Gyurko R, Han F, Scherrer-Crosbie M, Aretz TH, Hajjar R et al (2001) Accelerated atherosclerosis, aortic aneurysm formation, and ischemic heart disease in apolipoprotein E/endothelial nitric oxide synthase double-knockout mice. Circulation 104:448–454
22. Ponnuswamy P, Schröttle A, Ostermeier E, Grüner S, Huang PL, Ertl G et al (2012) ENOS protects from atherosclerosis despite relevant superoxide production by the enzyme in ApoE−/− mice. PLoS One 7:e30193
23. Kuhlencordt PJ, Chen J, Han F, Astern J, Huang PL (2001) Genetic deficiency of inducible nitric oxide synthase reduces atherosclerosis and lowers plasma lipid peroxides in apolipoprotein E-knockout mice. Circulation 103:3099–3104
24. Miyamoto Y, Saito Y, Nakayama M, Shimasaki Y, Yoshimura T, Yoshimura M, Harada M, Kajiyama N, Kishimoto I, Kuwahara K, Hino J, Ogawa E, Hamanaka I, Kamitani S, Takahashi N, Kawakami R, Kangawa K, Yasue H, Nakao K (2000) Replication protein A1 reduces transcription of the endothelial nitric oxide synthase gene containing a -786T-->C mutation associated with coronary spastic angina. Hum Mol Genet 9(18):2629–2637
25. Nakayama M, Yasue H, Yoshimura M, Shimasaki Y, Kugiyama K, Ogawa H, Motoyama T, Saito Y, Ogawa Y, Miyamoto Y, Nakao K (1999) T-786-->C mutation in the 5′-flanking region of the endothelial nitric oxide synthase gene is associated with coronary spasm. Circulation 99(22):2864–2870
26. Rai H, Parveen F, Kumar S, Kapoor A, Sinha N (2014) Association of endothelial nitric oxide synthase gene polymorphisms with coronary artery disease: an updated meta-analysis and systematic review. PLoS One 9(11):e113363
27. Piedrafita FJ, Molander RB, Vansant G, Orlova EA, Pfahl M, Reynolds WF (1996) An Alu element in the myeloperoxidase promoter contains a composite SP1-thyroid hormone-retinoic acid response element. J Biol Chem 271:14412–14420
28. Arslan S, Berkan Ö, Bayyurt B, Beton O, Şahin NLÖL, Aydemir EI (2017 Dec) Effects of MPO-463G/A and -129G/A polymorphisms on coronary artery disease risk and patient survival in a Turkish population. Biomed Rep 7(6):547–552
29. Wang Y, Chen XY, Wang K, Li S, Zhang XY (2017) Myeloperoxidase polymorphism and coronary artery disease risk: a meta-analysis. Medicine (Baltimore) 96(27):e7280

30. Chambers DE, Parks DA, Patterson G, Roy R, McCord JM, Yoshida S, Parmley LF, Downey JM (1985) Xanthine oxidase as a source of free radical damage in myocardial ischemia. J Mol Cell Cardiol 17(2):145–152
31. Singh TP, Skalina T, Nour D, Murali A, Morrison S, Moxon JV, Golledge J (2018) A meta-analysis of the efficacy of allopurinol in reducing the incidence of myocardial infarction following coronary artery bypass grafting. BMC Cardiovasc Disord 18(1):143
32. Scheepers LE, Wei FF, Stolarz-Skrzypek K, Malyutina S, Tikhonoff V, Thijs L, Salvi E, Barlassina C, Filipovský J, Casiglia E, Nikitin Y, Kawecka-Jaszcz K, Manunta P, Cusi D, Boonen A, Staessen JA, Arts IC (2016) Xanthine oxidase gene variants and their association with blood pressure and incident hypertension: a population study. J Hypertens 34(11):2147–2154
33. Yang J, Kamide K, Kokubo Y, Takiuchi S, Horio T, Matayoshi T, Yasuda H, Miwa Y, Yoshii M, Yoshihara F, Nakamura S, Nakahama H, Tomoike H, Miyata T, Kawano Y (2008) Associations of hypertension and its complications with variations in the xanthine dehydrogenase gene. Hypertens Res 31(5):931–940
34. Vane JR, Bakhle YS, Botting RM (1998) Cyclooxygenases 1 and 2. Annu Rev Pharmacol Toxicol 38:97–120
35. Schonbeck U, Sukhova GK, Graber P, Coulter S, Libby P (1999) Augmented expression of cyclooxygenase-2 in human atherosclerotic lesions. Am J Pathol 155(4):1281–1291
36. Liu H, Xu Z, Sun C, Gu D, Teng X, Zhao Y, Zheng Z (2017) A variant in COX-2 gene is associated with left main coronary artery disease and clinical outcomes of coronary artery bypass grafting. Biomed Res Int 2017:2924731
37. Rostoff P, Szczeklik W, Piwowarska W, Konduracka E, Sanak M, Nessler J (2014) Association of common cyclooxygenase-2 (COX-2) gene polymorphisms with clinical and angiographic characteristics of patients with coronary artery disease. Przegl Lek 71(6):314–318
38. Mehrabian M, Allayee H (2003) 5-lipoxygenase and atherosclerosis. Curr Opin Lipidol 14(5):447–457
39. Mehrabian M, Allayee H, Wong J, Shi W, Wang XP, Shaposhnik Z, Funk CD, Lusis AJ (2002) Identification of 5-lipoxygenase as a major gene contributing to atherosclerosis susceptibility in mice. Circ Res 91(2):120–126
40. Spanbroek R, Grabner R, Lotzer K, Hildner M, Urbach A, Ruhling K, Moos MP, Kaiser B, Cohnert TU, Wahlers T, Zieske A, Plenz G, Robenek H, Salbach P, Kuhn H, Radmark O, Samuelsson B, Habenicht AJ (2003) Expanding expression of the 5-lipoxygenase pathway within the arterial wall during human atherogenesis. Proc Natl Acad Sci U S A 100(3):1238–1243
41. Todur SP, Ashavaid TF (2012) Association of Sp1 tandem repeat polymorphism of ALOX5 with coronary artery disease in Indian subjects. Clin Transl Sci 5(5):408–411
42. Kaur N, Singh J, Reddy S (2018) Interaction between ALOX15 polymorphisms and coronary artery disease in North Indian population. Clin Exp Hypertens 40(4):398–405
43. Girelli D, Martinelli N, Trabetti E, Olivieri O, Cavallari U, Malerba G, Busti F, Friso S, Pizzolo F, Pignatti PF, Corrocher R (2007) ALOX5AP gene variants and risk of coronary artery disease: an angiography-based study. Eur J Hum Genet 15(9):959–966
44. Liu D, Liu L, Song Z, Hu Z, Liu J, Hou D (2017) Genetic variations of oxidative stress related genes ALOX5, ALOX5AP and MPO modulate ischemic stroke susceptibility through main effects and epistatic interactions in a Chinese population. Cell Physiol Biochem 43(4):1588–1602
45. Peng JR, Lu TT, Chang HT, Ge X, Huang B, Li WM (2016) Elevated levels of plasma superoxide dismutases 1 and 2 in patients with coronary artery disease. Biomed Res Int 2016:3708905
46. Neves AL, Mohammedi K, Emery N, Roussel R, Fumeron F, Marre M, Velho G (2012) Allelic variations in superoxide dismutase-1 (SOD1) gene and renal and cardiovascular morbidity and mortality in type 2 diabetic subjects. Mol Genet Metab 106(3):359–365

47. Möllsten A, Jorsal A, Lajer M, Vionnet N, Tarnow L (2009) The V16A polymorphism in SOD2 is associated with increased risk of diabetic nephropathy and cardiovascular disease in type 1 diabetes. Diabetologia 52(12):2590–2593
48. Tian C, Liu T, Fang S, Du X, Jia C (2012) Association of C47T polymorphism in SOD2 gene with coronary artery disease: a case-control study and a meta-analysis. Mol Biol Rep 39(5):5269–5276
49. Grammer TB, Renner W, Hoffmann MM, Kleber M, Winkelhofer-Roob BM, Boehm BO, Maerz W (2009) SOD3 R231G polymorphism associated with coronary artery disease and myocardial infarction. The Ludwigshafen Risk and Cardiovascular Health (LURIC) Study. Free Radic Res 43(7):677–684
50. Wickremasinghe D, Peiris H, Chandrasena LG, Senaratne V, Perera R (2016) Case control feasibility study assessing the association between severity of coronary artery disease with Glutathione Peroxidase-1 (GPX-1) and GPX-1 polymorphism (Pro198Leu). BMC Cardiovasc Disord 16:111
51. Ravn-Haren G, Olsen A, Tjønneland A, Dragsted LO, Nexø BA, Wallin H, Overvad K, Raaschou-Nielsen O, Vogel U (2006) Associations between GPX1 Pro198Leu polymorphism, erythrocyte GPX activity, alcohol consumption and breast cancer risk in a prospective cohort study. Carcinogenesis 27(4):820–825
52. Ramprasath T, Murugan PS, Kalaiarasan E, Gomathi P, Rathinavel A, Selvam GS (2012 Feb) Genetic association of Glutathione peroxidase-1 (GPx-1) and NAD(P)H:Quinone Oxidoreductase 1(NQO1) variants and their association of CAD in patients with type-2 diabetes. Mol Cell Biochem 361(1-2):143–150
53. Forsberg L, Lyrenäs L, de Faire U, Morgenstern R (2001) A common functional C-T substitution polymorphism in the promoter region of the human catalase gene influences transcription factor binding, reporter gene transcription and is correlated to blood catalase levels. Free Radic Biol Med 30(5):500–505
54. Rochu D, Chabrière E, Masson P (2007) Human paraoxonase: a promising approach for pre treatment and therapy of organophosphorus poisoning. Toxicology 233(1-3):47–59
55. Zielaskowska J, Olszewska-Słonina D (2006) The polymorphism of paraoxonase and its effects in physiological and pathological processes. Adv Clin Exp Med 15:1073–1078
56. Rosenblat M, Gaidukov L, Khersonsky O, Vaya J, Oren R, Tawfik DS, Aviram M (2006) The catalytic histidine dyad of high density lipoprotein-associated serum paraoxonase-1 (PON1) is essential for PON1-mediated inhibition of low density lipoprotein oxidation and stimulation of macrophage cholesterol efflux. J Biol Chem 281(11):7657–7665
57. La Du BN, Aviram M, Billecke S, Navab M, Primo-Parmo S, Sorenson RC, Standiford TJ (1999) On the physiological role(s) of the paraoxonases. Chem Biol Interact 119–120:379–388
58. Berrougui H, Loued S, Khalil A (2012) Purified human paraoxonase-1 interacts with plasma membrane lipid rafts and mediates cholesterol efflux from macrophages. Free Radic Biol Med 52(8):1372–1381
59. Shih DM, Lusis AJ (2015) The role of paraoxonase in cardiovascular diseases. Ann Clin Lab Sci 45(2):226–233
60. Garin MC, James RW, Dussoix P, Blanché H, Passa P, Froguel P, Ruiz J (1997) Paraoxonase polymorphism Met-Leu54 is associated with modified serum concentrations of the enzyme. A possible link between the paraoxonase gene and increased risk of cardiovascular disease in diabetes. J Clin Invest 99(1):62–66
61. Barbieri M, Bonafè M, Marfella R, Ragno E, Giugliano D, Franceschi C, Paolisso G (2002) LL-paraoxonase genotype is associated with a more severe degree of homeostasis model assessment IR in healthy subjects. J Clin Endocrinol Metab 87(1):222–225
62. Roest M, Rodenburg J, Wiegman A, Kastelein JJ, Voorbij HA (2006) Paraoxonase genotype and carotid intima-media thickness in children with familial hypercholesterolemia. Eur J Cardiovasc Prev Rehabil 13(3):464–466
63. Christidis DS, Liberopoulos EN, Kakafika AI, Miltiadous GA, Liamis GL, Kakaidi B, Tselepis AD, Cariolou MA, Elisaf MS (2007) Effect of paraoxonase 1 polymorphisms on

the response of lipids and lipoprotein-associated enzymes to treatment with fluvastatin. Arch Med Res 38(4):403–410
64. Mackness B, Mackness MI, Arrol S, Turkie W, Durrington PN (1998) Effect of the human serum paraoxonase 55 and 192 genetic polymorphisms on the protection by high density lipoprotein against low density lipoprotein oxidative modification. FEBS Lett 423:57–60
65. Doshi SN, McDowell IF, Moat SJ, Lang D, Newcombe RG, Kredan MB, Lewis MJ, Goodfellow J (2001) Folate improves endothelial function in coronary artery disease: an effect mediated by reduction of intracellular superoxide? Arterioscler Thromb Vasc Biol 21(7):1196–1202
66. Hogg N (1999) The effect of cyst(e)ine on the auto-oxidation of homocysteine. Free Radic Biol Med 27(1–2):28–33
67. Vinukonda G, Shaik Mohammad N, Md Nurul Jain J, Prasad Chintakindi K, Rama Devi Akella R (2009) Genetic and environmental influences on total plasma homocysteine and coronary artery disease (CAD) risk among South Indians. Clin Chim Acta 405(1-2):127–131
68. Ma Y, Peng D, Liu C, Huang C, Luo J (2017) Serum high concentrations of homocysteine and low levels of folic acid and vitamin B12 are significantly correlated with the categories of coronary artery diseases. BMC Cardiovasc Disord 17(1):37
69. Divyya S, Naushad SM, Addlagatta A, Murthy PV, Reddy CR, Digumarti RR, Gottumukkala SR, Kumar A, Rammurti S, Kutala VK (2012) Paradoxical role of C1561T glutamate carboxypeptidase II (GCPII) genetic polymorphism in altering disease susceptibility. Gene 497(2):273–279
70. Vijaya Lakshmi SV, Naushad SM, Seshagiri Rao D, Kutala VK (2013) Oxidative stress is associated with genetic polymorphisms in one-carbon metabolism in coronary artery disease. Cell Biochem Biophys 67(2):353–361
71. Vijaya Lakshmi SV, Naushad SM, Rupasree Y, Seshagiri Rao D, Kutala VK (2011) Interactions of 5′-UTR thymidylate synthase polymorphism with 677C → T methylene tetrahydrofolate reductase and 66A → G methyltetrahydrofolate homocysteine methyl-transferase reductase polymorphisms determine susceptibility to coronary artery disease. J Atheroscler Thromb 18(1):56–64
72. Patwari P, Higgins LJ, Chutkow WA et al (2006) The interaction of thioredoxin with Txnip. Evidence for formation of a mixed disulfide by disulfide exchange. J Biol Chem 281:21884
73. He X, Ma Q (2012) Redox regulation by nuclear factor erythroid 2-related factor 2: gatekeeping for the basal and diabetes-induced expression of thioredoxin-interacting protein. Mol Pharmacol 82:887
74. Wang XB, Han YD, Zhang S, Cui NH, Liu ZJ, Huang ZL, Li C, Zheng F (2016) Associations of polymorphisms in TXNIP and gene-environment interactions with the risk of coronary artery disease in a Chinese Han population. J Cell Mol Med 20(12):2362–2373
75. Mozzini C, Fratta Pasini A, Garbin U, Stranieri C, Pasini A, Vallerio P, Cominacini L (2014) Increased endoplasmic reticulum stress and Nrf2 repression in peripheral blood mononuclear cells of patients with stable coronary artery disease. Free Radic Biol Med 68:178–185
76. Lakshmi SV, Naushad SM, Saumya K, Rao DS, Kutala VK (2012 Oct) Role of CYP1A1 haplotypes in modulating susceptibility to coronary artery disease. Indian J Biochem Biophys 49(5):349–355
77. Bhat MA, Gandhi G (2018) CYP2D6 (C2850T, G1846A, C100T) polymorphisms, haplotypes and MDR analysis in predicting coronary artery disease risk in north-west Indian population: a case-control study. Gene 663:17–24
78. Zhang T, Yu K, Li X (2018) Cytochrome P450 family 4 subfamily F member 2 (CYP4F2) rs1558139, rs2108622 polymorphisms and susceptibility to several cardiovascular and cerebrovascular diseases. BMC Cardiovasc Disord 18(1):29
79. Zhang ZX, Zhang Y (2014) Glutathione S-transferase M1 (GSTM1) null genotype and coronary artery disease risk: a meta-analysis. Int J Clin Exp Med 7(10):3378–3384
80. Kariž S, Nikolajević Starčević J, Petrovič D (2012) Association of manganese superoxide dismutase and glutathione S-transferases genotypes with myocardial infarction in patients with type 2 diabetes mellitus. Diabetes Res Clin Pract 98(1):144–150

81. Moore-Morris T, van Vliet PP, Andelfinger G, Puceat M (2018) Role of epigenetics in cardiac development and congenital diseases. Physiol Rev 98(4):2453–2475
82. Schleithoff C, Voelter-Mahlknecht S, Dahmke IN, Mahlknecht U (2012) On the epigenetics of vascular regulation and disease. Clin Epigenetics 4(1):7
83. Matouk CC, Marsden PA (2008) Epigenetic regulation of vascular endothelial gene expression. Circ Res 102(8):873–887
84. Duthie SJ (2011) Epigenetic modifications and human pathologies: cancer and CVD. Proc Nutr Soc 70(1):47–56
85. Turunen MP, Aavik E, Yla-Herttuala S (2009) Epigenetics and atherosclerosis. Biochim Biophys Acta 1790(9):886–891
86. Baccarelli A, Rienstra M, Benjamin EJ (2010) Cardiovascular epigenetics: basic concepts and results from animal and human studies. Circ Cardiovasc Genet 3(6):567–573
87. Friso S, Pizzolo F, Choi SW, Guarini P, Castagna A, Ravagnani V, Carletto A, Pattini P, Corrocher R, Olivieri O (2008) Epigenetic control of 11 beta-hydroxysteroid dehydrogenase 2 gene promoter is related to human hypertension. Atherosclerosis 199(2):323–327
88. Loehrer FM, Angst CP, Haefeli WE, Jordan PP, Ritz R, Fowler B (1996) Low whole-blood S-adenosylmethionine and correlation between 5-methyltetrahydrofolate and homocysteine in coronary artery disease. Arterioscler Thromb Vasc Biol 16(6):727–733
89. Debreceni B, Debreceni L (2012) Why do homocysteine-lowering B vitamin and antioxidant E vitamin supplementations appear to be ineffective in the prevention of cardiovascular diseases? Cardiovasc Ther 30(4):227–233
90. Castro R, Rivera I, Struys EA, Jansen EE, Ravasco P, Camilo ME, Blom HJ (2003) Jakobs C, and Tavares de Almeida I. Increased homocysteine and S-adenosylhomocysteine concentrations and DNA hypomethylation in vascular disease. Clin Chem 49:1292–1296
91. Devlin AM, Arning E, Bottiglieri T, Faraci FM, Rozen R, Lentz SR (2004) Effect of Mthfr genotype on diet-induced hyperhomocysteinemia and vascular function in mice. Blood 103:2624–2629
92. Lund G, Andersson L, Lauria M, Lindholm M, Fraga MF, Villar-Garea A, Ballestar E, Esteller M, Zaina S (2004) DNA methylation polymorphisms precede any histological sign of atherosclerosis in mice lacking apolipoprotein E. J Biol Chem 279:29147–29154
93. Sharma P, Kumar J, Garg G, Kumar A, Patowary A, Karthikeyan G, Ramakrishnan L, Brahmachari V, Sengupta S (2008) Detection of altered global DNA methylation in coronary artery disease patients. DNA Cell Biol 27(7):357–365
94. Kim CS, Kim YR, Naqvi A, Kumar S, Hoffman TA, Jung SB, Kumar A, Jeon BH, McNamara DM, Irani K (2011) Homocysteine promotes human endothelial cell dysfunction via site-specific epigenetic regulation of p66shc. Cardiovasc Res 92(3):466–475
95. Lakshmi SV, Naushad SM, Reddy CA, Saumya K, Rao DS, Kotamraju S, Kutala VK (2013) Oxidative stress in coronary artery disease: epigenetic perspective. Mol Cell Biochem 374(1–2):203–211
96. Lacza Z, Pankotai E, Busija DW (2009) Mitochondrial nitric oxide synthase: current concepts and controversies. Front Biosci (Landmark Ed) 14:4436–4443
97. Kim GH, Ryan JJ, Archer SL (2013) The role of redox signaling in epigenetics and cardiovascular disease. Antioxid Redox Signal 18(15):1920–1936
98. Venkov CD, Rankin AB, Vaughan DE (1996) Identification of authentic estrogen receptor in cultured endothelial cells. A potential mechanism for steroid hormone regulation of endothelial function. Circulation 94:727–733
99. Miller VM, Duckles SP (2008) Vascular actions of estrogens: functional implications. Pharmacol Rev 60:210–241
100. Post WS, Goldschmidt-Clermont PJ, Wilhide CC, Heldman AW, Sussman MS, Ouyang P, Milliken EE, Issa JP (1999) Methylation of the estrogen receptor gene is associated with aging and atherosclerosis in the cardiovascular system. Cardiovasc Res 43(4):985–991
101. Kim J, Kim JY, Song KS, Lee YH, Seo JS, Jelinek J, Goldschmidt-Clermont PJ, Issa JP (2007) Epigenetic changes in estrogen receptor beta gene in atherosclerotic cardiovascular tissues and in-vitro vascular senescence. Biochim Biophys Acta 1772(1):72–80

102. Sedding DG, Tröbs M, Reich F, Walker G, Fink L, Haberbosch W, Rau W, Tillmanns H, Preissner KT, Bohle RM, Langheinrich AC (2009) 3-Deazaadenosine prevents smooth muscle cell proliferation and neointima formation by interfering with Ras signaling. Circ Res 104(10):1192–1200
103. Friso S, Lotto V, Choi SW, Girelli D, Pinotti M, Guarini P, Udali S, Pattini P, Pizzolo F, Martinelli N, Corrocher R, Bernardi F, Olivieri O (2012) Promoter methylation in coagulation F7 gene influences plasma FVII concentrations and relates to coronary artery disease. J Med Genet 49(3):192–199
104. Guay SP, Brisson D, Munger J, Lamarche B, Gaudet D, Bouchard L (2012) ABCA1 gene promoter DNA methylation is associated with HDL particle profile and coronary artery disease in familial hypercholesterolemia. Epigenetics 7(5):464–472
105. Zhuang J, Peng W, Li H, Wang W, Wei Y, Li W, Xu Y (2012) Methylation of p15INK4b and expression of ANRIL on chromosome 9p21 are associated with coronary artery disease. PLoS One 7(10):e47193
106. Chang PY, Chen YJ, Chang FH, Lu J, Huang WH, Yang TC, Lee YT, Chang SF, Lu SC, Chen CH (2013) Aspirin protects human coronary artery endothelial cells against atherogenic electronegative LDL via an epigenetic mechanism: a novel cytoprotective role of aspirin in acute myocardial infarction. Cardiovasc Res 99(1):137–145
107. Yang TC, Chen YJ, Chang SF, Chen CH, Chang PY, Lu SC (2014) Malondialdehyde mediates oxidized LDL-induced coronary toxicity through the Akt-FGF2 pathway via DNA methylation. J Biomed Sci 21:11
108. Kumar A, Kumar S, Vikram A, Hoffman TA, Naqvi A, Lewarchik CM, Kim YR, Irani K (2013) Histone and DNA methylation-mediated epigenetic downregulation of endothelial Kruppel-like factor 2 by low-density lipoprotein cholesterol. Arterioscler Thromb Vasc Biol 33(8):1936–1942
109. Jia L, Zhu L, Wang JZ, Wang XJ, Chen JZ, Song L, Wu YJ, Sun K, Yuan ZY, Hui R (2013) Methylation of FOXP3 in regulatory T cells is related to the severity of coronary artery disease. Atherosclerosis 228(2):346–352
110. Guay SP, Brisson D, Lamarche B, Gaudet D, Bouchard L (2014) Epipolymorphisms within lipoprotein genes contribute independently to plasma lipid levels in familial hypercholesterolemia. Epigenetics 9(5):718–729
111. Jiang D, Zheng D, Wang L, Huang Y, Liu H, Xu L, Liao Q, Liu P, Shi X, Wang Z, Sun L, Zhou Q, Li N, Xu L, Le Y, Ye M, Shao G, Duan S (2013) Elevated PLA2G7 gene promoter methylation as a gender-specific marker of aging increases the risk of coronary heart disease in females. PLoS One 8(3):e59752
112. Su J, Li X, Yu Q, Liu Y, Wang Y, Song H, Cui H, Du W, Fei X, Liu J, Lin S, Wang J, Zheng W, Zhong J, Zhang L, Tong M, Xu J, Chen X (2014) Association of P2Y12 gene promoter DNA methylation with the risk of clopidogrel resistance in coronary artery disease patients. Biomed Res Int 2014:450814
113. Niu PP, Cao Y, Gong T, Guo JH, Zhang BK, Jia SJ (2014) Hypermethylation of DDAH2 promoter contributes to the dysfunction of endothelial progenitor cells in coronary artery disease patients. J Transl Med 12:170
114. Breitling LP (2013) Current genetics and epigenetics of smoking/tobacco-related cardiovascular disease. Arterioscler Thromb Vasc Biol 33(7):1468–1472
115. Fiorito G, Guarrera S, Valle C, Ricceri F, Russo A, Grioni S, Mattiello A, Di Gaetano C, Rosa F, Modica F, Iacoviello L, Frasca G, Tumino R, Krogh V, Panico S, Vineis P, Sacerdote C, Matullo G (2014) B-vitamins intake, DNA-methylation of One Carbon Metabolism and homocysteine pathway genes and myocardial infarction risk: the EPICOR study. Nutr Metab Cardiovasc Dis 24(5):483–488
116. Baccarelli A, Bollati V (2009) Epigenetics and environmental chemicals. Curr Opin Pediatr 21:243–251
117. Breton CV, Byun HM, Wenten M, Pan F, Yang A, Gilliland FD (2009) Prenatal tobacco smoke exposure affects global and gene-specific DNA methylation. Am J Respir Crit Care Med 180:462–467

118. Sharma P, Garg G, Kumar A, Mohammad F, Kumar SR, Tanwar VS, Sati S, Sharma A, Karthikeyan G, Brahmachari V, Sengupta S (2014) Genome wide DNA methylation profiling for epigenetic alteration in coronary artery disease patients. Gene 541(1):31–40
119. Fernández-Sanlés A, Sayols-Baixeras S, Subirana I, Degano IR, Elosua R (2017) Association between DNA methylation and coronary heart disease or other atherosclerotic events: a systematic review. Atherosclerosis 263:325–333
120. Campos EI, Reinberg D (2009) Histones: annotating chromatin. Annu Rev Genet 43:559–599
121. Kaypee S, Sudarshan D, Shanmugam MK, Mukherjee D, Sethi G, Kundu TK (2016) Aberrant lysine acetylation in tumorigenesis: implications in the development oftherapeutics. Pharmacol Ther 162:98–119
122. Li B, Carey M, Workman J (2007) The role of chromatin during transcription. Cell 128:707–719
123. Jenuwein T, Allis C (2001) Translating the histone code. Science 293:1074–1080
124. Saha A, Wittmeyer J, Cairns BR (2006) Chromatin remodelling: the industrial revolution of DNA around histones. Nat Rev Mol Cell Biol 7:437–447
125. Zhang Y, Reinberg D (2001) Transcription regulation by histone methylation: interplay between different covalent modifications of the core histone tails. Genes Dev 15:2343–2360
126. O'Hagan HM, Wang W, Sen S, Destefano Shields C, Lee SS, Zhang YW, Clements EG, Cai Y, Van Neste L, Easwaran H, Casero RA, Sears CL, Baylin SB (2011) Oxidative damage targets complexes containing DNA methyltransferases, SIRT1, and polycomb members to promoter CpG Islands. Cancer Cell 20:606–619
127. Xu XF, Ma XL, Shen Z, Wu XL, Cheng F, Du LZ (2010) Epigenetic regulation of the endothelial nitric oxide synthase gene in persistent pulmonary hypertension of the newborn rat. J Hypertens 28:2227–2235
128. Gusterson R, Jazrawi E, Adcock I, Latchman D (2003) The transcriptional co-activators CREBbinding protein (CBP) and p300 play a critical role in cardiac hypertrophy that is dependent on their histone acetyltransferase activity. J Biol Chem 278:6838–6847
129. Haddad F, Bodell P, Qin A, Giger J, Baldwin K (2003) Role of antisense RNA in coordinating cardiac myosin heavy chain gene switching. J Biol Chem 278:37132–37138
130. Majumdar G, Johnson I, Kale S, Raghow R (2008) Epigenetic regulation of cardiac musclespecific genes in H9c2 cells by Interleukin-18 and histone deacetylase inhibitor m-arboxycinnamic acid bis-hydroxamide. Mol Cell Biochem 312:47–60
131. Zhang Q, Chen H, Wang L, Liu D, Hill J, Liu Z (2011) The histone trimethyllysine demethylase JMJD2A promotes cardiac hypertrophy in response to hypertrophic stimuli in mice. J Clin Invest 121:2447–2456
132. Bingham A, Ooi L, Kozera L, White E, Wood I (2007) The repressor element 1-silencing transcription factor regulates heart-specific gene expression using multiple chromatin-modifying complexes. Mol Cell Biol 27:4082–4092
133. Stein A, Jones T, Herron T, Patel S, Day S, Noujaim S, Milstein ML, Klos M, Furspan PB, Jalife J, Dressler GR (2011) Loss of H3K4 methylation destabilizes gene expression patterns and physiological functions in adult murine cardiomyocytes. J Clin Invest 121:2641–2650
134. Kaneda R, Takada S, Yamashita Y, Choi Y, Nonaka-Sarukawa M, Soda M, Misawa Y, Isomura T, Shimada K, Mano H (2009) Genome-wide histone methylation profile for heart failure. Genes Cells 14:69–77
135. Tsutsumi YM, Horikawa YT, Jennings MM, Kidd MW, Niesman IR, Yokoyama U, Head BP, Hagiwara Y, Ishikawa Y, Miyanohara A, Patel PM, Insel PA, Patel HH, Roth DM (2008) Cardiac-specific overexpression of caveolin-3 induces endogenous cardiac protection by mimicking ischemic preconditioning. Circulation 118:1979–1988
136. Carè A, Catalucci D, Felicetti F, Bonci D, Addario A, Gallo P, Bang ML, Segnalini P, Gu Y, Dalton ND, Elia L, Latronico MV, Høydal M, Autore C, Russo MA, Dorn GW 2nd, Ellingsen O, Ruiz-Lozano P, Peterson KL, Croce CM, Peschle C, Condorelli G (2007) MicroRNA-133 controls cardiac hypertrophy. Nat Med 13:613–618
137. Latronico M, Condorelli G (2009) MicroRNAs and cardiac pathology. Nat Rev Cardiol 6:419–429

138. Shilo S, Roy S, Khanna S, Sen CK (2008) Evidence for the involvement of miRNA in redox regulated angiogenic response of human microvascular endothelial cells. Arterioscler Thromb Vasc Biol 28:471–477
139. Garzon R, Liu S, Fabbri M, Liu Z, Heaphy CE, Callegari E, Schwind S, Pang J, Yu J, Muthusamy N, Havelange V, Volinia S, Blum W, Rush LJ, Perrotti D, Andreeff M, Bloomfield CD, Byrd JC, Chan K, Wu LC, Croce CM, Marcucci G (2009) MicroRNA-29b induces global DNA hypomethylation and tumor suppressor gene reexpression in acute myeloid leukemia by targeting directly DNMT3A and 3B and indirectly DNMT1. Blood 113:6411–6418
140. Wang YS, Chou WW, Chen KC, Cheng HY, Lin RT, Juo SH (2012) MicroRNA-152 mediates DNMT1-regulated DNA methylation in the estrogen receptor alpha gene. PLoS One 7:e30635
141. Noonan EJ, Place RF, Pookot D, Basak S, Whitson JM, Hirata H, Giardina C, Dahiya R (2009) miR-449a targets HDAC-1 and induces growth arrest in prostate cancer. Oncogene 28:1714–1724
142. Chan SY, Loscalzo J (2010) MicroRNA-210: a unique and pleiotropic hypoxamir. Cell Cycle 9:1072–1083
143. Sun D, Zhang J, Xie J, Wei W, Chen M, Zhao X (2012) MiR-26 controls LXR-dependent cholesterol efflux by targeting ABCA1 and ARL7. FEBS Lett 586:1472–1479
144. Kim J, Yoon H, Ramírez C, Lee S, Hoe H, Fernández-Hernando C, Kim J (2012) MiR-106b impairs cholesterol efflux and increases Aβ levels by repressing ABCA1 expression. Exp Neurol 235:476–483
145. Fang Y, Shi C, Manduchi E, Civelek M, Davies P (2010) MicroRNA-10a regulation of proinflammatory phenotype in athero-susceptible endothelium in vivo and in vitro. Proc Natl Acad Sci U S A 107:13450–13455
146. Sun X, Icli B, Wara A, Belkin N, He S, Kobzik L, Hunninghake GM, Vera MP, MICU Registry, Blackwell TS, Baron RM, Feinberg MW (2012) MicroRNA-181b regulates NF-κB-mediated vascular inflammation. J Clin Invest 122:1973–1990

Role of Monoamine Oxidases in Heart Diseases

6

Vinayak Gupta, Vikas Arige, and Nitish R. Mahapatra

Abstract

Monoamine oxidases (MAOs) are flavoenzymes that metabolize biogenic amines, dietary amines, and catecholamines in the brain and peripheral tissues. While MAOs are known to contribute to psychiatric and neurodegenerative (Parkinson's and Alzheimer's) diseases for a long time, recent studies have established their role in heart diseases as these enzymes potently generate reactive oxygen species (ROS) in cardiomyocytes via oxidative deamination of mainly norepinephrine and serotonin. Indeed, MAOs have emerged as important regulators of mitochondrial/endothelial/cardiac dysfunction, essential hypertension, ventricular hypertrophy, myocardial infarction, cardiomyocyte apoptosis, post-ischemic cardiac damage, and heart failure. Transcriptional and posttranscriptional regulation of MAOs (via certain transcription factors or microRNAs) may emerge as new therapeutic strategies for treatment of cardiovascular pathological conditions. The next-generation MAO inhibitors (that do not cause irreversible inhibition of MAOs) may also be useful for management of cardiovascular disease states involving dysregulated expression/activity of MAOs.

Keywords

Cardiovascular · Monoamine oxidase · Reactive oxygen species · Catecholamines · Transcription factors · microRNAs

V. Gupta
Cardiovascular Genetics Laboratory, Department of Biotechnology, Bhupat and Jyoti Mehta School of Biosciences, Indian Institute of Technology Madras, Chennai, Tamil Nadu, India

Department of Biotechnology, Bennett University, Greater Noida, Uttar Pradesh, India

V. Arige · N. R. Mahapatra (✉)
Cardiovascular Genetics Laboratory, Department of Biotechnology, Bhupat and Jyoti Mehta School of Biosciences, Indian Institute of Technology Madras, Chennai, Tamil Nadu, India
e-mail: nmahapatra@iitm.ac.in

S. Chakraborti et al. (eds.), *Modulation of Oxidative Stress in Heart Disease*,
https://doi.org/10.1007/978-981-13-8946-7_6

6.1 Introduction

Monoamine oxidases (MAOs) (EC 1.4.3.4) are flavin adenine dinucleotide (FAD)-dependent enzymes which metabolize biogenic amines, dietary amines, and catecholamines (viz., epinephrine, norepinephrine, and dopamine) in the brain and peripheral tissues. MAOs oxidatively deaminate these amines into corresponding aldehydes and generate hydrogen peroxide (H_2O_2) and ammonia (NH_3) during this reaction. Aldehydes generated in these reactions are further metabolized into corresponding organic acids by aldehyde dehydrogenases [1]. MAOs are expressed as integral proteins in the outer membrane of mitochondria. Based on the differences observed in substrate/inhibitor specificity and cell-/tissue-specific expression, MAOs are classified into two types, namely, MAOA and MAOB [2]. In brief, epinephrine, norepinephrine, and serotonin are preferentially metabolized by MAOA, while phenylalanine and benzylamine are mainly metabolized by MAOB. Dopamine, tyramine, and tryptamine are common substrates for both the MAOs [3]. Selective MAOA and MAOB inhibitors are clinically used to treat depression and Parkinson's disease [4].

Apart from sharing ~70% identity between their amino acid sequences, both MAOs have a conserved pentapeptide sequence (viz., Ser-Gly-Gly-Cys-Tyr), which serves as the FAD binding domain [1]. In several mammalian species including human, mouse, and rat, *MAOs* are mapped to the p arm of the X chromosome; these two genes are located next to each other in a tail-to-tail fashion. Identical exon-intron organization, equal number of exons, and high sequence similarity suggest that *MAOA* and *MAOB* are derived from a common ancestral gene (Fig. 6.1). Both MAOs are ubiquitously expressed in all cell types except red blood cells in a tissue-specific manner [5]. The human heart contains high levels of both isozymes; in the rat heart, MAOA is abundant, while MAOB is almost absent and the reverse is true for the mouse heart [6, 7]. *MAOA* expression is regulated by several transcription factors including circadian-clock components (via E-box elements), GATA2 (GATA binding protein-2), Krüppel-like factor-11 (Klf11), R1, sirtuin 1, Sp1 (specificity protein 1), SRY (sex-determining region gene on the Y chromosome), and TBP (TATA-binding protein), while *MAOB* expression is reported to be regulated by c-Jun, Egr1 (early growth response protein1), Klf-11, and Sp1 [8–12]. Interestingly, MAOs are also regulated by molecules of cardiovascular relevance such as androgen, glucocorticoid, retinoic acid (RA), forskolin, and tumor necrosis factor-α (TNF-α) [8, 9].

Besides the well-studied functions of MAOs in neuronal/behavioral disorders, cancer metastasis, and embryonic development [13–16], a lot of research has been performed in recent years to explore their possible roles in mitochondrial/endothelial/cardiac dysfunction, essential hypertension, ventricular hypertrophy, myocardial infarction, cardiomyocyte apoptosis, postischemic cardiac damage, and heart failure as discussed below [17–24]. This chapter aims to summarize our current understanding on the role of these enzymes in heart diseases.

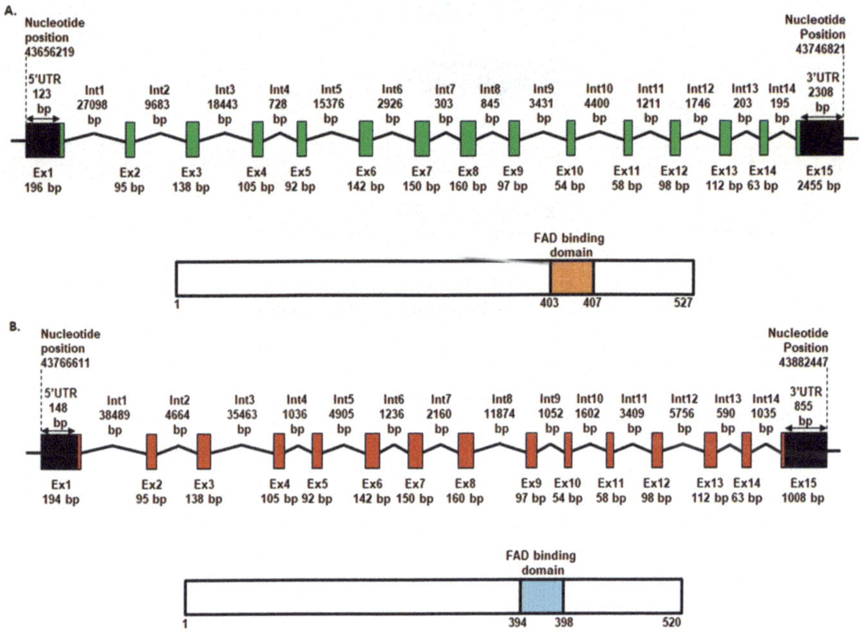

Fig. 6.1 Schematic representation of human MAOA and MAOB genes and their protein products. The human MAOA (panel **A**) and MAOB (panel **B**) genes consist of 15 exons seperated by 14 introns (UCSC Genome Browser refGenes NM_000240 and NM_000898). The lengths of UTRs, exons, and introns are mentioned. *FAD* flavin adenine dinucleotide, *UTR* untranslated region, *Ex* exon, *Int* intron, *bp* base pair. MAOA protein consists of 527 amino acids and the amino acids 403–407 serve as the FAD binding site. MAOB protein consists of 520 amino acids and the amino acids 394–398 serve as the FAD binding site

6.2 Role of MAOs in Cardiac Cell Death and Chronic Ventricular Dysfunction

MAOs are potent generators of reactive oxygen species (ROS) or oxidative stress due to oxidative deamination of mainly norepinephrine and serotonin in cardiac tissues [25–28]. Depending on the type of available substrate and ROS generated by MAOs, different signal transduction mechanisms lead to distinct phenotypes including cell proliferation/hypertrophy, basilar artery contraction, or apoptosis/necrosis [17, 18, 27–30]. For example, transgenic mice with cardiac-specific MAOA overexpression displays oxidative stress-induced p53 activation, which leads to downregulation of peroxisome proliferator-activated receptor gamma co-activator 1α (PGC1α) (a crucial regulator of mitochondrial biogenesis/function) that in turn causes mitochondrial dysfunction, cardiomyocyte necrosis, and chronic ventricular dysfunction [18] (Fig. 6.2). Moreover, ROS generated via MAOA can also block autophagic flux of lysosomes by reducing the lysosomal acidification and by preventing the nuclear translocation of transcription factor-EB (TF-EB) (that acts as a master regulator of lysosomal biogenesis and autophagy) [29] (Fig. 6.2).

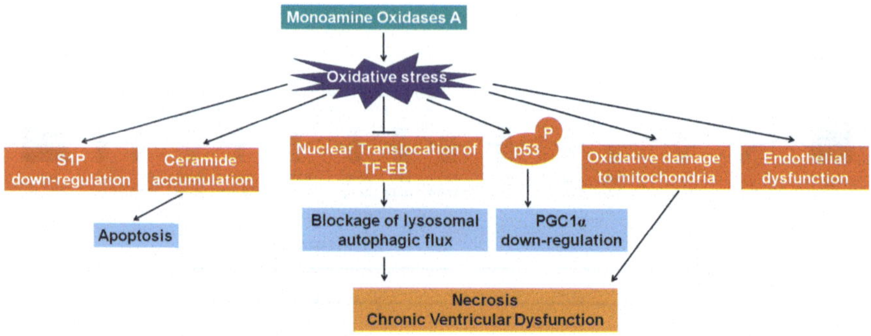

Fig. 6.2 Plausible mechanisms of MAOA-mediated apoptosis, necrosis, endothelial dysfunction, and ventricular dysfunction. MAOA-generated oxidative stress causes p53 activation and consequently downregulates PGC1α. Oxidative stress also impairs lysosome's function by blocking the nuclear translocation of TF-EB, which in turn leads to blockade of autophagic flux. p53- and TF-EB-mediated pathways lead to necrosis/chronic ventricular dysfunction. MAO-generated oxidative stress can also promote apoptosis via ceramide accumulation and downregulation of S1P in cardiomyocytes. Blunt-headed arrow indicates "inhibition" of nuclear translocation of TF-EB. *S1P* sphingosine 1-phosphate, *PGC1α* peroxisome proliferator-activated receptor-coactivator 1α, *P* phosphorylation

6.3 Role of MAOs in Cardiac Hypertrophy and Heart Failure

In contrast to cardiac cell death via apoptosis or necrosis, MAOs can lead to cardiac hypertrophy via different signaling pathways. In biomechanically stretched cardiomyocytes, *MAOA* has been reported to be upregulated (by ~four-fold) leading to cardiac hypertrophy and consequent heart failure [31]. These cellular changes are due to oxidative stress generated during oxidative deamination of serotonin and norepinephrine.

Serotonin (5-hydroxytyramine [5-HT]), a MAOA-specific substrate and a potent vasoactive amine, induces cardiomyocyte hypertrophy in a MAOA-dependent manner via activation of extracellular-regulated kinases (ERK1/2 that are essential signaling molecules for cell growth) [28] (Fig. 6.3). This cardiac hypertrophy is partly 5-HT_{2B} receptor dependent as reflected by cellular response following treatment with amine transporter inhibitors (imipramine) and MAO inhibitor (pargyline) [28]. In corroboration, MAOA contributes to oxidative stress in human heart valves following exposure to serotonin and dopamine [25]. In the circulatory system, the major source of 5-HT is platelets. Upon aggregation/activation, a large amount of 5-HT is released from the platelets into the circulation causing either vasorelaxation via endothelial cells or vasoconstriction via vascular smooth muscle cells [24]. In addition, 5-HT-dependent MAOA-mediated ROS also leads to basilar artery contraction in rats [26].

Norepinephrine stimulates the MAOA enzyme activity in neonatal and adult cardiomyocytes in vitro that leads to ROS production and maladaptive hypertrophy [27]. These in vitro changes may involve the transcription factor NFAT (nuclear

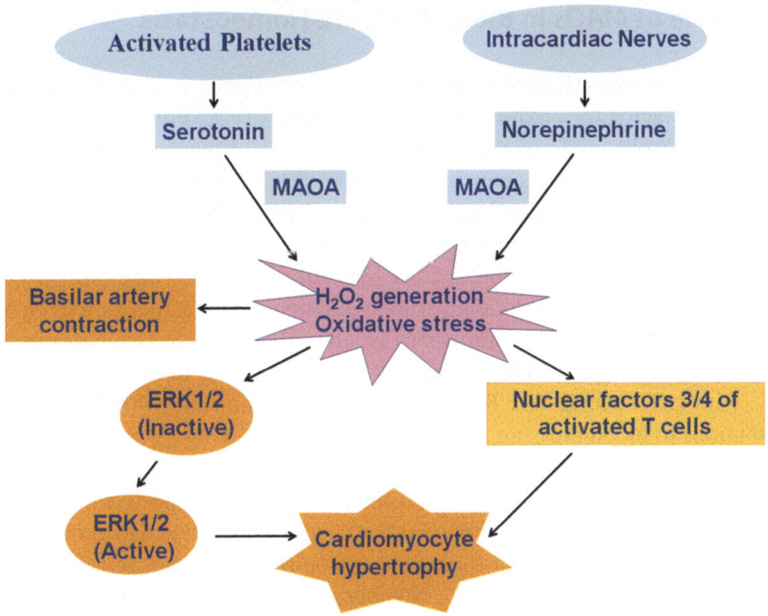

Fig. 6.3 Signaling pathways underlying development of cardiomyocyte hypertrophy via MAOA-mediated catabolism of serotonin and norepinephrine. Serotonin and norepinephrine are released from the activated platelets and intracardiac nerves, respectively. Following interaction with their respective receptors and signaling, they are sequestered into the cytoplasm via respective transporters present in the membrane. Serotonin and norepinephrine are degraded by MAOA-generating hydrogen peroxide/oxidative stress that, in turn, contributes to basilar artery contraction or cardiomyocyte hypertrophy. ERK1/2, extracellular signal-regulated kinase 1/2

factors of activated T cells) that contributes to maladaptive hypertrophic signaling (Fig. 6.3). In line with this finding, pharmacological or genetic inhibition of MAOA prevents the occurrence of heart failure in mice subjected to pressure overload [27]. Corroboratively, transcriptomic and proteomic analyses reveal that MAOA is one of the most upregulated proteins in a well-defined rat model of chronic heart failure (which has volume overload due to surgically created aorto-caval fistula) [32]. Similarly, enzyme activity and expression of both MAOs are significantly elevated in left and right ventricles of end-stage ischemic failing hearts in human [33].

In addition, *MAOB* knockout mice show compensated cardiac hypertrophy following pressure overload induced by transverse aortic constriction. They are also found to be resistant to adverse left ventricular (LV) dilation and dysfunction upon pressure overload. Thus, MAOB activity also contributes to oxidative stress and structural and functional derangements in the heart [19]. Moreover, oxidative stress also diminishes the activity of aldehyde dehydrogenase which may, in turn, cause the accumulation of toxic aldehydes. These accumulated aldehydes may induce mitochondrial dysfunction contributing to myocardial damage [19].

6.4 Role of *MAOs* in Blood Pressure Homeostasis

Essential hypertension (EH), a common, multifactorial/polygenic health problem, is the chief risk factor for cardiovascular/renal diseases (viz., myocardial infarction, heart failure, stroke, and end-stage renal disease) [34]. Catecholamines have been implicated to play an important role in the pathogenesis of EH. For example, dopamine modulates blood pressure via generation of ROS, interaction with the renin-angiotensin-aldosterone system (RAAS), regulation of epithelial sodium transport, and vascular smooth muscle contractility [35, 36] (Fig. 6.4). Therefore, *MAO*s are logical candidate genes for blood pressure regulation. Of note, there are three blood pressure QTLs (Bp65, Bp64, and Bp56) (source: Rat Genome Database) in the X chromosome of rat; both MAOA and MAOB are localized in the Bp65 and Bp64 QTLs (with LOD scores of 5.8 and 5.2, respectively) in line with their plausible contributions to blood pressure modulation (Fig. 6.5).

Several studies reported higher level of catecholamines in hypertensive individuals and in rodent models of hypertension compared to their respective normotensive controls [37–39]. This difference may, at least partly, be attributed to altered expression or enzyme activity of catecholamine catabolizing enzymes (e.g., MAOs and catecholamine-o-methyltransferase). Notably, two independent microarray studies on adrenal gland and kidney tissues of mouse models of human essential hypertension (viz., BPH (blood pressure high) and BPL (blood pressure low) mice) showed that *MAOA* expression was elevated by ~1.3- and ~3.3-fold, respectively, in BPL mice [40, 41]. Based on these observations, we speculate that higher *MAOA* levels

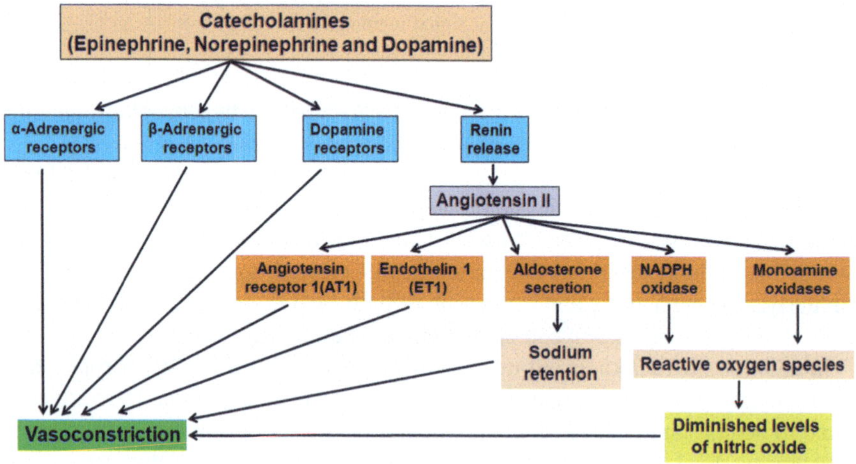

Fig. 6.4 Plausible molecular mechanisms of blood pressure regulation by catecholamines. Catecholamines alter the blood pressure homeostasis either through adrenergic/dopaminergic receptors or by increasing the release of renin from the adrenal cortex. Higher level of renin produces more angiotensin II which leads to vasoconstriction via angiotensin receptor 1, increasing endothelin-1, aldosterone secretion, and ROS generation (via enhancing the expression/activity of MAOs and NADPH oxidase)

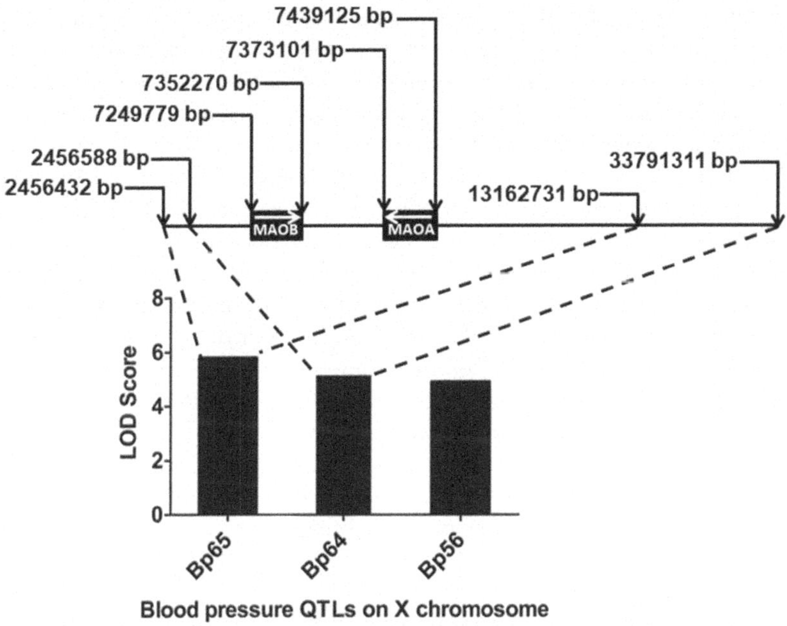

Fig. 6.5 Graphical representation of the blood pressure QTLs on the X chromosome of rat. Blood pressure QTLs (Bp65, Bp64, and Bp56) on the rat X chromosome and their respective LOD scores obtained from Rat Genome Database. Two of these three BP QTLs harbor the MAOA and MAOB genes, suggesting their important roles in BP regulation. The genomic positions of MAOA and MAOB genes in the BP QTLs are indicated

in BPL may contribute to lower catecholamine levels, which in turn could lead to low blood pressure phenotype. MAOs also inhibit nitric oxide synthase (NOS2) expression and consequently reduce the levels of the vasodilator nitric oxide (NO) (Fig. 6.4) [42]. Consistently, MAOA enzyme activity was ~1.4-fold higher in the kidneys of normotensive Wistar-Kyoto (WKY) rat than the spontaneously hypertensive rat (SHR) [43]. Surprisingly, some studies reported higher MAOA enzyme activity in the heart, aorta, femoral arteries, isolated cardiomyocytes, and brain of SHR compared to WKY rats [20, 42, 43]. SHR rats have also been reported to have higher MAOA protein level in their basilar arteries compared to WKY rats [26]. Similarly, MAOB enzyme activity was reported to be ~2.8-fold higher in isolated cardiomyocytes of SHR compared to WKY rats [21]. However, comparative microarray analysis showed that SHR adrenal gland tissues exhibited ~0.52-fold underexpression of *MAOB* than that of WKY [44]. The mechanism of such differential expression/activity pattern of MAOs across different tissues of SHR and WKY remains unclear. Of note, a recent study reported that in vivo administration of lipopolysaccharide and angiotensin II augments the vascular expression of both the MAOs leading to increased generation of H_2O_2 and subsequent endothelial dysfunction [22] (Fig. 6.4).

MAOB-specific substrates (phenylethylamine, tyramine, and tryptamine) are bioactive endogenous amines present in mammalian peripheral as well as central nervous system in low concentration (less than 1% of biogenic amines); therefore, they are called trace amines (TAs) [45]. These amines lack the catechol nucleus but are similar to biogenic amines in terms of structure and metabolism; these are described as "false neurotransmitters" or "sympathomimetic amines." TAs are present in food products like cheese, red wine, chocolates, etc. MAO inhibitor-treated patients consuming a TA-rich diet may develop complications such as tachycardia and hypertension. This hypertensive crisis is described as "cheese effect" irrespective of the nature of TA-rich food [46, 47]. The molecular mechanism of TA-induced hypertension is based on the fact that tyramine and phenylethylamine are structurally similar to norepinephrine. Therefore, these molecules enter sympathetic neurons by the same monoamine membrane transporter and displace norepinephrine. Consequently, norepinephrine is diffused from the cytoplasm into the synaptic cleft, leading to α-adrenoceptor-mediated vasoconstriction and the sudden rise in blood pressure [46, 48]. Thus, various studies support the role of *MAOs* in modulating blood pressure under pathophysiological conditions.

6.5 Mechanisms of Transcriptional Regulation of MAOs

6.5.1 Transcriptional Regulation of MAOA

Because transcription factors play crucial roles in gene regulation, a potential strategy for developing novel therapeutics against disease conditions can be attained by modulating the expression and/or activity of a specific transcription factor [49–51]. Regulatory mechanisms for both MAOs have been studied extensively. For example, previous reports showed that Sp1 and SRY synergistically enhanced the human *MAOA* (h*MAOA*) promoter activity in a dose-dependent manner. Of note, SRY plays a very important role in blood pressure homeostasis [52]. Indeed, apart from MAOA, promoters of several other key cardiovascular-regulatory genes including tyrosine hydroxylase (the rate-limiting enzyme in the catecholamine biosynthesis pathway) [53], chromogranin B [54], and genes in RAAS pathway [55] are responsive to Sry and influence blood pressure.

Sp/Klf family, Sp3, Sp4, and KLF11 are the other transcription factors which have also been reported to regulate h*MAOA* promoter. KLF11 and Sp4 are known to trans-activate the h*MAOA* gene expression; on the other hand, Sp3 and a novel transcription factor known as R1 (RAM2/CDCA7L/JPO2) repress h*MAOA* gene expression as they compete for the same binding site with Sp1 [9, 56]. KLF transcription factors, in general, interact with histone acetyltransferases (HATs), including p300, for gene regulation. Consistently, co-transfection of p300 and KLF11 expression plasmids with h*MAOA* promoter luciferase construct showed that activation of h*MAOA* by KLF11 was further augmented in the presence of p300 [56]. The mouse *MAOA* (m*MAOA*) promoter is also well-characterized; m*MAOA* gene expression is regulated by GATA2, Sp1, and TBP in a coordinated manner [8]. Of note, not only m*MAOA*, these transcription factors also enhance the MAOA protein levels in humans [8]. It is interesting to note that GATA2 may also hamper the inflammatory state in atherosclerosis and obesity [57], indicating a possible role of MAOA in these disease conditions. In addition, circadian-clock components (via E-box elements) and NAD-dependent deacetylase sirtuin 1 (SIRT1) have also been reported to regulate m*MAOA* gene expression [11, 12]. SIRT1-/GATA2-mediated *MAOA* gene regulation is critical because single-nucleotide polymorphisms (SNPs) present in both of these upstream regulators of *MAOA* are associated with cardiovascular/cardiometabolic disorders or their risk traits [58–61].

Dopamine, a common substrate for both the MAOs, regulates the expression and enzymatic activity of *MAOA* via D-2-like receptors in mesangial renal cells although such regulation has not been observed in proximal tubule renal cells [62]. Dexamethasone, a synthetic glucocorticoid hormone, has also been shown to augment *MAOA* gene expression in human skeletal myocytes via glucocorticoid receptor and Sp1. These results provide molecular mechanism for the pathogenesis of glucocorticoid-induced myopathy [63]. In addition, forskolin-mediated cAMP-PKA (protein kinase A) pathway and TNF-α also increase *MAOA* gene expression via Sp1 [8]. This observation has therapeutic implications since forskolin (a diterpene isolated from root of *Coleus forskohlii)* was reported to have beneficial effects in cardiovascular diseases including congestive heart failure and hypertension [64–67]. It may also be noted that a recent study established the role of GATA2, Sp1, and TBP in regulating *MAOA* gene expression under ischemia-like pathophysiological conditions [8].

6.5.2 Transcriptional Regulation of MAOB

Several studies reported the characterization of the human *MAOB* (h*MAOB*) promoter. Unlike the h*MAOA* promoter, the core h*MAOB* promoter contains a TATA box; it also harbors two Sp1 binding domains, which are separated by a CACCC element [68]. Egr1 also regulates h*MAOB* expression by binding to the distal Sp1 domain [69, 70]. Another transcription factor called Sp4 trans-activates h*MAOB* promoter activity via direct interaction with the Sp1 sites; this activation has been reported to be repressed by Sp3 and Krüppel-like zinc-finger transcription factor KLF5 (also called BTEB2) as Sp3/KLF5 compete for the Sp1-binding sites [9, 56]. Site-directed mutagenesis revealed that CACCC sequence (present between the two Sp1-binding sites) is a repressor element. It is important to note that the transforming growth factor-β-inducible early gene TIEG2 (also called KLF11) and Sp3 exhibit dual functions for the regulation of h*MAOB*. TIEG2 acts as a repressor at the CACCC element whereas it acts as an activator at the distal Sp1 site of h*MAOB* promoter. However, due to its higher affinity for the Sp1 site than the CACCC element, the overall effect of TIEG2 is activation of the h*MAOB* gene expression [68]. Egr1 and c-Jun can also regulate h*MAOB* gene expression by interacting with the overlapping Sp1/Egr-1/Sp1 sites [9]. Interestingly, phorbol 12-myristate 13-acetate enhances h*MAOB* gene expression by increasing the Egr1/c-Jun gene expression via activation of PKC (protein kinase C) and MAPK (mitogen-activated protein kinase) signaling pathways [9]. Our recent studies suggest that *MAOB* gene expression may also be regulated by cyclic AMP/PKA/CREB (cAMP response element binding protein) pathway (unpublished observation).

The roles of a number of hormones, such as androgen, glucocorticoid, estrogen, and RA, have been demonstrated in h*MAOB* gene regulation [9]. Of note, RA enhances *MAOB* expression through retinoic acid receptor α (RARα) and retinoid X receptor α (RXRα) transcription factors. RARα physically interacts with Sp1 to form a transcriptional regulatory complex and recruited to Sp1-binding sites at h*MAOB* promoter [9]. Of note, RAR/RXR have a crucial role in cardiovascular pathophysiology as evident from the fact that knockout of RAR/RXR in mice leads to the development of heart defects such as defects in the conduction system, heart malformations, and heart failure. On the other hand, elevated level of RAR or RXR leads to dilated cardiomyopathy and congestive heart failure [71, 72].

Similar to *MAOA* gene regulation, dexamethasone has been reported to stimulate h*MAOB* promoter activity via glucocorticoid response element (GRE) and Sp1-binding sites in vitro and in vivo. The molecular mechanism of this activation involves activation of glucocorticoid receptor by dexamethasone, which then translocates into the nucleus and binds to GRE [9, 56, 73]. Interestingly, glucocorticoids and their receptors have direct effects on the heart, blood vessels, and cardiometabolic risk factors which are discussed in detail elsewhere [74, 75]. Dopamine may also activate MAOB expression similar to the case of MAOA; the dopamine-mediated upregulation of MAOB seems to be modulated by cyclic AMP response element (CRE) in the proximal MAOB promoter (unpublished observation).

6.5.3 Potential Therapeutic Application of the Transcriptional Regulators of MAOs

As detailed above, some of the transcription factors (viz., Sp1, KLF11, possibly Egr1, and CREB) are common regulators of MAOA and MAOB. Regulation of these molecules as a new therapeutic strategy for management of cardiovascular diseases may be worth studying. Of note, mithramycin A, an antibiotic produced by *Streptomyces argillaceus*, is used to treat various diseases including testicular carcinoma and chronic myeloid leukemia by virtue of its ability to diminish binding of Sp1 and Egr1 to regulatory promoter elements (and thereby modulating gene expression) [76, 77]. Mithramycin A has also been reported to diminish the binding of Sp1 and Egr1 to the MAOB promoter, thereby offering neuroprotection in a mouse model of Parkinson's disease [78, 79]. Moreover, in endothelial cells, mithramycin A prevented the TNF-α-mediated fractalkine (a chemokine) expression suggesting that it could function as an anti-inflammatory agent [80]. In view of these reports, it will be interesting to evaluate therapeutic potential of mithramycin A and other agents that may regulate the expression of MAOs via interactions with the key transcription factors in the context of cardiovascular diseases.

6.6 Posttranscriptional Regulation of MAOs: Potential Role for Several microRNAs

MicroRNAs (miRNAs) are small (~22 nucleotides), noncoding RNAs which have emerged as important posttranscriptional regulators of gene expression either by inhibiting translation or by degrading mRNA [81]. They are involved in regulating various physiological processes including development, metabolism, and maintaining homeostasis [82, 83]. Dysregulated expression of miRNAs is associated with various complications including cardiovascular diseases. In addition to this, circulating miRNAs serve as excellent noninvasive biomarkers for diagnosis and prognosis of diseases [84]. Some miRNAs are also being evaluated for their therapeutic applications in various disease states. For example, miravirsen (a miR-122 inhibitor) is under clinical trials for the treatment of chronic hepatitis C infection [85]. Similarly, a few miRNAs are at various preclinical/clinical stages for the plausible treatment of various pathological conditions [86].

MiRNA-142 is reported to diminish *MAOA* expression in neuronal cells by downregulating SIRT1 [87]. Computational analysis of the *MAOA and MAOB* 3'-UTRs using ten miRNA prediction tools (DIANA-microT [88], miRanda [89], miRDB [90], miRWalk [91], RNAhybrid [92], PICTAR4, PICTAR5 [93], PITA [94], RNA22 [95], and Targetscan [93]) revealed putative binding sites for 641 and 297 miRNAs, respectively. MiRNAs predicted by at least three tools and based on the thermodynamic scores obtained using PITA ($\Delta\Delta G < -10$) and RNAhybrid ($\Delta G < -20$ kcal/mole) are presented in Table 6.1. Interestingly, miR-608 and miR-125a-3p harbor putative binding sites in the 3'-UTRs of both the MAOs representing these miRNAs as candidates for further studies. However, experimental validations of interactions between miR-608/miR-125a-3p and MAOA/MAOB are required for confirmation of their roles in regulating MAOA/MAOB expression.

An early increase in plasma levels of miRNA-133a and miR-133b in myocardial infarction (MI) and coronary artery disease is well-documented indicating that these miRNAs could serve as novel diagnostic markers for these diseases [96, 97]. Interestingly, in silico analysis using PITA and RNAhybrid revealed putative binding sites for miR-133a and miR-133b in the 3'-UTR of both the MAOs. Both MAOA and MAOB are also predicted by miRwalk (version 3.0) as putative targets of miR-1224. Besides this, a recent study reported the increase in miR-1224 levels in human hepatocytes and serum under acute liver failure. The levels of miR-1224 were also augmented in mice subjected to ischemia/reperfusion compared to control [98]. Furthermore, in mouse, lipopolysaccharide (LPS)-induced miR-1224 was shown to downregulate the expression of Sp1 [99]; this finding suggests that miR-1224 may also indirectly regulate MAOs, since Sp1 governs the expression of both the MAOs. miR-1224 may also regulate MAOs via modulation of the expression of CREB, a key regulator of catecholamine biosynthetic genes, since CREB is a target of miR-1224 [100] and forskolin/cAMP augments MAOA [8]/MAOB expression/activity [101]. Our in vitro experiments also provided evidence for regulation of MAOB by miR-1224 (unpublished observation). Of note, the expression of MAOs is augmented by LPS and angiotensin II (AngII) in murine aortic rings, which is mediated

Table 6.1 Putative microRNAs that may bind to the 3′-UTR of human MAOA and MAOB[a]

Gene	miRNA	Predicted by number of tools	PITA (ΔΔG)	RNAhybrid (ΔG), kcal/mole
MAOA	**hsa-miR-608**	**7**	**−16.35**	**−33.3**
MAOA	hsa-miR-449b	7	−14.23	−29.1
MAOA	hsa-miR-449a	7	−12.93	−28.8
MAOA	**hsa-miR-125a-3p**	**4**	**−11.81**	**−28.7**
MAOA	hsa-miR-412	7	−10.75	−22.8
MAOA	hsa-miR-769-5p	4	−10.45	−34.3
MAOA	hsa-miR-1262	4	−10.15	−24.9
MAOA	hsa-miR-34c-5p	7	−10.13	−25.6
MAOB	hsa-miR-1207-5p	3	−20.95	−36.9
MAOB	hsa-miR-485-5p	5	−15.11	−24.0
MAOB	hsa-miR-296-3p	4	−13.14	−30.2
MAOB	**hsa-miR-608**	**4**	**−12.64**	**−34.3**
MAOB	**hsa-miR-125a-3p**	**3**	**−12.35**	**−29.7**
MAOB	hsa-miR-1294	4	−11.38	−24.1
MAOB	hsa-miR-486-3p	7	−10.96	−27.6
MAOB	hsa-miR-641	6	−10.86	−27.3
MAOB	hsa-miR-630	9	−10.74	−27.4
MAOB	hsa-miR-654-5p	4	−10.14	−27.1
MAOB	hsa-miR-184	4	−10.02	−24.4

[a]Ten prediction tools (viz., DIANA-microT, miRanda, miRDB, miRWalk, RNAhybrid, PICTAR4, PICTAR5, PITA, RNA22, and Targetscan miRNA) were used to predict the putative miRNAs that may bind to the 3′-UTRs of human MAOA and MAOB. The number of programs predicting binding sites for a miRNA is shown. Some of the predicted miRNAs are common to both MAOA and MAOB (viz., miR-608 and miR-125a-3p); those are shown in bold. This table includes only those miRNAs that were predicted to have ΔΔG values of less than −10 (as per PITA program; https://genie.weizmann.ac.il/pubs/mir07/mir07_prediction.html) and ΔG values of less than −20 kcal/mole (as per the RNAhybrid program; https://bibiserv2.cebitec.uni-bielefeld.de/rnahybrid) since these values (i.e., ΔΔG < −10 and ΔG < −20 kcal/mole) indicate higher accessibility and affinity of miRNA/mRNA interaction

by phosphatidylinositol kinase and nuclear factor-κB [22]. Taken together, this increase in miR-1224 could be a compensatory mechanism to block MAOA/MAOB gene expression by binding to their 3′-UTRs and by targeting both Sp1 and CREB (Fig. 6.6). This is further substantiated by the evidence that MAO inhibitors are protective against oxidative stress [102]. All these observations indicate a complex interplay between miR-1224, Sp1 and CREB in regulating MAOA/MAOB expression which warrants further investigation.

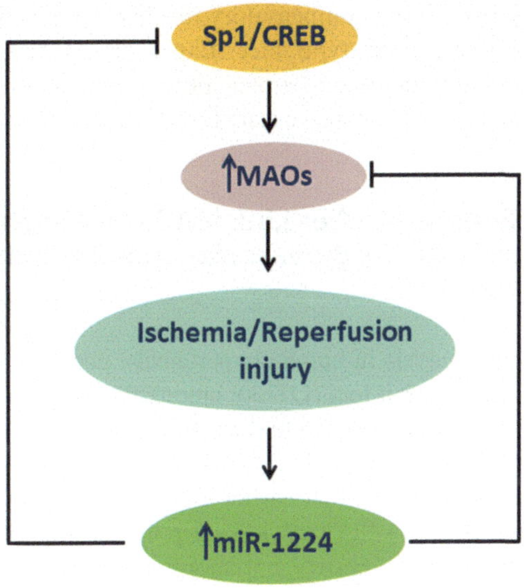

Fig. 6.6 Possible interplay of miR-1224, Sp1, and CREB in governing MAO gene regulation. The transcription factors Sp1 and CREB regulate MAOA/MAOB gene expression, which in turn may contribute to oxidative stress during ischemia/reperfusion (I/R) injury. The levels of miR-1224 are augmented under I/R condition which could be a compensatory mechanism to block Sp1, CREB, and MAOA/MAOB gene expressions. Upward arrows indicate "increase" and blunt-headed arrows indicate "inhibition" of function of the corresponding molecules. *Sp1* specificity protein 1, *CREB* cAMP response element binding protein, *miR-1224* microRNA-1224

6.7 Cardiovascular Implications of Systemic Ablation of MAOA/MAOB in Mouse

Generation of MAOA or MAOB knockout mice was carried out by inserting interferon β transgene or neomycin resistance gene into exon 2 and 6 of *MAOA* and *MAOB*, respectively [103]. As expected, *MAOA* knockout mice displayed higher levels of its substrates (catecholamines and serotonin) in the brain along with various neurochemical and physiological changes in comparison with the wild-type animals [103]. Similarly, adult *MAOB* knockout mice showed ~8.0-fold higher level of phenylethylamine in the brain while no statistically significant increase in serotonin, norepinephrine, and dopamine, due to the substrate specificity of MAOB. The most striking cardiovascular characteristic of *MAOA/MAOB* knockout mice was their hypotensive nature and reduced heart rate in the resting, restrained state [103]. This finding is in corroboration with the resting hypotension in Norrie disease patients who have deletions in *MAOA* gene [104, 105]. But this is in contrast to the fact that higher level of catecholamine is mostly associated with hypertension. This phenotype can be explained by the fact that high level of catecholamine may be a cause or effect, which can lead to higher/lower blood pressure. Most probably, these knockout mice developed some compensatory mechanism which leads to lowered blood pressure than that of wild-type mice. Expectedly, *MAOA/MAOB* knockout mice were found to have increased baroreceptor activity that serves to regulate blood pressure and leads to hypotensive state [106].

6.8 Human Genetic Studies Link MAOs and Their Upstream Regulators with Cardiovascular and Cardiometabolic Risk Factors

Genome-wide linkage analysis in human hypertensive population revealed several blood pressure quantitative trait loci (QTLs); among them, the blood pressure QTL on the X chromosome (Xp11.4-Xq11) harbors several genes of cardiovascular relevance including *MAOA* and *MAOB* [107]. This observation is in line with the identification of blood pressure QTLs that harbor these genes in the X chromosome in rats (Fig. 6.5). Some of the well-characterized polymorphisms (VNTR (variable number of tandem repeats) and EcoRV polymorphism) present in *MAOA* gene are also associated with cardiovascular or cardiometabolic risk factors including body mass index, lipid levels, and obesity [108–111]. In brief, the most widely studied polymorphism in h*MAOA* gene is a VNTR (30-bp repeat sequence present in 3, 3.5, 4, or 5 copies) present at ~1.2 kb upstream of the coding region in h*MAOA*. Several studies in the last few decades reported the functional role of this VNTR in the context of neuronal/behavioral traits. According to those studies, alleles with 3.5 or 4 copies of the repeat sequence displayed substantially higher (two- to tenfold) transcriptional activity when compared to alleles with 3 or 5 copies of the VNTR [112–114]. Another polymorphism present in *MAOA* gene, i.e., EcoRV polymorphism or T/C polymorphism (rs1137070) located within exon 14, has been associated with altered MAOA enzyme activity [115]. Briefly, MAOA gene with allele T harbors an *Eco*RV site and higher *MAOA* activity than that of MAOA gene with allele C and no *Eco*RV site. Interestingly, this T/C polymorphism causes a nucleotide substitution at the third position of a codon and does not affect the amino acid sequence (Asp to Asp). Perhaps, the polymorphism is in linkage disequilibrium with another genetic variation to regulate MAOA enzyme activity [115], or the rate of translation of the mRNA transcript could be altered due to this synonymous T/C polymorphism [116]. Moreover, this SNP was also associated with gout and hyperuricemia (another risk factor for cardiovascular disorders) [52, 117].

Several studies have associated SNPs in the upstream regions of the crucial transcriptional regulators of *MAOA* including SIRT1 and GATA2 with weight/body mass index/systolic blood pressure/diastolic blood pressure/hypertension/hyperglycemia in different populations across the world [58–61, 118, 119]. Sirtuin proteins (SIRT1–SIRT7) are nicotinamide adenine dinucleotide (NAD)-dependent deacetylases. The most conserved member of the sirtuin family, SIRT1, regulates the PGC1α activity via deacetylation, thereby protecting the cells against oxidative stress. In addition, SIRT1 deacetylates many other crucial transcription factors and cofactors including p53 [120], forkhead box class O (FOXO) proteins [121], and nuclear factor-κB [122]. Of note, MAOA upregulation leads to necrosis or chronic ventricular dysfunction via p53-PGC1α-mediated pathway as shown in Fig. 6.2. It is evident from human genetic studies that the SIRT1 SNP rs2273773 (C/T in exon 5, a silent mutation) is associated with seasonal variation in weight and diastolic blood pressure/hypertension in Finnish nationwide population [58]. Another study has also probed the association of SIRT1 SNPs (rs7895833 (A/G in the promoter region), rs7069102 (C/G in intron 4), and rs2273773 (C/T in the coding region)) with cardiovascular/cardiometabolic risk factors. For example, the mutant alleles for rs7069102 and rs2273773 were detected at significantly higher frequencies in cardiovascular disease patients compared to control subjects, increasing the disease risk by 2.4- and 1.9-fold, respectively, in mutant allele carriers than in wild-type allele carriers. In contrast, the allele frequency for rs7895833 did not differ between both groups [61]. Another study in a Japanese population showed the association of rs7895833, rs7069102, and rs2273773 with different cardiovascular/cardiometabolic phenotypes including fasting glucose/hyperglycemia/body fat ratio/systolic blood pressure/diastolic blood pressure/hypertension [60]. Thus, SIRT1 emerged as a potential therapeutic target for metabolic syndrome [123–125]. In addition to SIRT1, human genetic studies have identified *GATA2* as a novel susceptibility gene for coronary artery disease by showing the association of *GATA2* SNPs with cardiovascular/cardiometabolic risk traits [118, 119].

6.9 Conclusions and Perspectives

A growing body of research suggests that dysregulation of MAOs plays an important role in several cardiovascular pathophysiological conditions (including essential hypertension, LV remodeling, heart failure, cardiomyocyte hypertrophy, and I/R injury) possibly due to ROS generated by MAOs. Therefore, regulation of MAOs (perhaps, by tissue-specific regulation of some transcription factors) may emerge as a new therapeutic strategy for treatment of cardiovascular pathologies. Although, so far, conclusive studies on the applicability of MAO inhibitors with heart disease patients are lacking, general MAO inhibitors were previously used as therapeutics for cardiovascular diseases and have been reported to reduce blood pressure and intensity/frequency of anginal pain [126]. However, the main concern for the use of these irreversible MAO inhibitors is a phenomenon called "cheese effect" which, subsequently, causes hypertensive crisis. The efficiency of the next-generation reversible MAO inhibitors that are devoid of these harmful effects remains to be evaluated in cardiovascular pathologies. It is also important to note that although MAOB is highly abundant in the human myocardium, most of the studies focused on MAOA; therefore, future research should be designed to understand the contribution of MAOB to these complications. Systematic studies identifying posttranscriptional regulators (certain microRNAs or their inhibitors) of MAOs may also lead to identification of novel cardiovascular therapeutics. Based on human genetic studies, computational predictions, and regulatory mechanisms, certain common molecular factors may also emerge as potential therapeutic agents for dysregulated MAO expression/activity.

Acknowledgments The authors are thankful to the researchers who contributed to studies on the monoamine oxidases. This work was supported by a grant from the Department of Biotechnology, Government of India, to NRM (project number: BT/PR5017/MED/30/756/2012). VG and VA received research fellowships from the Ministry of Human Resource Development, Government of India, and Department of Science and Technology, Government of India, respectively.

Conflict of Interest Statement The authors declare that there is no conflict of interest.

References

1. Nagatsu T (2004) Progress in monoamine oxidase (MAO) research in relation to genetic engineering. Neurotoxicology 25:11–20
2. Grimsby J, Chen K, Wang LJ, Lan NC, Shih JC (1991) Human monoamine oxidase A and B genes exhibit identical exon-intron organization. Proc Natl Acad Sci U S A 88:3637–3641
3. Shih JC, Chen K, Ridd MJ (1999) Monoamine oxidase: from genes to behavior. Annu Rev Neurosci 22:197–217
4. Finberg J (2014) Update on the pharmacology of selective inhibitors of MAO-A and MAO-B: focus on modulation of CNS monoamine neurotransmitter release. Pharmacol Ther 143:133–152
5. Grimsby J, Lan NC, Neve R, Chen K, Shih JC (1990) Tissue distribution of human monoamine oxidase A and B mRNA. J Neurochem 55:1166–1169
6. Saura J, Richards JG, Mahy N (1994) Age-related changes on MAO in Bl/C57 mouse tissues: a quantitative radioautographic study. J Neural Transm Suppl 41:89–94
7. Strolin Benedetti M, Thomassin J, Tocchetti P, Dostert P, Kettler R, Da Prada M (1994) Species differences in changes of heart monoamine oxidase activities with age. J Neural Transm Suppl 41:83–87
8. Gupta V, Khan AA, Sasi BK, Mahapatra NR (2015) Molecular mechanism of monoamine oxidase A gene regulation under inflammation and ischemia-like conditions: key roles of the transcription factors GATA2, Sp1 and TBP. J Neurochem 134:21–38
9. Shih JC, Wu JB, Chen K (2011) Transcriptional regulation and multiple functions of MAO genes. J Neural Transm (Vienna) 118:979–986
10. Wong W, Ou X, Chen K, Shih J (2002) Activation of human monoamine oxidase B gene expression by a protein kinase C MAPK signal transduction pathway involves c-Jun and Egr-1. J Biol Chem 277:22222–22230
11. Libert S, Pointer K, Bell EL, Das A, Cohen DE, Asara J, Kapur K, Bergmann S, Preisig M, Otowa T et al (2011) SIRT1 activates MAO-A in the brain to mediate anxiety and exploratory drive. Cell 147:1459–1472
12. Hampp G, Ripperger J, Houben T, Schmutz I, Blex C, Perreau-Lenz S, Brunk I, Spanagel R, Ahnert-Hilger G, Meijer J et al (2008) Regulation of monoamine oxidase A by circadian-clock components implies clock influence on mood. Curr Biol 18:678–683
13. Bortolato M, Chen K, Shih J (2008) Monoamine oxidase inactivation: from pathophysiology to therapeutics. Adv Drug Deliv Rev 60:1527–1533
14. Wang C, Billett E, Borchert A, Kuhn H, Ufer C (2013) Monoamine oxidases in development. Cell Mol Life Sci 70:599–630
15. Wu J, Shao C, Li X, Li Q, Hu P, Shi C, Li Y, Chen Y, Yin F, Liao C et al (2014) Monoamine oxidase A mediates prostate tumorigenesis and cancer metastasis. J Clin Invest 124:2891–2908
16. Nicotra A, Pierucci F, Parvez H, Senatori O (2004) Monoamine oxidase expression during development and aging. Neurotoxicology 25:155–165

17. Pchejetski D, Kunduzova O, Dayon A, Calise D, Seguelas M, Leducq N, Seif I, Parini A, Cuvillier O (2007) Oxidative stress-dependent sphingosine kinase-1 inhibition mediates monoamine oxidase A-associated cardiac cell apoptosis. Circ Res 100:41–49
18. Villeneuve C, Guilbeau-Frugier C, Sicard P, Lairez O, Ordener C, Duparc T, De Paulis D, Couderc B, Spreux-Varoquaux O, Tortosa F et al (2013) p53-PGC-1α pathway mediates oxidative mitochondrial damage and cardiomyocyte necrosis induced by monoamine oxidase-A upregulation: role in chronic left ventricular dysfunction in mice. Antioxid Redox Signal 18:5–18
19. Kaludercic N, Carpi A, Nagayama T, Sivakumaran V, Zhu G, Lai E, Bedja D, De Mario A, Chen K, Gabrielson KL et al (2014) Monoamine oxidase B prompts mitochondrial and cardiac dysfunction in pressure overloaded hearts. Antioxid Redox Signal 20:267–280
20. Kaludercic N, Carpi A, Menabò R, Di Lisa F, Paolocci N (2011) Monoamine oxidases (MAO) in the pathogenesis of heart failure and ischemia/reperfusion injury. Biochim Biophys Acta 1813:1323–1332
21. Pino R, Failli P, Mazzetti L, Buffoni F (1997) Monoamine oxidase and semicarbazide-sensitive amine oxidase activities in isolated cardiomyocytes of spontaneously hypertensive rats. Biochem Mol Med 62:188–196
22. Sturza A, Leisegang M, Babelova A, Schröder K, Benkhoff S, Loot A, Fleming I, Schulz R, Muntean D, Brandes R (2013) Monoamine oxidases are mediators of endothelial dysfunction in the mouse aorta. Hypertension 62:140–146
23. Sturza A, Duicu O, Vaduva A, Dănilă M, Noveanu L, Varró A, Muntean D (2015) Monoamine oxidases are novel sources of cardiovascular oxidative stress in experimental diabetes. Can J Physiol Pharmacol 93:555–561
24. Kaludercic N, Mialet-Perez J, Paolocci N, Parini A, Di Lisa F (2014) Monoamine oxidases as sources of oxidants in the heart. J Mol Cell Cardiol 73:34–42
25. Peña-Silva R, Miller J, Chu Y, Heistad D (2009) Serotonin produces monoamine oxidase-dependent oxidative stress in human heart valves. Am J Physiol Heart Circ Physiol 297:H1354–H1360
26. Poon C, Seto S, Au A, Zhang Q, Li R, Lee WYW, Leung GPH, Kong SK, Yeung JHK, Ngai SM et al (2010) Mitochondrial monoamine oxidase-A-mediated hydrogen peroxide generation enhances 5-hydroxytryptamine-induced contraction of rat basilar artery. Br J Pharmacol 161:1086–1098
27. Kaludercic N, Takimoto E, Nagayama T, Feng N, Lai EW, Bedja D, Chen K, Gabrielson KL, Blakely RD, Shih JC et al (2010) Monoamine oxidase A-mediated enhanced catabolism of norepinephrine contributes to adverse remodeling and pump failure in hearts with pressure overload. Circ Res 106:193–202
28. Bianchi P, Pimentel DR, Murphy MP, Colucci WS, Parini A (2005) A new hypertrophic mechanism of serotonin in cardiac myocytes: receptor-independent ROS generation. FASEB J 19:641–643
29. Santin Y, Sicard P, Vigneron F, Guilbeau-Frugier C, Dutaur M, Lairez O, Couderc B, Manni D, Korolchuk VI, Lezoualc'h F et al (2016) Oxidative stress by monoamine oxidase-A impairs transcription factor EB activation and Autophagosome clearance, leading to Cardiomyocyte necrosis and heart failure. Antioxid Redox Signal 25:10–27
30. Maceyka M, Payne SG, Milstien S, Spiegel S (2002) Sphingosine kinase, sphingosine-1-phosphate, and apoptosis. Biochim Biophys Acta 1585:193–201
31. Frank D, Kuhn C, Brors B, Hanselmann C, Lüdde M, Katus HA, Frey N (2008) Gene expression pattern in biomechanically stretched cardiomyocytes: evidence for a stretch-specific gene program. Hypertension 51:309–318
32. Petrak J, Pospisilova J, Sedinova M, Jedelsky P, Lorkova L, Vit O, Kolar M, Strnad H, Benes J, Sedmera D et al (2011) Proteomic and transcriptomic analysis of heart failure due to volume overload in a rat aorto-caval fistula model provides support for new potential therapeutic targets – monoamine oxidase A and transglutaminase 2. Proteome Sci 9:69

33. Manni ME, Rigacci S, Borchi E, Bargelli V, Miceli C, Giordano C, Raimondi L, Nediani C (2016) Monoamine oxidase is Overactivated in left and right ventricles from ischemic hearts: an intriguing therapeutic target. Oxidative Med Cell Longev 2016:4375418
34. Deng AY (2007) Genetic basis of polygenic hypertension. Hum Mol Genet 16(Spec No. 2):R195–R202
35. Missale C, Nash SR, Robinson SW, Jaber M, Caron MG (1998) Dopamine receptors: from structure to function. Physiol Rev 78:189–225
36. Zeng C, Zhang M, Asico LD, Eisner GM, Jose PA (2007) The dopaminergic system in hypertension. Clin Sci 112:583–597
37. Andrew R, Best SA, Watson DG, Midgley JM, Reid JL, Squire IB (1993) Analysis of biogenic amines in plasma of hypertensive patients and a control group. Neurochem Res 18:1179–1182
38. Goldstein DS (1983) Plasma catecholamines and essential hypertension. An analytical review. Hypertension 5:86–99
39. Grobecker G, Roizen MF, Weise V, Saavedra JM, Kopin IJ (1975) Sympathoadrenal medullary activity in young, spontaneously hypertensive rats. Nature 258:267–268
40. Fries RS, Mahboubi P, Mahapatra NR, Mahata SK, Schork NJ, Schmid-Schoenbein GW, O'Connor DT (2004) Neuroendocrine transcriptome in genetic hypertension: multiple changes in diverse adrenal physiological systems. Hypertension 43:1301–1311
41. Puig O, Wang I-M, Cheng P, Zhou P, Roy S, Cully D, Peters M, Benita Y, Thompson J, Cai T-Q (2010) Transcriptome profiling and network analysis of genetically hypertensive mice identifies potential pharmacological targets of hypertension. Physiol Genomics 42A:24–32
42. Vega A, Chacón P, Monteseirín J, El Bekay R, Alvarez M, Alba G, Conde J, Martín-Nieto J, Bedoya FJ, Pintado E et al (2004) A new role for monoamine oxidases in the modulation of macrophage-inducible nitric oxide synthase gene expression. J Leukoc Biol 75:1093–1101
43. Yasuhara H, Tonooka M, Wada I, Oguchi K, Sakamoto K, Kamijo K (1983) Hemodynamics and monoamine oxidase activity in spontaneously hypertensive rats (SHR). Jpn J Pharmacol 33:1057–1064
44. Friese RS, Mahboubi P, Mahapatra NR, Mahata SK, Schork NJ, Schmid-Schönbein GW, O'Connor DT (2005) Common genetic mechanisms of blood pressure elevation in two independent rodent models of human essential hypertension. Am J Hypertens 18:633–652
45. Berry MD (2004) Mammalian central nervous system trace amines. Pharmacologic amphetamines, physiologic neuromodulators. J Neurochem 90:257–271
46. Knoll J, Miklya I, Knoll B, Markó R, Rácz D (1996) Phenylethylamine and tyramine are mixed-acting sympathomimetic amines in the brain. Life Sci 58:2101–2114
47. Blackwell B, Marley E, Ryle A (1964) Hypertensive crisis associated with monoamine-oxidase inhibitors. Lancet 1:722–723
48. Raiteri M, Del Carmine R, Bertollini A, Levi G (1977) Effect of sympathomimetic amines on the synaptosomal transport of noradrenaline, dopamine and 5-hydroxytryptamine. Eur J Pharmacol 41:133–143
49. Neef DW, Jaeger AM, Thiele DJ (2011) Heat shock transcription factor 1 as a therapeutic target in neurodegenerative diseases. Nat Rev Drug Discov 10:930–944
50. Altucci L, Leibowitz MD, Ogilvie KM, de Lera AR, Gronemeyer H (2007) RAR and RXR modulation in cancer and metabolic disease. Nat Rev Drug Discov 6:793–810
51. Evans WE, Guy RK (2004) Gene expression as a drug discovery tool. Nat Genet 36:214–215
52. Ely D, Underwood A, Dunphy G, Boehme S, Turner M, Milsted A (2010) Review of the Y chromosome, Sry and hypertension. Steroids 75:747–753
53. Milsted A, Serova L, Sabban EL, Dunphy G, Turner ME, Ely DL (2004) Regulation of tyrosine hydroxylase gene transcription by Sry. Neurosci Lett 369:203–207
54. Zhang K, Rao F, Wang L, Rana BK, Ghosh S, Mahata M, Salem RM, Rodriguez-Flores JL, Fung MM, Waalen J et al (2010) Common functional genetic variants in catecholamine storage vesicle protein promoter motifs interact to trigger systemic hypertension. J Am Coll Cardiol 55:1463–1475

55. Milsted A, Underwood AC, Dunmire J, DelPuerto HL, Martins AS, Ely DL, Turner ME (2010) Regulation of multiple renin-angiotensin system genes by Sry. J Hypertens 28:59–64
56. Duncan J, Johnson S, Ou X-M (2012) Monoamine oxidases in major depressive disorder and alcoholism. Drug Discov Ther 6:112–122
57. Menghini R, Marchetti V, Cardellini M, Hribal ML, Mauriello A, Lauro D, Sbraccia P, Lauro R, Federici M (2005) Phosphorylation of GATA2 by Akt increases adipose tissue differentiation and reduces adipose tissue-related inflammation: a novel pathway linking obesity to atherosclerosis. Circulation 111:1946–1953
58. Kovanen L, Donner K, Partonen T (2015) SIRT1 polymorphisms associate with seasonal weight variation, depressive disorders, and diastolic blood pressure in the general population. PLoS One 10:e0141001
59. Shimoyama Y, Suzuki K, Hamajima N, Niwa T (2011) Sirtuin 1 gene polymorphisms are associated with body fat and blood pressure in Japanese. Transl Res 157:339–347
60. Shimoyama Y, Mitsuda Y, Tsuruta Y, Suzuki K, Hamajima N, Niwa T (2012) SIRTUIN 1 gene polymorphisms are associated with cholesterol metabolism and coronary artery calcification in Japanese hemodialysis patients. J Ren Nutr 22:114–119
61. Kilic U, Gok O, Bacaksiz A, Izmirli M, Elibol-Can B, Uysal O (2014) SIRT1 gene polymorphisms affect the protein expression in cardiovascular diseases. PLoS One 9:e90428
62. Pizzinat N, Marchal-Victorion S, Maurel A, Ordener C, Bompart G, Parini A (2003) Substrate-dependent regulation of MAO-A in rat mesangial cells: involvement of dopamine D2-like receptors. Am J Physiol Renal Physiol 284:F167–F174
63. Manoli I, Le H, Alesci S, McFann KK, Su YA, Kino T, Chrousos GP, Blackman MR (2005) Monoamine oxidase-A is a major target gene for glucocorticoids in human skeletal muscle cells. FASEB J 19:1359–1361
64. Ammon HP, Müller AB (1985) Forskolin: from an ayurvedic remedy to a modern agent. Planta Med 51:473–477
65. Schlepper M, Thormann J, Mitrovic V (1989) Cardiovascular effects of forskolin and phosphodiesterase-III inhibitors. Basic Res Cardiol 84(Suppl 1):197–212
66. Dubey MP, Srimal RC, Nityanand S, Dhawan BN (1981) Pharmacological studies on coleonol, a hypotensive diterpene from Coleus forskohlii. J Ethnopharmacol 3:1–13
67. Wysham DG, Brotherton AF, Heistad DD (1986) Effects of forskolin on cerebral blood flow: implications for a role of adenylate cyclase. Stroke 17:1299–1303
68. Ou X-M, Chen K, Shih JC (2004) Dual functions of transcription factors, transforming growth factor-beta-inducible early gene (TIEG)2 and Sp3, are mediated by CACCC element and Sp1 sites of human monoamine oxidase (MAO) B gene. J Biol Chem 279:21021–21028
69. Chen K (2004) Organization of MAO A and MAO B promoters and regulation of gene expression. Neurotoxicology 25:31–36
70. Shih JC, Chen K (2004) Regulation of MAO-A and MAO-B gene expression. Curr Med Chem 11:1995–2005
71. Colbert MC, Hall DG, Kimball TR, Witt SA, Lorenz JN, Kirby ML, Hewett TE, Klevitsky R, Robbins J (1997) Cardiac compartment-specific overexpression of a modified retinoic acid receptor produces dilated cardiomyopathy and congestive heart failure in transgenic mice. J Clin Invest 100:1958–1968
72. Kotake D, Sato T, Hirasawa N (2014) Retinoid signaling in pathological remodeling related to cardiovascular disease. Eur J Pharmacol 729:144–147
73. Edelstein SB, Breakefield XO (1986) Monoamine oxidases A and B are differentially regulated by glucocorticoids and 'aging' in human skin fibroblasts. Cell Mol Neurobiol 6:121–150
74. Buttgereit F, Burmester G-R, Lipworth BJ (2009) Inflammation, glucocorticoids and risk of cardiovascular disease. Nat Clin Pract Rheumatol 5:18–19
75. Walker BR (2007) Glucocorticoids and cardiovascular disease. Eur J Endocrinol 157:545–559
76. Barceló F, Ortiz-Lombardía M, Martorell M, Oliver M, Méndez C, Salas JA, Portugal J (2010) DNA binding characteristics of mithramycin and chromomycin analogues obtained by combinatorial biosynthesis. Biochemistry 49:10543–10552

77. Choi ES, Nam JS, Jung JY, Cho NP, Cho SD (2014) Modulation of specificity protein 1 by mithramycin A as a novel therapeutic strategy for cervical cancer. Sci Rep 24:7162
78. Yao L, Dai X, Sun Y, Wang Y, Yang Q, Chen X, Liu Y, Zhang L, Xie W, Liu J (2018) Inhibition of transcription factor SP1 produces neuroprotective effects through decreasing MAO B activity in MPTP/MPP+ Parkinson's disease models. J Neurosci Res 96:1663–1676
79. Yu Q, Huang Q, Du X, Xu S, Li M, Ma S (2018) Early activation of Egr-1 promotes neuroinflammation and dopaminergic neurodegeneration in an experimental model of Parkinson's disease. Exp Neurol 302:145–154
80. Ahn SY, Cho CH, Park KG, Lee HJ, Park SK, Lee IK, Koh GY (2004) Tumor necrosis factor-alpha induces fractalkine expression preferentially in arterial endothelial cells and mithramycin A suppresses TNF-alpha-induced fractalkine expression. Am J Pathol 164:1663–1672
81. Orellana EA, Kasinski AL (2015) MicroRNAs in cancer: a historical perspective on the path from discovery to therapy. Cancers (Basel) 7:1388–1405
82. Reinhart BJ, Slack FJ, Basson M, Pasquinelli AE, Bettinger JC, Rougvie AE, Horvitz HR, Ruvkun G (2000) The 21-nucleotide let-7 RNA regulates developmental timing in Caenorhabditis elegans. Nature 403:901–906
83. Hartig SM, Hamilton MP, Bader DA, McGuire SE (2015) The miRNA interactome in metabolic homeostasis. Trends Endocrinol Metab 26:733–745
84. Barwari T, Joshi A, Mayr M (2016) MicroRNAs in cardiovascular disease. J Am Coll Cardiol 68:2577–2584
85. Titze-de-Almeida R, David C, Titze-de-Almeida SS (2017) The race of 10 synthetic RNAi-based drugs to the pharmaceutical market. Pharm Res 34:1339–1363
86. van Rooij E, Kauppinen S (2014) Development of microRNA therapeutics is coming of age. EMBO Mol Med 6:851–864
87. Chaudhuri AD, Yelamanchili SV, Fox HS (2013) MicroRNA-142 reduces monoamine oxidase A expression and activity in neuronal cells by downregulating SIRT1. PLoS One 8:e79579
88. Maragkakis M, Reczko M, Simossis VA, Alexiou P, Papadopoulos GL, Dalamagas T, Giannopoulos G, Goumas G, Koukis E, Kourtis K et al (2009) DIANA-microT web server: elucidating microRNA functions through target prediction. Nucleic Acids Res 37:W273–W276
89. Betel D, Wilson M, Gabow A, Marks DS, Sander C (2008) The microRNA.org resource: targets and expression. Nucleic Acids Res 36:D149–D153
90. Wong N, Wang X (2015) miRDB: an online resource for microRNA target prediction and functional annotations. Nucleic Acids Res 43:D146–D152
91. Dweep H, Gretz N, Sticht C (2014) miRWalk database for miRNA-target interactions. Methods Mol Biol 1182:289–305
92. Krüger J, Rehmsmeier M (2006) RNAhybrid: microRNA target prediction easy, fast and flexible. Nucleic Acids Res 34:W451–W454
93. Witkos TM, Koscianska E, Krzyzosiak WJ (2011) Practical aspects of microRNA target prediction. Curr Mol Med 11:93–109
94. Kertesz M, Iovino N, Unnerstall U, Gaul U, Segal E (2007) The role of site accessibility in microRNA target recognition. Nat Genet 39:1278–1284
95. Miranda KC, Huynh T, Tay Y, Ang Y-S, Tam W-L, Thomson AM, Lim B, Rigoutsos I (2006) A pattern-based method for the identification of MicroRNA binding sites and their corresponding heteroduplexes. Cell 126:1203–1217
96. D'Alessandra Y, Devanna P, Limana F, Straino S, Di Carlo A, Brambilla PG, Rubino M, Carena MC, Spazzafumo L, De Simone M et al (2010) Circulating microRNAs are new and sensitive biomarkers of myocardial infarction. Eur Heart J 31:2765–2773
97. D'Alessandra Y, Carena MC, Spazzafumo L, Martinelli F, Bassetti B, Devanna P, Rubino M, Marenzi G, Colombo GI, Achilli F et al (2013) Diagnostic potential of plasmatic MicroRNA signatures in stable and unstable angina. PLoS One 8:e80345

98. Roy S, Bantel H, Wandrer F, Schneider AT, Gautheron J, Vucur M, Tacke F, Trautwein C, Luedde T, Roderburg C (2017) miR-1224 inhibits cell proliferation in acute liver failure by targeting the antiapoptotic gene Nfib. J Hepatol 67:966–978
99. Niu Y, Mo D, Qin L, Wang C, Li A, Zhao X, Wang X, Xiao S, Wang Q, Xie Y et al (2011) Lipopolysaccharide-induced miR-1224 negatively regulates tumour necrosis factor-α gene expression by modulating Sp1. Immunology 133:8–20
100. Qian J, Li R, Wang Y-Y, Shi Y, Luan W-K, Tao T, Zhang J-X, Xu Y-C, You Y-P (2015) MiR-1224-5p acts as a tumor suppressor by targeting CREB1 in malignant gliomas. Mol Cell Biochem 403:33–41
101. Nakano T, Nagatsu T, Higashida H (1985) Expression of A and B types of monoamine oxidase in differentiated neuroblastoma hybrid cells. J Neurochem 44:755–758
102. Rațiu C, Uțu D, Petruș A, Norbert P, Olariu S, Duicu O, Sturza A, Muntean DM (2018) Monoamine oxidase inhibition improves vascular function and reduces oxidative stress in rats with lipopolysaccharide-induced inflammation. Gen Physiol Biophys 37:687–694
103. Holschneider DP, Chen K, Seif I, Shih JC (2001) Biochemical, behavioral, physiologic, and neurodevelopmental changes in mice deficient in monoamine oxidase A or B. Brain Res Bull 56:453–462
104. Sims KB, de la Chapelle A, Norio R, Sankila EM, Hsu YP, Rinehart WB, Corey TJ, Ozelius L, Powell JF, Bruns G (1989) Monoamine oxidase deficiency in males with an X chromosome deletion. Neuron 2:1069–1076
105. Murphy DL, Sims KB, Karoum F, de la Chapelle A, Norio R, Sankila EM, Breakefield XO (1990) Marked amine and amine metabolite changes in Norrie disease patients with an X-chromosomal deletion affecting monoamine oxidase. J Neurochem 54:242–247
106. Holschneider DP, Scremin OU, Roos KP, Chialvo DR, Chen K, Shih JC (2002) Increased baroreceptor response in mice deficient in monoamine oxidase A and B. Am J Physiol Heart Circ Physiol 282:H964–H972
107. Harrap SB, Wong ZYH, Stebbing M, Lamantia A, Bahlo M (2002) Blood pressure QTLs identified by genome-wide linkage analysis and dependence on associated phenotypes. Physiol Genomics 8:99–105
108. Brummett BH, Boyle SH, Siegler IC, Zuchner S, Ashley-Koch A, Williams RB (2008) Lipid levels are associated with a regulatory polymorphism of the monoamine oxidase-A gene promoter (MAOA-uVNTR). Med Sci Monit 14:CR57–CR61
109. Camarena B, Santiago H, Aguilar A, Ruvinskis E, González-Barranco J, Nicolini H (2004) Family-based association study between the monoamine oxidase A gene and obesity: implications for psychopharmacogenetic studies. Neuropsychobiology 49:126–129
110. Need AC, Ahmadi KR, Spector TD, Goldstein DB (2006) Obesity is associated with genetic variants that alter dopamine availability. Ann Hum Genet 70:293–303
111. Fuemmeler BF, Agurs-Collins TD, McClernon FJ, Kollins SH, Kail ME, Bergen AW, Ashley-Koch AE (2008) Genes implicated in serotonergic and dopaminergic functioning predict BMI categories. Obesity (Silver Spring) 16:348–355
112. Deckert J, Catalano M, Syagailo YV, Bosi M, Okladnova O, Di Bella D, Nöthen MM, Maffei P, Franke P, Fritze J et al (1999) Excess of high activity monoamine oxidase A gene promoter alleles in female patients with panic disorder. Hum Mol Genet 8:621–624
113. Sabol SZ, Hu S, Hamer D (1998) A functional polymorphism in the monoamine oxidase A gene promoter. Hum Genet 103:273–279
114. Wu Y-H, Fischer DF, Swaab DF (2007) A promoter polymorphism in the monoamine oxidase A gene is associated with the pineal MAOA activity in Alzheimer's disease patients. Brain Res 1167:13–19
115. Hotamisligil GS, Breakefield XO (1991) Human monoamine oxidase A gene determines levels of enzyme activity. Am J Hum Genet 49:383–392
116. Sauna ZE, Kimchi-Sarfaty C (2011) Understanding the contribution of synonymous mutations to human disease. Nat Rev Genet 12:683–691
117. Feig DI, Kang D-H, Johnson RJ (2008) Uric acid and cardiovascular risk. N Engl J Med 359:1811–1821

118. Muiya NP, Wakil S, Al-Najai M, Tahir AI, Baz B, Andres E, Al-Boudari O, Al-Tassan N, Al-Shahid M, Meyer BF et al (2014) A study of the role of GATA2 gene polymorphism in coronary artery disease risk traits. Gene 544:152–158
119. Connelly JJ, Wang T, Cox JE, Haynes C, Wang L, Shah SH, Crosslin DR, Hale AB, Nelson S, Crossman DC et al (2006) GATA2 is associated with familial early-onset coronary artery disease. PLoS Genet 2:e139
120. Haigis MC, Sinclair DA (2010) Mammalian sirtuins: biological insights and disease relevance. Annu Rev Pathol 5:253–295
121. Motta MC, Divecha N, Lemieux M, Kamel C, Chen D, Gu W, Bultsma Y, McBurney M, Guarente L (2004) Mammalian SIRT1 represses forkhead transcription factors. Cell 116:551–563
122. Yeung F, Hoberg JE, Ramsey CS, Keller MD, Jones DR, Frye RA, Mayo MW (2004) Modulation of NF-kappaB-dependent transcription and cell survival by the SIRT1 deacetylase. EMBO J 23:2369–2380
123. Guarente L (2006) Sirtuins as potential targets for metabolic syndrome. Nature 444:868–874
124. Westphal CH, Dipp MA, Guarente L (2007) A therapeutic role for sirtuins in diseases of aging? Trends Biochem Sci 32:555–560
125. Jiang W (2008) Sirtuins: novel targets for metabolic disease in drug development. Biochem Biophys Res Commun 373:341–344
126. Griffith GC (1960) Amine oxidase inhibitors; their current place in the therapy of cardiovascular diseases. Circulation 22:1156–1165

Key Cellular Effectors in ROS-Mediated Cardiac Diseases

7

Ratul Datta Chaudhuri, Santanu Rana, Kaberi Datta, and Sagartirtha Sarkar

Abstract

The "Oxygen Paradox" proposes that it is tough for aerobic organisms to live without oxygen, but it is difficult to live with oxygen as well. Assigned a job of incessant pumping, the heart, being an obligate aerobic organ, epitomizes the paradoxical effects of oxygen. Much of them are attributed to the reactive oxygen species (ROS) that mold the embryonic development and normal functioning of the heart under homeostatic conditions on one hand and the progression of cardiovascular diseases on the other. The ROS generation within the heart is equated at controlled physiological levels to the scavenger endogenous antioxidants that are employed to prevent their accumulation. A shift in the balance causes toxic levels of ROS to accumulate, self-accentuate, and inflict damage to cellular components, leading to myocardial oxidative injury. In addition, a number of pathophysiological signalling pathways are triggered by amassed ROS which culminate into enhanced myocardial apoptosis, fibrosis, inflammation, and contractile dysfunction—hallmarks of a failing heart. Adverse left ventricular remodeling as in pathological cardiac hypertrophy and myocardial infarction is intricately associated with oxidative stress, which prompts researchers to focus their attention on the redox biology of the heart in health and disease. This has been yielding far-reaching clinical implications in the field of antioxidant therapy and redox biomarker discovery. As cardiac disorders continue to be the highest contributor to the *Global Burden of Disease*, a molecular detailing of where, what, when, and how ROS is conducive to the remodeling of the cellular milieu in the heart would provide a holistic appreciation of cardiac disease biology.

Ratul Datta Chaudhuri and Santanu Rana contributed equally.

R. D. Chaudhuri · S. Rana · K. Datta · S. Sarkar (✉)
Department of Zoology, University of Calcutta, Kolkata, West Bengal, India

S. Chakraborti et al. (eds.), *Modulation of Oxidative Stress in Heart Disease*,
https://doi.org/10.1007/978-981-13-8946-7_7

Keywords
Pathological cardiac hypertrophy · Myocardial infarction · Antioxidants · Oxidative stress

7.1 Introduction

Reactive oxygen species (ROS) describe a group of reactive molecules and free radicals derived from molecular oxygen, viz., superoxide ($O_2{\bullet}^-$), hydrogen peroxide (H_2O_2), hydroxyl radical (OH•), hypochloride anion (OCl^-), hydroperoxide (ROOH), hydroperoxyl radical (HOO•), peroxyl radical (ROO•), and singlet oxygen ($^1O_2$) [1]. In living organisms, ROS are generated from the plasma membrane, the cytosol, the peroxisomes, and on membranes of mitochondria and the endoplasmic reticulum [2]. Such molecules have been implicated to be associated with a wide range of disease forms such as chronic obstructive pulmonary diseases, inflammatory bowel disease, neurodegenerative disorders, cardiovascular diseases, and aging. Research in the last few decades have established a link between free radicals and cardiac tissue injury from biochemical, physiological, and pharmacological data. The first published article in PubMed on the cellular role of ROS dated as late as 1945. "Reactive oxygen species" as keyword in PubMed fetches more than 2,20,661 English-written research articles, almost 23,262 of which are review articles, while use of "reactive oxygen species and cardiac" as keywords results in around 14,869 articles. Remarkably, a "free radical theory" proposed by Denham Harman in 1956 suggested endogenous oxygen radicals generated within cells over time result in cumulative cellular damage targeting DNA, protein, lipids, and other components of the cell [3]. Till date, redox biological research has successfully established the role of ROS as second messengers in cellular signalling and as mediators of pathophysiology within the diseased tissue. Accrued evidence suggests that oxidative damage directly or indirectly alters the cellular and molecular milieu of the diseased myocardium.

According to the World Health Organization (WHO), cardiovascular diseases have been designated as the major cause of mortality worldwide. Pathological stress involving the cardiovascular tissues is classified according to the anatomical structures affected (i.e., myocardium, valves, coronary arteries, aortic root, endocardium, pericardium), impaired physiological function (i.e., heart failure, pulmonary hypertension), and thirdly, the abnormality found in cardiac physiological parameters (such as systolic dysfunction, pulmonary valve stenosis, concentric/eccentric myocardial hypertrophy, arrhythmia, congenital anomalies). Herein, we focus upon cardiac diseases that are associated with an enlarged, thick, or rigid myocardium or with the dire loss of cardiomyocytes due to either chronic hemodynamic load or acute coronary insufficiency, both being manifested by reduced functional efficiency of the heart. Chronic heart failure is characterized by structural or functional cardiac abnormalities that lead to reduced cardiac output [4, 5]. Acute cardiac injury during myocardial infarction (MI) leads to loss of myocyte and increased myocardial strain post-MI, causing eccentric hypertrophy of the remaining myocytes through neurohormonal activation leading to fibrosis and progressive left ventricular dilatation, altogether altering the shape of the left ventricle from elliptical to spherical, thus bringing about left ventricular remodeling [6, 7].

Reportedly, ROS directly induce tissue damage along with the activation and expression of antioxidant enzymes-proteins that minimize the ROS-mediated perturbations. ROS are deleterious to organisms at high concentrations; when ROS levels cannot be mitigated by the antioxidant defense within a cell, a state of "oxidative stress" is triggered. Oxidative stress is a phenomenon which is related to the development of many pathological conditions [8]. Despite their destructive activity, ROS are a well-described species of second messengers in a variety of biological and physiological processes. Early in evolution, as it seems from their unique presence throughout the animal kingdom, nature selected ROS as a signal transduction mediator to allow for adaptation to changes in the microenvironment of tissues of aerobic organisms due to various endogenous signals.

The enhanced production of ROS during pathological stress can lead to lipid peroxidation, protein oxidation, nucleic acid damage, enzyme inhibition, activation of programmed cell death (PCD) pathway, and ultimately death of the cells. Whether ROS will act as damaging or signalling molecule depends on the delicate equilibrium between ROS production and scavenging. Reportedly, studies from our laboratory and others, over the last few decades, have shown that the enhanced production of ROS during pathological cardiac hypertrophy and MI contribute to a collaboration between mitochondria and endoplasmic reticulum, leading to disease progression toward heart failure [7, 9]. Moreover, over a decade, our laboratory has focused to decipher the various molecular regulators of chronic and acute forms of cardiovascular diseases and the therapeutic targets and strategies to reduce the myocardial oxidative stress. Herein, the sources and the role of ROS-mediated mechanisms would be elucidated within cardiac pathophysiological tissues to benefit clinicians and research scientists alike toward future discovery of pharmacological interventions in management of the morbid diseases.

7.2 Chemical Nature of ROS

Oxidative stress, as redox chemistry defines, is an "increase in the reduction potential or a large decrease in the reducing capacity of the cellular redox couples" [1]. Ground state molecular oxygen is itself a radical and is often known as a diradical; according to Pauli's exclusion principle, for two electrons to occupy the same atomic orbital, they must possess opposite spin. Electrons added to the oxygen diradical must be transferred one at a time when it gets reduced, with highly reactive oxygen-derived free radicals as intermediates in the reaction [10].

ROS can be defined as oxygen-containing chemically reactive molecules. Some of them are considered as free radicals since they contain unpaired electrons (i.e., radicals) and are capable of independent existence (hence called free) like superoxide anion ($O_2•^-$), hydroxyl radical (OH•), and lipid radicals. The unpaired electrons from these ROS in an attempt to complete their orbitals and gain stability steal electrons from other molecules and give rise to other stable ROS like hydrogen peroxide (H_2O_2), peroxynitrite ($ONOO^-$), and hypochlorous acid (HOCl) which are not considered to be free radicals yet result in an oxidizing effect within the cell [8].

Superoxide anion ($O_2\bullet^-$) from whichever source can either spontaneously dismutate (be reduced) to H_2O_2 or be converted to H_2O_2 catalyzed by superoxide dismutase (SOD). SOD decreases the half-life of $O_2\bullet^-$ from 10^{-9}–10^{-11} s to 10^{-15} s. Two $O_2\bullet^-$ radicals get converted to oxygen and H_2O_2 in the SOD-catalyzed reaction:

$$2O_2\bullet^- + 2H^+ \rightarrow H_2O_2 + O_2$$

H_2O_2 can be reduced to water by the action of peroxidase and catalase. Alternatively, the iron-catalyzed Haber-Weiss reaction produces hydroxyl radicals (OH•) from H_2O_2 and O_2- anion.

$$Fe^{3+} + O_2\bullet^- \rightarrow Fe^{2+} + O_2$$

The second step in the reaction is known as the Fenton reaction which makes use of reduced transition metals. The optimal pH for this reaction is 3.0–6.0.

$$Fe^{2+} + H_2O_2 \rightarrow Fe^{3+} + OH\bullet + OH^-$$
$$Fe^{3+} + H_2O_2 \rightarrow Fe^{2+} + OOH\bullet + H^+$$

The destructive OH• radical can attack unsaturated fatty acid side chains, generating a carbon radical in the process. This lipid radical undergoes subsequent rearrangement to form a conjugated diene which gives rise to free radicals in the presence of oxygen. The latter initiates a chain reaction by attacking additional fatty acid side chains, thereby producing lipid peroxide.

$$\text{Lipid}-\text{H} + \text{OH}\bullet \rightarrow \text{H}_2\text{O} + \text{Lipid}\bullet$$
$$\text{Lipid}\bullet + \text{O}_2 \rightarrow \text{Lipid OO}\bullet$$
$$\text{Lipid}-\text{H} + \text{Lipid OO}\bullet \rightarrow \text{Lipid}-\text{OOH} + \text{Lipid}\bullet$$

On the other hand, $O_2\bullet^-$ can also react with nitric oxide (NO) forming another relatively reactive molecule, peroxynitrite:

$$NO + O_2\bullet^- \rightarrow ONOO-$$

ONOO− is further oxidized or reacts with a hydrogen radical (H•), generating the stable HOONO which rapidly dismutates into OH• and free nitrogen species ($NO_2\bullet$). With a very short half-life of about 10^{-9}s, the highly reactive OH• radical cannot diffuse to large distances and thus causes damage to cellular components very close to its origin. But the less reactive $O_2\bullet^-$ radical can pass from the mitochondrion to the cytosol through anion channels like the voltage-dependent anion channel (VDAC). H_2O_2 on the other hand diffuses freely into the cytosol from the mitochondrial sites of its generation [11]. Thus, the term oxidative stress encompasses the overall reactions between the reactive oxygen and reactive nitrogen species that together produce the deleterious effects.

7.3 History of ROS Research in Cardiac Disease: A Twentieth-Century Perspective

The prevalent therapy for cardiovascular diseases was primarily limited to surgical interventions such as bypass surgeries and percutaneous balloon angioplasties which had been developing since the late 1960s. Incidentally, it was almost during this time that the earliest reports of ROS production in the heart revealed physiological amounts of peroxide generation within suspensions of electron transport particles within beef heart [12]. Pigeon heart mitochondrial suspensions also showed peroxide generation particularly with succinate, without detectable levels of glutathione peroxidase and catalase enzymes [13], whereas catalase was found to be present in the rat heart mitochondrial matrix [14].

Oxygen toxicity in the heart was a concept introduced in the late 1970s when initial evidences showed the efficacy of various molecules like alpha-tocopherol and sodium selenite in mitigating myocardial damage due to reperfusion post-ischemia [15, 16]. An exciting solution to the intriguing problem of hypoxic-reoxygenation damage to the heart was provided by studies showing hypoxia results in a proportional decline of SOD and glutathione peroxidase activity with time in isolated Langendorff perfused rat hearts which continues upon reoxygenation [17–20]. Another interesting study reported that cardiac myocytes from rabbit interventricular septal preparations were more susceptible to damage from hydroxyl radical than from superoxide, as in the former case, mitochondrial swelling and basement membrane blebbing were evident [21]. Interestingly, isolated mitochondria from hypoxic heart tissue showed lesser ROS generation compared to normoxic ones, which was increased only upon increasing the duration and concentration of calcium addition [22].

The perfect time for administration of free radical scavengers to obtain maximum myocardial protection was an intriguing question that had to draw attention. Although animals that received them prior to the ischemic episode showed enhanced myocardial function post-reperfusion [23], it was subsequently demonstrated that scavenging free radicals at the time of reflow decreases reperfusion injury [24, 25]. But the administration of SOD and catalase post-reperfusion did not alter the extent of the infarct size [26]. Finally, considering all these accounts, experimental proof acknowledged that ischemia reduced the intracellular glutathione pool and the initiation of reperfusion caused a burst of ROS production that led to cellular injury by dampening the residual antioxidant activity [27].

With the advent of biophysical tools, the application of electron paramagnetic resonance (EPR) spectroscopy to analyze free radical production following reperfusion of the ischemic rabbit heart was evaluated [28]. This study, along with others [29], reports the generation of ROS during ischemia too. However, direct evidence to this phenomenon surfaced as late as in 1997 when an investigation showed that the ROS which are generated prior to reperfusion during ischemia, cause little cell death until reperfusion and is related to the residual oxygen during ischemia, and this is a major contributor to reperfusion injury [30].

In vitro evidence showed that treatment with xanthine oxidase at pH 7.0 reduced sarcoplasmic reticular calcium uptake and ATPase activity that was in part reverted by glucose. As such, this was a slice of the novel concept that the glucose-insulin-potassium combination acts as a free radical scavenging system during global hypothermic ischemia-reperfusion [31]. Similar models reported an increase in ROS production and lipid peroxidation products and a decrease in SOD activity [32]. Subsequently, cultured murine myocardial cells exhibited loss of sarcolemmal integrity, indicative of sarcolemmal dysfunction due to oxygen-derived free radicals, the species of which were found to be dependent upon pH and concentration of iron salts. Interestingly, a combination of SOD, catalase, dimethyl sulfoxide and an alkaline pH proved to be beneficial to the membrane integrity [33]. Xanthine oxidase blockade by allopurinol also provided beneficial effects to the heart similar to SOD [34].

The first set of reports of the clinical role of ROS in pathological cardiac hypertrophy included the usage of butylated hydroxyanisole, vitamin E and catalase as antioxidants to mitigate the hypertrophic effects of tumor necrosis factor alpha (TNF-α) and angiotensin II (AngII) [35]. Graded hypoxia was reported to cause parallel increase in ROS and contractile dysfunction in cardiomyocytes which could be reverted by normoxic exposure; however, physiological levels of ROS signalling as second messengers were traced to the same mitochondrial origin. The decrease in V_{max} of mitochondrial cytochrome oxidase led to enhanced mitochondrial superoxide production [36]. Free radicals are also believed to take part in the peroxidative attack of the unsaturated fatty acid of the mitochondrial phospholipid cardiolipin, associated with a decline in mitochondrial cytochrome c oxidase activity during ischemia-followed reperfusion insult [37]. Besides this, mechanical strain-induced tenascin-C in cardiac myocytes could also be attenuated using antioxidants, suggesting the role of ROS in left ventricular remodeling [38]. This even extends to the effects of other pro-hypertrophic agents like ouabain which inhibit cardiac myocyte Na^+/K^+-ATPase that have also been shown to be reversed upon ROS inhibition [39].

The perpetual search of newer roles of cardiac ROS is still ongoing and has definitely garnered popular medical interest in antioxidant therapy in various experimental models of cardiac pathologies which range from naturally occurring phytochemicals like resveratrol [40] and fisetin [41] to intracellular signalling intermediaries like the glutathionylated SET and MYND domain containing 2 (SMYD2) protein [42], peroxisome proliferator activated receptor alpha (PPARα) [43], cytochrome oxidase subunit 6B1 (COX6B1) [44], mammalian Ste20-like kinase 1 (Mst1) [45], and GJA1-20k, an isoform of cardiac connexin 43 [46].

7.4 Biological Sources of ROS

There has been a large number of reported sources of ROS, enzymatic and otherwise, that are linked intricately to cardiovascular health and disease. Although the relative contributions of the enzymatic ROS-generating systems are not well known, the three principal cell types in the mammalian heart, *viz.*, cardiomyocytes,

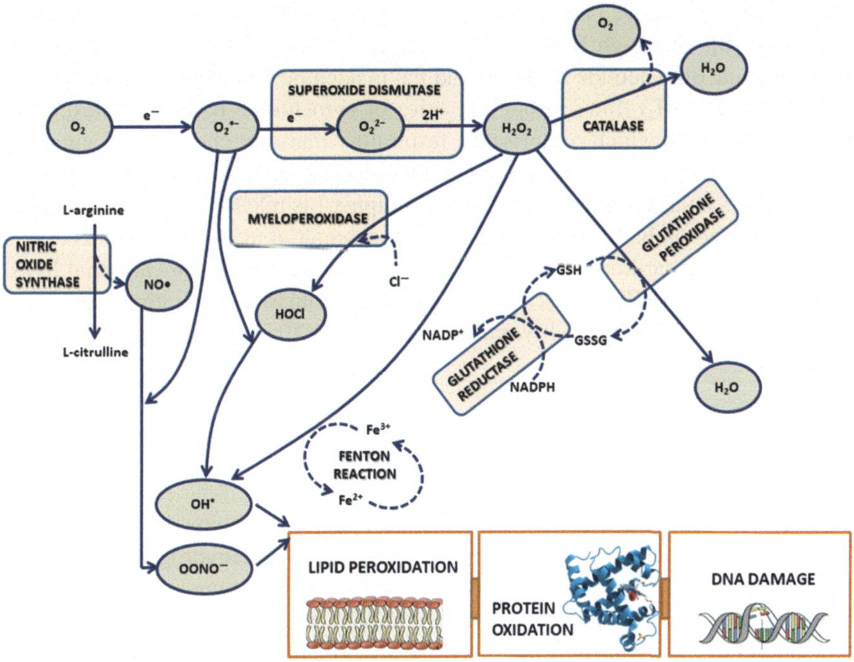

Fig. 7.1 Schematic representation of the different sources of myocardial ROS generation. The combined effects of the reactive forms are primarily inflicted upon cellular DNA, proteins, and lipids. Dashed lines indicate the fate of the reaction by-products

fibroblasts, and endothelial cells are known to possess all these sources. They mostly include mitochondrial respiratory chain, nicotinamide adenine dinucleotide phosphate (NADPH) oxidases, xanthine oxidase, lipoxygenase (LO), uncoupled nitric oxide synthase (NOS), and myeloperoxidase (MPO) (Fig. 7.1). Notably, it needs to be highlighted that several of the myocardial injuries stem from ischemic insults. With the decrease in the oxygen supply in itself, the ROS levels would be presumed to decline since oxygen is the prerequisite to ROS-generating chemical reactions. The most plausible explanation is that at a given level of oxygen availability, superoxide generation from the mitochondria is positively correlated with the factors that enhance the reduction state of their electron transport chain [36]. In other words, the strongly reduced mitochondrial redox state initiates ROS generation even under limiting oxygen supply by promoting electron donation to the residual oxygen.

7.4.1 Mitochondrial Respiratory Chain

Since cardiomyocytes require a huge amount of adenosine triphosphate (ATP) to fuel their relentless contractions, mitochondria comprise almost 35% of their volume in mammalian hearts which are in essence obligate aerobic organs.

Mitochondrial oxidative phosphorylation produces ROS as a by-product of aerobic respiration [8]. The high free energy of electrons derived from reduced nicotinamide adenine dinucleotide (NADH) and flavin adenine dinucleotide (FADH) (i.e., $NADH_2$ and $FADH_2$) flows through a series of cytochrome-based complexes like complexes I and III which comprise the respiratory transport chain to molecular O_2, giving rise to ATP, and in this process exhibit electron leak to finally convert about 1–2% of molecular O_2 to $O_2\bullet^-$. $O_2\bullet^-$ from complex I is released into the mitochondrial matrix and on both sides of the inner mitochondrial membrane from complex III. Isolated adult cardiomyocytes also show a phenomenon of mitochondrial ROS-induced ROS release (RIRR) as a burst of mitochondrial ROS generation is observed upon mitochondrial permeability transition pore (mPTP) induction caused by a trigger of ROS.

7.4.2 Xanthine Oxidase in the Cytosol

Xanthine oxidoreductase is a homodimeric molybdoflavin enzyme that exists in two functionally distinct and interconvertible forms, xanthine dehydrogenase and xanthine oxidase. Both of them are involved in the oxidation of hypoxanthine and xanthine during purine metabolism and catalyze the chemical conversion of hypoxanthine to xanthine and xanthine to uric acid, but only the oxidase form generates $O_2\bullet^-$ and H_2O_2. H_2O_2 is, however, the major oxidant product of xanthine oxidase activity. Under stressful situations, it is the oxidase form that predominates. Xanthine dehydrogenase can readily be converted to the oxidase form either reversibly by formation of disulfide bonds due to oxidation of cysteine residues or by irreversible proteolysis. In addition, it may serve to produce nitric oxide (NO) under hypoxic conditions. Xanthine oxidase thus serves a major contributor of ROS especially in ischemia-reperfusion models [2].

$$\text{Hypoxanthine} + H_2O + O_2 \leftrightharpoons \text{Xanthine} + H_2O_2$$
$$\text{Xanthine} + H_2O + O_2 \leftrightharpoons \text{Uric acid} + H_2O_2$$

7.4.3 NADPH Oxidase (Nox) in the Cell Membrane

The Nox family comprises of transmembrane enzymes mostly, but not always, across the cell membrane, each based on a core catalytic subunit. They utilize NADPH as the electron donor to catalyze the reduction of molecular oxygen to superoxide and hydrogen peroxide. Although specific cell types in the heart express specific predominant Nox isoforms, in general Nox2 and Nox4 are particularly highly expressed in the heart. Although known for their physiological roles, various Nox enzymes generate elevated ROS in the diseased heart [2].

$$\text{NADPH} + 2O_2 \leftrightarrow \text{NADP}^+ + 2O_2\bullet^- + H^+$$

Cardiomyocytes predominantly express Nox2 and Nox4, while Nox5 is expressed in the cardiovascular system of higher animals only. The main stimuli activating Nox2 in cardiomyocytes and endothelial cells seem to be G protein-coupled receptor (GPCR) agonists such as AngII (a known mediator of hypertension and hypertrophy) and endothelin-1, inflammatory cytokines such as TNFα, metabolites such as glucose, insulin (mediator of diabetic cardiomyopathy), oxidized low-density lipoprotein (LDL), glycated proteins, and mechanical forces like stretch and strain. These activate the inactive enzyme complexed with p22phox (phox: phagocyte NAPDH-oxidase) at the cell membrane (collectively called flavocytochrome b558) and the other dormant cytosolic regulatory subunits p47phox, p67phox, p40phox, and Rac1.

Nox4 has been found to be a major source of oxidative stress in the failing heart. It has been implicated in pressure overload cardiac condition as well as in hypertrophic stimuli, promoting apoptosis and mitochondrial dysfunction in cardiomyocytes by enhanced ROS production. It has also been known to mediate hypertrophy by the oxidation of histone deacetylase 4 (HDAC4), causing its nuclear exit. Unlike Nox2, it does not require activation by binding of various proteins as it is constitutively active, generating low levels of ROS in the cell, particularly H_2O_2.

7.4.4 Myeloperoxidase (MPO) in Azurophilic Granules

Abundant in polymorphonuclear neutrophils, MPO is a hemoprotein that catalyzes the formation of hypohalous acids from H_2O_2 and halides (Cl^-, Br^-, and I^-) or pseudohalide (SCN^-). Hypohalous acids act as the potent reactive oxidant that takes part in antimicrobial respiratory burst. Action of MPO is produced in tissues, in this case the heart muscle, during inflammatory infiltration of neutrophils after cardiac damages like ischemia, MI, and reperfusion injury [47]. They are also active in participating in the inflammatory stimulation of culprit atherosclerotic plaques in the vascular system, especially the coronary arteries, due to which they have been proposed as useful risk markers in acute coronary syndromes. Macrophages in atherosclerotic plaques also express MPO which is endogenously regulated by granulocyte macrophage colony-stimulating factor. Endothelial cells either capture soluble MPO released by activated neutrophils into the circulation or through direct cellular contact mediated by β2-integrin. This apart, MPO also contributes to atrial fibrillation and associated fibrosis. However, MPO-derived ROS do not affect myocardial tissue necrosis after MI but adversely affect left ventricular remodeling and function.

7.4.5 eNOS in Membrane Caveolae

A family of calmodulin-dependent homodimeric enzymes called nitric oxide synthases (NOS) is responsible for nitric oxide (NO) production in tissues. Each monomer within the dimeric state consists of an N-terminal oxygenase domain and a

C-terminal reductase domain. NOS in general produce NO upon catalyzing the chemical conversion of L-arginine to L-citrulline. Electrons derived from NADPH are transferred via the flavins [flavin adenine dinucleotide (FAD) and flavin mononucleotide (FMN)] to the haem, present in the N-terminal oxygenase domain of NOS. eNOS, originally identified in vascular endothelium, is also present in a number of other cell types within the heart and elsewhere [2] and is a potent source of $O_2^{\bullet -}$ besides NO. However, the reduction in the availability of its cofactor tetrahydrobiopterin (BH4) and substrates like L-arginine shapes the balance between NO and $O_2^{\bullet -}$ generation to predominate $O_2^{\bullet -}$ release. This condition is known as eNOS uncoupling when electrons are diverted towards molecular oxygen. Consequently, the increased $O_2^{\bullet -}$ rapidly reacts with NO to form peroxynitrite ($ONOO^-$) which further oxidizes BH4 leading to eNOS uncoupling and even more production of $O_2^{\bullet -}$. Therefore, the augmentation of BH4 is of considerable pharmacological interest to reduce the burden of various cardiovascular disorders.

7.4.6 Lipoxygenase (LO) in the Nucleus

Lipoxygenases are nonheme iron-containing dioxygenase that catalyze the oxygenation of polyunsaturated fatty acids such as linoleic acid and arachidonic acid to form corresponding hydroperoxides. ROS are usually produced as a by-product of this reaction [2]. Metabolites of 5-LO further generate ROS by stimulating Nox. Studies expose that 5-LO is a major contributor toward atherosclerotic susceptibility, and 15-LO is enzymatically active in young atherogenetic lesions but not in advanced plaques. 5-LO is abundantly expressed in arterial walls of patients afflicted with atherosclerosis. 5-LO localizes to macrophages, dendritic cells, foam cells, mast cells, and neutrophilic granulocytes, and the number of 5-LO-expressing cells markedly increases in advanced lesions. Genetic disruption of the 12/15-LO gene attenuates atherosclerosis in apolipoprotein E (ApoE)- and LDL receptor-deficient mice. A randomized trial study thus brings hope as it revealed suppression of MI biomarkers upon chemical inhibition of 5-LO-activating protein (FLAP) [48].

7.4.7 Endoplasmic Reticulum (ER) Stress: Protein Disulfide Isomerase (PDI)-ER Oxidoreductase (ERO-1) System

The ER lumen is a highly oxidizing environment which is conducive to proper protein folding. The active site of the chaperone and oxidoreductase PDI contains cysteine residues which accept electrons from substrate polypeptides of nascent proteins, causing reduction of PDI during disulfide bond formation. PDI activity is molecular oxygen-dependent, and this forms the basis of the overwhelming load of misfolded proteins accumulating in the cardiac ER since a large number of cardiac diseases stem from ischemic or hypoxic injury. The FAD-binding protein ER oxidoreductin 1 (ERO-1) then accepts electrons from the reduced PDI and transfers them to molecular O_2, generating H_2O_2 in the process [2]. Under conditions of an increased protein-folding load, the ER becomes particularly vulnerable to oxidative stress since it possesses limited enzymatic antioxidant capacity.

7.4.8 Detection of ROS and Redox Biomarkers in Cardiac Diseases

The central role of oxidative signalling in cardiac pathophysiology places molecular markers of redox as excellent biometric tools for research and clinical application. As a putative integrator of the cellular stress during cardiovascular pathophysiology, downstream redox biomarkers provide a quantitative estimate of disease processes, e.g., neurohormonal activation in heart failure. ROS quantification though performed by robust biochemical techniques becomes much more technically challenging in vivo due to the short half-life of ROS at the sites of generation. The following techniques, among many others, are widely used for the direct detection of ROS.

7.4.9 Fluorometric Detection of ROS Using H_2DCFDA

The nonfluorescent ester dihydrodichlorofluorescein diacetate (H_2-DCFDA) being lipophilic easily penetrates the plasma membrane into the cytosol where it is cleaved rapidly by unspecific esterases, producing the membrane-impermeable nonfluorescent alcohol derivative H_2DCF (2′,7′-dihydrodichlorofluorescein). This compound is then oxidized by mitochondrial/cellular ROS with the formation of highly fluorescent end product 2′,7′-dichlorofluorescein (DCF) [49] (Fig. 7.2). The newly developed carboxylated H_2DCFDA analog (carboxy-H_2DCFDA), carries two negative charges at physiological pH. Following cleavage of the acetate and ester groups by intracellular esterases and oxidation, it forms carboxydichlorofluorescein, with additional negative charges that impede its leakage out of the cell.

7.4.10 Fluorometric Detection of Mitochondrial ROS Using MitoSOX™ Red

MitoSOX™ Red is a fluorogenic dye which is highly selective for the detection of mitochondrial superoxide generation (Fig. 7.2). It is oxidized by superoxide once inside the mitochondria due to which it emits a red fluorescence. The oxidation product gets distinctly fluorescent upon binding to nucleic acids. The oxidation of the probe is highly specific as any other ROS or reactive nitrogen species (RNS) are not able to generate the red fluorescence. This is because the superoxide-mediated oxidation of the indicator dihydroethidium causes its hydroxylation at the 2-position, yielding 2-hydroxyethidium. The latter exhibits a fluorescence excitation peak at ~400 nm that is absent in the excitation spectrum of the ethidium oxidation product generated by ROS other than superoxide. Chemically, the dye comprises of

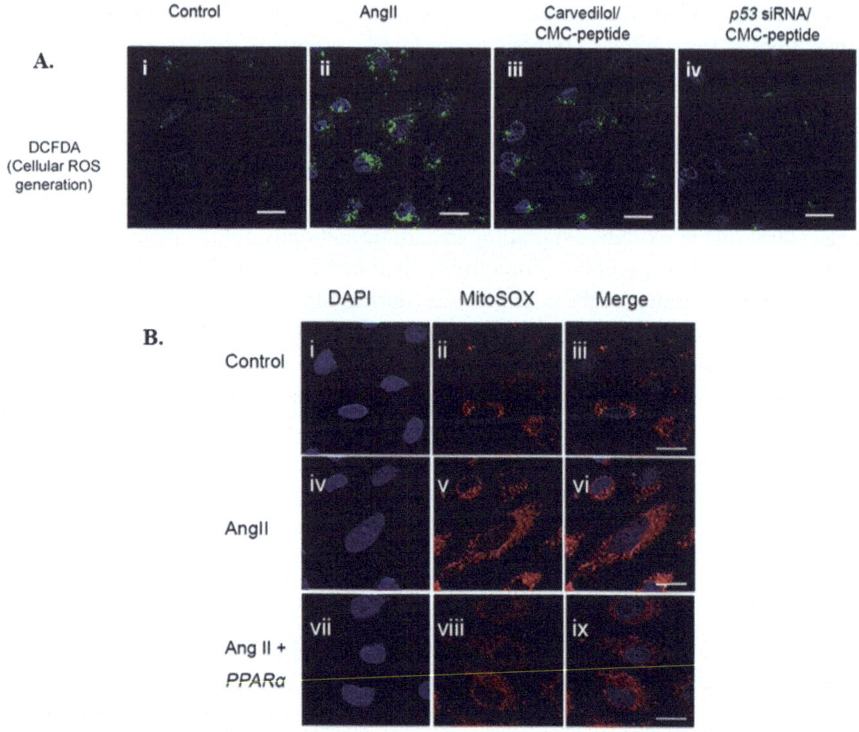

Fig. 7.2 Fluorometric detection of total cellular ROS generation using DCFDA (**A**) and ROS of mitochondrial origin by MitoSOX™ Red (**B**) in cultured neonatal cardiomyocytes. The blue fluorescence indicates DAPI (4′,6-diamidino-2-phenylindole)-stained nuclei.
(**A**) AngII-treated cardiomyocytes (panel ii) show an increase in green fluorescence signifying induced ROS generation compared to untreated myocytes (panel i). Amelioration of oxidative stress in hypertrophied myocytes by cardiomyocyte-targeted nanodelivery of carvedilol (panel iii) or of p53siRNA (panel iv).
(**B**) Reduction in mitochondrial ROS-induced fluorescence detected by MitoSOX™ Red (red) in PPARα overexpressed hypertrophied cardiomyocytes (panels viii, ix) compared to AngII-treated myocytes (panels v, vi).
(Reproduced from Rana et al., *Journal of Controlled Release*, 2015, and Rana et al., *Antioxidants & redox signaling*, 2018)

hydroethidine (HE), the two-electron reduced form of ethidium, which is a commonly used probe for $O_2•^-$, linked by a hexyl carbon chain to a triphenylphosphonium (TPP^+) group. The three lipophilic phenyl groups surrounding the positive charge on the phosphonium of the TPP^+ cations target molecules to mitochondria, thereby facilitating movement across phospholipid bilayers and accumulation within the mitochondrial matrix in response to the negative membrane potential [50].

7.4.11 Spectroscopic Detection of ROS Using EPR Spin Trapping

Electron paramagnetic resonance (EPR) or the electron spin resonance (ESR) spectroscopy is a tool to detect paramagnetic species having an unpaired electron, as in free radicals. In this technique, electron spins are excited, and the transition of the unpaired electrons in an applied magnetic field is detected by measuring the absorption of a quantum of microwave (MW) radiation that matches the energy gap between the two spin states (resonance conditions). However, if the half-life of radicals is too short to detect with EPR, compounds known as spin traps are used which convert them to stable long-lived radicals called spin adducts. Nitrones and nitroso compounds are highly acknowledged spin traps. Frequently used spin traps include alpha-phenyl N-tertiary-butyl nitrone (PBN) and 5,5-dimethyl-pyrroline N-oxide (DMPO) besides C-nitroso spin traps such as 3,5-dibromo-4-nitrosobenzenesulfonic acid (DBNBS). The spin-trapping reaction occurs with the covalent addition of the free radical to the double bond of the diamagnetic spin trap, with the resultant spin adduct having paramagnetic EPR spectrum. The EPR characteristics of spin adducts, like the g-value, hyperfine coupling constant (hfcc), and spin concentration, are obtained from their EPR spectra. This ultimately allows the qualitative and quantitative detection of ROS and RNS [51]. The unavailability of the robust apparatus and instrumentation required for this technique in most biochemical laboratories limit the use of this method.

7.4.12 Colorimetric Measurement of Nitric Oxide (NO) Production

Griess reagent is used for the indirect colorimetric determination of NO by spectrophotometric measurement of its stable decomposition products NO_3^- and NO_2^-. This method can only detect NO_2^-, and thus NO_3^- should first be reduced to NO_2^-. Griess reaction is a two-step diazotization reaction of sulfanilamide by the NO-derived nitrosating agent, dinitrogen trioxide (N_2O_3), which is generated from the acid-catalyzed formation of nitrous acid from nitrite (or autoxidation of NO) to produce a diazonium ion. The latter is then coupled to the naphthyl ring N-(1-naphthyl) ethylenediamine to form a chromophoric azo product that absorbs strongly at 540 nm [52].

The principles of the diazo coupling reaction method in a high-throughput combination with a dedicated high-performance liquid chromatography (HPLC) system developed by Ei Com Corporation, as well as certain other different fluorometric methods, allow sensitive and selective measurement of nitrite and nitrate in all biological matrices with ease and specificity [52]. One such fluorometric method employs aromatic diamino compound 2,3-diaminonaphthalene (DAN) as an indicator of NO formation. The relatively nonfluorescent DAN reacts rapidly with N_2O_3 generated from acidified nitrite (nitrous acid) or from the interaction of NO with oxygen to yield the highly fluorescent product 2,3-naphthotriazole. In addition to the DAN assay, more recent studies demonstrated that diaminofluoroscein-2 (DAF-2) may be used to determine the presence of NO in vitro and in situ.

7.4.13 Chemiluminescent Probes for ROS Detection

ROS detection by chemiluminescent probes rely on the fact that upon exposure to $O_2^{\bullet -}$, these probes release a photon, which, in turn, is detected by a scintillation counter or a luminometer. The most frequently used chemiluminescence technique in this regard is lucigenin-enhanced chemiluminescence. Chemically, lucigenin is bis-N-methylacridinium nitrate whose reactive radical form generated by univalent reduction is involved in the reaction with $O_2^{\bullet -}$ [53]. Therefore, lucigenin (LC^{2+}) is reduced by $O_2^{\bullet -}$ to its cationic radical ($LC^{\bullet +}$), which further reacts with a second $O_2^{\bullet -}$ generating the energy-rich dioxetane molecule (LCO_2) that emits a photon. However, this technique suffers from a limitation related to erroneous overestimation of $O_2^{\bullet -}$ due to the chemistry of the redox cycling phenomenon in which $O_2^{\bullet -}$ is generated as a result of the reaction of the lucigenin radical with oxygen. In contrast to lucigenin, another promising probe for $O_2^{\bullet -}$ is the cypradina luciferin-based molecule 2-Methyl-6-(4-methoxyphenyl)-3,7-dihydroimidazo[1,2-a]pyrazin-3-one, hydrochloride (MCLA) which does not undergo redox cycling. However, it is restricted to the detection of extracellular $O_2^{\bullet -}$ as it is cell impermeable [54].

In addition to the above biochemical techniques for the direct detection and measurement of ROS in biological systems, researchers taking interest in the pathophysiological aspects of ROS-mediated signalling in disease universally tend to look out for stable redox biomarkers that serve as an indirect yardstick of oxidative injury. They either may be detectable from the tissue or cell of interest or can be freely circulating ones. Since excessive ROS cause oxidative modifications in DNA, lipids, and proteins and cause increase in the antioxidant response system, these serve as important prognostive biomarkers of stress in preclinical and clinical studies. One factor that seems to question the potential of these biomarkers is their specificity, since the component of biological ROS content has an integral spatio-temporal variable built into it, in the sense that quite a few free radicals are unstable and are detectable within specific cellular microdomains. Therefore, analysis of other established parameters in assessing disease remains a sine qua non to the use of redox biomarkers.

7.4.14 Oxidized Proteins

ROS oxidize reduced glutathione (GSH), a small tripeptide (γ-L-glutamyl-L-cysteinyl-glycine), in the cell causing the sulfhydryl (thiol) group of its cysteine to get linked to a second GSH through a disulfide bridge. This results in the formation of the dimerized form of the protein, i.e., oxidized GSSG. The enzyme glutathione reductase can restore GSH by modification of GSSG. The increase in the GSSG/GSH ratio thus reflects oxidative stress. Detection can be achieved by an enzyme-recycling assay using the reagent 5,5-dithio-bis-(2-nitrobenzoic acid) (DTNB) which oxidizes GSH, resulting in the formation of GSSG and the chromophore 5-thio-2-nitrobenzoic acid (TNB). GSSG is again reduced to GSH by glutathione reductase using NADPH-derived reducing equivalent. The rate of TNB formation is determined by measuring TNB formation at 412 nm since it is proportional to the sum of GSH and GSSG present in the sample. Another reliable method is based on fluorometric detection post HPLC separation. Iodoacetic acid (IAA), as a thiol-alkylating agent, is used to form S-carboxymethyl derivatives with free thiols and fluorodinitrobenzene that reacts with amines allowing UV absorbance detection at 365 nm. Dansyl chloride derivatization to fluorescently tag amino groups as a modification to this protocol enhances assay sensibility to fluorometric detection.

Other oxidative protein modifications that can serve as redox biomarkers include analysis of nitration and S-glutathionylation of key proteins. RNS-mediated protein tyrosine nitration, like in those of the sarco(endo)plasmic reticulum Ca^{2+} ATPase 2a (SERCA2a) and MnSOD, is detected by immunostaining using antibodies to nitrotyrosine, although the most reliable method is based on tandem mass spectrometric experiments. Antibodies against nitrated proteins are also being developed. On the other hand, S-glutathionylation involves the formation of a stable yet reversible disulfide bridge between the cellular GSH and a reactive cysteine residue within the target protein, like SERCA2a, ryanodine receptor (RyR), eNOS, Na^+-K^+ pump, etc. Apart from highly sensitive mass spectrometric analysis, this modification is studied by Western blotting and enzyme-linked immunosorbance assays (ELISA) with monoclonal anti-glutathione antibody [55].

7.4.15 Lipid Peroxidation

Lipids are susceptible targets of oxidation by several free radicals because of abundant reactive double bonds they possess within their biochemical backbone. Lipid hydroperoxides are usually analyzed by HPLC, often followed by fluorescence detection. The most well-studied markers of lipid peroxidation are the unsaturated aldehydes which are the end products of lipid peroxidation. They include 4-hydroxy-trans-2-nonenal (HNE), malondialdehyde (MDA), isoprostanes, and acrolein. Among these, 4-HNE is the most abundant and hence the most studied product of arachidonic acid, the polyunsaturated fatty acids (PUFA) present in phospholipid cell membranes. HPLC and gas chromatography-mass spectrometric (GC/MS) techniques are suitable to detect these products. HPLC protocols typically make use of chemical aldehyde-reactive probes such as 2,4-dinitrophenylhydrazine and 1,3-cyclohexandione. For detection with negative chemical ionization using GC/MS, derivatizing reagents like pentafluorobenzyl oxime followed by silylation are also required. Newer techniques like liquid chromatography-mass spectrometry (LC/MS) are preferable since the derivatization step to enhance optical activity or to generate a volatile product is not required. Even the measurement of HNE-modified proteins along with their MS identification provides important ROS-related information in disease. Isoprostanes are another family of stable, arachidonic acid-derived prostaglandin-like compounds. They can be measured using GC/MS, LC/MS, ELISA, and radioimmunoassay in plasma and urine samples. Malondialdehyde (MDA) is another product of lipid peroxidation which is typically quantified from plasma samples by a colorimetric assay based on the reaction between MDA and thiobarbituric acid (TBA). This thiobarbituric acid reactive substances (TBARS) assay has been applied as an effective indicator of oxidative injury in cardiovascular disease models. ELISA and HPLC tools are also promising in measuring MDA [55].

However, in addition to the above methods, proteins modified by these lipid peroxidation-derived aldehydes can also be easily detected by Western blotting or immunohistochemistry with antibodies against aldehyde-modified proteins. Several anti-protein MDA and anti-protein HNE antibodies are now commercially available and have been found effective for immunohistochemistry, Western blotting, and ELISA. Polyclonal or monoclonal antibodies against HNE-treated keyhole limpet hemocyanin or bovine serum albumin are used to identify HNE-histidine adducts.

7.4.16 Measuring the Seral Antioxidant Capacity

The activity of circulating antioxidant enzymes such as catalase, glutathione peroxidase, and SOD has been quantified in plasma to measure antioxidant capabilities in the face of oxidative stress. Catalase activity can be measured by the non-spontaneous decomposition of H_2O_2 present in high concentration. On the contrary, peroxidase is known to possess a high affinity for H_2O_2 and can thus remove it even when it is present in low concentration. Measuring SOD activity is based on the principle of inhibition of the formation of nicotinamide adenine dinucleotide, phenazine methosulfate, and amino blue tetrazolium formazan. The commercial availability of these antioxidant enzyme-based assay kits allows them to be evaluated in large-scale high-throughput analysis [55].

7.5 Endogenous Myocardial Antioxidants

Unmitigated ROS produced within physiological ranges as well as their upregulated generation during pathological states can cause considerable damage to proteins, lipids, and DNA within the cell that produces it as well as in a paracrine fashion to adjacent cells. Therefore, an endogenous system of antioxidants plays a significant spatiotemporal role to quench ROS.

7.5.1 Superoxide Dismutases (SOD)

Superoxide dismutases are known to catalyze the conversion of $O_2^{\bullet -}$ to H_2O_2. SOD1 is a cytosolic Cu/ZnSOD, SOD2 is a mitochondrial MnSOD, whereas SOD3 is a membrane-bound Cu/ZnSOD. SOD infusion improves myocardial contraction after ischemia-reperfusion injury. Mice deficient in MnSOD exhibit early mitochondrial DNA damage during atherosclerotic initiation. Extracellular SOD also preserves pressure overload-induced cardiac dysfunction and adverse remodeling [56].

7.5.2 Catalase

Catalase is a tetrameric heme-containing enzyme that catalyzes the decomposition of H_2O_2 into oxygen and water. The enzyme is at first oxidized to a high-valent iron intermediate by reaction with H_2O_2, known as Compound I (Cpd I), which is rapidly reduced back to the resting state by further reacting with H_2O_2 to generate oxygen and a water molecule. Transgenic mice overexpressing catalase in the heart exhibit improved contractile force and reduced infarct size after ischemia-reperfusion insult compared to non-transgenic littermates. Cardiac-specific overexpression of catalase has been reported to prevent adverse myocardial remodeling and transition to overt heart failure [56].

7.5.3 Glutathione Peroxidase

Glutathione peroxidase is an enzyme having selenocysteine at its active site and is believed to be the predominant antioxidant defense system in the heart even under strong oxidative injury. Besides serving as a peroxynitrite reductase, it utilizes glutathione (GSH) to reduce H_2O_2 to water and lipid peroxides to their respective alcohols. An inverse association between glutathione peroxidase activity and cardiovascular events has been found in patients of coronary artery disease. Transgenic mice overexpressing glutathione peroxidase in the heart show resistance to myocardial ischemia-reperfusion injury shown to caused by downregulated cardiomyocyte apoptosis. Conversely, glutathione peroxidase deficiency accelerates cardiac hypertrophy and dysfunction in AngII-dependent hypertension [56].

7.5.4 Peroxiredoxins (Prx)

Peroxiredoxins are a diverse family of thiol-based antioxidant proteins that take part in redox signalling and act as peroxidases to detoxify H_2O_2, aliphatic and aromatic hydroperoxides, and peroxynitrite. The mammalian Prx family comprises six members (Prx 1–6), grouped as typical 2-Cys (Prx 1–4), atypical 2-Cys (Prx 5), and 1-Cys Prxs (Prx 6) classified on the basis of H_2O_2-sensitive catalytic cysteines. Being highly abundant in the heart, these enzymes are posttranslationally modified in response to oxidative stress in cardiomyocytes and undergo complex redox-dependent structural modifications. The human failing myocardium shows a decline in the protein expression of Prx 3–6 isoforms. Although another study reveals an increase in myocardial Prx 3 expression during MI, its cardioprotective effect has not been doubted. Oxidative stress reduces Prx 2 in cardiac myocytes, and overexpressing it prevents apoptotic cell death. While elevated serum levels of Prx 4 has been suggested to be a circulating biomarker in cardiovascular events, oxidation-induced dimerization of cardiac Prx 3 could be a specific potent biomarker of mitochondrial ROS production during ischemia. Prx 3 overexpressing transgenic mice are protected from cardiac failure and left ventricular remodeling after MI [56].

7.5.5 Thioredoxin

Thioredoxin is an oxidoreductase enzyme that catalyzes the reduction of other proteins by cysteine-thiol disulfide exchange. While thioredoxin 1 is cytosolic, thioredoxin 2 is mitochondrial. Thioredoxins have a highly conserved canonical CGPC catalytic motif, the cysteine residues of which they use to break disulfide bonds in oxidized substrate proteins. These two cysteine residues get oxidized to form a disulfide as a catalytic cycle ends. They are then converted back to the reduced state by thioredoxin reductase at the expense of the reduced form of NADP (i.e., NADPH). Thioredoxin reduces the oxidized form of thioredoxin peroxidase, and the reduced thioredoxin peroxidase acts as a ROS scavenger. The significance of the system is evident from studies that cardiac hypertrophy and oxidative stress is promoted upon inhibition of endogenous thioredoxin 1 even in baseline conditions besides pressure overload, when it upregulates the expression of miR-98/let-7. Elevated levels of circulating thioredoxin may also serve as a potent biomarker of ischemic heart disease. Transgenic mice overexpressing thioredoxin 1 in the heart exhibit reduced myocardial infarct size and improved ventricular recovery post-ischemia. Exogenously applied thioredoxin after ischemia-reperfusion protocols also showed improved infarct size, protection against arrhythmia, and antiapoptotic effects [56].

Genes for the abovementioned enzymes are known to possess a *cis*-acting antioxidant response element (ARE) in their promoter regions [56]. Gene expression through the ARE is mediated primarily by Nrf2 (nuclear factor E2-related factor 2). Basal low levels of Nrf2 (half-life of ~20 min) are due to its rapid degradation by proteasomes. Degradation of Nrf2 is triggered by polyubiquitination through the actin-associated Kelch-like ECH-associated protein 1 (Keap1) protein/Cullin3 ubiquitin ligase. Oxidation of Keap1 causes a disruption of this complex and translocation of increased levels of Nrf2 into the nucleus leading to an induction of the endogenous antioxidant defense system.

7.6 Pathophysiology of ROS-Induced Cardiac Diseases

ROS in the cardiac system are mostly studied for the obnoxious role they play in the progression of heart failure. But this preponderance does not preempt the appreciation of their physiological role in normal cellular homeostasis. Prior to the detailing of what damage they cause to the cardiovascular system, it must be emphasized that ROS can have salutary functions too. H_2O_2, among all the ROS, fits every criterion for being a second messenger in signalling pathways, in its properties of enzymatic production and degradation and specificity of thiol oxidation. That small nonlethal amounts of ROS in the heart per se evoke adaptive signalling and create compensatory responses are best described by the term "hormesis" [57].

One of the most distinguishing features of cardiomyocytes is the arrest in the cell cycle which is the primary postnatal change caused by an increase in ROS upon a transition to an oxygen-rich environment just after birth [58]. Also, a recent report has concluded exciting role of ROS in cardiac mechanotransduction. It has been suggested that in healthy cardiomyocytes, a burst of Ca^{2+} sparks—the elementary event associated with the release of free intracellular Ca^{2+} is triggered by ROS produced from Nox2. This Nox2 in turn is activated by physiologic mechanical stretch in myocytes. The ROS which is produced in the sarcolemmal and T-tubule membranes where Nox2 is located sensitizes the RyRs in their vicinity in the sarcoplasmic reticulum. However, this stretch-dependent increase in Ca^{2+} signalling sensitivity is deregulated during disease to trigger arrhythmia [59]. ROS find another important role in facilitating cardiac myocytes to distinguish calcium transients meant for the normal excitation-contraction coupling from those which result in excitation-transcription coupling during stress by suppressing the latter cascade in normal times [60]. In addition, physiological levels of NO and $O_2\bullet^-$ form intracellular peroxynitrite (ONOO–), which together with GSH activates SERCA by reversible S-glutathiolation of its Cys674 residue. In pathological states like atherosclerosis, chronic increase in ROS/RNS irreversibly oxidizes thiols that are critical for physiological function such as Cys674 thereby blocking S-glutathiolation [61]. On a different note, local mitochondrial ROS activate mitochondrial quality control system within the heart [62].

During ischemic preconditioning of the heart, free radicals act as second messengers potentiating tyrosine kinase phosphorylation which results in the activation of p38 mitogen-activated protein kinase (MAPK) and MAP kinase-activated protein kinase 2 (MAPKAPK2) leading to the activation of nuclear factor kappa-light-chain-enhancer of activated B cells (NF-κB) that provides cardioprotection [63]. NF-κB is a common target of multiple signals in the ischemic heart, like ROS, NO, tyrosine kinase, and protein kinase C (PKC), that play a significant role in late phase of ischemic preconditioning [64]. Preconditioning-derived oxygen radicals activate PKC which contribute to reduced infarct size [65]. The sarcoplasmic reticulum Ca^{2+} channel/ryanodine receptor is a possible target of sulfhydryl oxidation associated with the protective effect of ischemic preconditioning [66]. Low levels of ROS derived from either Nox2 or Nox4 protect the heart from ischemia-reperfusion injury by preventing the hypoxia-inducible factor (HIF)-1α inactivation and inhibition of PPARα [67]. In cultured cardiomyocytes, treatment with sublethal doses of 4-hydroxy-2-nonenal (4-HNE), a product of ROS-mediated lipid peroxidation, turns on Nrf2-mediated transcription and GSH biosynthesis which ultimately confers protection against ischemia-reperfusion stress [68].

Having an insight into the pathological effects of augmented ROS in the cardiovascular system without prior appreciation of its positive roles would thus strongly undermine the notion of myocardial ROS being a double-edged sword and a necessary evil. The significance of the adverse signalling set in by the imbalance between ROS and antioxidants (Fig. 7.3) is huge, as established by a large repertoire of existing literature, so much, so that the myocardial ROS content has, of late, been used as an important parameter for distinguishing infarct from non-infarct zones prior to global proteomic analysis in experimental post-MI models [69].

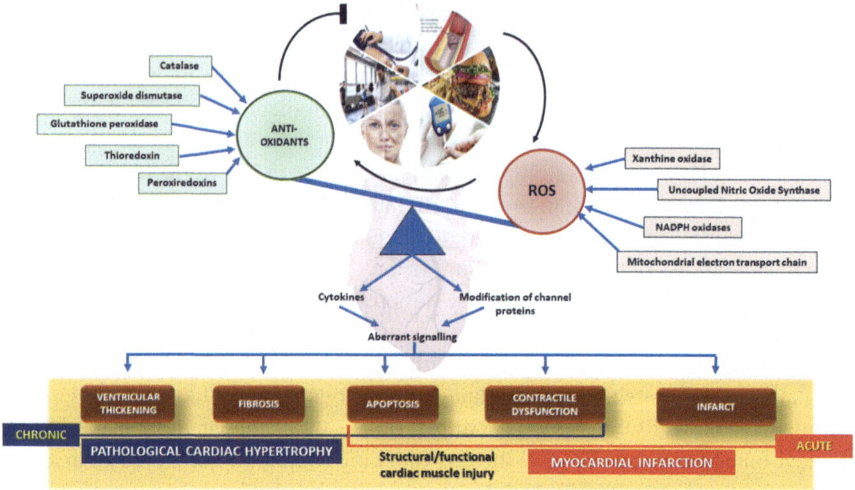

Fig. 7.3 Schematic diagram shows imbalance between the cellular ROS-generating systems and the ROS-scavenging antioxidant defense resulting in oxidative stress in the heart. In turn, aggravated ROS induce insufficient levels of antioxidant enzymes by the Nrf2 pathway. Thus, exogenous antioxidants have a huge potential in reducing the contribution of the deleterious factors that predispose the myocardium to oxidative injury. Acute oxidative stress results in MI, whereas chronic forms induce pathological left ventricular hypertrophy. The pie chart gives a snapshot summary of the factors that aggravate ROS generation in the heart

7.6.1 Cardiomyocyte Apoptosis

An increase in ROS culminates into cardiomyocyte apoptotic loss due to incident pathological stimulus. Apoptosis is a prime mode of cell death during heart failure [4]. Direct treatment of adult cardiac myocytes with ROS turns on the mitochondrial apoptotic pathway [70] or via lamin-A as in case of $O_2^{\bullet-}$ in particular, mostly associated with an increase in p53 protein content [71]. Upregulated p53 in turn leads to enhanced cardiomyocyte apoptosis [72]. Another report shows that treatment of cardiocytes with H_2O_2 activates the lectin-like oxidized low-density lipoprotein receptor-1 (LOX-1) and apoptosis [73]. However, the relative contribution of the various sources of ROS in the heart in etiologically different cardiac disease forms needs to be studied in detail, since a recent report has pointed out that the expression level of xanthine oxidase does not alter during pathological hypertrophy in comparison to control samples, unlike so during MI [7]. Whether the source of ROS generation molds the predominant route to apoptosis during such divergent disease forms therefore warrants further investigation. Kulisz et al. [74] in contrast have put forward evidence of pO_2-dependent mitochondrial ROS generation in hypoxic cardiomyocytes which they opine to contribute to adaptive responses. Inhibiting the Nox as a means of ROS scavenging upon AngII-induced cardiac hypertrophy has been shown to decrease the apoptotic burden [75]. In general, beta-adrenergic signalling leading to myocyte apoptosis is mediated by ROS [76]. However, Amin et al. [77]

also report the ROS dependence of hypertrophic phenotype brought about by alpha-adrenergic signalling due to norepinephrine treatment.

The downstream effectors of ROS-elicited apoptotic response are mostly attributed to the MAPK signalling cascade. While ROS generation due to endothelin and phenylepinephrine treatment to adult cardiac myocytes mediate cardiac hypertrophy via specific activation of the extracellular signal-regulated kinase (ERK) pathway [78], AngII-stimulated ROS leads to apoptosis by p38 activation downstream to Ca^{2+}/calmodulin-dependent protein kinase, even at low levels of Ca^{2+} [75]. Nonetheless, another study reports p38 to act upstream of ROS, and treatment with Vitamin C and E scavenged ROS and downregulated apoptosis and left ventricular function in failing rabbit hearts [79]. On the other hand, ROS from beta-adrenergic stimulation as well as xanthine oxidase activation culminates into mitochondrial and endoplasmic reticular stress-induced c-Jun N-terminal kinase (JNK)-mediated cell death [7, 76]. Treatment of adult rat cardiomyocytes with high concentration of H_2O_2 that induce apoptosis reveals that ERK 1/2 and Akt play a protective role against apoptosis while JNK is proapoptotic [80]. The serine-threonine kinase Akt, also known as protein kinase B (PKB) and acting downstream to the phosphoinositide 3-kinase (PI3K), is thus another effector in the cell fate determination under oxidative injury favoring cardioprotection [81]. Thus, pharmacological strategies to reduce cardiomyocyte apoptosis under oxidative stress target the Akt pathway. For example, the drug atorvastatin activates Akt phosphorylation and represses apoptosis in H_2O_2-treated cardiomyocytes [73], while the Chinese medicine QSKL reduces intracellular ROS, upregulates phosphorylation of PI3K and Akt, and attenuates myocardial apoptosis in in vitro and in vivo models of ischemia-reperfusion injury [82]. High glucose-induced Nox-derived ROS generation and resultant cardiomyocyte apoptosis can be rescued by treatment with resveratrol which specifically increases phosphorylation of 5′ adenosine monophosphate-activated protein kinase (AMPK) [83]. Similarly, overexpression of the trypsin-like serine protease corin protects cardiomyocytes against H_2O_2-induced injury by decreasing apoptosis and activating the PI3K/Akt and the nuclear factor kappa-light-chain-enhancer of activated B cells (NF-κB) signalling pathways and upregulating HIF-1α [84].

Related to cardiomyocyte death on a different note, oxidative stress caused by glucose deprivation in vitro results in autophagic induction. ROS production is simultaneously associated with decreased glutathione levels. Conversely, N-acetylcysteine treatment or overexpression of catalase or SOD disrupts autophagy [85]. Excess autophagy is a characteristic feature of failing hearts and is closely linked to cardiomyocyte death in pathological situations. At least upon ischemia-reperfusion insult, oxidative stress is strongly responsible for the induction of autophagy and subsequent myocardial injury [86]. Upon reperfusion post-ischemia, myocardial autophagosome accumulation is closely associated with increased ROS generation and resultant cardiomyocyte death, mainly due to decrease in lysosome-associated membrane protein-2 (LAMP-2) and Beclin-1 upregulation [87]. However, Beclin-1 is not upregulated during exogenous peroxide treatment to adult rat ventricular myocytes, which show an increase in AMPK activity and a decrease in the mechanistic target of rapamycin (mTOR)-ERK pathway, proteins related to autophagy. These H_2O_2-induced changes could be reverted by treatment with the adipokine adiponectin, underpinning its antioxidant potential [88]. Similarly, during cardiac hypertrophy, an increase in ROS production and the autophagosome formation is attenuated by overexpression of catalase specifically expressed in mitochondria in the heart [89]. The fine detailing of what ROS perpetrates on the heart can more be appreciated on comparing the two reports that Nox4 in the mitochondria promotes cardiomyocyte apoptosis during hypertrophy [90] but upon glucose deprivation, endoplasmic reticular Nox4 promotes cardiomyocyte autophagy that is related to enhanced survival [91]. However, in the contex of autophagy-related AMPK activity, it is interesting to note that besides being cardioprotective during hypertrophy, it shapes the pattern of hypertrophic response. In response to transverse aortic constriction (TAC) surgery, wild-type mice develop concentric hypertrophy, whereas AMPK-kinase dead mice develop eccentric hypertrophy and show enhanced ROS generation compared to the former group [92].

7.6.2 Cardiac Fibrosis

The failing myocardium is characterized by fibroblast proliferation and excess deposition of extracellular matrix proteins in the cardiac interstitium, leading to stiffness and loss of functional architecture of the ventricles. Fibroblasts respond both to direct stimulation by oxidative stressors and to cytokine signalling, viz., interleukin 6 (IL-6), TNF-α, and transforming growth factor beta (TGF-β) [6, 93], apart from paracrine and exosomal contribution from cardiomyocytes [6, 94]. The membrane-associated Nox complex is the predominant source of ROS in cardiac fibroblasts subjected to AngII treatment which leads to concomitant upregulation of collagen I and collagen III expression [95].

The outcome of excessive ROS generation on cardiac fibrosis remains a debatable issue. That ROS-induced oxidative stress leads to a decrease in collagen synthesis accompanied by enhanced activity of the matrix metalloproteinases (MMPs), MMP-2, MMP-9, and MMP-13 [96], is contradicted by a series of contrasting reports. Induction of cardiac fibroblasts by superoxide leads to their proliferation along with upregulated TGF-β gene expression [97], which has reportedly been a significant pro-fibrogenic cytokine that can act independent of AngII stimulation as well [98, 99].

On similar lines are other studies that indicate ROS as a fibrotic mediator in the stressed heart. AngII treatment to cardiac fibroblasts causes a JNK- and ERK-dependent upregulation of osteopontin expression which plays a prime role in post-MI remodeling by promoting collagen synthesis and accumulation [100]. Overexpression of mitochondria-targeted catalase, an antioxidant, reduces AngII-induced ventricular fibrosis [89]. Besides, the xanthine oxidase inhibitor allopurinol reduces ROS along with marked reduction in interstitial cardiac fibrosis after MI [101]. Rats orally fed with the phytomedicine curcumin exhibit attenuated oxidative stress, cardiac fibrosis, and ventricular dysfunction after reperfusion [102], and nanotized forms of the same compound increase its therapeutic efficacy within the diseased heart with improved bioavailability and retention [103].

Interestingly, the activation of fibroblasts in this context has been attributed to a seemingly increasing list of identified sources. A recent study has appreciated the role of Nox4 in AngII-mediated cardiac fibroblast proliferation and migration via AngII receptor type 1 (AT1)/Nox4 physical association. These are regulated by IL-18 and MMP-9 activities which are further dependent upon redox-sensitive modulation of p65 and c-Jun [104]. The increase in the protein expression of the intermediate-conductance Ca^{2+}-activated K^+ (KCa3.1) channels points toward an important role in the oxidative stress-induced proliferation and inflammatory reaction during the progression of cardiac fibrosis in hypertensive mice; this could, however, be reverted upon inhibition of ERK activity [105]. AngII infusion to cardiac-specific Nox4 transgenic mice worsened cardiac injury by robust elevation of Nox4 compared to AngII-infused control mice and upregulated fibrosis through a possible Akt-mTOR-NF-κB pathway [106]. Depletion of Nox4, but not Nox5, reduced TGF-β1-induced Smad 2/3 (homolog of SMA of *C. elegans* and of *Drosophila* protein mothers against decapentaplegic) phosphorylation and consequent activation of interstitial fibrosis [107]. In addition, Nox2 has been shown to be another significant mediator of profibrotic signalling after pressure overload as well as after MI by using murine models of Nox2-/- strains [108–110]. A recent study has projected the contribution of AngII-induced endothelial Nox2 activation to the profibrotic effect on the heart as an integration of phenomena like enhanced endothelial-to-mesenchymal transition and inflammation [111]. Yet another report has focused on the uncoupling of NOS3 as the prominent cause of myocardial ROS generation due to pressure overload that leads to maladaptive remodeling including fibrosis [112].

7.6.3 Contractile Dysfunction

The failing heart is characterized by contractile dysfunction primarily due to the depressed sarcomeric excitation-contraction coupling as a result of a deregulated calcium handling apparatus. For cardiomyocytes to activate contraction upon arrival of action potential, the intracellular Ca^{2+} concentration should rise from the resting 100 nM to 10 μM. The majority of this calcium is intracellularly released via the Ca^{2+}-sensitive RyRs in the sarcoplasmic reticulum. This is triggered by the external Ca^{2+} entering into the cell via sarcolemmal L-type Ca^{2+} channel (LTCC) [113]. ROS appear to function early during the development of cardiomyocyte contractile dysfunction upon exogenous H_2O_2 stimulation or hypoxia [36]. Importantly, cyclical mechanical stretch in cardiomyocytes causes hypertrophy and apoptosis due to amplitude-dependent enhanced ROS production [114]. Calcium channel currents are inhibited by oxidizing agents and hypoxia due to redox modifications on distinct cysteine residues on the α_{1C} subunit of recombinant human LTCC [115]. A reduction in cardiac LTCC activity induces cardiac hypertrophy and heart failure [116]. On the contrary, another study reports that H_2O_2-induced oxidation-dependent activation of Ca^{2+}/calmodulin (CaM)-dependent kinase II (CaMKII) facilitates rat ventricular LTCC and causes an increase in L-type Ca^{2+} current amplitude and slowed its inactivation [117]. H_2O_2 induces oxidation of methionine residues in the regulatory domain which sustains CaMKII activity even in the absence of Ca^{2+}/CaM, ultimately leading to cardiomyocyte apoptosis on AngII stimulation [118]. Oxidizing agents accelerate sarcoplasmic reticulum Ca^{2+} leak and decreases sarcoplasmic reticular Ca^{2+} content in normal hearts. ROS cause abnormal oxidative modification of key cysteine residues of RyR2 that enhances its activity and elevates Ca^{2+} leak from the sarcoplasmic reticulum during chronic heart failure [119]. Cardiac myocytes from post-MI dogs having ventricular flutter show increased ROS production and RyR oxidation [120]. Thus, the RyR open probability is favored by oxidizing conditions [121].

The increased cytosolic Ca^{2+} level during cardiomyocyte contraction has to return back to low levels during relaxation. The SERCA pump sequesters back Ca^{2+} into the sarcoplasmic reticulum followed by trans-sarcolemmal Ca^{2+} removal via the Na^+/Ca^{2+} exchanger (NCX) [113]. Hypoxic guinea pig ventricular myocytes have a significantly inhibited NCX activity which upon reoxygenation is reactivated at a time when ROS increase. Exogenous H_2O_2 application during the hypoxic protocol rapidly reactivates NCX, signifying that elevated ROS during reoxygenation rapidly reactivate NCX [122]. NADH/NAD^+ being an important redox couple, increasing cytosolic NADH inhibits NCX by accumulating ROS in adult cardiomyocytes [123]. On the other hand, hydroxyl radical inhibits SERCA function by directly attacking its ATP-binding site, and ATP binding to its active site prevents the loss of SERCA activity. This indicates that the ATP depletion during ischemia enhances the free radical-induced inhibition of SERCA activity on reperfusion [124]. In isolated perfused rat hearts, endothelin-1 treatment enhances ROS generation which partially is responsible for increase in contractility [125]. Following a challenge with hydroxyl radical, the maximum positive inotropic response to Ca^{2+} is significantly decreased due to inhibition of SERCA. Carvedilol could partly prevent such contractile dysfunction in atrial myocardial preparations [126]. Carvedilol treatment has also been shown to cause increase in SERCA expression accompanied by downregulation of ROS during cardiac hypertrophy [127]. Another compound called apocynin attenuates cardiac contractile dysfunction due to ischemia-reperfusion by downregulating Nox-derived ROS [128]. H_2O_2-mediated oxidation of SERCA at Cys674 also decreases its activity in adult ventricular myocytes. Similar oxidative modification downregulates SERCA activity and impairs myocyte relaxation in aged hearts [129]. The contribution of such posttranslational modification of SERCA toward impaired contractility is confronted since nitric oxide-derived nitroxyl in cardiomyocytes activates SERCA by its S-glutathiolation at Cys674 [130].

In addition to these changes that directly affect the function of the Ca^{2+}-handling apparatus, ROS also cause oxidative modifications on key sarcomeric proteins that lead to contractile dysfunction in the heart. The spectrum of affected proteins in this category is broad, ranging from high to low molecular weight ones. For example, oxidative stress modifies large proteins like titin and desmin that decreases the extensibility and increases passive resistance in titin and forms insoluble desmin aggregates which disrupt the sarcomeric lattice. Redox modifications of cysteine in myosin heavy chain (MHC) dampen myosin ATPase activity. Cysteine oxidation within actin alters actin filament sliding velocity and actomyosin ATPase activity. Redox modification of cardiac tropomyosin also alters its flexibility. Actin-myosin cross-bridge formation is thus hampered [131].

7.6.4 Inflammation

Activation of the immune system is a classic feature of failing hearts that leads to an increase in the production of pro-inflammatory cytokines, the most important ones within the cardiac perspective being TNF-α, IL-1, and IL-6 [132]. Interestingly, the regulation of the plethora of inflammatory genes involves the active participation of the transcription factor NF-κB which is a hallmark of the failing heart [133]. The "cytokine hypothesis" holds that a portfolio of biologically active cytokines is responsible for heart failure progression [134]. All the resident nucleated cells within the myocardium, including the cardiac myocyte, are potent sources of these inflammatory mediators [135]. However, it has been suggested that cardiac fibroblasts are more ideally suited than cardiomyocytes to initiate the inflammatory response to ischemia owing to their abundance, strategic interstitial location, and lesser susceptibility to oxidative stress-mediated death, as by H_2O_2 [136]. ROS production mediates the activation of the inflammasome by cardiac fibroblasts [137].

TNF-α has been found responsible for adverse myocardial remodeling, left ventricular dysfunction, and cardiomyocyte apoptosis [138]. H_2O_2 induces myocardial TNF-α production by a mechanism involving p38 MAPK-dependent cardiomyocyte death and myocardial dysfunction [139]. Conversely, TNF-α increases ROS production in cultured cardiomyocytes which result in mitochondrial DNA damage and dysfunction [140] in addition to hypertrophy [35]. This ROS production and resultant cardiomyocyte death occur due to caspase-8 activation [141]. Logically, TNF-α inhibition in experimental heart failure models ameliorates oxidative stress and mitochondrial dysfunction in the heart [142]. Furthermore, pro-inflammatory cytokines like TNF-α, IL-1β, and IFNγ enhance the production of superoxide and NO in the heart resulting in the production of peroxide which causes contractile failure [143]. Adverse cardiac remodeling due to a TNF-α trigger is caused due to faster and stronger MMP expression and activation in cardiomyocytes in comparison to fibroblasts, which is dependent upon superoxide generation and activation of PI3Kγ [144].

IL-1β induction has been shown in murine models of MI, and serum levels of IL-1β are elevated early in MI. It exerts pro-hypertrophic and proapoptotic effects in cardiomyocytes while depressing their contractility [145]. Treatment of neonatal rat ventricular myocytes with pathological concentration of IL-1β for 24 hours caused a significant rise in ROS and reduced the density of L-type Ca^{2+} current, which could be reverted by antioxidant treatment. Such decline in L-type Ca^{2+}

current density was traced to increased membrane translocation of PKCε downstream to IL-1β-mediated ROS activation [146].

IL-10 is one important anti-inflammatory cytokine whose mRNA levels have been detected in the failing heart. It is known to repress the levels of TNF-α, IL-1, and IL-6 [132]. Particularly, a reduction in IL-10 and in the IL-10/TNF-α ratio correlates well with a depressed cardiac function subsequent to MI [147]. IL-10 treatment to cardiac myocytes reversed all the effects of TNF-α treatment on cellular redox state including downregulating the elevated ROS levels caused by TNF-α [148]. In this phenomenon, p38 MAPK and ERK 1/2 play an interactive role in modulating oxidative stress and cardiomyocyte apoptosis [149].

The intrinsic stress response system to myocardial tissue injury is shaped up by the innate immune response system of the heart. While the pro-inflammatory cytokines serve as downstream molecular effectors, a family of receptors called Toll-like receptors (TLRs) and NOD-like receptors (NLRs) act as upstream molecular components of the system. These are, in essence, membrane-spanning and cytoplasmic pattern recognition receptors (PRRs), respectively. The heart expresses TLR-2 and TLR-4 in abundance, the latter being upregulated in failing human hearts [150]. In cardiac myocytes, TLR-2 takes part in the H_2O_2-induced activation of NF-κB and activator protein 1 (AP-1). Such oxidative stress-induced cardiomyocyte apoptosis was increased upon blockade of TLR-2 [151]. Furthermore, TLR-4 null mice show all signs of cardioprotection from doxorubicin-induced cardiomyopathy compared to wild-type mice, including a reduction in oxidative and inflammatory stress, cardiac apoptosis, and an improved cardiac function, which indicates a negative role of TLR-4 in cardiac remodeling [152]. Hyperglycemia upregulates TLR-4 in the heart, and its knockdown causes a decline in Nox activation, ROS production, and apoptosis, suggesting a role of TLR-4 in the diabetic heart [153]. Similar suppression of the myocardial ROS/TLR-4 axis in different models of diabetes using selenium [154] and the alkaloid matrine [155] has been reported.

Nucleotide-binding oligomerization domain-like receptor with a pyrin-domain (NLRP)3 activation mediates the release of IL-1β via the activation of the inflammasome. Following acute MI, inflammasome formation in the heart promotes loss of functional myocardium [156], cardiac fibroblasts being the active contributor to the process [157]. Therefore, the NLRP3 inflammasome mediates ischemia-reperfusion injury [158]. Although a recent study contradicts this conclusion as it reports that NLRP3 inflammasome activation in the heart is cardioprotective in ischemia-reperfusion [159], other compelling evidences show the opposite. For example, drugs like pirfenidone attenuate ROS levels and TAC-induced hypertrophy-associated fibrosis and inflammation by suppressing NLRP3 inflammasome formation [160]. The diabetic heart also exhibits oxidative stress-related NLRP3 inflammasome activity as contributors toward cardiac dysfunction [161]. A recent study also highlights that mitochondrial NLRP3 potentiates ROS to augment R-Smad activation during TGF-β stimulation to result in inflammasome-independent fibrosis [162].

7.7 Factors Modulating ROS-Induced Cardiac Pathologies

A large number of pathological stimuli to the heart exert cardiovascular injuries through the prodigious generation of ROS, yet their effects are reinforced by certain other factors like those discussed below, among others, which modify the effective functional phenotype of various heart failure models across population and in the laboratory (Fig. 7.3).

7.7.1 Age

The development of left ventricular hypertrophy, heart failure, and atrial fibrillation increases rapidly with age, underpinned by structural and functional alterations in the heart. The Framingham Heart Study and the Baltimore Longitudinal Study on Aging (BLSA) have shown this to hold for even healthy individuals [163]. With this is intertwined the concept of the deleterious side attacks of free radicals on cellular constituents as a contributor to aging [3]. This theory was later modified to specifically point out that mitochondria are the prime producers as well as the prime targets of the damage inflicted by ROS [164]. The increase in mitochondrial ROS generation from the aged heart was thereby reported by a large number of studies [165]. Mitochondria from aged hearts as a result of elevated ROS in mice display an increase in mitochondrial protein carbonylation and mtDNA mutation which culminate into upregulated signals for mitochondrial biogenesis via the peroxisome proliferator-activated receptor gamma coactivator-1 alpha (PGC-1α) pathway. These pathologies could successfully be reverted in mice overexpressing mitochondrial catalase [166]. In mice, the activity of thioredoxin is decreased but its expression level is increased in the aging heart, and this difference is further amplified after myocardial ischemia-reperfusion injury, indicating toward posttranslational modification of thioredoxin. As a result of increased production of RNS, thioredoxin nitration is increased in the aging heart which inhibits its activity and interaction with apoptosis signal-regulating kinase 1 (ASK1). This leads to upregulation of p38 MAPK activity, apoptosis, and a larger infarct size in postischemic aging hearts [167]. Another study also reports increase in ROS in both human and rats due to aging in MI, but RNS concentrations increased from young to middle age and decreased from middle age to the oldest in the murine model of ischemia-reperfusion [168].

7.7.2 Sex

The incidence of lower rates of cardiovascular diseases in premenopausal females compared to both age-matched males and postmenopausal females has been brought to focus through a number of anecdotal experimental configurations and meta-analyses across patient populations. This instigated research toward estrogen as a cardioprotective hormonal molecule. However, there were no direct evidences of reduced oxidative stress in the heart due to estrogen until it was reported that estrogen treatment to isolated cardiomyocytes subjected to hypoxia-reoxygenation activates p38β and PI3K, resulting in suppression of ROS and consequent p38α activation and apoptosis [169]. Female rats also show lower mitochondrial free radical production and cardiac oxidative stress than their male counterparts [170]. It was also reported that in female hearts, ROS generation is decreased due to increase in phosphorylation of α-ketoglutarate dehydrogenase following ischemia-reperfusion. The ROS-generated aldehyde adducts in such hearts were detoxified by increased activity of aldehyde dehydrogenase-2, which was dependent upon the PI3K pathway [171]. As such, immediate estradiol replacement to aged ovariectomized rats prevented NF-κB activation, cytokine overproduction, and increased ROS-handling capabilities leading to restoration of ventricular performance [172]. Increase in cystathionine-γ-lyase due to estradiol treatment in ovariectomized females diminishes oxidative stress in the heart which can be traced to the increased production of cardioprotective hydrogen sulfide [173]. Despite these reports, the appropriate dosage of estradiol required for reducing oxidative stress and promoting cardiac function was suggested to be lower than the standard pharmacological dose, with the benefit of avoidance of risks associated with estrogen replacement therapy [174]. Interestingly, basal ROS production from cardiac mitochondria is not affected due to ovariectomy but is enhanced only upon stimulation by high Ca^{2+} and antimycin-A, which further declines with estrogen and progesterone supplementation [175]. Thus, estradiol acting via estrogen receptors-α and estrogen receptors-β in cardiomyocytes increases mitochondrial p38β activation during ischemia-reperfusion injury. This in turn activates MnSOD by phosphorylation leading to downregulation of ROS production and consequent myocardial infarct size [176].

7.7.3 Exercise Training

A sedentary lifestyle is a major health concern for heart failure, for which exercise training proves beneficial in such subjects. Training decreases mitochondrial ROS in the heart through specific adaptations in the complex I and particular increase in catalase activity, without any improvement in the Ca^{2+}-induced mitochondrial dysfunction [177]. Free radicals, as such, do not play an important role in physiological hypertrophy, but they accentuate functional cardiac adaptations due to exercise via the PI3K/Akt signalling pathway. Therefore, the exercise regimen must be carefully considered if supplemented with dietary antioxidants to improve cardiac performance [178]. Swimming alleviates isoproterenol-induced ROS production and cardiac fibrosis in an AMPK-dependent manner through specific increases in myocardial antioxidants like SOD1, SOD2, and catalase and decreases in Nox4 protein expression [179]. Exercise preconditioning prior to ischemia-reperfusion provides adaptive signalling by Nox-derived ROS which replenish glutathione by redox-dependent modification in glutathione reductase [180]. Exercise-induced increase in left ventricular SOD and decrease in lipid peroxidation has been reported to improve myocardial tolerance to ischemia-reperfusion injury [181].

Downregulated antioxidant activities of glutathione peroxidase and catalase were observed in heart tissues from exercise-trained normotensive and hypertensive rats [182]. This contradiction particularly fosters attention to the exercise regimen used in the study which included a week of detraining after the exercise period, which as per recent reports is known to cause cardiomyocyte apoptosis and ventricular dysfunction in a PKC isoform switch-dependent fashion [183]. Long-term endurance training in rats improved the overall antioxidant protection system in cardiac homogenates in rats [184]. The duration and intensity of training seem to be an important parameter since short-term training does not yield antioxidant defense [185] whereas an acute bout of heavy exercise increases oxidative stress in the aged myocardium possibly due to limited antioxidant capacity [186]. Sprint-trained rats exhibit antioxidant defense in the heart by upregulation of glutathione peroxidase and glutathione reductase activity, but not SOD [187]. Moderate exercise in female rats, however, showed no lowering of cardiac lipid peroxide levels but a significant increase in antioxidants like SOD, catalase, glutathione peroxidase, and glutathione-S-transferase in heart homogenates compared to sedentary controls [188]. The beneficial effects of exercise-induced improvement in antioxidant activity in the heart, however, decline with age [189]. But upon application of an ischemia-reperfusion insult, exercise increases antioxidant activity and cardioprotection regardless of age [190]. Another dimension to this concept is that lifelong exercise can boost antioxidant activity in aged hearts [191].

7.7.4 High-Fat Diet

Obesity-associated metabolic syndrome stands to be an independent risk factor for cardiometabolic diseases leading to ventricular dysfunction. In this regard, a number of studies have made an attempt to define the role of ROS in such pathologies, mostly by feeding experimental models a high-fat diet (HFD). Mice fed with a HFD show decreased Sirtuin-3 (SIRT3) expression and increased ROS levels in the heart compared to normal diet-fed mice, the consequences of which are aggravated by SIRT3 loss [192]. SIRT3 in turn blocks Foxo3a-dependent antioxidant genes like MnSOD and catalase to increase ROS production in hypertrophied cardiomyocytes [193], and its increased expression protects them from oxidative stress-mediated apoptotic cell death [194]. Cardiac mitochondria from high-fat, high-sucrose-fed mice showed upregulated peroxide generation and mitochondrial dysfunction associated with oxidative posttranslational modification of mitochondrial complex I and II proteins, which was ameliorated in transgenic mice that express catalase in mitochondria [195, 196]. High-fat, high-sucrose-fed mice showed considerable decrease in ATP synthesis. Decreased myocardial ATP availability is a characteristic of advance heart failure [197], and preclinical strategies specifically targeted to cardiomyocytes to mitigate ROS in the diseased myocardium show an increase in ATP content [43]. Proteomic screening of global reversible cysteine oxidation in Western diet-fed mice has identified specific targets that perturb glycolysis, the tricarboxylic acid (TCA) cycle, beta-oxidation, and the mitochondrial electron transport chain

(ETC) that may be traced to enhanced free radical formation, resulting in cardiac dysfunction [198]. It is therefore of logic to derive antioxidant therapies to mitigate the HFD-induced cardiac dysfunction and fibrosis, as have been studied with a wide variety of such beneficial molecules as polyphenols like resveratrol and S17834 [199], catalase [200], folic acid [201], and mitoTempo [202].

7.7.5 Diabetes

Cardiovascular complications are a major cause of mortality in diabetic patients, the number of which has assumed epidemic proportions. Epidemiological data have suggested a positive correlation between diabetes and heart failure [203] characterized by adverse myocardial remodeling, inflammation, impaired excitation-contraction coupling, and ventricular dysfunction. Insulin resistance makes the heart more prone to ROS generation by the mitochondria. This is associated with the upregulation of the proton leak channels uncoupling protein (UCP)2/UCP3 and reduced mitochondrial membrane potential, leading to mitochondrial swelling, dysfunction, and consequent cardiomyocyte apoptosis. This, in turn, is further worsened by redox-sensitive KATP channels [204]. Increased production of ROS is involved in the pathology of diabetic hearts. Elevated cytosolic glucose oxidation due to diabetic blood glucose levels leads to an increase in the concentration of reducing equivalents like NADH, which participate in mitochondrial ROS generation [205]. Besides mitochondrial apoptosis, high glucose also induces necroptosis and inflammation in cardiocytes [206]. The diabetic heart is further worse as it possesses a low antioxidant capacity in comparison to other organs [205]. Therefore, pharmacological intervention to reduce ROS levels has resulted in decreased hyperglycemia-induced injuries in cardiomyocytes [207] and attenuated development of diabetic cardiomyopathy [208]. Hyperglycemia-induced cardiomyocyte apoptosis in rats is attenuated by antioxidant treatment [209]. Catalase [210] and MnSOD [211] have also been shown to provide cardioprotection in murine diabetic models.

7.7.6 Cardiotoxic Drugs

The effectiveness of a number of chemotherapeutic drugs is severely limited by the adversities they cause to the heart. For example, doxorubicin causes an increase in the oxidative injury to the heart as well as in cultured cardiomyocytes by activating the oxidative DNA damage-induced p53-based apoptotic pathway, which can be restored by pitavastatin treatment [212]. Carvedilol has also been shown to have antioxidant properties against doxorubicin-induced cardiomyopathy [213]. A reduction in reduced glutathione content and SOD activity and an increase in lipid peroxidation in the heart are caused by cisplatin treatment in rats, the effects of which are reversed by antioxidant treatment like y acetyl-L-carnitine, DL-α-lipoic acid, and silymarin [214] along with N-acetylcysteine [215]. N-acetylcysteine can completely

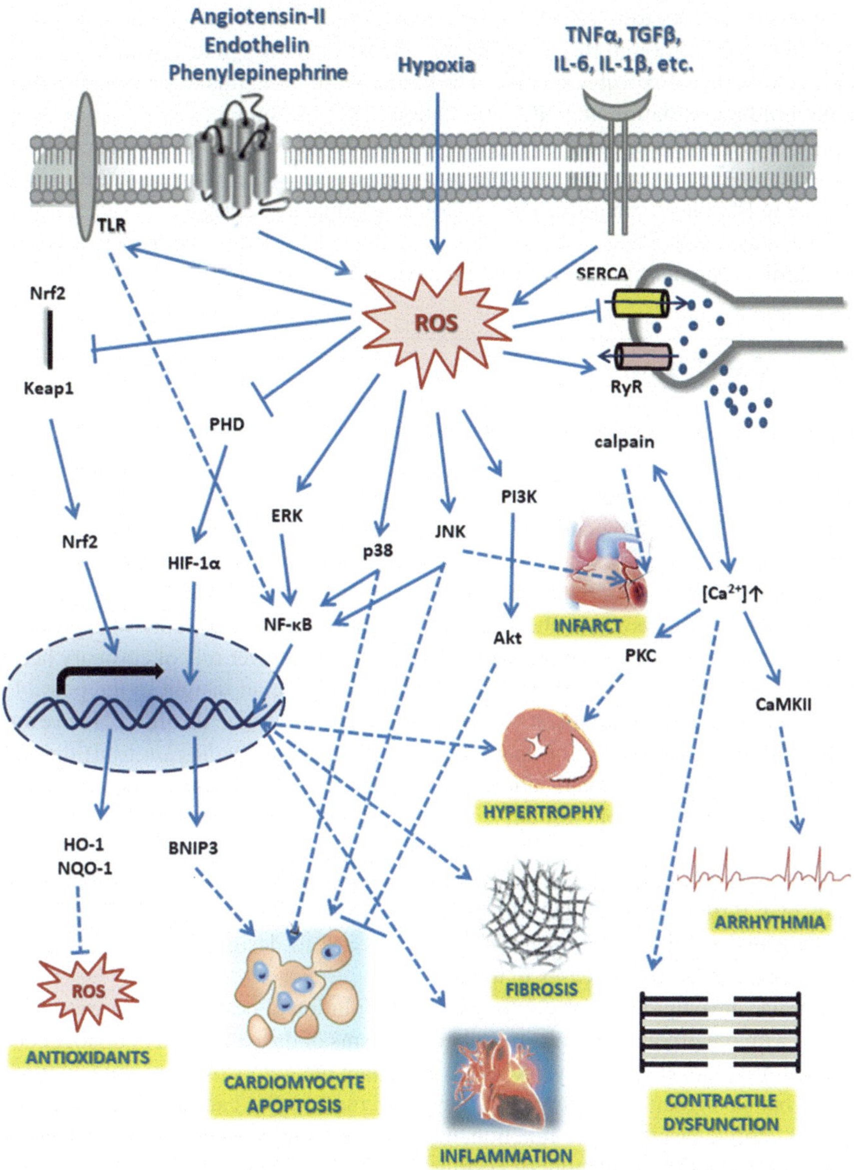

Fig. 7.4 Simplified flow diagram indicating the major signalling pathways that are affected by the redox imbalance in diseased heart. Heart failure encompasses increased cardiomyocyte apoptosis, fibrosis, inflammation, and contractile dysfunction, among others, that ultimately depress the myocardial performance. The dashed lines indicate the final role of the affected molecule in the signalling cascade. PHD, prolyl hydroxylase; NQO1, NAD(P)H quinone dehydrogenase 1; BNIP3, BCL2/adenovirus E1B 19kDa interacting protein 3

antagonize the apoptotic effects of a combination of 5-fluorouracil and levofolene on cardiocytes, although this combination of anticancer drugs results in lesser oxidative stress than doxorubicin [216]. The oxidative stress in this regard yields membrane lipid peroxidation and NO_2^- formation [217]. A complete exception to such drug-induced cardiotoxicity is represented by the action of the anticancer drug paclitaxel in case of myocardial ischemia-reperfusion injury whereby it results in a decline in ROS levels in a JNK-HO-1-dependent manner [218].

The major signalling pathways that are affected by ROS that lead to adverse myocardial remodeling in the failing heart are summarized in Fig. 7.4.

7.8 Conclusion

A long history of research in the redox biology of the heart supports an association between oxidative stress and myocardial dysfunction; accordingly, oxidative stress remains an attractive target for cardiovascular disease prevention and therapy. Reportedly, epidemiological data from Mediterranean populations showed lower cardiovascular mortality when compared with Northern European countries due to significant differences in the intake of antioxidant-rich foods and beverages [219]. In line, a meta-analysis of cohort including almost 400,000 patients showed high intake of vitamin E or vitamin C to be associated with a lower rate of coronary heart disease [220]. Collectively, these considerations provide an insight to the possible role of "traditional" antioxidants in therapy of cardiovascular disorders. However, interventional trials (HOPE, *Heart Outcomes Prevention Evaluation* trial) did not confirm the role of traditional antioxidant therapy in oxidative stress amelioration in clinical setting [221]. Interestingly, administration of statins resulted in an early antioxidant effect by enhancing plasma vitamin E level [222], thereby warranting a deeper understanding of the complex physiology of ROS. Robust epidemiological trials may further establish the role of ROS in progression of the pathophysiology. Nonetheless, significant clinical data are necessary to design treatment strategies against cardiovascular diseases in the future.

Cardiac developmental physiology and the evolution of heart disease are strongly linked with the production of ROS and mechanisms of dysregulation of endogenous oxidant-antioxidants pathways. ROS are important mediators of physiological functions, but in higher concentrations, they may modify the cardiac signalome and regress cardiac mechanics or function, thereby triggering a feedforward mechanism that leads to further worsening of systolic and diastolic function. Therefore, unraveling the molecular mechanisms of ROS accumulation behind disease pathology and progression might be critical in providing suitable targets for exploring innovative therapeutic avenues. Redox biology research in the last few decades have developed a greater understanding of ROS production. Innovative ROS detection strategies and novel targets for attenuation of ROS production have led to discovery of therapeutic strategies for cardiovascular diseases. Pharmacological strategies tested under preclinical setting toward amelioration of ROS production or betterment of ROS-mediated muscle injury have shown promising results [223]. However, better appreciation of new strategies may warrant adequately designed clinical trials, both from conceptual and methodological background.

To sum up, the existence of prodigious empiric data on the cardioprotective potential of ROS scavengers provides reasons enough to believe in the need for urgent advancement in antioxidant pharmacokinetics for healthcare, yet their failure in clinical trials remains a myth to be busted. Whether effectivity in such trials would stem from a combination therapy of antioxidants, or from more selectivity in ROS modulation to avoid scavenging of physiological ROS levels, or from better-controlled targeting of the ROS-generating cell, persists only of speculation until now. Toward this end, as a novel preclinical curative approach from our laboratory, the cardiomyocyte-targeted delivery platform with a stearic acid modified carboxymethyl chitosan conjugated to a 20-mer peptide has shown promising results toward betterment of oxidative stress and associated cardiac pathophysiology, by tinkering various novel molecular targets within cardiomyocytes [43, 94, 103, 127]. Interestingly, the designed nanoconstruct is instrumental in therapy of the diseased myocardium in a spatial scale by reducing bystander effects of the biomaterials; however, a greater degree of precision toward redox therapy of myocardium may be achieved if biomaterials are released from nanoconstruct when the ROS production-to-scavenging balance is dysregulated. Our laboratory has been working toward the controlled release of biomaterials that would be triggered on a spatiotemporal scale, specifically within cardiomyocytes that have an aberrant ROS production. Overall, we look forward to progress the field of cardiac therapeutic research and translate the preclinical results to clinical trials for betterment of human cardiac pathophysiology.

References

1. Genestra M (2007) Oxyl radicals, redox-sensitive signalling cascades and antioxidants. Cell Signal 19(9):1807–1819
2. Di Meo S, Reed TT, Venditti P, Victor VM (2016) Role of ROS and RNS sources in physiological and pathological conditions. Oxidative Med Cell Longev 2016:1–44
3. Harman D (1956) Aging: a theory based on free radical and radiation chemistry. J Gerontol 11(3):298–300
4. Sarkar S, Chawla-Sarkar M, Young D, Nishiyama K, Rayborn ME, Hollyfield JG, Sen S (2004) Myocardial cell death and regeneration during progression of cardiac hypertrophy to heart failure. J Biol Chem 279(50):52630–52642
5. Das B, Young D, Vasanji A, Gupta S, Sarkar S, Sen S (2010) Influence of p53 in the transition of myotrophin-induced cardiac hypertrophy to heart failure. Cardiovasc Res 87(3):524–534
6. Sarkar S, Vellaichamy E, Young D, Sen S (2004) Influence of cytokines and growth factors in ANG II-mediated collagen upregulation by fibroblasts in rats: role of myocytes. Am J Phys Heart Circ Phys 287(1):H107–H117
7. Mitra A, Basak T, Datta K, Naskar S, Sengupta S, Sarkar S (2013) Role of α-crystallin B as a regulatory switch in modulating cardiomyocyte apoptosis by mitochondria or endoplasmic reticulum during cardiac hypertrophy and myocardial infarction. Cell Death Dis 4(4):e582
8. Phaniendra A, Jestadi DB, Periyasamy L (2015) Free radicals: properties, sources, targets, and their implication in various diseases. Indian J Clin Biochem 30(1):11–26

9. Maulik SK, Kumar S (2012) Oxidative stress and cardiac hypertrophy: a review. Toxicol Mech Methods 22(5):359–366
10. Sen CK (1995) Oxidants and antioxidants in exercise. J Appl Physiol 79(3):675–686
11. Marín-García J (2011) Heart mitochondria: a receiver and integrator of signals. In: Signaling in the heart. Springer, Boston, pp 125–151
12. Jensen PK (1966) Antimycin-insensitive oxidation of succinate and reduced nicotinamide-adenine dinucleotide in electron-transport particles I. pH dependency and hydrogen peroxide formation. Biochim Biophys Acta (BBA)-Enzymol Biolo Oxid 122(2):157–166
13. Loschen G, Flohe L (1971) Respiratory chain linked H2O2 production in pigeon heart mitochondria. FEBS Lett 18(2):261–264
14. Nohl H, Hegner D (1978) Evidence for the existence of catalase in the matrix space of rat-heart mitochondria. FEBS Lett 89(1):126–130
15. Guarnieri C, Ferrari R, Visioli O, Caldarera CM, Nayler WG (1978) Effect of α-tocopherol on hypoxic-perfused and reoxygenated rabbit heart muscle. J Mol Cell Cardiol 10(10):893–906
16. Caldarera CM, Davalli P, Guarnieri C (1978) Effect of alpha-tocopherol and sodium selenite on post-anoxic re-oxygenated rat hearts. J Mol Cell Cardiol 10:16–16. 24–28 Oval Rd, London NW1 7DX, England: Academic Press Ltd-Elsevier Science Ltd
17. Guarnieri C, Flamigni F, Rossoni-Caldarera C (1979) Glutathione peroxidase activity and release of glutathione from oxygen-deficient perfused rat heart. Biochem Biophys Res Commun 89(2):678–684
18. Guarnieri C, Flamigni F, Caldarera CM (1980) Role of oxygen in the cellular damage induced by re-oxygenation of hypoxic heart. J Mol Cell Cardiol 12(8):797–808
19. Hearse DJ, Humphrey SM, Chain EB (1973) Abrupt reoxygenation of the anoxic potassium-arrested perfused rat heart: a study of myocardial enzyme release. J Mol Cell Cardiol 5(4):395–407
20. Hearse DJ, Humphrey SM, Nayler WG, Slade A, Border D (1975) Ultrastructural damage associated with reoxygenation of the anoxic myocardium. J Mol Cell Cardiol 7(5):315–324
21. Burton KP, McCord JM, Ghai GEETHA (1984) Myocardial alterations due to free-radical generation. Am J Phys Heart Circ Phys 246(6):H776–H783
22. Paraidathathu T, de Groot H, Kehrer JP (1992) Production of reactive oxygen by mitochondria from normoxic and hypoxic rat heart tissue. Free Radic Biol Med 13(4):289–297
23. Myers ML, Bolli R, Lekich RF, Hartley CJ, Roberts R (1985) Enhancement of recovery of myocardial function by oxygen free-radical scavengers after reversible regional ischemia. Circulation 72(4):915–921
24. Ambrosio G, Becker LC, Hutchins GM, Weisman HF, Weisfeldt ML (1986) Reduction in experimental infarct size by recombinant human superoxide dismutase: insights into the pathophysiology of reperfusion injury. Circulation 74(6):1424–1433
25. Ambrosio G, Weisfeldt ML, Jacobus WE, Flaherty JT (1987) Evidence for a reversible oxygen radical-mediated component of reperfusion injury: reduction by recombinant human superoxide dismutase administered at the time of reflow. Circulation 75(1):282–291
26. Jolly SR, Kane WJ, Bailie MB, Abrams GD, Lucchesi BR (1984) Canine myocardial reperfusion injury. Its reduction by the combined administration of superoxide dismutase and catalase. Circ Res 54(3):277–285
27. Fabiani R, Ceconi C, Curello S, Alfieri O, Visioli O (1993) Myocardial damage during ischaemia and reperfusion. Eur Heart J 14(suppl_G):25–30
28. Zweier JL, Flaherty JT, Weisfeldt ML (1987) Direct measurement of free radical generation following reperfusion of ischemic myocardium. Proc Natl Acad Sci 84(5):1404–1407
29. Arroyo CM, Kramer JH, Leiboff RH, Mergner GW, Dickens BF, Weglicki WB (1987) Spin trapping of oxygen and carbon-centered free radicals in ischemic canine myocardium. Free Radic Biol Med 3(5):313–315
30. Hoek TLV, Li C, Shao Z, Schumacker PT, Becker LB (1997) Significant levels of oxidants are generated by isolated cardiomyocytes during ischemia prior to reperfusion. J Mol Cell Cardiol 29(9):2571–2583

31. Hess ML, Okabe E, Poland J, Warner M, Stewart JR, Greenfield LJ (1983) Glucose, insulin, potassium protection during the course of hypothermic global ischemia and reperfusion: a new proposed mechanism by the scavenging of free radicals. J Cardiovasc Pharmacol 5(1):35–43
32. Rao PS, Mueller HS (1983) Lipid peroxidation and acute myocardial ischemia. In: Myocardial injury. Springer, Boston, pp 347–363
33. Scott JA, Khaw BA, Locke E, Haber E, Homcy C (1985) The role of free radical-mediated processes in oxygen-related damage in cultured murine myocardial cells. Circ Res 56(1):72–77
34. Chambers DE, Parks DA, Patterson G, Roy R, McCord JM, Yoshida S, Parmley LF, Downey JM (1985) Xanthine oxidase as a source of free radical damage in myocardial ischemia. J Mol Cell Cardiol 17(2):145–152
35. Nakamura K, Fushimi K, Kouchi H, Mihara K, Miyazaki M, Ohe T, Namba M (1998) Inhibitory effects of antioxidants on neonatal rat cardiac myocyte hypertrophy induced by tumor necrosis factor-α and angiotensin II. Circulation 98(8):794–799
36. Duranteau J, Chandel NS, Kulisz A, Shao Z, Schumacker PT (1998) Intracellular signaling by reactive oxygen species during hypoxia in cardiomyocytes. J Biol Chem 273(19):11619–11624
37. Paradies G, Petrosillo G, Pistolese M, Di Venosa N, Serena D, Ruggiero FM (1999) Lipid peroxidation and alterations to oxidative metabolism in mitochondria isolated from rat heart subjected to ischemia and reperfusion. Free Radic Biol Med 27(1–2):42–50
38. Yamamoto K, Dang QN, Kennedy SP, Osathanondh R, Kelly RA, Lee RT (1999) Induction of tenascin-C in cardiac myocytes by mechanical deformation role of reactive oxygen species. J Biol Chem 274(31):21840–21846
39. Xie Z, Kometiani P, Liu J, Li J, Shapiro JI, Askari A (1999) Intracellular reactive oxygen species mediate the linkage of Na+/K+-ATPase to hypertrophy and its marker genes in cardiac myocytes. J Biol Chem 274(27):19323–19328
40. Kuno A, Hosoda R, Sebori R, Hayashi T, Sakuragi H, Tanabe M, Horio Y (2018) Resveratrol Ameliorates Mitophagy Disturbance and Improves Cardiac Pathophysiology of Dystrophin-deficient mdx Mice. Sci Rep 8(1):15555
41. Dong B, Liu C, Xue R, Wang Y, Sun Y, Liang Z, Fan W, Jiang J, Zhao J, Su Q, Dai G (2018) Fisetin inhibits cardiac hypertrophy by suppressing oxidative stress. J Nutr Biochem 62:221
42. Godage DNM, VanHecke GC, Samarasinghe KT, Feng HZ, Hiske M, Holcomb J, Yang Z, Jin JP, Chung CS, Ahn YH (2018) SMYD2 glutathionylation contributes to degradation of sarcomeric proteins. Nat Commun 9(1):4341
43. Rana S, Datta R, Chaudhuri RD, Chatterjee E, Chawla-Sarkar M, Sarkar S (2018) Nanotized PPARα overexpression targeted to hypertrophied myocardium improves cardiac function by attenuating the p53-GSK3β-mediated mitochondrial death pathway. Antioxid Redox Signal 30(5):713–732
44. Zhang W, Wang Y, Wan J, Zhang P, Pei F (2018) COX6B1 relieves hypoxia/reoxygenation injury of neonatal rat cardiomyocytes by regulating mitochondrial function. Biotechnol Lett 41:1–10
45. Cheng Z, Zhang M, Hu J, Lin J, Feng X, Wang S, Wang T, Gao E, Wang H, Sun D (2018) Cardiac-specific Mst1 deficiency inhibits ROS-mediated JNK signalling to alleviate Ang II-induced cardiomyocyte apoptosis. J Cell Mol Med 23(1):543–555
46. Basheer WA, Fu Y, Shimura D, Xiao S, Agvanian S, Hernandez DM, Hitzeman TC, Hong T, Shaw RM (2018) Stress response protein GJA1-20k promotes mitochondrial biogenesis, metabolic quiescence, and cardioprotection against ischemia/reperfusion injury. JCI Insight 3(20). https://doi.org/10.1172/jci.insight.121900
47. Mullane KM, Kraemer R, Smith B (1985) Myeloperoxidase activity as a quantitative assessment of neutrophil infiltration into ischemic myocardium. J Pharmacol Methods 14(3):157–167
48. Hakonarson H, Thorvaldsson S, Helgadottir A, Gudbjartsson D, Zink F, Andresdottir M, Manolescu A, Arnar DO, Andersen K, Sigurdsson A, Thorgeirsson G (2005) Effects of a

5-lipoxygenase–activating protein inhibitor on biomarkers associated with risk of myocardial infarction: a randomized trial. JAMA 293(18):2245–2256
49. Karlsson M, Kurz T, Brunk UT, Nilsson SE, Frennesson CI (2010) What does the commonly used DCF test for oxidative stress really show? Biochem J 428(2):183–190
50. Robinson KM, Janes MS, Pehar M, Monette JS, Ross MF, Hagen TM, Murphy MP, Beckman JS (2006) Selective fluorescent imaging of superoxide in vivo using ethidium-based probes. Proc Natl Acad Sci 103(41):15038–15043
51. Suzen S, Gurer-Orhan H, Saso L (2017) Detection of reactive oxygen and nitrogen species by electron paramagnetic resonance (EPR) technique. Molecules, 22(1):181
52. Bryan NS, Grisham MB (2007) Methods to detect nitric oxide and its metabolites in biological samples. Free Radic Biol Med 43(5):645–657
53. Faulkner K, Fridovich I (1993) Luminol and lucigenin as detectors for O2ṡ−. Free Radic Biol Med 15(4):447–451
54. Nauseef WM (2014) Detection of superoxide anion and hydrogen peroxide production by cellular NADPH oxidases. Biochim Biophys Acta (BBA)-Gen Subj 1840(2):757–767
55. Ho E, Galougahi KK, Liu CC, Bhindi R, Figtree GA (2013) Biological markers of oxidative stress: applications to cardiovascular research and practice. Redox Biol 1(1):483–491
56. Brown DI, Griendling KK (2015) Regulation of signal transduction by reactive oxygen species in the cardiovascular system. Circ Res 116(3):531–549
57. Brown DA, Perry JB, Allen ME, Sabbah HN, Stauffer BL, Shaikh SR, Cleland JG, Colucci WS, Butler J, Voors AA, Anker SD (2017) Expert consensus document: mitochondrial function as a therapeutic target in heart failure. Nat Rev Cardiol 14(4):238
58. Puente BN, Kimura W, Muralidhar SA, Moon J, Amatruda JF, Phelps KL, Grinsfelder D, Rothermel BA, Chen R, Garcia JA, Santos CX (2014) The oxygen-rich postnatal environment induces cardiomyocyte cell-cycle arrest through DNA damage response. Cell 157(3):565–579
59. Prosser BL, Ward CW, Lederer WJ (2011) X-ROS signaling: rapid mechano-chemo transduction in heart. Science 333(6048):1440–1445
60. Morad M, Suzuki YJ (2000) Redox regulation of cardiac muscle calcium signaling. Antioxid Redox Signal 2(1):65–71
61. Adachi T, Weisbrod RM, Pimentel DR, Ying J, Sharov VS, Schöneich C, Cohen RA (2004) S-Glutathiolation by peroxynitrite activates SERCA during arterial relaxation by nitric oxide. Nat Med 10(11):1200
62. Song M, Chen Y, Gong G, Murphy E, Rabinovitch PS, Dorn GW (2014) Super-suppression of mitochondrial reactive oxygen species signaling impairs compensatory autophagy in primary mitophagic cardiomyopathy novelty and significance. Circ Res 115(3):348–353
63. Das DK, Maulik N, Sato M, Ray PS (1999) Reactive oxygen species function as second messenger during ischemic preconditioning of heart. Mol Cell Biochem 196(1–2):59–67
64. Xuan YT, Tang XL, Banerjee S, Takano H, Li RC, Han H, Qiu Y, Li JJ, Bolli R (1999) Nuclear factor-κB plays an essential role in the late phase of ischemic preconditioning in conscious rabbits. Circ Res 84(9):1095–1109
65. Baines CP, Goto M, Downey JM (1997) Oxygen radicals released during ischemic preconditioning contribute to cardioprotection in the rabbit myocardium. J Mol Cell Cardiol 29(1):207–216
66. Zucchi R, Yu G, Galbani P, Mariani M, Ronca G, Ronca-Testoni S (1998) Sulfhydryl redox state affects susceptibility to ischemia and sarcoplasmic reticulum Ca2+ release in rat heart: implications for ischemic preconditioning. Circ Res 83(9):908–915
67. Matsushima S, Kuroda J, Ago T, Zhai P, Ikeda Y, Oka S, Fong GH, Tian R, Sadoshima J (2013) Broad suppression of NADPH oxidase activity exacerbates ischemia/reperfusion injury through inadvertent downregulation of HIF-1 and upregulation of PPARα. Circ Res:CIRCRESAHA-112. https://doi.org/10.1161/CIRCRESAHA.111.300171
68. Zhang Y, Sano M, Shinmura K, Tamaki K, Katsumata Y, Matsuhashi T, Morizane S, Ito H, Hishiki T, Endo J, Zhou H (2010) 4-Hydroxy-2-nonenal protects against cardiac ischemia–reperfusion injury via the Nrf2-dependent pathway. J Mol Cell Cardiol 49(4):576–586

69. Datta K, Basak T, Varshney S, Sengupta S, Sarkar S (2017) Quantitative proteomic changes during post myocardial infarction remodeling reveals altered cardiac metabolism and Desmin aggregation in the infarct region. J Proteome 152:283–299
70. Aoki H, Kang PM, Hampe J, Yoshimura K, Noma T, Matsuzaki M, Izumo S (2002) Direct activation of mitochondrial apoptosis machinery by c-Jun N-terminal kinase in adult cardiac myocytes. J Biol Chem 277(12):10244–10250
71. von Harsdorf R, Li PF, Dietz R (1999) Signaling pathways in reactive oxygen species–induced cardiomyocyte apoptosis. Circulation 99(22):2934–2941
72. Chatterjee A, Mir SA, Dutta D, Mitra A, Pathak K, Sarkar S (2011) Analysis of p53 and NF-κB signaling in modulating the cardiomyocyte fate during hypertrophy. J Cell Physiol 226(10):2543–2554
73. Zhang L, Cheng L, Wang Q, Zhou D, Wu Z, Shen L, Zhang L, Zhu J (2015) Atorvastatin protects cardiomyocytes from oxidative stress by inhibiting LOX-1 expression and cardiomyocyte apoptosis. Acta Biochim Biophys Sin 47(3):174–182
74. Kulisz A, Chen N, Chandel NS, Shao Z, Schumacker PT (2002) Mitochondrial ROS initiate phosphorylation of p38 MAP kinase during hypoxia in cardiomyocytes. Am J Phys Lung Cell Mol Phys 282(6):L1324–L1329
75. Palomeque J, Rueda OV, Sapia L, Valverde CA, Salas M, Petroff MV, Mattiazzi A (2009) Angiotensin II–induced oxidative stress resets the Ca2+ dependence of Ca2+–calmodulin protein kinase II and promotes a death pathway conserved across different species. Circ Res 105(12):1204–1212
76. Remondino A, Kwon SH, Communal C, Pimentel DR, Sawyer DB, Singh K, Colucci WS (2003) β-Adrenergic receptor–stimulated apoptosis in cardiac myocytes is mediated by reactive oxygen species/c-Jun NH2-terminal kinase–dependent activation of the mitochondrial pathway. Circ Res 92(2):136–138
77. Amin JK, Xiao L, Pimental DR, Pagano PJ, Singh K, Sawyer DB, Colucci WS (2001) Reactive oxygen species mediate alpha-adrenergic receptor-stimulated hypertrophy in adult rat ventricular myocytes. J Mol Cell Cardiol 33(1):131–139
78. Tanaka K, Honda M, Takabatake T (2001) Redox regulation of MAPK pathways and cardiac hypertrophy in adult rat cardiac myocyte. J Am Coll Cardiol 37(2):676–685
79. Heusch P, Canton M, Aker S, Van De Sand A, Konietzka I, Rassaf T, Menazza S, Brodde OE, Di Lisa F, Heusch G, Schulz R (2010) The contribution of reactive oxygen species and p38 mitogen-activated protein kinase to myofilament oxidation and progression of heart failure in rabbits. Br J Pharmacol 160(6):1408–1416
80. Kwon SH, Pimentel DR, Remondino A, Sawyer DB, Colucci WS (2003) H2O2 regulates cardiac myocyte phenotype via concentration-dependent activation of distinct kinase pathways. J Mol Cell Cardiol 35(6):615–621
81. Matsui T, Rosenzweig A (2005) Convergent signal transduction pathways controlling cardiomyocyte survival and function: the role of PI 3-kinase and Akt. J Mol Cell Cardiol 38(1):63–71
82. Chang H, Li C, Wang Q, Lu L, Zhang Q, Zhang Y, Zhang N, Wang Y, Wang W (2017) QSKL protects against myocardial apoptosis on heart failure via PI3K/Akt-p53 signaling pathway. Sci Rep 7(1):16986
83. Guo S, Yao Q, Ke Z, Chen H, Wu J, Liu C (2015) Resveratrol attenuates high glucose-induced oxidative stress and cardiomyocyte apoptosis through AMPK. Mol Cell Endocrinol 412:85–94
84. Li Y, Xia J, Jiang N, Xian Y, Ju H, Wei Y, Zhang X (2018) Corin protects H 2 O 2-induced apoptosis through PI3K/AKT and NF-κB pathway in cardiomyocytes. Biomed Pharmacother 97:594–599
85. Marambio P, Toro B, Sanhueza C, Troncoso R, Parra V, Verdejo H, García L, Quiroga C, Munafo D, Díaz-Elizondo J, Bravo R (2010) Glucose deprivation causes oxidative stress and stimulates aggresome formation and autophagy in cultured cardiac myocytes. Biochim Biophys Acta (BBA)-Mol Basis Dis 1802(6):509–518

86. Hariharan N, Zhai P, Sadoshima J (2011) Oxidative stress stimulates autophagic flux during ischemia/reperfusion. Antioxid Redox Signal 14(11):2179–2190
87. Ma X, Liu H, Foyil SR, Godar RJ, Weinheimer CJ, Hill JA, Diwan A (2012) Impaired autophagosome clearance contributes to cardiomyocyte death in ischemia-reperfusion injury. Circulation:CIRCULATIONAHA-111. https://doi.org/10.1161/CIRCULATIONAHA.108.769331
88. Essick EE, Wilson RM, Pimentel DR, Shimano M, Baid S, Ouchi N, Sam F (2013) Adiponectin modulates oxidative stress-induced autophagy in cardiomyocytes. PLoS One 8(7):e68697
89. Dai DF, Johnson SC, Villarin JJ, Chin MT, Nieves-Cintrón M, Chen T, Marcinek DJ, Dorn GW, Kang YJ, Prolla TA, Santana LF (2011) Mitochondrial oxidative stress mediates angiotensin II–induced cardiac hypertrophy and Gαq overexpression–induced heart failure. Circ Res:CIRCRESAHA-110. https://doi.org/10.1161/CIRCRESAHA.110.232306
90. Ago T, Kuroda J, Pain J, Fu C, Li H, Sadoshima J (2010) Upregulation of Nox4 by hypertrophic stimuli promotes apoptosis and mitochondrial dysfunction in cardiac myocytes. Circ Res 106(7):1253–1264
91. Sciarretta S, Zhai P, Shao D, Zablocki D, Nagarajan N, Terada LS, Volpe M, Sadoshima J (2013) Activation of Nox4 in the endoplasmic reticulum promotes cardiomyocyte autophagy and survival during energy stress through the PERK/eIF-2α/ATF4 pathway. Circ Res:CIRCRESAHA-113. https://doi.org/10.1161/CIRCRESAHA.113.301787
92. Cates C, Rousselle T, Wang J, Quan N, Wang L, Chen X, Yang L, Rezaie AR, Li J (2018) Activated protein C protects against pressure overload-induced hypertrophy through AMPK signaling. Biochem Biophys Res Commun 495(4):2584–2594
93. Mir SA, Chatterjee A, Mitra A, Pathak K, Mahata SK, Sarkar S (2012) Inhibition of signal transducer and activator of transcription 3 (STAT3) attenuates interleukin-6 (IL-6)-induced collagen synthesis and resultant hypertrophy in rat heart. J Biol Chem 287(4):2666–2677
94. Datta R, Bansal T, Rana S, Datta K, Chaudhuri RD, Chawla-Sarkar M, Sarkar S (2017) Myocyte-derived Hsp90 modulates collagen upregulation via biphasic activation of STAT-3 in fibroblasts during cardiac hypertrophy. Mol Cell Biol 37(6):e00611–e00616
95. Lijnen P, Papparella I, Petrov V, Semplicini A, Fagard R (2006) Angiotensin II-stimulated collagen production in cardiac fibroblasts is mediated by reactive oxygen species. J Hypertens 24(4):757–766
96. Siwik DA, Pagano PJ, Colucci WS (2001) Oxidative stress regulates collagen synthesis and matrix metalloproteinase activity in cardiac fibroblasts. Am J Phys Cell Phys 280(1):C53–C60
97. Li PF, Dietz R, Von Harsdorf R (1999) Superoxide induces apoptosis in cardiomyocytes, but proliferation and expression of transforming growth factor-β1 in cardiac fibroblasts. FEBS Lett 448(2–3):206–210
98. Datta R, Bansal T, Rana S, Datta K, Chattopadhyay S, Chawla-Sarkar M, Sarkar S (2015) Hsp90/Cdc37 assembly modulates TGFβ receptor-II to act as a profibrotic regulator of TGFβ signaling during cardiac hypertrophy. Cell Signal 27(12):2410–2424
99. Bansal T, Chatterjee E, Singh J, Ray A, Kundu B, Thankamani V, Sengupta S, Sarkar S (2017) Arjunolic acid, a peroxisome proliferator-activated receptor alpha agonist regresses cardiac fibrosis by inhibiting non-canonical TGF-β signaling. J Biol Chem:jbc-M117. https://doi.org/10.1074/jbc.M117.788299
100. Xie Z, Singh M, Singh K (2004) ERK1/2 and JNKs, but not p38 kinase, are involved in reactive oxygen species-mediated induction of osteopontin gene expression by angiotensin II and interleukin-1β in adult rat cardiac fibroblasts. J Cell Physiol 198(3):399–407
101. Engberding N, Spiekermann S, Schaefer A, Heineke A, Wiencke A, Müller M, Fuchs M, Hilfiker-Kleiner D, Hornig B, Drexler H, Landmesser U (2004) Allopurinol attenuates left ventricular remodeling and dysfunction after experimental myocardial infarction: a new action for an old drug? Circulation 110(15):2175–2179
102. Wang NP, Wang ZF, Tootle S, Philip T, Zhao ZQ (2012) Curcumin promotes cardiac repair and ameliorates cardiac dysfunction following myocardial infarction. Br J Pharmacol 167(7):1550–1562

103. Ray A, Rana S, Banerjee D, Mitra A, Datta R, Naskar S, Sarkar S (2016) Improved bioavailability of targeted curcumin delivery efficiently regressed cardiac hypertrophy by modulating apoptotic load within cardiac microenvironment. Toxicol Appl Pharmacol 290:54–65
104. Somanna NK, Valente AJ, Krenz M, Fay WP, Delafontaine P, Chandrasekar B (2016) The Nox1/4 dual inhibitor GKT137831 or Nox4 knockdown inhibits angiotensin-II-induced adult mouse cardiac fibroblast proliferation and migration. AT1 physically associates with Nox4. J Cell Physiol 231(5):1130–1141
105. Wang LP, Fan SJ, Li SM, Wang XJ, Gao JL, Yang XH (2017) Oxidative stress promotes myocardial fibrosis by upregulating KCa3. 1 channel expression in AGT-REN double transgenic hypertensive mice. Arch Eur J Physiol 469(9):1061–1071
106. Zhao QD, Viswanadhapalli S, Williams P, Shi Q, Tan C, Yi X, Bhandari B, Abboud HE (2015) NADPH oxidase 4 induces cardiac fibrosis and hypertrophy through activating Akt/mTOR and NFκB signaling pathways. Circulation:CIRCULATIONAHA-114. https://doi.org/10.1161/CIRCULATIONAHA.114.011079
107. Cucoranu I, Clempus R, Dikalova A, Phelan PJ, Ariyan S, Dikalov S, Sorescu D (2005) NAD (P) H oxidase 4 mediates transforming growth factor-β1–induced differentiation of cardiac fibroblasts into myofibroblasts. Circ Res 97(9):900–907
108. Johar S, Cave AC, Narayanapanicker A, Grieve DJ, Shah AM, Johar S, Cave AC, Narayanapanicker A, Grieve DJ, Shah AM (2006) Aldosterone mediates angiotensin II-induced interstitial cardiac fibrosis via a Nox2-containing NADPH oxidase. FASEB J 20(9):1546–1548
109. Grieve DJ, Byrne JA, Siva A, Layland J, Johar S, Cave AC, Shah AM (2006) Involvement of the nicotinamide adenosine dinucleotide phosphate oxidase isoform Nox2 in cardiac contractile dysfunction occurring in response to pressure overload. J Am Coll Cardiol 47(4):817–826
110. Looi YH, Grieve DJ, Siva A, Walker SJ, Anilkumar N, Cave AC, Marber M, Monaghan MJ, Shah AM (2008) Involvement of Nox2 NADPH oxidase in adverse cardiac remodeling after myocardial infarction. Hypertension 51(2):319–325
111. Murdoch CE, Chaubey S, Zeng L, Yu B, Ivetic A, Walker SJ, Vanhoutte D, Heymans S, Grieve DJ, Cave AC, Brewer AC (2014) Endothelial NADPH oxidase-2 promotes interstitial cardiac fibrosis and diastolic dysfunction through proinflammatory effects and endothelial-mesenchymal transition. J Am Coll Cardiol 63(24):2734–2741
112. Takimoto E, Champion HC, Li M, Ren S, Rodriguez ER, Tavazzi B, Lazzarino G, Paolocci N, Gabrielson KL, Wang Y, Kass DA (2005) Oxidant stress from nitric oxide synthase–3 uncoupling stimulates cardiac pathologic remodeling from chronic pressure load. J Clin Invest 115(5):1221–1231
113. Marín-García J (2011) Rapid signaling pathways. In: Signaling in the heart. Springer, Boston, pp 49–68
114. Pimentel DR, Amin JK, Xiao L, Miller T, Viereck J, Oliver-Krasinski J, Baliga R, Wang J, Siwik DA, Singh K, Pagano P (2001) Reactive oxygen species mediate amplitude-dependent hypertrophic and apoptotic responses to mechanical stretch in cardiac myocytes. Circ Res 89(5):453–460
115. Fearon IM, Palmer ACV, Balmforth AJ, Ball SG, Varadi G, Peers C (1999) Modulation of recombinant human cardiac L-type Ca2+ channel α1C subunits by redox agents and hypoxia. J Physiol 514(3):629–637
116. Goonasekera SA, Hammer K, Auger-Messier M, Bodi I, Chen X, Zhang H, Reiken S, Elrod JW, Correll RN, York AJ, Sargent MA (2012) Decreased cardiac L-type Ca 2+ channel activity induces hypertrophy and heart failure in mice. J Clin Invest 122(1):280–290
117. Song YH, Cho H, Ryu SY, Yoon JY, Park SH, Noh CI, Lee SH, Ho WK (2010) L-type Ca2+ channel facilitation mediated by H2O2-induced activation of CaMKII in rat ventricular myocytes. J Mol Cell Cardiol 48(4):773–780
118. Erickson JR, Mei-ling AJ, Guan X, Kutschke W, Yang J, Oddis CV, Bartlett RK, Lowe JS, O'Donnell SE, Aykin-Burns N, Zimmerman MC (2008) A dynamic pathway for calcium-independent activation of CaMKII by methionine oxidation. Cell 133(3):462–474

119. Terentyev D, Gyorke I, Belevych AE, Terentyeva R, Sridhar A, Nishijima Y, Carcache de Blanco E, Khanna S, Sen CK, Cardounel AJ, Carnes CA (2008) Redox modification of ryanodine receptors contributes to sarcoplasmic reticulum Ca2+ leak in chronic heart failure. Circ Res 103(12):1466–1472
120. Belevych AE, Terentyev D, Viatchenko-Karpinski S, Terentyeva R, Sridhar A, Nishijima Y, Wilson LD, Cardounel AJ, Laurita KR, Carnes CA, Billman GE (2009) Redox modification of ryanodine receptors underlies calcium alternans in a canine model of sudden cardiac death. Cardiovasc Res 84(3):387–395
121. Niggli E, Ullrich ND, Gutierrez D, Kyrychenko S, Poláková E, Shirokova N (2013) Posttranslational modifications of cardiac ryanodine receptors: Ca2+ signaling and EC-coupling. Biochim Biophys Acta (BBA)-Mol Cell Res 1833(4):866–875
122. Eigel BN, Gursahani H, Hadley RW (2004) ROS are required for rapid reactivation of Na+/Ca2+ exchanger in hypoxic reoxygenated guinea pig ventricular myocytes. Am J Phys Heart Circ Phys 286(3):H955–H963
123. Liu T, O'Rourke B (2013) Regulation of na+/ca2+ exchanger by pyridine nucleotide redox potential in ventricular myocytes. J Biol Chem:jbc-M113. https://doi.org/10.1074/jbc.M113
124. Xu KY, Zweier JL, Becker LC (1997) Hydroxyl radical inhibits sarcoplasmic reticulum Ca2+-ATPase function by direct attack on the ATP binding site. Circ Res 80(1):76–81
125. Kubin AM, Skoumal R, Tavi P, Kónyi A, Perjés Á, Leskinen H, Ruskoaho H, Szokodi I (2011) Role of reactive oxygen species in the regulation of cardiac contractility. J Mol Cell Cardiol 50(5):884–893
126. Flesch M, Maack C, Cremers B, Baumer AT, Sudkamp M, Bohm M (1999) Effect of β-Blockers on free radical–induced cardiac contractile dysfunction. Circulation 100(4):346–353
127. Rana S, Datta K, Reddy TL, Chatterjee E, Sen P, Pal-Bhadra M, Bhadra U, Pramanik A, Pramanik P, Chawla-Sarkar M, Sarkar S (2015) A spatio-temporal cardiomyocyte targeted vector system for efficient delivery of therapeutic payloads to regress cardiac hypertrophy abating bystander effect. J Control Release 200:167–178
128. Chen Q, Parker WC, Devine I, Ondraskik R, Habtamu T, Bartol K, Casey B, Patel H, Chau W, Kuhn T (2016) Apocynin exerts dose-dependent cardioprotective effects by attenuating reactive oxygen species in ischemia/reperfusion. Cardiovasc Pharmacol 5(2). https://doi.org/10.4176/2329-6607.1000176
129. Qin F, Siwik DA, Lancel S, Zhang J, Kuster GM, Luptak I, Wang L, Tong X, Kang YJ, Cohen RA, Colucci WS (2013) Hydrogen peroxide–mediated SERCA cysteine 674 oxidation contributes to impaired cardiac myocyte relaxation in senescent mouse heart. J Am Heart Assoc 2(4):e000184
130. Lancel S, Zhang J, Evangelista A, Trucillo MP, Tong X, Siwik DA, Cohen RA, Colucci WS (2009) Nitroxyl activates SERCA in cardiac myocytes via glutathiolation of cysteine 674. Circ Res 104(6):720–723
131. Sumandea MP, Steinberg SF (2011) Redox signaling and cardiac sarcomeres. J Biol Chem:jbc-R110. https://doi.org/10.1074/jbc.R110.175489
132. Anker SD, von Haehling S (2004) Inflammatory mediators in chronic heart failure: an overview. Heart 90(4):464–470
133. Gordon JW, Shaw JA, Kirshenbaum LA (2011) Multiple facets of NF-κB in the heart: to be or not to NF-κB. Circ Res 108(9):1122–1132
134. Seta Y, Shan K, Bozkurt B, Oral H, Mann DL (1996) Basic mechanisms in heart failure: the cytokine hypothesis. J Card Fail 2(3):243–249
135. Kapadia S, Lee J, Torre-Amione G, Birdsall HH, Ma TS, Mann DL (1995) Tumor necrosis factor-alpha gene and protein expression in adult feline myocardium after endotoxin administration. J Clin Invest 96(2):1042–1052
136. Zhang X, Azhar G, Nagano K, Wei JY (2001) Differential vulnerability to oxidative stress in rat cardiac myocytes versus fibroblasts. J Am Coll Cardiol 38(7):2055–2062
137. Chen W, Frangogiannis NG (2013) Fibroblasts in post-infarction inflammation and cardiac repair. Biochim Biophys Acta (BBA)-Mol Cell Res 1833(4):945–953

138. von Haehling S, Jankowska EA, Anker SD (2004) Tumour necrosis factor-α and the failing heart. Basic Res Cardiol 99(1):18–28
139. Meldrum DR, Dinarello CA, Cleveland JC Jr, Cain BS, Shames BD, Meng X, Harken AH (1998) Hydrogen peroxide induces tumor necrosis factor α–mediated cardiac injury by a P38 mitogen-activated protein kinase–dependent mechanism. Surgery 124(2):291–297
140. Suematsu N, Tsutsui H, Wen J, Kang D, Ikeuchi M, Ide T, Hayashidani S, Shiomi T, Kubota T, Hamasaki N, Takeshita A (2003) Oxidative stress mediates tumor necrosis factor-α–induced mitochondrial DNA damage and dysfunction in cardiac myocytes. Circulation 107(10):1418–1423
141. Roberge S, Roussel J, Andersson DC, Meli AC, Vidal B, Blandel F, Lanner JT, Le Guennec JY, Katz A, Westerblad H, Lacampagne A (2014) TNF-α-mediated caspase-8 activation induces ROS production and TRPM2 activation in adult ventricular myocytes. Cardiovasc Res 103(1):90–99
142. Moe GW, Marin-Garcia J, Konig A, Goldenthal M, Lu X, Feng Q (2004) In vivo TNF-α inhibition ameliorates cardiac mitochondrial dysfunction, oxidative stress, and apoptosis in experimental heart failure. Am J Phys Heart Circ Phys 287(4):H1813–H1820
143. Ferdinandy P, Danial H, Ambrus I, Rothery RA, Schulz R (2000) Peroxynitrite is a major contributor to cytokine-induced myocardial contractile failure. Circ Res 87(3):241–247
144. Awad AE, Kandalam V, Chakrabarti S, Wang X, Penninger JM, Davidge ST, Oudit GY, Kassiri Z (2009) Tumor necrosis factor induces matrix metalloproteinases in cardiomyocytes and cardiofibroblasts differentially via superoxide production in a PI3Kγ-dependent manner. Am J Phys Cell Phys 298(3):C679–C692
145. Bujak M, Frangogiannis NG (2009) The role of IL-1 in the pathogenesis of heart disease. Arch Immunol Ther Exp 57(3):165–176
146. El Khoury N, Mathieu S, Fiset CE (2014) Interleukin-1β reduces L-type Ca2+ current through protein kinase C epsilon activation in mouse heart. J Biol Chem:jbc-M114. https://doi.org/10.1074/jbc.M114.549642
147. Kaur K, Sharma AK, Singal PK (2006) Significance of changes in TNF-α and IL-10 levels in the progression of heart failure subsequent to myocardial infarction. Am J Phys Heart Circ Phys 291(1):H106–H113
148. Kaur K, Sharma AK, Dhingra S, Singal PK (2006) Interplay of TNF-α and IL-10 in regulating oxidative stress in isolated adult cardiac myocytes. J Mol Cell Cardiol 41(6):1023–1030
149. Dhingra S, Sharma AK, Singla DK, Singal PK (2007) p38 and ERK1/2 MAPKs mediate the interplay of TNF-α and IL-10 in regulating oxidative stress and cardiac myocyte apoptosis. Am J Phys Heart Circ Phys 293(6):H3524–H3531
150. Mann DL (2011) The emerging role of innate immunity in the heart and vascular system: for whom the cell tolls. Circ Res 108(9):1133–1145
151. Frantz S, Kelly RA, Bourcier T (2001) Role of TLR-2 in the activation of nuclear factor κB by oxidative stress in cardiac myocytes. J Biol Chem 276(7):5197–5203
152. Riad A, Bien S, Gratz M, Escher F, Heimesaat MM, Bereswill S, Krieg T, Felix SB, Schultheiss HP, Kroemer HK, Tschöpe C (2008) Toll-like receptor-4 deficiency attenuates doxorubicin-induced cardiomyopathy in mice. Eur J Heart Fail 10(3):233–243
153. Zhang Y, Peng T, Zhu H, Zheng X, Zhang X, Jiang N, Cheng X, Lai X, Shunnar A, Singh M, Riordan N (2010) Prevention of hyperglycemia-induced myocardial apoptosis by gene silencing of Toll-like receptor-4. J Transl Med 8(1):133
154. Liu ZW, Zhu HT, Chen KL, Qiu C, Tang KF, Niu XL (2013) Selenium attenuates high glucose-induced ROS/TLR-4 involved apoptosis of rat cardiomyocyte. Biol Trace Elem Res 156(1–3):262–270
155. Liu ZW, Wang JK, Qiu C, Guan GC, Liu XH, Li SJ, Deng ZR (2015) Matrine pretreatment improves cardiac function in rats with diabetic cardiomyopathy via suppressing ROS/TLR-4 signaling pathway. Acta Pharmacol Sin 36(3):323
156. Mezzaroma E, Toldo S, Farkas D, Seropian IM, Van Tassell BW, Salloum FN, Kannan HR, Menna AC, Voelkel NF, Abbate A (2011) The inflammasome promotes adverse cardiac

remodeling following acute myocardial infarction in the mouse. Proc Natl Acad Sci 108(49):19725–19730
157. Kawaguchi M, Takahashi M, Hata T, Kashima Y, Usui F, Morimoto H, Izawa A, Takahashi Y, Masumoto J, Koyama J, Hongo M (2011) Inflammasome activation of cardiac fibroblasts is essential for myocardial ischemia/reperfusion injury. Circulation:CIRCULATIONAHA-110. https://doi.org/10.1161/CIRCULATIONAHA.110.982777
158. Sandanger Ø, Ranheim T, Vinge LE, Bliksøen M, Alfsnes K, Finsen AV, Dahl CP, Askevold ET, Florholmen G, Christensen G, Fitzgerald KA (2013) The NLRP3 inflammasome is up-regulated in cardiac fibroblasts and mediates myocardial ischaemia–reperfusion injury. Cardiovasc Res 99(1):164–174
159. Sandanger Ø, Gao E, Ranheim T, Bliksøen M, Kaasbøll OJ, Alfsnes K, Nymo SH, Rashidi A, Ohm IK, Attramadal H, Aukrust P (2016) NLRP3 inflammasome activation during myocardial ischemia reperfusion is cardioprotective. Biochem Biophys Res Commun 469(4):1012–1020
160. Wang Y, Wu Y, Chen J, Zhao S, Li H (2013) Pirfenidone attenuates cardiac fibrosis in a mouse model of TAC-induced left ventricular remodeling by suppressing NLRP3 inflammasome formation. Cardiology 126(1):1–11
161. Sharma A, Tate M, Mathew G, Vince JE, Ritchie RH, De Haan JB (2018) Oxidative stress and NLRP3-inflammasome activity as significant drivers of diabetic cardiovascular complications: therapeutic implications. Front Physiol 9:114
162. Bracey NA, Gershkovich B, Chun J, Vilaysane A, Meijndert HC, Wright JR, Fedak PW, Beck PL, Muruve DA, Duff HJ (2014) Mitochondrial NLRP3 induces reactive oxygen species to promote smad signalling and fibrosis independent from the inflammasome. J Biol Chem:jbc-M114. https://doi.org/10.1074/jbc.M114.550624
163. Lakatta EG, Levy D (2003) Arterial and cardiac aging: major shareholders in cardiovascular disease enterprises: Part II: the aging heart in health: links to heart disease. Circulation 107(2):346–354
164. Harman D (1972) The biologic clock: the mitochondria? J Am Geriatr Soc 20(4):145–147
165. Judge S, Leeuwenburgh C (2007) Cardiac mitochondrial bioenergetics, oxidative stress, and aging. Am J Phys Cell Phys 292(6):C1983–C1992
166. Dai DF, Chen T, Johnson SC, Szeto H, Rabinovitch PS (2012) Cardiac aging: from molecular mechanisms to significance in human health and disease. Antioxid Redox Signal 16(12):1492–1526
167. Zhang H, Tao L, Jiao X, Gao E, Lopez BL, Christopher TA, Koch W, Ma XL (2007) Nitrative thioredoxin inactivation as a cause of enhanced myocardial ischemia/reperfusion injury in the aging heart. Free Radic Biol Med 43(1):39–47
168. Fan Q, Chen M, Fang X, Lau WB, Xue L, Zhao L, Zhang H, Liang YH, Bai X, Niu HY, Ye J (2013) Aging might augment reactive oxygen species (ROS) formation and affect reactive nitrogen species (RNS) level after myocardial ischemia/reperfusion in both humans and rats. Age 35(4):1017–1026
169. Kim JK, Pedram A, Razandi M, Levin ER (2006) Estrogen prevents cardiomyocyte apoptosis through inhibition of reactive oxygen species and differential regulation of p38 kinase isoforms. J Biol Chem 281(10):6760–6767
170. Colom B, Oliver J, Roca P, Garcia-Palmer FJ (2007) Caloric restriction and gender modulate cardiac muscle mitochondrial H2O2 production and oxidative damage. Cardiovasc Res 74(3):456–465
171. Lagranha CJ, Deschamps A, Aponte A, Steenbergen C, Murphy E (2010) Sex differences in the phosphorylation of mitochondrial proteins result in reduced production of reactive oxygen species and cardioprotection in females. Circ Res 106(11):1681–1691
172. Stice JP, Chen L, Kim SC, Jung JS, Tran AL, Liu TT, Knowlton AA (2011) 17β-Estradiol, aging, inflammation, and the stress response in the female heart. Endocrinology 152(4):1589–1598
173. Zhu X, Tang Z, Cong B, Du J, Wang C, Wang L, Ni X, Lu J (2013) Estrogens increase cystathionine-γ-lyase expression and decrease inflammation and oxidative stress in the myocardium of ovariectomized rats. Menopause 20(10):1084–1091

174. Campos C, Casali KR, Baraldi D, Conzatti A, Araújo ASDR, Khaper N, Llesuy S, Rigatto K, Belló-Klein A (2014) Efficacy of a low dose of estrogen on antioxidant defenses and heart rate variability. Oxidative Med Cell Longev 2014:218749
175. Rattanasopa C, Phungphong S, Wattanapermpool J, Bupha-Intr T (2015) Significant role of estrogen in maintaining cardiac mitochondrial functions. J Steroid Biochem Mol Biol 147:1–9
176. Luo T, Liu H, Kim JK (2016) Estrogen protects the female heart from ischemia/reperfusion injury through manganese superoxide dismutase phosphorylation by mitochondrial p38β at threonine 79 and serine 106. PLoS One 11(12):e0167761
177. Starnes JW, Barnes BD, Olsen ME (2007) Exercise training decreases rat heart mitochondria free radical generation but does not prevent Ca2+-induced dysfunction. J Appl Physiol 102(5):1793–1798
178. Nelson MJ, Harris MB, Boluyt MO, Hwang HS, Starnes JW (2011) Effect of N-2-mercaptopropionyl glycine on exercise-induced cardiac adaptations. Am J Phys Regul Integr Comp Phys 300(4):R993–R1000
179. Ma X, Fu Y, Xiao H, Song Y, Chen R, Shen J, An X, Shen Q, Li Z, Zhang Y (2015) Cardiac fibrosis alleviated by exercise training is AMPK-dependent. PLoS One 10(6):e0129971
180. Frasier CR, Moukdar F, Patel HD, Sloan RC, Stewart LM, Alleman RJ, La Favor JD, Brown DA (2013) Redox-dependent increases in glutathione reductase and exercise preconditioning: role of NADPH oxidase and mitochondria. Cardiovasc Res 98(1):47–55
181. Powers SK, Demirel HA, Vincent HK, Coombes JS, Naito H, Hamilton KL, Shanely RA, Jessup J (1998) Exercise training improves myocardial tolerance to in vivo ischemia-reperfusion in the rat. Am J Phys Regul Integr Comp Phys 275(5):R1468–R1477
182. Hong H, Johnson P (1995) Antioxidant enzyme activities and lipid peroxidation levels in exercised and hypertensive rat tissues. Int J Biochem Cell Biol 27(9):923–931
183. Naskar S, Datta K, Mitra A, Pathak K, Datta R, Bansal T, Sarkar S (2014) Differential and conditional activation of PKC-isoforms dictates cardiac adaptation during physiological to pathological hypertrophy. PLoS One 9(8):e104711
184. Venditti P, Di Meo S (1996) Antioxidants, tissue damage, and endurance in trained and untrained young male rats. Arch Biochem Biophys 331(1):63–68
185. Moran M, Delgado J, Gonzalez B, Manso R, Megias A (2004) Responses of rat myocardial antioxidant defences and heat shock protein HSP72 induced by 12 and 24-week treadmill training. Acta Physiol Scand 180(2):157–166
186. Bejma J, Ramires P, Ji LL (2000) Free radical generation and oxidative stress with ageing and exercise: differential effects in the myocardium and liver. Acta Physiol Scand 169(4):343–351
187. Atalay M, Seene T, Hänninen O, Sen CK (1996) Skeletal muscle and heart antioxidant defences in response to sprint training. Acta Physiol Scand 158(2):129–134
188. Avula CR, Fernandes G (1999) Modulation of antioxidant enzymes and lipid peroxidation in salivary gland and other tissues in mice by moderate treadmill exercise. Aging Clin Exp Res 11(4):246–252
189. Navarro A, Gomez C, López-Cepero JM, Boveris A (2004) Beneficial effects of moderate exercise on mice aging: survival, behavior, oxidative stress, and mitochondrial electron transfer. Am J Phys Regul Integr Comp Phys 286(3):R505–R511
190. Starnes JW, Taylor RP, Park Y (2003) Exercise improves postischemic function in aging hearts. Am J Phys Heart Circ Phys 285(1):H347–H351
191. Gunduz F, Senturk UK, Kuru O, Aktekin B, Aktekin MR (2004) The effect of one year swimming exercise on oxidant stress and antioxidant capacity in aged rats. Physiol Res 53(2):171–176
192. Zeng H, Vaka VR, He X, Booz GW, Chen JX (2015) High-fat diet induces cardiac remodelling and dysfunction: assessment of the role played by SIRT 3 loss. J Cell Mol Med 19(8):1847–1856
193. Sundaresan NR, Gupta M, Kim G, Rajamohan SB, Isbatan A, Gupta MP (2009) Sirt3 blocks the cardiac hypertrophic response by augmenting Foxo3a-dependent antioxidant defense mechanisms in mice. J Clin Invest 119(9):2758–2771

194. Sundaresan NR, Samant SA, Pillai VB, Rajamohan SB, Gupta MP (2008) SIRT3 is a stress-responsive deacetylase in cardiomyocytes that protects cells from stress-mediated cell death by deacetylation of Ku70. Mol Cell Biol 28(20):6384–6401
195. Sverdlov AL, Elezaby A, Behring JB, Bachschmid MM, Luptak I, Tu VH, Siwik DA, Miller EJ, Liesa M, Shirihai OS, Pimentel DR (2015) High fat, high sucrose diet causes cardiac mitochondrial dysfunction due in part to oxidative post-translational modification of mitochondrial complex II. J Mol Cell Cardiol 78:165–173
196. Sverdlov AL, Elezaby A, Qin F, Behring JB, Luptak I, Calamaras TD, Siwik DA, Miller EJ, Liesa M, Shirihai OS, Pimentel DR (2016) Mitochondrial reactive oxygen species mediate cardiac structural, functional, and mitochondrial consequences of diet-induced metabolic heart disease. J Am Heart Assoc 5(1):e002555
197. Doenst T, Nguyen TD, Abel ED (2013) Cardiac metabolism in heart failure: implications beyond ATP production. Circ Res 113(6):709–724
198. Behring JB, Kumar V, Whelan SA, Chauhan P, Siwik DA, Costello CE, Colucci WS, Cohen RA, McComb ME, Bachschmid MM (2014) Does reversible cysteine oxidation link the Western diet to cardiac dysfunction? FASEB J 28(5):1975–1987
199. Qin F, Siwik DA, Luptak I, Hou X, Wang L, Higuchi A, Weisbrod RM, Ouchi N, Tu VH, Calamaras TD, Miller EJ (2012) The polyphenols resveratrol and S17834 prevent the structural and functional sequelae of diet-induced metabolic heart disease in mice. Circulation:CIRCULATIONAHA-111. https://doi.org/10.1161/CIRCULATIONAHA.111.067801
200. Liang L, Shou XL, Zhao HK, Ren GQ, Wang JB, Wang XH, Ai WT, Maris JR, Hueckstaedt LK, Ma AQ, Zhang Y (2015) Antioxidant catalase rescues against high fat diet-induced cardiac dysfunction via an IKKβ-AMPK-dependent regulation of autophagy. Biochim Biophys Acta (BBA)-Mol Basis Dis 1852(2):343–352
201. Li W, Tang R, Ouyang S, Ma F, Liu Z, Wu J (2017) Folic acid prevents cardiac dysfunction and reduces myocardial fibrosis in a mouse model of high-fat diet-induced obesity. Nutr Metab 14(1):68
202. Gutiérrez-Tenorio J, Marín-Royo G, Martínez-Martínez E, Martín R, Miana M, López-Andrés N, Jurado-López R, Gallardo I, Luaces M, San Román JA, González-Amor M (2017) The role of oxidative stress in the crosstalk between leptin and mineralocorticoid receptor in the cardiac fibrosis associated with obesity. Sci Rep 7(1):16802
203. Erqou S, Lee CTC, Suffoletto M, Echouffo-Tcheugui JB, de Boer RA, van Melle JP, Adler AI (2013) Association between glycated haemoglobin and the risk of congestive heart failure in diabetes mellitus: systematic review and meta-analysis. Eur J Heart Fail 15(2):185–193
204. Wilson AJ, Gill EK, Abudalo RA, Edgar KS, Watson CJ, Grieve DJ (2018) Reactive oxygen species signalling in the diabetic heart: emerging prospect for therapeutic targeting. Heart 104(4):293–299
205. Huynh K, Bernardo BC, McMullen JR, Ritchie RH (2014) Diabetic cardiomyopathy: mechanisms and new treatment strategies targeting antioxidant signaling pathways. Pharmacol Ther 142(3):375–415
206. Liang W, Chen M, Zheng D, He J, Song M, Mo L, Feng J, Lan J (2017) A novel damage mechanism: Contribution of the interaction between necroptosis and ROS to high glucose-induced injury and inflammation in H9c2 cardiac cells. Int J Mol Med 40(1):201–208
207. Zhong P, Wu L, Qian Y, Fang Q, Liang D, Wang J, Zeng C, Wang Y, Liang G (2015) Blockage of ROS and NF-κB-mediated inflammation by a new chalcone L6H9 protects cardiomyocytes from hyperglycemia-induced injuries. Biochim Biophys Acta (BBA)-Mol Basis Dis 1852(7):1230–1241
208. Rajesh M, Mukhopadhyay P, Bátkai S, Mukhopadhyay B, Patel V, Haskó G, Szabó C, Mabley JG, Liaudet L, Pacher P (2009) Xanthine oxidase inhibitor allopurinol attenuates the development of diabetic cardiomyopathy. J Cell Mol Med 13(8b):2330–2341
209. Fiordaliso F, Bianchi R, Staszewsky L, Cuccovillo I, Doni M, Laragione T, Salio M, Savino C, Melucci S, Santangelo F, Scanziani E (2004) Antioxidant treatment attenuates hyperglycemia-induced cardiomyocyte death in rats. J Mol Cell Cardiol 37(5):959–968

210. Ye G, Metreveli NS, Donthi RV, Xia S, Xu M, Carlson EC, Epstein PN (2004) Catalase protects cardiomyocyte function in models of type 1 and type 2 diabetes. Diabetes 53(5):1336–1343
211. Shen X, Zheng S, Metreveli NS, Epstein PN (2006) Protection of cardiac mitochondria by overexpression of MnSOD reduces diabetic cardiomyopathy. Diabetes 55(3):798–805
212. Yoshida M, Shiojima I, Ikeda H, Komuro I (2009) Chronic doxorubicin cardiotoxicity is mediated by oxidative DNA damage-ATM-p53-apoptosis pathway and attenuated by pitavastatin through the inhibition of Rac1 activity. J Mol Cell Cardiol 47(5):698–705
213. Matsui H, Morishima I, Numaguchi Y, Toki Y, Okumura K, Hayakawa T (1999) Protective effects of carvedilol against doxorubicin-induced cardiomyopathy in rats. Life Sci 65(12):1265–1274
214. El-Awady ESE, Moustafa YM, Abo-Elmatty DM, Radwan A (2011) Cisplatin-induced cardiotoxicity: Mechanisms and cardioprotective strategies. Eur J Pharmacol 650(1):335–341
215. Rosic G, Selakovic D, Joksimovic J, Srejovic I, Zivkovic V, Tatalović N, Orescanin-Dusic Z, Mitrovic S, Ilic M, Jakovljevic V (2016) The effects of N-acetylcysteine on cisplatin-induced changes of cardiodynamic parameters within coronary autoregulation range in isolated rat hearts. Toxicol Lett 242:34–46
216. Lamberti M, Porto S, Marra M, Zappavigna S, Grimaldi A, Feola D, Pesce D, Naviglio S, Spina A, Sannolo N, Caraglia M (2012) 5-Fluorouracil induces apoptosis in rat cardiocytes through intracellular oxidative stress. J Exp Clin Cancer Res 31(1):60
217. Lamberti M, Porto S, Zappavigna S, Addeo E, Marra M, Miraglia N, Sannolo N, Vanacore D, Stiuso P, Caraglia M (2014) A mechanistic study on the cardiotoxicity of 5-fluorouracil in vitro and clinical and occupational perspectives. Toxicol Lett 227(3):151–156
218. Cao H, Wang Y, Wang Q, Wang R, Guo S, Zhao X, Zhang Y, Tong D, Yang Z (2016) Taxol prevents myocardial ischemia-reperfusion injury by inducing JNK-mediated HO-1 expression. Pharm Biol 54(3):555–560
219. Tuomilehto J, Kuulasmaa K, Torppa J (1987) WHO MONICA Project: geographic variation in mortality from cardiovascular diseases. Baseline data on selected population characteristics and cardiovascular mortality. World Health Stat Q 40(2):171–184
220. Ye Z, Song H (2008) Antioxidant vitamins intake and the risk of coronary heart disease: meta-analysis of cohort studies. Eur J Cardiovasc Prev Rehabil 15(1):26–34
221. Yusuf S, Dagenais G, Pogue J, Bosch J, Sleight P (2000) Vitamin E supplementation and cardiovascular events in high-risk patients. N Engl J Med 342(3):154–160
222. Cangemi R, Loffredo L, Carnevale R, Perri L, Patrizi MP, Sanguigni V, Pignatelli P, Violi F (2007) Early decrease of oxidative stress by atorvastatin in hypercholesterolaemic patients: effect on circulating vitamin E. Eur Heart J 29(1):54–62
223. Kornfeld OS, Hwang S, Disatnik MH, Chen CH, Qvit N, Mochly-Rosen D (2015) Mitochondrial reactive oxygen species at the heart of the matter: new therapeutic approaches for cardiovascular diseases. Circ Res 116(11):1783–1799

Reactive Oxygen Species Generation in Neutrophils: Modulation by Nitric Oxide

8

Sachin Kumar and Madhu Dikshit

Abstract

Neutrophils, the phagocytic and short-lived cells, were initially noticed for powerful microbicidal action; however, their specific depletion helped in gaging their unidentified importance in myocardial ischemia-reperfusion injury. Moreover, change in the number of circulating/or migrating neutrophils at the inflammatory site attracted scientists to investigate their significance in various pathologies. Importantly, inhibition of neutrophil recruitment and reactive oxygen species (ROS) generation ability improved cardiac function including cardiac hypertrophy and remodeling in diverse conditions. ROS and protease release from neutrophils have been associated with tissue damages including myocarditis, myocardial infarction, and ischemia-reperfusion injury. Nitric oxide (NO), a pleotropic molecule, modulates various physiological functions including vascular tone and cardiac homeostasis. NO controls most of neutrophil functions including ROS generation that influence release of several inflammatory mediators. Neutrophil ROS that depends on NADPH oxidase (NOX-2) system is regulated by diverse mechanisms including posttranslational modifications, protein interactions, and cofactors. In this chapter, we discuss various regulatory mechanisms involved in NO-mediated modulation of neutrophil reactive oxygen and nitrogen species (RONS) generation, and also NO production by neutrophils,

S. Kumar (✉)
Pharmacology Division, CSIR-Central Drug Research Institute, Lucknow, India
e-mail: Sachin.ku@cdri.res.in

M. Dikshit (✉)
Drug Discovery Research Centre, Translational Health Science and Technology Institute, NCR Biotech Science Cluster, Faridabad, Haryana, India
e-mail: madhudikshit@thsti.res.in

S. Chakraborti et al. (eds.), *Modulation of Oxidative Stress in Heart Disease*,
https://doi.org/10.1007/978-981-13-8946-7_8

which has impacted our understanding of the inflammatory diseases including cardiovascular disorders.

Keywords
Neutrophils · ROS · Nitric oxide · Inflammation · NADPH oxidase · Inducible nitric oxide synthase · Neuronal nitric oxide synthase

8.1 Introduction

Neutrophils are first cells to respond to infection/injury and play crucial role in inflammation [1–3]. Furthermore, inflammation is an important pathological component in cardiovascular diseases including myocarditis, dilated cardiomyopathy, myocardial infarction (MI), and ischemia-reperfusion injury [4–6]. Diverse models have revealed increased accumulation of neutrophils in first hour of injury and reperfusion that causes exacerbation of the inflammatory response and MI injury [7, 8]. Several early studies using antibodies-mediated specific depletion of neutrophils have demonstrated reduction of myocardial infarct size by neutrophil depletion [9–11]. Interestingly, studies from our group demonstrated that the number of neutrophils decreased after thrombosis, and antioxidants were found to be protective against thrombosis [12, 13]. Further studies have identified role of neutrophil activation, release of extracellular traps, and reactive oxygen species in thrombosis [14, 15]. The severity of heart disease including myocarditis and ischemia-reperfusion injury uniquely correlated with the proportion of neutrophils both in mice models and human patients [6, 16, 17]. Indeed, inhibition of neutrophil recruitment and activity improves cardiac function [6, 9, 18, 19]. Similarly, neutrophils also identified to play important role in neuro-inflammation, ischemic stroke, and neurodegeneration as depletion or inhibition of neutrophils infiltration led to protection to neuropathology [20, 21]. Furthermore, all of these pathologies are often accompanied with high oxidative stress that is associated with collateral damage in ischemic stroke and neuro-disorders. The contribution of neutrophils in these pathologies seems crucial as these cells produce high amount of ROS through NADPH oxidase system and subsequent release of NETs coincides with these pathologies. Thus, understanding of various mechanisms involved in ROS generation in PMNs is of great importance. Interestingly, the early accumulation of neutrophils to myocardial injury exacerbates tissue injury through the reactive oxygen species or proteases that are major contributors of inflammation and tissue damage. Furthermore, mice deficient in neutrophil-derived ROS have shown protection in cardiac remodeling, contractile dysfunction, and cardiac hypertrophy in diverse conditions [22–24]. On the similar note, antioxidants have mixed success in myocardial Infarction [25]. Neutrophils also degranulate an array of matrix metalloproteases that regulate cardiac remodeling in combination with ROS production [26]. Unraveling of ROS and nitric oxide (NO) interaction and subsequent formation of several toxic species made researchers inquisitive about the modulatory effect of these species on the intracellular signaling network, which introduced the science of redox biology. ROS and NO both are present in atmosphere and are also produced by all the live forms, suggesting them to be the major players of cellular fate and function. NO regulates vascular tone and cardiac homeostasis. Interestingly, NO donors or drugs enhancing NO levels protect against ischemia-reperfusion injury [27, 28]. In addition, NO also controls neutrophil degranulation and migration and ROS generation to regulate neutrophil-mediated inflammation and possibly ROS-mediated cardiac dysfunction. Here we will overview neutrophil function specifically ROS generation and NO status in neutrophils and then describe NO-mediated regulation of ROS in neutrophils in association to cardiac homeostasis.

8.2 Neutrophils

Elie Metchnikoff's (1883) micro-phagocytes, popularly known as polymorphonuclear leukocytes (PMNs/neutrophils), are the most abundant leukocyte (60–70%) in circulation. Neutrophils initially considered as terminally differentiated, postmitotic cellular entity present with tremendous ability of recognition, chemotaxis, phagocytosis, and production of highly reactive oxygen species and microbicidal proteases to attack and kill the invading pathogens [3, 29, 30], while sustained activation of neutrophils is also associated with more or less collateral tissue damage in inflammatory conditions [3]. Interestingly, neutrophils have gone through a paradigm shift from an innate phagocyte to driver of acquired immunity with time [1, 2, 29]. Though in recent decades participation of neutrophil in adaptive immunity has also been observed by activation and regulation of T, B, and dendritic cells via expression of MHCs and cytokines [1, 2, 31–33], still neutrophils are major player to control infection and inflammation through phagocytic function in association with diverse proteases and ROS production ability via NADPH oxidase activation. In the following section, we discuss the respiratory burst or ROS generation in neutrophils and its modulation by NO.

8.2.1 Respiratory Burst/ROS Generation in Neutrophils

Early studies have demonstrated that neutrophils during phagocytosis produce large amounts of H_2O_2 through "respiratory burst" phenomenon [34, 35]. Interestingly, in neutrophils, this is associated with extra respiration that mainly depended on hexose monophosphate (HMP) shunt but was independent of mitochondrial activity [35]. Further research identified microbicidal function of cytoplasmic granules that were discharged into the phagocytic vacuole [36, 37]. Later, distinct neutrophil granules were characterized, and several granule proteins were found to exhibit microbicidal activity [38, 39]. Soon after the discovery of superoxide dismutase by McCord and Fridovich in 1969, activated neutrophils were found to generate superoxide, and this leukocyte oxidase activity was lacking in chronic granulomatous disease (CGD) patients [40, 41]. In addition, antimicrobial function of myeloperoxidase (MPO)

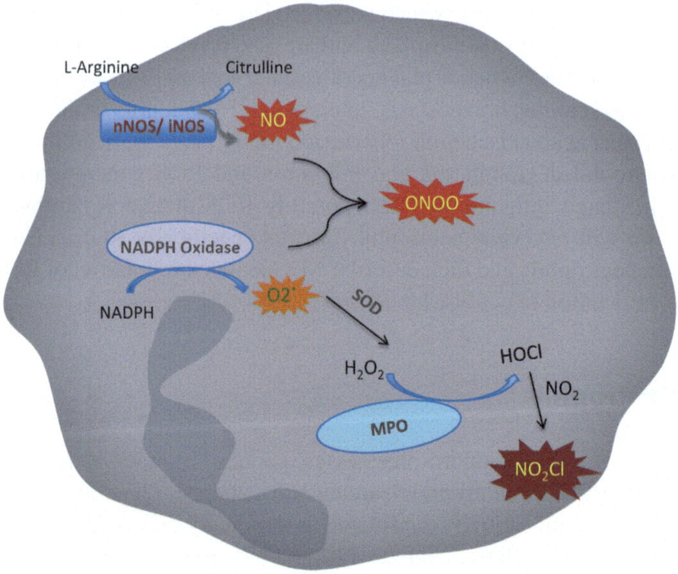

Fig. 8.1 Key enzymes involved in ROS, NO and RONS generation in neutrophils

through halogenation of targets was also described [42, 43]. Thus neutrophils exhibit killing of pathogens through both oxygen-dependent (including superoxide-, hydrogen peroxide-, and myeloperoxidase-dependent hypohalous acids) and oxygen-independent (involving granular proteins such as lysozyme, lactoferrin, and proteases such as elastase, cathepsins) mechanisms [29, 44]. Besides these beneficial functions, neutrophils have also been involved in tissue damage (Fig. 8.1).

Thus respiratory burst was reported by Baldridge and Gerald [34] during the process of phagocytosis in neutrophils, due to the activity of NADPH oxidase, a multi-subunit enzymatic complex [34]. Importantly, respiratory burst is responsible for more than 90% of the total oxygen consumption by neutrophils [45]. This leads to generation of O_2^- into the phagosome or to the exterior milieu. Superoxide anions are relatively less noxious but form additional toxic oxygen species, in particular H_2O_2, by spontaneous dismutation which may then oxidize halides, in particular Cl^-, to hypohalous acid, e.g., HOCl, catalyzed by myeloperoxidase released from azurophil granules during degranulation. In phagocytosis, neutrophils uptake the invading organism into vacuolar structure or phagosome. Importantly, both oxygen-dependent and oxygen-independent antimicrobial functions take place in phagosome. Upon stimulation, multi-subunit of NADPH oxidase system get assembled at membrane. $p47^{Phox}$, $p67^{Phox}$, and a Rac-related GTP protein that in steady-state condition reside in cytoplasm move to the plasma membrane, where it forms complex with b-type hemoprotein and cytochrome b_{558} consisting of dimer of $gp91^{Phox}$ and $p22^{Phox}$ and binds to FAD and NADPH [29, 46]. Further transfer of an electron from the cytochrome to oxygen leads to the superoxide generation [29, 46]. These superoxide radicals induce a series of oxidative events generating other strong oxidants, which result in microbial killing. In CGD patients, mutation of NADPH oxidase subunits causes failure in superoxide generation. Importantly, these patients face recurrent infections due to dysfunction of oxidant-dependent antimicrobial mechanism in phagocytic cells [46]. Interestingly, in addition to superoxide generation, neutrophils also produce NO through nitric oxide synthase. Importantly, NO has been defined to exhibit both superoxide scavenging and modulation of NADPH oxidase activity that are discussed in following sections.

8.3 Nitric Oxide

NO, initially known as endothelium-derived relaxing factor (EDRF), is a tiny lipophilic reactive radical gas that mediates both regulatory and cytotoxic functions [47]. Further studies revealed involvement of enzyme nitric oxide synthase (NOS) in NO synthesis in various cells including vascular endothelial cells [48], macrophages [49], and neurons [47]. Importance of NO in diverse system was well appreciated by defining this as "molecule of the year" (*Science* journal in 1992) and Nobel Prize to Furchgott, Ignarro, and Murad in 1998. Immune cells such as eosinophils, platelets, neutrophils, monocytes, and macrophages also synthesize NO. Among them, neutrophils, the most abundant leukocytes, have been observed to participate in diverse pathological conditions. Interestingly, deregulated NO synthesis and signaling in neutrophils are suggested to involve in these pathologies (Fig. 8.2).

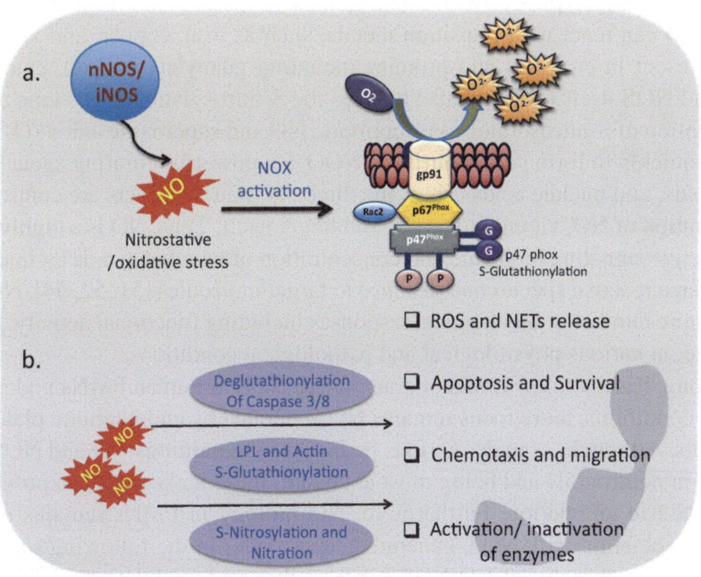

Fig. 8.2 (**a**) NO-NOS mediated regulation of NADPH Oxidase, (**b**) NO mediated modulation of other proteins involved in diverse neutrophil functions

NO synthases (NOSs) catalyze the conversion of L-arginine to L-citrulline and NO in nicotinamide adenine dinucleotide phosphate (NADPH)-dependent manner. There are three isoforms of NOS including neuronal NOS (nNOS), endothelial NOS (eNOS), and inducible NOS (iNOS). eNOS and nNOS, the constitutive NOS (cNOS), are considered to produce low level of NO and depend on calcium for its activity; however, inducible iNOS that can be augmented by inflammatory milieu is calcium independent and produce high NO for prolonged time [50]. NOS enzyme contains two domains: the C-terminal reductase domain contains binding sites for NADPH, FMN, and FAD, while the N-terminal oxygenase domain binds with biopterin (BH_4), heme, and L-Arg. Calmodulin (CaM or calcium-modulated protein) binding triggers flow to electrons from the reductase to the oxygenase domain [47]. Heme and biopterin play essential role in enzyme dimerization, a stable conformation for electron transport. iNOS tightly interact with calmodulin and thus functions in calcium-independent manner [50].

NO diffuses actively in aqueous as well as hydrophobic environments; further its biological concentration is defined partially by vicinity and cellular redox environment. Furthermore, its target cell specificity depends on its concentration, compartmentalization, exposure time, chemical reactivity, vicinity, and priming of target cells. NO functions can be divided into direct and indirect categories [51, 52]. In direct effect, NO at low concentration react directly with a biological target molecule for a short period. In contrast, NO at much higher concentrations interacts with oxygen or superoxide to generate RONS and thus indirectly regulates biological processes. At molecular level, NO modulates the biological activities by reaction with transition metal (direct effect) and nitration and nitrosylation of tyrosine and cysteine residues, respectively, of the proteins and modify their functions (indirect effect). NO can react with transition metals, such as iron, copper, and zinc, abundantly present in enzymes and proteins including guanylate cyclase, cytochrome P450, and NOS itself [53]. NO also induces the S-nitrosylation at cysteine residues via formation of S-nitrosothiols. Furthermore, NO and superoxide anion (O_2^-) interact very quickly to form peroxynitrite ($ONOO^{\cdot-}$), a powerful nitrating agent for proteins, lipids, and nucleic acids. Thus, the direct vs indirect effects are controlled by concentration of NO, vicinity to target, and target itself. Thus, NO is a highly diffusible, reactive signaling molecule and concentration of which depends on interaction with diverse reactive species and distance to target molecules [51, 52, 54]. NO regulates infinite number of biochemical responses including functional activity, growth, death, etc. in various physiological and pathological conditions.

For long, based on low levels of constitutive NOS in human PMNs under physiological conditions, more focus remains NO generated by endothelium, platelets, or other cells. Although, as under various pathological conditions NO and NOS levels increase in neutrophils and being most abundant, these leukocytes can provide significant NO in circulation. Furthermore, NO biology in PMNs remains complex with its association with NOX-generated superoxide. In the following section, we will emphasize on status of NO/NOS in neutrophils and modulation of PMNs function with special focus on ROS.

8.4 NO/NOS in Neutrophils

Presence of NOS in PMNs was first recognized in a study conducted by using rat peritoneal neutrophils and vascular ring preparation [55]. Both peritoneal and peripheral rat PMNs elicited a relaxation response in rat aortic rings when added to endothelium-denuded aortic rings [55, 56]. Human PMNs produce NO at a rate of 0.75 ± 0.2 ρmoles/min/10^6 cells in the presence of SOD [57, 58]. Furthermore, neutrophils exhibited platelet aggregation inhibitory response that was enhanced in the presence of superoxide dismutase but abolished by NO scavenger, suggesting NO-mediated inhibitory effect of PMNs on platelets [56]. Interestingly, N^G-monomethyl L-arginine and N^G nitro-L-arginine methyl ester also inhibited PMNs mediated response, indicating presence of an NOS in PMNs [56, 59]. Neutrophil-mediated inhibition of platelet aggregation was thus due to NO, which stimulated the formation of cGMP in the platelets [59]. Role of NO and neutrophils in regulation of hemostasis was revealed by our early studies in rat thromboembolism model [56]. Malawista et al. suggested functional importance of NO in the microbicidal activity of neutrophils [60]. In other study, phagocytosis-induced NO generation was implied in peroxynitrite-mediated formation of nitrotyrosine [61]. Consistent to this down the road, presence of NOS in rat and human PMNs have been established at protein and transcript level, while NO production has been revealed by measurement of nitrite (NO_2^-), DAF-2DA, and NOS activity by conversion of radiolabeled L-arginine to radiolabeled L-citrulline [62–66]. Subsequent studies were addressed to characterize the enzymatic source responsible for NO generation in PMNs.

In leukocytes, Wheeler et al. (1997) identified neutrophils as the primary source of iNOS in patients with urinary tract infections [67]. Furthermore, increase in neutrophil iNOS was observed after bacterial infection [67]. While Miles et al. [63] failed to detect iNOS mRNA, protein, or enzymatic activity in circulating PMNs from rat and human, that was increased after cultured for 4–6 h in vitro [63]. Presence of iNOS was also observed in cytokine-stimulated [61] or bacterial-infected human neutrophils that was evident in the primary granules [67]. Wallerath et al. [68] were successful in detecting NOS isoform transcripts expression in bone marrow human neutrophil granulocytes, megakaryocytes, and platelets. Cedergren et al. noticed constitutive expression of iNOS in human PMNs and suggested that failure of iNOS observation in resting PMNs by others could be because of incomplete release of the membrane-bound enzyme and inadequate proteinase inhibition [66], as more than 90% of iNOS is tightly bound to membrane in human PMNs [67]. Human PMNs present in oral cavity [69] and isolated from patients with sepsis syndrome [70] have possessed iNOS mRNA and protein and exhibit enzyme activity. In addition, human and rat neutrophils were shown to express nNOS mRNA constitutively [71, 72].

Biochemical characterization of NO formation in our lab using radioactive L-arginine demonstrated the utilization of L-arginine to synthesize NO formation by control and LPS-treated PMNs [73]. We also showed distribution of functionally active nNOS and iNOS in cytosolic and nuclear compartments of rat PMNs [65]. Both nNOS and iNOS were found to colocalize and interact with caveolin-1 in rat PMNs [65]. Interestingly, caveolins functioned as negative regulators of NOS by regulating post-transcription proteasomal degradation and direct interaction [74, 75]; however, signal for eNOS was not detectable [65]. While investigating the importance of neutrophil-derived NO generation in various experimental models, we found augmented NO formation in PMNs isolated from spontaneously hypertensive rat (SHR) in comparison to normotensive Wistar rats [76]. Furthermore, expression of iNOS was significantly more in the neutrophils obtained from SHR, while nNOS expression remained unaffected. We also found importance of high ascorbate in sustaining neutrophil NOS expression, catalysis, and oxidative burst [77]. Ascorbate helped in maintaining the redox-sensitive tetrahydrobiopterin content to support NO synthesis by neutrophils. Recent findings from our lab also demonstrated constitutive expression of iNOS in human PMNs, and it was evident in plasma membrane, cytosol, nucleus, mitochondria, as well as elastase and gelatinase containing granules [78]. Immunofluorescence staining further documented the presence of nNOS and iNOS in human PMNs, while eNOS was not detected in agreement with previous report [68]. Similarly, in RT-PCR, transcripts for nNOS and iNOS but not for eNOS were identified [68]. Presence of eNOS in neutrophils is still controversial as only one group published so far has proposed the presence of eNOS in resting human neutrophils [79] and suggested decreased eNOS expression during acute myocardial infarction or TNFα treatment. In yet another study, it was observed that neutrophil nNOS expression during their maturation in the bone marrow remains unchanged, while iNOS levels were augmented in the rats [80]. Importantly, eNOS expression was attenuated leading to undetectable levels in mature neutrophils. RT-PCR and Western blot analysis further demonstrated low expression of NOS isoform (nNOS and iNOS) in human PMNs in comparison to mice circulating PMNs [63, 66, 71]. Together, research so far unequivocally identified presence of both nNOS and iNOS in rat, mice, and human neutrophils [65, 66, 72, 77, 81, 82]. Thus, neutrophils possess functional iNOS and nNOS under tight regulation that provide substantial amount of NO and further can modulate ROS generation and vascular homeostasis.

8.5 NO and Neutrophil ROS

Exogenous NO is known to affect most of neutrophil functions in a concentration-dependent manner [83]. Studies in rodent and human PMNs also demonstrated that NOS enzymes were present though in variable amounts right from promyelocytes to the segmented stage. NOS primarily catalyzes NO generation, but it can also produce $O_2^{\cdot-}$, $ONOO^-$, and NO_3^- depending on the environment [84]. In normal productive cycle in the presence of all cofactors, NOS synthesizes NO. All the three

NOS isoforms (nNOS, eNOS, and iNOS) can also generate O_2^- under L-arg or BH_4 limitation, by mechanism known as uncoupling [84]. When the NO concentration accumulates in the range of micromolar, $ONOO^{\cdot -}$ and NO_3^- are generated from NOS known as futile cycle [84].

Further, NO regulates ROS levels by modulating superoxide levels by scavenging, regulating NADPH oxidase system, and producing additional oxidant species. NO has been extensively investigated to modulate oxidative burst in neutrophils in our laboratory and is also supported by others [73, 77, 85–90]. The observations convincingly indicated toward NO-mediated augmentation of free radical generation from PMNs [73, 82, 85–87, 91]. Thus, several studies from this laboratory proposed role of NO in the modulation of neutrophil free radical generation by using multiple fluorescent probes and other methodologies [73, 82, 85–87, 91–93]. Interestingly, suppression of luminol-dependent chemiluminescence (LCL) response was observed in the presence of NO in micromole concentration [86], while by using other probes, formation of RONS was also evident [82, 87, 91, 93]. Coherent to an early study, that utilized D and L- enantiomers of NOS inhibitors to test neutrophil ROS and found differential effect on LCL based on enantiomers interaction and advice to be cautious for NO-mediated effects on LCL responses [92]. This NO-mediated superoxide scavenging and distinct effect was further explored. Interestingly, Kumar et al. demonstrated contrasting effect on DCF and DHE adducts and advocated a precaution while using these probes, based on superoxide-scavenging ability of NO [87]. In other study, ascorbate has been shown to regulate tetrahydrobiopterin levels and thus control NOS activity and NO content [88, 89], while ascorbate deficiency led to decrease in NO and free radical generation in neutrophils [77]. Together, most of these studies using diverse probes for reactive oxidant detection suggest increase in ROS generation in NO-treated PMNs. Hereafter, direct biochemical association and regulation of NADPH oxidase by NO are described.

Clancy et al. revealed a direct interaction of NO with the membrane subunit of the NADPH oxidase complex [90], suggesting NO-mediated regulation of NOX system, while another study demonstrated an inhibitory association of NO with both membranous and cytosolic subunits [94]. Further, Lee et al. reported an inhibitory effect also at a higher concentration of NO [95]. In an intriguing study, NOS was found to interact with Rac2 to regulate activation of NADPH oxidase system that was translocated to phagosomes during phagocytosis [82]. Together, NO was demonstrated to contribute to ROS/RNS generation and microbial killing. $ONOO^-$ also exhibited a biphasic effect like NO, being stimulatory at lower concentrations through the MEK/ERK/MAPK pathway but inhibitory at very higher concentrations [85]. NO was also observed to regulate neutrophil-derived free radical generation following hypoxia-reoxygenation [85] that was blocked in the presence of NO scavenger. Moreover, hypoxic neutrophils following oxygenation exhibited a significant increase in the respiratory burst in a NO-dependent manner [85]. Further study suggested involvement of protein kinase C and calcium signaling in neutrophil ROS generation following hypoxia reoxygenation [96]. These observations are of significance in explaining the damaging effects of neutrophils at the hypoxic environment of the inflammatory loci. Klink et al. have reported that NO donors decrease PMA- and/or fMLP-induced p47 phosphorylation on tyrosine and serine/threonine residues and PKC on serine residues and ROS production with MAPK phosphorylation [97]. In a time-dependent study, Nagarkoti et al. demonstrated NO donors itself induces phosphorylation and glutathionylation of p47 that led to increase in ROS generation [91]. These differential outcomes might be a result of probe selection of ROS detection [87]. This probe-based dilemma was further resolved by using biochemistry approach, in which NO was found to interact with Rac2 resulting in translocation of NADPH oxidase subunits to the plasma membrane [82, 91]. Furthermore, intracellular and extracellular calcium levels also have a modulatory impact on NOS activity and free radical generation [98]. Another study has suggested the involvement of K^+ channels and kinases in NO-mediated augmentation of respiratory burst [99]. Our laboratory has also revealed NO-dependent NETs release from human neutrophils for the first time through ROS-dependent mechanisms, as NOX and myeloperoxidase inhibitors reduce NETs release [93]. In a mechanistic study, PMA-induced ROS has led to activation of ERK and p38 MAPK that has regulated NETs release from human neutrophils [100]. Recent study from this lab has shown that sustained ROS generation by NOX-2 was due to the glutathionylation of the important cysteine residues in the P47phox, and if glutathionylation was blocked, PMA-induced respiratory burst could not be sustained even in the presence of phosphorylation. Similarly, FMLP-induced short burst of ROS formation was converted to sustained generation by promoting glutathionylation of phosphorylated P47phox [91].

It is intriguing to note that factors instigating oxidative burst might also simultaneously trigger NOS in neutrophils or vice versa. For example, lipopolysaccharide (LPS), a membrane component of gram-positive bacteria, potentially induced iNOS and L-arginine uptake. In addition, it also induced free radical generation with arachidonic acid from peripheral and peritoneal neutrophils [73]. Nitrite treatment was found to elevate free radical generation and myeloperoxidase (MPO) release from neutrophils [73]. Interestingly, NOS inhibitors, aminoguanidine and 7-nitroindazole, have inhibitory effect on arachidonic acid-induced free radical generation and MPO-derived ROS in PMNs [73, 87]. This together suggests that NO-mediated regulation of free radical generation from PMNs is dependent on NOX and MPO activity. In a recent study, mechanism of NO-induced sustained ROS generation in neutrophil was found to be dependent on S-glutathionylation of p47phox unit of NOX system [91]. Interestingly, NO treatment induced p47phox glutathionylation responsible for sustained ROS generation, which was not the case in fMLP stimulation [91]. In other study, prolonged treatment with NO was found to induce apoptosis of neutrophils through ROS-mediated caspase-8 and caspase-3 activation and mitochondrial death pathway [101]. NO thus affect ROS generation by modulating multiple mechanisms in a time- and concentration-dependent manner, leading to NETs release or apoptosis. Understanding of the molecular mechanisms involved is of immense importance, as these might help in identifying the signaling networks associated with infective and noninfective inflammatory conditions.

In addition, nitric oxide as well as ROS regulates neutrophil migration and chemotaxis [102–105]. An early report identified role of NO and NOS signaling in neutrophil migration in carrageenan- (Cg) or LPS-induced inflammation [104]. NOS inhibitors, NG-nitro-L-arginine and aminoguanidine, enhanced neutrophil migration in Cg model that was reversed by co-treatment with L-arginine, suggesting involvement of NOS pathway [104]. Furthermore, iNOS-deficient mice exhibited more neutrophil migration upon Cg challenge than wild-type mice. This increase in neutrophils at inflammatory sites was further observed due to increase in adhesion and mitigation of apoptosis [104]. In other study, L-NAME-mediated NOS inhibition led to induction of microparticles from neutrophils [105]. Interestingly, these L-NAME-induced microparticles were found to enhance neutrophil migration to IL-8 [105]. Another study from our group revealed that NO modulates actin polymerization to regulate neutrophil migration and polarization. More specifically, NO treatment led to S-glutathionylation of cytoskeleton proteins actin and L-plastin (LPL) in neutrophils [106]. Further experiments identified that S-glutathionylation of LPL cause impaired neutrophil migration and bactericidal activity [106]. Furthermore, elevated LPL S-glutathionylation was evident in neutrophils from diabetic patients and *db/db* mice that exhibit neutrophil dysfunctions [106]. Interestingly, ROS also regulate S-glutathionylation of actin [102]; this suggests an intriguing cross talk of ROS-RNS through NADPH oxidase and NO synthases. Further, NO can modulate proliferation and cytostasis of diverse cells including promyelocytic HL-60s depending on concentration [107]. In other study, NO was found to suppress proliferation of murine and human Th17 cells. NO also inhibited IL-22 and IL-23 receptor signaling in Th17 cells [108]. Interestingly, IL-17-IL-23 axis also regulates neutrophil homeostasis [109, 110]. Thus, NO-mediated regulation of neutrophil ROS and IL-17 axis might be responsible for distinct characteristics of neutrophil subsets and generation of neutrophils through granulopoiesis.

Neutrophils, being reservoirs of oxidants and proteases, possess strong antioxidant defense mechanisms which remain on constant vigil to maintain the redox balance. Neutrophils are protected against self-destruction by intracellular superoxide dismutase, ascorbate, GSH, and catalase [44]. Furthermore, to counter against ROS-mediated killing by neutrophils, *Escherichia coli* expressed enterobactin (Ent), a catecholate siderophore for iron that inhibited PMA-induced ROS and NETosis [111], thus identified a novel microbial antiradical defense to mitigate neutrophil responses. Further, it would be important to investigate the effect of these siderophore on NO-induced neutrophil ROS generation, as free iron might affect NOS activity. Extent of neutrophils with pro-inflammatory characteristics, also described as aged neutrophils, was dependent on microbiota [112]. Interestingly, gut microbiota have been suggested to impact cardiovascular function [113]. Thus, microbiota-mediated regulation of NO and neutrophil ROS is important to understand and warrants future investigation.

8.6 Neutrophils and NO/NOS in Diseases

Neutrophil-dependent inflammation through high levels of NO and ROS has been involved in various pathological conditions. High levels of plasma nitrate concentration have been reported in patients with septicemia with normal or elevated number of neutrophils in peripheral blood than to those with neutropenia [114]. Furthermore, during hypoxia-reoxygenation, neutrophil-derived NO induces free radical generation [85]. An increase in the release of NO from PMNs after thrombosis [56] suggests its role in the regulation of homeostasis. Circulating neutrophils from hypertensive patients were present with high oxidative stress than normotensive counterparts [115]. Consistently, neutrophils from spontaneously hypertensive rats exhibited high iNOS and augmented NO generation that might be responsible for oxidative and inflammatory stress in hypertension [116]. Circulating neutrophils via suppression of bacteria and IFNγ-dependent iNOS expression have also been shown to maintain physiological blood pressure [117] as neutrophil depletion led to low blood pressure and suggested to maintain the optimal vascular tone.

In addition, dysregulation of neutrophils NOS levels and NO signaling has been reported in neurological disorders. Increase in the neutrophil nitrite content and its role in Parkinson's disease has been detected [118], while no change was observed in plasma and platelets nitrite levels. Interestingly, PMNs catalase activity was decreased, while SOD and GPx were unaffected in PD patients, while a significant decrease (>70%) of nitrite level and NOS activity was observed in neutrophils from patients of depression [119]. Moreover, Gatto et al. [81] have reported overexpression of neutrophil nNOS in Parkinson's disease. In contrast, in schizophrenia patients, a specific and significant decrease in NO levels in PMNs has been reported, while antioxidant enzyme activities remain unaffected in the PMN of schizophrenia patients [120].

Interestingly, in yet another study, we observed a reduction in the circulating levels of NO/nitrite, which was also found in the cellular fractions of bone marrow in the patients of myeloid leukemia [121]. A study focused on bio-antioxidants has revealed alterations of GSH-redox cycle, total thiol groups, and vitamins E and C in blood, platelets, neutrophils, heart, and lung in thrombosis [122]. Clinical relevance of NO and its metabolite nitrite and nitrate increases to high level in systemic inflammatory response syndrome (SIRS), sepsis, and septic shock. Data has revealed a direct association between NO metabolites and progression of septic shock [123]. In addition, neutrophils and inflammation also control myocarditis, dilated cardiomyopathy, myocardial infarction, and ischemia-reperfusion injury [4–6]. Decisive role of neutrophil ROS has been well demonstrated in cardiac remodeling, contractile dysfunction, and cardiac hypertrophy using mice deficient in neutrophil-derived ROS that have exhibited protection in diverse conditions [22–24]. Neutrophils being abundant leukocytes might thus add substantial amount of NO in circulation with a potential widespread impact on vascular homeostasis [124]. Furthermore, knowledge of NO-mediated regulation of neutrophil ROS generation is important for our better understanding of the pathological conditions. It is also desired to conduct studies to monitor the changes in some of the abovementioned parameters for diagnostic, prognostic, and therapeutic purposes.

8.7 Conclusions

Together, NO as a pleiotropic molecule regulates homeostasis and neutrophil-driven inflammation through modulation of ROS production. NO also exhibits differential effects on NADPH oxidase, and MPO system in neutrophils and on quenching of superoxide forms more toxic peroxynitrite. Being these cells abundant in circulation, cells can provide sufficient ROS levels to cause tissue damage including heart dysfunction. Further recent studies suggested role of NO in the modulation of neutrophil NOX- and MPO-mediated RONS generation by multiple mechanisms, thus providing unique approaches to specifically target enzymes/pathways to intercept inflammation. Neutrophil ROS also participates in cardiac remodeling and cardiac hypertrophy that might be targeted therapeutically through further understanding of regulatory mechanisms.

References

1. Mayadas TN et al (2014) The multifaceted functions of neutrophils. Annu Rev Pathol 9:181–218
2. Mantovani A et al (2011) Neutrophils in the activation and regulation of innate and adaptive immunity. Nat Rev Immunol 11(8):519–531
3. Nathan C (2006) Neutrophils and immunity: challenges and opportunities. Nat Rev Immunol 6(3):173–182
4. Afanasyeva M et al (2004) Quantitative analysis of myocardial inflammation by flow cytometry in murine autoimmune myocarditis: correlation with cardiac function. Am J Pathol 164(3):807–815
5. Nemoto S et al (2002) Escherichia coli LPS-induced LV dysfunction: role of toll-like receptor-4 in the adult heart. Am J Physiol Heart Circ Physiol 282(6):H2316–H2323
6. Guasti L et al (2011) Neutrophils and clinical outcomes in patients with acute coronary syndromes and/or cardiac revascularisation. A systematic review on more than 34,000 subjects. Thromb Haemost 106(4):591–599
7. Vinten-Johansen J (2004) Involvement of neutrophils in the pathogenesis of lethal myocardial reperfusion injury. Cardiovasc Res 61(3):481–497
8. Carbone F et al (2013) Pathophysiological role of neutrophils in acute myocardial infarction. Thromb Haemost 110(3):501–514
9. Jordan JE et al (1999) The role of neutrophils in myocardial ischemia-reperfusion injury. Cardiovasc Res 43(4):860–878
10. Romson JL et al (1983) Reduction of the extent of ischemic myocardial injury by neutrophil depletion in the dog. Circulation 67(5):1016–1023
11. Jolly SR et al (1986) Reduction of myocardial infarct size by neutrophil depletion: effect of duration of occlusion. Am Heart J 112(4):682–690
12. Dikshit M et al (1989) Role of free radicals in pulmonary thromboembolism in mice. Thromb Res 55(5):549–557
13. Dikshit M et al (1992) Effect of pulmonary thromboembolism on circulating neutrophils in mice. Thromb Res 66(2–3):133–139

14. van Montfoort ML et al (2013) Circulating nucleosomes and neutrophil activation as risk factors for deep vein thrombosis. Arterioscler Thromb Vasc Biol 33(1):147–151
15. Brill A et al (2012) Neutrophil extracellular traps promote deep vein thrombosis in mice. J Thromb Haemost 10(1):136–144
16. Knuefermann P et al (2002) Cardiac inflammation and innate immunity in septic shock: is there a role for toll-like receptors? Chest 121(4):1329–1336
17. Hiroi T et al (2012) Neutrophil TRPM2 channels are implicated in the exacerbation of myocardial ischaemia/reperfusion injury. Cardiovasc Res 97(2):271–281
18. Abdel-Rahman U et al (2007) Inhibition of neutrophil activity improves cardiac function after cardiopulmonary bypass. J Inflamm (Lond) 4:21
19. Garcia-Prieto J et al (2017) Neutrophil stunning by metoprolol reduces infarct size. Nat Commun 8:14780
20. Zenaro E et al (2015) Neutrophils promote Alzheimer's disease-like pathology and cognitive decline via LFA-1 integrin. Nat Med 21(8):880–886
21. Jickling GC et al (2015) Targeting neutrophils in ischemic stroke: translational insights from experimental studies. J Cereb Blood Flow Metab 35(6):888–901
22. Bendall JK et al (2002) Pivotal role of a gp91(phox)-containing NADPH oxidase in angiotensin II-induced cardiac hypertrophy in mice. Circulation 105(3):293–296
23. Grieve DJ et al (2006) Involvement of the nicotinamide adenosine dinucleotide phosphate oxidase isoform Nox2 in cardiac contractile dysfunction occurring in response to pressure overload. J Am Coll Cardiol 47(4):817–826
24. Looi YH et al (2008) Involvement of Nox2 NADPH oxidase in adverse cardiac remodeling after myocardial infarction. Hypertension 51(2):319–325
25. Rodrigo R et al (2013) Molecular basis of cardioprotective effect of antioxidant vitamins in myocardial infarction. Biomed Res Int 2013:437613
26. Ma Y et al (2013) Neutrophil roles in left ventricular remodeling following myocardial infarction. Fibrogenesis Tissue Repair 6(1):11
27. Schulz R et al (2004) Nitric oxide in myocardial ischemia/reperfusion injury. Cardiovasc Res 61(3):402–413
28. Hansen PR (1995) Role of neutrophils in myocardial ischemia and reperfusion. Circulation 91(6):1872–1885
29. Segal AW (2005) How neutrophils kill microbes. Annu Rev Immunol 23:197–223
30. Kolaczkowska E, Kubes P (2013) Neutrophil recruitment and function in health and inflammation. Nat Rev Immunol 13(3):159–175
31. Sandilands GP et al (2005) Cross-linking of neutrophil CD11b results in rapid cell surface expression of molecules required for antigen presentation and T-cell activation. Immunology 114(3):354–368
32. Beauvillain C et al (2007) Neutrophils efficiently cross-prime naive T cells in vivo. Blood 110(8):2965–2973
33. Neuman E et al (1992) Regulation of MHC class I synthesis and expression by human neutrophils. J Immunol 148(11):3520–3527
34. Bladridge CW, Gerard RW (1932) The extra respiration of phagocytes. Am J Phys 103(3):235–236
35. Karnovsky ML (1962) Metabolic basis of phagocytic activity. Physiol Rev 42:143–168
36. Hirsch JG, Cohn ZA (1960) Degranulation of polymorphonuclear leucocytes following phagocytosis of microorganisms. J Exp Med 112:1005–1014
37. Cohn ZA, Hirsch JG (1960) The isolation and properties of the specific cytoplasmic granules of rabbit polymorphonuclear leucocytes. J Exp Med 112:983–1004
38. Bainton DF et al (1971) The development of neutrophilic polymorphonuclear leukocytes in human bone marrow. J Exp Med 134(4):907–934
39. Borregaard N et al (2007) Neutrophil granules: a library of innate immunity proteins. Trends Immunol 28(8):340–345
40. McCord JM, Fridovich I (1969) Superoxide dismutase. An enzymic function for erythrocuprein (hemocuprein). J Biol Chem 244(22):6049–6055

41. Baehner RL, Nathan DG (1967) Leukocyte oxidase: defective activity in chronic granulomatous disease. Science 155(764):835–836
42. Klebanoff SJ (1968) Myeloperoxidase-halide-hydrogen peroxide antibacterial system. J Bacteriol 95(6):2131–2138
43. Klebanoff SJ, White LR (1969) Iodination defect in the leukocytes of a patient with chronic granulomatous disease of childhood. N Engl J Med 280(9):460–466
44. Roos D, Winterbourn CC (2002) Lethal weapons. Science 296(5568):669–671
45. Babior BM et al (1973) Biological defense mechanisms. The production by leukocytes of superoxide, a potential bactericidal agent. J Clin Invest 52(3):741–744
46. Babior BM et al (2002) The neutrophil NADPH oxidase. Arch Biochem Biophys 397(2):342–344
47. Bredt DS, Snyder SH (1994) Nitric oxide: a physiologic messenger molecule. Annu Rev Biochem 63:175–195
48. Moncada S et al (1989) Biosynthesis of nitric oxide from L-arginine. A pathway for the regulation of cell function and communication. Biochem Pharmacol 38(11):1709–1715
49. Gross SS et al (1991) Cytokine-activated endothelial cells express an isotype of nitric oxide synthase which is tetrahydrobiopterin-dependent, calmodulin-independent and inhibited by arginine analogs with a rank-order of potency characteristic of activated macrophages. Biochem Biophys Res Commun 178(3):823–829
50. Alderton WK et al (2001) Nitric oxide synthases: structure, function and inhibition. Biochem J 357(Pt 3):593–615
51. Thomas DD et al (2008) The chemical biology of nitric oxide: implications in cellular signaling. Free Radic Biol Med 45(1):18–31
52. Wink DA et al (1996) Chemical biology of nitric oxide: regulation and protective and toxic mechanisms. Curr Top Cell Regul 34:159–187
53. Ignarro LJ (1991) Signal transduction mechanisms involving nitric oxide. Biochem Pharmacol 41(4):485–490
54. Knowles RG, Moncada S (1994) Nitric oxide synthases in mammals. Biochem J 298:249–258
55. Rimele TJ et al (1988) Interaction of neutrophils with vascular smooth muscle: identification of a neutrophil-derived relaxing factor. J Pharmacol Exp Ther 245(1):102–111
56. Dikshit M et al (1993) Pulmonary thromboembolism-induced alterations in nitric oxide release from rat circulating neutrophils. J Pharmacol Exp Ther 265(3):1369–1373
57. Wright CD et al (1989) Generation of nitric oxide by human neutrophils. Biochem Biophys Res Commun 160(2):813–819
58. Salvemini D et al (1989) Human neutrophils and mononuclear cells inhibit platelet aggregation by releasing a nitric oxide-like factor. Proc Natl Acad Sci U S A 86(16):6328–6332
59. Faint RW et al (1991) Platelet aggregation is inhibited by a nitric oxide-like factor released from human neutrophils in vitro. Br J Haematol 77(4):539–545
60. Malawista SE et al (1992) Evidence for reactive nitrogen intermediates in killing of staphylococci by human neutrophil cytoplasts. A new microbicidal pathway for polymorphonuclear leukocytes. J Clin Invest 90(2):631–636
61. Evans TJ et al (1996) Cytokine-treated human neutrophils contain inducible nitric oxide synthase that produces nitration of ingested bacteria. Proc Natl Acad Sci U S A 93(18):9553–9558
62. Rodenas J et al (1995) Simultaneous generation of nitric oxide and superoxide by inflammatory cells in rats. Free Radic Biol Med 18(5):869–875
63. Miles AM et al (1995) Nitric oxide synthase in circulating vs. extravasated polymorphonuclear leukocytes. J Leukoc Biol 58(5):616–622
64. Chen LY, Mehta JL (1996) Further evidence of the presence of constitutive and inducible nitric oxide synthase isoforms in human platelets. J Cardiovasc Pharmacol 27(1):154–158
65. Saini R et al (2006) Nitric oxide synthase localization in the rat neutrophils: immunocytochemical, molecular, and biochemical studies. J Leukoc Biol 79(3):519–528
66. Cedergren J et al (2003) Inducible nitric oxide synthase (NOS II) is constitutive in human neutrophils. APMIS 111(10):963–968

67. Wheeler MA et al (1997) Bacterial infection induces nitric oxide synthase in human neutrophils. J Clin Invest 99(1):110–116
68. Wallerath T et al (1997) Identification of the NO synthase isoforms expressed in human neutrophil granulocytes, megakaryocytes and platelets. Thromb Haemost 77(1):163–167
69. Nakahara H et al (1998) Biochemical properties of human oral polymorphonuclear leukocytes. Free Radic Res 28(5):485–495
70. Tsukahara Y et al (1998) Expression of inducible nitric oxide synthase in circulating neutrophils of the systemic inflammatory response syndrome and septic patients. World J Surg 22(8):771–777
71. Greenberg SS et al (1998) Human and rat neutrophils constitutively express neural nitric oxide synthase mRNA. Nitric Oxide 2(3):203–212
72. Greenberg SS et al (1996) An in vivo cytokine and endotoxin-independent pathway for induction of nitric oxide synthase II mRNA, enzyme, and nitrate/nitrite in alveolar macrophages. Biochem Biophys Res Commun 227(1):160–167
73. Sethi S et al (2001) Nitric oxide- and oxygen-derived free radical generation from control and lipopolysaccharide-treated rat polymorphonuclear leukocyte. Nitric Oxide 5(5):482–493
74. Sato Y et al (2004) Identification of caveolin-1-interacting sites in neuronal nitric-oxide synthase. Molecular mechanism for inhibition of NO formation. J Biol Chem 279(10):8827–8836
75. Felley-Bosco E et al (2000) Caveolin-1 down-regulates inducible nitric oxide synthase via the proteasome pathway in human colon carcinoma cells. Proc Natl Acad Sci U S A 97(26):14334–14339
76. Chatterjee M et al (2007) Biochemical and molecular evaluation of neutrophil NOS in spontaneously hypertensive rats. Cell Mol Biol (Noisy-le-Grand) 53(1):84–93
77. Chatterjee M et al (2008) Ascorbate sustains neutrophil NOS expression, catalysis, and oxidative burst. Free Radic Biol Med 45(8):1084–1093
78. Jyoti A et al (2014) Interaction of inducible nitric oxide synthase with rac2 regulates reactive oxygen and nitrogen species generation in the human neutrophil phagosomes: implication in microbial killing. Antioxid Redox Signal 20(3):417–431
79. de Frutos T et al (2001) Expression of an endothelial-type nitric oxide synthase isoform in human neutrophils: modification by tumor necrosis factor-alpha and during acute myocardial infarction. J Am Coll Cardiol 37(3):800–807
80. Kumar S et al (2010) Functional and molecular characterization of NOS isoforms in rat neutrophil precursor cells. Cytometry A 77(5):467–477
81. Gatto EM et al (2000) Overexpression of neutrophil neuronal nitric oxide synthase in Parkinson's disease. Nitric Oxide 4(5):534–539
82. Jyoti A et al (2014) Interaction of inducible nitric oxide synthase with rac2 regulates reactive oxygen and nitrogen species generation in the human neutrophil phagosomes: implication in microbial killing. Antioxid Redox Signal 20(3):417–431
83. Sethi S, Dikshit M (2000) Modulation of polymorphonuclear leukocytes function by nitric oxide. Thromb Res 100(3):223–247
84. Stuehr DJ et al (2004) Update on mechanism and catalytic regulation in the NO synthases. J Biol Chem 279(35):36167–36170
85. Sethi S et al (1999) Nitric oxide-mediated augmentation of polymorphonuclear free radical generation after hypoxia-reoxygenation. Blood 93(1):333–340
86. Seth P et al (1994) Modulation of rat peripheral polymorphonuclear leukocyte response by nitric oxide and arginine. Blood 84(8):2741–2748
87. Kumar S et al (2010) Nitric oxide-mediated augmentation of neutrophil reactive oxygen and nitrogen species formation: critical use of probes. Cytometry A 77(11):1038–1048
88. Sharma P et al (2004) Ascorbate-mediated enhancement of reactive oxygen species generation from polymorphonuclear leukocytes: modulatory effect of nitric oxide. J Leukoc Biol 75(6):1070–1078
89. Sharma P et al (2003) Role of ascorbate in the regulation of nitric oxide generation by polymorphonuclear leukocytes. Biochem Biophys Res Commun 309(1):12–17

90. Clancy RM et al (1992) Nitric oxide, an endothelial cell relaxation factor, inhibits neutrophil superoxide anion production via a direct action on the NADPH oxidase. J Clin Invest 90(3):1116–1121
91. Nagarkoti S et al (2018) S-Glutathionylation of p47phox sustains superoxide generation in activated neutrophils. Biochim Biophys Acta 1865(2):444–454
92. Dikshit M et al (1996) Interaction of nitric oxide synthase inhibitors and their D-enantiomers with rat neutrophil luminol dependent chemiluminescence response. Br J Pharmacol 119(3):578–582
93. Patel S et al (2010) Nitric oxide donors release extracellular traps from human neutrophils by augmenting free radical generation. Nitric Oxide 22(3):226–234
94. Fujii H et al (1997) Nitric oxide inactivates NADPH oxidase in pig neutrophils by inhibiting its assembling process. J Biol Chem 272(52):32773–32778
95. Lee C et al (2000) Biphasic regulation of leukocyte superoxide generation by nitric oxide and peroxynitrite. J Biol Chem 275(50):38965–38972
96. Sethi S et al (2000) Mechanisms involved in the augmentation of arachidonic acid-induced free-radical generation from rat neutrophils following hypoxia-reoxygenation. Thromb Res 98(5):445–450
97. Klink M et al (2009) Effect of nitric oxide donors on NADPH oxidase signaling pathway in human neutrophils in vitro. Immunobiology 214:692
98. Dikshit M, Sharma P (2002) Nitric oxide mediated modulation of free radical generation response in the rat polymorphonuclear leukocytes: a flow cytometric study. Methods Cell Sci 24(1–3):69–76
99. Patel S et al (2009) Ion channel modulators mediated alteration in NO-induced free radical generation and neutrophil membrane potential. Free Radic Res 43(5):514–521
100. Keshari RS et al (2013) Reactive oxygen species-induced activation of ERK and p38 MAPK mediates PMA-induced NETs release from human neutrophils. J Cell Biochem 114(3):532–540
101. Dubey M et al (2016) Nitric oxide-mediated apoptosis of neutrophils through caspase-8 and caspase-3-dependent mechanism. Cell Death Dis 7(9):e2348
102. Sakai J et al (2012) Reactive oxygen species-induced actin glutathionylation controls actin dynamics in neutrophils. Immunity 37(6):1037–1049
103. Hattori H et al (2010) Reactive oxygen species as signaling molecules in neutrophil chemotaxis. Commun Integr Biol 3(3):278–281
104. Dal Secco D et al (2003) Neutrophil migration in inflammation: nitric oxide inhibits rolling, adhesion and induces apoptosis. Nitric Oxide 9(3):153–164
105. Nolan S et al (2008) Nitric oxide regulates neutrophil migration through microparticle formation. Am J Pathol 172(1):265–273
106. Dubey M et al (2015) L-Plastin S-glutathionylation promotes reduced binding to beta-actin and affects neutrophil functions. Free Radic Biol Med 86:1–15
107. Kumar S et al (2010) Cdk2 nitrosylation and loss of mitochondrial potential mediate NO-dependent biphasic effect on HL-60 cell cycle. Free Radic Biol Med 48(6):851–861
108. Niedbala W et al (2011) Regulation of type 17 helper T-cell function by nitric oxide during inflammation. Proc Natl Acad Sci U S A 108(22):9220–9225
109. Stark MA et al (2005) Phagocytosis of apoptotic neutrophils regulates granulopoiesis via IL-23 and IL-17. Immunity 22(3):285–294
110. Smith E et al (2007) IL-23 is required for neutrophil homeostasis in normal and neutrophilic mice. J Immunol 179(12):8274–8279
111. Saha P et al (2017) Bacterial Siderophores hijack neutrophil functions. J Immunol 198(11):4293–4303
112. Zhang D et al (2015) Neutrophil ageing is regulated by the microbiome. Nature 525(7570):528–532
113. Tang WH et al (2017) Gut microbiota in cardiovascular health and disease. Circ Res 120(7):1183–1196

114. Neilly IJ et al (1995) Plasma nitrate concentrations in neutropenic and non-neutropenic patients with suspected septicaemia. Br J Haematol 89(1):199–202
115. Pontremoli S et al (1989) Enhanced activation of the respiratory burst oxidase in neutrophils from hypertensive patients. Biochem Biophys Res Commun 158(3):966–972
116. Chatterjee M et al (2009) Augmented nitric oxide generation in neutrophils: oxidative and pro-inflammatory implications in hypertension. Free Radic Res 43(12):1195–1204
117. Morton J et al (2008) Circulating neutrophils maintain physiological blood pressure by suppressing bacteria and IFNgamma-dependent iNOS expression in the vasculature of healthy mice. Blood 111(10):5187–5194
118. Barthwal MK et al (1999) Polymorphonuclear leukocyte nitrite content and antioxidant enzymes in Parkinson's disease patients. Acta Neurol Scand 100(5):300–304
119. Srivastava N et al (2002) A study on nitric oxide, beta-adrenergic receptors and antioxidant status in the polymorphonuclear leukocytes from the patients of depression. J Affect Disord 72(1):45–52
120. Srivastava N et al (2001) Nitrite content and antioxidant enzyme levels in the blood of schizophrenia patients. Psychopharmacology 158(2):140–145
121. Jain M et al (2017) Cellular and plasma nitrite levels in myeloid leukemia - a pathogenetic decrease. Biol Chem 398
122. Kumari R et al (1994) Alterations in the bioantioxidants following thrombosis. Free Radic Biol Med 17(5):481–484
123. Kothari N et al (2012) Role of active nitrogen molecules in progression of septic shock. Acta Anaesthesiol Scand 56(3):307–315
124. Horckmans M et al (2017) Neutrophils orchestrate post-myocardial infarction healing by polarizing macrophages towards a reparative phenotype. Eur Heart J 38(3):187–197

Heat Shock Proteins and Their Associated Oxidative Stress-Induced Heart Disease

9

Sangeeta Mitra, Rakhi Dasgupta, and Angshuman Bagchi

Abstract

Heat shock proteins, apart from having a strong impactful function as molecular chaperones, are involved in a variety of diseases including cardiovascular diseases. Various studies have reported that there is an elevation of concentration gradients of circulating heat shock protein antibodies. These HSP antibodies have a strong alliance in case of extremity and advancement of cardiovascular diseases. A major stress factor, such as oxidative stress, contributes largely to endothelial dysfunction through several mechanisms, hence leading to the development of associated cardiovascular diseases. During this time, the heart accumulates misfolded proteins and chaperones/co-chaperone network function for preventing misfolding, refolding denatured proteins, and targeting them for further degradation. In this review, the cardioprotective roles of these chaperones, co-chaperones, and heat shock factors (HSF) will be discussed in correlation with oxidative stress, inflammatory cytokines, and others which are said to be acquainted with the evolution and advancement of cardiovascular diseases.

Keywords

Heat shock proteins · Small Hsps · ROS · Oxidative stress · Induced cardiovascular diseases

S. Mitra · R. Dasgupta (✉) · A. Bagchi (✉)
Department of Biochemistry and Biophysics, University of Kalyani, Kalyani, West Bengal, India

S. Chakraborti et al. (eds.), *Modulation of Oxidative Stress in Heart Disease*, https://doi.org/10.1007/978-981-13-8946-7_9

9.1 Introduction

The discovery of heat shock proteins is one of the most evolutionary findings in the history of molecular biology. It has shaped the present scenario of research in the particular field and has given it a completely new dimension. In the year 1962, the heat shock proteins were first pioneered by Ferruccio Ritossa and his coworkers. They observed that heat shock generates a different kind of puffing pattern and gene expression profile in the salivary gland of *Drosophila melanogaster* larva. Heat shock proteins are a certain type of multigene family proteins which are induced due to application of excess heat and temperature. They are highly conserved throughout different species and can be induced in prokaryotes, eukaryotes as well as in plant cells [1–4].

9.2 "Heat Shock" Proteins: The "Stress" Proteins

Heat stress helps in upregulation of the spontaneous conflation of a multigene group of heat shock proteins. This specific kind of retaliation is often known as the heat shock response. The sublethal heat stress response apparently promotes the capability of a cell to bear with a comparatively lethal subsequent heat stress challenge. This occurrence is known as thermotolerance, which deals with an important role in studies of in vitro as well as in vivo experimental models. This phenomenon shows a certain similarity in the heat shock responses and also gives protectional coverage against either simulated ischemia or hypoxia. Several multigene families consist of some stress proteins which show a range of 10 to 160 kDa in molecular size, and those are available in various membrane-enclosed regions and in significant cellular compartments (Hsp 27, Hsp 70, Hsp 90, etc.). The expression of heat shock protein genes is often induced by several significant stress activities like inflammation, amino acid analogues, puromycin, ethanol, heavy metal analogues, ischemic stress, and oxidative stress. Besides this, there are some broad varieties of stressors like arsenicals, tissue explanation, and infections caused by certain viruses. This particular term "stress proteins" also include glucose regulatory proteins (GPRs) which are localized in the rough endoplasmic reticulum (ER), and being a member of HSP family, they are also considered as "stress proteins." GRPs are induced by glucose deprivation and also by some stress factors. They also include certain HSP inducers (amino acid analogues, heat, etc.) and also induce glycosylation inhibitors that are said to be stressors selective for GRPs and Ca'+ ionophores. From the studies of both eukaryotic and prokaryotic responses, it is considered that the stress factors that induce GPRs as well as HSPs do share a nature that causes cells to synthesize proteins in both ER compartments and the compartments of the nucleoplasm. They also have a nature of damaging the aberrant proteins directly. Through several experiments, it is proved that the cells can properly make a distinctive comparison between local and denatured forms of the similar protein. There is an experimental evidence of 70 kd heat shock cognate protein (Hsc70), and it was said that confiscating Hsc70 may readily initiate the process of induction. Such a sequestration is responsible for the release of heat shock transcription factor (HSF) from which a reversible association can be seen with that of Hsc70. This ultimately releases HSF and readily triggers the heat shock genes. Moreover, the particular complexes that carries both HSF and Hsc70 were found in the extracts of Hela cell (Baler, University of Miami). Another experiment proves that in normal cells, Hsc70 shows an association with advancing chains of polypeptide. However, in a condition of stress, this kind of transient association stays for a long time. This eventually helps to exhaust the pool of Hsc70 and thus creates a high deficient functions of Hsc70. Heat shock response can activate the human hsp70 gene. This activation is readily reduced by cycloheximide. The evidence shows that the reduction in the synthesized polypeptides can be considered to be tactful targets for thermal damage. This releases the bound Hsc70 into an available pool. Structurally, it is evident that for translocation throughout the cell membrane, polypeptides start unfolding themselves which can be considered as transient proteins in cellular organisms [2, 3, 5–14].

9.3 The Classification

The classification of heat shock proteins is done based on the molecular weight they possess, for example, 60kDa molecular weight; HSp60. Molecular chaperones vary from 10 to over 100 kDa in range. According to shape, size, and molecular weight, the range of molecular chaperones varies within the cell. The physiological roles including specific locations of molecular chaperones of heat shock proteins can be present in different intracellular compartments. The entire families of HSPs are located in various locations within the cell; for example, HSP10, HSP60, and HSP75 are sited in cell organelle, mitochondria. Other Hsp families are located in the nucleus, cytoplasm, cytosol, and endoplasmic reticulum. In the process of embryonic development, the molecular weight of small heat shock proteins that ranges from 15 to 43 kDa are regarded as heat shock protein β (HSPBs); these HSPBs also possess chaperone activity. The chaperons exhibit different kinds of functions, and they are stabilization, translocation, and assembly of oligomeric proteins and proteases such as ubiquitin-dependent proteasomes which facilitate the indignity of partially or completely damaged proteins.

9.4 The Expression and Function

The expression and function of heat shock proteins is apparently constitutive in nature. The HSPs involve both chaperones and protease activities that are essential for overcoming the changes in involving protein degradation and denaturation. It is observed that in *Escherichia coli* organisms, the heat shock response is controlled by a specified factor named sigma-32 (σ32). This σ32 factor is triggered by the rpoH gene, which eventually interacts with heat shock promoters. These heat shock promoters are localized in the upstream of heat shock genes. Evidences show that the intracellular concentrations of these proteins can be increased by two- to threefold by application of excess heat which results in protein misfolding, aggregation, and protein unfolding. HSPs also function in cellular mechanism, such as receptor maturation, signal transduction, and protein trafficking. These actions are exhibited by the HSPs by depicting the term "heat shock" as a misstatement. Members of this protein family are Hsp100, Hsp90, Hsp70, Hsp60, and Hsp40, including other small heat shock proteins in the family [13–19].

9.5 Heat Shock Proteins: A Molecular Chaperon

Several studies prove the concept of HSPs playing an important role as a molecular chaperon. Molecular chaperons are those proteins which helps in folding of other polypeptides and also helps in the association of their specific oligomeric forms. Hsps have a high molecular weight in different vertebrates and plants (fungi, *Drosophila*, yeast, etc.). Small heat shock proteins usually maintain a molecular mass between 15 and 30kDa. Mammalian Hsps are said to be localized in the cytosol of the cell as well as in the tissues and consist of 32 subunits. They show an oligomeric confirmation and can be present without the presence of any external stress factors like elevated temperature. A major role of heat shock protein is to prevent the aggregation of polypeptides and proteins. They also play an important role in the unfolding of cellular proteins especially in stressed conditions. This chaperone-resembling activity was first observed in the eye lens protein α-crystallin and in the hetero-oligomer products of two genes, αA- and αB-crystallin, against the heat-persuaded aggregation of β-crystallin, γ-crystallin, and alcohol dehydrogenase. Experimental evidences say that aggregation of α-glucosidase is prevented by human Hsp 27 and murine Hsp 25. Bovine αB-crystallin also functions in aggregation of citrate synthase. Homo-oligomers of αA- and αB-crystallin also prevent aggregating a few of the target proteins, thus exhibiting a chaperon-like activity. There are several Hsps which show temperature-dependent chaperone-like activity. For example, Hsp B2 and Hsp 22 help in aggregation of target proteins. Hsp 27 proceeds with thermal induction and thus helps to increase the size of oligomeric form. Thus, this small heat shock protein correlates with elevating the molecular chaperon-like activity. Sometimes heat-induced conformational changes can be seen in rat Hsp22. An increased exposure of hydrophobic surfaces is eventually exhibited by Hsp22. This shows a strong presence of chaperone-like activity [6, 20–25].

9.6 Chaperones and Proteases

Heat shock proteins may act as molecular chaperons as well as exhibit protease activities. The chaperons exhibit different kinds of functions such as stabilization, translocation, and assembly of oligomeric proteins. On the other hand, proteases such as ubiquitin-dependent proteasomes facilitate the degradation of damaged proteins.

9.7 A Brief on Chaperon Network

In case of stressed conditions, there are several structural proteins as well as some enzymes that are responsible for some lethal changes both structurally and functionally. These proteins thus prevent the accumulation of nonlocalized proteins and help in refolding the disaggregated and denatured proteins so that they can retain their functional conformation. They also remove potentially harmful polypeptides that are nonfunctional which appears from denaturation, aggregation or in misfolding condition. These structural proteins and enzymes play a major role in cell survival under typical stressed conditions. There are several classes of Hsps that show chaperon-like activity and thus cooperate in cellular protection. These heat shock proteins sometimes play an overlapping function to protect the proteins from stressed conditions. In the case of thermosensitive *E. coli* mutants, there is a coordination between several classes of chaperones which shows plasmid-controlled chaperone expression. There are also some plant heat shock proteins which exhibit chaperon-like activities. It is observed that Hsp18.1 from pea (*Pisum sativum*) shows an interaction with heat-denatured protein. This plHsp 18.1 folds in such a competent way so that it can help in further refolding by the members of Hsp100 or Hsp 70 chaperon families. In the case of Hsp16.6 from *Synechocystis* sp., there is a binding evidence of nonnative proteins. They eventually protect them from getting aggregated and thus provide a complete pool of substrates for successive refolding. The complexes of Hsp70 and Hsp100 families are mainly taking part in it [26–29].

9.8 Hsp as Cardiovascular Chaperon

Heat shock proteins are stated as well-conserved stress proteins, and one of the primary functions of Hsp is exhibiting a chaperon-like activity. Under triggered conditions, they are enhanced by several cellular stress stimuli. There are several molecular chaperons which act as an important factor and begin a spontaneous expression in normal cardiac function with a very minimal level. These molecular chaperon components of cardiomyocyte eventually increase the target expression in connection with cardiac stress. In the case of reperfusion or ischemia, HSPs usually show upregulation in the heart. In congestive failure and hemodynamic overload, these HSPs show a chaperon-like activity and thus undergo upregulation in the heart. Hsps have a major role in various cardiovascular diseases and also have several major implications in cardiac-pathological conditions like myocardial hypertrophy, cardiomyopathy, and ischemic preconditioning. These cardiac-pathological conditions upregulate HSPs for cardioprotection. Some of the stress proteins having low molecular weight along with their major HSP families having predominant functions in cardiovascular diseases are presented below in the table [30, 31] (Table 9.1).

Table 9.1 Some Hsps and their source and importance in the cardiovascular system [14, 32–35]

Hsp families (stress proteins)	Organelle/ tissue	Protein targets	Cellular functions	Cardiovascular significance
Hsp20	Mainly cytoplasm	Hsp27, αB-crystallin	Vasorelaxation	Cardiac dysfunction, apoptosis
Hsp27	Nucleus/ cytoplasm	p38 kinase, αB-crystallin	Actin dynamics, thermotolerance	Atherosclerosis, heart attack, and stroke causes ischemia, cross-tolerance (obligatory condition)
Hsp32	Cytoplasm	αB-crystallin	Thermotolerance, cleaves heme to get carbon monoxide and the antioxidant molecule biliverdin	Regulation of ischemia/ reperfusion-induced cardiac injury
Hsp47	Endoplasmic reticulum	Procollagen I Procollagen III	Quality control of caring Procollagen synthesis	Reactive and reparative interstitial fibrosis
[Small HsP] Heme oxygenase (HO-1, HO-2)	Mainly cytoplasm	—	Hypoxia	Antioxidant activity
αB-Crystallin/ Hsp22	Cytoplasm	Hsp27	Stabilization of cytoskeleton near Z bands	Cross-tolerance Ischemia (obligatory conditioning)
Hsp60	Mitochondria	Chaperonin 10	Protein import/ folding	Heat stress, ischemia (cytochrome c and apoptosis)
Hsp70	Cytoplasm	—	Active in protein folding in endoplasmic reticulum (ER)	—
mHsp75	Mitochondria	Translocation and protein folding	—	Cytochrome c and apoptosis
Grp78/BiP	Endoplasmic reticulum	Grp94	Protein folding (unfolded proteins)	Cystic fibrosis transmembrane conductance regulator (CFTR binding)
Hsc70 (cognate)	Cytoplasm	Peroxisome	Folding of Hsp40 protein	Binding of cystic fibrosis transmembrane conductance regulator

(continued)

Table 9.1 (continued)

Hsp families (stress proteins)	Organelle/ tissue	Protein targets	Cellular functions	Cardiovascular significance
Hsp70	Cytoplasm/ nucleus	Platelets and RAD46	Folding of Hsp40 protein (heat stress/ unfolded proteins)	Heat stress, ischemia/reperfusion injury, cytoprotection (ischemic events)
Hsp90α (Hsp86)	Cytoplasm	p23, Hsp70, Hsp56,	Functional activity as aporeceptor due to heat stress	Atherosclerosis Estrogens
Hsp90β (Hsp84)	Cytoplasm	Immunophilins, steroid receptors	—	Immunosuppressive therapy, cardiac transplantation
Grp94	Endoplasmic reticulum	94-Kinase	Calcium-binding chaperone	Ischemia reperfusion
Osp94	Renal medulla	Osmotic and heat stress	—	Dehydration, hyperosmolar stress
Hsp110 (human)	Nucleolus/ cytoplasm	Hsc70, Hsp40	Thermotolerance	Ischemic cross-tolerance

9.9 Regulation and Expression of Heat Shock Protein

The induction and organization of heat shock proteins are regulated by a special kind of transcription factor, namely, heat shock protein factor (HSF), localized at the promoter region of the heat shock gene. In the case of vertebrates, four HSFs have been identified, and among these, two are ubiquitously present and conserved throughout all species of vertebrates; these are HSF1 and HSF2. Among these two, HSF1 plays the major role in vertebrates during stress conditions; on the other hand, it is found that during differentiation and early development processes, HSF2 shows higher activity. HSF1 is a monomeric molecule which is present in the cytoplasm in latent state, and also it is incompetent to bind and interact with the DNA molecule. During stress conditions, HSF1 is activated by an intracellular deluge of newly synthesized nonnative proteins, and these proteins are hyperphosphorylated in a Ras-dependent mitogen-activated protein kinase. After that, HSF1 is being transmuted to phosphorylated trimers having the capacity to get interacted with the DNA, and subsequently they are being translocated from the cytoplasm to the nucleus.

On the other hand, HSF2 proteins are said to be temperature sensitive. During increased temperature, it gets inactivated and also sequestered to the cytoplasm. Thus, it is being prevented from interacting with HSF1 in stressed cells.

The mechanisms by which heat shock proteins are being induced are highly regulated; otherwise their continuous production will affect the homeostasis and intracellular cell functions and thus leads to cell death or apoptosis. One notable mechanism which is responsible for the regulation as well as the expression of the heat shock protein is facilitated by the interaction of Hsp70 to the transactivation domain of HSF1, which leads to restraining the transcription process of the heat shock gene. DNA binding as well as the HSF1 in stress-induced phosphorylation remains unaltered by the interaction between Hsp70 and HSF1. Another mechanism which regulates heat shock protein synthesis is the interaction between heat shock protein binding factor 1 (HSBP1), the active trimeric form of HSF1, and Hsp70, which results in inhibition of the capacity of HSF1 to bind to DNA. The localization of HSBP1 is mainly in the nucleus, whereas HSBP1 mRNA can be found in higher concentrations in different cell lines and animal tissues that remain unaffected by heat shock [14, 15, 18, 19, 34–37].

9.10 Heat Shock Protein in the Cardiovascular System

In recent studies, it has been shown that a large number of people from the western world is suffering from acute and chronic ischemic heart disease which even leads to fatal death, although various exogenous pharmacological protective measures have been taken, such as coronary vasodilators, calcium antagonists, and blocking agents of the angiotensin, which later coverts into b-adrenoreceptors and enzymes too. It is reported that the heart might get some positive advantages from an endogenous source if only some protective measures are possessed. In this context, a phenomenon like elevation and upregulation of heat shock protein synthesis can improve the level of endurance to ischemic heart disease, e.g., in humans.

9.11 Heat Shock Proteins Play Active Roles in the Recovery of Ischemic Heart Disease

HSP70 is a particular heat shock protein that is found to have a significant function in preventing the onset of ischemic heart disease. Research conducted on animal models has shown that:

1. Forced overexpression of HSP70 confers a cytoprotective effect in cultured cells which includes myocytes that eventually simulate ischemia
2. Myocardial function is seen to be improved in transgenic mice due to overexpression of HSP70
3. Other than HSP70, HSP27 and αβ-crystallin are found to be effective in protecting cardiomyocytes against ischemic damage

Although results are pretty convincing, the proper mechanism of heat shock proteins in promoting cardioprotection has not been deciphered completely yet [38–44].

9.12 Biochemical Activities of HSPs in Response to Myocardial Infarction

In an experiment conducted by Benjamin and McMillan, it was observed that in cultured myogenic cells, highly acidic condition (pH 6.70) was ineffective in binding HSF1 with exposed DNA to simulate ischemia (if ATP stores were being preserved). On the other hand, when ATP molecules were severely depleted (65%), the binding of HSF1 to DNA was stimulated, in spite of normal pH range. In the case of intact ischemic heart, approximately 15 minutes of ischemia generates reversible injury and is also related to decrease in ATP stores (65%). On the contrary, lethal injury is related to prolonged ischemia (>40%) and high depletion of ATP stores (>90%). It can be concluded that biochemical features of HSPs as well as the ATP-dependent HSF1 regulatory pathway are both improbable to be unfavorably affected during the entire span of transient ischemia as well as myocardial ischemic injury. It also confirms the cardioprotective properties of heat shock proteins. In spite of these favorable outcomes, few salient features regarding HSPs and cardiovascular diseases have to be understood, and they are as follows:

1. How they are related to other intrinsic pathways which are involved in cardioprotection against reactive oxygen species
2. Specific functions of the different members of the HSP70 multigene family
3. How these pathways are related during acute ischemia and other physiological processes which induce heat shock protein production [2, 5, 6]

9.13 Relation Between Infections, Hsps, and Cardiovascular Diseases

Development of atherosclerosis appears to be a balanced event between regulatory and proinflammatory immune responses. The presence of concurrent infection is a decisive factor in immunogenic responses during the development of atherosclerosis. Accumulated evidences gathered from the infection of *Chlamydia pneumonia* indicate that infective pathogens take part in the development of atherosclerosis. The organism *C. pneumonia* is found in atherosclerosis plaque, and it is able to induce foam cell formation in macrophages. Researchers have isolated T-cells specific for *C. pneumonia* from atherosclerosis plaque, which concludes the fact that the organism is able to elicit T-cell-mediated immunity. Substantial work has been done on isolated Hsp60 from *C. pneumonia*. It shows that HSP 60 may have the potential to induce macrophage production of TNF-α and matrix metalloproteinase activity. In spite of these convincing results, the definitive correlation between atherosclerosis and infection has not been elucidated clearly. Still some researchers have deciphered a positive correlation between antibodies against *C. pneumoniae* and the antibody levels of antimycobacterial hsp65 [45–50].

Table 9.2 Table showing some hands-on laboratory experimental details along with their clinical manifestation in cardiovascular disease [5, 6, 51]

Disease name	Laboratory studied data	Clinical manifestation
Cardiovascular diseases	(i) Overexpression of SOD gives a nature of protection against several injuries	Alpha-Tocopherol, Beta-Carotene Cancer Prevention (ATBC) Study says that no overall benefit on cardiovascular disease (CVD) rate with vitamin E or beta-carotene is found but increase in CVD deaths with beta-carotene 10 is found as evidence
	(ii) Interference and disarrangement of SOD often lead to heart failure	
	Vitamin E provides protection against development and growth of atherosclerosis 87	Cambridge Heart Antioxidant Study (CHAOS) says that vitamin E eventually lowers the rate of nonfatal myocardial infarct
	Pre-atherosclerotic blood vessels show the elevation of functional ROS86	Physicians' Health Study I (PHS I) says that there is strong evidence in high-risk subgroups

Some laboratory data clearly depict about the oxidative functions in the pathogenesis of cardiovascular diseases. Several proceedings of antioxidant vitamins such as vitamin A and E describe their potential outcomes along with some marginal results (Table 9.2).

9.14 Pathogens Involved in Cardiovascular Diseases

A study conducted by Haraszthy et al. [52] identified 22 different pathogens which take active part in human carotid atheromas. The identification was done based on 50 carotid atheromas that found at endarterectomy due to the presence of 16S rDNA via polymerase chain reaction. Over here, specific probes are used for periodontal pathogens like *B. forsythus*, *Actinobacillus actinomycetemcomitans*, *P. gingivalis*, and *Prevotella intermedia*. The result showed that 13 (26 %) were positive for *P. gingivalis*, 15 (30%) of the specimen were positive for *B. forsythus*, 9 (18%) were positive for *A. actinomycetemcomitans*, and 7 (14 %) were positive for *P. intermedia*. Additionally, 9 (18%) of these atheromas were detected in *C. pneumoniae* DNA. It is evident from these studies that pathogens like *C. pneumonia* can have a major role in the succession of atherosclerosis. In other studies, *P. gingivalis* was found to be proficient to invade both the carotid and coronary endothelium during cell culture (Table 9.3).

9.15 Disease Mechanism

A proposal of direct mechanism is made by Herzberg and colleagues and Herzberg and Meyer [53, 54], which suggests that dental infection bacteria is able to trigger cardiovascular diseases. In another study, it was revealed that an oral gram-positive

Table 9.3 Causative agents incriminated in atherogenesis

Bacteria	Virus
Helicobacter pylori	Cytomegalovirus (CMV)
Porphyromonas gingivalis	Herpes simplex virus (HSV)
Streptococcus mutans	Coxsackie B
Chlamydia pneumoniae	Hepatitis A
Enterobacteriaceae	Influenza A

bacteria (*Streptococcus sanguis*) and a gram-negative periodontal pathogen *P. gingivalis* promote the production of platelet aggregation-associated proteins. These might induce atheroma production. Endotoxin lipopolysaccharide (LPS) is on such stimuli, which is able to activate the inflammatory cytokine cascade. These cytokines can be a potent source of atherosclerosis stimulation, which induces the liver so that it can produce some acute-phase proteins. Acute-phase proteins like C-reactive protein (CRP) and fibrinogen can affect coagulation, platelet aggregation, and activation. The expression of leukocyte adhesion molecules such as ICAM and VCAM can also be increased by inflammatory cytokines, and LPS is seen in periodontal disease. This often leads to atheroma formation [55, 56].

9.16 Cardiovascular Disease and Heat Shock Protein Antibodies

In context with the functional activity as molecular chaperones, the proteins of HSP family performs immunodominance-like phenomenon which describes that immune responses are depicted against only a few antigenic peptides among thousands produced, and as a result, only a distinct element of the immune response to pathogenic microorganisms is forwarded toward heat shock protein peptides. They are obtained from the phylogenetic similarity between mammalian and microbial forms of these molecules (45–75% identical residues in the case of the 60 kDa family). A proposal said that those elements could behave like harmful autoantigens, and moreover, from those particular infectious agents that cause immune responses to heat shock protein, determinant might play an important role in cross-reaction having similarity with “self” molecules. As a result, a connection between several autoimmune diseases and their respective infection conditions is highly in demand.

There is a strong association observed between the progressive engrossment of antibodies to the 65 kDa mycobacterial heat shock protein (HSP65) that is said to be 75% homologous in nature to human Hsp60 with effectiveness, severity, and elevation of cardiovascular disease. As a consequence, anti-Hsp65 antibody values predict the 5-year mortality of the corresponding patients having carotid atherosclerosis and the several prevalence of cardiovascular properties. Therefore, the proposal comes that immunity forwarding to heat shock proteins may have a high chance to influence the elevation, progression, and development of various severe cardiovascular diseases. As a result, those antibodies via cross-reactivity to Hsp65 mediate endothelial cytotoxicity expressed with Hsp60, and on the surface of human endothelial cells after heat treatment, this can guide the evidence that suggests that such particular interactions can result in endothelial injury [57–60].

9.17 Oxidative Stress and Cardiovascular Diseases

Around 1950, the eminent scientist Denham Harman observed the "free radical theory" of aging. According to his words, endogenous oxygen radicals were procured in cells, and thus a consecutive pattern of cumulative damage is observed. After a decade, an enzyme was evident whose one of the main functions is removal of superoxide anions. This is vividly known as superoxide dismutase (SOD). SOD has been classified into three enzymatic types: (i) Cu/Zn SOD, (ii) Mn SOD, and (iii) extracellular SOD. This SOD spontaneously dismutases $O_2–$ to H_2O_2. As a result, H_2O_2 eliminates glutathione peroxidase (GPx) and catalase to water. SOD is also responsible for providing other essential mechanistic details for Harman's hypothesis. Oxidative stress plays a major role in the development of cardiovascular diseases as well as in pathogenesis. Several oxidase enzymes such as cyclooxygenase, xanthine oxidase, and nicotinamide adenine dinucleotide phosphate (NADPH) are responsible for the generation of ROS. Uncoupled endothelial NO synthase (eNOS) and mitochondrial electron transport are also involved in the process of ROS generation. ROS undergoes involvement of a variety of cell types, like endothelial cells, mononuclear cells, and vascular smooth muscle cells (VSMCs). HOCl, $ONOO^-$, and H_2O_2 are classified as non-free radicals, and OH, O_2^-, and NO are classified as free radicals; both groups have potent oxidation ability [61–70] (Fig. 9.1).

9.18 Oxidative Stress in Response to Pathogenesis

Various patterns of evidence demonstrate that several physiological manifestations like atherosclerosis, myocardial infarction, hypertension, heart failure, dyslipidemia, diabetes mellitus, and angina pectoris include a major role played by oxidative stress. It is said that protective antioxidant mechanisms show a multifactorial as well as a complex nature. There are such antioxidant defense systems like catalase, SOD, scavenge ROS, and GPx. These antioxidant defense systems play an important role in the inhibition of NO degradation. Several sensitive transcriptional factors such as hypertrophy, cell proliferation, and apoptosis are induced by oxidative stress. Moreover, oxidative stress also induces activation of various signaling cascades. Lipid peroxidation induces overexpression of redox genes, which eventually damages the endothelial or myocardial cells. Oxidative stress-induced atherosclerosis leads to the development

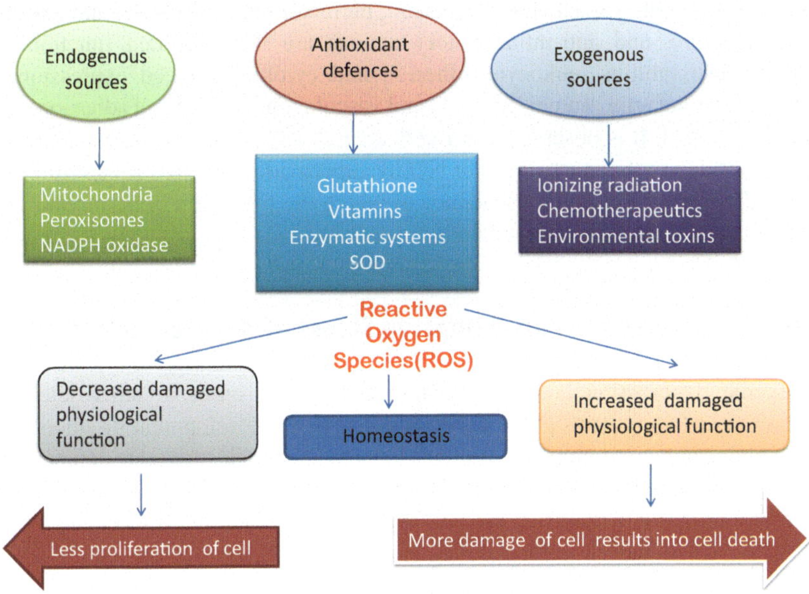

Fig. 9.1 Generation of ROS and its corresponding cellular response

of atherosclerosis. Excess ROS such as free radicals involves oxidation of various molecules. It is also evident that protein oxidation induces overexpression of redox genes and results in the damage of vascular smooth muscle cell (VSMCs) as well as plays a major role in damaging endothelial cells.

9.19 Effect of Oxidative Stress on Endothelium

There are different kinds of bioactive layers found in our system. Among them, the endothelium acts as the inner layer of blood vessels. The most important function of the endothelium is to control vascular tone permeability. Extracellular matrices (collagens), adhesion molecules, and other regulatory mediators like endothelin-1 (ET-1), NO, angiotensin II (Ang II), prostanoids, and von Willebrand factor (VWF) are being produced by the endothelium.

Research shows that in the case of atherosclerotic disease, endothelial dysfunction (ED) is an early sign which leads to subsequent clinical manifestations as well as complications. Accumulated evidences indicate the fact that ED can be termed as a strong predictor of future onset of cardiovascular diseases. It is observed that reactive oxygen species (ROS) plays a key function in promoting ED and thus may act as intracellular messengers. These intracellular messengers (ROS) eventually tend to modulate other signaling pathways. If production of ROS species gets increased, it can severely damage the endothelium and can lead to atherosclerosis. NO produced by endothelial cells is one of the most important chemical modulators, and it acts as a formidable vasodilator. Decreasing permeability, antiplatelet and antiproliferative nature, and anti-inflammatory actions are the important functions of NO. NO also inhibits leukocyte adhesion and cytokine-induced expression of monocyte chemotactic protein (MCP-1) and vascular endothelial cell adhesion molecule (VCAM-1). It is observed that ED decreases NO production and availability to an extent due to the inactivation of NO by superoxide. Superoxides are found to react with NO rapidly, as a consequence of which the formation of peroxynitrite is readily observed with the bioavailability loss of NO [71–74].

9.20 Relation Between Oxidative Stress and Atherosclerosis

Endothelial dysfunction is a major cause behind the onset of atherosclerosis. Generation of free radicals plays an integral part in the development of atherosclerosis which can lead to myocardial infarction and sudden death in due course. Free radicals can severely damage the inner layer of the blood vessels. The development of atherosclerosis is a multistep process where plasma cholesterol level and proliferation of smooth muscle cells play key roles [70].

9.21 Cardioprotective Effects of Heat Shock Proteins

9.21.1 Stress Induced

The concept of heat stress preconditioning was first observed by Currie et al. [75]. It was treated as a strategy for myocardial infarction. It also gets several benefits for the ischemic myocardium. Cardiovascular function often tends to get protected by pre-induction of HSP70 following trauma hemorrhage [76]. According to Dillman et al. [77], the increased level of inducible Hsp70 plays a major role in hearts subjected to prolonged ischemia and necrosis. Overexpression of HSP70 usually protects the heart against the lethal damaging effects of ischemia. Correction of metabolic acidosis and recovery of high energy phosphate stores act as competent determinants as a protecting component of the heart. According to the description by Latchman [78], the cardioprotective mechanistic details of HSPs inhibit the mitochondrial caspase 9 pathway by HSP27, HSP70, and HSP90. The overexpression of HSP70 enhances NO production in response to cytokine stimulation [79]. HSP70 has also been known as a cytoprotective therapeutic agent [80]. It basically reduces I/R by decreasing the huge risk of postoperative atrial fibrillation [81]. It is observed that in any kind of oxidative stress, free radicals are formed in myocardial I/R injury that results in a phenomena named “myocardial stunning.” This is also known as ventricular dysfunction, or arrhythmias. Molecular chaperones also have some intense effect on cardioprotection in connection with the reactive oxygen species (ROS). The cardioprotection by these molecular chaperones involves some of the protective mechanism of superoxide dismutase (SOD) and glutathione peroxidase (GPx and catalase) [82]. Various models of cell culture exhibit their activity in expressing the mechanistic path of cardiac protection of HSPs. In pediatric patients, during cardiac surgery heat shock proteins play a major role in cardioprotective effect [83].

9.21.2 In Vascular Endothelial Cells

In context with maintaining the hemodynamic stress of the vascular cells [84], heat shock proteins play certain important functions by acting as a molecular chaperon. The reperfusion does not extend to the coronary endothelium [85]. As a result, the endothelial cells, which is also known as cardiomyocyte, also express heat shock proteins in response to congestive heart failure. Coronary endothelial cells are considered as a foremost localization of induction of HSP70 family which eventually provides protection to heat stress on the recovery of cellular and endothelial function [86]. It has also been observed that if the levels of anti-HSP antibodies are elevated, it can actually enhance ischemic stroke [87].

9.21.3 HSP32 and Its Role in Cardioprotection

HSP32 is a 32 KDa protein which induced different kinds of stress situations. Some of the stressed conditions are ischemia/reperfusion, hypoxia, heavy metals (e.g., selenium, cobalt, cadmium, stannous ions) and hydrogen peroxide. Hsp32 plays a key function in cardioprotection by mediating vasodilation of VSMCs and guanylyl cyclase-dependent platelet inhibition. Angiotensin II treatment conducted on mice model shows a decrease in Hsp32 mRNA expression. Evidences also show that it has the potential to modulate oxidative stress during ischemia and vascular tone as well as inhibit platelet aggregation thus promoting cardioprotection [88–90].

9.21.4 Inducible Cardiac Synthesis of Hsps

Heat shock proteins show their presence in vascular compartments as well as in cardiac compartments and can be induced by some specific stressors. It is noted that the type of proteins expressed in the heart is in some way different from the vascular compartment. In the case of adult mice, those in nonstressed conditions such as Hsps (Hsp27, Hsp70, and Hsp84) are constitutively expressed in several tissues, including the heart. These four Hsps usually retain a comparative low level in the heart compared with the other tissues. In contrast, cardiac Hsp70 levels are similar to those in other tissues in the case of the adult rabbit. It is observed that levels are much higher in unstressed rats, whereas intermediate levels of this B-crystallin show their presence for Hsp27.

9.22 The Proteomic Aspect of Heat Shock Proteins Related to Cardiac Response and Stress

Proteomic analysis in cardiovascular biology is a hot topic of interest nowadays. Endothelial cells and smooth muscles are said to be responsible for the changes that occurred in the cardiovascular system. Proteomic approaches can unite these particular changes in response to cardiovascular stress. It is observed that in terms of cardiovascular diseases, differential proteomic approaches consist of mitochondrial HSP70 precursor, mitochondrial stress protein (HSP70), protein changes of HSP72, HSP70, mitochondrial matrix protein p1 (membrane-bound HSP60), and HSP27. Changes in cardiac protein expression determine the posttranslational modifications of cardiac proteins in response to cardiomyopathy.

As the process of CHF induction is over, the cardiac sHSP tends to be expressed in an elevated way. This shows a kind of proteomic analysis of cardiac sHSP expression in the case of congestive heart failure (CHF). A large number of HSPs function as molecular chaperones of the cardiovascular system. In the pathophysiology of CHF [91], αB-crystallin, HSP20, and HSP27 significantly increase the level of CHF compared to a normal heart and also perform a censorious redeeming role. Almost 50 HSP27 protein species revealed the analysis of dilated cardiomyopathy-diseased human myocardial tissue by immunoblotting mechanism [92].

9.23 Conclusions

Heat shock proteins (HSPs) principally persuade stress stimuli, specifically in cardiovascular diseases. By inhibiting cellular apoptotic mechanisms, they eventually protect the cardiac tissue from further damage and recurring injuries. Heat shock proteins are said to be significant in the prevention of apoptosis. By regulating cells under normal conditions, most of the intrinsic HSPs play a vital role in biochemistry. A majority of HSPs are induced by stress stimuli and thus promote the resistance mechanism of tissues upon initial stress. These Hsps act like the first line of defense. Heat shock proteins undergo a mechanism of self-induced cardioprotection as well as in vascular endothelial cells. Hsps are also involved in constitutive and inducible effect of cardiac synthesis. This cardioprotective nature of the heat shock protein could be indispensable to reclaim cardiac tissues during consecutive cardiac stress. Thus, comprehension of the physiological functions of HSPs would definitely be beneficial for the preparation and development of synthetic drugs and also exhibit effectiveness against cardiovascular prophylaxis. Therefore, heat shock proteins are said to be the "guardian" components in response to cardiovascular stresses.

References

1. Benjamin IJ, McMillan DR (1998) Stress (heat shock) proteins molecular chaperones in cardiovascular biology and disease. Circ Res 83:117–132. https://doi.org/10.1161/01.RES.83.2.117
2. Lindquist S, Craig EA (1988) The heat-shock proteins. Annu Rev Genet 22:631–677
3. Ritossa F (1996) Discovery of the heat shock response. Cell Stress Chaperones 1:97–98
4. Tissieres A, Mitchell HK, Tracy UM (1974) Protein synthesis in salivary glands of Drosophila melanogaster: relation to chromosome puffs. J Mol Biol 84:389–398
5. Hartl F (1996) Molecular chaperones in cellular protein folding. Nature 381:571–579
6. Gething MJ, Sambrook J (1992) Protein folding in the cell. Nature 355:33–45
7. Hutter JJ, Mestril R, Tam EK, Sievers RE, Dillmann WH, Wolfe CL (1996) Overexpression of heat shock protein 72 in transgenic mice decreases infarct size in vivo. Circulation 94:1408–1411
8. Hightower LE, Sadis SE, Takenaka IM (1994) Interactions of vertebrate hsc70 and hsp70 with unfolded proteins and peptides. In: Morimoto RI, Tissieres A, Georgopoulos C (eds) The biology of heat shock proteins and molecular chaperones. Cold Spring Harbor Press, Cold Spring Harbor
9. Hightower LE, Hendershot LM (1997) Molecular chaperones and the heat shock response at Cold Spring Harbor. Cell Stress Chaperones 2:1–11
10. Currie RW, Karmazyn M, Kloc M, Mailer K (1988) Heat-shock response is associated with enhanced postischemic ventricular recovery. Circ Res 63:543–549
11. Marber MS, Latchman DS, Walker JM, Yellon DM (1993) Cardiac stress protein elevation 24 hours after brief ischemia or heat stress is associated with resistance to myocardial infarction. Circulation 88:1264–1272
12. Arrigo AP, Suhan JP, Welch WJ (1988) Dynamic changes in the structure and intracellular locale of the mammalian low-molecular-weight heat shock protein. Mol Cell Biol 8:5059–5071
13. Abravaya K, Phillips B, Morimoto RI (1991) Attenuation of the heat shock response in HeLa cells is mediated by the release of bound heat shock transcription factor and is modulated by changes in growth and in heat shock temperatures. Genes Dev 5:2117–2127
14. Satyal SH, Chen D, Fox SG et al (1998) Negative regulation of the heat shock transcriptional response by HSBP1. Genes Dev 12:1962–1974
15. Abravaya K, Myers MP, Murphy SP, Morimoto RI (1992) The human heat shock protein Hsp70 interacts with HSF, the transcription factor that regulates heat shock gene expression. Genes Dev 6:1153–1164
16. Amici C, Sistonen L, Santoro MG, Morimoto RI (1992) Antiproliferative prostaglandins activate heat shock transcription factor. Proc Natl Acad Sci 89:6227–6231
17. Gamer J, Multhaup G, Tomoyasu T, McCarty JS, Rudiger S, Schonfeld H-J, Schirra C, Bujard H, Bukau B (1996) A cycle of binding and release of the DnaK, DnaJ and GrpE chaperones regulates activity of the Escherichia coli heat shock transcription factor s32. EMBO J 15:607–617
18. Mosser DD, Duchaine J, Massie B (1993) The DNA-binding activity of the human heat shock transcription factor is regulated in vivo by Hsp70. Mol Cell Biol 13:5427–5438
19. Mosser DD, Caron AW, Bourget L, Denis-Larose C, Massie B (1997) Role of the human heat shock protein Hsp70 in protection against stress-induced apoptosis. Mol Cell Biol 17:5317–5532
20. Gupta SC, Sharma A, Mishra M, Mishra RK, Chowdhuri DK (2010) Life Sci 86(11–12):377–384
21. Boston RS, Viitanen PV, Vierling E (1996) Complexity of the heat stress response in plants. Plant Mol Biol 32(1–2):191–222
22. Li Z, Menoret A, Srivastava P (2002) Roles of heat-shock proteins in antigen presentation and cross-presentation. Curr Opin Immunol 14(1):45–51
23. Chisholm ST, Coaker G, Day B, Staskawicz BJ (2006) Host-microbe interactions: shaping the evolution of the plant immune response. Cell 124(4):803–814

24. Chen W, Syldath U, Bellmann K, Burkart V, Kolb H (1999) Human 60-kDa heat-shock protein: a danger signal to the innate immune system. J Immunol 162(6):3212–3219
25. Williams JH, Ireland HE (2008) Sensing danger--Hsp72 and HMGB1 as candidate signals. J Leukoc Biol 83(3):489–492
26. Veinger L et al (1998) The small heat-shock protein IbpB from E. coli stabilizes stress-denatured proteins for subsequent refolding by a multichaperone network. J Biol Chem 273:11032–11037
27. Lee GJ, Vierling E (2000) A small heat shock protein cooperates with heat shock protein 70 systems to reactivate a heat-denatured protein. Plant Physiol 122:189–198
28. Mogk A et al (2003) Refolding of substrates bound to small Hsps relies on a disaggregation reaction mediated most efficiently by ClpB/DnaK. J Biol Chem 278:31033–31042
29. Mogk A et al (2003) Small heat shock proteins, ClpB and the DnaK system form a functional triade in reversing protein aggregation. Mol Microbiol 50:585–595
30. Tiefenbrunn AJ, Sobel BE (1992) Timing of coronary recanalization: paradigms, paradoxes, and pertinence. Circulation 85:2311–2315
31. Rogers WJ, Bowlby LJ, Chandra NC, French WJ, Gore JM, Lambrew CT, Rubison RM, Tiefenbrunn AJ, Weaver WD (1994) Treatment of myocardial infarction in the United States (1990 to 1993): observations from the National Registry of Myocardial Infarction. Circulation 90:2103–2114.7
32. Denis M, Gustafsson J (1989) The Mr approximately 90,000 heat shock protein: an important modulator of ligand and DNA-binding properties of the glucocorticoid receptor. Cancer Res 49:2275s–2281s
33. Pratt W (1993) The role of heat shock proteins in regulating the function, folding, and trafficking of the glucocorticoid receptor. J Biol Chem 268:21455–21458
34. Wu C, Clos J, Giorgi G, Haroun RI, Kim S-J, Rabindran SK, Westwood JT, Wisniewski J, Yim G (1994) Structure and regulation of heat shock transcription factor. In: Morimoto RI, Tissieres A, Georgopoulos C (eds) The biology of heat shock proteins and molecular chaperones. Cold Spring Harbor Laboratory Press, Cold Spring Harbor, pp 395–416
35. Borkovich KA, Farrelly FW, Finkelstein DB, Taulien J, Lindquist S (1989) hsp82 is an essential protein that is required in higher concentrations for growth of cells at higher temperatures. Mol Cell Biol 9:3919–3930
36. Wu BJ, Kingston RE, Morimoto RI (1986) Human HSP70 promoter contains at least two distinct regulatory domains. Proc Natl Acad Sci 83:629–633
37. Morimoto RI, Jurivich DA, Kroeger PE, Mathur SK, Murphy SP, Nakai A, Sarge K, Abravaya K, Sistonen LT (1994) Regulation of heat shock gene transcription by a family of heat shock factors. In: Morimo RI (ed) The biology of heat shock proteins and molecular chaperones. Cold Spring Harbor Laboratory Press, Cold Spring Harbor
38. Plumier JC, Ross BM, Currie RW, Angelidis CE, Kazlaris H, Kollias G, Pagoulatos GN (1995) Transgenic mice expressing the human heat shock protein 70 have improved post-ischemic myocardial recovery. J Clin Invest 95:1854–1860
39. Marber MS, Mestril R, Chi SH, Sayen MR, Yellon DM, Dillmann WH (1995) Over expression of the rat inducible 70-kD heat stress protein in atransgenic mouse increases the resistance of the heart to ischemic injury. J Clin Invest 95:1446–1456
40. Radford NB, Fina M, Benjamin IJ, Moreadith RW, Graves KH, Zhao P, Gavva S, Wiethoff A, Sherry AD, Malloy CR, Williams RS (1996) Cardio protective effects of 70-kDa heat shock protein transgenic mice. Proc Natl Acad Sci U S A 93:2339–2342
41. Williams RS, Thomas JA, Fina M, German Z, Benjamin IJ (1993) Human heat shock protein 70 (hsp70) protects murine cells from injury during metabolic stress. J Clin Invest 92:503–508
42. Mestril R, Chi SH, Sayen MR, O'Reilly K, Dillmann WH (1994) Expression of inducible stress protein 70 in rat heart myogenic cells confers protection against simulated ischemia-induced injury. J Clin Invest 93:759–767
43. Martin JL, Mestril R, Hilal-Dandan R, Brunton LL, Dillmann WH (1997) Small heat shock proteins and protection against ischemic injury in cardiac myocytes. Circulation 96:4343–4348

44. Benjamin IJ, Kroger B, Williams RS (1990) Activation of the heat shock transcription factor by hypoxia in mammalian cells. Proc Natl Acad Sci U S A 87:6263–6267
45. Graham Pockley A (2002) Heat shock proteins, inflammation, and cardiovascular disease. Circulation 105:1012–1017. https://doi.org/10.1161/hc0802.103729
46. Kuo CC, Shor A, Campbell LA et al (1993) Demonstration of Chlamydia pneumoniae in atherosclerotic lesions of coronary arteries. J Infect Dis 167:841–849
47. Jackson LA, Campbell LA, Kuo CC et al (1997) Isolation of Chlamydia pneumoniae from a carotid endarterectomy specimen. J Infect Dis 176:292–295
48. Laitinen K, Laurila A, Pyhala L et al (1997) Chlamydia pneumoniae infection induces inflammatory changes in the aortas of rabbits. Infect Immun 65:4832–4835
49. Muhlestein JB, Anderson JL, Hammond EH et al (1998) Infection with Chlamydia pneumoniae accelerates the development of atherosclerosis and treatment with azithromycin prevents it in a rabbit model. Circulation 97:633–636
50. Caligiuri G, Rottenberg M, Nicoletti A et al (2001) Chlamydia pneumoniae infection does not induce or modify atherosclerosis in mice. Circulation 103:2834–2838
51. Strickland E, Qu BH, Millen L, Thomas PJ (1997) The molecular chaperone Hsc70 assists the in vitro folding of the N-terminal nucleotide-binding domain of the cystic fibrosis trans membrane conductance regulator. J Biol Chem 272:25421–25424
52. Haraszthy VI, Zambon JJ, Trevisan M, Zeid M, Genco RJ (2000) Identification of periodontal pathogens in atheromatous plaques. J Periodontol 71:1554–1560
53. Herzberg M, Brintzenhofe K, Clawson C (1983) Aggregation of human platelets and adhesion of Streptococcus sanguis. Infect Immun 39:1457–1469
54. Herzberg MC, Meyer MW (1996) Effects of oral flora on platelets: possible consequences in cardiovascular disease. J Periodontol 67(supplement 10):1138–1142
55. Castell JV, Andus T, Kunz D, Heinrich P (1989) Interleukin-6: the major regulator of acute-phase protein synthesis in man and rat. Ann N Y Acad Sci 557:87–99
56. Yamauchi-Takihara K, Ihara Y, Ogata A, Yoshizaki K, Azuma J, Kishimoto T (1995) Hypoxic stress induces cardiac myocyte-derivedinterleukin-6. Circulation 91:1520–1524
57. McLaughlin MM, Kumar S, McDonnell PC, Van Horn S, Lee JC, Livi GP, Young PR (1996) Identification of mitogen-activated protein (MAP) kinase activated protein kinase-3, a novel substrate of CSBP p38 MAP kinase. J Biol Chem 271:8488–8492
58. Huot J, Lambert H, Lavoie JN, Guimond A, Houle F, Landry J (1995) Characterization of 45-kDa/54-kDa HSP27 kinase, a stress-sensitive kinase which may activate the phosphorylation-dependent protective function of mammalian 27-kDa heat-shock protein HSP27. Eur J Biochem 227:416–427
59. Landry J, Chretien P, Lambert H, Hickey E, Weber LA (1989) Heat shock resistance conferred by expression of the human HSP27 gene in rodent cells. J Cell Biol 109:93–101
60. Zu YL, Wu F, Gilchrist A, Ai Y, Labadia ME, Wang CK (1994) The primary structure of a human MAP kinase activated protein kinase 2. Biochem Biophys Res Commun 200:1118–1124
61. Harman D (1957) Aging: a theory based on free radical and radiation chemistry. J Gerontol 2:298–300
62. McCord JM, Fridovich I (1969) Superoxide dismutase. An enzymatic function for erythrocuperin (hemocuperin). J Biol Chem 244:6049–6055
63. Ku HH, Brunk UT, Sohal RS (1993) Relationship between mitochondrial superoxide and hydrogen peroxide production and longevity of mammalian species. Free Radic Biol Med 15:621–627
64. Finkel T (1998) Oxygen radicals and signaling. Curr Opin Cell Biol 10:248–253
65. Nishikawa T et al (2000) Normalizing mitochondrial superoxide production blocks three pathways of hyperglycaemic damage. Nature 404:787–790
66. Nemoto S, Takeda K, Yu ZX, Ferrans VJ, Finkel T (2000) A role for mitochondrial oxidants as regulators of cellular metabolism. Mol Cell Biol 20:7311–7318
67. Suh YA et al (1999) Cell transformation by the superoxide-generating oxidase Mox1. Nature 401:79–82

68. Geiszt M, Kopp JB, Varnai P, Leto TL (2000) Identification of Renox, an NAD(P)H oxidase in kidney. Proc Natl Acad Sci U S A 97:8010–8014
69. Shi Y, Mosser DD, Morimoto RI (1998) Molecular chaperones as HSF1-specific transcriptional repressors. Genes Dev 12:654–666
70. Lakshmi SVV (2009) Oxidative stress and cardiovascular disease. Indian J Biochem Biophys 46:421–440
71. Vidal-Puig AJ et al (2000) Energy metabolism in uncoupling protein 3 gene knockout mice. J Biol Chem 275:16258–16266
72. Vanhoutte PM (1989) Endothelium and control of vascular function: state of the art lecture. Hypertension 13:658–667
73. Raij L (1993) Nitric oxide and the kidney. Circulation 87(Suppl V):V-26–V-29
74. Higashi Y, Oshima T, Ozono R, Watanabe M, Matsuura H, Kajiyama G (1995) Effects of L-arginine infusion on renal hemodynamics in patients with mild essential hypertension. Hypertension 25:898–902
75. Currie RW, Karmazyn M, Kloc M, Mailer K (1988) Heat-shock response is associated with enhanced postischemic ventricular recovery. Circ Res 63(3):543–549
76. Mizushima Y, Wang P, Jarrar D et al (2000) Preinduction of heat shock proteins protects cardiac and hepatic functions following trauma and hemorrhage. Am J Physiol Regulatory Integrative Comp Physiol 278:R352–R359
77. Dillmann WH, Mehta HB, Barrieux A, Guth BD, Neeley WE, Ross J (1986) Ischemia of the dog heart induces the appearance of a cardiac mRNA coding for a protein with migration characteristics similar to heatshock/stress protein 71. Circ Res 59(1):110–114
78. Latchman D (2001) Heat shock proteins and cardiac protection. Cardiovasc Res 51(4):637–646
79. Bellmann K, Burkart V, Bruckhoff J, Kolb H, Landry J (2000) p38-dependent enhancement of cytokine-induced nitric-oxide synthase gene expression by heat shock protein 70. J Biol Chem 275(24):18172–18179
80. Rokutan K, Hirakawa T, Teshima S et al (1998) Implications of heat shock/stress proteins for medicine and disease. J Med Investig 44:137–147
81. Rammos K (2002) Low preoperative HSP70 atrial myocardial levels correlate significantly with high incidence of postoperative atrial fibrillation after cardiac surgery. Cardiovasc Surg 10(3):228–232
82. Benjamin IJ, McMillan DR (1998) Stress (heat shock) proteins. Circ Res 83(2):117–132
83. Giannessi D (2003) A possible cardioprotective effect of heat shock proteins during cardiac surgery in pediatric patients. Pharmacol Res 48(5):519–529
84. Xu Q, Wick G (1996) The role of heat shock proteins in protection and pathophysiology of the arterial wall. Mol Med Today 2(9):372–379
85. Laude K (2002) Endothelial protective effects of preconditioning. Cardiovasc Res 55(3):466–473
86. Amrani M, Latif N, Morrison K, Gray CC, Jayakumar J, Corbett J, Goodwin A, Dunn M, Yacoub M (1998) Relative induction of heat shock protein in coronary endothelial cells and cardiomyocytes: implications for myocardial protection. J Thorac Cardiovasc Surg 115(1):200–209
87. Gromadzka G, Zielińska J, Ryglewicz D, Fiszer U, Członkowska A (2001) Elevated levels of anti-heat shock protein antibodies in patients with cerebral ischemia. Cerebrovasc Dis 12(3):235–239
88. Cruse I, Maines MD (1988) Evidence suggesting that the two forms of heme oxygenase are products of different genes. J Biol Chem 263:3348–3353
89. Ewing JF, Maines MD (1991) Rapid induction of heme oxygenase 1 mRNA and protein by hyperthermia in rat brain: heme oxygenase 2 is not a heat shock protein. Proc Natl Acad Sci 88(12):5364–5368
90. Mccoubrey WK, Huang TJ, Maines MD (1997) Isolation and characterization of a cDNA from the rat brain that encodes Hemoprotein Heme Oxygenase-3. Eur J Biochem 247(2):725–732

91. Dohke T, Wada A, Isono T, Fujii M, Yamamoto T, Tsutamoto T, Horie M (2006) Proteomic analysis reveals significant alternations of cardiac small heat shock protein expression in congestive heart failure. J Card Fail 12(1):77–84
92. Scheler C, Müller E-C, Stahl J, Müller-Werdan U, Salnikow J, Jungblut P (1997) Identification and characterization of heat shock protein 27 protein species in human myocardial two-dimensional electrophoresis patterns. Electrophoresis 18(15):2823–2831

Modulation of Oxidative Stress in Cardiovascular Diseases

10

Jay C. Jha, Madhura Bose, and Karin Jandeleit-Dahm

Abstract

Cardiovascular disease (CVD), the leading cause of morbidity and mortality, represents a major global health and economic burden worldwide [1, 2]. The World Health Organization report has projected that approximately half of all deaths in developed countries will be due to CVD by 2020. CVD is a multifactorial disorder, which encompasses a broad range of injuries of the vasculature and heart including atherosclerosis, coronary heart disease leading to myocardial infarction, peripheral vascular disease, stroke, aneurysms, and cardiomyopathy [3–6] (Fig. 10.1). There is no single cause for CVD, but there are a range of risk factors, which increase the likelihood for clinical manifestations of cardiovascular disease. These risk factors for CVD include obesity, dyslipidemia, diabetes, hypertension, smoking, and aging as well as a positive family history and environmental factors [5, 7–9]. A significant number of studies have shown a close association among these cardiovascular risk factors. Indeed, hypertension, dyslipidemia, obesity, insulin resistance, and chronic hyperglycemia often coexist and synergistically enhance the risk for CVD-related deaths [1, 5, 7, 8]. Reports suggest that diabetes increases the risk of stroke and myocardial infarction with diabetic patients demonstrating a 1.7 times higher risk of CVD death than nondiabetic individuals [1]. In addition, the risk for CVD including coronary disease and stroke is elevated with a rise in blood pressure [8, 10]. Smoking is an avoidable risk factor of CVD, and a person's risk of CVD mortality can be reduced by 36% over 2 years upon cessation of smoking [11]. The burden of CVD risk increases with age and can be decreased partly by modifying and monitoring other coexisting CVD risk factors [12]. CVD can also result from environmental and demographic factors. The high prevalence

J. C. Jha · M. Bose · K. Jandeleit-Dahm (✉)
Department of Diabetes, Monash University, Melbourne, VIC, Australia
e-mail: karin.jandeleit-Dahm@monash.edu

S. Chakraborti et al. (eds.), *Modulation of Oxidative Stress in Heart Disease*,
https://doi.org/10.1007/978-981-13-8946-7_10

of CVD and its risk factors among the general population have motivated research investigating the pathological mechanisms of CVD and to develop novel approaches to prevent the progression of this disease. This has led to a better understanding of the underlying pathogenic mechanisms for the development and progression of CVD.

Keywords
Obesity · Diabetes · Oxidative stress · Hypertention · Cardiovascular disease

10.1 Introduction

Cardiovascular disease (CVD), the leading cause of morbidity and mortality, represents a major global health and economic burden worldwide [1, 2]. The World Health Organization report has projected that approximately half of all deaths in developed countries will be due to CVD by 2020. CVD is a multifactorial disorder, which encompasses a broad range of injuries of the vasculature and heart including atherosclerosis, coronary heart disease leading to myocardial infarction, peripheral vascular disease, stroke, aneurysms, and cardiomyopathy [3–6] (Fig. 10.1). There is no single cause for CVD, but there are a range of risk factors, which increase the likelihood for clinical manifestations of cardiovascular disease. These risk factors for CVD include obesity, dyslipidemia, diabetes, hypertension, smoking, and aging as well as a positive family history and environmental factors [5, 7–9]. A significant number of studies have shown a close association among these cardiovascular risk factors. Indeed, hypertension, dyslipidemia, obesity, insulin resistance, and chronic hyperglycemia often coexist and synergistically enhance the risk for CVD-related deaths [1, 5, 7, 8]. Reports suggest that diabetes increases the risk of stroke and myocardial infarction with diabetic patients demonstrating a 1.7 times higher risk of CVD death than nondiabetic individuals [1]. In addition, the risk for CVD including coronary disease and stroke is elevated with a rise in blood pressure [8, 10]. Smoking is an avoidable risk factor of CVD, and a person's risk of CVD mortality can be reduced by 36% over 2 years upon cessation of smoking [11]. The burden of CVD risk increases with age and can be decreased partly by modifying and monitoring other coexisting CVD risk factors [12]. CVD can also result from environmental and demographic factors. The high prevalence of CVD and its risk factors among the general population have motivated research investigating the pathological mechanisms of CVD and to develop novel approaches to prevent the progression of this disease. This has led to a better understanding of the underlying pathogenic mechanisms for the development and progression of CVD.

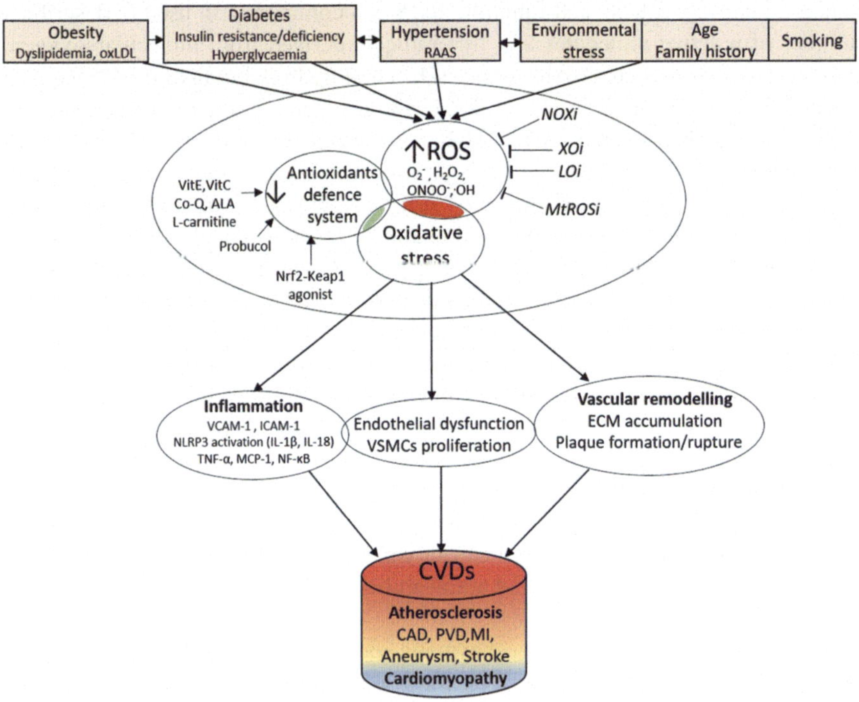

Fig. 10.1 Schema represents the modulation of oxidative stress in cardiovascular diseases and potential avenues for intervention. Reactive oxygen species (ROS); superoxide anion (O_2^-); hydrogen peroxide (H_2O_2); hydroxyl radical (OH^-); peroxynitrite ($ONOO^-$); oxidized low-density lipoprotein (Ox-LDL); renin-angiotensin-aldosterone system (RAAS); vascular cell adhesion molecule-1 (VCAM-1); intracellular adhesion molecule-1 (ICAM-1); tumor necrosis factor-α (TNF-α); interleukin-1-β (IL1-β); interleukin-18 (IL-18); monocyte chemoattractant protein-1 (MCP-1); nuclear transcription factor-κB (NFκB); vascular smooth muscle cells (VSMCs); extracellular matrix (ECM); NOX inhibitor (*NOXi*); xanthine oxidase inhibitor (*XOi*); lipoxygenase inhibitor (*LOi*); mitochondrial ROS inhibitor (*mtROSi*); coronary heart disease (CAD); myocardial infarction (MI); and peripheral vascular disease (PVD)

10.2 Oxidative Stress in CVD

A plethora of both experimental and clinical studies suggest that oxidative stress, a pathological state as a result of imbalance between reactive oxygen species (ROS) production and antioxidant defense systems, plays a crucial role in the pathogenesis of CVD [13]. ROS act as a double-edged sword in cellular processes; at a low level, it participates in the cell signaling process, whereas at a high level, it has cytotoxic effects by interacting with macromolecules including DNA, proteins, and lipids leading to cell death [14–16]. It has been suggested that the cellular redox potential is an important determinant of cell function, and interruption of redox balance adversely affects cell function. ROS is a common term used to describe a number of reactive molecules and free radicals derived from molecular oxygen. Free radicals such as superoxide anion ($O_2^{\cdot -}$) and hydroxyl radical ($OH^{\cdot -}$) have an extremely high chemical reactivity due to the presence of unpaired free electron. Other ROS like hydrogen peroxide (H_2O_2), peroxynitrite ($ONOO^-$), and hypochlorous acid (HOCl) are not free radicals as they lack the free unpaired electron but have oxidizing effects instead of reactive effects resulting in oxidant stress [14–16]. Generation of ROS requires a series of chain reactions which also leads to formation of more ROS through a vicious cycle [15, 16]. In the physiological system, ROS promotes cellular activities, regulates hormone level, preserves chemical balance, enhances synaptic plasticity, and induces enzymes. Moreover, ROS also helps to combat against invading pathogens and induce an immune response against the pathogens [15]. To some extent, ROS are neutralized and kept at homeostatic levels by intracellular antioxidant enzymes such as superoxide dismutase (SOD), glutathione peroxidase (GPx), and catalase. Other nonenzymatic antioxidants consumed as supplements include β-carotene, ascorbic acid, and tocopherol supplements which balance ROS at normal levels. The disruption of the balance between ROS production and antioxidant defense capacity results in oxidative damage of the cell membrane integrity causing altered permeability and changes in expression of proteins [16, 17].

A variety of enzymatic and nonenzymatic sources of ROS exists in cardiovascular cells. Therefore, ROS produced by these sources play a crucial role in normal vascular physiology but also in cardiovascular disease. The sources of ROS in the cardiovascular system includes the mitochondrial respiratory chain, NADPH oxidases (NOX), lipoxygenase (LO), xanthine oxidase (XO), cytochrome P450 (CYP), and uncoupled nitric oxide synthases (NOS) [15–17]. Among these enzymatic sources of ROS, NADPH oxidases are the only known enzymes solely dedicated to ROS generation [13, 18, 19]. In addition to these, a number of external mediators also contribute to the production of ROS including ionizing radiation, heavy metals, metal complexes, nanoparticles, cigarette smoke, various drugs, and certain types of other chemical compounds [20].

The increase in cardiovascular ROS due to overproduction and/or a decrease in degradation as well as a decline in the level of antioxidants result in oxidative damage to various cellular components of the heart and blood vessels [13, 15]. The common risk factors for CVD include chronic hyperglycemia, hypertension, hypercholesterolemia, smoking, and aging which further enhance the level of ROS generation in CVD [13, 15–17]. Enhanced levels of cardiovascular ROS mediated by these risk factors can activate the key elements of proinflammatory and profibrotic pathways causing endothelial dysfunction; upregulation of adhesion molecules (VCAM-1, ICAM-1), cytokines (TNF-α), and chemokines (MCP-1); activation of transcription factors (NFκB), the NLRP3 inflammasome (IL-1β, IL-18), and metalloproteinases (MMPs); accumulation of extracellular matrix proteins (collagens and fibronectin) and proteoglycans; as well as induction of proliferation and migration of smooth muscle cells, lipid peroxidation, and change in vasomotor functions, which collectively lead to CVD [15, 17, 21] (Fig. 10.1).

10.3 Antioxidants in Reducing CVD

Several clinical studies have been conducted to investigate the use of dietary antioxidant supplements containing vitamins (e.g., vitamin E, vitamin C, coenzyme Q10, L-carnitine, and a-lipoic acid) in combination or alone to alleviate the burden of oxidative stress-induced cardiovascular tissue damage in CVD including diabetes and hypertension. However, the effectiveness of antioxidant interventions in clinical trials still remains a controversy and thereby remains a challenge for future therapeutic approaches. Indeed, a meta-analysis showed that many of the trials failed to improve the disease outcome and in fact some of them had adverse effects [22]. Therefore, the use of these dietary antioxidants is arguable as their mechanism of action and function remain poorly understood. Since exogenous antioxidants have largely failed to improve disease outcomes in clinical trials, new approaches to combat dysregulation of redox status are necessary to attenuate the progression of CVD.

Nrf2 Activators Endogenous antioxidants play an important role in combating oxidative stress in various diseases, including CVD. The nuclear factor E2-related factor-2 (Nrf2) plays a pivotal role as a key regulator of the expression of several critical antioxidant and detoxification genes. Activation of antioxidant response element by Nrf2 in combination with its negative regulator, Kelch-like ECH-associated protein 1 (Keap1), leads to increased expression and activity of several antioxidants, including glutathione S-transferase, hemeoxigenase-1 (HO-1), and c-glutamylcysteine synthetase, and NADPH quinone oxidoreductase [23, 24]. Nrf2-deficient mice showed decreased expression of antioxidant genes in association with increased tissue oxidative stress and toxicity in the vasculature with activation of inflammatory pathways [25, 26]. Therefore, Nrf2 activation or its downstream genes can be targeted by therapeutics to lower the impact of CVD. The Keap1-Nrf2 and Nox pathways have been shown to regulate both mitochondrial and cytosolic ROS production [27]. Indeed, the deficiency of Nrf2 in cardiomyocytes showed ROS (hydrogen peroxide, peroxynitrite, and 4-hydroxy-2-nonenal) induced cell injury [28]. However, Nrf2 overexpression in endothelial cells resulted in decreased expression of inflammatory mediators such as TNF-α, IL-1β, MCP1, and VCAM1, suggesting an anti-inflammatory potential of Nrf2 [29]. Activators of Nrf2 vary in their chemical properties and the mechanism of action. Nrf2 activation has been shown to be critical in the defense against a variety of cardiovascular diseases, including high glucose-induced oxidative damage to heart tissue [30]. Indeed, Protandim, an Nrf2 activator, showed a robust increase in heme oxygenase-1 (HO-1) in coronary artery endothelial cells and cardiac myocytes [30, 31]. In addition, in a clinical study, Protandim showed decreased levels of ROS in association with a significant increase in uric acid and the activity of erythrocyte SOD and catalase [30, 31]. In addition, administration of a novel Nrf2 activator, bardoxolone methyl derivative, dh404, in diabetic mice showed upregulation of Nrf2-responsive genes including HO-1 and downregulation of proinflammatory mediators such as VCAM-1 and the p65 subunit of NF-κB in association with attenuation of endothelial dysfunction and atherosclerosis via reduction in both systemic and vascular ROS [32, 33]. Furthermore, in a clinical trial, despite an initial benefit in patients with diabetic nephropathy, bardoxolone methyl in the subsequent phase 3 trial was prematurely terminated because of higher cardiovascular mortality in the treated group (ClinicalTrials.gov NCT01351675) [34, 35]. In addition, the use of other Nrf2 activators including sulforaphanes and MG132 demonstrated attenuation of cardiac hypertrophy, fibrosis, and inflammation as well as reduced aortic wall thickness and structural derangement of the aorta via reduction in ROS production and upregulation of Nrf2-dependent antioxidative function in diabetic mice [36–38]. However, the study also revealed that chronic use of these Nrf2 activators could be detrimental to cardiac function. Findings from these studies suggest that Nrf2 activators may have a significant therapeutic potential against CVD; however, further detailed studies are necessary to unravel the role of Nrf2 in the pathogenesis of cardiovascular diseases prior to its therapeutic use for the treatment of CVD [39, 40].

Probucol The antioxidant, Probucol, a clinically used lipid-lowering drug, has reported to possess cardioprotective effects in experimental models of CVD by enhancing the endogenous antioxidant reserve [41]. The use of Probucol showed alleviation of atherosclerosis through accelerating the process of reverse cholesterol transport by lowering the level of ox-LDL and improving HDL function in association with improving anti-inflammatory and antioxidant functions [42, 43]. In addition, a study using WHHL rabbits as a model of human familial hypercholesterolemia showed that administration of Probucol reduced plasma cholesterol levels and resulted in a prominent reduction of aortic en face lesions, reduced coronary artery stenosis, and increased plaque stability as well as reduced macrophages and increased smooth muscle cells [44]. These results suggest that Probucol treatment may have beneficial effects on the plaque stability of hypercholesterolemic patients. In addition, Probucol also suppressed hydrogen peroxide-induced ROS in human endothelial cells. AGI-1067, a stable analog for Probucol, has similar antioxidant and vasculoprotective properties [45]. In ApoE-deficient mice, AGI-1067 exhibited lipid-lowering and anti-inflammatory functions, thereby reducing the progression of atherosclerosis [45]. A phase III clinical trial of AGI-1067 has shown reduced morbidity from stroke and myocardial infarction in patients with atherosclerosis [43, 46].

TPCD NP Because conventional antioxidant therapies have shown limited clinical outcomes, it has been suggested that a broad-spectrum ROS-scavenging nanoparticle could function as a potent therapy in atherosclerosis [47, 48]. A broad-spectrum ROS-eliminating material was synthesized and named TPCD and its nanoparticle, TPCD NP. The nanoparticle was rapidly and effectively internalized by macrophages and vascular smooth muscle cells and inhibited inflammation and apoptosis in macrophages, by reducing intracellular ROS production. In ApoE$^{-/-}$ mice, TPCD

NPs inhibited development of atherosclerosis and decreased systemic and local oxidative stress and inflammation [48]. It also reduced inflammatory cell infiltration in atherosclerotic plaques. The study indicated that TPCD NPs were safe after long-term treatment and have the potential to be developed as an anti-atherosclerotic nanotherapy [48].

10.4 Agents Targeting the Source of ROS in Reducing the Risk of CVD

Evidence suggests that the use of dietary antioxidant supplements failed to improve the health of patients with cardiovascular diseases. This is partly because of the lack of information about the specificity and the mechanism of action of these antioxidants. This raises many questions in relation to our current knowledge of the molecular processes involved in ROS formation and downstream effects. Based on previous experimental studies, it appears to be more beneficial to directly target the source of ROS in order to maintain the cellular antioxidant-redox homeostasis, thereby alleviating the oxidative damage of cardiovascular tissues [16, 18, 49, 50].

NADPH Oxidases Significant progress has been made to better understand the role of pro-oxidant enzymes, NADPH oxidases (NOX), in cardiovascular pathophysiology and the contribution of individual NOX isoforms in the pathogenesis of CVD [21, 51–53]. Under physiological conditions, most NOX isoforms have very low or no constitutive activity, but the enzyme can be activated in disease states such as hypertension and diabetes. In these situations, increased NOX-derived ROS surpasses the handling capacity of the endogenous antioxidant system, thus leading to increased oxidative stress and ultimately tissue injury [16, 21]. Nox 1, 2, 4, and 5 isoforms of NADPH oxidase are expressed in cardiomyocytes and vascular cells [21, 54–56]. Based on experimental evidence, it is postulated that NOX-generated ROS modulates both vascular physiology and pathology [21, 54, 56]. Studies in experimental animals revealed that Nox isoforms, particularly Nox2 and Nox4, are upregulated in heart disease and play crucial roles in cardiac hypertrophy, fibrosis, and cardiac remodeling [57–62]. However, Nox1 has been shown to be a potential target in atherosclerosis [21, 52, 63]. Certain compounds including apocynin [64], plumbagin [65], GLX351322 [66], and probucol [67] are shown to be associated with partial NOX inhibition and reduced ROS formation and showed a certain degree of protection against renal and cardiovascular injury in experimental studies. However, these compounds lack specificity toward the individual NOX isoform. Furthermore, angiotensin-converting enzyme inhibitors (ACEi), vitamins, angiotensin receptor antagonists, calcium channel blockers, as well as statins were found to inhibit NOX activity and reduce oxidative stress [19].

The advancement in NOX/ROS pathobiology has shown some degree of progress in developing NOX-specific agents. Some of these compounds, considered NOX-specific inhibitors, named GKT136901 and GKT137831, developed by

Genkotex (www.genkyotex.com/), have shown promising results at the preclinical level. Both GKT136901 and GKT137831 are dual inhibitors for the NOX1 and NOX4 isoforms, but they also confer a lesser degree of inhibitory action on NOX5 and an almost negligible effect on NOX2 [68–71]. However, the inhibitory mechanism of action of these compounds is poorly understood. Experimental data demonstrated that application of GKT136901 partially reduced atherosclerosis and renal injury in diabetes via reduction in ROS production [72–74]. In addition, we and others have demonstrated more pronounced athero- and renoprotective effects of GKT137831 in animal models of insulin-deficient diabetes (STZ-ApoE KO and OV26 mice and Akita mouse) in association with reduced renal and vascular ROS formation and inflammation [50, 52, 63, 75, 76], suggesting that concomitant NOX1 and NOX4 inhibition can provide simultaneous athero- and renoprotection. In addition, the antiatherosclerotic effect of GKT13781 was associated with a decrease in MCP-1 expression and reduced macrophage accumulation within vascular wall [52]. Studies in patients with GKT137831 have demonstrated excellent tolerability and reduction of various markers of chronic inflammation. A phase IIb clinical trial of GKT137831 in patients with type 2 diabetic kidney disease showed significant reduction in markers of inflammation and ROS; however, reduction in albuminuria was not observed [21]. In addition, a longer-duration and higher-dosage clinical trial of this drug is under investigation in patients with type 1 diabetic nephropathy. Since Nox5 is absent in rodents and present in humans in addition to other NOX isoforms, the function of NOX5 requires further extensive investigation. This will provide impetus for the development of isoform-specific NOX inhibitors. It needs to be shown if isoform-specific NOX inhibitors such as NOX5 inhibition alone or in combination with other NOX isoforms will have superior protection against end-organ tissue damage.

Xanthine Oxidase Xanthine oxidase (XO) is predominantly found in mammalian tissues with elevated expression in capillary endothelium [77]. This oxidase catalyzes the conversion of hypoxanthine to xanthine and generates $O_2^{\cdot-}$ and H_2O_2 [78]. Pharmacological inhibition of XO by allopurinol or oxypurinol or by inactivating XO by a tungsten-rich, molybdenum-deficient diet to experimental animals has been shown to reduce atherosclerosis [79]. It has been revealed that NADPH oxidase upregulates XO expression and thereby $O_2^{\cdot-}$ generation implying that factors controlling NADPH oxidase may also affect XO [80]. Among the XO inhibitors, allopurinol has proven to be effective as second-line drugs in patients suffering from chronic stable ischemic heart disease and has been recommended in this setting by current evidence-based guidelines [81]. It has been shown that inhibition of XO with allopurinol reduced ROS production and intracellular Ca^{2+} overload in both ischemia-reperfusion-injured rat hearts and hypoxia-reoxygenation-injured cardiomyocytes [82]. The use of allopurinol in patients with type 1 diabetic nephropathy is under investigation in the prevention of early loss of renal function (PERL-study, NCT 02017171). Furthermore, a clinical study showed reduction in the progression of kidney disease and cardiovascular risk in patients with gout and diabetes treated with allopurinol [83, 84]. In addition, febuxostat, a nonpurine inhibitor of XO, has been found to reduce stress-induced ROS production and adipose tissue inflammation in experimental animals [85]. Febuxostat has been shown to have pharmacological advantages over the commonly used doses of allopurinol as it has higher serum urate-lowering efficacy, lower reported hypersensitivity reactions, and no requirement for adjustment of doses in patients with moderate renal impairment [86, 87]. However, long-term studies are required to demonstrate any beneficial effect of febuxostat on cardiovascular outcomes in CVD [88].

Lipoxygenases Lipoxygenases (LOs) are nonheme, iron-containing enzymes that catalyze insertion of molecular oxygen into fatty acids [89]. Previous studies have depicted the role of LOs in cardiovascular pathophysiology. Studies have indicated that LOs promote atherosclerosis by producing ROS and oxidized low-density lipoprotein (Ox-LDL) [90, 91]. It has been shown in several experimental models that leukocyte-type 12-lipoxygenase- and 15-lipoxygenase-1 (12/15-LO)-deficient mice demonstrated decreased atherosclerosis [92–94]. Deficiency of 12/15-LO has also been shown to reduce interleukin production and the adhesion of monocytes to endothelial cells in atherosclerosis [92]. The number of 5-lipoxygenase (5-LO) positive cells has been reported to rise in human atherosclerotic plaque specimens, and increased 5-LO activity has been associated with plaque instability [95, 96]. Interestingly, inhibition of 5-LO decreased production of leukotrienes and coronary plaque burden (as assessed by CT scan) in patients suffering from acute coronary syndrome [96]. Another compound named DG031 which inhibits 5-LO activating protein (FLAP) has been found to decrease biomarkers of cardiovascular risk in patients [97]. Furthermore, it has been shown that targeting the 5-LO pathway could effectively lower coronary heart disease and abdominal aortic aneurysms. In addition, another 5-lipoxygenase inhibitor, VIA-2291 (atreleuton), reduced leukotriene production and coronary plaque burden in patients with recent acute coronary syndrome [96]. However, the results require to be confirmed in a larger scale trial for a longer duration. In addition to XO and LO, the cytochrome P450 (CYP) system has been shown to have an indirect role in ROS production in the cardiovascular system. It has been shown that cytochrome P450 produces ROS via activation of NOX, causing renal and vascular cell injury in diabetic mice [98, 99]. Upregulation of CYP expression has been shown to be associated with risk factors for MI, cardiomyopathy, and heart failure [99]. A cardioprotective role for CYP2C9 inhibitors including chloramphenicol, cimetidine, and sulfaphenazole was demonstrated in an ischemia-reperfusion model of MI [100, 101].

There is evidence to suggest that uncoupled eNOS-mediated ROS generation leads to low levels of endothelial NO bioavailability which results in vascular endothelial dysfunction in cardiovascular diseases [102]. Studies have shown that arterial segments from humans with atherosclerosis exhibit eNOS uncoupling [103]. In addition, eNOS uncoupling can be reversed by treating with 5-methyl tetrahydrofolate (5-MTHF; the active form of folic acid), which elevates intracellular BH4 levels, and thereby reduces vascular superoxide production [104].

Mitochondrial Respiratory Chain There is increasing evidence suggesting that mitochondrial respiratory chain-derived ROS are crucial mediators in the progression of CVDs [105, 106]. Though most of the electron flux through mitochondria is used to reduce cellular oxygen to water, 1–2% of the electrons are leaked from the redox centers in the electron transport chain to oxygen [107, 108]. As a result, oxygen is reduced to $O_2^{\cdot-}$ and acts as a precursor of most ROS and mediates oxidative chain reactions. $O_2^{\cdot-}$ is further dismuted by superoxide dismutase or spontaneously to H_2O_2 which in turn may be completely reduced to water or partially reduced to one of the powerful oxidants hydroxyl radical [107, 108]. It has been shown that there is a higher number of mitochondria in the cardiomyocytes in comparison to other cell types [107]. It is unknown whether mitochondrial deregulation is a cause or consequence of cellular dysfunction, but it appears to be part of a vicious cycle perpetuating ROS generation and end-organ injury. Under pathological conditions such as CVD, higher levels of mitochondrial ROS (mtROS) produced by increased flux through oxidative phosphorylation induce adverse effects, including cell apoptosis, hypertrophy, and inflammation leading to abnormal cardiac and vascular function [109–111]. Various studies have shown that mtROS regulate important vascular function in physiological conditions and activate inflammatory pathways in response to CVD risk factors [110, 112].

The importance of eliminating enhanced mitochondrial $O_2^{\cdot-}$ has been demonstrated in a study using animals which were depleted for the manganese superoxide dismutase (Mn-SOD) allele. It has been found that the Mn-SOD-deficient mice exhibited perinatal lethality due to cardiac dysfunction [113]. In addition, a cardiac-specific Mn-SOD deletion in animals showed progressive congestive heart failure with morphological changes of mitochondria [112]. On the other hand, overexpression of the mitochondrial antioxidant peroxiredoxin-3 (Prx-3) has been shown to prevent left ventricular (LV) remodeling and heart failure after myocardial infarction [114]. Overall these studies imply that overproduction and less utilization of mitochondrial ROS in the heart and vasculatures play a key role in the development and progression of CVD and heart failure.

Recent progress in characterizing mtROS has led to the generation of a new paradigm, in which blockade of mtROS production may serve as a promising therapy for inhibiting proinflammatory cytokine production and in turn CVD, including atherosclerosis and heart diseases. The outer mitochondrial membrane protein voltage-dependent anion channel 1 (VDAC1) has been considered a promising target for therapeutic intervention related to mitochondrial dysfunction in CVD. The compounds VBIT-3 and VBIT-4 have been shown to inhibit VDAC1 apoptosis-associated mitochondrial dysfunction, reestablishing mitochondrial membrane potential and ultimately lowering ROS production [115]. In addition, inhibition of mtROS by MitoTEMPO suppressed LPS-induced endothelial cell activation and aortic monocyte recruitment in ApoE-deficient mice [116]. Furthermore, treatment with MitoQ (mitochondria-targeted ubiquinone) in Akita mice showed decreased nuclear accumulation of profibrotic transcription factors, phospho-Smad2/3, and beta-catenin, indicating reduced TGF-β/Smad signaling in these mice [117]. These results indicate that targeting of mitochondrial ROS can be a promising therapy for vascular inflammation and cardiovascular diseases.

10.5 Conclusion

The precise underlying redox mechanism of CVD and consequences of elevated ROS in cardiovascular tissue are intricate and have not yet been fully elucidated. But it is well established that ROS play a critical role in the pathogenesis and development of CVD including atherosclerosis and heart diseases. Future studies are required to understand the mechanisms that control the activation of individual sources of ROS in cardiovascular cells, particularly the isoform-specific role of NOX enzymes in CVD, the crosstalk between NOX isoforms and other sources of ROS, and their involvement in cardiovascular diseases.

References

1. Benjamin EJ, Blaha MJ, Chiuve SE, Cushman M, Das SR, Deo R et al (2017) Heart disease and stroke statistics-2017 update: a report from the American Heart Association. Circulation 135(10):e146–e603
2. Collaborators GBDRF (2018) Global, regional, and national comparative risk assessment of 84 behavioural, environmental and occupational, and metabolic risks or clusters of risks for 195 countries and territories, 1990–2017: a systematic analysis for the global burden of disease study 2017. Lancet 392(10159):1923–1994
3. Haffner SM, Lehto S, Ronnemaa T, Pyorala K, Laakso M (1998) Mortality from coronary heart disease in subjects with type 2 diabetes and in nondiabetic subjects with and without prior myocardial infarction. N Engl J Med 339(4):229–234
4. Reaven GM, Chen YD (1996) Insulin resistance, its consequences, and coronary heart disease. Must we choose one culprit? Circulation 93(10):1780–1783
5. Reaven GM, Lithell H, Landsberg L (1996) Hypertension and associated metabolic abnormalities–the role of insulin resistance and the sympathoadrenal system. N Engl J Med 334(6):374–381
6. Selby JV, Newman B, Quiroga J, Christian JC, Austin MA, Fabsitz RR (1991) Concordance for dyslipidemic hypertension in male twins. JAMA 265(16):2079–2084
7. Haffner SM (1998) Epidemiology of type 2 diabetes: risk factors. Diabetes Care 21(Suppl 3):C3–C6
8. Oparil S, Zaman MA, Calhoun DA (2003) Pathogenesis of hypertension. Ann Intern Med 139(9):761–776
9. Reed T, Quiroga J, Selby JV, Carmelli D, Christian JC, Fabsitz RR et al (1991) Concordance of ischemic heart disease in the NHLBI twin study after 14–18 years of follow-up. J Clin Epidemiol 44(8):797–805
10. Dinesh Shah A, Langenberg C, Rapsomaniki E, Denaxas S, Pujades-Rodriguez M, Gale CP et al (2015) Type 2 diabetes and incidence of a wide range of cardiovascular diseases: a cohort study in 1.9 million people. Lancet 385(Suppl 1):S86
11. Rigotti NA, Clair C (2013) Managing tobacco use: the neglected cardiovascular disease risk factor. Eur Heart J 34(42):3259–3267
12. Dhingra R, Vasan RS (2012) Age as a risk factor. Med Clin North Am 96(1):87–91
13. Panth N, Paudel KR, Parajuli K (2016) Reactive oxygen species: a key hallmark of cardiovascular disease. Adv Med 2016:9152732

14. Droge W (2002) Free radicals in the physiological control of cell function. Physiol Rev 82(1):47–95
15. He F, Zuo L (2015) Redox roles of reactive oxygen species in cardiovascular diseases. Int J Mol Sci 16(11):27770–27780
16. Jha JC, Banal C, Chow BS, Cooper ME, Jandeleit-Dahm K (2016) Diabetes and kidney disease: role of oxidative stress. Antioxid Redox Signal 25(12):657–684
17. Taverne YJ, Bogers AJ, Duncker DJ, Merkus D (2013) Reactive oxygen species and the cardiovascular system. Oxidative Med Cell Longev 2013:862423
18. Jha JC, Watson AMD, Mathew G, de Vos LC, Jandeleit-Dahm K (2017) The emerging role of NADPH oxidase NOX5 in vascular disease. Clin Sci 131(10):981–990
19. Paravicini TM, Touyz RM (2008) NADPH oxidases, reactive oxygen species, and hypertension: clinical implications and therapeutic possibilities. Diabetes Care 31(Suppl 2):S170–S180
20. Galano A, Castaneda-Arriaga R, Perez-Gonzalez A, Tan DX, Reiter RJ (2016) Phenolic melatonin-related compounds: their role as chemical protectors against oxidative stress. Molecules 21(11):1442
21. Jha JC, Ho F, Dan C, Jandeleit-Dahm K (2018) A causal link between oxidative stress and inflammation in cardiovascular and renal complications of diabetes. Clin Sci 132(16):1811–1836
22. Akbar S, Bellary S, Griffiths HR (2011) Dietary antioxidant interventions in type 2 diabetes patients: a meta-analysis. Br J Diab Vasc Dis 11(2):62–68
23. Kaspar JW, Niture SK, Jaiswal AK (2009) Nrf2:INrf2 (Keap1) signaling in oxidative stress. Free Radic Biol Med 47(9):1304–1309
24. Kensler TW, Wakabayashi N, Biswal S (2007) Cell survival responses to environmental stresses via the Keap1-Nrf2-ARE pathway. Annu Rev Pharmacol Toxicol 47:89–116
25. Chan K, Kan YW (1999) Nrf2 is essential for protection against acute pulmonary injury in mice. Proc Natl Acad Sci U S A 96(22):12731–12736
26. Thimmulappa RK, Lee H, Rangasamy T, Reddy SP, Yamamoto M, Kensler TW et al (2006) Nrf2 is a critical regulator of the innate immune response and survival during experimental sepsis. J Clin Invest 116(4):984–995
27. Kovac S, Angelova PR, Holmström KM, Zhang Y, Dinkova-Kostova AT, Abramov AY (2015) Nrf2 regulates ROS production by mitochondria and NADPH oxidase. Biochim Biophys Acta Gen Subj 1850(4):794–801
28. Zhu H, Jia Z, Misra BR, Zhang L, Cao Z, Yamamoto M et al (2008) Nuclear factor E2-related factor 2-dependent myocardiac cytoprotection against oxidative and electrophilic stress. Cardiovasc Toxicol 8(2):71–85
29. Chen LH, Huang Q, Wan L, Zeng LY, Li SF, Li YP et al (2006) Expression, purification, and in vitro refolding of a humanized single-chain Fv antibody against human CTLA4 (CD152). Protein Expr Purif 46(2):495–502
30. Reuland DJ, McCord JM, Hamilton KL (2013) The role of Nrf2 in the attenuation of cardiovascular disease. Exerc Sport Sci Rev 41(3):162–168
31. Donovan EL, McCord JM, Reuland DJ, Miller BF, Hamilton KL (2012) Phytochemical activation of Nrf2 protects human coronary artery endothelial cells against an oxidative challenge. Oxidative Med Cell Longev 2012:132931
32. Sharma A, Rizky L, Stefanovic N, Tate M, Ritchie RH, Ward KW et al (2017) The nuclear factor (erythroid-derived 2)-like 2 (Nrf2) activator dh404 protects against diabetes-induced endothelial dysfunction. Cardiovasc Diabetol 16(1):33
33. Tan SM, Sharma A, Stefanovic N, Yuen DY, Karagiannis TC, Meyer C et al (2014) Derivative of bardoxolone methyl, dh404, in an inverse dose-dependent manner lessens diabetes-associated atherosclerosis and improves diabetic kidney disease. Diabetes 63(9):3091–3103
34. Pergola PE, Krauth M, Huff JW, Ferguson DA, Ruiz S, Meyer CJ et al (2011) Effect of bardoxolone methyl on kidney function in patients with T2D and stage 3b-4 CKD. Am J Nephrol 33(5):469–476
35. Pergola PE, Raskin P, Toto RD, Meyer CJ, Huff JW, Grossman EB et al (2011) Bardoxolone methyl and kidney function in CKD with type 2 diabetes. N Engl J Med 365(4):327–336

36. Bai Y, Cui W, Xin Y, Miao X, Barati MT, Zhang C et al (2013) Prevention by sulforaphane of diabetic cardiomyopathy is associated with up-regulation of Nrf2 expression and transcription activation. J Mol Cell Cardiol 57:82–95
37. Miao X, Cui W, Sun W, Xin Y, Wang B, Tan Y et al (2013) Therapeutic effect of MG132 on the aortic oxidative damage and inflammatory response in OVE26 type 1 diabetic mice. Oxidative Med Cell Longev 2013:879516
38. Wang Y, Zhang Z, Sun W, Tan Y, Liu Y, Zheng Y et al (2014) Sulforaphane attenuation of type 2 diabetes-induced aortic damage was associated with the upregulation of Nrf2 expression and function. Oxidative Med Cell Longev 2014:123963
39. Howden R (2013) Nrf2 and cardiovascular defense. Oxidative Med Cell Longev 2013:104308
40. Mann GE, Bonacasa B, Ishii T, Siow RC (2009) Targeting the redox sensitive Nrf2-Keap1 defense pathway in cardiovascular disease: protection afforded by dietary isoflavones. Curr Opin Pharmacol 9(2):139–145
41. El-Demerdash E, Awad AS, Taha RM, El-Hady AM, Sayed-Ahmed MM (2005) Probucol attenuates oxidative stress and energy decline in isoproterenol-induced heart failure in rat. Pharmacol Res 51(4):311–318
42. Tardif JC, Gregoire J, Schwartz L, Title L, Laramee L, Reeves F et al (2003) Effects of AGI-1067 and probucol after percutaneous coronary interventions. Circulation 107(4):552–558
43. Tardif JC, McMurray JJ, Klug E, Small R, Schumi J, Choi J et al (2008) Effects of succinobucol (AGI-1067) after an acute coronary syndrome: a randomised, double-blind, placebo-controlled trial. Lancet 371(9626):1761–1768
44. Li S, Liang J, Niimi M, Bilal Waqar A, Kang D, Koike T et al (2014) Probucol suppresses macrophage infiltration and MMP expression in atherosclerotic plaques of WHHL rabbits. J Atheroscler Thromb 21(7):648–658
45. Sundell CL, Somers PK, Meng CQ, Hoong LK, Suen KL, Hill RR et al (2003) AGI-1067: a multifunctional phenolic antioxidant, lipid modulator, anti-inflammatory and antiatherosclerotic agent. J Pharmacol Exp Ther 305(3):1116–1123
46. Tardif JC, Gregoire J, L'Allier PL, Ibrahim R, Anderson TJ, Reeves F et al (2008) Effects of the antioxidant succinobucol (AGI-1067) on human atherosclerosis in a randomized clinical trial. Atherosclerosis 197(1):480–486
47. Li L, Guo J, Wang Y, Xiong X, Tao H, Li J et al (2018) A broad-spectrum ROS-eliminating material for prevention of inflammation and drug-induced organ toxicity. Adv Sci (Weinh) 5(10):1800781
48. Wang Y, Li L, Zhao W, Dou Y, An H, Tao H et al (2018) Targeted therapy of atherosclerosis by a broad-spectrum reactive oxygen species scavenging nanoparticle with intrinsic anti-inflammatory activity. ACS Nano 12(9):8943–8960
49. Jha JC, Banal C, Okabe J, Gray SP, Hettige T, Chow BSM et al (2017) NADPH oxidase Nox5 accelerates renal injury in diabetic nephropathy. Diabetes 66(10):2691–2703
50. Jha JC, Gray SP, Barit D, Okabe J, El-Osta A, Namikoshi T et al (2014) Genetic targeting or pharmacologic inhibition of NADPH oxidase nox4 provides renoprotection in long-term diabetic nephropathy. J Am Soc Nephrol 25(6):1237–1254
51. Gray SP, Di Marco E, Kennedy K, Chew P, Okabe J, El-Osta A et al (2016) Reactive oxygen species can provide atheroprotection via NOX4-dependent inhibition of inflammation and vascular remodeling. Arterioscler Thromb Vasc Biol 36(2):295–307
52. Gray SP, Di Marco E, Okabe J, Szyndralewiez C, Heitz F, Montezano AC et al (2013) NADPH oxidase 1 plays a key role in diabetes mellitus-accelerated atherosclerosis. Circulation 127(18):1888–1902
53. Sirker A, Zhang M, Shah AM (2011) NADPH oxidases in cardiovascular disease: insights from in vivo models and clinical studies. Basic Res Cardiol 106(5):735–747
54. Di Marco E, Gray SP, Chew P, Kennedy K, Cooper ME, Schmidt HH et al (2016) Differential effects of NOX4 and NOX1 on immune cell-mediated inflammation in the aortic sinus of diabetic ApoE−/− mice. Clin Sci 130(15):1363–1374
55. Dikalov SI, Dikalova AE, Bikineyeva AT, Schmidt HH, Harrison DG, Griendling KK (2008) Distinct roles of Nox1 and Nox4 in basal and angiotensin II-stimulated superoxide and hydrogen peroxide production. Free Radic Biol Med 45(9):1340–1351

56. Martyn KD, Frederick LM, von Loehneysen K, Dinauer MC, Knaus UG (2006) Functional analysis of Nox4 reveals unique characteristics compared to other NADPH oxidases. Cell Signal 18(1):69–82
57. Amann B, Tinzmann R, Angelkort B (2003) ACE inhibitors improve diabetic nephropathy through suppression of renal MCP-1. Diabetes Care 26(8):2421–2425
58. Borchi E, Bargelli V, Stillitano F, Giordano C, Sebastiani M, Nassi PA et al (2010) Enhanced ROS production by NADPH oxidase is correlated to changes in antioxidant enzyme activity in human heart failure. Biochim Biophys Acta 1802(3):331–338
59. Johar S, Cave AC, Narayanapanicker A, Grieve DJ, Shah AM (2006) Aldosterone mediates angiotensin II-induced interstitial cardiac fibrosis via a Nox2-containing NADPH oxidase. FASEB J: official publication of the Federation of American Societies for Experimental Biology 20(9):1546–1548
60. Kuroda J, Ago T, Matsushima S, Zhai P, Schneider MD, Sadoshima J (2010) NADPH oxidase 4 (Nox4) is a major source of oxidative stress in the failing heart. Proc Natl Acad Sci U S A 107(35):15565–15570
61. Kuroda J, Sadoshima J (2010) NADPH oxidase and cardiac failure. J Cardiovasc Transl Res 3(4):314–320
62. Stevenson MD, Canugovi C, Vendrov AE, Hayami T, Bowles DE, Krause KH et al (2018) NADPH oxidase 4 regulates inflammation in ischemic heart failure: role of soluble epoxide hydrolase. Antioxid Redox Signal
63. Gray SP, Jha JC, Kennedy K, van Bommel E, Chew P, Szyndralewiez C et al (2017) Combined NOX1/4 inhibition with GKT137831 in mice provides dose-dependent Reno- and atheroprotection even in established micro- and macrovascular disease. Diabetologia 60(5):927–937
64. Susztak K, Raff AC, Schiffer M, Böttinger EP (2006) Glucose-induced reactive oxygen species cause apoptosis of podocytes and podocyte depletion at the onset of diabetic nephropathy. Diabetes 55(1):225–233
65. Ding Y, Chen ZJ, Liu S, Che D, Vetter M, Chang CH (2005) Inhibition of Nox-4 activity by plumbagin, a plant-derived bioactive naphthoquinone. J Pharm Pharmacol 57(1):111–116
66. Anvari E, Wikstrom P, Walum E, Welsh N (2015) The novel NADPH oxidase 4 inhibitor GLX351322 counteracts glucose intolerance in high-fat diet-treated C57BL/6 mice. Free Radic Res 49(11):1308–1318
67. Zhou G, Wang Y, He P, Li D (2013) Probucol inhibited Nox2 expression and attenuated podocyte injury in type 2 diabetic nephropathy of db/db mice. Biol Pharm Bull 36(12):1883–1890
68. Gaggini F, Laleu B, Orchard M, Fioraso-Cartier L, Cagnon L, Houngninou-Molango S et al (2011) Design, synthesis and biological activity of original pyrazolo-pyrido-diazepine, -pyrazine and -oxazine dione derivatives as novel dual Nox4/Nox1 inhibitors. Bioorg Med Chem 19(23):6989–6999
69. Sedeek M, Callera G, Montezano A, Gutsol A, Heitz F, Szyndralewiez C et al (2010) Critical role of Nox4-based NADPH oxidase in glucose-induced oxidative stress in the kidney: implications in type 2 diabetic nephropathy. Am J Physiol Ren Physiol 299(6):F1348–F1358
70. Laleu B, Gaggini F, Orchard M, Fioraso-Cartier L, Cagnon L, Houngninou-Molango S et al (2010) First in class, potent, and orally bioavailable NADPH oxidase isoform 4 (Nox4) inhibitors for the treatment of idiopathic pulmonary fibrosis. J Med Chem 53(21):7715–7730
71. Aoyama T, Paik YH, Watanabe S, Laleu B, Gaggini F, Fioraso-Cartier L et al (2012) Nicotinamide adenine dinucleotide phosphate oxidase in experimental liver fibrosis: GKT137831 as a novel potential therapeutic agent. Hepatology 56(6):2316–2327
72. Sedeek M, Gutsol A, Montezano AC, Burger D, Cat AND, Kennedy CR et al (2013) Renoprotective effects of a novel Nox1/4 inhibitor in a mouse model of type 2 diabetes. Clin Sci 124(3):191–202
73. Vendrov AE, Hakim ZS, Madamanchi NR, Rojas M, Madamanchi C, Runge MS (2007) Atherosclerosis is attenuated by limiting superoxide generation in both macrophages and vessel wall cells. Arterioscler Thromb Vasc Biol 27(12):2714–2721

74. Vendrov AE, Madamanchi NR, Niu XL, Molnar KC, Runge M, Szyndralewiez C et al (2010) NADPH oxidases regulate CD44 and hyaluronic acid expression in thrombin-treated vascular smooth muscle cells and in atherosclerosis. J Biol Chem 285(34):26545–26557
75. Gorin Y, Cavaglieri RC, Khazim K, Lee DY, Bruno F, Thakur S et al (2015) Targeting NADPH oxidase with a novel dual Nox1/Nox4 inhibitor attenuates renal pathology in type 1 diabetes. Am J Physiol Ren Physiol 308(11):F1276–F1287
76. You YH, Quach T, Saito R, Pham J, Sharma K (2016) Metabolomics reveals a key role for fumarate in mediating the effects of NADPH oxidase 4 in diabetic kidney disease. J Am Soc Nephrol: JASN 27(2):466–481
77. Jarasch ED, Grund C, Bruder G, Heid HW, Keenan TW, Franke WW (1981) Localization of xanthine oxidase in mammary-gland epithelium and capillary endothelium. Cell 25(1):67–82
78. Harrison R (2002) Structure and function of xanthine oxidoreductase: where are we now? Free Radic Biol Med 33(6):774–797
79. Pacher P, Nivorozhkin A, Szabo C (2006) Therapeutic effects of xanthine oxidase inhibitors: renaissance half a century after the discovery of allopurinol. Pharmacol Rev 58(1):87–114
80. McNally JS, Davis ME, Giddens DP, Saha A, Hwang J, Dikalov S et al (2003) Role of xanthine oxidoreductase and NAD(P)H oxidase in endothelial superoxide production in response to oscillatory shear stress. Am J Physiol Heart Circ Physiol 285(6):H2290–H2297
81. Zdrenghea M, Sitar-Taut A, Cismaru G, Zdrenghea D, Pop D (2017) Xanthine oxidase inhibitors in ischaemic heart disease. Cardiovasc J Afr 28(3):201–204
82. Kang SM, Lim S, Song H, Chang W, Lee S, Bae SM et al (2006) Allopurinol modulates reactive oxygen species generation and Ca2+ overload in ischemia-reperfused heart and hypoxia-reoxygenated cardiomyocytes. Eur J Pharmacol 535(1–3):212–219
83. Goicoechea M, Garcia de Vinuesa S, Verdalles U, Verde E, Macias N, Santos A et al (2015) Allopurinol and progression of CKD and cardiovascular events: long-term follow-up of a randomized clinical trial. Am J Kidney Dis: The Official Journal of the National Kidney Foundation 65(4):543–549
84. Singh JA, Ramachandaran R, Yu S, Curtis JR (2017) Allopurinol use and the risk of acute cardiovascular events in patients with gout and diabetes. BMC Cardiovasc Disord 17(1):76
85. Yisireyili M, Hayashi M, Wu H, Uchida Y, Yamamoto K, Kikuchi R et al (2017) Xanthine oxidase inhibition by febuxostat attenuates stress-induced hyperuricemia, glucose dysmetabolism, and prothrombotic state in mice. Sci Rep 7(1):1266
86. White WB, Gunawardhana L (2018) Cardiovascular safety of febuxostat. N Engl J Med 379(16):1584
87. White WB, Saag KG, Becker MA, Borer JS, Gorelick PB, Whelton A et al (2018) Cardiovascular safety of febuxostat or allopurinol in patients with gout. N Engl J Med 378(13):1200–1210
88. Lytvyn Y, Har R, Locke A, Lai V, Fong D, Advani A et al (2017) Renal and vascular effects of uric acid lowering in normouricemic patients with uncomplicated type 1 diabetes. Diabetes 66(7):1939–1949
89. Brash AR (1999) Lipoxygenases: occurrence, functions, catalysis, and acquisition of substrate. J Biol Chem 274(34):23679–23682
90. Kuhn H, Heydeck D, Hugou I, Gniwotta C (1997) In vivo action of 15-lipoxygenase in early stages of human atherogenesis. J Clin Invest 99(5):888–893
91. Shappell SB, Boeglin WE, Olson SJ, Kasper S, Brash AR (1999) 15-lipoxygenase-2 (15-LOX-2) is expressed in benign prostatic epithelium and reduced in prostate adenocarcinoma. Am J Pathol 155(1):235–245
92. Cyrus T, Pratico D, Zhao L, Witztum JL, Rader DJ, Rokach J et al (2001) Absence of 12/15-lipoxygenase expression decreases lipid peroxidation and atherogenesis in apolipoprotein e-deficient mice. Circulation 103(18):2277–2282
93. Funk CD, Cyrus T (2001) 12/15-lipoxygenase, oxidative modification of LDL and atherogenesis. Trends Cardiovasc Med 11(3–4):116–124

94. George J, Afek A, Shaish A, Levkovitz H, Bloom N, Cyrus T et al (2001) 12/15-lipoxygenase gene disruption attenuates atherogenesis in LDL receptor-deficient mice. Circulation 104(14):1646–1650
95. Cipollone F, Mezzetti A, Fazia ML, Cuccurullo C, Iezzi A, Ucchino S et al (2005) Association between 5-lipoxygenase expression and plaque instability in humans. Arterioscler Thromb Vasc Biol 25(8):1665–1670
96. Tardif JC, L'Allier PL, Ibrahim R, Gregoire JC, Nozza A, Cossette M et al (2010) Treatment with 5-lipoxygenase inhibitor VIA-2291 (Atreleuton) in patients with recent acute coronary syndrome. Circ Cardiovasc Imaging 3(3):298–307
97. Hakonarson H, Thorvaldsson S, Helgadottir A, Gudbjartsson D, Zink F, Andresdottir M et al (2005) Effects of a 5-lipoxygenase-activating protein inhibitor on biomarkers associated with risk of myocardial infarction: a randomized trial. JAMA 293(18):2245–2256
98. Eid AA, Gorin Y, Fagg BM, Maalouf R, Barnes JL, Block K et al (2009) Mechanisms of podocyte injury in diabetes: role of cytochrome P450 and NADPH oxidases. Diabetes 58(5):1201–1211
99. Rendic S, Di Carlo FJ (1997) Human cytochrome P450 enzymes: a status report summarizing their reactions, substrates, inducers, and inhibitors. Drug Metab Rev 29(1–2):413–580
100. Granville DJ, Tashakkor B, Takeuchi C, Gustafsson AB, Huang C, Sayen MR et al (2004) Reduction of ischemia and reperfusion-induced myocardial damage by cytochrome P450 inhibitors. Proc Natl Acad Sci U S A 101(5):1321–1326
101. Hunter AL, Cruz RP, Cheyne BM, McManus BM, Granville DJ (2004) Cytochrome p450 enzymes and cardiovascular disease. Can J Physiol Pharmacol 82(12):1053–1060
102. Grobe AC, Wells SM, Benavidez E, Oishi P, Azakie A, Fineman JR et al (2006) Increased oxidative stress in lambs with increased pulmonary blood flow and pulmonary hypertension: role of NADPH oxidase and endothelial NO synthase. Am J Physiol Lung Cell Mol Physiol 290(6):L1069–L1077
103. Antoniades C, Shirodaria C, Leeson P, Antonopoulos A, Warrick N, Van-Assche T et al (2009) Association of plasma asymmetrical dimethylarginine (ADMA) with elevated vascular superoxide production and endothelial nitric oxide synthase uncoupling: implications for endothelial function in human atherosclerosis. Eur Heart J 30(9):1142–1150
104. Verhaar MC, Wever RM, Kastelein JJ, van Dam T, Koomans HA, Rabelink TJ (1998) 5-methyltetrahydrofolate, the active form of folic acid, restores endothelial function in familial hypercholesterolemia. Circulation 97(3):237–241
105. Brownlee M (2005) The pathobiology of diabetic complications: a unifying mechanism. Diabetes 54(6):1615–1625
106. Widlansky ME, Wang J, Shenouda SM, Hagen TM, Smith AR, Kizhakekuttu TJ et al (2010) Altered mitochondrial membrane potential, mass, and morphology in the mononuclear cells of humans with type 2 diabetes. Transl Res 156(1):15–25
107. Crespo FL, Sobrado VR, Gomez L, Cervera AM, McCreath KJ (2010) Mitochondrial reactive oxygen species mediate cardiomyocyte formation from embryonic stem cells in high glucose. Stem Cells 28(7):1132–1142
108. Turrens JF (2003) Mitochondrial formation of reactive oxygen species. J Physiol 552(Pt 2):335–344
109. Handy DE, Loscalzo J (2012) Redox regulation of mitochondrial function. Antioxid Redox Signal 16(11):1323–1367
110. Zorov DB, Filburn CR, Klotz LO, Zweier JL, Sollott SJ (2000) Reactive oxygen species (ROS)-induced ROS release: a new phenomenon accompanying induction of the mitochondrial permeability transition in cardiac myocytes. J Exp Med 192(7):1001–1014
111. Zorov DB, Juhaszova M, Sollott SJ (2014) Mitochondrial reactive oxygen species (ROS) and ROS-induced ROS release. Physiol Rev 94(3):909–950
112. Nojiri H, Shimizu T, Funakoshi M, Yamaguchi O, Zhou H, Kawakami S et al (2006) Oxidative stress causes heart failure with impaired mitochondrial respiration. J Biol Chem 281(44):33789–33801

113. Li Y, Huang TT, Carlson EJ, Melov S, Ursell PC, Olson JL et al (1995) Dilated cardiomyopathy and neonatal lethality in mutant mice lacking manganese superoxide dismutase. Nat Genet 11(4):376–381
114. Matsushima S, Ide T, Yamato M, Matsusaka H, Hattori F, Ikeuchi M et al (2006) Overexpression of mitochondrial peroxiredoxin-3 prevents left ventricular remodeling and failure after myocardial infarction in mice. Circulation 113(14):1779–1786
115. Ben-Hail D, Begas-Shvartz R, Shalev M, Shteinfer-Kuzmine A, Gruzman A, Reina S et al (2016) Novel compounds targeting the mitochondrial protein VDAC1 inhibit apoptosis and protect against mitochondrial dysfunction. J Biol Chem 291(48):24986–25003
116. Li X, Fang P, Li Y, Kuo YM, Andrews AJ, Nanayakkara G et al (2016) Mitochondrial reactive oxygen species mediate lysophosphatidylcholine-induced endothelial cell activation. Arterioscler Thromb Vasc Biol 36(6):1090–1100
117. Chacko BK, Reily C, Srivastava A, Johnson MS, Ye Y, Ulasova E et al (2010) Prevention of diabetic nephropathy in Ins2(+/)(−)(AkitaJ) mice by the mitochondria-targeted therapy MitoQ. Biochem J 432(1):9–19

Part II

Pathophysiology of Oxidative Stress

Oxidative Stress and Heart Failure

11

Bodh I. Jugdutt and Bernadine A. Jugdutt

Abstract

Heart failure (HF) remains a major cause of disability, suffering, and death worldwide. The prevalence of HF increases with age and at an alarming pace in the elderly population aged 65 years or more. Importantly, the increase in HF prevalence, first seen in developed countries and currently in developing countries as well, has taken place despite tremendous advances in HF therapy and efforts to encourage implementation of management guidelines. The magnitude of this HF pandemic is staggering, affecting nearly 26 million people across the world. There are several reasons for this continued increase in HF prevalence despite optimal therapy; of these, two that stand out include (i) the aging-induced cardiovascular (CV) remodeling that modifies disease expression and response to therapy and aging-related increase in reactive oxygen species (ROS) and oxidative stress (OXS) that augment adverse left ventricular remodeling after myocardial injury; ii) the lifelong exposure to CV disease (CVD) risk factors that increase ROS and OXS, as well as inflammation. Other pathways and mechanisms leading to HF that are yet to be addressed may also involve OXS and inflammation. This chapter focuses on the evidence for ROS-induced myocardial damage during HF progression and some potential pharmacological interventions and strategies for reducing the damage. In addition, some key issues facing translation of experimental successes with antioxidant therapy into successes in clinical practice on the real-world stage are addressed.

Keywords

Aging · Healing · Infarct size · Hypertrophy · Heart failure with preserved ejection fraction · Heart failure with reduced ejection fraction · Hypertension ·

The authors have nothing to disclose.

B. I. Jugdutt (✉) · B. A. Jugdutt
Cardiology Division, Department of Medicine, Faculty of Medicine, University of Alberta, Edmonton, Alberta, Canada
e-mail: bjugdutt@ualberta.ca

S. Chakraborti et al. (eds.), *Modulation of Oxidative Stress in Heart Disease*,
https://doi.org/10.1007/978-981-13-8946-7_11

Inflammation · Myocardial infarction · Mitochondria · Oxidative stress · Prevention · Remodeling · Reperfusion injury

Abbreviations

ACS	acute coronary syndrome
ACE	angiotensin-converting enzyme
ACEIs	angiotensin-converting-enzyme inhibitors
ADAM	a disintegrin and metalloproteinase
AMP	adenosine monophosphate
Ang II	angiotensin II
ARB	angiotensin II type 1 receptor blocker
ARNI	angiotensin receptor neprilysin inhibitor
ATP	adenosine triphosphate
BH4	tetrahydrobiopterin
BNP	N-terminal B-type natriuretic peptide
CABG	coronary artery bypass surgery
CAD	coronary artery disease
CANTOS	Canakinumab Anti-Inflammatory Thrombosis Outcomes Study
CARE	cholesterol and recurrent events
cGMP	cyclic guanosine monophosphate
CKD	chronic kidney disease
CMR	cardiac magnetic resonance
CPB	cardiopulmonary bypass
CRP	C-reactive protein
CVD	cardiovascular disease
DM2	type 2 diabetes mellitus
ECG	electrocardiogram
ECM	extracellular matrix
eNOS	endothelial nitric oxide synthase
EPR	electron paramagnetic resonance
ESR	electron spin resonance
ET	endothelin
ETC	electron transfer chain
GDF	growth differentiation factor
GLP-1	antidiabetic glucagon-like peptide-1
GLP-1RA	glucagon-like peptide/receptor agonist
H2O2	hydrogen peroxide
HDL	high-density lipoprotein
HF	heart failure
HFmrEF	heart failure with midrange ejection fraction
HFpEF	heart failure with preserved ejection fraction

HFrEF	heart failure with reduced ejection fraction
hs-CRP	high-sensitivity C-reactive protein
IGF	insulin-like growth factor
IL	interleukin
IL-1ra	recombinant IL-1 receptor antagonist
iNOS	inducible nitric oxide synthase
I/R	ischemia-reperfusion
IRA	infarct-related artery
IZ	infarct zone
LDL	low-density lipoprotein
LV	left ventricular
LVAD	LV assist device
MACE	major adverse cardiovascular events
MI	myocardial infarction
MMP	matrix metalloproteinase
MPO	myeloperoxidase
MRA	mineralocorticoid receptor antagonist
MRI	magnetic resonance imaging
MUGA	multigated acquisition scan
NDEA	N-nitrosodiethylamine
NEP	neprilysin
NADPH	nicotinamide adenine dinucleotide phosphate
NDMA	N-nitrosodimethylamine
NIZ	noninfarct zone
•NO	nitric oxide
NOO^-	NO-derived peroxynitrite
NOS	nitric oxide synthase
NOX	NADPH oxidase
NSTEMI	non-ST-segment elevation MI
O2	oxygen
OFRs	oxygen free radicals
OPN	osteopontin
OXS	oxidative stress
•OH	hydroxyl radical
PCI	percutaneous coronary intervention
PDGF	platelet-derived growth factor
PKG	phosphokinase G
PPCI	primary PCI
$O2^{\bullet -}$	superoxide anion radical
RAAS	renin-angiotensin-aldosterone system
RCT	randomized clinical trial
RIPC	remote ischemic preconditioning
ROS	reactive oxygen species
SGLT2	sodium glucose cotransporter-2
SLPI	secretory leucocyte protease inhibitor

SOD	superoxide dismutase
SPARC	secreted protein acidic and rich in cysteine
STEMI	ST-segment elevation MI
TIMP	tissue inhibitor of metalloproteinase
TGF	transforming growth factor
TNF	tumor necrosis factor
VSMC	vascular smooth muscle cell

11.1 Introduction

In this second decade of the twenty-first century, heart failure (HF) remains a major cause of disability, suffering, and death worldwide [1–13]. The prevalence of HF increases with age and reaches alarming proportions in elderly people aged ≥65 years, a group that is growing steadily [14–19]. Over the last two decades alone, HF has grown into a serious pandemic, affecting nearly 26 million people in developed and developing countries across the world [3–6]. In Europe, both the prevalence and risk of HF in the adult and aging population are significant; it is reported that HF occurs in about 1–2% of adults and rises to more than 10% in those who are older than 70 years [1]; the lifetime risk of HF in people aged 55 years is reported at 33% for men and 28% for women [20]. In the United States, HF prevalence in people aged ≥20 years is reported to have increased from 5.7 million over 2009–2012 to 6.5 million over 2011–2014, with a projected increase by 46% over 2012–2030, which equates to over 8 million people aged ≥18 years with HF [6]; HF incidence in people aged >65 years approached 21 per 1000 or 2.1%, and HF risk was highest among African Americans [6]; the lifetime risk of HF was 20–45% for people aged 45–95 years and was higher for people with hypertension (HTN) and obesity irrespective of age [6]. Across Asia, estimates of HF prevalence ranged between 1.26 and 6.7% [6, 21]. Taken together, the global burden of HF in the aging population is clearly and undeniably staggering [3–6, 14–16, 22, 23]. Importantly, this increase in HF prevalence, first in developed countries and currently in developing countries as well, has taken place despite tremendous advances in HF therapy and efforts to encourage implementation of management guidelines [1, 2, 7–13].

There are many causes for HF (Fig. 11.1) and these have been extensively reviewed [1, 8]. New pathophysiological mechanisms leading to HF continue to be elucidated, and potential targets are being identified for therapeutic intervention [24, 25]. The two leading causes of HF remain myocardial infarction (MI) and HTN [1–19]. In his 2015 Lancet lecture, Braunwald boldly stated that it is “time to declare war on HF” and nicely underpinned several targets for urgent action [26]. However, as depicted in Fig. 11.2, one additional target that needs urgent action concerns the lifelong onslaught from reactive oxygen species (ROS), oxygen free radicals (OFRs), and oxidative stress (OXS) associated with exposure to cardiovascular disease (CVD) risk factors, such as hyperlipidemia, obesity, type 2 diabetes mellitus (DM2), HTN, and aging; although OXS has been recognized as an important

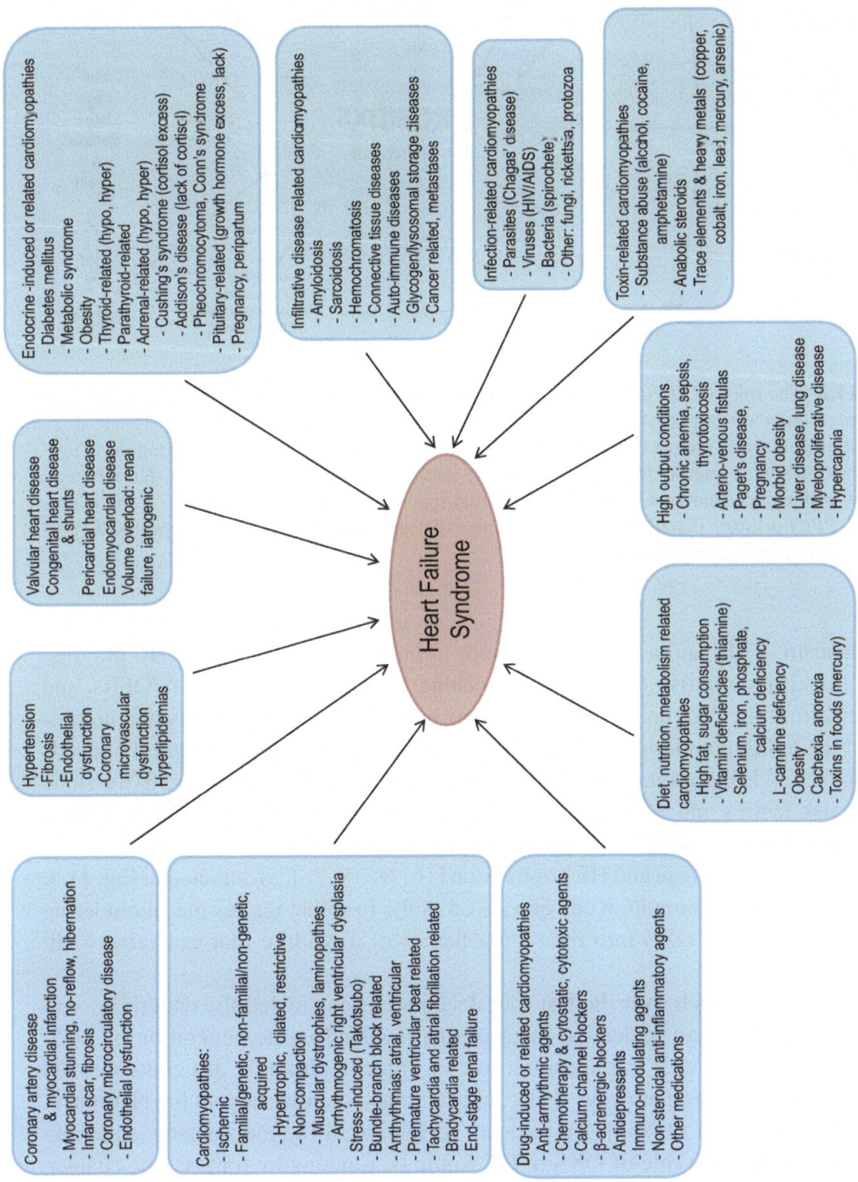

Fig. 11.1 Known etiologies leading to heart failure
Updated, adapted, and modified from Dickstein et al. 2008 [8] & Ponikowski et al. 2016 [1]

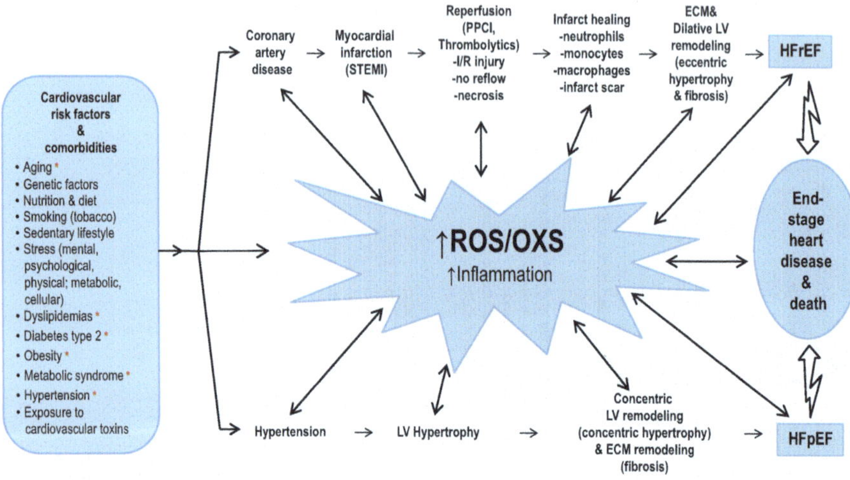

Fig. 11.2 The roles of ROS/OXS and inflammation in the cardiovascular disease continuum and the progression to heart failure. Adapted and modified from Jugdutt (2014) [18, 19]
*Cardiovascular risk factors tied to ↑ ROS, OXS, and inflammation. Other abbreviations as in text: *ECM* extracellular matrix, *HFpEF* heart failure with preserved ejection fraction, *HFrEF* heart failure with reduced ejection fraction, *I/R* ischemia-reperfusion, *LV* left ventricular, *OXS* oxidative stress, *PPCI* primary percutaneous coronary intervention, *ROS* reactive oxygen species, *STEMI* ST-segment elevation myocardial infarction

mechanism of HF and a major contributor to both aging [16, 18] and HF progression [27, 28], definitive therapy for quenching increased levels of ROS/OFRs, and thereby limiting OXS and its potential contribution to excess morbidity and mortality in the aging population with HF, is still lacking [22]. Another target that needs urgent action is inflammation, which is also related to lifelong exposure to the same CVD risk factors and is exacerbated by acute triggers such as ischemia, ischemia-reperfusion (I/R), and MI, and appears to interact synergistically with OXS to intensify myocardial damage and HF progression [16, 18, 19, 28], as depicted in Fig. 11.3. In that context, inflammation can be viewed as the fuse that ignites the smouldering background fire of OXS into the raging flames of a wildfire that exacerbates HF progression.

Extensive research over the last four decades has elucidated the underlying cellular, subcellular, and molecular mechanisms involved in the generation of ROS/OFRs and OXS as well as the central role of mitochondria [29, 30] and identified several key signaling pathways that could serve as potential targets for pharmacological intervention [31–33]. An in-depth review of all the mechanisms and pathways illustrating the role of OXS in HF would be too lengthy for this one chapter, and several of them are addressed in other chapters of this book. The chapter here focuses on some potential future pharmacological interventions aimed at reducing ROS-induced damage during HF progression in the clinical setting. In addition, some key issues facing the translation of experimental successes with antioxidant

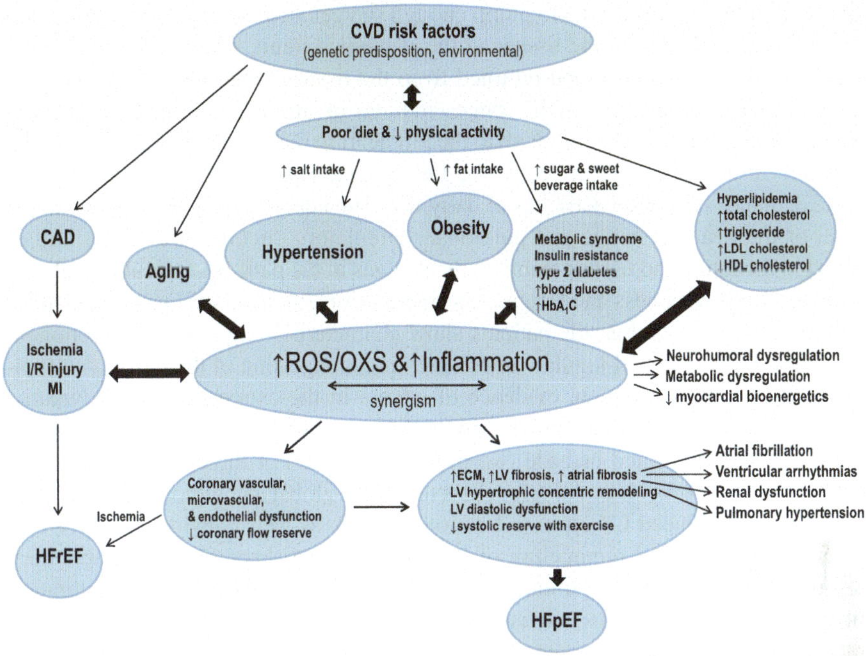

Fig. 11.3 Schematic of putative interactions between CVD risk factors and aging, comorbidities, increased ROS and OXS, increased inflammation, fibrosis, vascular, and LV dysfunction in the march to HFpEF

Abbreviations as in text: *CAD* coronary artery disease, *CVD* cardiovascular disease, *ECM* extracellular matrix, *HDL* high-density lipoprotein, *HFpEF* heart failure with preserved ejection fraction, *HFrEF* heart failure with reduced ejection fraction, *I/R* ischemia-reperfusion, *LDL* low-density lipoprotein, *LV* left ventricular, *MI* myocardial infarction, *OXS* oxidative stress, *ROS* reactive oxygen species

therapy into successes in clinical practice on the real-world stage are addressed, and the importance of proper validation in carefully designed randomized clinical trials (RCTs) is discussed.

11.2 Pathophysiology of Progressive Remodeling in Heart Failure

Six points about the pathophysiology of the progressive adverse remodeling leading to progression in HF severity need emphasis.

11.2.1 Multiple Causes of the HF Syndrome

It is important to appreciate that multiple etiologies (Fig. 11.1), via diverse pathophysiological mechanisms, converge to produce the final clinical syndrome of HF [1, 8]. In the final analysis, HF represents a failure of homeostasis on several fronts;

simply put, the left heart fails to pump enough oxygenated blood forward and maintain optimal perfusion of the tissues in the systemic circuit, whereas the right heart fails to pump all venous blood returned from the tissues to the lungs for reoxygenation in the pulmonary circuit; the congested lungs in turn fail to optimally reoxygenate the venous blood and return it all to the left heart to maintain circulatory flow. The reduced cardiac output generated by the failing heart can no longer match the metabolic demands of normal activities of daily life. The net effect of these failures at the level of the heart, systemic and pulmonary circulatory circuits, and the backup of blood in the lungs and tissues manifest themselves in the patient's typical complaints of generalized weakness and fatigue, shortness of breath, and swelling of the ankles and legs; on examination, the patients show characteristic signs such as prominent neck veins with elevated jugular venous pressure, congestion of the lungs with typical crackles on auscultation, evidence of edema in the extremities, hemodynamic evidence of reduced cardiac output and/or elevated intracardiac pressures at rest or with exercise stress, and bedside two-dimensional (2D) or three-dimensional (3D) echocardiographic evidence of left ventricular (LV) global and regional systolic dysfunction with reduced LV ejection fraction (LVEF), as well as evidence of LV diastolic dysfunction and remodeling of LV structure and shape [35–46]. Laboratory test panels often display metabolic abnormalities related to the severity and duration of stress on different organs and tissues, such as reduced mixed venous oxygen saturation and arteriovenous oxygen difference, metabolic and respiratory acidosis with elevated lactate levels, altered blood profile with anemia, electrolyte imbalance with changes in serum sodium (Na+) and potassium (K+) levels, changes in serum creatinine (kidney stress), serum albumin and liver enzymes (liver stress), and biomarkers such as N-terminal B-type natriuretic peptide (BNP) and NT-proBNP (cardiac stress) and C-reactive protein (CRP; for inflammation). Metabolic panels reveal evidence of neurohormonal activations such as the sympathetic nervous system, renin-angiotensin-aldosterone system (RAAS), and endothelin (ET) system. Evidence of hypothyroidism is often present in the blood test. Other tests, such as electrocardiogram (ECG), chest X-ray, multigated acquisition (MUGA) scan, nuclear stress test, magnetic resonance imaging (MRI) or cardiac magnetic resonance (CMR), contrast echo with 2D and 3D imaging, pharmacologic stress test, and cardiac catheterization with angiography, provide clues regarding the precise etiology. No routine blood tests are currently done to screen for OXS.

11.2.2 Two Main Subsets of HF with Divergent LV Remodeling

Notwithstanding the multiplicity of etiologies leading to HF (Fig. 11.1), the two commonest aforementioned causes are MI and HTN, and they account for the majority of cases of HF seen in clinical practice and show nearly equal distribution (about 50% with MI and about 50% with HTN) [1, 2, 7–13]. The type and degree of adverse cardiac remodeling in HF depend on the underlying cause; dilative and eccentric remodeling with eccentric LV hypertrophy develops after ST-elevation MI (STEMI) and volume overload conditions; and concentric remodeling with concentric LV hypertrophy develops in HTN and other pressure overload conditions (Fig. 11.2). The progressive, adaptive, and maladaptive remodeling of cardiac structure, geometric shape, and function that occurs over time in survivors of the different types of CVD insults takes different paths, as illustrated by MI [34–44] and HTN [45, 46], as well as various cardiomyopathies, including those associated with DM2 and obesity.

11.2.3 Different Subsets of HF Based on LVEF

Stratification of HF on the basis of LVEF has unmasked three categories with distinct phenotypes that are now recognized in the latest HF management guidelines [1, 2]: (i) HF with reduced LVEF <40% (HFrEF); (ii) HF with preserved LVEF ≥50% (HFpEF); and (iii) most recently, an intermediate group with mid-range LVEF 40–49% (HFmrEF). Besides the difference in LVEFs, the criteria for diagnosis of HFpEF and HFmrEF both include the presence of elevated BNP levels and either LV diastolic dysfunction or abnormal structure reflected in LV hypertrophy and/or left atrial enlargement [1]. In survivors of the acute phase of the insults, MRI or CMR can be used, where available, not only to quantify ventricular volumes, LVEF, and LV mass but also to assess fibrosis and scar size, and provide clues as to the precise etiology in various cardiomyopathies [1, 2]. In the context of ROS and OXS, the most data currently available is for these two main categories of HF, namely, HFrEF and HFpEF.

11.2.4 Divergent Types of Remodeling in the Two Subsets of HF Based on LVEF

Over the last four decades, the underlying structural, biochemical, cellular, subcellular, molecular, and metabolic derangements in HF following MI and HTN have been extensively researched, and key pathways and molecules have been identified and targeted by therapeutic interventions that have undisputedly reduced the number of patients dying and suffering from this debilitating chronic disease [1, 2, 25–35]. It is well appreciated that the nature of the insults and rates of progression of remodeling after MI and HTN are very different (Figs. 11.2 and 11.3). Typically, remodeling after anterior transmural MI or STEMI has two components; an early, dramatic, and rapidly developing regional expansion of the infarct zone with thinning and dilation (infarct expansion), followed by more gradual dilative global LV remodeling of both the IZ and NIZ during the healing and repair phases. Both these two components of remodeling are associated with HFrEF and poor outcome [34, 39–43, 52–58], and cumulative evidence suggests that both OXS and inflammation play distinct pathophysiologic roles in the two components of remodeling during the development and progression of HFrEF [27–44, 48–58]. In contrast, remodeling in HTN progresses at a much slower pace, in parallel with progression of the hypertensive disease and increasing blood pressure; this results in progressive concentric

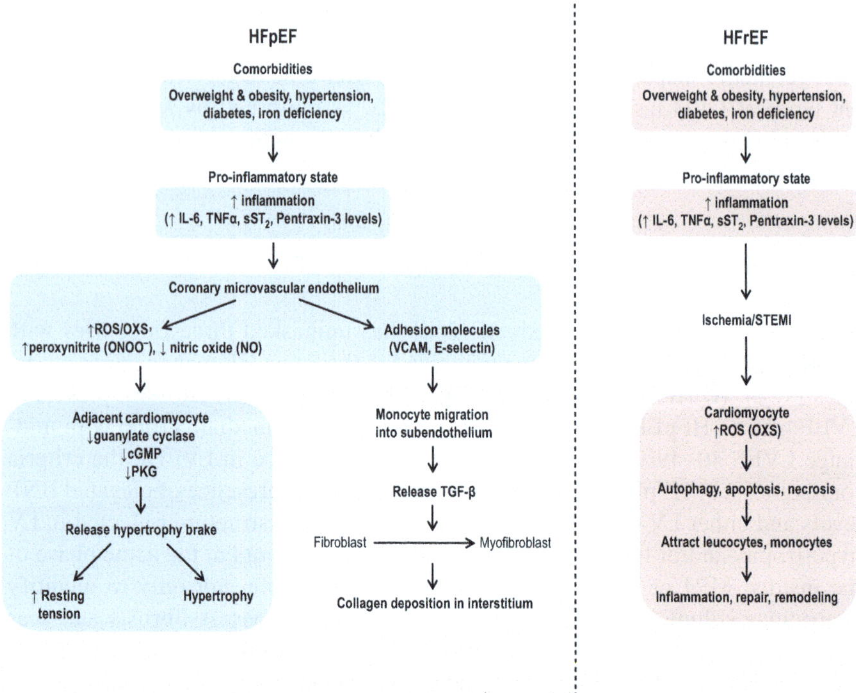

Fig. 11.4 Schematic showing steps in myocardial remodeling in HFpEF versus HFrEF. Adapted from Paulus and Tschöpe (2013) [46]
Abbreviations: as in text

LV remodeling with development of HFpEF and poor outcome [45, 46]. Cumulative evidence over the last two decades indicates that OXS and inflammation also play important pathophysiologic roles in the development and progression of HFpEF, often with significant contribution from various comorbidities [27–33, 59–69]. These current concepts are summarized in Figs. 11.3 and 11.4. Furthermore, research has shed light on the various mechanisms leading to OXS and inflammation and their roles in the pathophysiology of CVD and the progression to HF via the chains of ischemia-MI-I/R-/HFrEF and HTN-HFpEF, as illustrated in Figs. 11.2, 11.3, and 11.4. Research has also suggested novel potential therapeutic targets.

11.2.5 Evidence on the Roles of OXS and Inflammation in HFrEF and HFpEF

Evidence for the roles of OXS and/or inflammation in the progression of HFrEF and HFpEF were addressed in several studies, and some key reports are summarized below.

11.2.5.1 Multiple Biomarkers

In a clinical study of the different biological pathways that characterize HFrEF and HFpEF, Tromp et al. [59] analyzed 92 biomarkers in a cohort of 804 elderly HF patients (47% HFrEF, 27% HFpEF; mean age 74 years); they found that key markers of HFrEF were BNP, growth differentiation factor-15 (GDF-15), IL-1 receptor-like 1, and activating transcription factor 2, whereas key markers in HFpEF were integrin subunit beta-2 and catenin beta-1. They concluded that biomarker profiles in HFrEF are related to regulation of sequence-specific DNA-binding transcription, smooth muscle cell proliferation, and nitric oxide (NO) biosynthesis and metabolism, whereas those for HFpEF were related to cell adhesion, leucocyte migration, cytokine response and inflammation, neutrophil degranulation, and ECM organization, and the profile for HFmrEF was intermediate between those of HFrEF and HFpEF [59]. However, OXS was not addressed.

11.2.5.2 Multiple Comorbidities

In a small experimental study of multiple comorbidities in swine with chronic streptozotocin-induced diabetes, high-fat diet, and HTN induced by renal artery embolization over 6 months, Sorop et al. [60] documented that increased blood glucose and triglyceride, kidney dysfunction, and HTN were associated with evidence of systemic inflammation (increased IL-6 and tumor necrosis factor-α [TNF-α]), myocardial OXS (with increased superoxide or $O2^{\bullet -}$, NADPH oxidase or NOX activity, and endothelial NO synthase or eNOS uncoupling), coronary microvascular dysfunction (with decreased NO and impaired endothelial-dependent vasodilation), as well as increased myocardial collagen, decreased capillary/fiber ratio, with hemodynamic evidence of increased passive myocardial stiffness, LV diastolic dysfunction, and HFpEF [60]. However, OXS was not directly addressed in that study.

11.2.5.3 Microvascular Dysfunction

In a small clinicopathologic study of LV biopsies in 3 groups of patients (HFpEF, n = 36; aortic stenosis, n = 67; and HFrEF, n = 43), van Heerebeek et al. [65] found that protein kinase G (PKG) activity was lower in patients with HFpEF than in patients with aortic stenosis or HFrEF; importantly, the lower PKG level was associated with lower cyclic guanosine monophosphate (cGMP) concentration and higher nitrosative-OXS as well as a higher cardiomyocyte resting tension which was normalized by in vitro exogenous PKG administration [65]. They also noted that higher nitrosative-OXS, reflected in higher myocardial nitrotyrosine content in HFpEF than in HFrEF (and aortic stenosis) and known to impair NO-cGMP-PKG signaling, might result from the higher prevalence of associated comorbidities that increase OXS and inflammation (such as HTN, obesity, and DM2) in the HFpEF group. The patients in that study were older adults with a mean age of 60–65 years [65]. The overall findings suggested that microvascular dysfunction and PKG might be useful targets in HFpEF, but these need validation.

11.2.5.4 Coronary Flow Reserve and Microvascular Dysfunction

In a prospective observational study of coronary flow reserve in 202 patients with HFpEF and without unvascularized large vessel CAD, Shah et al. [62] documented a high prevalence of coronary microvascular dysfunction reflected in endothelial dysfunction (low reactive hyperemic index), higher albumin/creatinine ratio, NT-proBNP, and right ventricular dysfunction [62]. The patients in that study were elderly (mean age 72–75 years) with several comorbidities (such as HTN, obesity, DM2, hyperlipidemia, chronic kidney disease [CKD], cigarette smoking) known to increase OXS and inflammation.

In another study of coronary flow reserve using phase contrast cine-MRI of the coronary sinus to assess flow as an index of LV microvascular function in elderly patients (25 HFpEF, mean age 73 years; 13 hypertensive LV hypertrophy, age 67; 18 controls, age 65), Kato et al. documented that coronary flow reserve was lower in 76% of the HFpEF patients compared to that in patients with hypertensive LV hypertrophy and the controls; in addition, they found that coronary flow reserve correlated with serum BNP levels [64]. The findings suggested that impairment of coronary flow reserve might be related to the severity of HFpEF [64]. The patients in that study had several comorbidities (such as HTN, DM2, hyperlipidemia, cigarette smoking) that are known to increase OXS and inflammation.

11.2.5.5 OXS Markers

Despite these compelling reports of the role of OXS and inflammation in HFpEF, controversy exists. In one small study of 50 patients with and without HFpEF, Negi et al. [69] measured various OXS markers (such as derivatives of reactive oxidative metabolites, F2-isoprostanes, ratios of oxidized to reduced glutathione and cysteine) and angiotensin-converting-enzyme (ACE) levels and activity; while they found an association between HFpEF and male gender and higher body mass index (BMI), they did not find significant evidence of systemic renin-angiotensin system (RAS) activation or OXS, leading them to conclude that their finding may explain the failure of RAS inhibitors to alter outcomes in HFpEF [69].

11.2.5.6 Unexplained Mode of Death in HFpEF

In a systematic review of 1608 papers on HFpEF from 1985 to 2015, the authors found that about 25% of deaths were sudden, calling for a longitudinal multicenter global registry [70]. Whether there is an arrhythmic and/or ischemic contribution to the mode of death in HFpEF is not currently established; however, this appears very likely in view of the increased OXS [60–63, 65, 67–69], inflammation [59–61, 63], ECM and fibrosis [59–69], microvascular disease [62, 63], impaired coronary flow reserve [64, 66], and impaired bioenergetics [67], as depicted in Fig. 11.3.

11.2.5.7 Merits of RCT Versus Observational Data in HF

From another review of electronic databases on observational nonrandomized studies versus those in RCTs until the end of 2017 for the association between drug therapy and mortality in HF patients, the authors concluded that "treatment effects cannot be estimated from observational data" [71]. That finding supports the need for RCTs to guide therapy.

In summary, cumulative evidence supports the role of CVD risk factors in promoting OXS and inflammation (Figs. 11.2, 11.3, and 11.4), thereby leading to coronary microvascular dysfunction, subendocardial ischemia, regional and diffuse fibrosis, cardiac steatosis, vascular stiffness, and adverse LV remodeling.

11.2.6 Impact of Multiple Etiologies and Comorbidities on HF Therapy

The distinction between the two HF categories based on severity of systolic dysfunction is logical from a treatment perspective because the specific category might dictate the specific therapeutic approach to be recommended in compliance with the management guidelines [1, 2, 10–13]. Since the two main categories of HF (HFrEF and HFpEF) based on LVEF are characterized by two divergent types of LV remodeling (Fig. 11.2), it is not surprising that therapies recommended for the management of the two distinct categories of HF in the updated published guidelines should be quite different [1, 2, 10–13], as reflected in the summaries shown in Tables 11.1, 11.2, and 11.3. However, it should be noted that, whereas therapies for HFrEF are fairly well-defined (Tables 11.1 and 11.2), those for HFpEF still remain to be defined (Table 11.3); this is partly due to the heterogeneous causations and the presence of multiple CV comorbidities in HFpEF [66], such as HTN, obesity, metabolic syndrome, DM2, hyperlipidemias, CAD, atrial fibrillation, ventricular arrhythmias, renal dysfunction, and pulmonary hypertension (Fig. 11.3), and these require different specific therapies (Table 11.3). They also have several non-CV comorbidities, including osteoarthritis, hypothyroidism, chronic obstructive lung disease, sleep apnea, CKD, iron-deficiency anemia, anxiety, and depression, that require separate therapies. In addition, RCTs have not distinguished between HFpEF and HFmrEF because the phenotypes are still being characterized [72]; as a result, the recommendations have tended to lump the two categories into a single HFpEF category at this time [1, 2]. Furthermore and as mentioned before, at least 50% of HF patients have HFpEF, and they tend to be elderly, and aging is an important CV risk factor as shown in Figs. 11.3, 11.4, and 11.5.

11.2.7 Management Guidelines for HF Therapy and HF Pathophysiology

It should also be noted that management guidelines that are updated by the major CV societies worldwide represent the consensus opinion of experts based on available evidence mainly from RCTs, are meant to guide therapy, and do not address all pertinent issues [1, 2]. Guideline-driven management of HF centers around improving clinical status, functional capacity, and quality of life and reducing

Table 11.1 Recommended pharmacologic therapies during acute STEMI and reperfusion

During acute STEMI and reperfusion
Thrombolytics
Antiplatelet agents
$P2Y_{12}$ inhibitor (prasugrel, ticagrelor), clopidogrel
Aspirin
$GPII_b/III_a$ inhibitors
Cangrelor
Anticoagulants
Unfractionated heparin
Bivalirudin
Enoxaparin
Fondaparinux
Fibrinolytics
Fibrin-specific agent (tenecteplase, alteplase, reteplase)
Antiplatelet agents (aspirin, clopidogrel, $P2Y_{12}$ inhibitor)
Anticoagulants (enoxaparin, unfractionated heparin)
During maintenance after STEMI
Antithrombotic agents
Aspirin
Antiplatelet agents
Proton pump inhibitors
Oral anticoagulants
Beta-blockers
Metoprolol
Lipid-lowering agents
Statins
Ezetimide
PCSK9 inhibitors
Nitrates
Calcium antagonists
Angiotensin-converting-enzyme inhibitors (ACEIs)
Angiotensin II receptor blockers (ARBs)
Mineralocorticoid/aldosterone receptor antagonists (MRAs)
Eplerenone

Summarized and adapted from Ponokowski et al. (2016) [1]

Table 11.2 Recommended pharmacologic therapies for HFrEF

HFrEF
Angiotensin-converting-enzyme inhibitors (ACEIs)
Captopril, enalapril, lisinopril, ramipril, trandolapril
Beta-blockers
Bisoprolol, carvedilol, metoprolol, nebivolol
Angiotensin type I receptor blockers (ARBs)
Candesartan, valsartan, losartan
Mineralocorticoid/aldosterone receptor antagonists (MRAs)
Eplerenone, spironolactone
Angiotensin receptor neprilysin inhibitors (ARNIs)
Sacubitril/valsartan
I*f*-channel blocker
Ivabradine
Diuretics
Loop diuretics
Furosemide, bumetanide, torasemide
Thiazides
Bendroflumethiazide, hydrochlorothiazide, metolazone, indapamide
Potassium-sparing diuretics
Spironolactone/eplerenone, amiloride, triamterene
Other drugs
Hydralazine and isosorbide dinitrate
Digoxin
N-3 PUFA (polyunsaturated fatty acid)
Drugs of unproven benefit
3-Hydroxy-3-methylglutaryl-coenzyme A reductase inhibitors (statins)
Oral anticoagulants and antiplatelet therapy
Renin inhibitors
Drugs not recommended/ contraindicated
Non-dihydropyridine calcium-channel blockers (CCBs)
Verapamil and diltiazem unsafe
Amlodipine and felodipine safe, only if compelling indication

Summarized and adapted from Ponokowski et al. (2016) [1]

Table 11.3 Recommended and other pharmacologic therapies for HFpEF

In the absence of irrevocably positive RCT evidence that any drug treatment reduces mortality or morbidity in HFpEF, 2016 management guidelines recommend the following:
Treatment of cardiovascular comorbidities
Treatment of noncardiovascular comorbidities
Treatment of HF symptoms and signs of congestion with diuretics
Treatment effects on "symptom relief" reported as follows:
Diuretics: usually positive, as for HFrEF
ACEIs: evidence inconsistent
ARBs: evidence inconsistent except for candesartan (improve NYHA class)
Beta-blockers and MRAs: no evidence
Treatment effects on "HF hospitalizations" as follows:
ACEIs, ARBs
Patients in sinus rhythm: some evidence of benefit with nebivolol, digoxin, spironolactone, and candesartan
Patients in atrial fibrillation: evidence inconclusive for beta-blockers, absent for digoxin
Treatment effects on "mortality" as follows:
No RCT evidence for ACEIs, ARBs, beta-blockers, and MRAs except
Nebivolol reduced the "combined death and cardiovascular hospitalization endpoint"
Other drugs for specific comorbidities:
Atrial fibrillation:
Anticoagulants: positive evidence for reducing risk of thromboembolism
Antiplatelet agents: no proven benefit
Non-vitamin K oral anticoagulants (NOACS), such as apixaban, dabigatran, rivaroxaban: increase risk of hemorrhage in renal dysfunction, use contraindicated
Rate control: aggressive control might be deleterious
Digoxin, beta-blockers or rate-limiting CCBs, monotherapy or used in combination
Verapamil or diltiazem should not be combined with a beta-blocker
Hypertension:
Non-RCT evidence favors treatment of systolic blood pressure in HFpEF with
Diuretics, ACEIs, ARBs, and MRAs all effective; patients on ACEIs and beta-blockers should not receive the ARB olmesartan based on one study
Beta-blockers may be less effective
Type 2 diabetes:
Metformin, considered first-line drug
Type 2 diabetes and cardiovascular risk prevention:
Empagliflozin, sodium glucose cotransporter-2 (SGLT2): induces diuresis, natriuresis, weight loss; lowers blood pressure, preload/afterload; RCT evidence on efficacy of this drug for reducing blood glucose, HbA_1C, and body weight (probably by increased glucose excretion and osmotic diuresis); reducing HF hospitalization and cardiovascular mortality; reduce cardiovascular events and heart failure. Warning: aggressive management of dysglycemia may be harmful
Liraglutide, glucagon-like peptide/receptor agonist (GLP1RA): improve glycemic control, reduce risk of myocardial infarction, cerebrovascular accident, cardiovascular death in adults with type 2 diabetes and cardiovascular disease. Warning: stop drug if pancreatitis is suspected or confirmed

(continued)

Table 11.3 (continued)

Myocardial ischemia:
Treatment as per published management guidelines for angina and HFrEF
Exercise intolerance:
Combined endurance/resistance training
Obesity:
Diet and exercise
Various weight-loss drugs (FDA approved and non-FDA approved)
Hyperlipidemias:
Lipid-lowering agents as in myocardial ischemia, infarction, and HFrEF
Statins
Ezetimide
PCSK9 inhibitors
Other drugs proposed for HFpEF
LCZ696 (sacubitril/valsartan), angiotensin receptor neprilysin inhibitor: approved for treatment of HFrEF in 2015; trials in HFpEF are in progress and results are pending

Summarized and adapted from Ponokowski et al. (2016) [1], and updated from 2018 Congresses of the American College of Cardiology, European Society of Cardiology, and American Heart Association
Abbreviations as in text

hospitalization, morbidity, and mortality [1, 2]. It is recognized that several drugs that show efficacy in the short term may prove to be harmful in the long term [1, 2].

In HFrEF, pharmacotherapy consists of diuretics and neurohumoral antagonists such as ACE inhibitors (ACEIs), mineralocorticoid receptor antagonists (MRAs), and beta-blockers in the absence of contraindications or intolerance [Table 11.2]. Based on the results of a single RCT showing that the new compound LCZ696, which combines moieties of the angiotensin II (Ang II) type 1 receptor blocker (ARB) valsartan and the neprilysin (NEP) inhibitor sacubitril in a single molecule that is an angiotensin receptor neprilysin inhibitor (ARNI) and acts on both the RAAS and the neutral endopeptidase system, was superior to the ACEI enalapril in reducing HF mortality and hospitalization [73], it was recommended to replace ACEIs in HFrEF patients who continue to be symptomatic despite optimal therapy [1, 2, 74–76]. Several reports supported the replacement of ACEI with sacubitril/valsartan in patients with HFrEF [75, 76]. Moreover, since ARBs do not consistently reduce mortality, they are only used in ACEI-intolerant patients [1, 2]. The I_f-channel blocker ivabradine is used to control elevated heart rate above 70 beats per minute [1, 2].

Some safety issues with the ARNI sacubitril/valsartan include hypotension and angioedema [1, 2]. Recent reports in 2018 of contamination with human carcinogens N-nitrosodiethylamine (NDEA) in irbesartan and both N-nitrosodimethylamine (NDMA) and NDEA in valsartan supplied by certain pharmaceutical firms have raised additional concern that needs to be addressed. Of note, the recently updated recall list of the Food and Drug Administration (FDA) in the United States includes the ARB losartan for the same reason.

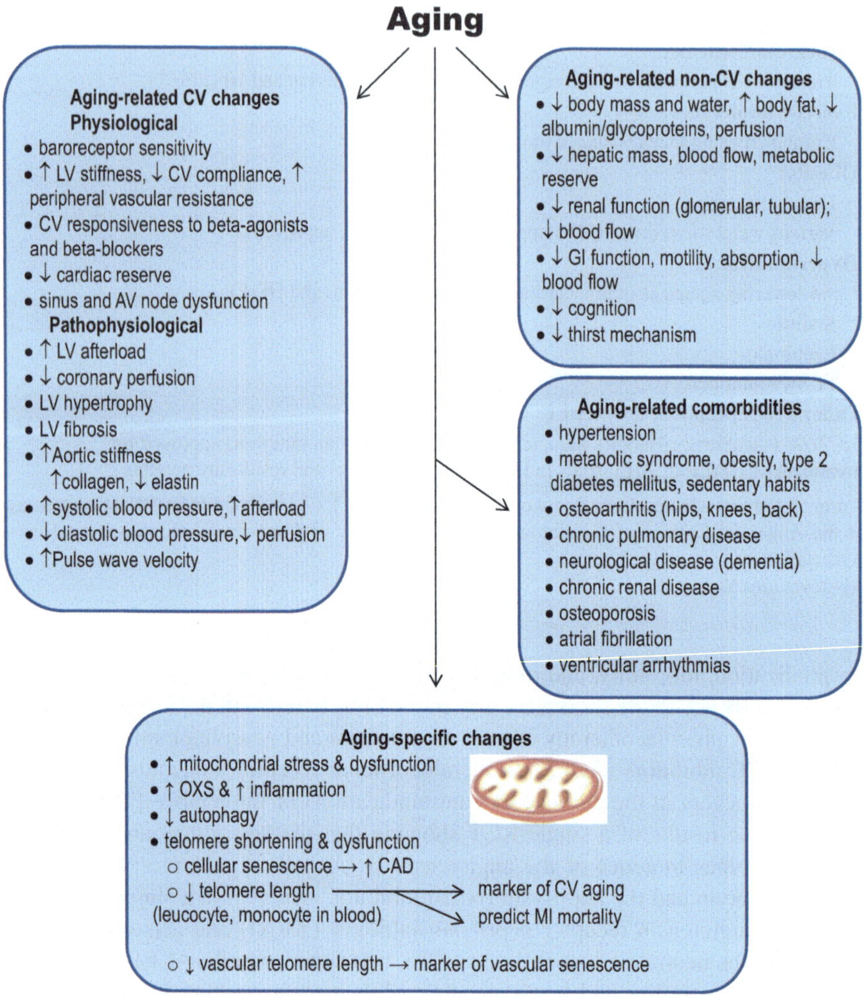

Fig. 11.5 Some aging-related cardiovascular and noncardiovascular physiological and pathophysiological changes and associated comorbidities
Adapted from Jugdutt [14–19, 34–36, 49]
Abbreviations: ↑ increased, enhanced, ↓ decreased, *AV* atrioventricular, *CV* cardiovascular, *GI* gastrointestinal, *HF* heart failure, *LV* left ventricular
Other abbreviations as in text

In the absence of firm and specific recommendations for HFpEF in the 2016 management guidelines [1, 2], several studies have been evaluating the efficacy of LCZ696 in HFpEF; while the final RCT results are still pending, there is considerable enthusiasm and hope of a positive outcome [76–81]. While awaiting RCT results in HFpEF, physicians in current clinical practice have tended to use therapies recommended for HFrEF to treat the HF component, including diuretics, beta-blockers, MRAs, ACEIs, and ARBs [1, 2], in addition to other specific therapies for common comorbidities such as HTN and DM2, an approach that has resulted in partial benefit [1, 2]. As summarized in Table 11.3, newer therapies, such as the sodium glucose cotransporter-2 (SGLT2) inhibitor empagliflozin and the glucagon-like peptide/receptor agonist (GLP-1RA) liraglutide, which were recently shown to improve control of DM2 as well as CV events and CV risks [80, 82, 83], are being implemented. However, there have been several negative studies; a large RCT of spironolactone in HFpEF failed to show benefit [84]. Although in patients with HFpEF an RCT with the phosphodiesterase-5 inhibitor sildenafil on exercise capacity [85] and another randomized double-blind crossover study with isosorbide mononitrate on daily activity [86] were negative, a subsequent meta-analysis of RCTs showed the benefit of exercise in HFpEF [87].

11.2.8 Studies on the Role of Inflammation in the Pathophysiology of HF Progression

The critical roles of acute and chronic inflammation during acute MI and the subsequent healing processes, and the progressive remodeling that spans these processes during remote MI and well beyond over years into the chronic HF stage, have been well documented, and various anti-inflammatory strategies have been proposed over the last five decades [34–36, 39, 41–44, 48–55, 58, 88–102]. These studies and others have underscored several pertinent points that need consideration when developing therapy for HFrEF after MI. The main points include the following.

11.2.8.1 Timing of Events Post-MI

The events that follow an acute MI all take place in tandem fashion (Table 11.4); early damage of muscle, matrix, and microvasculature triggers the healing process, which through a timed sequence of acute and chronic inflammation and associated biochemical, molecular, cellular, and subcellular reactions lead to formation of a fibrotic scar in the IZ, followed by fibrosis and hypertrophy in the NIZ and significant remodeling of structure, shape, and function [34–36, 41, 42, 49, 51–55, 101, 103]. Considering a single factor, such as the time interval needed for collagen deposition to reach a plateau during healing and repair after MI, this varies from weeks to months depending on the species, infarct size, reperfusion, and other factors; it usually takes a few days in mice, about 1 week in rats, 6 weeks in dogs, and 3 to 6 months in humans [51]. As reviewed before [51], this timing of the various events during the progression of fibrosis after MI is clearly important when deciding on timing and duration of therapies for HFrEF post-MI [34, 42, 99].

11.2.8.2 Multiple Cellular and Molecular Processes

An additional consideration with respect to therapy for HFrEF post-MI is the diversity and multiplicity of the cellular and molecular processes, cell types, and changes involved in the four main stages after STEMI or reperfused STEMI [50], as summarized in Table 11.4. Briefly, the early infarction phase over the first few hours involves damage to cardiomyocytes by apoptosis and necrosis, extracellular matrix

Table 11.4 Cell types and processes during stages of healing after MI and the march to HFrEF

Infarction phase, damage stage (hours)
Cardiomyocyte apoptosis and necrosis
ECM damage and vascular damage
Cardiomyocyte autophagy
Early healing phase, inflammation stage (days/weeks)
Different cell types:
Neutrophils
Monocytes: Ly-6C^{high}, Ly-6C^{low}
Macrophages: M1, M2
Mast cells
Lymphocytes: Treg
Dendritic cells
Late healing phase, proliferation stage (weeks/months)
Monocytes: Ly-6C^{low}
Macrophages: M2
Dendritic cells
Fibroblasts and myofibroblasts
Collagen/ECM deposition and ECM remodeling
Angiogenesis and vascular remodeling
Pericyte/endothelial cell
Maturation phase and scar formation stage (weeks/months)
Myofibroblasts
ECM remodeling
Collagen cross-link formation
Scar remodeling: contraction, late thinning; scar compaction; late scar expansion
Structural remodeling, during and after healing (days/months/years)
Early remodeling with infarct expansion (stretching, thinning, dilatation)
Scar remodeling (persistent myofibroblasts)
Structural remodeling
Infarct and noninfarct zones
Infarcted left ventricular chamber (dilatation, hypertrophy)
Other cardiac chambers: left atrium, right ventricle, right atrium
Left ventricular systolic and diastolic dysfunction
Left ventricular volume overload
Heart failure (HFrEF)
End-stage heart disease

Adapted and updated from Jugdutt (2013) [51]
Abbreviations as in text

(ECM) by matrix metalloproteinases (MMPs), and vascular cells by apoptosis and necrosis; the early healing phase with inflammation over days involves neutrophils, monocytes, macrophages, and mast cells; late healing with proliferation over weeks involves fibroblasts, myofibroblasts, collagen/ECM deposition, ECM remodeling, angiogenesis and vascular remodeling, and maturation with further ECM remodeling by cross-link formation, and scar formation, structural remodeling of the IZ and NIZ, and LV systolic and diastolic dysfunction [50]. The key modulators and mediators involved during the healing, repair, and fibrosis phases after MI and in the march to HFrEF, several of which can be targeted, have been reviewed before [51, 103] and are summarized in Table 11.5. Over all the phases leading to HFrEF, the three neurohumoral systems (RAAS, ET, and adrenergic) and OXS exert important modulating effects.

11.2.8.3 Persistent Inflammation Post-MI

The trickle of evidence over the last three decades indicates that low-grade inflammation and ECM remodeling both continue beyond the MI and collagen plateau phases during healing into the later phase of progression to the chronic HF phase, as reviewed elsewhere [101–108]; this is also an important consideration for timing and duration of therapy. During healing post-MI, the release of key factors that modulate healing, such as chemokines, cytokines, matrikines, growth factors including transforming growth factor-β (TGF-β), MMPs, and other matrix proteins, is quite precisely timed to orchestrate the sequence of acute and chronic inflammation with formation of granulation tissue, tissue repair with proliferation of fibroblasts, deposition of ECM, formation of myofibroblasts and scars, structural and functional remodeling of IZ and NIZ myocardium with cardiomyocyte hypertrophy and very little regeneration, and some angiogenesis [34, 51, 101–103, 108].

11.2.8.4 Timing of Remodeling Post-MI

The remodeling in post-MI survivors also occurs in a timed sequence, is progressive, and spans the infarction (first 24–48 h in humans) and healing (6 weeks to 3 months in humans) phases and far beyond (months to years) [34, 35, 40–44, 49, 51]; it is also modulated by multiple factors that orchestrate post-MI remodeling of myocardium, vascular tissue, ECM, and over time other cardiac chambers, tissues, cells, and molecules, resulting in a vicious cycle leading to end-stage HF. The additional role of exposure to CV risk factors that exacerbate OXS and inflammation in aging post-MI survivors during the march to HFrEF, just as in aging HTN victims in the march toward HFpEF as depicted in Figs. 11.2 and 11.3, in the progression of adverse remodeling needs to be recognized and addressed. Furthermore, the added contribution of non-CV risk factors in exacerbating OXS and inflammation and thereby promoting progressive adverse remodeling in the marches to both HFrEF and HFpEF need to be addressed.

Table 11.5 Some key modulators and mediators during healing, repair, and fibrosis phases after MI in the cascade to HFrEF

Infarction phase	RAAS, Ang II, ROS, ET, adrenergic overtone
Cardiomyocyte damage	
Necrosis	Nuclear factor κB (NFκB), Toll-like receptor 4 (TLR4)
Apoptosis	Caspases, etc.
Autophagy	AMP-activated kinase (AMPK), Beclin-1, NR4A2 (NR4A orphan nucleus receptor family member)
Vascular damage	Neutrophils, nuclear factor NF-kB (transcription factor family); TNFα, IL-1, chemokines (IL8), adhesion molecule ICAM-1, etc.
ECM damage	Proteases: matrix metalloproteinases (MMPs); stromelysin MMP-13; membrane-type MMP-14, MTP-MMP; collagenases: MMP-1, MMP-8, MMP-18; gelatinases: MMP-2, MMP-9; chymase/tryptase; MMP inhibitors, TIMP-1, TIMP-2, TIMP-3, TIMP-4
Early healing phase	Complement cascade
Acute inflammation	Cytokines: Tumor necrosis factor-α (TNF- α); interleukin (IL)-6; IL-1β; IL-18
Different cell types:	Chemokines: monocyte chemoattractant protein-1 (MCP-1)
Neutrophils, monocytes, macrophages, mast cells, lymphocytes: Treg, dendritic cells	Selectins, integrins, histamine
	Monocyte subsets Ly-6C^{high}, Ly-6C^{low}
	Macrophage receptors, macrophage subsets M1, M2
Late healing phase	
Chronic inflammation	Monocyte subset Ly-6C^{low}, macrophages: M2
Proliferation	Growth factors:
Monocytes, macrophages, dendritic cells	Transforming growth factor-β (TGFβ) → Smad signaling
Fibroblasts and myofibroblasts	Basic fibroblast growth factor (bFGF)
Collagen/ECM deposition	Platelet-derived growth factor (PDGF)
Pericytes, endothelial cells	Connective tissue growth factor (CTGF)
	Myocardial-related transcription factor-A
Maturation phase	Matricellular proteins/ECM remodeling:
Myofibroblasts	Secreted protein acidic and rich in cysteine (SPARC); osteopontin (OPN); Thrombospondin (TSP)-1
Collagen/ECM remodeling	Matrixins: A disintegrin and metalloproteinase (ADAM)-10, ADAM-17
Collagen cross-link formation	Protease inhibitors: Secretory leucocyte protease inhibitor (SLPI)
	Growth differentiation factor (GDF)-5; bone morphogenetic protein (BMP-2); bone morphogenetic protein (BMP)/growth differentiation factor-5 or BMP-14/GDF-5 (↑ scarring post-MI)
Fibrosis	Secreted frizzled-related protein-2 (sFRP-2); galectin-3; myocardin-related transcription factor (MRTF)-A

(continued)

Table 11.5 (continued)

Infarction phase	RAAS, Ang II, ROS, ET, adrenergic overtone
Angiogenesis and vascular remodeling	Vascular endothelial growth factor (VEGF), etc.
Repair	Stem cells: Akt-mesenchymal stem cell (MSC); Wnt signaling
	Blood flow; metabolism; mitochondrial pathways
Structural LV remodeling (infarct/noninfarct zone)	
Scar formation, contraction, compaction, expansion	
Structural remodeling of other cardiac chambers	
Left atrium, right atrium, right ventricle	
Development of LV systolic and diastolic dysfunction	
Development of heart failure/HFrEF	

Adapted and updated from Jugdutt (2013) [51]

11.2.8.5 Addressing Residual Inflammation in the Progression of Atherosclerosis Post-MI

It is well-known that (i) inflammation plays major roles in the initiation as well as the progression of atherosclerosis, as indicated by increased levels of the marker of inflammation, namely, high-sensitivity CRP (hs-CRP); and (ii) statins, through a pleotropic effect, reduce the residual inflammatory risk after MI [109–113]. However, the importance of "residual inflammation" in the progression of atherosclerosis in post-MI patients was only underscored in two recent RCTs [114, 115]. In the first report of the Canakinumab Anti-Inflammatory Thrombosis Outcomes Study (CANTOS) group, anti-inflammatory therapy with canakinumab to target IL-1β in 10,061 stable post-MI patients with evidence of residual inflammation, assessed by elevated hs-CRP levels, resulted in reducing the hs-CRP level and recurrence of CV events, independent of lipid-lowering [114]. In the second report of the CANTOS group, canakinumab therapy in 4833 patients with atherosclerosis showed that those patients in whom the IL-6 levels were lowered benefited from a reduction in major adverse CV events (MACE), hospitalization for unstable angina, and lower CV mortality as well as lower all-cause mortality independent of lipid-lowering. These findings suggested that modulation of the IL-6 proinflammatory pathway is beneficial for limiting vascular and CV events [115]. In the previous cholesterol and recurrent events (CARE) trial in MI survivors, levels of CRP increased over 5 years, indicating an increase in residual inflammation. Importantly, in that study, while therapy with the lipid-lowering agent pravastatin prevented the increase in CRP levels after MI, this benefit was not related to the extent of lipid-lowering, suggesting that a pleiotropic effect of the drug was involved [113]. In the more recent report of a cohort of 385 patients with hospitalization for HF from the original 10,061 patients with prior MI and elevated hs-CRP in CANTOS, the IL-1β inhibitor canakinumab showed a dose-dependent reduction in HF hospitalization and the composite endpoint of hospitalization for HF or HF-related mortality [116].

11.2.8.6 Addressing Inflammation in the ACS: Shift in Focus to Prevention of MI

In the last two decades, the thinking relating to the role of inflammation in acute coronary syndromes (ACS) and MI has shifted from the focus on treatment of the atherosclerotic plaque and epicardial coronary artery thrombosis, and the use of statins for lipid reduction and CV prevention, to the use of anti-inflammatory strategies for atherosclerosis and other strategies for ACS without thrombosis [117]. The CANTOS group continues to study IL-1 blockade with the IL-1β inhibitor canakinumab for interrupting the IL-1/IL-6 cascade and thereby reducing inflammatory risk in different settings [114, 115, 118, 119]. A small phase II RCT of 182 patients with non-ST elevation ACS showed that the recombinant IL-1 receptor antagonist (IL-1ra) anakinra reduced the elevated CRP, suggesting that IL-1 may be driving CRP elevation in ACS [120]. Interestingly, in that study, the CRP level rose again 16 days after treatment was stopped [120], suggesting that the persistent inflammation requires longer-term therapy. Furthermore, although the patients in CANTOS and the other studies with interleukin inhibitors were mostly older with mean ages of 60–64 years [114–116, 118, 120], and several factors in aging hearts are known to lead to increased ROS, $O2^{\bullet -}$, and myocardial Ang II which in turn trigger increased proinflammatory cytokines, MMPs, and OXS markers and thereby modulate post-MI healing and repair and progression of HF, OXS markers were not measured in those studies.

11.2.8.7 Role of IL-8 in Enhanced Damage After STEMI

While the canine aging study of Jugdutt et al. drew attention to the role of IL-6 as one important mediator of adverse LV remodeling after STEMI [58], other chemokines and proinflammatory cytokines may also be involved [121]. In a recent study of 258 patients with STEMI undergoing percutaneous coronary intervention (PCI) and who were followed for a median of 70 months, high levels of IL-8 in serially drawn blood samples were associated with large infarct size, impaired LV functional recovery, and adverse clinical outcome [122]. Interestingly, in that study, levels of IL-8 remained higher in nonsurvivors compared to survivors at 4 months [122]. While that study supports targeting of IL-8 for suppressing post-MI inflammation [122], the authors did not measure markers of OXS. Of note, the patients in that study were older adults, with an average age of 60 years (range 53–66 years) [122].

11.2.8.8 Surge of ROS and OXS After Reperfused STEMI

It is known that after STEMI, the extent of myocardial injury is massive and with reperfusion after a time delay; both the intensity of the inflammatory response and the extent of damage caused by the initial MI and the subsequent delayed reperfusion can be significant [48, 58]. Myocardial damage after STEMI is further exacerbated by the burst of OFR release and OXS with reperfusion as shown by Bolli et al. [123]. In that study, Bolli et al. showed that the levels of ROS, measured by electron paramagnetic resonance (EPR) and a spin trap, also called electron spin resonance (ESR), spectroscopy in the venous effluent from the stunned zone of myocardium in the dog model and sampled via a catheter positioned in the anterior interventricular vein, increased significantly to a peak over the first 20 minutes and persisted for several hours after reperfusion [116].

11.2.8.9 Role of Dysregulation of Immune Pathways in Adverse Post-MI Remodeling and HF

In a review on the topic, Prabhu and Frangogiannis [124] summarized the innate immune mechanisms involved in the four main steps of the cascade between MI and scar formation: (i) danger signals from necrotic cells in MI lead to activation of innate immune pathways that trigger inflammation; (ii) increased expression of pro-inflammatory cytokines (such as IL-1 and TNF-α) and chemokines (such as monocyte chemoattractant protein-1/CCL2) in response to stimulation of Toll-like receptor (TLR) signaling and complement promotes adhesive interactions between leukocytes and endothelial cells that lead to extravasation of neutrophils and monocytes; (iii) activation of repair mechanisms with suppression of inflammatory response cells leads to fibroblast proliferation and differentiation into myofibroblasts (driven by the RAAS and TGF-β), increase in ECM proteins, and scar formation; and (iv) scar maturation follows, with cross-linking of the collagen matrix, and removal of granulation tissue by apoptosis. The authors suggested that the combination of dysregulation of the immune pathways, impaired suppression or resolution of post-MI inflammation, failure of spatial containment of the inflammatory response, and excessive fibrosis contributes to adverse post-MI remodeling and HF [124]; however, they did not mention the possible contributions of persistent OXS, residual inflammation, the interaction between OXS, and inflammation in the progression from MI to HFrEF. While there is a vast body of compelling research evidence that supports therapeutic modulation of the inflammatory and repair responses after MI [124], Granger and Kochar pointed out that targeting inflammation in acute MI with a specific inflammatory agent might be "an elusive goal" [125]. In fact, the management guidelines up to 2016 recommend that anti-inflammatory agents such as steroids and nonsteroidal anti-inflammatory agents should be avoided after STEMI [1, 2, 109, 110].

11.2.8.10 Role of Different Monocyte Subsets in Post-MI Healing and HF

In another provocative review of innate immune mechanisms during healing after MI, Nahrendorf et al. [102] underscored the importance of two populations of monocytes involved in post-MI healing and found in both mice and humans [102, 126]; briefly, they noted that there is a biphasic response post-MI in the mouse, with Ly-6C^{high} monocytes (resembling CD16^{-} monocytes in humans) that are dominant in the early inflammatory phase and Ly-6C^{low} monocytes (resembling CD16^{+} monocytes in humans) that are dominant in the subsequent reparative phase [102]. They also noted that monocyte numbers are increased in atherosclerosis and increased recruitment of Ly-6C^{high} monocytes after plaque rupture impairs healing of MI and promotes HF via a cascade of increased chemokines (such as TNF-α), increased protease activity (such as MMPs, cathepsins), resolution of inflammation (with decreased TGF-β), and decreased collagen synthesis, based on previous findings in apoE$^{-/-}$ mice [102, 127]; importantly, they postulated a bell-shaped parabolic relation between monocyte numbers and healing after MI, in which both too little or too many of the monocytes can impair healing [102]. The authors proposed "shifting the monocyte response to a hypothetical vertex that denotes 'optimal' healing" as a goal of therapy for preventing HF [102]. They also suggested tailoring therapy to modulate the recruitment of monocyte subsets [102].

In a more recent study, Ruparelia et al. found that the patterns of gene expression associated with monocytes in inflammation and proliferation are switched on before they infiltrate the injured myocardium, suggesting that early therapy might be beneficial [126]. Of note, in that study, the average ages of the control and STEMI groups of human subjects were 63 and 60 years, respectively [126]. It should also be noted that during the early inflammatory response, the recruited neutrophils, macrophages, and monocytes release OFRs and contribute to OXS.

Together, these studies underscore the need to address the possible contributions of continued inflammation and OXS as well as ECM remodeling during healing after MI and the progression to HFrEF in post-MI survivors.

11.2.9 Role of Aging on the Pathophysiology of HF Progression

The importance of aging in the progression of adverse ventricular remodeling and HF (Fig. 11.4) is slowly becoming appreciated since the publication of several critical reviews and books on the topic [1–4, 14–19, 128]. The main points are summarized below.

11.2.9.1 Aging-Related Changes and HFpEF

As reviewed before [14, 15, 18, 19, 46–51, 103, 128], the progressive physiological, biological, and structural changes that characterize CV aging lead to increased ECM deposition and fibrosis (driven by increased Ang II, aldosterone, and TGFβ) in the heart, thereby increasing ventricular-arterial stiffening and LV diastolic dysfunction resulting in HFpEF (Fig. 11.5). In addition, increased ECM deposition and fibrosis plus decreased elastin (associated with increased MMP-9, MMP-12, cysteine proteinases cathepsins S, K, L, and serine proteinase neutrophil elastase from inflammatory cells) lead to increased aortic stiffness which in turn leads to the chain of increased aortic pulse wave velocity, systolic blood pressure, and afterload, which results in systolic HTN, LVH, and fibrosis (driven by increased Ang II, aldosterone, and TGFβ). As pulse pressure widens, a decrease in diastolic blood pressure below 70 mmHg can lead to the J-curve effect and thereby decrease diastolic myocardial perfusion, which in turn can trigger subendocardial ischemia and increase CV risk [128]. In that construct, the physiological changes during aging predispose the subject to HFpEF (Fig. 11.4); the added insults from HTN and the associated combined effects of LV pressure overload, LV hypertrophy, excess LV and aortic ECM and fibrosis, and decrease in aortic elastin collaborate to drive the concentric LV remodeling in the initial stage of HFpEF. Exposure to other CVD risk factors cooperates in exacerbating OXS, inflammation, and microvascular dysfunction in the survivors, thereby contributing to the progressive adverse remodeling and the march toward increasing severity of HFpEF progression (Figs. 11.2, 11.3, 11.4, and 11.5). Later, as a result of acute ischemia or other aggravating factors, a mixture of the two types of LV remodeling may develop and lead to HFrEF superimposed on HFpEF, thereby accelerating the march toward end-stage disease (Figs. 11.2, 11.3, and 11.4).

11.2.9.2 Aging-Related Changes and HFrEF

In the aging patient who develops acute STEMI, the background physiological changes of CV aging that predispose the patient to HFpEF alter and augment the responses to acute injury; the combination of enhanced damage to the myocardium, ECM, microcirculation, and endothelium with STEMI drives the development of dilative LV remodeling and the shift toward HFrEF (Fig. 11.4). The continued exposure to risk factors that exacerbate OXS and inflammation in the survivor contributes to progressive adverse remodeling (Figs. 11.2, 11.3, and 11.4). Furthermore, reperfusion therapy with restoration of coronary blood flow in the infarct-related artery (IRA) after STEMI leads to an acute surge in OXS, which, in combination with the intense acute inflammation, exacerbates the damage [19, 31, 34, 48–52, 58, 101, 102].

11.2.9.3 Aging-Related Changes in Healing and Post-STEMI Remodeling

As mentioned before, progressive adverse remodeling that takes place during the subsequent healing, repair, and beyond the healing/repair phases in survivors of STEMI spurs on the progression to HFrEF (Fig. 11.2). Expanding on that general theme, it is now established that with aging (Fig. 11.5), adverse post-MI remodeling is more severe [14–19]. This is due to impaired healing and repair of the damaged tissue resulting from a dysregulation of the pathways that are involved [15, 18]. The augmented adverse remodeling affects the entire left ventricle and, in time, leads to remodeling of the left atrium and right ventricle as well (Fig. 11.3). Of note, cardiomyocyte hypertrophy and fibrosis after STEMI develop in both the spared myocardium within the IZ and the myocardium in the NIZ [51]. In the aging mouse model of reperfused MI, Bujak et al. showed that enhanced adverse remodeling was associated with suppression of the inflammatory response, delayed granulation tissue formation, and reduced collagen deposition [48]. In the aging canine model of reperfused STEMI, with reperfusion after 90 minutes of ischemia after coronary occlusion, Jugdutt et al. [58] showed age-dependent early increases in markers of damage (increased ischemic injury, infarct size, cardiomyocyte apoptosis, blood flow impairment and no-reflow), structural remodeling (increased LV dilation and dysfunction) and matrix remodeling (increased expression of secretory leucocyte protease inhibitor [SLPI], secreted protein acidic and rich in cysteine [SPARC], osteopontin [OPN], a disintegrin and metalloproteinase [ADAM]-10 and ADAM-17, and MMP-9 and MMP-2), and inflammation (with increased inducible NO synthase [iNOS], proinflammatory cytokines IL-6 and TNF-α, and TGF-β1, and decreased anti-inflammatory cytokine IL-10). Importantly in that study, early therapy with the ARB candesartan, initiated at the time of reperfusion, attenuated these adverse age-dependent changes [58].

As reviewed previously [50], the main changes in molecular and cellular responses during healing and repair after STEMI/reperfused STEMI in older subjects include (i) increased RAAS, Ang II, and ROS activity; (ii) impaired or defective healing, with amplification of damage in the infarction phase (reflected in increased apoptosis, necrosis, and ECM degradation), dysregulation of inflammation in the early inflammation phase (with decreased antioxidant response), and defective healing and repair in the late phase (reflected in decreased myofibroblasts, collagen/ECM, angiogenesis, and telomere lengths) resulting in a defective scar. The ECM changes during repair include increased collagen type III (more elastic and pliable) and decreased collagen type I (more rigid) and cross-linking [50].

11.2.9.3.1 Aging-Related Changes in Mitochondrial OXS and Inflammation During HF Progression

The roles of aging, mitochondrial OXS, inflammation, and CVD risk factors during HF progression have been underlined here in Figs. 11.2, 11.3, 11.4, 11.5, 11.6 and 11.7, and the pertinent advances recently reviewed by Paneni et al. [128] are summarized below under 14 subheadings.

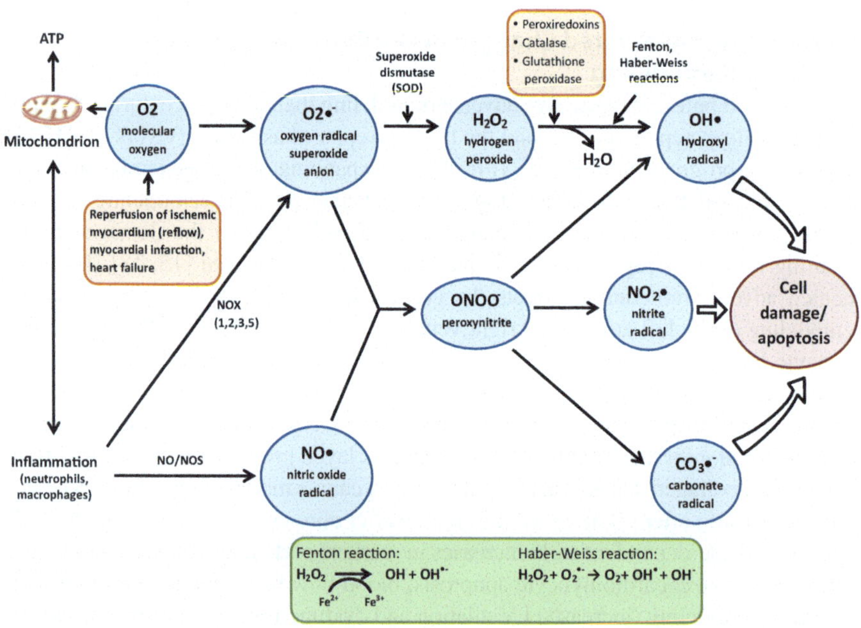

Fig. 11.6 Schematic of main sources of oxidants, superoxide, and peroxynitrite. Adapted from Bartesaghi and Radi (2018) [98]
Peroxynitrite (ONOO-) is cytotoxic and enhances cell damage during reperfusion following ischemia and can cause protein tyrosine nitration which can serve as a biomarker of oxidative stress. The nitrite radical NO2• is a strong oxidizing and nitrating agent. Protein tyrosine nitration is pertinent in reperfusion and inflammation. The carbonate radical CO3•- is involved mainly in nitro-oxidative damage
Abbreviations: *NOX* NADPH oxidases, *NOS* nitric oxide synthases

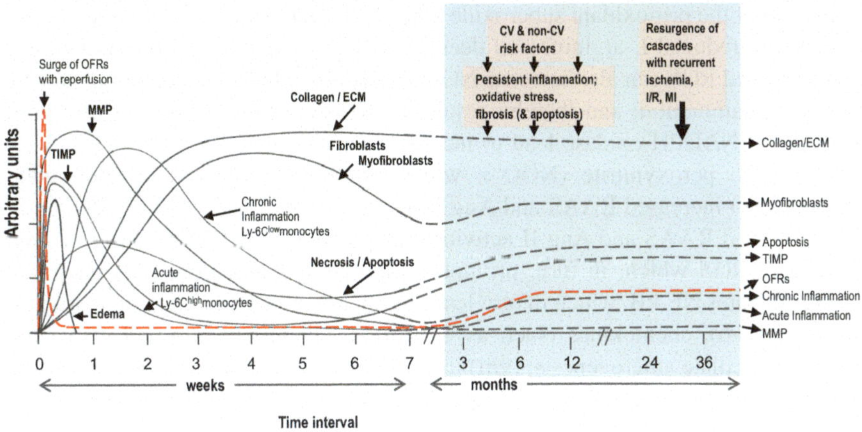

Fig. 11.7 Schematic depicting the postulated temporal evolution of changes in myocardial damage, extracellular matrix, inflammation, and oxidative stress after myocardial infarction during progression of HFrEF
Data based on studies in the canine model by Jugdutt BI et al. using histopathology for myocardial damage (necrosis and apoptosis), inflammation, collagen and fibrosis, and biochemistry for collagen, MMP, and TIMP [18, 34–36, 42, 44, 49–51, 55, 58, 92, 94, 98, 103, 128, 129] and Bolli et al. using electron paramagnetic resonance and a spin trap for reactive oxygen species [123], and a review of inflammation and repair postinfarction by Nahrendorf et al. based initially on data in mouse [102]
Abbreviations: *MMP* matrix metalloproteinase, *TIMP* tissue inhibitor of metalloproteinase
Other Abbreviations as in text

11.2.9.3.2 The Problem of Changing Demographics of HF with Aging

Expanding on previous reviews on aging and HF [14–19, 50, 51], Paneni et al. reemphasized the impact of lifelong vascular remodeling on the development of CVD risk and HF [128]. Updating previous demographic data on aging [14], they underline the projected doubling of the population aged >65 years from 12% in 2010 to 22% in 2040 [128]. Importantly, they point out that the prevalence of CVD is increasing in people aged >65 years, more so in those aged >80 years, and will increase by 10% over the next 20 years [128]. More alarming, the projected increases between 2010 and 2030 are for 27 million more people with HTN, 8 million more with CHD, 4 million more with stroke, and 3 million more with HF, and these will be mainly in the expanding elderly group [128]. They also underscore several important points that are pertinent for therapeutic interventions [128] and are outlined below.

11.2.9.3.3 Molecular Mechanisms in Aging-Related Vascular Remodeling

The evidence for how the changes associated with aging-related vascular remodeling contribute to CVD risk and adverse CV events and what key cellular mechanisms and pathways are involved in endothelial dysfunction [128] are as follows: (i) decreased cofactor tetrahydrobiopterin (BH4), in eNOS-mediated generation of NO from L-arginine, leads to eNOS uncoupling with decreased NO release and increased formation of the pro-oxidant superoxide anion ($O2•^-$); (ii) increased arginase activity leads to reduced L-arginine and decreased NO production and bioavailability; (iii) increased ROS can also increase NO degradation, which is mediated in part by chronic inflammation, and thereby results in NO depletion [128]; (iv) increased TNF-α and NADPH oxidase lead to increased superoxide ($O2•^-$) that reacts with NO to form peroxynitrite (NOO^-), which nitrosylates eNOS and antioxidant enzymes; (v) increased RAAS and Ang II activities contribute to NO inactivation; (vi) increased RAAS and Ang II activities also activate NADPH oxidase, thereby increasing ROS which, in turn, promotes vascular inflammation; (vii) increased H2O2 activates NF-kB, which in turn leads to release of proinflammatory cytokines (such as IL-6), chemokines (such as TNF-α), and adhesion molecules that are known to mediate atherogenesis; (viii) increased ET-1 levels in the blood and aortic wall promote vasoconstriction and impair endothelium-dependent dilatation; ix) increased cyclooxygenase (COX)-derived eicosanoids (such as prostaglandin [PG]-H2, thromboxane [Tx]-A2, and PGF2α) that are known to promote vasoconstriction and thrombosis, combined with decreased prostacyclin ($PG\text{-}I_2$) that is known to prevent these effects, result in increased vasoconstrictor tone and thrombogenicity; (x) increased collagen (from increased collagen deposition and decreased breakdown, as well as increased advanced glycation end-products [AGEs]) and decreased elastin (via the aforementioned increase in MMP-9, MMP-12, cysteine proteinases cathepsins S, K, L, and serine proteinase neutrophil elastase from inflammatory cells with elastolytic activity) lead to increased arterial stiffness (especially the thoracic aorta) and reduced distensibility of large elastic arteries; and (xi) increased TGF-β activity, which induces increased synthesis of interstitial collagen by the adjacent vascular smooth muscle cells (VSMCs) and increased RAAS, Ang II, and TGFβ, which increase both synthesis of collagen and lysis of elastin in the arterial wall, contribute further to arterial stiffness [128]. Taken together, vascular aging is associated with arterial stiffening, endothelial dysfunction with decreased endothelial-dependent vasodilation and antithrombotic property, and increased OXS and proinflammatory cytokines, which act in concert to increase the predisposition to atherosclerosis, thrombosis, and CVD [128].

11.2.9.3.4 Impact of Lack of Evidence-Based Therapy for HFpEF in the Elderly

As mentioned before, whereas aging-related adverse myocardial and vascular remodeling are known to lead to HTN and HFpEF in the elderly, the lack of evidence-based effective therapies for HFpEF further predisposes the patients to increased risk of myocardial ischemia, MI, ischemic cardiomyopathy, and a switch to HFrEF [128]. Furthermore, while aging is associated with decreased skeletal calcium, the elderly develop increased risk of calcific aortic stenosis that is partly due to increased inflammation and can contribute to increase in afterload, LVH, and subendocardial ischemia. Paradoxically, statins appear to hasten coronary artery calcification [128]. In addition, aging-related frailty with sarcopenia (i.e., loss of muscle mass and function) results in increased sensitivity to drugs used for treating HTN in the elderly [15, 19, 128]. The diagnosis of frailty is complicated by the associated osteoporosis and obesity (that is partly due to the proinflammatory state) [128]. Frailty also appears to interact with the increased prevalence of CVD in the elderly, thereby resulting in increased vulnerability to stressors and OXS.

11.2.9.3.5 Aging-Related Increased Incidence of Cardiac Amyloidosis

OXS and inflammation may play a role in cardiac amyloidosis. Increase in light-chain amyloidosis is associated with the increase in multiple myeloma with aging, whereas cardiac amyloidosis is associated with wild-type transthyretin (wtTTR) that is more common in older men. Imaging detected wtTTR cardiac amyloidosis in about 13% of patients with HFpEF aged ≥60 years, whereas autopsy showed it in about 20% of people who died at age >80 years [128]. Although there is no proven treatment for wtTTR cardiac amyloidosis, several trials are under way. Chemotherapy is used in light-chain amyloidosis.

11.2.9.3.6 Aging-Related Telomere Shortening During Cellular Senescence

As previously reviewed [19], evidence suggests that telomere shortening is associated with increased risk of CVD during aging. It is well established that aging-related increase in senescent cells in the vascular wall and heart contributes to adverse remodeling [128]. Telomeres consist of repetitive nucleotide sequences (TTAGGG on one strand and AATCCC on the other strand) at the ends of mammalian chromosomes. These act as caps that preserve chromosome stability and integrity by preventing deterioration or fusion with neighboring chromosomes; every cell division shortens telomeric DNA. When a critical length is reached, the capping function is lost, leading to DNA damage and apoptosis. Studies have shown an association between decreased leukocyte telomere length (TL) and vascular cell senescence, aortic stenosis, CV risk factors (such as HTN, DM2, obesity, and smoking), and risk of atherothrombotic events. Other studies showed correlations between leukocyte TL and atherosclerosis, ischemic and hemorrhagic stroke, and between reduced leukocyte TL and risk of plaque and its progression [128]. A meta-analysis (43,725 participants; CVD, 8400) showed that patients with the shortest leukocyte TL had a higher risk of coronary heart disease and cerebrovascular disease [129].

11.2.9.3.7 Mitochondrial Oxidative Stress and Cardiovascular Aging: Role of the $p66^{Shc}$ Gene

The molecular events in CV aging, mitochondrial OXS, chromatin remodeling, and genomic instability may all be linked [128]. Six lines of evidence from translational studies support the idea that mitochondrial OXS contributes to cellular senescence through a chain of $O2•^-$ or H2O2 formation, ROS overload, senescence, DNA damage, inflammation, and cell death pathways including apoptosis [128]. The evidence supporting the role of the mitochondrial adaptor $p66^{Shc}$ gene in this chain of cellular events is as follows: (i) cells lacking $p66^{Shc}$ show reduced intracellular ROS, whereas mice lacking $p66^{Shc}$ have reduced ROS upon exposure to high OXS, and mice with $p66^{Shc}$ deletion show increased longevity; (ii) aging $p66^{Shc}$-deficient mice have reduced ROS and preserved NO bioavailability and are protected from both systemic and cerebral endothelial dysfunctions; (iii) p66^{Shc}-deficient mice with brain injury from I/R show reduced ROS production in the brain and reduced stroke size, and in vivo postischemic silencing of p66^{Shc} prevents I/R brain injury in mice; (iv) increased p66^{Shc} expression found in stroke patients correlates with neurological deficits; (v) increased p66^{Shc} is present in peripheral blood mononuclear cells of patients with ACS and DM2; and (vi) p66^{Shc} protein activation is found in patients with CV risk factors, including hyperglycemia, oxidized low-density lipoprotein, smoking, and HTN. Together, these findings suggest that p66^{Shc} may be a potential therapeutic target for age-related CVD and increased mitochondrial ROS [128].

11.2.9.3.8 Mitochondrial Oxidative Stress and Cardiovascular Aging: Role of JunD

Evidence suggests that the activated protein-1 (AP-1) transcription factor JunD mediates aging-related OXS [128]. JunD, which is formed from dimeric complexes from three main families of DNA-binding proteins (Jun, Fos, and ATF/CREB), is involved in regulation of cell growth and survival, as well as in protection against OXS by modulating genes involved in antioxidant defense and ROS production [128]. Five lines of evidence that support the role of JunD are as follows: (i) JunD levels are lower in the aorta of the aging mouse and in peripheral blood mononuclear cells from old compared to those in young healthy humans; (ii) young mice lacking JunD show endothelial dysfunction and vascular senescence similar to that found in old wild-type mice; (iii) JunD null mice show increased aging markers p53 and p16INK4a, reduced telomerase activity, and mitochondrial DNA damage in their aortas, whereas overexpression of JunD rescues vascular aging features in old mice; (iv) the age-associated decrease in JunD leads to an imbalance between the oxidant NADPH oxidase and scavenger enzymes manganese superoxide dismutase (MnSOD) and aldehyde dehydrogenase 2, which results in early redox changes, mitochondrial dysfunction, and vascular senescence; and (v) mice lacking JunD develop less hypertrophy after mechanical pressure overload whereas mice with cardiomyocyte-specific expression of JunD develop LV dilatation and contractile dysfunction; moreover, fra-1 transgenic mice overexpressing the AP-1 member fos-related antigen and lacking JunD develop dilated cardiomyopathy associated with defective mitochondria and increased cardiomyocyte apoptosis; the findings suggested that JunD promotes adaptive or maladaptive hypertrophy depending on its level [130].

11.2.9.3.9 Mitochondrial Oxidative Stress and Cardiovascular Aging: Role of Sirtuin-1

The evidence supporting the beneficial role of the sirtuins (silent information regulator [SIR] genes), which are NAD$^+$-dependent enzymes of the big nicotinamide adenine dinucleotide (NAD)-dependent protein family, in human aging [128] is as follows: (i) endogenous sirtuin-1 (SIRT1) expression in VSMCs correlates inversely with donor age, and age-related loss of SIRT1 correlates with reduced stress response and increased senescence; (ii) endothelial-specific SIRT1 overexpression or chronic exposure to a SIRT1 activator in hypercholesterolemic mice decreases atherogenesis, whereas reduced SIRT1 increases atherosclerosis; (iii) inhibition of SIRT1 by immunosuppressant drugs (such as sirolimus and everolimus) leads to endothelial senescence; (iv) inhibition of SIRT1 with sirtinol impairs eNOS function, while activation of SIRT1 improves endothelial NO availability; (v) inhibition of SIRT1 with the endogenous inhibitor, microRNA-217, suppresses SIRT1-dependent eNOS function and promotes endothelial senescence; (vi) SIRT1 has been shown to regulate p66^{Shc} transcription, whereas reduced SIRT1 leads to NF-kB p65 acetylation, resulting in increased inflammatory genes; (vii) SIRT1 has been shown to repress pathways of arterial aging, thereby preventing DNA damage, cell cycle arrest, and oxidative stress; and (vii) SIRT1 has been shown to activate the energy regulator enzyme 5′-adenosine monophosphate (AMP)-activated protein kinase involved in glucose homeostasis, thereby maintaining cellular ATP levels and maintaining endothelial integrity via regulation of eNOS activity and autophagy [128]. Together, the findings suggest that activation of SIRT1 may preserve endothelial function during aging and contribute to prevention of CVD and progression to HF.

11.2.9.3.10 Mitochondrial Oxidative Stress and Cardiovascular Aging: Role of Klotho

Recent evidence suggests that Klotho is an important antiaging gene and Klotho protein acts as a circulating hormone, which binds to a cell-surface receptor and thereby suppresses intracellular signals of insulin and IGF-1that are known to favor longevity [128]. Studies show that (i) Klotho deletion in mice induces premature aging and reduces life span, whereas overexpression increases life span and protects against age-related CV and kidney dysfunction; (ii) high plasma levels of Klotho in patients are associated with a reduced risk of CVD, whereas low serum Klotho concentrations predict CAD and arterial stiffness [128]. Together, these findings suggest that Klotho may be a useful biomarker and a potential target for limiting age-related CVD and stalling progression to HF.

11.2.9.3.11 Aging-Related Dysregulation of DNA Repair and Damaged DNA Overload

Evidence suggests that the buildup of damaged genetic material throughout life and defects in the DNA repair mechanism after damage (caused by chemicals, mutations, and epigenetic alterations [i.e., "change in gene activity without change in DNA sequence"]) might result in genomic instability and cellular senescence and thereby promote CV aging and dysfunction [128]. Studies have shown that (i) extensive nuclear DNA damage in the Hutchinson-Gilford progeria syndrome is associated with premature atherosclerosis and CVD that lead to early MI or stroke; (ii) mice with genomic instability from defective repair genes develop premature aging that is associated with endothelial cell senescence, vascular stiffness, and HTN, thought to be due to dysregulation of eNOS and sirtuin and to increased NADPH oxidase; and (iii) humans develop sporadic genomic mutations, as supported by findings of (a) DNA damage in circulating cells and plaques of patients with atherosclerosis, (b) chromosomal damage and mitochondrial DNA deletions in peripheral blood mononuclear cells of patients with coronary heart disease that correlate with severity of the disease, (c) association between variation in nucleotide excision repair components and carotid-femoral pulse wave velocity, and (d) increased phosphodiesterase type 1 (PDE1A) expression and impaired NO-cGMP signaling and endothelial dysfunction in senescent VSMCs, and association between PDE1A polymorphisms and diastolic blood pressure and carotid intima-media thickness in patients [128]. Together, these findings suggest that preserving genome stability may limit age-related CVD that contributes to HF progression.

11.2.9.3.12 Aging-Related Defects in Angiogenesis

Evidence suggests that defects in angiogenesis during CV aging lead to (i) increased stroke, peripheral artery disease, and MI in elderly, with worse outcomes compared to younger patients; (ii) increased mortality and rate of limb amputation after acute limb ischemia; (iii) reduced capillary density, associated with microvascular disease, defective eNOS functionality, and impaired insulin sensitivity in elderly men; (iv) decreased proliferative capacity, telomerase activity, and production of angiogenic growth factors such as VEGF-A in senescent endothelial cells; (v) impaired endothelial migration leading to reduced tube formation; (vi) reduced hypoxia-inducible factor 1α (HIF1α) activity, due mainly to increased degradation and reduced nuclear translocation by importin α; (vii) reduced activity of the transcriptional coactivator PGC-1α that is associated with hypoxia-driven angiogenesis; and (viii) dysregulation of angiogenic pathways associated with an age-dependent decrease in the number and function of stem and progenitor cells [128]. These findings are pertinent for the development of therapies for the elderly.

11.2.9.3.13 Aging-Related Vascular Remodeling via Nongenomic Regulation of Changes in Chromatin

Recent evidence suggests that, besides the well-recognized gene-driven regulation of aging, nongenomic regulation of aging via epigenetic modifications of transcription programs in OXS, inflammation, angiogenesis, and metabolism can also promote maladaptive pathways leading to vascular aging [128]. The epigenetic modifications that are acquired throughout life appear to be quite stable and explain how environmental factors may interact with genomic DNA and change gene expression. Importantly, epigenetic changes may be transmitted down or inherited and thereby contribute to senescent traits and CVD in young adults [128]. Epigenetic processes that modify chromatin include DNA methylation, acetylation, phosphorylation, ubiquitination, and sumoylation. Studies have shown that (i) DNA methylation, which is the most widely studied, decreases with age, whereas the rate of demethylation correlates inversely with age in mice and humans; (ii) unmethylated or partially methylated CpG (cytosine-guanine) islands are present in atherosclerotic plaques and leukocytes from patients and atherosclerosis-prone mice; (iii) changes in DNA methylation localize at promoter sites of several regulatory genes, including NOS and the vascular endothelial growth factor receptor, that are involved in atherosclerosis and aging; (iv) histone methylation has been implicated in regulating life span and vascular homeostasis that involves endothelial NF-kB, SIRT1, and other genes, including methyltransferase Set7 expression which is increased in peripheral blood mononuclear cells from DM2 patients and correlates with NF-kB-mediated inflammation, OXS, and endothelial dysfunction; (v) histone deacetylation by SIRT1 may influence age-related CVD and transgenic overexpression of SIRT1 was shown to improve metabolic efficiency and endothelial function in old mice; and (vi) SIRT6 can prevent endothelial dysfunction and atherosclerosis by epigenetic modulation of multiple atherosclerosis-related genes, including the pro-atherogenic gene of the TNF family [128]. Together, these findings implicate chromatin modifications in aging-related CVD.

11.2.9.3.14 Aging-Related Changes and Stem Cell-Based Therapies for Vascular Repair in the Elderly

With regard to the status of vascular repair and of autologous bone marrow-derived stem cell transplantation in the elderly, the conflicting results of RCTs of stem cell-based therapies in MI may be explained by the fact that, in most cases, the aged bone marrow cells have been down the senescence pathways and the substrate in the elderly involved activation of pathways that may impair the healing ability of transplanted stem cells. Therapies that can potentially modulate epigenetic modifications of chromatin and thereby restore gene expression may be useful [128]. Another consideration is the supplementation with adjunctive therapy for targeting both increased OXS and inflammation in elderly patients. Other aging-related molecular remodeling including impaired adrenergic signaling and calcium handling are discussed in detail elsewhere [16, 19].

11.2.9.4 Aging-Related Increase in Mitochondrial ROS and Autophagy

A prevalent theory is that aging is general, including CV aging, and involves increased generation of ROS from dysfunctional mitochondria leading to ROS-induced cumulative cellular damage [131, 132]; the dysfunctional mitochondria are thought to accumulate during cellular aging due to decreased autophagy [131, 133]. Autophagy, which performs cellular housekeeping and maintains homeostasis in the cell by renewing or recycling cytoplasmic materials and organelles (including mitochondria), removing toxic protein aggregates and harmful ROS, and providing essential energy and biomolecules to cells, is a cytoprotective function that has been shown to decline with aging [133, 134]. With aging, evidence for both the increase in basal ROS levels [135] and the increase in dysfunctional mitochondria from reduced or dysregulated autophagy [134] have been documented. Since cardiomyocytes are not frequently replaced, they are subjected to OXS and oxidative damage during aging [135]. Dai et al. have reviewed the interactions among mitochondrial ROS, redox, and other cellular signaling pathways as well as various therapeutic strategies and drugs (such as mitochondrial antioxidants MitoQ, SkQ1, and the mitochondrial protective peptide SS-31) that can potentially improve mitochondrial function during aging [132].

Evidence shows that homeostasis by autophagy is maintained through cues from "danger-associated molecular patterns" (DAMPs), such as ROS, mitochondrial

DNA, and extracellular ATP that are detected by "pattern recognition receptors" (PRRs) comprised of several families, including "Toll-like receptors" (TLRs) and "NOD-like receptors" (NLRs), which control autophagy and are also involved in innate and adaptive immune response [136]. DAMPs activate the NLRP3 inflammasome in response to ROS and other stimuli, such as extracellular ATP and lysosomal disruption [136]. Evidence suggests that autophagy and inflammation are interdependent [137]. As mentioned before, impaired autophagy is associated with an increase in defective mitochondria and levels of ROS and leads to increased inflammatory cytokine levels; evidence suggests that increased ROS activates inflammasomes which process IL-1β.

While preclinical studies are consistent with decreased autophagy during aging (Fig. 11.5) and autophagy-driven regulation of the degree of inflammation, whether the decrease in autophagy with aging leads to increased basal levels of inflammation and OXS in the human heart or the CV system during HF needs further study. The studies reviewed by Linton et al. did not specifically address this question but the results were interesting [136]. In one small study of 9 HF patients with idiopathic ischemic cardiomyopathy maintained on an LV assist device (LVAD), LV biopsies showed a downregulation of the autophagy gene mRNA and protein [138]. In another study of 170 patients undergoing coronary artery bypass surgery (CABG), 22% who developed postoperative atrial fibrillation showed evidence of impaired autophagy in their atrial biopsies, while levels of hs-CRP, inflammation, fibrosis, and other markers were similar for patients with or without postoperative atrial fibrillation [139]. In a further study of 19 patients undergoing surgery with cardiopulmonary bypass (CPB), the test results suggested depletion of autophagy proteins [140]. In another study of patients undergoing heart surgery with I/R, biopsies of the right atrial appendage showed evidence of upregulation of autophagy-related genes in 13% and downregulation in 4% [141]. A small study of patients undergoing CABG and remote ischemic preconditioning (RIPC) failed to show changes in autophagy in LV biopsies [142], probably due to issues with study design [136]. In a cohort of older cardiac surgery patients (mean age 62 years, range 33–87 years) who had right atrial biopsies before and after CPB, Linton's group found evidence for a "robust" autophagic response to the ischemic stress that was age-independent, albeit as assessed by lipidation of the autophagy protein LC3 which is somewhat controversial [143]. Another pertinent finding with respect to exposure to comorbidities is that animals with metabolic syndrome and DM2 show suppression of autophagy; in a cohort of their surgical patients, Linton's group found evidence for impaired autophagy in those DM2 patients with poor glycemic control and HbA_1C level > 7% [136].

Several studies addressed the role of autophagy in postremodeling and HF; in one study in young mice with chronic MI, the autophagy inhibitor bafilomycin A1 worsened remodeling, whereas the autophagy enhancer rapamycin attenuated remodeling, suggesting that autophagy might protect against adverse post-MI remodeling [144]; another more recent study in young mice with chronic MI showed that TLR3 upregulation contributes to persistent autophagy which promotes HF and death, whereas TLR3 deletion inhibits autophagy, reduces MI size, attenuates HF, and improves survival, and the autophagy inducer rapamycin abolishes these benefits [145].

11.2.9.5 Role of AGEs and RAGEs in OXS and Heart Failure

AGEs are condensates of glucose that nonenzymatically form cross-links between collagen molecules, thereby rendering them resistant to enzymatic degradation [128]. AGEs are known to accumulate in vessel walls and contribute to both microvascular and macrovascular complications through formation of cross-links. They exert adverse effects through either receptor or nonreceptor mechanisms; in the receptor mechanism, they bind to specific cell-surface receptors for AGEs (RAGEs) thereby activating RAGE and leading to a chain of NF-kB upregulation, proinflammatory cytokine (such as TNF-α, IL-1β) activation, growth factors (such as PDGF, IGF-1), adhesion molecules (such as VCAM-1), and increased ROS which converge on tissue damage. While soluble AGEs activate monocytes, basement membrane AGEs inhibit monocytes; AGE-bound RAGE increases endothelial permeability; AGEs block NO activity and induce ROS production; AGEs have also been implicated in diabetic vascular injury [146, 147]. Prasad et al. [147] referred to the adverse effects of AGE and AGE-RAGE as "AGE-RAGE stress" and its endogenous defense mechanisms as "antistressors"; the endogenous antistressor mechanism involves enzyme-mediated (through glyoxalase 1 and 2) and AGE receptor-mediated (AGER-1 and AGER-2) degradation of AGE, and increased soluble receptor of AGE (sRAGE); exogenous defense strategies include decreasing AGE consumption, preventing AGE formation, and downregulating RAGE. Pharmacological agents known to reduce AGE formation include those used to treat HFrEF and HTN (ACEIs, ARBs), DM2 (anti-DM2 drugs; aminoguanidine) and CVD risk (statins), and others [147]. Drugs that elevate sRAGE include ACEIs, statins, and antidiabetic drugs [147].

11.2.9.6 Role of Synergism Between OXS and Inflammation in HF Progression

The topic of stress and disease has raised concern since ancient times. Selye studied the role of stress in health and disease, focusing mainly on its role in the brain and related disorders [148]. Interest in the role of OXS as an important mediator of CVD was rekindled in the 1980s [31–33]. The vast literature on the biology, physiology, and biochemistry of OFRs, ROS, and OXS that stemmed from extensive basic, translational, and clinical research over nearly 5 decades since the 1970s has been recently reviewed elsewhere [31–33]. This has improved our understanding of the pathobiology, biochemistry, and pharmacology of OFRs (Fig. 11.6) and unveiled the mechanisms leading to OXS and its contribution to the pathophysiology of CVD, including myocardial ischemia, I/R and MI, and subsequent initiation and progression to HFrEF [27, 28, 31–33, 128]. In addition, the realization that both OXS and inflammation contribute to the progression of HFpEF and HFrEF has perked interest in the critical roles of OXS and inflammation in HF progression and underscored the importance of OXS and inflammation as potential targets for stalling HF progression [28–33]. The additional contributions of CVD risk factors in increasing both OXS and inflammation have also been underscored [31–33]. With the passage of time during aging, OXS and inflammation appear to propel the progression of HF forward toward end-stage heart disease (Figs. 11.2 and 11.4). However, more research is needed to unravel the mechanisms of how exactly CVD risk factors lead to increased OXS and inflammation and thereby mediate HF progression.

We hypothesize that the synergism between OXS and inflammation is an important factor in accelerating HF progression toward end-stage heart disease, and both OXS and inflammation need to be targeted for optimal results.

11.2.9.6.1 OXS and Inflammation in Post-STEMI Remodeling and HF Progression

The OXS-inflammation interaction during post-STEMI cardiac remodeling and its contribution during HF progression have been a topical issue for the last two decades (Figs. 11.2, 11.3, and 11.4). Although persistent inflammation has been addressed recently in several aforementioned RCTs showing the benefit of some anti-inflammatory strategies during post-MI HF progression [114, 115], RCTs that target persistent OXS alone or the OXS-inflammation interaction have not been launched.

11.2.9.6.2 Development of ROS Overload and OXS Post-STEMI and HF

The development of ROS overload and OXS in HF represents a breakdown of homeostasis that occurs whenever the production of OFRs far exceeds the capacity of endogenous antioxidant defense mechanisms to neutralize them; when that happens, the result is damage to the myocardium and other CV tissues. It is known that during aerobic respiration, molecular oxygen (O2) acts as the final acceptor of electrons in the ETC, thereby leading to generation of adenosine triphosphate (ATP) through the Krebs cycle in the mitochondria. The ATP supplies the energy needed for maintaining contractile function, cardiac output, and organ perfusion. Under physiological conditions, oxygen molecules are reduced to the ROS superoxide ($O2•^-$) which is converted to hydrogen peroxide (H2O2) by the endogenous antioxidant superoxide dismutase (SOD); H2O2 is then broken down into H2O by other antioxidants (such as catalase or glutathione peroxidase), as shown in Fig. 11.6. However, under pathological conditions (such as myocardial ischemia, I/R, MI, and HF), the formation of ROS or OFRs including $O2•^-$, H2O2, hydroxyl radical (•OH), hypochlorite (HOCl), and NO-derived peroxynitrite (NOO^-) increases; the scavenging ability of the antioxidants is exceeded, thereby resulting in ROS overload and OXS (Fig. 11.6) [30, 31, 149]. H2O2 is converted to the reactive (•OH) via the Fenton or Haber-Weiss reactions (Fig. 11.6). Increased NO from inducible NOS (iNOS) leads to increase in the reactive NOO^- (Fig. 11.6). Whereas most ROSs are lipid insoluble and remain within the cell, H2O2 can cross cell membranes and thereby cause damage at remote regions [149]. Endogenous antioxidants include SOD, glutathione peroxidases, catalase, and peroxiredoxins [149] (Fig. 11.6). Sources of ROS include enzymes in mitochondrial ETC, plasma membrane, peroxisomes, endoplasmic reticulum and nuclear membrane, and other enzymes such as xanthine oxidase, MPO, and cytochrome P450 enzymes (monooxygenases), some of which are found in macrophages and neutrophils that accompany inflammation, NADPH oxidases, and soluble heme proteins [149]. ROS produced by cardiomyocytes and infiltrating inflammatory cells leads to cellular damage through disruption of membranes, proteins, and nucleic acids and activation of cell death pathways that trigger apoptosis [150, 151] (Fig. 11.6). During postischemic reperfusion after MI, ROS may be produced by both enzymatic and nonenzymatic systems and in both cardiomyocytes and infiltrating inflammatory cells. Reperfusion stimulates neutrophil activation and lipid peroxidation in membranes which result in increased ROS. Although NO is nonreactive, iNOS generates reactive •NO from NO; NOXs generate reactive superoxide ($O2•^-$) from O2; the •NO and $O2•^-$ interact to yield highly reactive peroxynitrite (NOO^-) (Fig. 11.6). Evidence suggests that peroxynitrite leads to the formation of hydroxyl, nitrite, and carbonate radicals, which mediate cell damage (Fig. 11.6), and tyrosine nitration which serves as a biomarker of OXS [149]. Besides protein nitration, increased ROS and NOO^- also lead to lipid peroxidation, single-strand nucleic acid breaks, and chromosomal changes [149]. Some pertinent harmful effects of ROS overload during I/R include (i) stimulation of platelets to release platelet activating factor, thereby attracting more neutrophils that exacerbate damage; (ii) attenuation of NO function, thereby enhancing endothelial dysfunction through NOO^- formation and augmenting damage; and (iii) blunting of endothelium-dependent vasodilation and enhanced ET-1-induced vasoconstriction, which together aggravate the decrease in reflow.

11.2.9.6.3 Reasons for Failure to Reduce ROS Overload and OXS Post-STEMI and HF in Patients

Despite experimental animal studies suggesting that administration of exogenous antioxidants including SOD, xanthine oxidase inhibitors (such as allopurinol), N-acetyl cysteine, vitamin E, and vitamin C during after I/R and MI yields favorable results, overall clinical experience has been disappointing [1, 2].

There are several possible reasons for the lack of clinical benefit of antioxidants during and after I/R and MI at the bedside. These include the following: (i) in the setting of I/R and MI, the presence of intense inflammation can have many harmful effects as mentioned before, including increased ROS and activation of platelets which aggregate and provide the matrix scaffold for thrombi to form and plug blood vessels, thereby augmenting cardiomyocyte damage; (ii) the acute inflammation after STEMI also induces vascular damage, which (as mentioned before) is augmented by reperfusion, and involves sequential activation of resident macrophages, release of proinflammatory cytokines (such as IL-1 and TNF-α), upregulation of vascular adhesion molecules (such as selectins, integrins, and ICAM-1) and CXC chemokines (such as IL-8, macrophage inflammatory protein-2 [MIP-2]), neutrophil adhesion and transmigration, damage to endothelial cells, basement membrane and matrix, and increased permeability with microvascular hemorrhage; importantly, the intensity of the inflammatory response is regulated by the balance between proinflammatory cytokines (such as TNF-α, IL-1, and CXC chemokines) and anti-inflammatory cytokines (such as IL-10 and IL-13); (iii) the intense inflammatory

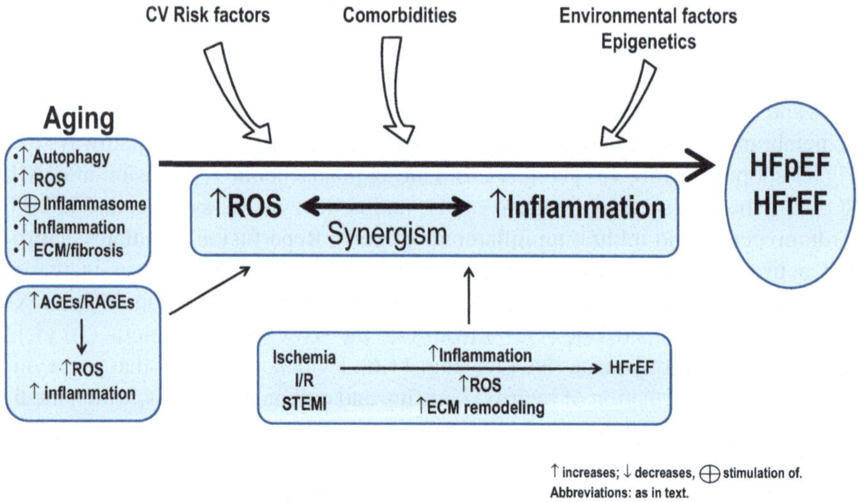

Fig. 11.8 Schematic depicting concept of synergism between persistent ROS and inflammation in heart failure progression
↑ increases, ↓ decreases, ⊕ stimulation of
Abbreviations: as in text

response with I/R and reperfused STEMI appears to collaborate in the early surge in ROS in the first 2–10 min and the persistence for several hours thereafter (Figs. 11.7 and 11.8) [123]; since inflammation extends far beyond reperfusion and generates ROS as discussed before (Fig. 11.7), it can be expected to contribute to the ROS pool far beyond the early postreperfusion phase into the subsequent phase of healing and repair, with obvious pathophysiological and therapeutic implications; of note, SOD plus catalase, given over 15 min before and continued for 30 min after reflow, blocked ROS production in the stunned zone, and improved functional recovery in the dog model [152]; (iv) intracellular remodeling after I/R injury is associated with increased proteolytic enzyme activity and alterations in gene expression and translation mechanisms which can have persistent effects [153]; (v) humans with STEMI are middle-aged or older adults, or elderly, with a host of aging-related issues as discussed before (Fig. 11.5); for example, aging may blunt the response to therapy, as found with an ARB after reperfused STEMI in dogs [58] and with post-ischemic conditioning in mice [97]; aging may also modify inflammation, healing, and repair [48, 58, 124]; (vi) in the context of aging, age equivalence should be considered when assessing therapies in animal models with the goal of translation to humans. For example, a 6-week-old mouse or rat would be equivalent to a young human child by age, and not even to a young adult [99]; (vii) persistent ROS production over the healing phases of acute and chronic inflammation [48, 50, 97, 101–103], and far beyond, during progressive remodeling and HF as discussed below; (viii) the rate of reperfusion differs between animal models and humans, being usually more abrupt in most animal models than in humans who undergo PCI and/or thrombolysis after STEMI [97]; (ix) besides SOD and catalase, other protective endogenous mechanisms against ROS include adenosine, opening of ATP-sensitive potassium (K-ATP) channels, and release of NO [154]; (x) ROS overload and OXS-induced damage after reperfusion of STEMI and thereafter in humans may be enhanced by comorbidities such as hyperlipidemia, HTN, and DM2 (Figs. 11.2 and 11.3) that aggravate endothelial dysfunction and lead to adverse vascular remodeling [110, 111, 128]; (xi) targeting I/R injury and OXS in humans is further complicated by the multiple factors, players, mechanisms, mediators, signaling pathways, background drugs, pathologies, approaches, and timings involved [109–111] (Figs. 11.3, 11.4, 11.5, 11.6, and 11.7); and xii) despite successful primary percutaneous coronary intervention (PPCI) after STEMI and achieving TIMI grade 3 flow in IRAs, reperfusion at the tissue level is often incomplete in as many as 9–15 % of patients due to a combination of microvascular damage and distal embolization of bits of thrombi and debris from atherosclerotic plaques [155, 156]. Taken together, these points should be considered in designing RCTs to demonstrate the benefit of therapies targeting ROS overload and alleviate OXS in HFrEF and HFpEF. Importantly, rather than targeting ROS alone, both ROS and inflammation need to be targeted.

11.2.9.6.4 ROS-Driven Progression of Early and Late Remodeling Post-STEMI and HF in Patients

The mechanisms that modulate post-STEMI cardiac remodeling have been addressed above and elsewhere before [34, 39–43, 52–58], as has the contribution of inflammation [27–44, 48–58, 102, 124, 126]. However, the molecular and cellular mechanisms that modulate ROS-driven progression of early and late remodeling after STEMI and thereby contribute to chronic HF were not addressed. Grieve et al. very nicely underlined some of the evidence implicating ROS, OXS, and redox signaling in the progression of early and late remodeling after STEMI [157]. They noted that under physiological conditions, ROS can (i) act as second messengers in several cellular functions, including reacting with cysteine residues in sulfhydryl groups of proteins, and thereby induce conformational changes that mediate signal transduction; (ii) stimulate DNA synthesis and induce expression of growth-related genes (such as c-fos, c-jun, and c-myc); and (iii) alter the activity of redox-sensitive phosphatases and kinases such as mitogen-activated protein kinases (MAPK), thereby leading to changes in gene transcription and cell phenotype. Under pathophysiological conditions, they suggested a chain of increased intracellular ROS, ROS-induced activation of kinases (such as p38MAPK, extracellular signal-related kinase [ERK]-1/2, c-Jun N-terminal kinase [JNK], protein kinase B/C, and protein tyrosine phosphatase/tyrosine protein kinase c-Src), redox-sensitive gene transcription, and the development and progression of early and late ventricular remodeling [157].

Increased ROS and OXS have been documented in animal models of IR, MI, and HF and in patients with chronic HF in early studies almost two decades ago [27, 123, 152, 158–162]. Hill et al. showed that rats with MI have marginal increases in SOD, glutathione peroxidase and catalase activities, and vitamin E levels during the nonfailure stage at 1 week but develop increased redox state (reduced/oxidized glutathione ratio) and lipid peroxidation associated with decreased SOD, glutathione peroxidase and catalase activities, and vitamin E levels during moderate to severe HF at 16 weeks, suggesting that HF post-MI is associated with an antioxidant-deficit and increased OXS [158]. Kinugawa et al. showed, using ESR spectroscopy in mice with MI, that the hydroxyl radical (•OH) scavenger dimethylthiourea (DMTU) attenuated the increase in •OH as well as MMP-2 in the NIZ [159]. Singh et al. suggested that antioxidant vitamins reduced OXS in patients with suspected MI [160]. Dhalla et al. showed that aortic banding in guinea pigs induced hypertrophy associated with decreased OXS (increased redox state) at 10 weeks followed by HF with increased OXS at 20 weeks; importantly, chronic vitamin E therapy improved the reserve of endogenous vitamin E, reduced OXS (decreased glutathione/oxidized glutathione ratio), and ultrastructural damage at 20 weeks, suggesting that long-term antioxidant therapy can potentially attenuate or prevent HF [161]. Ide et al. showed, using the dog model of congestive HF induced by 4 weeks of rapid ventricular pacing, that ROS (using ESR spectroscopy for measuring ROS superoxide anion) increased nearly threefold in HF; this was due to a functional block of electron transport at complex I (reflected in decreased complex I enzymatic activity) and uncoupling of the respiratory chain, leading to increased production of mitochondrial ROS and resulting in contractile dysfunction and structural damage to the myocardium during HF [162]. Kim et al. showed, using adenoviral-mediated gene transfer of a rac1 gene product in VSMCs in culture, that inhibition of rac1 blocks the release of ROS upon reoxygenation (as with reperfusion) and protects against reoxygenation-induced cell damage [163]. Talukder et al. used cardiomyocyte-specific overexpression of active rac to show that increased myocardial rac leads to increased injury, contractile dysfunction, and MI after I/R [164].

It is clear that there are multiple sources of ROS during HF, including inflammatory cells such as neutrophils and monocytes [49, 51, 157], cardiomyocyte mitochondria [29–31, 162, 165], xanthine oxidases [157], phagocytic-type NADPH oxidases [157, 166, 167], and dysfunctional NO synthase [157, 168, 169]. Vascular injury can occur not only during the initial MI but also with recurrent ischemia, I/R, and MI during progression of both HFrEF and HFpEF (Figs. 11.3 and 11.7). As mentioned before, acute inflammation appears to be regulated by anti-inflammatory mediators (such as IL-10, IL-13, and SLPI) that in turn can regulate NF-kB activation and dampen the release of proinflammatory mediators (such as TNFα and IL-1), thereby attenuating the damaging effects of oxidants and proteases derived from recruited neutrophils [170]. Lentsch et al. suggested that therapy to limit vascular injury should be based on anti-inflammatory mediators [170]. Many other studies have implicated other specific pathways involved in ROS overload and potential signaling pathways and molecules that could be targeted to quench ROS overload and reduce OXS during the adverse remodeling and progression to HF after MI [27–33].

11.2.9.6.5 Role of ECM Remodeling in Progression of LV Remodeling in HFrEF After STEMI

The evidence for the role of the ECM in cardiac remodeling has been reviewed elsewhere [18, 34–37, 42, 44, 49–58]. Several points need emphasis: (i) dramatic ECM remodeling in early STEMI is driven by the sharp and early rise in MMP levels (Fig. 11.7), causing a chain of MMP/TIMP imbalance, ECM degradation, decreased collagen, adverse LV remodeling, LV dysfunction, HFrEF, and adverse outcome; (ii) the disturbed ECM homeostasis after acute injury continues during healing and repair phases, and beyond initial scar formation in survivors, and contributes to further progressive ECM and LV remodeling and HF progression; whereas the MMP and TIMP levels subside over weeks as inflammation subsides and the early rise in ROS falls (Fig. 11.7); chronically high MMP/TIMP ratios may promote continued ECM degradation and contribute to the commonly observed progressive LV dilation during healing and remote MI and progressive increase in severity of HFrEF; (iii) continued low-grade inflammation and low level of ROS production (Figs. 11.7 and 11.8), in part due to the continued inflammation itself, contribute further to the progression of HFrEF; (iv) in contrast to that construct, a chronically low MMP/TIMP ratio can contribute to increased ECM and fibrosis in both the IZ and NIZ and thereby lead to increased stiffness and diastolic dysfunction and aggravate LV systolic dysfunction over time; (v) defective ECM and fibrosis together with increased cross-linking can also augment adverse LV remodeling and systolic/diastolic dysfunction; in addition, decreased ECM deposition and fibrosis and/or defective ECM (with increased type III collagen and decreased or abnormal cross-linking) can further exacerbate adverse LV remodeling, LV systolic dysfunction, and LV rupture; (vi) whereas LV remodeling is a major mechanism for LV enlargement leading to HF after STEMI, it is the ECM disruption that mediates the initial step in LV dilatation and is the key mechanism underlying LV structural remodeling and Ang II (a primary effector molecule of the RAAS) that drives both ECM and LV remodeling [34, 35]; (vii) the aforementioned mix of cells (inflammatory cells, fibroblast, and vascular cells) together with a mix of molecules (growth factors, cytokines, chemokines, and matrikines) in the healing post-MI substrate act in concert to further modulate LV remodeling [36]; (viii) fibroblasts regulate ECM synthesis/deposition and mediate ECM degradation/turnover through MMP/TIMP balance, and various MMPs (including MMP-1, MMP-2, and MMP-9) modulate ECM remodeling; concurrently, growth factors (such as TGF-β) and proinflammatory cytokines (such as Ang II, IL-6, and TNF-α) that are released after MI modulate MMP/TIMP imbalance, ECM degradation or interstitial fibrosis, and remodeling [36]; (ix) diffusion of proteins and migration of cells from the IZ to borders and the NIZ may extend fibrotic remodeling to those areas, whereas MMP-9 can interact with inflammatory response proteins in the post-MI inflammatory response (such as activator protein-1, specificity protein-1 and NF-κB) and, in fact, shows correlation with various inflammatory markers (such as IL-6, hs-CRP, and fibrinogen), LV hypertrophy, adverse remodeling with LV dilation, dysfunction in HFrEF and HFpEF as well as CV mortality [49, 50, 96]; (x) TIMP-3 also regulates inflammation and inhibits ADAM-17 and ADAM-10, which in turn can alter integrins (cell-surface matrix receptors) and thereby disrupt cell-matrix interactions, degrade ECM, and contribute to LV dilation; xi) ADAMS also interact with inflammatory cytokines and alter MMPs and thereby impact LV remodeling and/or injury; these interactions between the matrix proteins and inflammatory cytokines may modulate ECM damage; (xii) since inflammation is a key modulator of healing/repair after MI and impacts both ECM and cardiac remodeling, tight regulation of the inflammatory response is essential not only for adequate healing/repair and scar formation but also during the subsequent progression to HF and beyond; it follows that dysregulation of the inflammatory response can result in defective scars, increased adverse ECM and LV remodeling, and more rapid progression to HFrEF and end-stage disease; (xiii) the inflammatory cells also release MMPs and oxidants; infiltrating neutrophils release MMPs such as MMP-9; activated neutrophils and monocytes produce MPO which enhances remodeling through generation of oxidants and MMP activation; the aforementioned Ly-6C^{high} monocytes secrete inflammatory cytokines, ROS, and matrix-degrading proteases, whereas the Ly-6C^{low} monocytes trigger collagen/ECM synthesis by myofibroblasts and promote healthy infarct scar formation. Myofibroblasts persist in the infarct scar and maintain ECM; xiv) ROS modulates fibroblast proliferation and collagen synthesis and activates MMPs [171–173] which are redox-sensitive [172–174]; and various cytokines (such as Ang II, cytokines, and cyclic load) stimulate both ECM remodeling and intracellular ROS generation [175–177]; and (xv) ACEIs and ARBs attenuate both adverse post-MI remodeling and OXS [178–182]; chronic treatment with the ROS scavenger DMTU was shown to inhibit adverse LV remodeling and HF, and reduce collagen deposition, MMP-2 activity, LV hypertrophy, and dilatation in mice [159]; the antioxidant probucol was shown to improve LV function, prevent LV dilatation, and reduce cardiac fibrosis post-MI in rats [183]. Although a time-course analysis showed nuclear-mitochondrial cross-talk in global myocardial ischemia, the effect of HF therapies on the cross-talk needs further study [184].

Taken together, the findings support the idea that interactions between ROS and inflammation after STEMI contribute to adverse ECM and cardiac remodeling and progression to HFpEF.

11.3 Conclusion and Future Directions

The rising trend in disability and death from HF has persisted despite adherence to therapies recommended in the published management guidelines from major societies including the American Heart Association, American College of Cardiology, and European Society of Cardiology [1–5, 7]. Clearly, this trend burdens the available healthcare resources [9]. As mentioned before, HF prevalence in Europe is reported to be about 1–2% in adults, rising to over 10% in those older than 70 years [1], and the lifetime risk of HF in people aged 55 years is 33% in men and 28% for women [20]. In the United States, the lifetime risk of HF was 20–45% for people aged 45–95 years and was higher for people with HTN and obesity irrespective of age [6]. The recommended pharmacologic therapies in the guidelines for ST-segment elevation MI (STEMI) and HF are listed in Tables 11.1, 11.2, and 11.3, respectively.

Whereas mortality and morbidity after ACS and MI have improved dramatically over the last four decades due to adherence to guideline-driven timely reperfusion and adjunctive therapies [1, 2, 74], there remains a 7% mortality and 22% morbidity in patients with STEMI at 1 year after prompt reperfusion and PPCI [1, 110, 111]. Although several pharmacological interventions were shown to be effective for preventing and limiting the ravages of ROS and OXS during myocardial ischemia, I/R, MI, and HF in animal models, they did not prove to be effective in patients [1, 2, 7, 12, 109, 110]. The guidelines therefore did not recommend the use of antioxidants in the setting of myocardial ischemia, I/R, MI, and HF [1, 2, 7–10, 12, 109, 110].

As is clear from the discussions in this chapter, the area of prevention of OXS on the clinical front is complex. While there are several reasons for the continued increase in HF prevalence despite optimal approved therapy, three reasons that stand out include (i) the aging-induced CV remodeling that modifies disease expression and response to therapy; (ii) the lifelong exposure to CVD risk factors that increase ROS and OXS as well as inflammation (Fig. 11.4); and (iii) the contribution of synergism between ROS overload/OXS and inflammation as proposed here (Fig. 11.8) and other yet to be addressed pathways and mechanisms leading to HF.

The multiplicity of CVD risk factors, comorbidities, and pathophysiological mechanisms and pathways involved in HFpEF, coupled with the lack of guideline-approved therapy for HFpEF in 2018, actually provides unique opportunities for drug development in this area. As shown in Fig. 11.8, whereas OXS and inflammation appear to show synergism in mediating damage, this aspect of the background mechanisms still remains to be fully characterized and addressed. Whereas there are recommended therapies for HFrEF, the persistence of background levels of inflammation as well as ROS and MMPs, as shown in Fig. 11.7, also provides unique opportunities for drug development in this area. Already, the CANTOS trial group has tested effective adjunctive anti-inflammatory therapy for limiting the residual inflammation on top of background therapies in post-STEMI/HFrEF patients [114, 115]. A similar approach can be used to test a carefully selected antioxidant or pathway in a well-designed RCT for HFrEF and HFpEF. It is also important to select a biomarker or panel of biomarkers for OXS that can be used to assess response to therapy and to predefine realistic endpoints.

In this war against OXS, the matter of background therapy is an important consideration when undertaking RCTs. Background drugs used by both HFrEF and HFpEF patients include ACEIs, ARBs, MRAs, beta-blockers, and other drugs that already decrease OXS. In addition, in spite of the lack of support for the use of antioxidants in the management guidelines, off-the-counter vitamins and other related drugs that are advertised to reduce OXS are widely used, especially in the elderly. Furthermore, the health-conscious patients are already on diet and exercise programs that reduce OXS.

Thus, the selection of the appropriate placebo arm of an OXS-RCT becomes a difficult task. It is essential for investigators designing the RCTs to limit OXS to have a clear appreciation of the key basic mechanisms and problems facing the translation of strategies that are found to be effective when tested in research studies carried out in appropriate animal models. Well-designed preliminary clinical studies in humans are also essential before launching equally well-designed RCTs and before moving to application at the bedside of patients suffering from HF. What, when, and exactly how best to target OXS and translate promising experimental therapy to the real-world by well-designed RCTs before application at the bedside by practicing cardiologists and physicians remains an unanswered question. Waging war on OXS is a worthy goal but it can be challenging.

Acknowledgement None.

Funding Sources None.

References

1. Ponikowski P, Voors AA, Anker SD, Bueno H, Cleland JG, Coats AJS et al (2016) 2016 ESC guidelines for the diagnosis and treatment of acute and chronic heart failure. The Task Force for the diagnosis and treatment of acute and chronic heart failure of the European Society of Cardiology (ESC). Developed with the special contribution of the Heart Failure Association (HFA) of the ESC. Eur Heart J 37:2129–2200
2. Yancy CW, Jessup M, Butler J, Casey DE, Colvin MM, Drazner MH et al (2016) 2016 ACC/AHA/HFSA focused update on new pharmacological therapy for heart failure: an update of the 2013 ACCF/AHA guideline for the management of heart failure. A report of the American College of Cardiology/American Heart Association Task force on clinical practice guidelines and the Heart Failure Society of America. Circulation 134:e282–e293
3. Lloyd-Jones D, Adams RJ, Brown TM, Carnethon MR, Dai S, de Simone G et al (2010) Heart disease and stroke statistics – 2010 update: a report from the American Heart Association Statistics Committee and Stroke Statistics Subcommittee. Circulation 121:e46–e215
4. Roger VL, Go AS, Lloyd-Jones DM, Benjamin EJ, Berry JD, Borden WB et al (2012) Heart disease and stroke statistics-2012 update: a report from the American Heart Association. Circulation 125:e2–e220
5. Benjamin EJ, Blaha MJ, Chiuve SE, Cushman M, Das SR, Deo R et al (2017) Heart disease and stroke statistics-2017 update. A report from the American Heart Association. Circulation 35:e146–e603
6. Benjamin EJ, Virani SS, Callaway CW, Chamberlain AH, Chang AR, Cheng S et al (2018) Heart disease and stroke statistics- 2018 update. A report from the American Heart Association. Circulation 137:e67–e492
7. Hunt SA, Abraham WT, Chin MH, Feldman AM, Francis GS, Ganiats TG et al (2009) 2009 focused update incorporated into the ACC/AHA 2005 guidelines for the diagnosis and Management of Heart Failure in adults: a report of the American College of Cardiology Foundation/American Heart Association Task Force on practice guidelines: developed in collaboration with the International Society for Heart and Lung Transplantation. Circulation 119:e391–e479
8. Dickstein K, Cohen-Solal A, Filippatos G et al (2008) ESC guidelines for the diagnosis and treatment of acute and chronic heart failure 2008: the Task Force for the diagnosis and treatment of acute and chronic heart failure 2008 of the European Society of Cardiology. Eur J Heart Fail 10:933–989
9. Jessup M, Abraham WT, Casey DE, Feldman AM, Francis GS, Ganiats TG et al (2009) Focused update: ACCF/AHA guidelines for the diagnosis and management of heart failure

in adults: a report of the American College of Cardiology Foundation/American Heart Association Task Force on practice guidelines. Circulation 119:1977–2016
10. McMurray J, Adamopoulos S, Anker S, Auricchio A, Böhm M, Dickstein K et al (2012) ESC guidelines for the diagnosis and treatment of acute and chronic heart failure 2012- the task force for the diagnosis and treatment of acute and chronic heart failure of the European Society of Cardiology. Developed in collaboration with the Heart Failure Association (HFA) of the ESC. Eur J Heart Fail 14:803–869
11. Butler J, Fonarow GC, Zile MR, Lam CS, Roessig L, Schelbert EB et al (2014) Developing therapies for heart failure with preserved ejection fraction: current state and future directions. JACC Heart Fail 2:97–112
12. Yancy CW, Jessup M, Bozkurt B, Butler J, Casey DE, Drazner MH et al (2013) 2013 ACCF/AHA guideline for the Management of Heart Failure: executive summary: a report of the American College of Cardiology Foundation/American Heart Association Task Force on practice guidelines. Circulation 128:1810–1852
13. Filippatos G, Khan SS, Ambrosy AP, Cleland JGF, Collins SP, Lam CSP et al (2015) International registry to assess medical practice with longitudinal observation for treatment of heart failure (REPORT-HF): rationale for and design of a global registry. Eur J Heart Fail 17:527–533
14. Jugdutt BI (2010) Aging and heart failure: changing demographics and implications for therapy in the elderly. Heart Fail Rev 15:401–405
15. Jugdutt BI (2010) Heart failure in the elderly: advances and challenges. Expert Rev Cardiovasc Ther 8:695–715
16. Jugdutt BI (ed) (2014) Aging and heart failure: mechanisms and management. Springer, New York
17. Jugdutt BI (2014) Changing demographics of the aging population with heart failure and implications for therapy. Chapt 1. In: Jugdutt BI (ed) Aging and heart failure: mechanisms and management. Springer, New York, pp 1–14
18. Jugdutt BI (2014) Aging and remodeling of the RAS and RAAS and related pathways: implications for heart failure therapy. Chapt 18. In: Jugdutt BI (ed) Aging and heart failure: mechanisms and management. Springer, New York, pp 259–290
19. Jugdutt BI (2014) Biology of aging and implications for heart failure therapy and prevention. Chapt 2. In: Jugdutt BI (ed) Aging and heart failure: mechanisms and management. Springer, New York, pp 14–34
20. Bleumink GS, Knetsch AM, Sturkenboom MCJM, Straus SMJM, Hofman A, Deckers JW et al (2004) Quantifying the heart failure epidemic: prevalence, incidence rate, lifetime risk and prognosis of heart failure The Rotterdam Study. Eur Heart J 25:1614–1619
21. Sakata Y, Shimokawa H (2013) Epidemiology of heart failure in Asia. Circ J 77:2209–2217
22. Roger VL (2013) Epidemiology of heart failure. Circ Res 113:646–659
23. Savarese G, Lund LH (2017) Global public health burden of heart failure. Card Fail Rev 3:7–11
24. Braunwald E (2012) Paul Lichtlen lecture 2011: the rise of cardiovascular medicine. Eur Heart J 33:838–846
25. Braunwald E (2013) Research advances in heart failure. A compendium. Circ Res 113:633–645
26. Braunwald E (2015) The war against heart failure: the Lancet lecture. Lancet 385:812–824
27. Tsutsui H, Kinugawa S, Matsushima S (2011) Oxidative stress and heart failure. Am J Physiol Heart Circ Physiol 301:H2181–H2190
28. Münzel T, Gori T, Keaney JF Jr, Maack C, Daiber A (2015) Pathophysiological role of oxidative stress in systolic and diastolic heart failure and its therapeutic implications. Eur Heart J 36:2555–2564
29. Murphy MP (2009) How mitochondria produce reactive oxygen species. Biochem J 417:1–13
30. Murphy E, Ardehali H, Balaban RS, DiLisa F, Dorn GW, Kitsis RN et al (2016) American Heart Association council on basic cardiovascular sciences, council on clinical cardiol-

ogy, and council on functional genomics and translational biology. Mitochondrial function, biology, and role in disease: a scientific statement from the American Heart Association. Circ Res 118:1960–1991
31. Sack MN, Fyhrquist FY, Saijonmaa OJ, Fuster V, Kovacic JC (2017) Basic biology of oxidative stress and the cardiovascular system: part 1 of 3-part series. J Am Coll Cardiol 70:196–211
32. Münzel T, Camici GG, Maack C, Bonetti NR, Fuster V, Kovacic JC (2017) Impact of oxidative stress on the heart and vasculature: part 2 of a 3-part series. J Am Coll Cardiol 70:212–229
33. Niemann B, Rohrback S, Miller MR, Newby DE, Fuster V, Kovacic JC (2017) Oxidative stress and cardiovascular risk: obesity, diabetes, smoking, and pollution: part 3 of a 3-part series. J Am Coll Cardiol 70:230–251
34. Jugdutt BI (2003) Ventricular remodeling post-infarction and the extracellular collagen matrix. When is enough enough? Circulation 108:1395–1403
35. Jugdutt BI (2003) Remodeling of the myocardium and potential targets in the collagen degradation and synthesis pathways. Curr Drug Targets Cardiovasc Haematol Disord 3:1–30
36. Jugdutt BI (2008) Aging and remodeling during healing of the wounded heart: current therapies and novel drug targets. Curr Drug Targets 9:325–344
37. Jugdutt BI, Michorowski B (1987) Role of infarction expansion in rupture of the ventricular septum after acute myocardial infarction. A two-dimensional echocardiography study. Clin Cardiol 10:641–652
38. Jugdutt BI, Warnica JW (1988) Intravenous nitroglycerin therapy to limit myocardial infarct size, expansion and complications: effect of timing, dosage and infarct location. Circulation 78:906–919. Erratum in (1989) Circulation 79:1151
39. Jugdutt BI, Basualdo CA (1989) Myocardial infarct expansion during indomethacin and ibuprofen therapy for symptomatic post-infarction pericarditis: effect of other pharmacologic agents during early remodelling. Can J Cardiol 5:211–221
40. Jugdutt BI (1990) Identification of patients prone to infarct expansion by the degree of regional shape distortion on an early two-dimensional echocardiogram after myocardial infarction. Clin Cardiol 13:28–40
41. Pfeffer MA, Braunwald E (1990) Ventricular remodelling after myocardial infarction. Circulation 81:1161–1172
42. Jugdutt BI (1993) Prevention of ventricular remodelling post myocardial infarction: timing and duration of therapy. Can J Cardiol 9:103–114
43. Gaudron P, Eilles C, Kugler I, Ertl G (1993) Progressive left ventricular dysfunction and remodeling after myocardial infarction. Potential mechanisms and early predictors. Circulation 87:755–763
44. Jugdutt BI (1996) Prevention of ventricular remodeling after myocardial infarction and in congestive heart failure. Heart Fail Rev 1:115–129
45. Carabello BA (2002) Concentric versus eccentric remodeling. J Cardiac Fail 8(6 Suppl):S258–S263
46. Paulus WJ, Tschöpe C, Sanderson JE, Rusconi C, Flachskampf FA, Rademakers FE et al (2007) How to diagnose diastolic heart failure: a consensus statement on the diagnosis of heart failure with normal left ventricular ejection fraction by the Heart Failure and Echocardiography Associations of the European Society of Cardiology. Eur Heart J 28:2539–2550
47. Lakatta EG, Levy D (2003) Arterial and cardiac aging: major shareholders in cardiovascular disease enterprises. Part I. Aging arteries: a "set up" for vascular disease. Circulation 107:139–146
48. Bujak M, Kweon HJ, Chatila K, Li N, Taffet G, Frangogiannis NG (2008) Aging-related defects are associated with adverse cardiac remodeling in a mouse model of reperfused myocardial infarction. J Am Coll Cardiol 51:1384–1392
49. Jugdutt BI (2011) Modulators of remodeling after myocardial infarction. In: Dhalla NS, Nagano M, Ostadal B (eds) Molecular defects in cardiovascular disease. Springer Media, Inc, New York, pp 231–242

50. Jugdutt BI, Jelani A (2013) Aging and markers of adverse remodeling after myocardial infarction, chapter 27. In: Jugdutt BI, Dhalla NS (eds) Cardiac remodeling: molecular mechanisms. Springer, New York, pp 487–512
51. Jugdutt BI (2013) Regulation of fibrosis after myocardial infarction: implications for ventricular remodeling, chapter 29. In: Jugdutt BI, Dhalla NS (eds) Cardiac remodeling. Molecular mechanisms. Springer, New York, pp 525–545
52. Jugdutt BI, Amy RW (1986) Healing after myocardial infarction in the dog: changes in infarct hydroxyproline and topography. J Am Coll Cardiol 7:91–102
53. Jugdutt BI, Khan MI (1992) Impact of increased infarct transmurality on remodeling and function during healing after anterior myocardial infarction in the dog. Can J Physiol Pharmacol 70:949–958
54. Jugdutt BI, Tang SB, Khan MI, Basulado CA (1992) Functional impact on remodeling during healing after non-Q-wave versus Q-wave anterior myocardial infarction in the dog. J Am Coll Cardiol 20:722–731
55. Jugdutt BI, Joljart MJ, Khan MI (1996) Rate of collagen deposition during healing after myocardial infarction in the rat and dog models: mechanistic insights into ventricular remodeling. Circulation 94:94–101
56. Spinale FG (2002) Matrix metalloproteinases: regulation and dysregulation in the failing heart. Circ Res 90:520–530
57. Spinale FG (2007) Myocardial matrix remodeling and the matrix metalloproteinases: influence on cardiac form and function. Physiol Rev 87:1285–1342
58. Jugdutt BI, Jelani A, Palaniyappan A, Idikio H, Uweira RE, Menon V et al (2010) Aging-related changes in markers of ventricular and matrix remodelling after reperfused ST-segment elevation myocardial infarction in the canine model. Effect of early therapy with an angiotensin II type 1 receptor blocker. Circulation 122:341–351
59. Tromp J, Westenbrink BD, Ouwerkerk W, van Veldhuisen DJ, Samani NJ, Ponikowski P et al (2018) Identifying pathophysiological mechanisms in heart failure with reduced versus preserved ejection fraction. J Am Coll Cardiol 72:1081–1090
60. Sorop O, Heinonen I, van Kranenburg M, van de Wouw J, de Beer VJ, Nguyen TN et al (2018) Multiple comorbidities produce left ventricular diastolic dysfunction associated with coronary microvascular dysfunction, oxidative stress, and myocardial stiffening. Cardiovasc Res 114:954–964
61. O'Gallagher K, Shah AM (2018) Modelling the complexity of heart failure with preserved ejection fraction. Cardiovasc Res 114:919–921
62. Shah SJ, Lam CSP, Svedlund S, Saraste A, Hage C, Tan R-S et al (2018) Prevalence and correlates of coronary microvascular dysfunction in heart failiure with preserved ejection fraction: PROMIS-HFpEF. Eur Heart J 39:3439–3450
63. Wei J, Nelson MD, Sharif B, Scufelt C, Merz CNB (2018) Why do we care about coronary microvascular dysfunction and heart failure with preserved ejection fraction: addressing knowledge gaps for evidence-based guidelines. Eur Heart J 39:3451–3453
64. Kato S, Saito N, Kirigaya H, Gyotoku D, Linuma N, Kusakawa Y et al (2016) Impairment of coronary flow reserve evaluated by phase contrast cine-magnetic resonance imaging in patients with heart failure with preserved ejection fraction. J Am Heart Assoc 5:e002649. https://doi.org/10.1161/JAHA.115.002649)
65. van Heerebeek L, Hamdani N, Falcão-Pires I, Leite-Moreira AF, Begieneman MP, Bronzwaer JG et al (2012) Low myocardial protein kinase G activity in heart failure with preserved ejection fraction. Circulation 126:830–839
66. Paulus WJ, Tschöpe C et al (2013) A novel paradigm for heart failure with preserved ejection fraction: comorbidities drive myocardial dysfunction and remodeling through coronary microvascular endothelial inflammation. J Am Coll Cardiol 62:263–271
67. Ter Maaten JM, Damman K, Verhaar MC, Paulus WJ, Duncker DJ, Cheng C et al (2016) Connecting heart failure with preserved ejection fraction and renal dysfunction: the role of endothelial dysfunction and inflammation. Eur J Heart Fail 18:588–598

68. Hiebert JB, Shen Q, Thimmesch A, Pierce J (2017) Impaired myocardial bioenergetics in HFpEF and the role of antioxidants. Open Cardiovasc Med J 10:158–162
69. Negi SI, Jeong EM, Shukrullah I, Veleder E, Jones DP, Fan TH et al (2015) Renin-angiotensin activation and oxidative stress in early heart failure with preserved ejection fraction. Biomed Res Int 2015:825027. https://doi.org/10.1155/2015/825027
70. Vaduganathan M, Patel RB, Michel A, Shah SJ, Senni MGheorgiade M et al (2017) Mode of death in heart failure with preserved ejection fraction. J Am Coll Cardiol 69:556–569
71. Rush CJ, Campbell RT, Jhund PS, Petrie MC, McMurray JJV (2018) Association is not causation: treatment effects cannot be estimated from observational data in heart failure. Eur Heart J 39:3417–3438
72. Shah SJ, Katz DH, Selvaraj S, Burke MA, Yancy CW, Gheorghiade M et al (2015) Phenomapping for novel classification of heart failure with preserved ejection fraction. Circulation 131:269–279
73. McMurray JJ, Packer M, Desai AS, Gong J, Lefkowitz MP, Rizkala AR, PARADIGM-HF Investigators and Committees et al (2014) Angiotensin-neprilysin inhibition versus enalapril in heart failure. N Engl J Med 371:993–1004
74. Maggioni AP, Anker SD, Dahlström U, Filippatos G, Ponikowski P, Zannad F et al (2013) Are hospitalized or ambulatory patients with heart failure treated in accordance with European Society of Cardiology guidelines? Evidence from 12,440 patients of the ESC Heart Failure Long-Term Registry. Eur J Heart Fail 15:1173–1184
75. Mernendez JT (2016) The mechanism of action of LCZ696. Card Fail Rev 2:40–46
76. Hayman S, Atherton JJ (2016) Should angiotensin receptor neprilysin inhibitors replace angiotensin converting enzyme inhibitors in heart failure with a reduced ejection fraction? Card Fail Rev 2:47–50
77. Solomon SD, Zile M, Pieske B, Voors A, Shah A, Kraigher-Krainer E et al (2012) The angiotensin receptor neprilysin inhibitor LCZ696 in heart failure with preserved ejection fraction: a phase 2 double-blind randomised controlled trial. Lancet 380:1387–1395
78. Senni M, Paulus WJ, Gavazzi A, Fraser AG, Diez J, Solomon SD et al (2014) New strategies for heart failure with preserved ejection fraction: the importance of targeted therapies for heart failure phenotypes. Eur Heart J 35:2797–2815
79. Lim SL, Lam CSP (2016) Breakthrough in heart failure with preserved ejection fraction: are we there yet? Korean J Intern Med 31:1–14
80. Owens AT, Brozena SC, Jessup M (2016) New management strategies in heart failure. Circ Res 118:480–495
81. Ilieşiu AM, Hodorogea AS (2018) Treatment of heart failure with preserved ejection fraction. Adv Exp Med Biol 1067:67–87
82. Zheng SL, Roddick AJ, Aghar-Jaffar R, Shun-Shin MJ, Francis D, Oliver N et al (2018) Association between use of sodium-glucose cotransporter 2 inhibitors, glucagon-like peptide 1 agonists, and dipeptidyl peptidase 4 inhibitors with all-cause mortality in patients with type 2 diabetes: a systematic review and meta-analysis. JAMA 319:1580–1591
83. Das SR, Everett BM, Birtcher KK, Brown JM, Cefalu WT, Januzzi JL Jr et al (2018) 2018 ACC expert consensus decision pathway on novel therapies for cardiovascular risk reduction in patients with type 2 diabetes and atherosclerotic cardiovascular disease pathways. A report of the American College of Cardiology task force on expert consensus decision pathways Endorsed by the American Diabetes Association Writing Committee. J Am Coll Cardiol. https://doi.org/10.1016/j.jacc.2018.09.020
84. Pitt B, Pfeffer MA, Assmann SF, Boineau R, Anand IS, Claggett B (2014) Spironolactone for heart failure with preserved ejection fraction. N Engl J Med 370:1383–1392
85. Redfield MM, Chen HH, Borlaug BA, Semigran MJ, Lee KL, Lewis G et al (2013) Effect of phosphodiesterase-5 inhibition on exercise capacity and clinical status in heart failure with preserved ejection fraction: a randomized clinical trial. JAMA 309:1268–1277
86. Redfield MM, Anstrom KJ, Levine JA (2015) Isosorbide mononitrate in heart failure with preserved ejection fraction. N Engl J Med 373:2314–2324

87. Fukuta H, Goto T, Wakami K, Ohte N (2016) Effects of drug and exercise intervention on functional capacity and quality of life in heart failure with preserved ejection fraction: a meta-analysis of randomized controlled trials. Eur J Prev Cardiol 23:78–85
88. Jugdutt BI (1985) Delayed effects of early infarct-limiting therapies on healing after myocardial infarction. Circulation 72:907–914
89. Jugdutt BI (1996) Pharmacological intervention in post-infarction wound healing. EXS 76:501–512
90. Jugdutt BI, Lucas A, Khan MI (1997) Effect of angiotensin-converting enzyme inhibition on infarct collagen deposition and remodelling during healing after transmural canine myocardial infarction. Can J Cardiol 13:657–668
91. Jugdutt BI, Menon V (2002) Beneficial effects of therapy on the progression of structural remodeling during healing after reperfused and nonreperfused myocardial infarction: different effects on different parameters. J Cardiovasc Pharmacol Ther 7:95–107
92. Jugdutt BI, Palaniyappan A, Uwiera RR, Idikio H (2009) Role of healing-specific-matricellular proteins and matrix metalloproteinases in age-related enhanced early remodeling after reperfused STEMI in dogs. Mol Cell Biochem 322:25–36
93. Jugdutt BI (2008) Pleiotropic effects of cardiac drugs on healing post-MI. The good, bad, and ugly. Heart Fail Rev 13:439–452
94. Jugdutt BI, Idikio H, Uwiera RR (2007) Therapeutic drugs during healing after myocardial infarction modify infarct collagens and ventricular distensibility at elevated pressures. Mol Cell Biochem 304:79–91
95. Jugdutt BI (2007) Cyclooxygenase inhibition and adverse remodeling during healing after myocardial infarction. Circulation 115:288–291
96. Jugdutt BI (2006) Matrix metalloproteinases as markers of adverse remodeling after myocardial infarction. J Card Fail 12:73–76
97. Jugdutt BI, Jelani A (2008) Aging and defective healing, adverse remodeling, and blunted post-conditioning in the reperfused wounded heart. J Am Coll Cardiol 51:1399–1403
98. Palaniyappan A, Uwiera RR, Idikio H, Menon V, Jugdutt C, Jugdutt BI (2013) Attenuation of increased secretory leukocyte protease inhibitor, matricellular proteins and angiotensin II and left ventricular remodeling by candesartan and omapatrilat during healing after reperfused myocardial infarction. Mol Cell Biochem 376:175–188
99. Jugdutt BI (2011) Optimizing pharmacotherapy for limiting cardiovascular remodeling a matter of timing therapy to match biology. J Am Coll Cardiol 57:2029–2030
100. Manhenke C, Ueland T, Jugdutt BI, Godang K, Aukrust P, Dickstein K et al (2014) The relationship between markers of extracellular cardiac matrix turnover: infarct healing and left ventricular remodelling following primary PCI in patients with first-time STEMI. Eur Heart J 35:395–402
101. Frangogiannis NG, Smith CW, Entman ML (2002) The inflammatory response in myocardial infarction. Cardiovasc Res 53:31–47
102. Nahrendorf M, Pittet MJ, Swirski FK (2010) Monocytes: protagonists of infarct inflammation and repair after myocardial infarction. Circulation 121:2437–2445
103. Jugdutt BI (2015) Aging-related changes in cardiac extracellular matrix: implications for heart failure in older patients. J Cardiol Curr Res 3(3):00101. https://doi.org/10.15406/jccr.2015.03.00101
104. Weber KT (1997) Extracellular matrix remodeling in heart failure. A role for De Novo Angiotensin II generation. Circulation 96:4065–4082
105. Jugdutt BI (2005) Extracellular matrix and cardiac remodeling, chapter 3. In: Villareal FJ (ed) Interstitial fibrosis in heart failure. Springer, New York, pp 23–55
106. Deschamps AM, Spinale FG (2006) Pathways of matrix metalloproteinase induction in heart failure: bioactive molecules and transcriptional regulation. Cardiovasc Res 69:666–676
107. Spinale FG, Zile MR (2013) Integrating the myocardial matrix into heart failure recognition and management. Circ Res 113:300–309

108. Jugdutt BI (2014) Cardiac matrix remodeling and heart failure. In: Li R (ed) Cardiac regeneration and repair: pathology and therapies, Vol 1, Part 1. Woodhead Publishing Ltd, Oxford, pp 3–26
109. O'Gara PT, Kushner FG, Ascheim DD, Casey DE Jr, Chung MK, de Lemos JA et al (2013) 2013 ACCF/AHA guideline for the management of ST-elevation myocardial infarction: executive summary: a report of the American college of cardiology foundation/American heart association task force on practice guidelines. J Am Coll Cardiol 61:485–510
110. Ibanez B, James S, Agewall S, Antunes MJ, Bucciarelli-Ducci C, Bueno H et al (2018) 2017 ESC guidelines for the management of acute myocardial infarction in patients presenting with ST-segment elevation: the task force for the management of acute myocardial infarction in patients presenting with ST-segment elevation of the European Society of Cardiology (ESC). Eur Heart J 39:119–177
111. Hausenloy DJ, Botker HE, Engstrom T, Erlinge D, Heusch G, Ibanez B et al (2017) Targeting reperfusion injury in patient with ST-segment elevation myocardial infarction: trials and tribulations. Eur Heart J 38:935–941
112. Libby P (2001) Current concepts of the pathogenesis of the acute coronary syndromes. Circulation 104:365–372
113. Ridker PM, Rifai N, Pfeffer MA, Sacks F, Braunwald E (1999) Long-term effects of pravastatin on plasma concentration of C-reactive protein: the cholesterol and recurrent events (CARE) investigators. Circulation 100:230–235
114. Ridker PM, Everett TT, Thuren T, MacFadyen JG, Chang WH, Ballantyne C, for the CANTOS trial group et al (2017) Antiinflammatory therapy with canakinumab for atherosclerotic disease. N Engl J Med 377:1119–1131
115. Ridker PM, Libby P, MacFadyen JG, Thuren T, Ballantyne C, Fonseca F, on behalf of the CANTOS Trial Group et al (2018) Modulation of the interleukin-6 signalling pathway and incidence rates of atherosclerotic events and all-cause mortality: analyses from the canakinumab anti-inflammatory thrombosis outcomes study (CANTOS). Eur Heart J 39:3499–3507
116. Everett BM, Cornel JH, Lainscak M, Anker SD, Abbate A, Thuren T et al (2018) Anti-inflammatory therapy with canakinumab for the prevention of hospitalization for heart failure. Circulation. https://www.ahajournals.org/. https://doi.org/10.1161/CIRCULATIONAHA.118.038010
117. Crea F, Libby P (2017) Acute coronary syndromes. The way forward from mechanisms to precision treatment. Circulation 136:1155–1166
118. Ridker PM (2017) Canakinumab for residual inflammatory risk. Implications of CANTOS for clinical practice and drug development. Eur Heart J 38:3545–3548. https://doi.org/10.1093/eurheartj/ehx723
119. Abbate A, Dinarello CA (2015) Anti-inflammatory therapies in acute coronary syndromes: is IL-1 blockade a solution? Eur Heart J 36:337–339
120. Morton AC, Rothman AM, Greenwood JP, Gunn J, Chase A, Clarke B et al (2015) The effect of interleukin-1 receptor antagonist therapy on markers of inflammation in non-ST elevation acute coronary syndromes: the MRC-ILA Heart Study. Eur Heart J 36:377–384
121. Frangogiannis NG (2004) Chemokines in the ischemic myocardium: from inflammation to fibrosis. Inflamm Res 53:585–595
122. Shetelig C, Limalanathan S, Hoffmann P, Seljeflot I, Gran JM, Eritsland J et al (2018) Association of IL-8 with infarct size and clinical outcomes in patients with STEMI. J Am Coll Cardiol 72:187–198
123. Bolli R, Jeroudi MO, Patel BS, DuBose CM, Lai EK, Roberts R et al (1989) Direct evidence that oxygen-derived free radicals contribute to postischemic myocardial dysfunction in the intact dog. Proc Natl Acad Sci U S A 86:4695–4599
124. Prabhu SD, Frangogiannis NG (2016) The biological basis for cardiac repair after myocardial infarction. From inflammation to fibrosis. Circ Res 119:91–112
125. Granger CB, Kochar A (2018) Understanding and targeting inflammation in acute myocardial infarction. An elusive goal. Am Coll Cardiol 72:199–200

126. Ruparelia N, Godec J, Lee R, Chai JT, Dall'Armellina E, McAndrew D et al (2015) Acute myocardial infarction activates distinct inflammation and proliferation pathways in circulating monocytes, prior to recruitment, and identified through conserved transcriptional responses in mice and humans. Eur Heart J 36:1923–1934
127. Panizzi P, Swirski FK, Figueiredo JL, Waterman P, Sosnovik D, Aikawa E et al (2010) Impaired infarct healing in atherosclerotic mice with Ly-6Chi monocytosis. J Am Coll Cardiol 55:1629–1638
128. Paneni F, Diaz Cañestro C, Libby P, Lüscher TF, Camici GG (2017) The aging cardiovascular system: understanding it at the cellular and clinical levels. J Am Coll Cardiol 69:1952–1967
129. Haycock PC, Heydon EE, Kaptoge S, Butterworth AS, Thompson A, Willeit P et al (2014) Leucocyte telomere length and risk of cardiovascular disease: systematic review and meta-analysis. BMJ 349:g4227. https://doi.org/10.1136/bmj.g4227
130. Ricci R, Eriksson U, Oudit GY, Eferl R, Akhmedov A, Sumara I et al (2005) Distinct functions of junD in cardiac hypertrophy and heart failure. Genes Dev 19:208–213
131. Kroemer G, Marino G, Levine B (2010) Autophagy and the integrated stress response. Mol Cell 40:280–293
132. Dai DF, Chiao YA, Marcinek DJ, Szeto HH, Rabinovitch PS (2014) Mitochondrial oxidative stress in aging and healthspan. Longevity & Healthspan 3:6. https://doi.org/10.1186/2046-2395-3-6
133. Terman A, Brunk UT (2005) Autophagy in cardiac myocyte homeostasis, aging, and pathology. Cardiovasc Res 68:355–365
134. Terman A, Kurz T, Navratil M, Arriaga EA, Brunk UT (2010) Mitochondrial turnover and aging of long-lived postmitotic cells: the mitochondrial-lysosomal axis theory of aging. Antioxid Redox Signal 12:503–535
135. Judge S, Jang YM, Smith A, Hagen T, Leeuwenburgh C (2005) Age-associated increases in oxidative stress and antioxidant enzyme activities in cardiac interfibrillar mitochondria: implications for the mitochondrial theory of aging. FASEB J 19:419–421
136. Linton P-J, Gurney M, Sengstock D, Mentzer RM Jr, Gottlieb RA (2016) This old heart: cardiac aging and autophagy. J Mol Cell Cardiol 83:44–54
137. Levine B, Mizushima N, Virgin HW (2011) Autophagy in immunity and inflammation. Nature 469:323–335
138. Kassiotis C, Ballal K, Wellnitz K, Vela D, Gong M, Salazar R et al (2009) Markers of autophagy are downregulated in failing human heart after mechanical unloading. Circulation 120:S191–S197
139. Garcia L, Verdejo HE, Kuzmicic J, Zalaquett R, Gonzalez S, Lavandero S et al (2012) Impaired cardiac autophagy in patients developing postoperative atrial fibrillation. J Thorac Cardiovasc Surg 143:451–459
140. Jahania SM, Sengstock D, Vaitkevicius P, Andres A, Ito BR, Gottlieb RA et al (2013) Activation of the homeostatic intracellular repair response during cardiac surgery. J Am Coll Surg 216:719–726
141. Singh KK, Yanagawa B, Quan A, Wang R, Garg A, Khan R et al (2014) Autophagy gene fingerprint in human ischemia and reperfusion. J Thorac Cardiovasc Surg 147:1065–1072
142. Gedik N, Thielmann M, Kottenberg E, Peters J, Jakob H, Heusch G et al (2014) No evidence for activated autophagy in left ventricular myocardium at early reperfusion with protection by remote ischemic preconditioning in patients undergoing coronary artery bypass grafting. PLoS One 9:e96567
143. Sengstock D, Jahania S, Andres A, Ito BR, Gottlieb RA, Mentzer RM Jr (2014) Homeostatic intracellular repair response (HIR2) is increased in older adults and is upregulated by ischemia. J Am Geriatr Soc 62:S108
144. Kanamori H, Takemura G, Goto K, Maruyama R, Tsujimoto A, Ogino A et al (2011) The role of autophagy emerging in postinfarction cardiac remodeling. Cardiovasc Res 91:330–339
145. Gao T, Zhang S-P, Wang J-F, Liu L, Wang Y, Cao Z-Y et al (2018) TLR3 contributes to persistent autophagy and heart failure in mice after myocardial infarction. J Cell Mol Med 22:395–408

146. Goldin A, Beckman JA, Schmidt AM, Creager MA (2006) Advanced glycation end products. Sparking the development of diabetic vascular injury. Circulation 114:597–605
147. Prasad K, Mishra M (2018) AGE-RAGE stress, stressors, and antistressors in health and disease. Int J Angiol 27:1–12
148. Selye H (1976) Stress in health and disease, 1st edn. Butterworths, Boston, pp 762–807
149. Bartesagh S, Radi (2018) Fundamentals on the biochemistry of peroxynitrite and protein tyrosine nitration. Redox Biol 14:618–625
150. Kumar D, Jugdutt BI (2003) Apoptosis and oxidants in the heart. J Lab Clin Med 142:288–297
151. Jugdutt BI, Idikio HA (2005) Apoptosis and oncosis in acute coronary syndromes: assessment and implications. Mol Cell Biochem 270:177–200
152. Bolli R, Jeroudi M, Patel B, Aruoma O, Halliwell B, Lai E et al (1989) Marked reduction of free radical generation and contractile dysfunction by antioxidant therapy begun at the time of reperfusion. Evidence that myocardial "stunning" is a manifestation of reperfusion injury. Circ Res 65:607–622
153. Ducas A, Bartekova M, Dhalla NS (2015) Ischemia-reperfusion injury of the heart: moving forward with our knowledge. J Heart Health 1:1–10. https://doi.org/10.16966/2379-769X.110
154. Jugdutt BI (2002) Nitric oxide and cardioprotection during ischemia-reperfusion. Heart Fail Rev 7:391–405
155. Neumann F-J, Gick M (2018) Direct stenting in ST-segment elevation myocardial infarction: convenient, but not improving outcomes. Eur Heart J 39:2480–2483
156. Mahmood KO, Jolly SS, James S, Dzavik V, Cairns JA, Olivecrona CK et al (2018) Clinical impact of direct stenting and interaction with thrombus aspiration in patients with ST-segment elevation myocardial infarction undergoing percutaneous coronary intervention: thrombectomy trialists collaboration. Eur Heart J 39:2472–2479
157. Grieve DJ, Byrne JA, Cave AC, Shah A (2004) Role of oxidative stress in cardiac remodelling after myocardial infarction. Heart Lung Circ 13:132–138
158. Hill MF, Singal PK (1996) Antioxidant and oxidative stress changes during heart failure subsequent to myocardial infarction in rats. Am J Pathol 148:291–300
159. Kinugawa S, Tsutsui H, Hayashidani S, Ide T, Suematsu N, Satoh S et al (2000) Treatment with dimethylthiourea prevents left ventricular remodelling and failure after experimental myocardial infarction in mice: role of oxidative stress. Circ Res 87:392–398
160. Singh RB, Niaz MA, Rastogi SS, Rastogi S (1996) Usefulness of antioxidant vitamins in suspected acute myocardial infarction (the Indian experiment of infarct survival-3). Am J Cardiol 77:232–236
161. Dhalla AK, Hill MF, Singal PK (1996) Role of oxidative stress in transition of hypertrophy to heart failure. J Am Coll Cardiol 28:506–514
162. Ide T, Tsutsui H, Kinugawa S, Utsumi H, Kang D, Hattori N et al (1999) Mitochondrial electron transport complex I is a potential source of oxygen free radicals in the failing myocardium. Circ Res 85:357–363
163. Kim K-S, Takeda K, Sethi R, Pracyk JB, Tanaka K, Zhou YF et al (1998) Protection from reoxygenation injury by inhibition of rac1. J Clin Invest 101:1821–1826
164. Talukder MAH, Elnakish MT, Yang F, Nishijima Y, Alhai MA, Velayutham M et al (2013) Cardiomyocyte-specific overexpression of an active form of Rac predisposes the heart to increased myocardial stunning and ischemia-reperfusion injury. Am J Physiol Heart Circ Physiol 304:H294–H302
165. Ide T, Tsutsui H, Kinugawa S, Suematsu N, Hayashidani S, Ichikawa K et al (2000) Direct evidence for increased hydroxyl radicals originating from superoxide in the failing myocardium. Circ Res 86:152–157
166. Griendling KK, Sorescu D, Ushio-Fukai M (2000) NAD(P)H oxidase. Role in cardiovascular biology and disease. Circ Res 86:494–501
167. Heymes C, Bendall JK, Ratajczak P, Cave AC, Samuel JL, Hasenfuss G et al (2003) Increased myocardial NADPH oxidase activity in human heart failure. J Am Coll Cardiol 41:2164–2171
168. Bauersachs J, Bouloumie A, Fraccarollo D, Hu K, Busse R, Ertl G (1999) Endothelial dysfunction in chronic myocardial infarction despite increased vascular endothelial nitric oxide

synthase and soluble guanylate cyclase expression: role of enhanced vascular superoxide production. Circulation 100:292–298
169. Saavedra WF, Paolocci N, St. John ME, Skaf MW, Stewart GC, Xie JS et al (2002) Imbalance between xanthine oxidase and nitric oxide synthase signaling pathways underlies mechanoenergetic uncoupling in the failing heart. Circ Res 90:297–304
170. Lentsch AB, Ward PA (2000) Regulation of inflammatory vascular damage. J Pathol 190:343–348
171. Irani K, Xia Y, Zweier JL, Sollott SJ, Der CJ, Fearon ER et al (1997) Mitogenic signaling mediated by oxidants in Ras-transformed fibroblasts. Science 275:1649–1652
172. Arnold RS, Shi J, Murad E, Whalen AM, Sun CQ, Polavarapu R et al (2001) Hydrogen peroxide mediates the cell growth and transformation caused by the mitogenic oxidase Nox1. Proc Natl Acad Sci U S A 98:5550–5555
173. Siwik DA, Pagano PJ, Colucci WS (2001) Oxidative stress regulates collagen synthesis and matrix metalloproteinase activity in cardiac fibroblasts. Am J Phys 280:C53–C60
174. Morin I, Li WQ, Su S, Ahmad M, Zafarullah M (1999) Induction of stromelysin gene expression by tumor necrosis factor alpha is inhibited by dexamethasone, salicylate, and N-acetylcysteine in synovial fibroblasts. J Pharmacol Exp Ther 289:1634–1640
175. Sano M, Fukuda K, Sato T, Kawaguchi H, Suematsu M, Matsuda S et al (1999) ERK and p38 MAPK, but not NF-kB, are critically involved in reactive oxygen species-mediated induction of IL-6 by angiotensin II in cardiac fibroblasts. Circ Res 89:661–669
176. Kunsch C, Medford RM (1999) Oxidative stress as a regulator of gene expression in the vasculature. Circ Res 85:753–766
177. Campbell SE, Katwa LC (1997) Angiotensin II stimulated expression of transforming growth factor-beta1 in cardiac fibroblasts and myofibroblasts. J Mol Cell Cardiol 29:1947–1958
178. Anthonio RL, van Veldhuisen DJ, van Gilst WH (1998) Left ventricular dilatation after myocardial infarction: ACE inhibitors, beta-blockers, or both? J Cardiovasc Pharmacol 32(Suppl 1):S1–S8
179. Mankad S, d'Amato TA, Reichek N, McGregor WE, Lin J, Singh D et al (2001) Combined angiotensin II receptor antagonism and angiotensin-converting enzyme inhibition further attenuates postinfarction left ventricular remodelling. Circulation 103:2845–2850
180. Diez J, Querejeta R, Lopez B, Gonzalez A, Larman M, Martinez Ubago JL (2002) Losartan-dependent regression of myocardial fibrosis is associated with reduction of left ventricular chamber stiffness in hypertensive patients. Circulation 105:2512–2517
181. Kuno A, Miura T, Tsuchida A, Hasegawa T, Miki T, Nishino Y et al (2002) Blockade of angiotensin II type 1 receptors suppressed free radical production and preserved coronary endothelial function in the rabbit heart after myocardial infarction. J Cardiovasc Pharmacol 39:49–57
182. Moe G, Konig A, Liu P, Jugdutt BI (2008) Selective type 1 angiotensin II receptor blockade attenuates oxidative stress and regulates angiotensin II receptors in the canine failing heart. Mol Cell Biochem 17:97–104
183. Sia YT, Lapointe N, Parker TG, Tsoporis JN, Deschepper CF, Calderone A et al (2002) Beneficial effects of long-term use of the antioxidant probucol in heart failure in the rat. Circulation 105:2549–2555
184. Marín-García J, Damle S, Jugdutt BI, Moe GW (2012) Nuclear-mitochondrial cross-talk in global myocardial ischemia. A time-course analysis. Mol Cell Biochem 364:225–234

Oxidative Stress in Metabolic Syndrome: Experimental Model of Biomarkers

12

María del Carmen Baez, Mariana Tarán, Mónica Moya, and María de la Paz Scribano Parada

Abstract

Metabolic syndrome (MS) is characterized by the convergence of several risk factors at the same time in which each individual contributes to cardiovascular risk. Although the factors that establish the relationship between metabolic alterations and vascular changes that predispose to cardiovascular events are not fully understood, it is likely that endothelial dysfunction has decisive importance in this regard. We implemented the use of fibrinogen, nitric oxide, adiponectin, and superoxide dismutase, to evaluate the implication of these phenomena in mitochondrial function and morphology in a MS model. It was demonstrated that the sustained oxidative stress situation induces histological alterations at the aortic level. This pathological and oxidative state leads to a mitochondrial dysfunction with repercussion in the morphology of this organelle.

Due to the intimate link to insulin resistance (IR), obesity, and MS, the importance of studying the implication of the inflammatory phenomenon and associated oxidative stress is understood, in order to establish the probable physiopathogenic mechanisms with the aim of generating strategies that prevent the incidence and prevalence of this pathology, given that it has huge consequences in health system. For this it is necessary to identify the determinants of the disease in order to implement preventive measures for control and monitoring as well as to study therapeutic strategies that can be implemented to reduce the incidence of this multisyndromic pathology.

M. del Carmen Baez (✉) · M. Tarán · M. de la Paz Scribano Parada
Cardiovascular Biomarkers Laboratory, Cátedra de Física Biomédica, Facultad de Ciencias Médicas, Universidad Nacional de Córdoba, Córdoba, Argentina

M. Moya
Cardiovascular Biomarkers Laboratory, Cátedra de Física Biomédica, Facultad de Ciencias Médicas, Universidad Nacional de Córdoba, Córdoba, Argentina

Física Biomédica, Ciencias Médicas, Universidad Nacional de La Rioja, La Rioja, Argentina

S. Chakraborti et al. (eds.), *Modulation of Oxidative Stress in Heart Disease*,
https://doi.org/10.1007/978-981-13-8946-7_12

Keywords
Metabolic syndrome · Oxidative stress · Mitochondria

12.1 Introduction

Metabolic syndrome (MS) is characterized by the convergence of several risk factors at the same time in which each individual contributes to cardiovascular risk and its association increases exponentially and not only additively [1–3]. In addition diabetes mellitus type 2 (T2DM), obesity, and hypertension are associated with prothrombotic, hypofibrinolytic, and proinflammatory alterations that contribute individually to cardiovascular risk (Fig. 12.1) [4–6].

The biochemical parameters that accompany syndrome are insulin resistance and compensatory hyperinsulinism associated with a metabolism of hydrocarbon metabolism disorders, lipid alterations (such as hypertriglyceridemia, lowering of high-density lipoprotein cholesterol (c-HDL), presence of low cholesterol lipoprotein density type B, increase of free fatty acids, postprandial lipemia), obesity, and high blood pressure [7].

Although the factors that establish the relationship between metabolic alterations and vascular changes that predispose to cardiovascular events could elicit, a pivotal role in endothelial dysfunction has been suggested to be decisive importance in this regard [8].

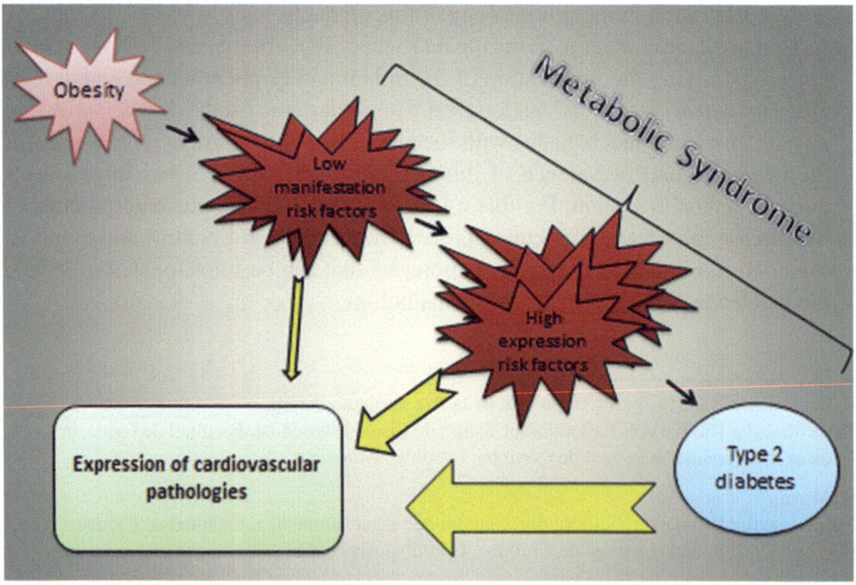

Fig. 12.1 Evolution of metabolic syndrome

Currently, there is increased evidence about that MS pathophysiology is related to a low-grade inflammatory condition, given that the patients exhibit high level of cytokines and inflammatory markers. All these would be the result of hypercaloric diets with low-energy expenditure, associating with an increase of fat body tissue, increasing abdominal visceral fat [9].

There are also primary triggers that induce the inflammatory cascade, such as the presence of high levels of oxidized LDL and early inducers of atherogenesis. Other secondary triggering factors that are involved in the maintenance and amplification of cytokine production include the presence of mechanical factors, angiotensin II levels, free radicals, and metalloproteinases, among others [10].

Proinflammatory stimuli increase acute-phase reactants by activating specific cytokine production patterns for different proteins. As a consequence of the activation, some markers such as plasma fibrinogen (FP), C-reactive protein (CRP), cytokines (interleukin-6 (IL-6)), tumor necrosis factor-alpha (TNF-α), adhesion molecules (intercellular adhesion molecule, ICAM I), and other coagulation factors may be increased in plasma and have prognostic value for future cardiovascular events [11, 12].

Although these stimuli increase the concentrations of acute-phase reactants, not all of them increase uniformly in the same or different diseases, indicating that they are individually regulated (Fig. 12.2) [13]. This behavior of inflammatory markers along with insulin could play an important role as a cardiovascular risk factor and oxidative stress have been proposed to be as a potential inducer of inflammation, susceptibility to obesity, and comorbid states [12, 14, 15]. However, interaction is not clearly elucidated among them in the pathogenic mechanism of this multisyndromic disease.

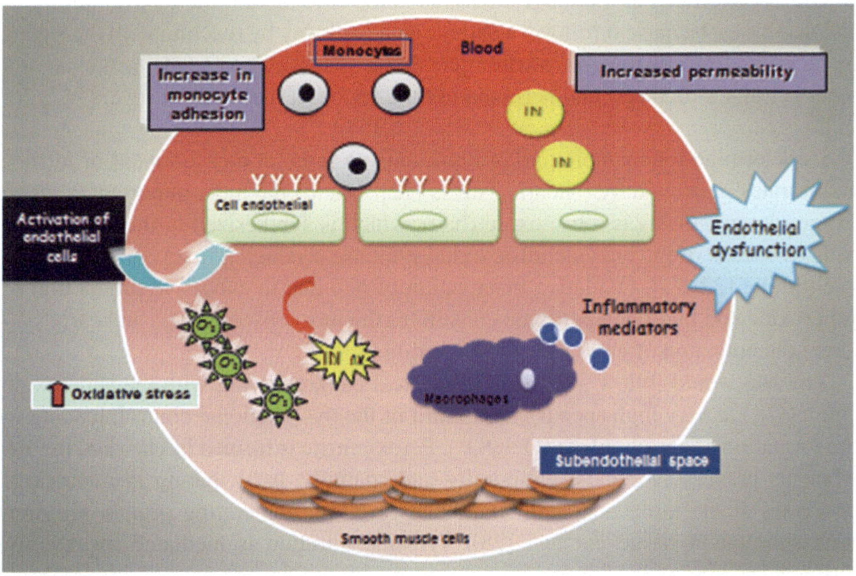

Fig. 12.2 Vascular inflammatory process

In fact, in each of the disorders that characterize MS (diabetes, obesity, hypertension, and hyperlipidemia), there is an alteration in the function of the endothelium, which responds poorly to the stimuli of relaxation [16]. These dysfunctional endothelia lose the ability to regulate their vital functions acquiring procoagulant properties instead of anticoagulants and are likely to modify the synthesis of reactive oxygen intermediates.

The deterioration of health due to some pathology can be evaluated through the use of biomarkers, parameters that allow knowing the normal or abnormal state of an individual and express the probability that an undesired effect will occur as a result of an exposure [17]. The existence of a transversal and prospective association between the plasma measurement of oxidative stress and the risk of presenting MS would be independent, since the prospective association persists even after including in the analysis of the classic risk factors [18]. In this context, we implemented the use of fibrinogen, nitric oxide, adiponectin, and superoxide dismutase, to evaluate the implication of these phenomena in mitochondrial function and morphology in a MS model.

12.1.1 Nitric Oxide

NO is a relatively stable gas, which has an unpaired electron, thus giving it the property of free radical. This molecule exerts various physiological and pathological effects [19]. As a free radical, NO easily undergoes chemical reactions of addition, substitution, and oxidation; these reactions constitute the molecular basis of its different biological effects.

Under conditions of oxidative stress, as in the case of MS, the bioavailability of NO decreases, because it follows a pathological pathway by reacting with the superoxide radical forming peroxynitrite; generating oxidations in lipids, proteins, and other cellular structures; and perpetuating the lesion in the vascular layers [12, 20, 21].

Its determination as a biomarker represents an indirect measurement of nitrites and nitrates generated by oxidative stress [12]. One of the most important reactive oxygen intermediates is nitric oxide (NO), which is synthesized in the endothelial cells by means of the enzyme nitric oxide synthase producing in an equimolar way L-citrulline [22, 23]. Oxidative inactivation of NO due to excessive production of superoxide and hydrogen peroxide constitutes the most characteristic and early systemic phenomenon of endothelial dysfunction [24–27].

The NO would follow a pathophysiological path, coupling with the superoxide anion (O_2^-) at very high speeds, at the limit of the dysfunctional control, leading to the formation of peroxynitrite ($ONOO^-$). Peroxynitrite is formed in vitro and in vivo when its precursors coexist temporally and spatially, participating in a series of processes at the molecular and subcellular level and generating protein nitration with subsequent tissue damage [28]. Tyrosine nitration is mediated by reactive nitrogen species.

On the other hand, this nitration of the tyrosine residues in proteins prevents their functional interactions and jeopardizes the cell viability, being able to alter the conformation and structure of the proteins, their catalytic activity, and/or their susceptibility for digestion [29]. There are researches indicating that the vasculature per se produces significant amounts of superoxide, which reduces the bioavailability of NO through the degradation of it and the formation of peroxynitrite [30].

It is argued that the existence of a pro-oxidative state triggered by oxidative stress can induce insulin resistance by causing phosphorylation of insulin receptors [31].

The excess in the formation of these highly reactive molecules alters the activity of antioxidant enzymes that catalyze the decomposition of these harmful oxidants, neutralize their toxicity, and prevent their concentrations from becoming pathological [32, 33]. Some have postulated that antioxidant mechanisms would be responsible for the deterioration of insulin action [34]. On the other hand, the uncontrolled production of ROS would also damage DNA and induce lipid peroxidation of the cell membrane, affecting its permeability and functionality, generating accumulation of lipid peroxides that pass into the bloodstream, and increasing peroxidation of lipoproteins promoting endothelial dysfunction in SM [35].

12.1.2 Fibrinogen

Fibrinogen is a soluble glycoprotein found in plasma synthesized mainly in the liver and has a half-life of 100 h. As a coagulation factor, fibrinogen is a precursor of fibrin. It also participates in processes of inflammation, atherogenesis, and thrombogenesis [36].

On the other hand, it is an acute-phase reactant whose concentrations increase in plasma in response to a tissue injury; therefore it is postulated that it exists a close relationship between fibrinogen and atherosclerotic vascular disease. Likewise, it is considered that fibrinogen is not only a cardiovascular risk factor, but also it can be implemented as a biological marker in pathologies with inflammatory components such as the metabolic syndrome [37].

It is about the complex process of endothelial dysfunction, where hyperfibrinogenemia could reflect the proinflammatory state of MS probably through the physiopathological pathway of NO and the modified enzymatic activity of SOD [38].

12.1.3 Superoxide Dismutase

The human organism has a system of oxidant detoxification by the enzyme superoxide dismutase (SOD), whose plasma levels are essential to preserve the availability of NO.

This enzyme is part of the natural defense mechanism that the human body possesses. Its main function is to neutralize the free radicals produced physiologically in the organism by the metabolic processes. It is mainly responsible for the dismutation of the superoxide anion into hydrogen peroxide, a less harmful compound for the cell avoiding cellular oxidative stress [12, 39, 40]. The normal functioning of superoxide dismutase would be inhibited in a situation of oxidative stress, due to the excess of free radicals that saturate the enzymatic activity, a situation that leads to the generation of tissue lesions in pathologies such as MS [36, 38]. Its determination would be useful to establish the cellular antioxidant status in multifactor pathologies.

SOD neutralizes naturally the superoxide anion and catalyzes its enzymatic conversion to hydrogen peroxide (H_2O_2) at a much higher rate of spontaneous conversion; evidently the antioxidant chain is continued with the definitive reduction to water, by peroxidases or catalases of the peroxide formed, but the importance of SOD reaction is that at the level of the endothelial wall, the massive production of superoxide decreases the bioavailability of NO.

Experimental studies have shown an inhibition in the enzymatic activity of SOD, and this mechanism could be relevant in the endothelial dysfunction observed in pathologies such as MS [32, 43]. This endogenous enzyme establishes a biochemical "surveillance," since the superoxide anion is generally formed as an intermediary in the oxygenation reactions of substrates, thus protecting the tissues from the deleterious action of the superoxide radical [40]. In SM there would be a direct relationship between the production of free radicals by induction of oxidative stress and the endogenous antioxidant system specifically SOD [41, 42]; however there are no conclusive data in the literature regarding the binomial SM and SOD and so with the importance of studying the behavior of SOD in experimental models [44, 45].

12.1.4 Mitochondria: Function and Morphology

The production of mitochondrial superoxide is an important mediator of cellular oxidative injury; likewise, succinate-cytochrome c reductase (SCR) present in the electron transport chain could be involved as a mediator in the generation and as an alternative target of NO in the regulation of mitochondrial respiration [46].

The control of mitochondrial functions depends on two variables, the concentration of NO and the oxygen level; each one acts on a range of concentrations and gradients, and there are critical points in which both variables are intercepted generating peroxynitrite. In this way, NO, superoxide anion, and peroxynitrite play a crucial role in the regulation of mitochondrial function [47, 48]. This process would irreversibly inhibit different components of the mitochondrial respiratory chain. This effect would be produced through different enzymatic ways of inactivation of dehydrogenated NADH (complex I) and dehydrogenated succinate (complex II), as well as through the inhibition of ATP synthase [49]. The generation of ROS in mitochondria can be triggered by several factors such as efficiency of the electron transport chain, the concentration of oxygen, the availability of electron donors such as NADH and FADH2 of cytokines, and the activity of antioxidant defenses [50].

The ability of the mitochondrial matrix to produce NO has several implications. On the one hand, NO would act as a physiological messenger that modulates the speed of electron flow under physiological conditions by binding to the active site of oxygen in cytochrome oxidase in a competitive manner. This binding regulates the oxygen consumption representing a beneficial effect. The interruption of cytochrome oxidative phosphorylation and, consequently, the alteration of the cellular metabolic activity could be produced by NO that reacts with heme's iron (Fe) forming a nitrosylated inclusion complex [51].

In addition, the endothelial NOS (eNOS) requires the presence of the Ca^{2+} ion for its catalytic activity and in the cells involved in the systemic inflammatory response to the intracellular Ca^{2+} homeostasis. The cellular breakdown would generate a massive influx into the ion cell of Ca^{+} and Na^{+} and an increase in cell volume due to the entry of water, inducing alterations at the mitochondrial level and initiating an increase in ROS inside these organelles, a process that could contribute to the initiation of vascular pathologies in which mitochondrial antioxidant defenses are overcome. This plays an important role in mitochondrial energy regulation, modulated by the release of hydrogen peroxide, which is converted to hydroxyl radical. In situations of oxidative stress, the hydroxyl radical would inhibit the ATPase activity of the Na^{+}/Ca^{2+} pump and causes damage to the cell membrane, which would increase its permeability. The continuous flow of uncontrolled Ca^{2+} activates different enzymes (proteases, endonucleases, and phospholipases) creating a vicious circle, which would contribute to a higher production of free radicals affecting oxidative phosphorylation and, therefore, the production of antioxidants in a process that requires energy [51, 52].

The changes observed at the mitochondrial level also involve the Krebs cycle. The oxidation of fatty acids, aerobic respiration, related to oxidative mitochondrial phosphorylation and the synthesis of ATP. For example, peroxynitrite would carry out the nitration of tyrosines in several mitochondrial proteins, such as the ferrosulfurated cycle of Krebs aconitase and the dehydrogenated glyceraldehyde-3P involved in glycolysis, leading to an inhibition of mitochondrial respiration and a fall in the synthesis of ATP.

When mitochondrial respiration is inhibited, electrons would accumulate in the respiratory chain, which would lead to the complex NADH-ubiquinone reductase (complex I) and ubiquinol-cytochrome c reductase (complex III) that possess ubiquinone as a common component and increase the generation of superoxide and consequently peroxynitrite and hydrogen peroxide, perpetuating the lesions generated by oxidative stress [53, 54]. Therefore, the reduction of electron transport and oxidative stress could represent an important objective for the modulation of the function of adipocytes in the SM [55, 56]. DM associated with metabolic disorders can lead to mitochondrial dysfunction, activation of NADPH oxidase (NOX), and excess ROS production, which results in oxidative stress and promotes cardiovascular disorders [57, 58].

The modifications generated would be responsible for the morphofunctional mitochondrial alteration, changes to which probably contributed the modified inflammatory biomarkers, variations in the concentration of adiponectin, insulin resistance, and hyperinsulinism. It is demonstrated that mitochondrial damage is related to the pathogenesis of several metabolic disorders, especially those related to insulin resistance, such as obesity and DM2 [59].

From the aforementioned, we understand the importance of studying the implication of the inflammatory phenomenon and oxidative stress in pathologies with vascular components to establish on the one hand the probable physiopathogenic mechanisms with the aim of generating strategies that prevent the appearance of events by decreasing or controlling the risk factors. If the concentration of mitochondrial ROS were reduced, endothelial dysfunction and its consequences in the initial stages of MS would probably be controlled or reversed [60].

The clinical importance of an experimental model is the identification of early markers that would help find ways to prevent or slow the development of DM2 and the progression of atherosclerosis since glucose intolerance is a proinflammatory, proatherogenic, prothrombotic, and facilitating condition of diabetes mellitus [61, 62].

That is why we designed SM research from the basic sciences in order to implement biomarkers of oxidative stress and also to analyze the probable histological alterations in the vascular endothelium and liver. In addition, we can assess whether morphofunctional modifications of mitochondria of smooth aortic muscle cells are generated.

12.2 Metabolic Syndrome Validation

The investigation was carried out according to the guide for care and use of laboratory animals published by the US National Institute of Health, NIH publication (N°58–23, revised 1996), and the Ethic Committee from the Medicine School (National University of Cordoba) Res. N°49/17 has also approved the experimental animal procedures. The model of SM was performed in male Wistar rats, divided into two groups: (A) control group ($n = 12$) and (B) group with SM ($n = 12$) [63, 40]. After the SM induction, the blood and tissue samples were processed for further analysis. Plasma level was analyzed: blood glucose [64], insulinemia [65], and lipid profile [64], complementing with homeostatic model assessment (HOMA) [66] to validate this pathology.

In the experimental SM model, the results presented in Table 12.1 showed that rats that receive fructose in drinking water in a chronic manner provide a useful model for the diagnosis of the factors that make up SM; this is induced by changes in the intake and expresses numerous alterations similar to the human subject with SM [40, 63]. Unlike glucose, diets with high levels of fructose, being proinflammatory molecules, induce in rodents glucose intolerance, dyslipidemia by lipolysis, angiogenesis, endothelial dysfunction, vasoconstriction, fibrinolysis, and insulin resistance [67, 68] as demonstrated in our results. The administration of 10% fructose generated in the SM group experimentally induced hyperglycemia, hypertriglyceridemia, decreased HDL levels, and increased circulating levels of total cholesterol together with hyperinsulinemia, confirming that the experimental model presents the characteristic manifestations of MS [40, 69].

Table 12.1 Plasma levels of glycemia, insulinemia, and lipid profile in control group and group SM

	Control (A)	SM (B)
Glycemia (mg/dL)	115 ± 1,1	176 ± 17,3
Insulinemia (uU/mL)	4 ± 0,82	29,5 ± 4,52
HDL (mg/dL)	61 ± 0,01	28,3 ± 1,14
Total cholesterol (mg/dL)	69,7 ± 1,6	133 ± 9,6
Triglycerides (mg/dL)	46,2 ± 6	75 ± 12,9
HOMA	3 ± 0,38	11 ± 1,3

Mean ± ES: glycemia: (A) vs. (B): $p < 0.001$. Insulinemia: (A) vs. (B): $p < 0.001$. HDL (high-density lipoprotein): (A) vs. (B): p < 0.001. Total cholesterol: (A) vs. (B): $p < 0.001$. Triglycerides: (A) vs. (B): $p < 0.001$. HOMA: (A) vs. (B): $p < 0.001$. ($n = 12$ per group) SM (metabolic syndrome) [35]

In addition, the calculation of HOMA is a useful model for the quantification of insulin resistance, reflecting the function of beta cells requiring only a fasting serum sample. It is calculated by multiplying fasting plasma insulin (FPI) by fasting plasma glucose (FPG) and then dividing by the constant 22.5 [66, 70].

In our results, an increase in the HOMA value was obtained in the SM group, validating the presence of insulin resistance in the experimental SM model. Arterial hypertension, diabetes, and obesity are common but not independent pathologies, and their combination is reflected in the SM [71]. Several inflammatory mediators produced by adipose tissue and the interrelationship between immune and metabolic cells affect insulin signaling for ROS generation and subsequently endothelial dysfunction leading to insulin resistance (IR) and heart diseases [72].

Inflammation of the hypothalamus can be induced experimentally by a diet high in fat or by the administration of fructose [73] causing hyperphagia, and it has been documented that it impairs the release of insulin from the β cells and the peripheral action of insulin. Therefore, the chronic excess of nutrients, such as lipids and glucose, can simultaneously trigger inflammatory responses, which disrupt the metabolic function, generating oxidative stress, inflammation, and MS.

The significant hypertriglyceridemia observed in the group with MS would be demonstrating a concentration of triglycerides rich in abnormal lipoprotein (lipoprotein-a) or modified lipoprotein associated with prothrombotic states.

On the other hand, inhibitory dysfunction of lipolysis in adipocytes produces an activation of triglyceride lipolysis and the release to the peripheral circulation of free fatty acids. Both hyperinsulinemia and said fatty acids decrease the action in the adipose tissue of the catalytic enzyme lipoprotein lipase, generating an increase in the production of triglycerides [74], so that their plasma values are increased (Table 12.1).

The results of another lipid profile parameter such as HDL were analyzed, and significant decrease was observed in the group with MS. This is due to the generalized metabolic effect of IR that increases lipolysis, with greater availability of

FFA. When this mechanism becomes dysfunctional due to the decrease in circulating plasma HDL, it translates into an increase in proinflammatory properties, endothelial imbalance, and mechanisms of atherothrombosis and fibrinolysis.

On the other hand, normally the initial formation of HDL cholesterol is secreted in discoidal particles of the liver. The lipoprotein cholesterol-HDL has a protective role in promoting cellular cholesterol efflux and reverse cholesterol transport [75, 76]. When EHNA-type alterations are present, the synthesis of this lipoprotein and its precursors is affected [77], decreasing expression of cell adhesion molecules (CAMs) and the increasing endothelial NOS expression modifying the activation, release, and bioavailability of NO [78, 79].

12.2.1 Pathological Evaluation in Liver Preparations of Rats with MS

The excessive accumulation of fat (steatosis) is linked to diseases such as diabetes mellitus type 2, IR, central obesity, hyperlipidemia with low levels of high-density lipoprotein (HDL cholesterol), hypertriglyceridemia, and hypertension, becoming an increasingly common liver problem. In addition, nonalcoholic steatohepatitis (NASH) is considered to be the prevalent liver disease in obese people and with MS [80]. Due to its intimate link to IR, obesity, and MS, the importance of studying the implication of the inflammatory phenomenon and associated oxidative stress is understood, in order to establish the probable physiopathogenic mechanisms with the aim of generating strategies that prevent the incidence and prevalence of this pathology [81].

To confirm the presence of NASH, a distinctive hepatic seal of the metabolic syndrome, 15 liver sections of 3–5 μm each were analyzed by optical microscopy in all the batches studied, objectifying in the group with SM-induced alterations consistent with cholestasis, sinusoidal congestion, binucleation, and periportal inflammatory infiltrate (Fig. 12.3a, b), compatible with NASH, which corroborates the hepatic lesions characteristic of MS, compared to the control group that did not show histological changes (Fig. 12.4) [82]. It should be noted that there are also histological changes of NASH that include steatohepatitis (fatty liver in addition to parenchymal inflammation with or without accompanying focal necrosis), steatosis (fatty liver), and varying degrees of fibrosis, including cirrhosis, although they were not visualized in the tissues analyzed [83].

The hypertriglyceridemia measured in animals with MS would be due to the accumulation of triglycerides in the hepatocytes as a result of the entry of FFA to the liver at higher levels than necessary; therefore the remainder is exported to the blood circulation, being an adjuvant factor in the increase in cardiovascular risk [45].

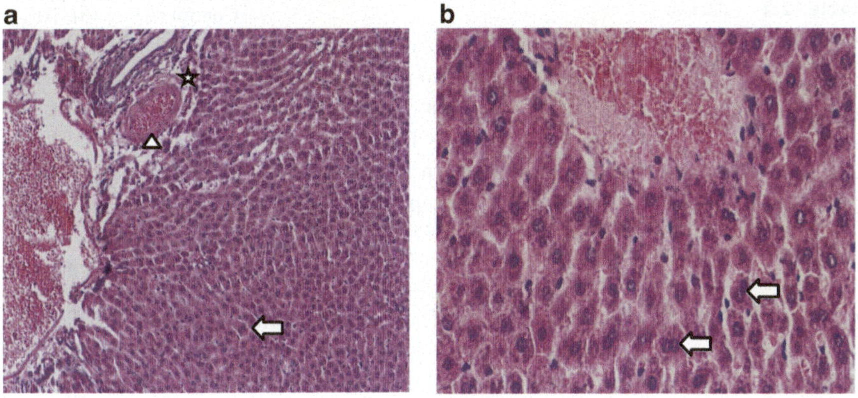

Fig. 12.3 Liver histological section corresponding to the group of animals with induced SM (B). (**a**): The MO shows vacuolization (star), congestive vasculature (triangle) and binucleation (arrowhead). (H/EX40). (**b**): The MO shows binucleation and periportal inflammatory infiltrate (arrowhead)

Fig. 12.4 Liver histological section corresponding to the group of control animals (A): the MO shows normal hepatic histology (H/EX40)

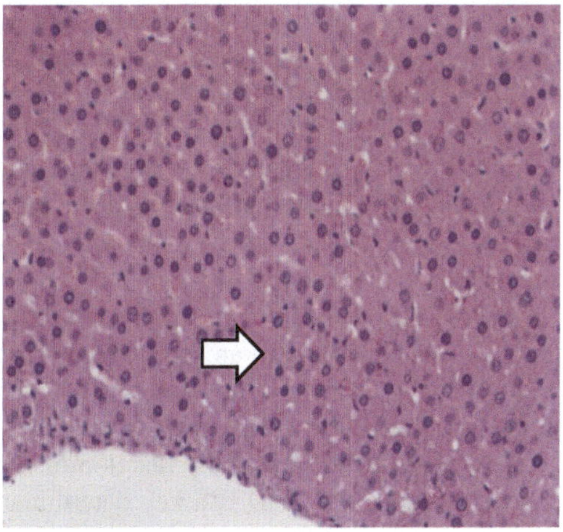

12.2.2 Analysis of Inflammatory Biomarkers and Oxidative Stress

In animals with NASH, it can be proven that when inflammatory responses are simultaneously triggered, the plasma concentrations of these acids and glycerol increase, interrupting the metabolic function of insulin and triggering oxidative stress with an increase in the concentrations of reactive oxygen species (ROS), decreasing the amount of bioactive NO by chemical inactivation when toxic peroxynitrite is formed, showing a relationship with the decreased plasma NO values

Table 12.2 Values of fibrinogen, adiponectin, nitric oxide, and enzymatic activity of superoxide dismutase in rats with experimentally induced SM

	Control (A)	SM (B)
Fibrinogen (mg/dL)	203 ± 9	292 ± 11
Adiponectin (μg/dL)	11,17 ± 0,11	8,34 ± 0,2
NO (μM)	23,58 ± 1,4	8,7 ± 1,2
SOD (U/mL)	138,5 ± 3,6	181 ± 6

ME ± ES: fibrinogen: (A) vs. (B): $p < 0,00.1$. Adiponectin: (A) vs. (B): $p < 0,01$. NO: (A) vs. (B): $p < 0,001$. SOD: (A) vs. (B): $p < 0,01$. ($n = 12$ per group) SM (metabolic syndrome) [35]

observed in animals with SM induction. The increased concentrations of fibrinogen show the inflammatory state present in SM induced, verifying that this would behave as an independent indicator in SM and establishes a close relationship with the oxidative stress process reflected by the decrease in the bioavailability of NO, which quantifies indirectly the production of superoxide anion (O_2^-).

Consequently, when we corroborated the experimental SM model, we proceeded to study the plasma concentrations of fibrinogen [84], adiponectin [16] (inflammatory markers), nitric oxide [85] (oxidative marker), and superoxide dismutase activity [86] (marker) antioxidant in red blood cell lysate (Table 12.2).

In this way it could be established that in this proinflammatory process, fibrinogen would generate oxidative stress with probable loss of the capacity of endothelial cells to regulate their vital functions in the production of vasoactive substances such as NO [87]. Therefore, the inflammatory state and the endothelial disbalance described would demonstrate a dysregulation in the circulating levels of fibrinogen, forming an important proinflammatory marker being predictive of the evolution of cardiovascular diseases such as myocardial infarction, cerebrovascular accident, and venous thromboembolism [88].

The increase in fibrinogen values observed in the induced SM group evidences that this would behave as an independent predictor in SM and establishes a close relationship with the oxidative stress reflected by NO modifications, which indirectly quantifies the production of superoxide anion ($O_2^{\bullet-}$).

Due to the efficiency of the reaction of the superoxide with the NO, the local concentration of the SOD is a key determinant of the bioactivity of the NO. The increased values of SOD in the experimental model is possibly an adaptive response, characteristic of biological systems tending to compensate oxidative stress, which can be interpreted as a situation of increased redox environment.

This increase in SOD would allow us to infer that the endothelial dysfunction of the vascular wall would initially be a stimulus for its synthesis and later descend by enzymatic saturation [89]. The increase in the activity of the enzyme SOD observed in the results associated with changes in fasting blood glucose levels, dyslipidemia due to decreased HDL and hypertriglyceridemia, IR, hypoadiponectinemia, and increased inflammatory and oxidative components are indicators of oxidative stress present in the experimental model. This would reflect the importance of SOD as an endogenous antioxidant mechanism that tries to compensate the increase of free radicals; however, it is sometimes insufficient either due to low productivity or the excess of the production rate of this enzyme, which would allow alterations to progress in both models [40, 90].

The histological changes visualized at the hepatic level, consisting of cholestasis, sinusoidal congestion, binucleation, and periport inflammatory infiltrate, are compatible with the first changes of NASH. It should be remembered that it is the most common liver disease in obese patients with MS [91, 92].

The interaction of these mechanisms would be responsible for the hyperfibrinogenemia, the decrease in the bioavailability of NO, and the increase in SOD activity demonstrated in the study results associated with the observed liver changes. The hypertriglyceridemia measured in animals with MS is due to the accumulation of triglycerides in the hepatocytes as a result of the entry of FFA to the liver at levels higher than necessary; therefore the remainder is exported to the blood circulation [93, 94].

Authors suggest that inflammation of the liver and IR could play an important role in the development of endothelial dysfunction and atherosclerosis in patients with NASH, especially young and middle-aged, who could benefit from early prevention strategies to help decrease the risk of developing manifest cardiovascular disease [38, 95].

Consequently oxidative stress, lipoperoxidation, and the abnormal production of proinflammatory adipocytokines have been linked to hepatocyte damage and apoptosis. Experimental studies have found that adiponectin, one of the adipocytokines that decreases in the SM as observed in the results (Table 12.2), antagonizes the excess storage of lipids in the liver protecting against inflammation and fibrosis, so that its decrease is associated with the severity of hepatic steatosis, necroinflammation, and fibrosis.

The massive arrival of free fatty acids to the liver via portal, mainly from the lipolysis of visceral adipose tissue, generates a decrease in insulin clearance, expressing hyperinsulinism, and increases in gluconeogenesis reflected as hyperglycemia and in the synthesis of VLDL responsible for quantified dyslipidemia [96]. The lipolysis of abdominal fat has special importance in the pathogenesis of NASH [97]; in fact, insulin resistance, peripheral levels of adiponectin, the presence of other cytokines such as TNF-α and IL-6, CRP, insulin, and glucose remain unchanged after the removal of subcutaneous fat. The reduction of visceral fat improves the resistance to insulin and the remaining metabolic disorders associated with NASH [98]; it is particularly resistant to the action of insulin [99], and, consequently, it is more easily hydrolyzed.

In addition, the liver, by occupying a strategic place in the portal circulation, directly receives the free fatty acids (FFA) released during the lipolysis of abdominal fat. In animals with NASH, it can be proven that the plasma concentrations of these acids and glycerol are greatly increased and that insulin has a reduced capacity to prevent the release of n of these lipolysis products [100, 101]. The AGL that reach the liver activate the nuclear receptor PPARα (peroxisome proliferator-activated receptor) inducing the transcription of numerous genes involved in the catabolism and elimination of fatty acids [102, 103]. This protein complex intervenes in the use of FFA, in the synthesis of triglycerides inducing steatosis, and in the phospholipids, influencing gluconeogenesis, which is expressed in hyperglycemia; they also modify the oxidation in mitochondria, in peroxisomes, or in microsomes, and these three oxidations are of great importance since they can contribute to cellular oxidative stress.

12.2.3 Pathological Assessment of Thoracic Aorta in Rats with MS

Several clinical studies recognize MS as responsible for endothelial dysfunction triggered by insulin resistance and hyperinsulinism; even in patients with MS, an association between an increase in the thickness of the intima and media of the carotid has been observed, this image being also a marker of subclinical atherosclerosis [104]. These lesions would lead to an increased risk of cardiovascular events, especially in pathologies that involve multiple risk factors such as MS.

Beside the biochemical parameters and biomarkers analyzed, anatomopathological studies by MO in thoracic aorta showed that the state of MS (Fig. 12.5) generates endothelial denudation, thickening of the intima, increase in the extracellular matrix, myxoid changes in the subendothelium and protrusion of the vascular wall toward the light of the aorta, and changes that objectify the repercussion at vascular level and whose functional expression would be endothelial dysfunction [20].

The inflammatory signals elicited by biomarkers activate endothelium via an increase in ROS production, which have been supported by the histopathology studies of thoracic aorta [40, 103].

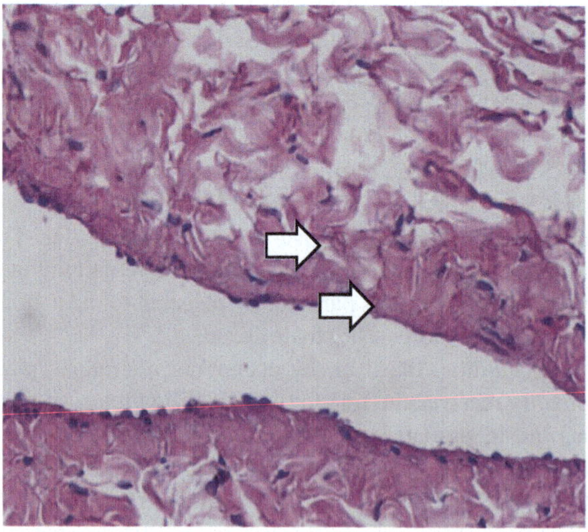

Fig. 12.5 Optical microscopy of rat thoracic aorta with induced MS showing endothelial denudation with red blood cells adhered, intimal thickening, myxoid changes in the extracellular matrix and disorganization of the internal muscular lamina in most cuts (arrow head) (H/EX60) [35]

12.2.4 Study of the Mitochondrial Morphology of Smooth Muscle Cells of the Thoracic Aorta of Rats with MS

Due to the biochemical and pathological changes observed, the analysis of the mitochondrial morphology and functionality of the mitochondrial respiratory chain was carried out, given that it is the final target of oxidative stress.

The plasma results obtained from the SM indicators and histopathological lesions demonstrate a common final pathway, which is oxidative stress and endothelial dysfunction. These processes are closely related to morphological and functional alterations of the mitochondria in the smooth muscle cells of the vascular wall (Figs. 12.6 and 12.7), expressing an important role both in the beginning and in the progress of these multisyndromic pathologies [53, 105, 106]. Numerous investigations provided evidence to suggest that mitochondrial dysfunction is one of the main causes of IR and related cardiometabolic diseases [107].

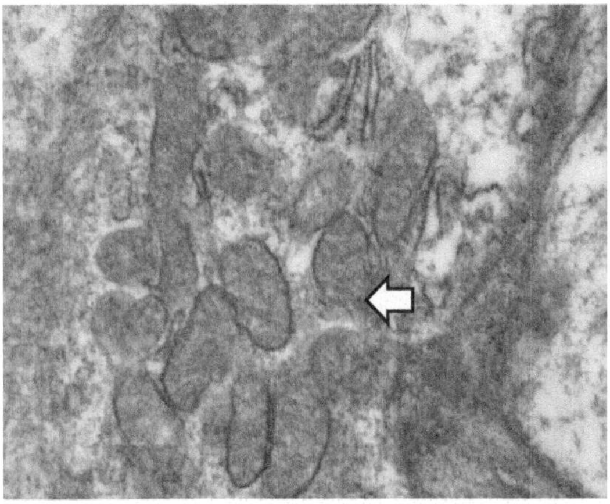

Fig. 12.6 Microphotograph of mitochondria in control group, structure of membranes, and crests without changes and maintaining normal shape and size (arrow), 27,800X [34].

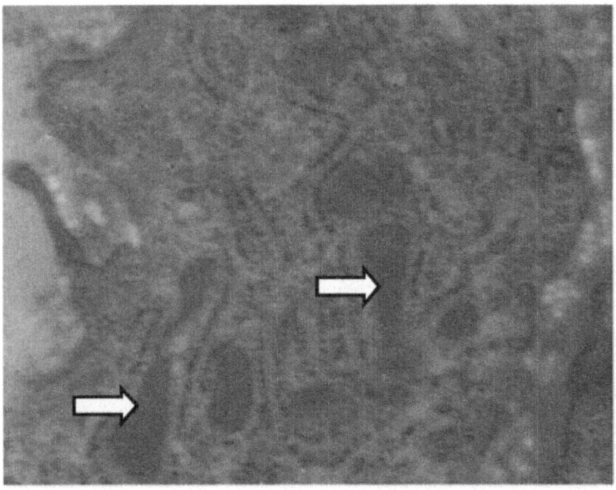

Fig. 12.7 Microphotograph of mitochondria in SM group. Mitochondrial groupings with structural deformations and areas of swelling (head of arrows) are observed, 27,800X

The mitochondrial oxidative capacity in the SM is totally correlated with the number and size of the same [108]; these modifications would be due to the action of the AGL and the enzymes that generate or catabolize the regulation of the morphology, that is why the importance of control of the plasma lipid profile as well as the synthesis of adiponectin, which, being diminished in the experimental SM, conditions the appearance of proinflammatory and pro-oxidative products that increase the peroxidation of HDL, generating repercussions in the development of metabolic alterations [109].

Another probable cause of these morphological modifications is due to the fact that this organelle constitutes a complex, interconnected, and highly dynamic network, maintained by permanent, opposite, and balanced fusion and mitochondrial fission events [110].

Both the number of tubules and their connections, as well as the subcellular distribution of the organelle, are actively controlled. In this way, the term "mitochondrial dynamics" has been coined to encompass at least three different processes:

(a) The remodeling of the mitochondrial reticulum through fusion/fission processes, which is closely linked to the cellular metabolic state and is controlled by the activity of a group of guanosine triphosphate hydrolases proteins (GTPases) related to the dynamin family [111, 112].
(b) Subcellular mitochondrial motility, particularly relevant in polarized cells and corresponding to the mitochondrial displacement dependent on the kinesin 1 and 3 motors and the Milton and Miro adapters [113], which ensures the local supply of ATP in biological processes with high-energy requirements and the use of these organelles as calcium buffer [111, 112, 114].
(c) The remodeling of the mitochondrial ultrastructure and the condensation of its matrix, processes classically considered as a reflection of the mitochondrial metabolic state. Both the different functional states of the mitochondria and their ultrastructure play a role in remodeling of mitochondrial crests as observed in the results [111, 112, 115].

Since mitochondria are organized to form intricate mitochondrial networks and these networks work as electrical units to transmit the mitochondrial membrane potential, a normal mitochondrial function is required for the correct homeostasis of oxidative substrates, since a mitochondrial dysfunction can lead to metabolic diseases or contribute to the pathophysiology of obesity. In addition to the described morphological alterations of the mitochondria, modifications were observed in the functionality of these organelles.

12.2.5 Analysis of the Enzymatic Activity of the Mitochondrial Respiratory Chain in Smooth Aortic Muscle Cells from Rats with MS

A depreciation of the Krebs cycle would have been indicated when the measurements of the activity of the electron chain were analyzed, because a progressive decrease in the activity of citrate synthase was observed (Fig. 12.8). This is reflected by decreasing the activity of the mitochondrial respiratory chain reflected in the depletion of the activity of the enzymes NADH dehydrogenase (IC), succinate ubiquinone reductase (CII), cytochrome C reductase (CIII), and cytochrome C oxidase (CIV). We must emphasize that this deterioration of mitochondrial function is progressive and accentuated, manifesting via oxidative stress that proinflammatory and pro-oxidative indicators in both experimental models have an impact on mitochondrial morphofunctionality.

Pathological processes such as inflammation, hypoxia/ischemia, and oxidative stress exert harmful effects on the structure and function of mitochondria [116]. As a result, there is a decrease in the phosphorylation potential that directly leads to the depression of cellular energy, and consequently there is a decrease in the ability of the cell to control homeostasis. Mitochondrial lesions are reversible in short periods of time and irreversible when harmful stimuli persist, since oxidative stress in the mitochondria through the formation of peroxynitrites ($ONOO^-$) inhibits many mitochondrial proteins including the subunits of complexes I and II of the respiratory chain (Figs. 12.9 and 12.10), which results in the depletion of oxidative phosphorylation as observed in our results [116–118].

In some pathological situations such as hypoxia, complex II would also be a producer of physiologically relevant mitochondrial ROS [119]. In the absence of ADP, the electrons derived from succinate (substrate FADH2-linked to complex II)

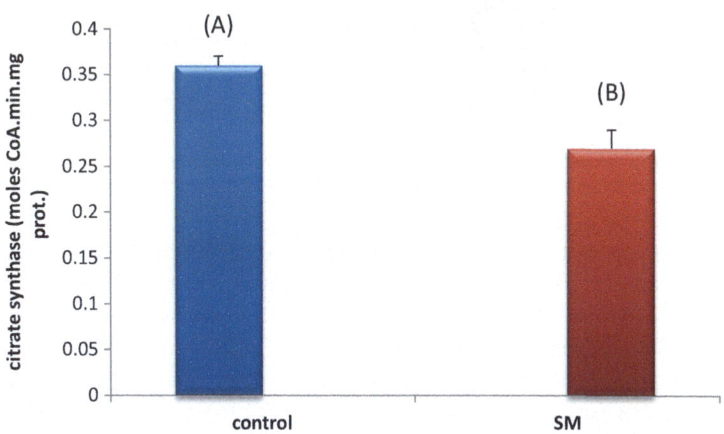

Fig. 12.8 Enzymatic activity of citrate synthase in rats with experimental SM. (n = 12 per group) SM (metabolic syndrome). ME ± ES: (A) vs. (B): $p < 0{,}01$

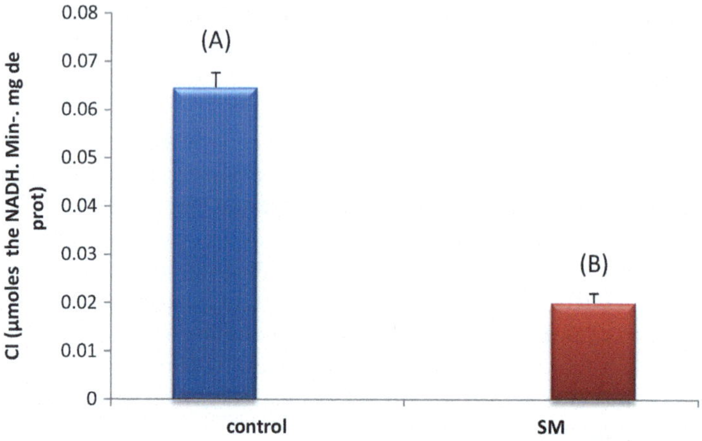

Fig. 12.9 Enzymatic activity of complex I in rats with experimental SM. ($n = 12$ per group) SM (metabolic syndrome). ME ± ES: (A) vs. (B): $p < 0{,}01$

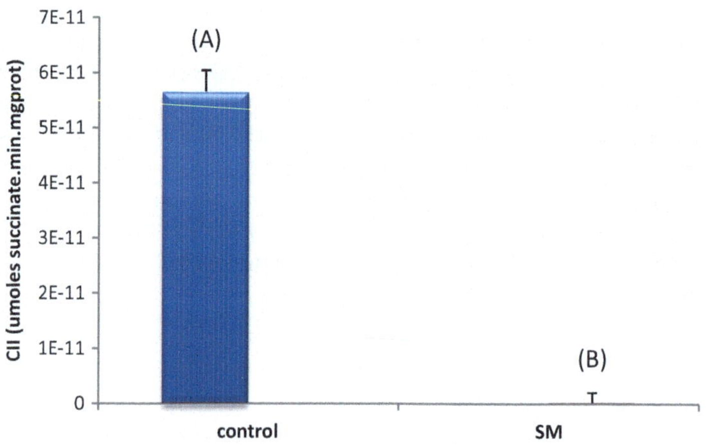

Fig. 12.10 Enzymatic activity of complex II in rats with experimental SM. ($n = 12$ per group). SM (metabolic syndrome). ME ± ES: (A) vs. (B): $p < 0{,}0001$

can flow in the opposite direction increasing the production of O_2 by complex I; for this reason, complex I is considered the main physiological and pathologically generating site of ROS in mitochondria [109, 120, 121]. This increased production of ROS would alter the mitochondrial morphological function, and it is known that the main sites of production of mitochondrial superoxide derive mainly from complexes I and III; these two are the main electron leakage sites in the electron transfer chain, and complex I is the most vulnerable (Fig. 12.11) [122].

In the experimental model of MS, the activity of citrate synthase and complex I decreased could be by increased oxidation and reduced glucose storage together

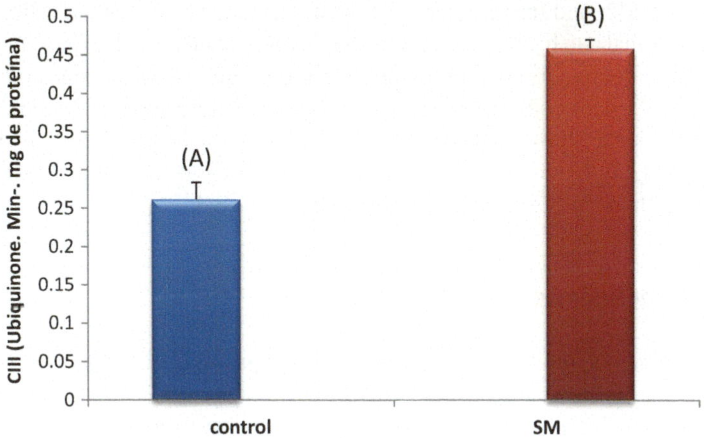

Fig. 12.11 Enzymatic activity of complex III in rats with experimental SM. ($n = 12$ per group). SM (metabolic syndrome). ME ± ES: (A) vs. (B): $p < 0.01$

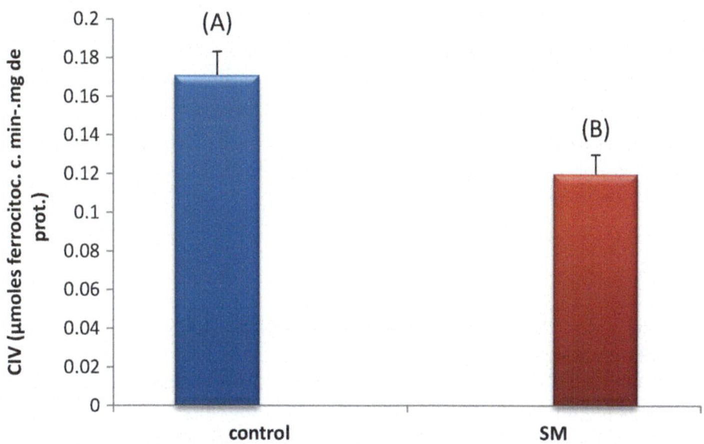

Fig. 12.12 Enzymatic activity of complex IV in rats with experimental SM. ($n = 12$ per group). SM (metabolic syndrome). ME ± ES: (A) vs. (B): $p < 0.01$

with the reduced activity of the tricarboxylic acid cycle, β oxidation, and the decrease in the chain of electron transfer, as has been described in pathologies with obesity and DM2 [123]. In other experimental models of obesity and IR, reduction was found in the expression of eNOS, in mtDNA, and in respiratory proteins such as cytochrome C oxidase, complex IV of CRM, and cytochrome c, as well as in oxygen consumption and ATP production along with changes in mitochondrial morphology. Similar behavior was observed in complex IV (Fig. 12.12), due to the modifications in the bioavailability of NO, a molecule that modulates mitochondrial O_2 consumption through mitochondrial inhibition of complex IV, which regulates mitochondrial biogenesis [124].

Obesity, DM2, and insulin resistance would produce a chronic elevation of circulating FFA that can become cytotoxic; the increase in electron leakage and uncoupling in the mitochondria is a serious problem in these conditions since fatty acids would cause oxidative stress and alterations in the structure and mitochondrial function. The evidence shows that oxidative damage to mitochondrial proteins leads to progressive dysfunction and that the deterioration of mitochondrial function is a unifying mechanism of several risk factors such as MS. [125]

12.3 Conclusions

In our experimental model, a proinflammatory and pro-oxidative state was observed at the vascular level, verified by modified levels of the analyzed biomarkers. In addition, it was demonstrated that the sustained oxidative stress situation induces histological alterations at the aortic level. This pathological and oxidative state leads to a mitochondrial dysfunction with repercussion in the morphology of this organelle.

It is relevant to inquire about inflammation and oxidative stress, since these would seem to be the physiopathological origin of phenomena such as IR, MS, and obesity. Corroborating the existence of this link would allow establishing therapeutic guidelines that would modify the impact of these diseases on society, given that it has huge consequences in health system. For this it is necessary to identify the determinants of the disease in order to implement preventive measures for control and monitoring as well as to study therapeutic strategies that can be implemented to reduce the incidence of this multisyndromic pathology.

References

1. Grundy SM (2007) Metabolic syndrome: a multiplex cardiovascular risk factor. J Clin Endocrinol Metab 92:399–404
2. Grandl G, Wolfrum C (2018) Hemostasis, endothelial stress, inflammation, and the metabolic syndrome. Semin Immunopathol 40(2):215–224. https://doi.org/10.1007/s00281-017-0666-5
3. Gómez M, Manuel A, Patino Alonso MP et al (2013) Confirmatory factor analysis to assess the measure of adiposity that best fits the diagnosis of metabolic syndrome and relationship to physical activity in adults. European Journal of Nutrition, relationship to physical activity in adults. Eur J Nutr 52(5):1451–1459
4. Palomo IG, Gutiérrez CL, Alarcón ML et al (2009) Increased concentration of plasminogen activator inhibitor-1 and fibrinogen in individuals with metabolic syndrome. Mol Med Rep 2(2):253–257
5. Palomo IG, Moore-Carrasco R, Alarcon M et al (2010) Pathophysiology of the proatherothrombotic state in the metabolic syndrome. Front Biosci 2:194–208
6. Fernández-García JC, Cardona F, Tinahones FJ (2013) Inflammation, oxidative stress and metabolic syndrome: dietary modulation. Curr Vasc Pharmacol 11(6):906–919
7. Alberti KG, Eckel RH, Grundy SM, Zimmet PZ, Cleeman JI, Donato KA et al (2009) Harmonizing the metabolic syndrome: a joint interim statement of the international Diabetes Federation Task force on epidemiology and prevention; National Heart, Lung, and Blood Institute; American heart Association; World heart Federation; International Atherosclerosis Society; and International Association for the study of obesity. Circulation 20:1640–1645

8. Ghezzi P, Bonetto V, Fratelli M (2013) Thiol–disulfide balance: from the concept of oxidative stress to that of redox regulation. Antioxid Redox Signal 7(7–8):964–972
9. Nikolopoulou A, Kadoglou NP (2012) Obesity and metabolic syndrome as related to cardiovascular disease. Expert Rev Cardiovasc Ther 10(7):933–939
10. Krintus M, Kozinski M, Kubica J, Sypniewska G (2014) Critical appraisal of inflammatory markers in cardiovascular risk stratification. Crit Rev Clin Lab Sci 51:263–279. https://doi.org/10.3109/10408363.2014.913549
11. Stoner L, Lucero AA, Palmer BR, Jones LM, Young JM, Faulkner J (2013) Inflammatory biomarkers for predicting cardiovascular disease. Clin Biochem 46(15):1353–1371
12. Lowe F (2014) Biomarkers of oxidative stress. Syst Biol Free Radicals Antioxid:65–87
13. Packard RR, Libby P (2008) Inflammation in atherosclerosis: from vascular biology to biomarker discovery and risk prediction. Clin Chem 54(1):24–38
14. Monteiro R, Azevedo I (2010) Chronic inflammation in obesity and the metabolic syndrome. Mediat Inflamm 2010:1–10. Article ID 289645. https://doi.org/10.1155/2010/289645
15. Vazzana N, Santilli F, Sestili S, Cuccurullo C, Davi G (2011) Determinants of increased cardiovascular disease in obesity and metabolic syndrome. Curr Med Chem 18(34):5267–5280
16. Esposito K, Ciotola M, Giugliano D et al (2006) Oxidative stress in the metabolic syndrome. J Endocrinol Investig 29(9):791–795
17. Sandra S, Arango V (2012) Biomarkers for the evaluation of human health risks. Rev Fac Nac Salud Pública 30(1):75–82
18. Roos CJ, Quax PH, Jukema JW (2012) Cardiovascular metabolic syndrome: mediators involved in the pathophysiology from obesity to coronary heart disease. Biomark Med 6(1):35–52
19. Duarte MJ, Espinosa López RF, Diaz MS, Sánchez Rojas G (2008) Nitric oxide: metabolism and clinical implications. Med Int 24(6):397–406
20. Baez MC, Tarán M, Scribano MP, Balceda A, Buonanotte C, Blencio S, Fonseca S, Moya M (2017) Inflammatory and oxidative stress markers as indicator of atherogenesis in rats: antioxidants as preventive pharmacological methods. Anti-Inflammatory Anti-Allergy Agents Med Chem 16:1–7
21. Carrillo Calvillo J, Bear Sandoval IR (2004) Biomarkers, inflammation, oxidative stress, lipids and atherothrombosis atherosclerosis: an inflammatory process. Arch Cardiol Mex 74(2):S379–S384
22. Bonomini F, Tengattini S, Fabiano A, Bianchi R, Rezzani R (2008) Atherosclerosis and oxidative stress. Histol Histopathol 23(3):381–390
23. Zweier JL, Li H, Samouilov A, Liu X (2010) Mechanisms of nitrite reduction to nitric oxide in the heart and vessel wall. Nitric Oxide 22(2):83–90
24. Gleissner CA, Leitinger N, Ley K (2007) Effects of native and modifi ed low-density lipoproteins on monocyte recruitment in atherosclerosis. Hypertension 50(2):276–283
25. Thomas SR, Witting PK, Drummond GR (2008) Redox control of endothelial function and dysfunction: molecular mechanisms and therapeutic opportunities. Antioxid Redox Signal 10(10):1713–1766
26. Victor VM, Rocha M, Solá E, Bañuls C, Garcia-Malpartida K, Hernández-Mijares A (2009) Oxidative stress, endothelial dysfunction and atherosclerosis. Curr Pharm Des 15(26):2988–3002
27. Brand MD (2016) Mitochondrial generation of superoxide and hydrogen peroxide as the source of mitochondrial redox signaling. Free Radic Biol Med 100:14–31
28. Battes LC, Cheng JM, Oemrawsingh RM, Boersma E, Garcia-Garcia HM, de Boer SP et al (2014) Circulating cytokines in relation to the extent and composition of coronary atherosclerosis: results from the ATHEROREMO-IVUS study. Atherosclerosis 236(1):18–24
29. Abello N et al (2009) Protein tyrosine nitration: selectivity, physicochemical and biological consequences, denitration, and proteomics methods for the identification of tyrosine-nitrated proteins. J Proteome Res 8(7):3222–3238

30. Surmeli NB, Litterman NK, Miller AF, Groves JT (2010) Peroxynitrite mediates active site tyrosine nitration in manganese superoxide dismutase. Evidence of a role for the carbonate radical anion. J Am Chem Soc 132(48):17174–17185
31. Li LF, Li J (2007) Link between oxidative stress and insulin resistance. Chin Med Sci J 22(4):254–259
32. Puddu P, Puddu G, Cravero E, De Pascalis S, Muscari A (2009) The emerging role of cardiovascular risk factor-induced mitochondrial dysfunction in atherogenesis. J Biomed Sci 16:112. https://doi.org/10.1186/1423-0127-16-112
33. Karbach S, Wenzel P, Waisman A, Münzel T, Daiber A (2014) eNOS uncoupling in cardiovascular diseases – the role of oxidative stress and inflammation. Curr Pharm Des 20(22):3579–3594
34. Ježek P, Dlasková A, Plecitá-Hlavatá L (2012) Redox homeostasis in pancreatic β cells. Oxidative Med Cell Longev 2012:932838. https://doi.org/10.1155/2012/932838
35. Bakhtiari A, Hajian-Tilaki K, Omidvar S, Nasiri Amiri F (2017) Association of lipid peroxidation and antioxidant status with metabolic syndrome in Iranian healthy elderly women. Biomed Rep 7(4):331–336
36. Kamath S, Lip GY (2003) Fibrinogen: biochemistry, epidemiology and determinants. QJM 96:711–729
37. Moya M, Scribano MP, Baez M (2017) Nuevo Rol para una vieja proteína: Fibrinógeno. Editorial Académica Española
38. Jiménez-Rosales A, Amaya-Chávez A, Domínguez García MV, Camarillo-Romero E, Huitrón Bravo GG, Cruz AM (2013) Association of inflammatory and oxidative stress biomarkers in subjects with cardiovascular risk. Am J Ther 20(4):422–431
39. Scribano MP, Baez MC, Becerra F, Tarán M, Signorini F, Balceda AG, Moya M (2014) Effects of atorvastatin on oxidative stress biomarkers and mitochondrial morphofunctionality in hyperfibrinogenemia-induced atherogenesis. Adv Med 2014. Article ID 947258:1–6
40. Garagiola ML, Tarán M, Scribano MP, Balceda A, García E, Fonseca I, Moya M, Baez MC (2016) Myeloperoxidase as an indicator of oxidative stress in metabolic syndrome. Argent J Cardiol 84(6):514–518
41. Aouacheri O, Saka S, Krim M, Messaadia A, Maidi I (2015) The investigation of the oxidative stress-related parameters in type 2 diabetes mellitus. Can J Diabetes 39(1):44–49
42. Wang Y-X (ed) (2017) Pulmonary vasculature redox signaling in health and disease, vol 967. Springer, Cham
43. Sena CM, Pereira AM, Seiça R (2013) Endothelial dysfunction – a major mediator of diabetic vascular disease. Biochim Biophys Acta 1832(12):2216–2231
44. Yubero-Serrano EM, Delgado-Lista J, Peña-Orihuela P, Perez-Martinez P, Fuentes F, Marin C et al (2013) Oxidative stress is associated with the number of components of metabolic syndrome: LIPGENE study. Exp Mol Med 45:e28. https://doi.org/10.1038/emm.2013.53
45. Mahendra JV, Kumar SD, Anuradha TS, Prashanth T, Nagaraj RS, Vishali V (2015) Plasma fibrinogen in type 2 diabetic patients with metabolic syndrome and its relation with ischemic heart disease (IHD) and retinopathy. J Clin Diagn Res 9(1):BC18–BC21
46. Chen YR, Zweier JL (2014) Cardiac mitochondria and reactive oxygen species generation. Circ Res 114(3):524–537
47. Giovanni B, Rodriguez R (2008) Ultrastructure of the mitochondrion and its bearing on function and bioenergetics. Antioxid Redox Signal 10(8):1313–1342
48. Chistiakov DA, Shkurat TP, Melnichenko AA, Grechko AV, Orekhov AN (2018) The role of mitochondrial dysfunction in cardiovascular disease: a brief review. Ann Med 50(2):121–127
49. Hulsmans M, Van Dooren E, Holvoet P (2012) Mitochondrial reactive oxygen species and risk of atherosclerosis. Curr Atheroscler Rep 14(3):264–276
50. Vichova T, Motovska Z (2013) Oxidative stress: predictive marker for coronary artery disease. Exp Clin Cardiol 18(2):e88–e91
51. Chong SJ, Low IC, Pervaiz S (2014) Mitochondrial ROS and involvement of Bcl-2 as a mitochondrial ROS regulator. Mitochondrion 19(Pt A):39–48

52. Takeuchi A, Kim B, Matsuoka S (2015) The destiny of Ca2+ released by mitochondria. J Physiol Sci 65(1):11–24
53. Radi R (2013) Protein tyrosine nitration: biochemical mechanisms and structural basis of functional effects. Acc Chem Res 46(2):550–559
54. Oliveira PJ (ed) (2018) Mitochondrial biology and experimental therapeutics. Springer Nature, Cham
55. Newton BW, Cologna SM, Moya C, Russell DH, Russell WK, Jayaraman A (2011) Proteomic analysis of 3T3-L1 adipocyte mitochondria during differentiation and enlargement. J Proteome Res 10(10):4692–4702
56. Sil R, Chakraborti AS (2016) Oxidative inactivation of liver mitochondria in high fructose diet-induced metabolic syndrome in rats: effect of glycyrrhizin treatment. Phytother Res 30(9):1503–1512
57. Shen GX (2012) Mitochondrial dysfunction, oxidative stress and diabetic cardiovascular disorders. Cardiovasc Hematol Disord Drug Targets 12(2):106–112
58. Faria A, Persaud SJ (2017) Cardiac oxidative stress in diabetes: mechanisms and therapeutic potential. Pharmacol Ther 172:50–62
59. Peinado JR, Diaz-Ruiz A, Frühbeck G, Malagon MM (2014) Mitochondria in metabolic disease: getting clues from proteomic studies. Proteomics 14(4–5):452–466
60. Zorov DB, Juhaszova M, Sollott SJ (2014) Mitochondrial reactive oxygen species (ROS) and ROS-induced ROS release. Physiol Rev 94(3):909–950
61. Yayan J (2013) Emerging families of biomarkers for coronary artery disease: inflammatory mediators. Vasc Health Risk Manag 9:435–456
62. Voudris KV, Chanin J, Feldman DN, Charitakis K (2015) Novel inflammatory biomarkers in coronary artery disease: potential therapeutic approaches. Curr Med Chem 22(22):2680–2689
63. Renna N, Vázquez M, González S, Lama C, Cruzado M, Miatello R (2007) Expresión vascular de factores de transcripción proinflamatorios en un modelo de Síndrome Metabólico. Rev Argent Cardiol 75:36–41
64. Evaluation and Treatment of High Blood Cholesterol in Adults (2001) Executive summary of the third report of the National Cholesterol Education Program (NCEP) Expert Panel on Detection, Evaluation and Treatment of High Blood Cholesterol in Adults (Adult Treatment Panel III). JAMA 285:2486–2497
65. National Diabetes Data Group (1979) Classification and diagnosis of diabetes and other categories of glucose intolerance. Diabetes 28:1039–1057
66. Wallace TM, Levy JC, Matthews DR (2004) Use and abuse of HOMA modeling. Diabetes Care 27:1487–1495
67. Nakagawa T, Hu H, Zharikov S, Tuttle KR, Short RA, Glushakova O, Ouyang X, Feig DI, Block ER, Herrera-Acosta J, Patel JM, Johnson RJ (2006) A causal role for uric acid in fructose-induced metabolic syndrome. Am J Phys Renal Phys 290(3):625–631
68. Stanhope KL (2012) Role of fructose-containing sugars in the epidemics of obesity and metabolic syndrome. Annu Rev Med 63:329–343
69. Bursać BN, Vasiljević AD, Nestorović NM, Veličković NA, Vojnović Milutinović DD, Matić GM, Djordjevic AD (2014) High-fructose diet leads to visceral adiposity and hypothalamic leptin resistance in male rats–do glucocorticoids play a role? J Nutr Biochem 25(4):446–455
70. Qu HQ, Li Q, Rentfro AR, Fisher-Hoch SP, McCormick JB (2011) The definition of insulin resistance using HOMA-IR for Americans of Mexican descent using machine learning. PLoS One 6(6):e21041
71. Simmons RK, Alberti KG, Gale EA, Colagiuri S, Tuomilehto J, Qiao Q, Ramachandran A, Tajima N, Brajkovich MI, Ben Nakhi A, Reaven G, Hama SB, Mendis S, Roglic G (2010) The metabolic syndrome: useful concept or clinical tool? Report of a WHO expert consultation. Diabetologia 53(4):600–605
72. Maury E, Brichard SM (2010) Adipokine dysregulation, adipose tissue inflammation and metabolic syndrome. Mol Cell Endocrinol 314(1):1–16
73. Thaler JP, Yi C-X, Schur EA et al (2012) Obesity is associated with hypothalamic injury in rodents and humans. J Clin Invest 122(1):153–162

74. Caimi G, Lo Presti R, Montana M, Noto D, Canino B, Averna MR, Hopps E (2014) Lipid peroxidation, nitric oxide metabolites, and their ratio in a group of subjects with metabolic syndrome. Oxidative Med Cell Longev 824756:2014
75. Rosenson RS, Brewer HB Jr, Davidson WS, Fayad ZA, Fuster V (2012) Cholesterol efflux and atheroprotection: advancing the concept of reverse cholesterol transport. Circulation 125(15):1905–1919
76. Gomaraschi M, Ossoli A, Pozzi S, Nilsson P, Cefalù A, Averna M, Kuivenhoven J, Hovingh G, Veglia F, Franceschini G, Calabresi L (2014) 133 eNOS activation by HDL is impaired in genetic CETP deficiency. PLoS One 9(5):e95925
77. Rodriguez AD (2014) The fixed combination of pravastatin and fenofibrate: what can it provide? Clin Investig Arterioscler 26(1):12–16
78. Besler C, Heinrich K, Rohrer L, Doerries C, Riwanto M et al (2011) Mechanisms underlying adverse effects of HDL on eNOS-activating pathways in patients with coronary artery disease. J Clin Invest 121(7):2693–2708
79. Gomaraschi M et al (2014) eNOS activation by HDL is impaired in genetic CETP deficiency. PLoS One 9(5):e95925
80. Gaggini M, Morelli M, Buzzigoli E, DeFronzo RA, Bugianesi E, Gastaldelli A (2013) Non-alcoholic fatty liver disease (NAFLD) and its connection with insulin resistance, dyslipidemia, atherosclerosis and coronary heart disease. Nutrients 5(5):1544–1560
81. Bhaswant M, Poudyal H, Mathai ML, Ward LC, Mouatt P, Brown L (2015) Green and black cardamom in a diet-induced rat model of metabolic syndrome. Nutrients 7(9):7691–7707
82. Ozturk ZA, Kadayifci A (2014) Insulin sensitizers for the treatment of non-alcoholic fatty liver disease. World J Hepatol 6(4):199–206
83. Abd El-Kader SM, El-Den Ashmawy EM (2015) Non-alcoholic fatty liver disease: the diagnosis and management. World J Hepatol 7(6):846–858
84. Ratnoff OD, Menzie AC (1957) A new method for the determination of fibrinogen in small samples of plasma. J Lab Clin Med 37:316–320
85. Moncada S, Palmer RM, Higgs EA (1989) The biological significance of nitric oxide formation from L-arginine. Biochem Soc Trans 17(4):642–644
86. Woolliams JA, Wiener G, Anderson PH, McMurray CH (1983) Variation in the activities of glutathione peroxidase and superoxide dismutase and in the concentration of copper in the blood in various breed crosses of sheep. Res Vet Sci 34:253–256
87. Vanhoutte PM, Shimokawa H, Feletou M, Tang EH (2017) Endothelial dysfunction and vascular disease – a 30th anniversary update. Acta Physiol (Oxford) 219(1):22–96
88. Sabater-Lleal M, Huang J, Chasman D, Naitza S, Dehghan A, Johnson AD (2013) Multiethnic meta-analysis of genome-wide association studies in >100 000 subjects identifies 23fibrinogenassociated loci but no strong evidence of a causal association between circulating fibrinogen and cardiovascular disease. Circulation 128(12):1310–1324
89. El Assar M, Ruiz de Adana JC, Angulo J, Pindado Martínez ML, Hernández MA, Rodríguez-Mañas L (2013) Preserved endothelial function in human obesity in the absence of insulin resistance. J Transl Med 11:263. https://doi.org/10.1186/1479-5876-11-263
90. Feoli AM, Macagnan FE, Piovesan CH, Bodanese LC, Siqueira IR (2014) Xanthine oxidase activity is associated with risk factors for cardiovascular disease and inflammatory and oxidative status markers in metabolic syndrome: effects of a single exercise session. Oxidative Med Cell Longev 2014:1–8. 587083. https://doi.org/10.1155/2014/587083
91. Smith BW, Adams LA (2011) Non-alcoholic fatty liver disease. Crit Rev Clin Lab Sci 48(3):97–113
92. Manco M (2011) Metabolic syndrome in childhood from impaired carbohydrate metabolism to nonalcoholic fatty liver disease. J Am Coll Nutr 30(5):295–303
93. Castro GS, Cardoso JF, Vannucchi H, Zucoloto S, Jordão AA (2011) Fructose and NAFLD: metabolic implications and models of induction in rats. Acta Cir Bras 26(2):45–50
94. Tarán Mariana D, Baez María C, de La Paz Scribamo Parada M, Balceda A, Sergio B, Miriam B, Mónica M (2018) Experimental model of oxidative stress markers in subclinical atherogenesis associated with metabolic syndrome. SM J Cardiol Cardiovasc Dis 4(1):1021

95. Cao YX, Li LP (2014) Relationship of non-alcoholic steatohepatitis with arterial endothelial function and atherosclerosis. Zhonghua Gan Zang Bing Za Zhi 22(3):205–208
96. Van der Poorten D, Samer CF, Ramezani-Moghadam M, Coulter S, Kacevska M, Schrijnders D, Wu LE, McLeod D, Bugianesi E, Komuta M et al (2013) Hepatic fat loss in advanced non-alcoholic steatohepatitis: are alterations in serum adiponectin the cause? Hepatology (Baltim) 57:2180–2188
97. Weiß J, Rau M, Geier A (2014) Non-alcoholic fatty liver disease: epidemiology, clinical course, investigation, and treatment. Dtsch Arztebl Int 111(26):447–452
98. Nassir F, Ibdah JA (2014) Role of mitochondria in nonalcoholic fatty liver disease. Int J Mol Sci 15(5):8713–8742
99. Wong R, Aijaz Ahmed A (2014) Obesity and non-alcoholic fatty liver disease: disparate associations among Asian populations. World J Hepatol 6(5):263–273
100. Tilg H, Moschen AR (2010) Evolution of inflammation in nonalcoholic fatty liver disease: the multiple parallel hits hypothesis. Hepatology 52(5):1836–1846
101. Torre-Villalvazo I, Bunt AE, Alemán G, Marquez-Mota CC, Diaz-Villaseñor A, Noriega LG, Estrada I, Figueroa-Juárez E et al (2018) Adiponectin synthesis and secretion by subcutaneous adipose tissue is impaired during obesity by endoplasmic reticulum stress. J Cell Biochem 119(7):5970–5984
102. Bellafante E, Murzilli S, Salvatore L, Latorre D, Villani G, Moschetta A (2013) Hepatic-specific activation of peroxisome proliferator-activated receptor γ coactivator-1β protects against steatohepatitis. Hepatology 57(4):1343–1356
103. Silva AKS, Peixoto CA (2018) Role of peroxisome proliferator-activated receptors in non-alcoholic fatty liver disease inflammation. Cell Mol Life Sci 75:2951–2961. https://doi.org/10.1007/s00018-018-2838-4. [Epub ahead of print]
104. Novo S, Peritore A, Guarneri FP, Corrado E, Macaione F, Evola S, Novo G (2012) Metabolic syndrome (MetS) predicts cardio and cerebrovascular events in a twenty years follow-up. A prospective study. Atherosclerosis 223(2):468–472
105. Bernabé García J, Zafrilla Rentero P, Mulero Cánovas J, Gómez Jara P, Leal Hernández M, Abellán Alemán J (2014) Biochemical and nutritional markers and antioxidant activity in metabolic syndrome. Endocrinol Nutr 61(6):302–308
106. Chistiakov DA, Sobenin IA, Bobryshev YV, Orekhov AN (2012) Mitochondrial dysfunction and mitochondrial DNA mutations in atherosclerotic complications in diabetes. World J Cardiol 4(5):148–156
107. Rocha M, Apostolova N, Herance JR, Rovira-Llopis S, Hernandez-Mijares A, Victor VM (2014) Perspectives and potential applications of mitochondria-targeted antioxidants in cardiometabolic diseases and type 2 diabetes. Med Res Rev 34(1):160–189
108. Lim S, Rashid MA, Jang M, Kim Y, Won H, Lee J, Woo JT, Kim YS, Murphy MP, Ali L, Ha J, Kim SS (2011) Mitochondria-targeted antioxidants protect pancreatic β-cells against oxidative stress and improve insulin secretion in glucotoxicity and glucolipotoxicity. Cell Physiol Biochem 28(5):873–886
109. Han D, Canali R, Rettori D, Kaplowitz N (2003) Effect of glutathione depletion on sites and topology of superoxide and hydrogen peroxide production in mitochondria. Mol Pharmacol 64(5):1136–1144
110. Parra V, Verdejo H, del Campo A, Pennanen C, Kuzmicic J, Iglewski M et al (2011) The complex interplay between mitochondrial dynamics and cardiac metabolism. J Bioenerg Biomembr 43:47–51
111. Marín-García J, Goldenthal MJ (2002) La mitocondria y el corazón. Rev Esp Cardiol 55(12):1293–1310
112. Kuzmicic J, Del Campo A, López-Crisosto C, Morales PE, Pennanen C, Bravo-Sagua R et al (2011) Mitochondrial dynamics: a potential new therapeutic target for heart failure. Rev Esp Cardiol (English Edition) 64(10):916–923
113. Hirokawa N, Takemura R (2005) Molecular motors and mechanisms of directional transport in neurons. Nat Rev Neurosci 6:201–214

114. Hollenbeck PJ, Saxton WM (2005) The axonal transport of mitochondria. J Cell Sci 118(23):5411–5419
115. Germain M, Mathai JP, McBride HM, Shore GC (2005) Endoplasmic reticulum BIK initiates DRP1-regulated remodelling of mitochondrial cristae during apoptosis. EMBO J 24(8):1546–1556
116. Seppet E et al (2009) Mitochondria and energetic depression in cell pathophysiology. Int J Mol Sci 10(5):2252–2303
117. Stowe DF, Camara AK (2009) Mitochondrial reactive oxygen species production in excitable cells: modulators of mitochondrial and cell function. Antioxid Redox Signal 11(6):1373–1414
118. Quinlan CL, Perevoshchikova IV, Hey-Mogensen M, Orr AL, Branda MD (2013) Sites of reactive oxygen species generation by mitochondria oxidizing different substrates. Redox Biol 1(1):304–312
119. Quinlan CL, Orr AL, Perevoshchikova IV, Treberg JR, Ackrell BA, Brand MD (2012) Mitochondrial complex II can generate reactive oxygen species at high rates in both the forward and reverse reactions. J Biol Chem 287(32):27255–27264
120. Fisher-Wellman KH, Neufer PD (2012) Linking mitochondrial bioenergetics to insulin resistance via redox biology. Trends Endocrinol Metab 23(3):142–153
121. Koju N et al (2019) Pharmacological strategies to lower crosstalk between nicotinamide adenine dinucleotide phosphate (NADPH) oxidase and mitochondria. Biomed Pharmacother 111:1478–1498
122. Yan LJ (2014) Pathogenesis of chronic hyperglycemia: from reductive stress to oxidative stress. J Diabetes Res 2014:1–11. 137919. https://doi.org/10.1155/2014/137919
123. Rector RS, Uptergrove GM, Borengasser SJ, Mikus CR, Morris EM, Naples SP, Laye MJ, Laughlin MH, Booth FW, Ibdah JA, Thyfault JP (2010) Changes in skeletal muscle mitochondria in response to the development of type 2 diabetes or prevention by daily wheel running in hyperphagic OLETF rats. Am J Physiol Endocrinol Metab 298(6):1179–1187
124. Miller MW, Knaub LA, Olivera-Fragoso LF, Keller AC, Balasubramaniam V, Watson PA, Reusch JE (2013) Nitric oxide regulates vascular adaptive mitochondrial dynamics. Am J Physiol Heart Circ Physiol 304(12):H1624–H1633
125. Madamanchi NR, Runge MS (2007) Mitochondrial dysfunction in atherosclerosis. Circ Res 100(4):460–473

Oxidative Stress as a Critical Determinant of Adult Cardiac Progenitor Cell-Fate Decisions

13

Diego Herrero, Susana Cañón, Guillermo Albericio, Susana Aguilar, Rosa María Carmona, Adrián Holguín, and Antonio Bernad

Abstract

Tissue homeostasis and the response to injury require a tight regulation of the balance between self-renewal and differentiation of adult stem/progenitor cells. Recent evidence obtained in several tissues suggests that this balance is regulated, at least in part, by the cellular redox status via the control of reactive oxygen species (ROS) levels and cellular metabolism. In this chapter, we consider the main sources and the relevance of oxidative stress in adult stem turnover and the key signaling pathways involved, with a particular focus on cardiac progenitor cell turnover. While it is generally accepted that the mammalian heart has high physiological levels of ROS and an oxidative metabolism, few studies have explored the importance of redox signaling in cardiac progenitor cells. We propose that low-ROS areas in the heart are permissive niches for adult cardiac progenitor cells. Accordingly, manipulation of ROS-related signaling pathways in the adult heart might open new horizons for stem cell therapy by enhancing their heretofore limited cardiac regenerative potential.

Keywords

Cardiac progenitor cell · CPC · ROS · Bmi1 · Turnover · Cardiomyocyte · Low-ROS niche

D. Herrero · S. Cañón · G. Albericio · S. Aguilar · R. M. Carmona · A. Holguín
A. Bernad (✉)
Department of Immunology and Oncology, Spanish National Center for Biotechnology (CNB-CSIC), Universidad Autónoma de Madrid, Madrid, Spain
e-mail: abernad@cnb.csic.es

S. Chakraborti et al. (eds.), *Modulation of Oxidative Stress in Heart Disease*,
https://doi.org/10.1007/978-981-13-8946-7_13

13.1 Introduction

Reactive oxygen species (ROS) are a heterogeneous group of highly reactive molecules that derive from the partial reduction of molecular oxygen. Intracellular ROS are mainly observed as hydrogen peroxide (H_2O_2), hydroxyl free radicals (HO·), and superoxide anion (O_2^-) (reviewed in [1]). Originally considered as exclusively harmful subproducts of oxidative respiration generated as a direct consequence of metabolic cellular processes, ROS are increasingly recognized as critical contributors to the regulation of cellular dynamics [2, 3].

When produced in low levels, ROS are involved in the regulation of several signaling pathways that control cell-fate determination, proliferation, differentiation, senescence, and programmed cell death [4–7]. By contrast, excessive ROS formation, which occurs following the deregulation of ROS balance as a consequence of transient or constitutive oxidative stress situations, can promote an appreciable alteration of the cellular dynamics and cause oxidative damage to DNA, lipids, and proteins [8].

The effects of ROS on cell biology processes are especially important in the regulation of adult stem cell compartments, as the cellular pathways they modulate are directly involved in the regulation of stem cell self-renewal and differentiation [9]. Whereas low levels of ROS are required to maintain a controlled proliferation rate, which is necessary for the long-term maintenance of stem cells [10], disproportionate ROS levels induce overproliferation and differentiation, with the consequent reduction of life span and premature exhaustion of the stem cell pool [11, 12]. In this review, we summarize our current understanding of redox regulation in adult stem cell biology, with a particular focus on cardiac progenitor cells (CPC). A better understanding of redox-regulated mechanisms might foster more rational therapeutic approaches that directly target adult stems cells and their niches.

13.2 Maintenance of Intracellular Redox Balance

Given the above, it is not surprising that stem cells are equipped with an efficient antioxidant system to exert a tight control over the ROS balance [13, 14]. Accordingly, a better understanding of the mechanisms involved in the generation and degradation of ROS might yield valuable insights into how stem cells regulate these cellular processes (Fig. 13.1).

13.2.1 Sources of ROS Production

In homeostasis, ROS are produced at stable levels and are tightly controlled to prevent premature cell differentiation [15, 16]. Mitochondria are considered the main endogenous producers of ROS, which are generated through leakage of electrons during normal oxidative phosphorylation (reviewed in [17]). However, ROS can also be formed in other cellular compartments, including peroxisomes (Fig. 13.1).

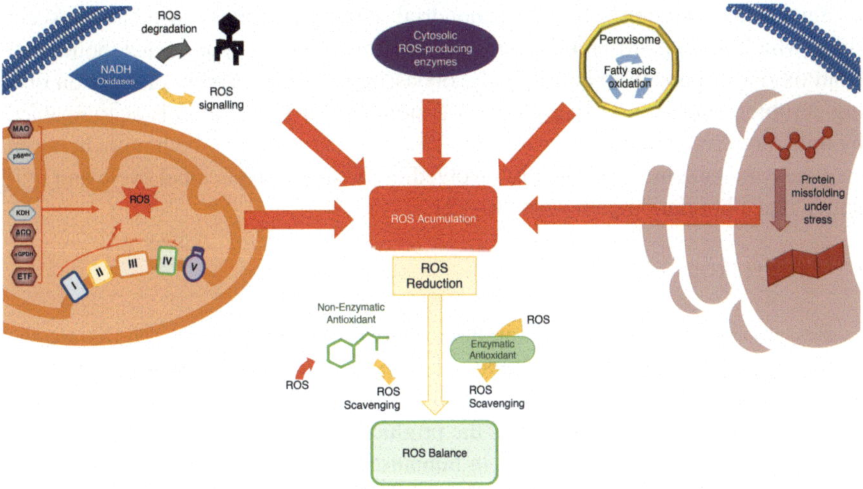

Fig. 13.1 Integration of main cellular ROSgenic and antioxidant systems, working coordinately to maintain redox homeostasis in adult stem cells

- Mitochondria.

Under normal conditions, up to 2% of the oxygen consumed by mitochondria is converted into ROS, with variable levels depending on cell type, environment, and the metabolic activity of the organism/tissue [18, 19]. The main intracellular source of ROS originates from the mitochondrial electron transport chain, which channels electrons across different respiratory protein complexes to ultimately generate a proton gradient used for ATP formation at complex V [20]. ROS are formed when this electron flow is inadvertently interrupted, aborting the complete reduction of O_2 to H_2O (reviewed in [17]).

In addition to respiration, other reactions inside mitochondria can lead to ROS production, although their contribution to total ROS generation is small and depends on the specific tissue and the oxidative stress condition. Reactions catalyzed by α-glycerophosphate dehydrogenase and flavoproteins, which transfer electrons to coenzyme Q [21, 22], and aconitase, which participates in the Krebs cycle by transforming citrate to isocitrate [23], can lead to ROS production. Furthermore, monoamine oxidases, which oxidize different amines, are able to generate H_2O_2 [24]. Some proteins can contribute to ROS formation under conditions of oxygen or nutrient deprivation. For instance, when oxidized, the adaptor protein p66^{shc} translocates from the cytosol to the mitochondrial intermembrane space and generates H_2O_2 [25]. Interestingly, p66^{shc} is critically involved in the hypoxia survival response and self-renewal regulation in normal [26] and transformed adult stem cells [27]. Similarly, α-ketoglutarate dehydrogenase has been proposed as a source of ROS under conditions of low intracellular nicotinamide adenine dinucleotide (NAD) levels [28].

Stem cells maintain a low mitochondrial activity and mass in their quiescent state, which inhibits differentiation [29]. Moreover, they execute a metabolic shift from oxidative phosphorylation to glycolysis, which reduces ROS production during aerobic respiration [13]. By contrast, stem cell differentiation is accompanied by mitochondrial remodeling, which increases the expression of both mitochondrial enzymes and enzymes specifically involved in the tricarboxylic acid cycle and oxidative phosphorylation pathways [30–32].

- NADPH Oxidases.

Beyond the mitochondrial respiratory chain, NADPH oxidases (NOX) are the major alternative intracellular sources of ROS within cells. NOX are a family of transmembrane proteins that reduce O_2 to O_2^- or H_2O_2 by accepting electrons from NADPH and whose sole function is the production of ROS. Seven different members of the NOX family are known in humans: NOX1, NOX2, NOX3, and NOX5 produce O_2^-, whereas NOX4, DUOX1, and DIOX2 generate H_2O_2. The O_2^- generated by NOX is usually reduced through a dismutation reaction to generate H_2O_2, the main ROS signaling species (reviewed in [33]).

NOX-generated ROS are especially important in the regulation of stem cell differentiation and proliferation. For example, in cardiac stem cells, they regulate differentiation to the three main cardiac lineages [34, 35], and in neural stem cells, they maintain self-renewal capacity [36]. In addition, it has been described that ROS produced by NOX are relevant in the degradation of pathogenic molecules by oxidation (reviewed in [33]).

- Other ROS-Producing Organelles.

Peroxisomes and the endoplasmic reticulum also generate ROS at different rates depending on the cell type (Fig. 13.1). Peroxisomes produce ROS during long-chain fatty acid metabolism, the preferred metabolic pathway in the heart [37], and the endoplasmic reticulum produces significant amounts of ROS under stress, which are derived from the unfolded protein response oxidative machinery [38].

- Cytosolic Enzymes.

Several proteins in the cytosol that are charged with the metabolism of toxic compounds represent another appreciable source of ROS. Chief among these are the cytochrome p450 enzymes, which belong to a family of monooxygenases specialized in the degradation of xenobiotics such as alcohol and function to reduce the levels of toxic compounds through the consumption of NADPH and the subsequent generation of ROS (reviewed in [39]). Other ROS-producing enzymes in the cytoplasm include lipoxygenases and xanthine oxidases, which catalyze the dioxygenation of polyunsaturated fatty acids in lipids containing a cis,cis-1,4- pentadiene and the oxidation of hypoxanthine to xanthine, respectively [40, 41].

13.2.2 ROS Scavenging Systems

Cells have developed a diverse range of antioxidant mechanisms to transform ROS into less reactive molecules [42]. These scavenging mechanisms are particularly important for stem cells as their cell biology is strongly influenced by ROS [43]. ROS scavenging can be performed directly by specialized antioxidant enzymes or, alternatively, through spontaneous reactions involving nonenzymatic antioxidants (Fig. 13.1) [42].

- Antioxidant Enzymes.

The main intracellular processing of ROS is performed by a variety of antioxidant enzymes specialized in the neutralization of ROS. Among them, superoxide dismutase (SOD), glutathione peroxidase (GPx), catalase, peroxiredoxins, and the thioredoxin system are considered the more relevant [42].

SOD are metalloproteins that catalyze the dismutation of O_2^- into O_2 and H_2O_2. As the first defensive barrier against ROS, SOD maintain ROS levels inside mitochondria, thereby preventing the activation of proapoptotic pathways by cytochrome C release [44]. Moreover, SOD activity regulates essential cellular processes through H_2O_2 production, the main intracellular ROS second messenger [45]. Catalase and GPx act coordinately to decompose H_2O_2 into O_2 and H_2O, which can abolish downstream signaling. Catalases and peroxiredoxins directly interact with H_2O_2, whereas GPx uses glutathione as a reductant. The thioredoxin system is essential for maintaining the functionality of antioxidant enzymes, as ROS scavenging by GPx or peroxiredoxin results in their oxidation, which can be reversed through thioredoxin-driven reduction. Oxidized thioredoxin can then be reduced by thioredoxin reductase, which maintains a stable antioxidant system [42]. The expression of these enzymes is significantly upregulated in several types of stem cells protecting them from harmful ROS increases [46–48].

- Nonenzymatic Antioxidants.

Nonenzymatic antioxidants, which can be produced by endogenous metabolism or through the exogenous intake of foods or supplements, are also ROS detoxifiers and include vitamins, carotenoids, flavonoids, or different minerals [42]. Indeed, several metabolites and other molecules have reductant capacity to generate less oxidative molecules.

13.3 ROS Protein Regulation in Stem/Progenitor Cells

Stem and progenitor cell niches [49] comprise cellular components (stem/progenitor cells, fibroblasts, endothelial cells of blood vessels, etc.) and noncellular factors, such as extracellular matrix and metabolic factors, and also biophysical signals including temperature, shear forces cell-cell contact signaling, and O_2 tension. The

Table 13.1 Main ROS regulatory proteins in adult stem cells

ROS regulatory proteins	Low-ROS levels: redox homeostasis	High ROS levels: oxidative stress	Refs.
HIF1-α	Glycolytic metabolism	Increased cell sensitivity to stress	[64–74]
	Self-renewal, pluripotency, and maintenance		
	Quiescence and low proliferation	Increased apoptosis	
	Migration and homing		
	Survival		
	Blockage of differentiation and cell death		
FoxO	Glycolytic metabolism	Cell cycle arrest	[79–84]
	Self-renewal, pluripotency, and maintenance	Loss of quiescence and self-renewal	
	Quiescence and low proliferation	Terminal differentiation	
	Survival	Increased apoptosis	
	Blockage of differentiation and cell death		
Nrf2	Cell death protection	Enhanced cell cycle entry	[88–93]
	Self-renewal and pluripotency		
	Low proliferation	Promoted differentiation	
	Blockage of differentiation and cell death		
Bmi1	Self-renewal and pluripotency	Enhanced cell differentiation programs	[96–102]
	Low proliferation		
	Blockage of differentiation	Loss of self-renewal and stemness	
P53	**Modulated P53 and P38 levels allow:**	**Damage-activated P53 and P38 trigger:**	[103–107]
	Proliferation and survival	Senescence and altered proliferation	
	Self-renewal and multipotency	Loss of quiescence and self-renewal	
P38	Quiescence and maintenance	Enhanced terminal differentiation	
ATM	Cell cycle delay and quiescence	Altered proliferation and self-renewal	[104]
	Self-renewal and maintenance	Loss of quiescence and maintenance	
APE1/ref-1	Low proliferation	Altered proliferation	[108]
	Cell death protection and survival		
	Self-renewal	Altered self-renewal, quiescence, and maintenance	
	Preserved differentiation potential		

integration of all these components regulates the maintenance, self-renewal, function, multipotency, and differentiation capacities of stem cells. Early studies revealed that stem cell niches were more hypoxic than initially expected and that O_2 was a key regulator of stem cell biology (reviewed in [50]). Accordingly, low O_2 levels favored the survival of hematopoietic stem cells (HSC), neural stem cells (NSC) and human embryonic stem cells (hES), and also some tumor cells [51–53].

The advantages of hypoxic niches reside in the protection of stem cells from detrimental oxidative stress through reducing the source of ROS itself (O_2) and the fine regulation of ROS production. Redox balance is known to regulate both embryonic and adult stem cell functions [54], and thus stem cell niches have developed a protective strategy to maintain low-ROS production principally through low mitochondrial biogenesis and/or through expression and activation of ROS scavenging proteins (Fig. 13.1). Several antioxidant proteins can regulate ROS levels in stem cells by inducing hypoxia-inducible factors (HIF), FoxO transcription factors, nuclear factor erythroid 2-related factor 2 (Nrf-2), and polycomb family member B lymphoma Mo-MLV insertion region 1 homolog (Bmi1). DNA repair proteins, such as ataxia-telangiectasia mutated (ATM), the apyrimidinic endonuclease1/redox factor-1 (APE1/Ref-1), and p53, are also involved in ROS responses [55]. Below, we summarize some important properties of these factors in adult stem cells (Table 13.1).

13.3.1 Hypoxia-Inducible Factors (HIF)

The HIF family of transcription factors are heterodimeric proteins composed of an oxygen-responsive alpha subunit (1α, 2α, or 3α) and a constitutively expressed beta (β) subunit. HIF proteins are highly regulated transcriptional factors that activate downstream hypoxia signaling in response to O_2 reduction. Hypoxic stem cell niches maintain stem cells at low proliferative rates preventing their exhaustion or senescence. Because HIF were reported to trigger cell cycle arrest in several cell types [56–58], it was tempting to assume that they could play similar roles in the hypoxic stem cell niches. Indeed HIF regulate several stem cell functions including cell fate, proliferation, survival, and pluripotency [59–62].

Cellular adaptation to hypoxia includes a metabolic shift from oxidative phosphorylation to glycolysis, accompanied by the upregulation of angiogenesis and alterations in cell cycle and apoptosis. HIF-1α activation mediates these shifts and governs stem cell maintenance and proliferation [63]. HIF proteins are also known to control pluripotency and differentiation through direct interactions with downstream mediators including NOTCH, OCT4, or WNT, among others [64–67]. Adult human and mouse HSC express high levels of HIF-1α, which regulates their quiescence, maintenance, and expansion. In this context, HIF-1α must also be tightly regulated. Inhibition of HIF-1α function negatively affects hematopoiesis, and HSC become more stress sensitive with increased apoptosis, leading ultimately to embryonic lethality, whereas excess levels of HIF-1α destabilize HSC function, causing premature cell exhaustion [68]. Low O_2 levels also regulate NSC and mesenchymal stem cell (MSC) niches. Thus, HIF-1α regulates NSC maintenance, proliferation, differentiation, and maturation in the adult brain [69–71], and hypoxia influences adult MSC in a similar manner with regard to cell fate or stemness. For example, when cultured in low O_2, MSC show increased genetic stability and decreased differentiation potential [72, 73]. In ES cultures, the synergy between hypoxia and ROS can promote the expression of differentiated cardiogenic markers through HIF-1α regulation [74].

13.3.2 FOXO Subfamily of Transcription Factors

The FOXO transcription factor family is represented by four isoforms in mammals (FOXO1, FOXO3, FOXO4, and FOXO6), and they regulate diverse cellular functions including cell cycle arrest, apoptosis, glucose metabolism, DNA damage repair, and stress resistance [75]. FOXO factors limit the expansion of stem/progenitor cell pools in several tissues and are critical regulators of oxidative stress through the fine-tuning of ROS to maintain self-renewal, quiescence, and multipotency [76–78]. Under damage conditions such as oxidative stress and ROS accumulation, FOXO factors enter the nucleus where they activate proapoptotic signals and downstream antioxidant targets that mediate cell cycle arrest and dormancy. By contrast, inactivation of FOXO in HSC by PI3K/AKT-mediated phosphorylation results in its exclusion from the nucleus, leading to loss of quiescence and also terminal differentiation [26]. In HSC and NSC populations, FOXO activation (FOXO3 or FOXO1/3/4, respectively) promotes the loss of self-renewal and also transient cell cycle entry, which ultimately triggers apoptosis and cell exhaustion [79–81]. In MSC, oxidative stress-mediated FOXO activation induces cell cycle arrest, quiescence, antioxidant gene activation, and loss of differentiation of osteogenic or adipogenic lineages [78, 82, 83]. Similarly, in hES, FOXO1 has been related to pluripotency maintenance through interaction with OCT4 and NANOG and, in oxidative conditions, can induce cell cycle arrest and apoptosis [84].

13.3.3 Nuclear Factor (Erythroid-Derived 2)-Like 2 (NRF2)

Cells express NRF2 ubiquitously. In homeostasis, NRF2 forms a complex with KEAP1 protein, which regulates NRF2 by tethering it in the cytoplasm and facilitating ubiquitination, leading to its proteosomal degradation. Damage-induced ROS increases modify cysteine residues on KEAP1, and the resulting conformational change releases NRF2, which enters the nucleus and activates genes encoding antioxidant proteins and detoxification enzymes, protecting cells from redox disturbance [85–87]. NRF2 is an essential regulator in HSC, and, in human NSC, it can protect against oxidative-induced cell death [88]. Mice deficient for *Nrf2* have an expanded HSC pool, suggesting a regulatory role for NRF2 in HSC migration and retention in the niche [89]. In response to oxidative damage, stabilized NRF2 enhances HSC cell cycle entry and progenitor contribution during hematopoietic regeneration [90] and also controls both pluripotency potential and self-renewal capacities in hES [91]. Interestingly, enforced NRF2 expression in MSC in vitro mimicked the beneficial effects of the hypoxic MSC niche, protecting MSC from oxidative stress and enhancing self-renewal while inhibiting differentiation potential [92, 93].

13.3.4 Bmi1

BMI1 is member of polycomb repressive complex 1 (PRC1), which acts as an epigenetic repressor of a large number of target genes [94]. The polycomb group of proteins (PcG) and, in particular, BMI1 is implicated in the maintenance of self-renewal capacity in several adult stem cell populations [95–99]. *Bmi1*-deficient mice present mitochondrial dysfunction and elevated levels of ROS, which suggested a relevant function of BMI1 and PcG as regulators of mitochondrial function and redox balance [100]. Indeed, BMI1 confers protection from oxidative stress and premature senescence and maintains the multipotency and proliferation capacities of both HSC and NSC populations. Accordingly, HSC and NSC deficient in *Bmi1* showed impaired self-renewal and proliferation arrest that culminated in premature aging and postnatal cell depletion [96, 97]. The impact of ROS regulation in cardiac stem cell functionality, however, has been less explored. We recently identified a multipotent and self-renewal population of cardiac progenitors ($Bmi1^{+}$CPC) expressing high levels of Bmi1 [99, 101], which presented low ROS [9]. In homeostatic $Bmi1^{+}$CPC, BMI1 repressed cell-fate genes, including the cardiogenic differentiation program. Oxidative stress nonetheless modified BMI1 activity in vivo, derepressing canonical target genes in favor of Bmi1 antioxidant and anticlastogenic functions. This redox-mediated regulatory mechanism is not restricted to damage situations; ROS-associated differentiation of cardiac progenitors in steady state was also apparent, albeit at lower levels. These findings illustrate a ROS-dependent BMI1 role in adult cardiac progenitor cell-fate decisions [102] (see also Sect. 13.5.2).

13.3.5 DNA Damage Response

There is a wealth of evidence showing the critical association between ROS control and the DNA damage repair machinery in healthy homeostasis. High levels of ROS in MSC lead to the activation of p38 MAPK, which in turn activates the cell cycle modulator $p16^{INK4A}$, thus promoting the exit from quiescence, reducing self-renewal, and favoring senescence [103].

ATM is a protein kinase with an important role in activating the damage response pathway following DNA injury, leading to cell cycle delay or apoptosis. ATM is also a redox sensor and can be activated by elevated ROS levels. Accordingly, when ATM is dysfunctional in HSC, ROS production is deregulated, provoking exhaustion of the stem cell compartment [104].

The central tumor suppressor gene, *p53*, has been well characterized in adult stem cells. p53 can be stabilized by hypoxia, and its transcriptional activity can be modulated in a redox-dependent fashion. Specifically, under steady-state (or low stress) conditions, p53 promotes stem cell quiescence [105, 106]. Basal p53 expression is necessary to maintain MSC multipotency, whereas oxygen-induced increases in p53 expression modulate cell-fate and survival decisions [107].

Finally, the multifunctional protein complex APE1/REF-1 has been implicated in redox signaling in stem cells and also in cardiovascular cells. Independent studies have identified APE1, the main AP-endonuclease of the base excision repair pathway, as a factor able to stimulate the binding activity of several transcription factors, suggesting a redox-dependent chaperone activity for APE1 in the regulation of stem cell renewal (reviewed in [108]).

13.4 ROS and Cardiovascular Disease

To satisfy its exceptionally high-energy demands, the heart relies largely on fatty acid oxidation, which drives oxidative phosphorylation in mitochondria. It is estimated that each day, the human heart produces ~6 kg of ATP to sustain optimal function [109]. Fatty acid oxidation has been associated with high rates of mitochondrial ROS production that can impair mitochondrial and cellular functions. Under normal physiological conditions, however, moderate ROS levels are important signaling mediators in the cardiovascular system.

13.4.1 The Physiologically Delicate Equilibrium

A fine balance between ROS production and antioxidant activity is essential in the heart; otherwise, there would be a high risk to evolve toward cardiac hypertrophy, arrhythmia, myocardial ischemia/reperfusion damage, and/or heart failure [110].

Cardiovascular disease (CVD) is a major cause of mortality in the developed world, and, consequently, it has been extensively studied over the past decade. However, its multifactorial nature has hindered the full elucidation of the underlying pathogenic mechanisms. ROS play a pivotal role in the progression of CVD; in particular, ROS are key participants in endothelial dysfunction and atherosclerosis, which are prominent in CVD [111], and aberrant redox regulation is known to be associated with CVD progression [112]. In addition, it has been recently established that ROS-mediated regulation during post-myocardial infarction repair and remodeling could be affected by the distortion of circadian mechanisms, likely affecting brain functions [118].

NRF2 has emerged as a transcriptional factor critically involved in redox-sensitive signaling and promotion of the antioxidant response in the heart [112] through the regulation of expression of several endogenous antioxidants and detoxification enzymes (see also Sect. 13.3.3). Loss or deregulation of *Nrf2* in mice is linked to various manifestations of CVD [113]; for example, NRF2 function is involved in the control of hypertension, the protection against cardiac hypertrophy, and the manifestation of several comorbidities that course with CVD, including diabetes, all of which are associated with the prolonged exposure to increased ROS levels. *Nrf2* has also been implicated in ischemia-reperfusion injury, correlating with the cardioprotective effects associated with ischemic preconditioning [114]. Finally, deregulation of several endogenous sources of ROS is related with heart senescence and aging, with a potential role for APE1 (see Sect. 13.3.5) in cardiovascular pathophysiology [108].

13.4.2 Chronic Cardiac Oxidative Stress Derived from Exogenous Sources

As mentioned earlier, the main sources of ROS are endogenous, but in certain situations, these can be supplemented through acute or sustained external interventions. Breast radiation therapy has become a critical component in managing patients who receive breast-conserving surgery or have certain high-risk features after mastectomy [115]. Unfortunately, radiotherapy to the breast and chest can be associated with radiation-related morbidity and mortality that may offset some of its benefits. Moreover, because of increased life expectancy, late-appearing adverse effects in cancer patients are becoming more common.

In the context of CVD, radiation directly provokes DNA damage and ROS generation and also activates TGF-β1 and pro-inflammatory signaling, promoting myofibroblast accumulation and extracellular matrix production, thrombin generation, and platelet activation. This pathological process leads to pericardial disease, myocardial fibrosis, coronary artery disease, valvular lesions, and cardiac conduction system injury, which are collectively considered as radiation-induced heart disease (RIHD) (reviewed in [116]).

Although the inflammatory response pathway is very likely the predominant mediator of pro-fibrotic effects, other factors also contribute significantly, such as the establishment of chronic ROS production. Oxidative stress, in turn, increases the levels of inflammatory mediators, proteases, and adhesion molecules and decreases the levels of nitric oxide. Ultimately, these long-term stressful conditions promote epigenetic modifications in heart tissue [117].

There remains, nevertheless, scarce information about the eventual alterations affecting cardiac cell turnover in any of these pathological conditions. A deeper understanding of the conceivable effect of deregulated oxidative stress for the functional maintenance of CPC populations might be helpful to better comprehend the long-term effects of specific cardiac diseases.

13.5 Redox Signaling in Cardiac Renewal

Several studies over the last few years have provided solid evidence for a low but continuous cell turnover in the mammalian heart during adult life, where cardiomyocyte turnover is limited [118]. After injury, however, the adult heart has a very weak capacity to regenerate, and a fibrotic scar mostly replaces lost cardiomyocytes, leading to a pathological cardiac remodeling. The reasons for the inability of adult heart to respond efficiently to injury, nonetheless, remain poorly understood (reviewed in [119]).

13.5.1 Heart Regeneration in Lower Vertebrates

Teleost fish can effectively regenerate several organs, including the heart [120] (Fig. 13.2). Following heart injury and fibrin deposition, the zebrafish heart does not develop an intense collagen scar, which occurs in mammals, but instead preexisting cardiomyocytes proliferate to replace those lost after injury [121]. Indeed, genetic cell ablation studies show that zebrafish survive after depletion of more than 60% of their ventricular cardiomyocytes [122]. This regenerative capacity has been linked to the intrinsic mononuclear and monoploid state of their mature cardiomyocytes [123, 124] and also with the oxygenation state of the environment [125]. The zebrafish warm aquatic environment has a 1/30th oxygen capacitance compared with air, which explains the remarkable tolerance of zebrafish to hypoxia [126].

Heart regeneration studies have not been limited to zebrafish. Newt and salamander can also regenerate lost cardiac tissue and recover cardiac function within 3 months of injury, without evidence of scarring [127, 128].

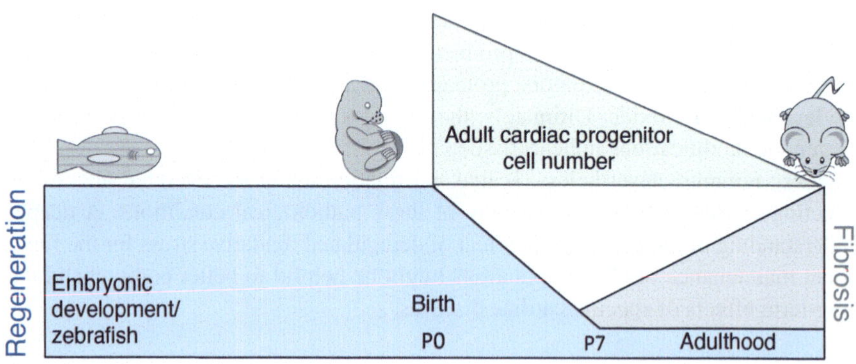

Fig. 13.2 Zebrafish and mammalian hearts demonstrate a high regenerative capacity during development. After birth (P0), this capacity is progressively substituted by a strong and preferential healing response (pivoting around P7), generating a nonfunctional fibrotic scar. In addition, numbers, and probably also their regeneration capacity, of cardiac progenitors diminish in an age-dependent manner

13.5.2 Heart Regeneration in Mice

In the mouse, embryonic, neonatal, and adulthood stages correlate with different cardiac regeneration capacities (Fig. 13.2). During intrauterine life, the blood ejected from the mammalian heart is only 65% saturated with a partial pressure of oxygen (PaO_2) of 25–28 mm Hg as compared with 90% of saturation with a PaO_2 of 35 mm Hg in the umbilical vein (maternal blood) [129]. Therefore, the mammalian fetal heart resides in a relatively hypoxic environment, similar to that found for zebrafish. In these environmental conditions, the fetal heart demonstrates a robust regenerative capacity. Genetic ablation of up to 50–60% of cardiac progenitor cells or immature cardiomyocytes is well tolerated during mouse embryonic development and does not disrupt normal heart organogenesis [130].

The neonatal mammalian heart partially maintains regenerative capacity and is capable of substantial regeneration following apex resection during the first week of life, which proceeds through cardiomyocyte proliferation [131]. However, a recent study described that resected neonatal hearts displayed thickening of the left ventricle wall and local fibrosis in adulthood [132]. During the early postnatal phase, the contribution of CPC to heart regeneration is negligible [101]. Embryonic-postnatal transition drastically modifies the oxygenation and metabolic state of cardiac cells. Accordingly, the arterial PaO_2 increases from 28 mm Hg to 100 mm Hg [133], and the mitochondrial mass and total ROS increase three-to-four fold [134]. These changes lead to the upregulation of the DNA damage response pathway (see 3.5) and a metabolic shift in energy metabolism in cardiac cells, from glycolysis during embryonic development to oxygen-dependent mitochondrial oxidative phosphorylation in adulthood [135]. Overall, these changes induce cell cycle exit and terminal differentiation of cardiomyocytes, which are mainly (80%) binucleated.

In the adult mammalian heart, there is a failure in the regeneration of the large majority of the cardiomyocytes lost after infarction, and tissue resident fibroblasts form a nonfunctional scar [136]. Over the last decade, several studies have demonstrated that resident CPC are present in the adult heart (Fig. 13.3a), including cardiosphere-derived cells (CDC), c-Kit^+, $Gli1^+$, $Sca1^+$, $Pdgfr\alpha^{+}$, and $Bmi1^+$ cells [99, 137–140]. However, in contrast to other tissues, the consensus on the existence of a definitive cardiac cell population(s) responsible for cardiac homeostasis or regeneration is still intensely debated [119]. Although a $CD45^-$ c-Kit^+ resident population was the first candidate proposed and has been studied extensively [137, 141], recent results have raised concerns about its real contribution to cardiomyocyte turnover [142, 143]. In mice, it has been described that the representation or functionality of cardiac progenitor cells such as $Bmi1^+$CPC [102] and c-Kit^+ cells [144] is reduced during physiological aging, consistent with the loss of regenerative capacity. Indeed, a correlation between aging and loss of CSC functionality has been established in a rat model of spontaneous hypertension, associated with a progressive increase in ROS levels and enhanced by hypertension [145].

It has recently been described that a rare population of cycling cardiomyocytes exists in the adult heart that expresses HIF-1α and associates with the intrinsic cardiac turnover capacity (Fig. 13.3b). These hypoxic cardiomyocytes display

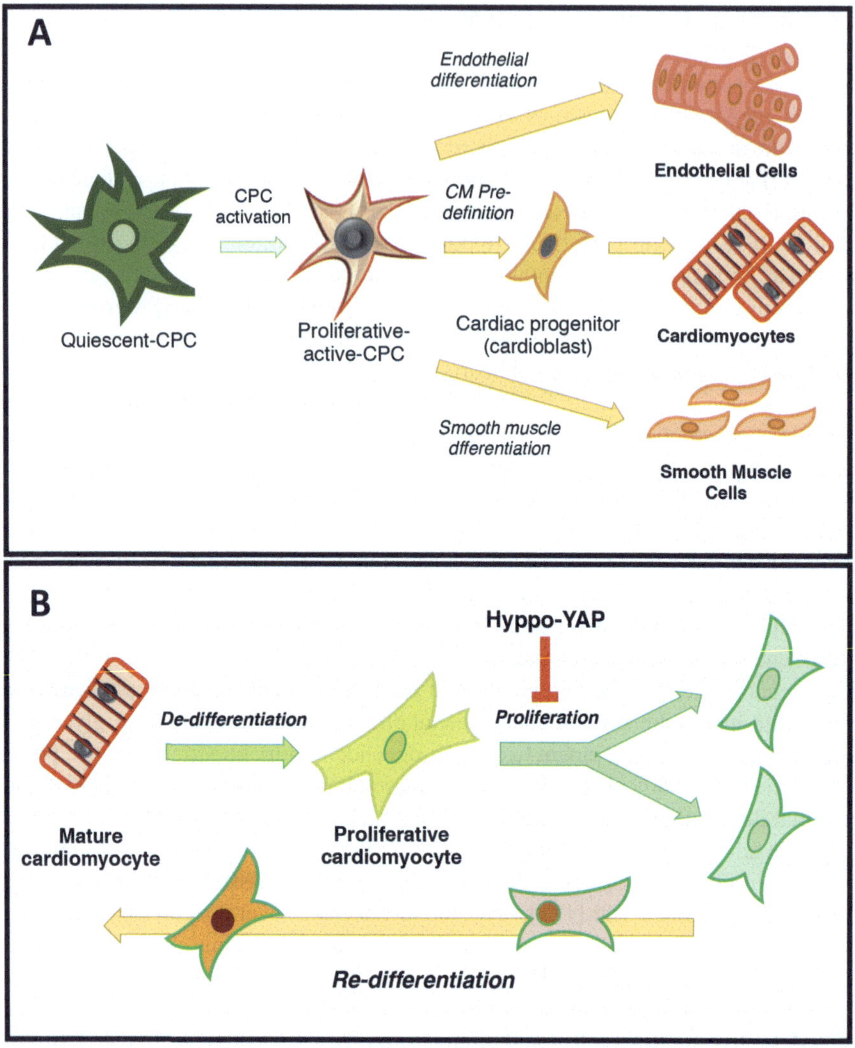

Fig. 13.3 Source of cardiac cells in adult heart. (**a**) Resident cardiac progenitor cells (CPC) are mainly maintained in a quiescent state. In response to unknown specific signals (both in homeostasis and after several damages), CPC are activated, become proliferative, and contribute significantly to the major heart cell lineages (endothelial cells > > smooth muscle cells > cardiomyocytes); for the cardiomyocyte differentiation pathway, an intermediate specific progenitor (cardioblast) has been defined [172]. (**b**) In concert, it has been proposed that a poorly defined subpopulation of mature cardiomyocytes might dedifferentiated in response to unknown signals and progress to a proliferative state. In a final step, their progeny would be able to re-differentiate to new cardiomyocytes. Hippo-YAP pathway seems to be a critical regulator of cell cycle in immature cardiomyocytes [148]

characteristics very similar to those of proliferative neonatal cardiomyocytes, such as smaller size, mononucleation, and low oxidative DNA damage [146]. In this regard, the Hippo-YAP signaling pathway, which is involved in cardiomyocyte responses to oxidative stress [147], also seems to be a critical regulator of cell cycle in immature cardiomyocytes being redundantly repressed by multiple miRNAs [148].

Studies exploring the role of metabolism in the homeostasis/differentiation of adult cardiac cells are few, but pioneering work in ES cells showed that redox status is a key regulator of cardiomyocyte differentiation [149] and that antioxidant treatment impaired cardiomyocyte differentiation from embryoid bodies [150]. In isolated CPC, several genetic (sulfiredoxin-1; $p16^{INK4A}$ knockdown) or chemical (i.e., bergamot fractions) approaches have been used to enhance their survival capacity, decreasing ROS levels and mitochondrial membrane potential concomitant with an upregulation of the primary antioxidant machinery and improved regeneration [151–153]. The scaffold protein β-arrestin2, which regulates multiple signaling pathways by desensitization and internalization of G-protein-coupled receptors, has been demonstrated to be important for c-Kit^+CPC survival, particularly in hypoxic cultures [154]. Finally, it has been shown that hypoxic preconditioning prior to transplantation enhances neonatal CPC invasion ability, activates pro-survival pathways [155], and favors pro-angiogenesis [156].

In adult murine hearts, hypoxemia alleviates oxidative DNA damage, favoring heart healing after myocardial infarction [157]. As in other adult tissues (reviewed in [158]), accumulating evidence suggests that cardiac cells with progenitor-related characteristics display low-ROS levels and, therefore, low oxidative damage:

- c-Kit^+ cardiac progenitor cells are nested in hypoxic niches and express high levels of SOD [159, 160].
- $Pdgfr\alpha^+$ cardiac progenitor cells are enriched in side population (SP) cells [140], which are identified by ATP-binding cassette transport proteins that contribute to their survival against oxidative stress [161].
- $Bmi1^{high+}$ CPC are cells with low levels of ROS whose differentiation status is directly related to ROS levels [9] (Fig. 13.4).

Although $Bmi1^+$CPC seem to be quite resistant to short-term oxidative damage-induced apoptosis in vivo, there remains cellular damage that provokes a clear differential depletion of these cells in the long term [102]. It has been demonstrated in vitro that human and murine CDC and $Bmi1^+$CPC are particularly sensitive to O_2 tension during their isolation and expansion [162, 163]; low O_2 tension favors the preservation of genetic stability and reduces the senescence rate during the required expansion prior to transplantation [164]. All these findings clearly underscore the fundamental role that redox status plays in cardiac progenitor cell responses.

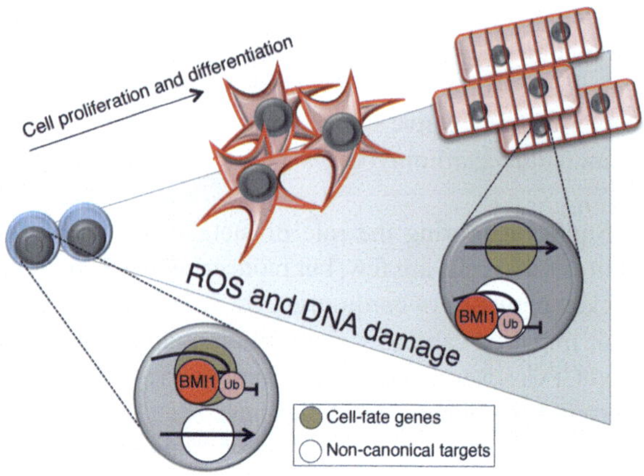

Fig. 13.4 Under low oxidative stress conditions (low ROS; cells in blue); BMI1 inhibits cardiogenic differentiation program (cell-fate genes; in brown). ROS increment (intermediates or mature cells; reddish cells) provokes a shift in BMI1 function to be mainly dedicated to mitochondrial regulation and DNA repair (noncanonical genes; in white)

13.5.3 Identification of Areas with Low-ROS Levels in the Adult Heart

There is increasing evidence that the niche plays a crucial role in adult stem cell maintenance and differentiation (reviewed in [158]). The adult stem cell niche reduces cellular stress principally through regulation of ROS levels and O_2 tension [165]. The HSC niche, undoubtedly one of the best-characterized niches, maintains HSC in a low-ROS [166] and low-PaO_2 [167] environment. Accordingly, it was reasonable to hypothesize that adult cardiac progenitor cell niches should be situated in hypoxic areas in the heart. Because the epicardium and subepicardium are areas of low vascularization, the first characterization studies focused on glycolytic epicardial cells, leading to the identification of a heterogeneous progenitor-like multipotent population that expressed HIF-1α and relied on glycolytic metabolism [168]. More recently, an elegant fate mapping study in transgenic mice expressing a hypoxia-inducible protein showed that ~0.01% of total cardiomyocytes in the murine adult heart were hypoxic and were responsible for the majority of proliferating cardiomyocytes [169]. However, the authors did not find a biased cell distribution of these hypoxic cardiomyocytes, suggesting that the subepicardium is not the unique adult cardiac progenitor cell niche.

Because several proposed adult cardiac progenitor cell populations are found perivascularly [138, 170], the coronary vasculature has also been postulated as a putative niche-like structure in the adult heart [171]. Similar to what is found in the HSC niche [166], Bmi1$^+$CPC are preferentially located close to perivascular low-ROS areas in the murine adult heart (Fig. 13.5). Indeed, functional interaction with and/or proximity to endothelial cells seems to promote quiescence of Bmi1$^+$CPC. Several chemical or genetic approaches to manipulate ROS levels, as well as several forms of cardiac damage, disrupt the Bmi1$^+$CPC-endothelium

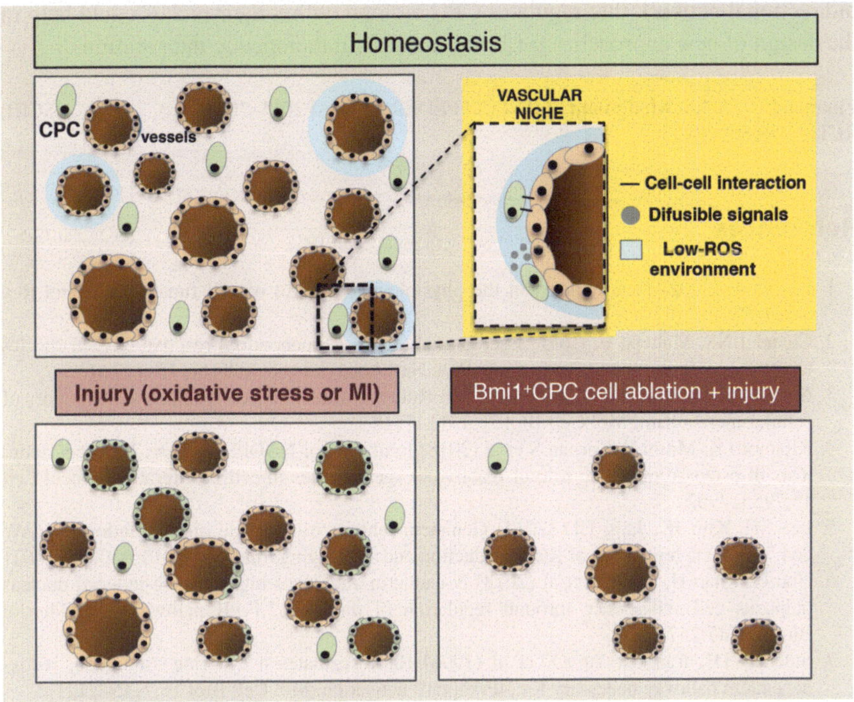

Fig. 13.5 The adult mouse heart presents perivascular areas with low-ROS levels (blue) that lodge the majority of the progenitor cells (CPC), including $Bmi1^+$ CPC. In addition, coronary vasculature regulates CPC behavior through both cell-cell contacts and secreted factors. Genetic ablation of $Bmi1^+$ CPC in vivo provokes a deficient angiogenic response after myocardial infarct (MI)

crosstalk [173], suggesting an important regulatory role for ROS in $Bmi1^+$CPC biology (Fig. 13.5). Moreover, ablation (diphtheria toxin-mediated) of $Bmi1^+$CPC in vivo confirmed their essential role in normal heart recovery after acute myocardial infarction (Fig. 13.5). Accordingly, $Bmi1^+$CPC-depleted animals presented substantially deteriorated heart angiogenesis and ejection fraction, resulting in an ischemic-dilated cardiac phenotype [102].

Overall, these arguments strongly support that low-ROS perivascular areas might define a physiological niche for $Bmi1^+$CPC.

13.6 Conclusion

Cellular ROS levels and the appropriate function of antioxidant systems orchestrate, at least in part, stem cell self-renewal and differentiation capacities. Low-ROS areas provide a safe haven for cardiac progenitor cells. The existing low-ROS regions around coronary vessels constitute a favorable vascular niche that protects CPC from the detrimental effects of oxidative stress. A deeper understanding of the bidirectional network that regulates CPC biology within their niches could help in the design of new approaches for future successful therapeutic interventions.

Financial Acknowledgment MINECO/FEDER; SAF2015–70882-R and ISCIII; RETICS-RD12/001.

References

1. Dröge W (2002) Free radicals in the physiological control of cell function. Physiol Rev 82:47–95
2. Chandel NS, Maltepe E, Goldwasser E et al (1998) Mitochondrial reactive oxygen species trigger hypoxia-induced transcription. Proc Natl Acad Sci U S A 95:11715–11720
3. Nemoto S, Takeda K, Yu ZX et al (2000) Role for mitochondrial oxidants as regulators of cellular metabolism. Mol Cell Biol 20:7311–7318
4. Kitamoto K, Miura Y, Karnan S et al (2018) Inhibition of NADPH oxidase 2 induces apoptosis in osteosarcoma: the role of reactive oxygen species in cell proliferation. Oncol Lett 15:7955–7962
5. Lee SH, Kim JK, Jang HD (2014) Genistein inhibits osteoclastic differentiation of RAW 264.7 cells via regulation of ROS production and scavenging. Int J Mol Sci 15:10605–10621
6. Hou G, Zhao H, Teng H et al (2018) N-cadherin attenuates high glucose-induced nucleus pulposus cell senescence through regulation of the ROS/NF-κB pathway. Cell Physiol Biochem 47:257–265
7. Sulciner DJ, Irani K, Yu ZX et al (1996) Rac1 regulates a cytokine-stimulated, redox-dependent pathway necessary for NF-kappaB activation. Mol Cell Biol 16:7115–7121
8. Mathew R, Karp CM, Beaudoin B et al (2009) Autophagy suppresses tumorigenesis through elimination of p62. Cell 137:1062–1075
9. Herrero D, Tomé M, Cañón S et al (2018) Redox-dependent BMI1 activity drives in vivo adult cardiac progenitor cell differentiation. Cell Death Differ 25:807–820
10. Kim JH, Song SY, Park SG et al (2012) Primary involvement of NADPH oxidase 4 in hypoxia-induced generation of reactive oxygen species in adipose-derived stem cells. Stem Cells Dev 21:2212–2221
11. Borodkina A, Shatrova A, Abushik P et al (2014) Interaction between ROS dependent DNA damage, mitochondria and p38 MAPK underlies senescence of human adult stem cells. Aging 6:481–495
12. Shi B, Wang Y, Zhao R et al (2018) Bone marrow mesenchymal stem cell-derived exoso-mal miR-21 protects C-kit+ cardiac stem cells from oxidative injury through the PTEN/PI3K/Akt axis. PLoS One 13:e0191616
13. Varum S, Rodrigues AS, Moura MB et al (2011) Energy metabolism in human pluripotent stem cells and their differentiated counterparts. PLoS One 6:e20914
14. Saretzki G, Armstrong L, Leake A et al (2004) Stress defense in murine embryonic stem cells is superior to that of various differentiated murine cells. Stem Cells 22:962–971
15. Schmelter M, Ateghang B, Helmig S et al (2006) Embryonic stem cells utilize reactive oxygen species as transducers of mechanical strain-induced cardiovascular differentiation. FASEB J 20:1182–1184
16. Ji AR, Ku SY, Cho MS et al (2010) Reactive oxygen species enhance differentiation of human embryonic stem cells into mesendodermal lineage. Exp Mol Med 42:175–186
17. Brand MD (2010) The sites and topology of mitochondrial superoxide production. Exp Gerontol 45:466–472
18. Aon MA, Stanley BA, Sivakumaran V et al (2012) Glutathione/thioredoxin systems modulate mitochondrial H2O2 emission: an experimental-computational study. J Gen Physiol 139:479–491

19. Tahara EB, Navarete FD, Kowaltowski AJ (2009) Tissue-, substrate-, and site-specific characteristics of mitochondrial reactive oxygen species generation. Free Radic Biol Med 46:1283–1297
20. Lapuente-Brun E, Moreno-Loshuertos R, Acín-Pérez R et al (2013) Supercomplex assem-bly determines electron flux in the mitochondrial electron transport chain. Science 340:1567–1570
21. Drahota Z, Chowdhury SK, Floryk D et al (2002) Glycerophosphate-dependent hydrogen peroxide production by brown adipose tissue mitochondria and its activation by ferricyanide. J Bioenerg Biomembr 34:105–113
22. Frerman FE (1987) Reaction of electron-transfer flavoprotein ubiquinone oxidoreductase with the mitochondrial respiratory chain. Biochim Biophys Acta 893:161–169
23. Vasquez-Vivar J, Kalyanaraman B, Kennedy MC (2000) Mitochondrial aconitase is a source of hydroxyl radical. An electron spin resonance investigation. J Biol Chem 275:14064–14069
24. Hauptmann N, Grimsby J, Shih JC et al (1996) The metabolism of tyramine by monoamine oxidase A/B causes oxidative damage to mitochondrial DNA. Arch Biochem Biophys 335:295–304
25. Giorgio M, Migliaccio E, Orsini F et al (2005) Electron transfer between cytochrome c and p66Shc generates reactive oxygen species that trigger mitochondrial apoptosis. Cell 122:221–233
26. Tothova Z, Kollipara R, Huntly BJ et al (2007) FoxOs are critical mediators of hematopoietic stem cell resistance to physiologic oxidative stress. Cell 128:325–339
27. Sansone P, Storci G, Giovannini C et al (2007) p66Shc/Notch-3 interplay controls self-renewal and hypoxia survival in human stem/progenitor cells of the mammary gland expanded in vitro as mammospheres. Stem Cells 25:807–815
28. Starkov AA, Fiskum G, Chinopoulos C et al (2004) Mitochondrial alpha-ketoglutarate dehydrogenase complex generates reactive oxygen species. J Neurosci 24:7779–7788
29. Prigione A, Fauler B, Lurz R et al (2010) The senescence-related mitochondrial/oxidative stress pathway is repressed in human induced pluripotent stem cells. Stem Cells 28:721–733
30. Chung S, Dzeja PP, Faustino RS et al (2007) Mitochondrial oxidative metabolism is required for the cardiac differentiation of stem cells. Nat Clin Pract Cardiovasc Med 4(Suppl 1):S60–S67
31. St John JC, Ramalho-Santos J, Gray HL et al (2005) The expression of mitochondrial DNA transcription factors during early cardiomyocyte in vitro differentiation from human embryonic stem cells. Cloning Stem Cells 7:141–153
32. Chung S, Arrell DK, Faustino RS et al (2010) Glycolytic network restructuring integral to the energetics of embryonic stem cell cardiac differentiation. J Mol Cell Cardiol 48:725–734
33. Skonieczna M, Hejmo T, Poterala-Hejmo A et al (2017) NADPH oxidases: insights into selected functions and mechanisms of action in cancer and stem cells. Oxidative Med Cell Longev 2017:9420539
34. Li J, Stouffs M, Serrander L et al (2006) The NADPH oxidase NOX4 drives cardiac differentiation: role in regulating cardiac transcription factors and MAP kinase activation. Mol Biol Cell 17:3978–3988
35. Nadworny AS, Guruju MR, Poor D et al (2013) Nox2 and Nox4 influence neonatal c-kit(+) cardiac precursor cell status and differentiation. Am J Physiol Heart Circ Physiol 305:H829–H842
36. Topchiy E, Panzhinskiy E, Griffin WS (2013) Nox4-generated superoxide drives angiotensin II-induced neural stem cell proliferation. Dev Neurosci 35:293–305
37. Moruno-Manchon JF, Uzor NE, Kesler SR et al (2018) Peroxisomes contribute to oxidative stress in neurons during doxorubicin-based chemotherapy. Mol Cell Neurosci 86:65–71
38. Jiang S, He R, Zhu L et al (2018) Endoplasmic reticulum stress-dependent ROS production mediates synovial myofibroblastic differentiation in the immobilization-induced rat knee joint contracture model. Exp Cell Res S0014-4827(18):30316–30311
39. Zangar RC, Davydov DR, Verma S (2004) Mechanisms that regulate production of reactive oxygen species by cytochrome P450. Toxicol Appl Pharmacol 199:316–331

40. Chuang DY, Simonyi A, Kotzbauer PT et al (2015) Cytosolic phospholipase A2 plays a crucial role in ROS/NO signaling during microglial activation through the lipoxygenase pathway. J Neuroinflammation 12:199
41. Huang CC, Chen KL, Cheung CH et al (2013) Autophagy induced by cathepsin S inhibition induces early ROS production, oxidative DNA damage, and cell death via xanthine oxidase. Free Radic Biol Med 65:1473–1486
42. He L, He T, Farrar S et al (2017) Antioxidants maintain cellular redox homeostasis by elimination of reactive oxygen species. Cell Physiol Biochem 44:532–553
43. Dernbach E, Urbich C, Brandes RP et al (2004) Antioxidative stress-associated genes in circulating progenitor cells: evidence for enhanced resistance against oxidative stress. Blood 104:3591–3597
44. Bognar Z, Kalai T, Palfi A et al (2006) A novel SOD-mimetic permeability transition inhibitor agent protects ischemic heart by inhibiting both apoptotic and necrotic cell death. Free Radic Biol Med 41:835–848
45. Saretzki G, Walter T, Atkinson S et al (2008) Downregulation of multiple stress defense mechanisms during differentiation of human embryonic stem cells. Stem Cells 26:455–464
46. Solari C, Vázquez Echegaray C, Cosentino MS et al (2015) Manganese superoxide dismutase gene expression is induced by nanog and Oct4, essential pluripotent stem cells' transcription factors. PLoS One 10:e0144336
47. Miao W, Xufeng R, Park MR et al (2013) Hematopoietic stem cell regeneration enhanced by ectopic expression of ROS-detoxifying enzymes in transplant mice. Mol Ther 21:423–432
48. Kwon T, Bak Y, Park YH et al (2016) Peroxiredoxin II is essential for maintaining stemness by redox regulation in liver cancer cells. Stem Cells 34:1188–1197
49. Spradling A, Drummond-Barbosa D, Kai T (2001) STem cells find their niche. Nature 414:98–104
50. Simon MC, Keith B (2008) The role of oxygen availability in embryonic development and stem cell function. Nat Rev Mol Cell Biol 9:285–296
51. Panchision DM (2009) The role of oxygen in regulating neural stem cells in development and disease. J Cell Physiol 220:562–568
52. Eliasson P, Jönsson JI (2010) The hematopoietic stem cell niche: low in oxygen but a nice place to be. J Cell Physiol 222:17–22
53. Silván U, Díez-Torre A, Arluzea J et al (2009) Hypoxia and pluripotency in embryonic and embryonal carcinoma stem cell biology. Differentiation 78:159–168
54. Cho YM, Kwon S, Pak YK et al (2006) LeeDynamic changes in mitochondrial biogenesis and antioxidant enzymes during the spontaneous differentiation of human embryonic stem cells. Biochem Biophys Res Commun 348:1472–1478
55. Wang K, Zhang T, Dong Q et al (2013) Redox homeostasis: the linchpin in stem cell self-renewal and differentiation. Cell Death Dis 4:e537
56. Gardner LB, Li Q, Park MS et al (2001) Hypoxia inhibits G1/S transition through regulation of p27 expression. J Biol Chem 276:7919–7926
57. Iida T, Mine S, Fujimoto H et al (2002) Hypoxia-inducible factor-1alpha induces cell cycle arrest of endothelial cells. Genes Cells 7:143–149
58. Koshiji M, Kageyama Y, Pete EA et al (2004) HIF-1alpha induces cell cycle arrest by functionally counteracting Myc. EMBO J 23:1949–1956
59. Adelman DM, Gertsenstein M, Nagy A et al (2000) Placental cell fates are regulated in vivo by HIF-mediated hypoxia responses. Genes Dev 14:3191–3203
60. Lee SW, Jeong HK, Lee JY et al (2012) Hypoxic priming of mESCs accelerates vascular-lineage differentiation through HIF1-mediated inverse regulation of Oct4 and VEGF. EMBO Mol Med 4:924–938
61. Francis KR, Wei L (2010) Human embryonic stem cell neural differentiation and enhanced cell survival promoted by hypoxic preconditioning. Cell Death Dis 1:e22
62. Ng KM, Lee YK, Chan YC et al (2010) Exogenous expression of HIF-1 alpha promotes cardiac differentiation of embryonic stem cells. J Mol Cell Cardiol 48:1129–1137

63. De Miguel MP, Alcaina Y, de la Maza DS et al (2015) Cell metabolism under microenvironmental low oxygen tension levels in stemness, proliferation and pluripotency. Curr Mol Med 15:343–359
64. Gustafsson MV, Zheng X, Pereira T et al (2005) Hypoxia requires notch signaling to maintain the undifferentiated cell state. Dev Cell 9:617–628
65. Mutoh TS (2012) Oxygen levels epigenetically regulate fate switching of neural precursor cells via hypoxia-inducible factor 1α-notch signal interaction in the developing brain. Stem Cells 30:561–569
66. Covello KL, Kehler J, Yu H et al (2006) HIF-2alpha regulates Oct-4: effects of hypoxia on stem cell function, embryonic development, and tumor growth. Genes Dev 20:557–570
67. Kaidi A, Williams AC, Paraskeva C (2007) Interaction between beta-catenin and HIF-1 promotes cellular adaptation to hypoxia. Nat Cell Biol 9:210–217
68. Takubo K, Goda N, Yamada W et al (2010) Regulation of the HIF-1alpha level is essential for hematopoietic stem cells. Cell Stem Cell 7:391–402
69. Zhang K, Zhou Y, Zhao T et al (2015) Reduced cerebral oxygen content in the DG and SVZ in situ promotes neurogenesis in the adult rat brain in vivo. PLoS One 10:e0140035
70. Lange C, Turrero-Garcia M, Decimo I et al (2016) Relief of hypoxia by angiogenesis promotes neural stem cell differentiation by targeting glycolysis. EMBO J 35:924–941
71. Li L, Candelario KM, Thomas K et al (2014) Hypoxia inducible factor-1α (HIF-1α) is required for neural stem cell maintenance and vascular stability in the adult mouse SVZ. J Neurosci 34:16713–16719
72. Estrada JC, Albo C, Benguría A et al (2012) Culture of human mesenchymal stem cells at low oxygen tension improves growth and genetic stability by activating glycolysis. Cell Death Differ 19:743–755
73. Valorani MG, Montelatici E, Germani A et al (2012) Pre-culturing human adipose tissue mesenchymal stem cells under hypoxia increases their adipogenic and osteogenic differentiation potentials. Cell Prolif 45:225–238
74. Ateghang B, Wartenberg M, Gassmann M et al (2006) Regulation of cardiotrophin-1 expression in mouse embryonic stem cells by HIF-1α and intracellular reactive oxygen species. J Cell Sci 119:1043–1052
75. Salih DA, Brunet A (2008) FoxO transcription factors in the maintenance of cellular homeostasis during aging. Curr Opin Cell Biol 20:126–136
76. van der Horst A, Burgering BM (2007) Stressing the role of FoxO proteins in lifespan and disease. Nat Rev Mol Cell Biol 8:440–450
77. Essers MA, de Vries-Smits LM, Barker N et al (2005) Functional interaction between beta-catenin and FOXO in oxidative stress signaling. Science 308:1181–1184
78. Kops GJ, Dansen TB, Polderman PE et al (2002) Forkhead transcription factor FOXO3a protects quiescent cells from oxidative stress. Nature 419:316–321
79. Paik JH, Ding Z, Narurkar R et al (2009) FoxOs cooperatively regulate diverse pathways governing neural stem cell homeostasis. Cell Stem Cell 5:540–553
80. Miyamoto K, Araki KY, Naka K et al (2007) Foxo3a is essential for maintenance of the hematopoietic stem cell pool. Cell Stem Cell 1:101–112
81. Miyamoto K, Miyamoto T, Kato R et al (2008) FoxO3a regulates hematopoietic homeostasis through a negative feedback pathway in conditions of stress or aging. Blood 112:4485–4493
82. Iyer S, Ambrogini E, Bartell SM et al (2013) FOXOs attenuate bone formation by suppressing Wnt signaling. J Clin Invest 123:3409–3419
83. Higuchi M, Dusting GJ, Peshavariya H et al (2013) Differentiation of human adipose -derived stem cells into fat involves reactive oxygen species and Forkhead box O1 mediated upregulation of antioxidant enzymes. Stem Cells Dev 22:878–888
84. Zhang X, Yalcin S, Lee DF et al (2011) FOXO1 is an essential regulator of pluripotency in human embryonic stem cells. Nat Cell Biol 13:1092–1099
85. Nguyen T, Nioi P, Pickett CB (2009) The Nrf2-antioxidant response element signaling pathway and its activation by oxidative stress. J Biol Chem 284:13291–13295

86. Itoh K, Wakabayashi N, Katoh Y et al (1999) Keap1 represses nuclear activation of antioxidant responsive elements by Nrf2 through binding to the amino-terminal Neh2 domain. Genes Dev 13:76–86
87. Kobayashi A, Kang MI, Okawa H et al (2004) Oxidative stress sensor Keap1 functions as an adaptor for Cul3-based E3 ligase to regulate proteasomal degradation of Nrf2. Mol Cell Biol 24:7130–7139
88. Li J, Johnson D, Calkins M et al (2005) Stabilization of Nrf2 by tBHQ confers protection against oxidative stress-induced cell death in human neural stem cells. Toxicol Sci 83:313–328
89. Tsai JJ, Dudakov JA, Takahashi K et al (2013) Nrf2 regulates haematopoietic stem cell function. Nat Cell Biol 15:309–316
90. Murakami S, Suzuki T, Harigae H et al (2017) NRF2 activation impairs quiescence and bone marrow reconstitution capacity of hematopoietic stem cells. Mol Cell Biol 37:e00086–e00017
91. Jang J, Wang Y, Kim HS et al (2014) Nrf2, a regulator of the proteasome, controls self-renewal and pluripotency in human embryonic stem cells. Stem Cells 32:2616–2625
92. Mohammadzadeh M, Halabian R, Gharehbaghian A et al (2012) Nrf-2 overexpression in mesenchymal stem cells reduces oxidative stress-induced apoptosis and cytotoxicity. Cell Stress Chaperones 17:553–565
93. Yoon DS, Choi Y, Lee JW (2016) Cellular localization of NRF2 determines the self-renewal and osteogenic differentiation potential of human MSCs via the P53–SIRT1 axis. Cell Death Dis 7:e2093
94. Bhattacharya R, Mustafi SB, Street M et al (2015) Bmi-1: at the crossroads of physiological and pathological biology. Genes Dis 2:225–239
95. Pietersen AM, van Lohuizen M (2008) Stem cell regulation by polycomb repressors: postponing commitment. Curr Opin Cell Biol 20:201–217
96. Molofsky AV, Pardal R, Iwashita T et al (2003) Bmi-1 dependence distinguishes neural stem cell self-renewal from progenitor proliferation. Nature 425:962–967
97. Park IK, Qian D, Kiel M et al (2003) Bmi-1 is required for maintenance of adult self-renewing haematopoietic stem cells. Nature 423:302–325
98. López-Arribillaga E, Rodilla V, Pellegrinet L et al (2015) Bmi1 regulates murine intestinal stem cell proliferation and self-renewal downstream of notch. Development 142:41–50
99. Valiente-Alandi I, Albo-Castellanos C, Herrero D et al (2015) Cardiac Bmi1+ cells contribute to myocardial renewal in the murine adult heart. Stem Cell Res Ther 6:205
100. Liu L, Cao L, Chen J et al (2009) Bmi1 regulates mitochondrial function and the DNA damage response pathway. Nature 459:387–392
101. Valiente-Alandi I, Albo-Castellanos C, Herrero D et al (2016) Bmi1 (+) cardiac progenitor cells contribute to myocardial repair following acute injury. Stem Cell Res Ther 7:100
102. Herrero D, Cañón S, Pelacho B et al (2018) Bmi1-progenitor cell ablation impairs the angiogenic response to myocardial infarction. Arterioscler Thromb Vasc Biol 38:2160–2173. https://doi.org/10.1161/ATVBAHA.118.310778
103. Brodkina A, Shatrova A, Abushik P, Nikolsky N, Burova E (2014) Interaction between ROS dependent DNA damage, mitochondria and p38 MAPK underlies senescence of human adult stem cells. Aging (Albany NY) 6:481–495
104. Kong Y, Song Y, Hu Y et al (2016) Increased reactive oxygen species and exhaustion of quiescent CD34-positive bone marrow cells may contribute to poor graft function after allotransplants. Oncotarget 7:30892–30906
105. Brien GL, Healy E, Jerman E et al (2015) A chromatin-independent role of Polycomb-like 1 to stabilize p53 and promote cellular quiescence. Genes Dev 29:2231–2243
106. Mohrin M, Bourke E, Alexander D et al (2010) Hematopoietic stem cell quiescence promotes error-prone DNA repair and mutagenesis. Cell Stem Cell 7:174–185
107. Boregowda SV, Krishnappa V, Strivelli J (2018) Basal p53 expression is indispensable for mesenchymal stem cell integrity. Cell Death Differ 25:677–690
108. Cesselli D, Aleksova A, Sponga S et al (2017) Cardiac cell senescence and redox signaling. Front Cardiovasc Med 4:38

109. Khaper N, Bailey CDC, Ghugre NR et al (2018) Implications of disturbances in circadian rhythms for cardiovascular health: a new frontier in free radical biology. Free Radic Biol Med 119:85–92
110. Kanaan GN, Harper ME (2017) Cellular redox dysfunction in the development of cardiovascular diseases. Biochim Biophys Acta 1861:2822–2829
111. He F, Zuo L (2015) Redox roles of reactive oxygen species in cardiovascular diseases. Int J Mol Sci 16:27770–27780
112. Barančík M, Grešová L, Barteková M et al (2016) Nrf2 as a key player of redox regulation in cardiovascular diseases. Physiol Res 65(Suppl 1):S1–S10
113. Erkens R, Kramer CM, Lückstädt W et al (2015) Left ventricular diastolic dysfunction in Nrf2 knock out mice is associated with cardiac hypertrophy, decreased expression of SERCA2a, and preserved endothelial function. Free Radic Biol Med 89:906–917
114. Xu B, Zhang J, Strom J et al (2014) Myocardial ischemic reperfusion induces de novo Nrf2 protein translation. Biochim Biophys Acta 1842:1638–1647
115. Taunk NK, Haffty BG, Kostis JB et al (2015) Radiation-induced heart disease: pathologic abnormalities and putative mechanisms. Front Oncol 5:39
116. Ahamed J, Laurence J (2017) Role of platelet-derived transforming growth factor-β1 and reactive oxygen species in radiation-induced organ fibrosis. Antioxid Redox Signal 27:977–988
117. Weigel C, Schmezer P, Plass C et al (2015) Epigenetics in radiation-induced fibrosis. Oncogene 34:2145–2155
118. Bergmann O, Zdunek S, Felker A et al (2015) Dynamics of cell generation and turnover in the human heart. Cell 161:1566–1575
119. Uygur A, Lee RT (2016) Mechanisms of cardiac regeneration. Dev Cell 36:362–374
120. Poss KD, Wilson LG, Keating MT (2002) Heart regeneration in Zebrafish. Science 298:2188–2190
121. Jopling C, Sleep E, Raya M et al (2010) Zebrafish heart regeneration occurs by cardiomyocyte dedifferentiation and proliferation. Nature 464:606–609
122. Wang J, Panáková D, Kikuchi K et al (2011) The regenerative capacity of zebrafish reverses cardiac failure caused by genetic cardiomyocyte depletion. Development 138:3421–3430
123. Kikuchi K, Holdway JE, Werdich AA et al (2010) Primary contribution to zebrafish heart regeneration by Gata4+ cardiomyocytes. Nature 464:601–605
124. González-Rosa JM, Sharpe M, Field D et al (2018) Myocardial polyploidization creates a barrier to heart regeneration in zebrafish. Dev Cell 44:433–446
125. Roesner A, Hankeln T, Burmester T (2006) Hypoxia induces a complex response of globin expression in zebrafish (Danio rerio). J Exp Biol 209:2129–2137
126. Rees BB, Sudradjat FA, Love JW (2001) Acclimation to hypoxia increases survival time of zebrafish, Danio rerio, during lethal hypoxia. J Exp Zool 289:266–272
127. Flink IL (2002) Cell cycle reentry of ventricular and atrial cardiomyocytes and cells within the epicardium following amputation of the ventricular apex in the axolotl, Amblystoma mexicanum: confocal microscopic immunofluorescent image analysis of bromodeoxyuridine-label. Anat Embryol 205:235–244
128. Oberpriller JO, Oberpriller JC (1974) Response of the adult newt ventricle to injury. J Exp Zool 187:249–253
129. Dawes GS, Mott JC, Widdicombe JG (1954) The foetal circulation in the lamb. J Physiol 126:563–587
130. Suturzu AC, Rajarajan K, Passer D et al (2014) The fetal mammalian heart generates a robust compensatory response to cell loss. Circulation 132:109–121
131. Porrello ER, Mahmoud AI, Simpson E et al (2011) Transient regenerative potential of the neonatal mouse heart. Science 331:1078–1080
132. Sampaio-Pinto V, Rodrigues SC, Laundos TL et al (2018) Neonatal apex resection triggers cardiomyocyte proliferation, neovascularization and functional recovery despite local fibrosis. Stem Cell Rep 10:860–874
133. Webster WS, Abela D (2007) The effect of hypoxia in development. Birth Defects Res C Embryo Today 81:215–228

134. Puente BN, Kimura W, Muralidhar SA et al (2014) The oxygen rich postnatal environment induces cardiomyocyte cell cycle arrest through DNA damage response. Cell 157:565–579
135. Millis RJ, Titmarsh DM, Koenig X et al (2017) Functional screening in human cardiac organoids reveals a metabolic mechanism for cardiomyocyte cell cycle arrest. Proc Natl Acad Sci U S A 114:E8372–E8381
136. Yang F, Liu YH, Yang XP et al (2002) Myocardial infarction and cardiac remodelling in mice. Exp Physiol 87:547–555
137. Beltrami AP, Barlucchi L, Torella D et al (2003) Adult cardiac stem cells are multipotent and support myocardial regeneration. Cell 114:763–776
138. Kramann R, Schneider RK, DiRocco DP et al (2015) Perivascular Gli1+ progenitors are key contributors to injury-induced organ fibrosis. Cell Stem Cell 16:51–66
139. Uchida S, De Gaspari P, Kostin S et al (2013) Sca1-derived cells are a source of myocardial renewal in the murine adult heart. Stem Cell Rep 1:397–410
140. Noseda M, Harada M, McSweeney S et al (2015) PDGFRα demarcates the cardiogenic clonogenic Sca1+ stem/progenitor cell in adult murine myocardium. Nat Commun 6:6930
141. Vicinanza C, Aquila I, Scalise M et al (2017) Adult cardiac stem cells are multipotent and robustly myogenic: c-kit expression is necessary but not sufficient for their identification. Cell Death Differ 24:2101–2116
142. van Berlo JH, Molkentin JD (2016) Most of the dust has settled: cKit+ progenitor cells are an irrelevant source of cardiac myocytes in vivo. Circ Res 118:17–19
143. He L, Li Y, Li Y et al (2017) Enhancing the precision of genetic lineage tracing using dual recombinases. Nat Med 23:1488–1498
144. Castaldi A, Dodia RM, Orogo AM et al (2017) Decline in cellular function of aged mouse c-kit^{+} cardiac progenitor cells. J Physiol 595:6249–6262
145. Saheera S, Nair RR (2017) Accelerated decline in cardiac stem cell efficiency in spontaneously hypertensive rat compared to normotensive wistar rat. PLoS One 12:e0189129
146. Kimura W, Xiao F, Canseco DC et al (2015) Hypoxia fate mapping identifies cycling cardiomyocytes in the adult heart. Nature 523:226–230
147. Shao D, Zhai P, Del Re P et al (2014) A functional interaction between hippo-YAP signalling and FoxO1 mediates the oxidative stress response. Nat Commun 5:3315
148. Diez-Cuñado M, Wei K, Bushway PJ et al (2018) miRNAs that induce human cardiomyocyte proliferation converge on the hippo pathway. Cell Rep 23:2168–2174
149. Crespo FL, Sobrado VR, Gomez L et al (2010) Mitochondrial reactive oxygen species mediate cardiomyocyte formation from embryonic stem cells in high glucose. Stem Cells 28:1132–1142
150. Sauer H, Rahimi G, Hescheler J et al (2000) Role of reactive oxygen species and phosphatidylinositol 3-kinase in cardiomyocyte differentiation of embryonic stem cells. FEBS Lett 476:218–223
151. Li X, He P, Wang XL et al (2018) Sulfiredoxin-1 enhances cardiac progenitor cell survival against oxidative stress via the upregulation of the ERK/NRF2 signal pathway. Free Radic Biol Med 123:8–19
152. Khatiwala RV, Zhang S, Li X et al (2018) Inhibition of p16^{INK4A} to rejuvenate aging human cardiac progenitor cells via the upregulation of anti-oxidant and NFκB signal pathways. Stem Cell Rev 14:612–625. https://doi.org/10.1007/s12015-018-9815-z
153. Carresi C, Musolino V, Gliozzi M et al (2018) Anti-oxidant effect of bergamot polyphenolic fraction counteracts doxorubicin-induced cardiomyopathy: role of autophagy and c-kit^{pos}CD45^{neg}CD31^{neg} cardiac stem cell activation. J Mol Cell Cardiol 119:10–18
154. Seo SK, Kim N, Lee JH et al (2018) β-arrestin2 affects cardiac progenitor cell survival through cell mobility and tube formation in severe hypoxia. Korean Circ J 48:296–309
155. Hernandez I, Baio JM, Tsay E et al (2018) Short-term hypoxia improves early cardiac progenitor cell function *in vitro*. Am J Stem Cells 7:1–17
156. Amirrasouli MM, Shamsara M (2017) Comparing the *in vivo* and *in vitro* effects of hypoxia (3% O_2) on directly derived cells from murine cardiac explants versus murine cardiosphere derived cells. J Stem Cells Regen Med 13:35–44

157. Nakada Y, Canseco DC, Thet SW et al (2017) Hypoxia induces heart regeneration in adult mice. Nature 541:222–227
158. Bigarella CL, Li J, Rimmelé P et al (2014) Stem cells and the impact of ROS signaling. Development 141:4206–4218
159. Sanada F, Kim J, Czarna A et al (2014) c-Kit-positive cardiac stem cells nested in hypoxic niches are activated by stem cell factor reversing the aging myopathy. Circ Res 114:41–55
160. Seshadri G, Che PL, Boopathy AV et al (2012) Characterization of superoxide dismutases in cardiac progenitor cells demonstrates a critical role for manganese superoxide dismutase. Stem Cells Dev 21:3136–3146
161. Krishnamurthy P, Ross DD, Nakanishi T et al (2004) The stem cell marker Bcrp/ABCG2 enhances hypoxic cell survival through interactions with heme. J Biol Chem 279:24218–22425
162. Li TS, Cheng K, Malliaras K (2011) Expansion of human cardiac stem cells in physiological oxygen improves cell production efficiency and potency for myocardial repair. Cardiovasc Res 89:157–165
163. Moscoso I, Tejados N, Barreiro O et al (2016) Podocalyxin-like protein 1 is a relevant marker for human c-kit(pos) cardiac stem cells. J Tissue Eng Regen Med 10:580–590
164. Sanz-Ruiz R, Casado-Plasencia A, Borlado LR et al (2017) Rationale and design of a clinical trial to evaluate the safety and efficacy of intracoronary infusion of allogeneic human cardiac stem cells in patients with acute myocardial infarction and left ventricular dysfunction: the randomized multicenter double-blind controlled CAREMI trial (cardiac stem cells in patients with acute myocardial infarction). Circ Res 121:71–80
165. Morrison SJ, Spradling C (2008) Stem cells and niches: mechanisms that promote stem cell maintenance throughout life. Cell 132:598–611
166. Itkin T, Gur-Cohen S, Spencer JA et al (2016) Distinct bone marrow blood vessels differentially regulate haematopoiesis. Nature 532:323–328
167. Spencer J, Ferraro F, Roussakis E et al (2014) Direct measurement of local oxygen concentration in the bone marrow of live animals. Nature 508:269–273
168. Kocabas F, Mahmoud AI, Sosic D et al (2012) The hypoxic epicardial and subepicardial microenvironment. J Cardiovasc Transl Res 5:654–665
169. Kimura W, Muralidhar S, Canseco DC et al (2014) Redox signaling in cardiac renewal. Antioxid Redox Signal 21:1660–1673
170. Fioret BA, Heimfeld JD, Paik DT et al (2014) Endothelial cells contribute to generation of adult ventricular myocytes during cardiac homeostasis. Cell Rep 8:229–241
171. Gómez-Gaviro MV, Lovell-Badge R, Fernández-Avilés F et al (2012) The vascular stem cell niche. J Cardiovasc Transl Res 5:618–630
172. Malliaras K, Ibrahim A, Tseliou E et al (2014) Stimulation of endogenous cardioblasts by exogenous cell therapy after myocardial infarction. EMBO Mol Med 6:760–777
173. Herrero D, Cañón S, Albericio G, Carmona RM, Aguilar S, Mañes S, Bernad A, (2019) Age-related oxidative stress confines damage-responsive Bmi1+ cells to perivascular regions in the murine adult heart. Redox Biology 22:101156

Role of Oxidative Stress in Hyperhomocysteinemia-Induced Heart Diseases

14

Nevena Jeremic, Jovana Bradic, Anica Petkovic, and Gregory Weber

Abstract

Evidence suggests that HHcy is closely related with risk of unwanted cardiovascular events. In state of excessively high levels of Hcy, metabolism of Hcy is disrupted and vascular tissue is exposed to its adverse effects. Based on epidemiological, retrospective, and prospective studies, hyperhomocysteinemia is considered as an independent risk factor for coronary heart and cerebrovascular and peripheral artery diseases. A considerable number of studies have been conducted in order to reveal the mechanisms through which Hcy contributes to endothelial injury. Endothelial dysfunction is characterized by impaired endothelium-dependent relaxation due to a decrease in available nitric oxide (NO). Hcy exerts harmful effects on vascular endothelium and smooth muscle cells, leading to impairment of arterial structure and function. The underlying mechanisms involve an increase in coagulation, synthesis of collagen, proliferation of vascular smooth muscle cells, initiation of inflammatory response, and elevated generation of pro-oxidants. Redox homeostasis is regulated by several intermediates involved in the methionine cycle, such as glutathione, hydrogen sulfide (H2S), and S-adenosyl methionine (SAM). Glutathione and H2S are responsible

N. Jeremic (✉)
Department of Physiology, School of Medicine, Uniuversity of Louisville, Louisville, KY, USA

Department of Pharmacy, Faculty of Medical Sciences, University of Kragujevac, Kragujevac, Serbia
e-mail: nevena.jeremic@louisville.edu

J. Bradic · A. Petkovic
Department of Pharmacy, Faculty of Medical Sciences, University of Kragujevac, Kragujevac, Serbia

G. Weber
Department of Physiology, School of Medicine, Uniuversity of Louisville, Louisville, KY, USA

S. Chakraborti et al. (eds.), *Modulation of Oxidative Stress in Heart Disease*, https://doi.org/10.1007/978-981-13-8946-7_14

for regulation of cellular redox state, while SAM is a main methyl donor in organisms, and is involved in the methylation pathway of Hcy. The exact mechanism(s) of HHcy-induced endothelial dysfunction has(ve) not been fully clarified. However, it's been proposed that endothelial dysfunction may be mediated by initiation of ROS production and reduction in capacity of antioxidant defense system. Therefore, in this chapter, we tried to consolidate current findings regarding role of oxidative stress in hyperhomocysteinemia.

Keywords

Homocysteine · Oxidative stress · Cardiovascular system · Heart

14.1 Introduction

Homocysteine (Hcy) is a sulfhydryl-containing amino acid which is synthesized during metabolism of methionine (Met) [1]. There are two major metabolic routes for elimination of Hcy: remethylation and transsulfuration pathways [2]. Remethylation refers to the transfer of methyl groups from 5-methyltetrahydrofolate (MTHF) to Hcy to form Met and occurs during Met deficiency. Less than 50% of Hcy is catabolized to cysteine and taurine (as final urinary products) by transsulfuration pathway [3]. This route is significant for removing excess Hcy, which is not necessary for methyl transfer. Furthermore, metabolism of Hcy requires the presence of cofactors, such as vitamin B12, B6, and folate [1]. Cellular export mechanisms exist to help maintain low levels of plasma Hcy (tHcy) (10 μmol/L) and together with transsulfuration pathway are responsible for the presence of low intracellular concentration [4]. The most reactive form of Hcy, known as cyclic thioester—Hcy thiolactone, forms during the metabolism of Hcy by methionyl-tRNA synthetase [5].

The total level of plasma tHcy may be altered by diet, lifestyle, and genetic factors [1]. It has been established that a diet deficient in folate, vitamin B6, and B12 as well as alcohol intake, smoking habits, and lack of physical activity may lead to an increase in tHcy. Hyperhomocysteinemia (HHcy) is a condition defined as having abnormally high levels of Hcy in plasma (above 15 μmol/L). Depending on the level of tHcy, hyperhomocysteinemia is classified into mild (between 16 and 30 μmol/L), intermediate (31–100 μmol/L), and severe (above 100 μmol/L) [1]. Excessively high levels of Hcy (above 500 μmol/L) are associated with accumulation of homocysteine and its metabolites in the urine-homocystinuria [6]. In this condition, metabolism of Hcy is disrupted, and vascular tissue is exposed to its adverse effects.

14.2 Homocysteine and Cardiovascular Diseases

Cardiovascular disease (CVD) is characterized as a class of diseases related to the heart or blood vessels and remains the leading cause of mortality and morbidity worldwide [7]. Therefore, significant effort has been invested in identifying new risk factors for CVD, as well as prevention. Mounting evidence suggests an association between HHcy and vascular diseases [1]. Based on epidemiological, retrospective, and prospective studies, hyperhomocysteinemia is considered as an independent risk factor for coronary heart, cerebrovascular, and peripheral artery diseases [8]. Data suggests that for every 4 μM increase in the level of Hcy, the relative risk for cardiovascular diseases rises by 1.3–1.4 [9]. It has been reported that in patients with Hcy levels below 9 μM/L, overall mortality was 3.8%, while patients with Hcy levels above 15 μM/L had a 24.7% increase in mortality [10]. Pronounced and sustained oxidative stress is proposed as one of the mechanisms underlying Hcy-induced CVD.

14.3 Hcy and Endothelial Dysfunction: Role of Oxidative Stress

Hcy may affect any component of the arterial wall; however, the most prominent effects and resulting damage are found in the endothelium. A considerable number of studies have been conducted in order to reveal the mechanisms through which Hcy contributes to endothelial injury, a critical event in the pathogenesis of atherosclerosis. Endothelial dysfunction is characterized by impaired endothelium-dependent relaxation due to a decrease in available nitric oxide (NO). This pathological state is associated with HHcy and is involved in pathogenesis of hypertension, diabetes, atherosclerosis, and renal and cardiac failure [1, 11].

McCully was the first to propose the "homocysteine hypothesis of arteriosclerosis" about 50 years ago [12]. He described the case of atherothrombosis in children with elevated Hcy, cystathionine, and Hcy-cysteine disulfide concentrations in plasma and urine. Since those discoveries, a substantial amount of scientific work has focused on evaluating the effects of Hcy on cardiovascular tissues. Numerous data support a role for Hcy involvement in the development of atherosclerosis through direct damage of the endothelium or by altering redox status. Hcy exerts harmful effects on vascular endothelium and smooth muscle cells, leading to impairment of arterial structure and function. The underlying mechanisms involve an increase in coagulation, synthesis of collagen, proliferation of vascular smooth muscle cells, initiation of inflammatory response, and elevated generation of pro-oxidants [13].

14.4 Homocysteine-Induced Oxidative Stress

Reactive oxygen species (ROS) are generated mostly as by-products of mitochondrial respiration, and under physiological conditions, there is a balance between production and elimination of free radicals [14]. An increase in ROS levels may affect mitochondrial gene expression, leading to DNA damage, and impair mitochondrial function. Redox homeostasis is regulated by several intermediates involved in the methionine cycle, such as glutathione, hydrogen sulfide (H_2S), and S-adenosyl methionine (SAM). Glutathione and H_2S are responsible for regulation of cellular redox state, while SAM is a main methyl donor in organisms and is involved in there-methylation pathway of Hcy [15]. Investigations have reported that intracellular accumulation of Hcy alters mitochondrial structure responsible for oxidative phosphorylation and ATP synthesis. Therefore, it's been suggested that Hcy-induced mitochondrial damage may be mediated via enhanced oxidative stress [1, 16].

The exact mechanism(s) of HHcy-induced endothelial dysfunction have not been fully clarified. However, it's been proposed that endothelial dysfunction may be mediated by initiation of ROS production and reduction in capacity of antioxidant defense system [11]. The free thiol group of Hcy binds, via disulfide bond, to plasma proteins or other Hcy, thus promoting generation of ROS. Prooxidants, such as strong oxidizing agent hydrogen peroxide (H_2O_2), ROS, and reactive nitrogen species (RNS), are produced as a result of Hcy binding. Hydroxyl radicals have the potential to induce further oxidation of lipids, proteins, carbohydrates, and nucleic acids, thus contributing to endothelial dysfunction. ROS and RNS are highly reactive molecules that can with almost all cellular components and thereby compromise cellular integrity [2].

Hcy-induced oxidative stress is caused by several mechanisms. In addition to Hcy autooxidation, hyperhomocysteinemia causes a decrease in antioxidant potential of the cell due to inhibition of activity of enzymatic antioxidants, such as superoxide dismutase (SOD) from endothelial surfaces [17]. Furthermore, this pathological condition affects enzymes such as nicotinamide adenine dinucleotide phosphate oxidase (NADPH oxidase, NOX) and nitric oxide synthase (NOS), which are involved in production of ROS [1].

The major non-mitochondrial source of ROS is NOX [18]. The impact of Hcy on various forms of NOX has been reported. Several genes responsible for variations in plasma Hcy levels have been identified so far. One of the gene loci associated with Hcy metabolism is NADPH oxidase 4 (NOX 4), an isoform of NOX [19]. Investigations have shown that incubation of tubular cells with Hcy was connected with an increased expression of NOX4, resulting in a high generation of superoxide anion radical (O_2^-) [20]. On the other hand, overproduction of O_2^- may be achieved by activation of angiotensin-converting enzyme (ACE) by Hcy, which is involved in the activation of NOX [21]. The NOX 2 isoform is present in the endothelium, and it has been confirmed that Hcy may also induce apoptosis of endothelial cells via NADPH oxidase-related oxidative stress. One research study has shown that incubation of human umbilical vein endothelial cells (HUVECs) with Hcy resulted in increased expression of the NOX 2 isoform. As a result, there was increased production of O_2^- and accumulation of nitrotyrosine residues. In addition to endothelial cells, a role for Hcy in programmed cell death of cardiomyocytes is proposed as well through activation of NOX 2-induced oxidative damage [22]. Moreover, Hcy may lead to necrosis and membrane flip-flop, due to inactivation of flippase [23].

Activation of endothelium under pathophysiological conditions is associated with suppression of NO signaling [11]. Normally, there are three isoforms of enzyme nitric oxide synthase: endothelial NOS (eNOS or NOS1), inducible NOS (iNOS or NOS2), and neural NOS (nNOS or NOS3) [1]. Nitric oxide (NO) can be generated by all NOS isoforms, which utilize L-arginine as the substrate in the presence of molecular oxygen and NADPH. Other significant cofactors in the NO synthase reaction are flavin mononucleotide (FMN), flavin adenine dinucleotide (FAD), and tetrahydrobiopterin (BH4) [24].

It has been established that when endothelial cells are damaged, bioavailability of NO is lowered, and various studies have reported that homocysteine may alter the NOS pathway. At physiological concentrations, tHcy reacts with NO in the presence of oxygen and forms S-nitroso-homocysteine, resulting in inhibition of sulfhydryl dependent generation of H_2O_2 [25]. S-nitroso-homocysteine possesses vasodilatory and antiplatelet properties, but in comparison to Hcy, it is not converted to Hcy thiolactone [26]. In hyperhomocysteinemia, one of the possible explanations for decreased NO bioavailability is oxidative degradation of NO due to an increase in the production of H_2O_2 and O_2^- [27]. In fact, NO interacts with ROS, particularly O_2^-, resulting in generation of very toxic peroxynitrite ($ONOO^-$) [28]. Peroxynitrite induces tyrosine nitration, which alters protein function, thus leading to endothelial dysfunction [29].

Hydroxyl radical formation in the presence of high Hcy promotes lipid peroxidation, leading to a decrease in eNOS expression and NO degradation [30, 31]. Additionally, HHcy may inhibit dimethylarginine dimethylaminohydrolase 2 (DDAH), an enzyme that catalyzes the conversion of asymmetric dimethylarginine (ADMA) to L-arginine, citrulline, and dimethylamine. Therefore, accumulation of ADMA inhibits activity of eNOS and iNOS, thus decreasing NO bioavailability [32, 33]. On the other hand, Hcy can induce pro-inflammatory cytokines, which subsequently upregulate iNOS, leading to an increase in NO synthesis [1].

A correlation between high Hcy levels and the concentration of malondialdehyde (MDA), a marker of lipid peroxidation, was confirmed in patients with stable or unstable angina [30]. On the other hand, it was reported that moderate hyperhomocysteinemia did not worsen oxidative status in these patients. However, other factors apart from Hcy are involved in impaired redox state in patients with coronary artery disease. Lipid peroxidation initiated by ROS during HHcy results in oxidation of LDL, a very potent pro-atherosclerotic mediator [31]. Several enzyme systems are involved, such as NOX, mitochondrial electron transport enzymes, xanthine oxidase, lipoxygenase, cyclo-oxygenase, myeloperoxidase, cytochrome P450 enzymes, and eNOS. Formation of oxidized LDL cholesterol is responsible for the presence of a prooxidant state in all phases of atherosclerosis, from beginning to the acute thrombotic events [1].

14.5 Homocysteine and Atherothrombosis: Role of Oxidative Stress

Since the early 1990s, it has been known that elevated levels of homocysteine may modulate functions and properties of endothelial cells, causing changes in the hemostasis system. As a result, there is an imbalance between procoagulant and anticoagulant factors. Data suggests that subjects with high levels of tHcy have an increased chance to develop venous thrombosis. Reasons for formation and propagation of thrombus are abnormalities of blood flow, blood vessel wall, and blood-clotting components [34]. Furthermore, increased levels of homocysteine leads to platelet adhesion to endothelial cells, as well as increment of β-thromboglobulin, tissue plasminogen activator, and factor VIIc [8].

In addition to the abovementioned pathways, elevation of platelet activation represents another risk factor in the pathogenesis of various cardiovascular diseases. An important reason for platelet activation and coagulation can be endothelial injury induced by an imbalance in oxygen concentration, oxidative stress, cytokines, and thrombin. Blood platelets can generate different reactive oxygen/nitrogen species (ROS/RNS) which can be produced by several pathways, such as arachidonic acid pathway (via cyclooxygenase or 12-lipoxygenase) stimulated by different agonists, the glutathione cycle and metabolism of phosphoinositides [35]. This production of ROS in platelets is due to activation of NOX [36–38] and xanthine oxidase [39]. Homocysteine is involved in release of arachidonic acid, formation of thromboxane A2, and protein tyrosine phosphorylation in blood platelets [35].

One of the main molecules that plays a role in thrombus formation is NO [8]. Previous investigators were focused on antithrombogenic properties of NO in arterioles rather than in venules. It was established that homocysteine-induced platelet activity may be due to promotion of oxidative stress and inhibition of NO formation in platelets [35]. Reduced levels of NO may be due to nitrosation of Hcy (reaction of HCy with NO to form S-nitroso-homocysteine). Furthermore, generation of peroxynitrite in blood platelets by Hcy causes cell death in cardiomyocyte cell line H9C2 [39].

Fibrinogen as the main substrate for coagulation cascade can be covalently modified by Hcy. Modification of lysine residues in fibrinogen, major binding sites for fibrinolytic enzymes, may alter fibrinolysis by the N-homocysteinylation reaction. Important functional consequences of homocysteinylation are (1) alteration of clot and fibrin structure and (2) increased resistance to fibrinolysis [40]. Elevated levels of coagulation factor VIII further increase the risk of development of CVD during hyperhomocysteinemia [41].

14.6 Homocysteine and Hypertension

A role for Hcy in the pathogenesis of essential hypertension has been reported. Incremental concentrations of Hcy directly affects the value of blood pressure. However, this impact is gender specific, with the stronger effects noticed in women. Elevation of 5 μM/L in concentration of Hcy induces a rise of 0.5 mmHg and 0.7 mmHg in diastolic and 0.7 and 1.2 mmHg in men and women, respectively. The presence of hyperhomocysteinemia attenuates vasodilatory effects of NO, enhances production of ROS, and alters vascular wall elasticity. Additionally, biosynthesis and function of vasodilator factors in vascular wall is strongly affected by high levels of Hcy. Consequently, there is impairment in formation of extracellular matrix components and intense myocyte proliferation and migration [8].

Hcy induces oxidative stress by changing the redox thiol status of smooth muscle vascular cells, which have redox-sensitive homocysteine receptor responsible for collagen expression. Therefore, there is activation of nuclear factor kappa-B (NF-kB) which contributes to enhanced vascular smooth muscle proliferation [11, 42]. Moreover, Hcy may increase tumor necrosis factor-alpha (TNFα) expression in coronary arteries, resulting in upregulation of NOX and iNOS [43]. These observations support a role for HHcy exerting harmful effects via oxidant injury of the endothelium. As a consequence of enhancement of collagen synthesis and accumulation, vascular structure is worsened, and systemic vascular resistance is increased. Additionally, diastolic dysfunction of the vessels and loss in flexibility occur. Abovementioned deleterious effects of HHcy are responsible for Hcy-induced elevation of blood pressure and development of hypertension [8, 11].

14.7 Positive Effects of Homocysteine

It was found that in homocystinuric patients, level of measured Hcy positively correlated with activity of extracellular superoxide dismutase (SOD) and GSHPx which are an important antioxidant in vascular tissue. This can be explained as protective response to overproduction of free radicals induced by Hcy. Consequently risk of vascular events in these patients is not extremely high [1, 44]. On the other hand, another protective effect of Hcy is reflected in reduction of endothelin-1 production, powerful vasoconstrictor, via oxidative stress products [45]. Furthermore, it was reported that Hcy at micromolar concentrations produces negligible quantities of H_2O_2 and does not elevate peroxynitrite formation but blocks dihydrorhodamine oxidation. Therefore, Hcy is acting as antioxidant on cellular and chemical system [46].

14.8 Therapeutic Strategies in Hyperhomocysteinemia

Therapies for lowering levels of Hcy are safe and relatively inexpensive. As mentioned earlier, folic acid and B vitamins are necessary for remethylation of homocysteine to methionine. Folic acid is considered to be more effective than vitamins B6 and B12, but its daily dose in clinical studies varies [47]. Levels of Hcy can be reduced about 25% by supplementation with 0.5–5.0 mg of folic acid. Vitamin B_{12} supplementation of at least 0.4 mg daily further lowers levels by 7%. B6 supplements can lower Hcy after loading of methionine [48]. A prospective, randomized clinical trial examined impact of prolonged administration of folate combined with vitamins B_6 and B_{12} on cardiovascular risk. They revealed that daily administration reduced Hcy levels significantly; however, the risk of death from cardiovascular causes, myocardical infarction, and stroke was not changed [49]. On the other hand, a few smaller trials showed no benefit of treatment with folate and vitamin B complex [50]. This discrepancy between results of clinical trials can be due to inherent limitations of observational studies.

It has been reported that administration of vitamins C and E to mild HHcy subjects disturbs activation of hemocoagulation and reduces potential for endothelial adhesion. In addition, supplementation with only vitamin C prevented endothelial cells deterioration due to high Hcy concentration [47, 51].

The effect of folate on pro-oxidative impact of Hcy has not been fully investigated. Nevertheless, folate supplementation inhibits proliferation of smooth muscle cells of the vessel wall, thus abolishing adverse effects induces by free radicals. However, administration of vitamins B6 and B12 did not lead to this effect. Another experiment showed that folate can improve endothelial function in HHcy. After 3 months of folate administration with B6 vitamin, makers of endothelial damage (soluble thrombomodulin and von Willebrand factor) were lowered, while the concentration of glutathione was not affected. Furthermore, concentrations of Hcy and LDL oxidation can be reduced by administration of vitamin cocktail with 0.65 mg folate [52–54].

References

1. Djuric D, Jakovljevic V, Zivkovic VI, Srejovic IM (2018) Homocysteine and homocysteine-related compounds: an overview of the roles in the pathology of the cardiovascular and nervous systems. Can J Physiol Pharmacol 96(10):991–1003
2. Škovierová H, Vidomanová E, Mahmood S, Sopková J, Drgová A, Červeňová T, Halašová E, Lehotský J (2016, Oct 20) The molecular and cellular effect of homocysteine metabolism imbalance on human health. Int J Mol Sci 17(10). pii: E1733. Review. PubMed PMID: 27775595; PubMed Central PMCID: PMC5085763
3. Cacciapuoti F (2018) Poor re-methylation of homocysteine and trans-methylation of methionine: cause and effect of hyper-homocysteinemia: which role for folic acid and Vitamins B-6-12 supplementation? Ann Clin Exp Metabol 3(1):1026
4. Ciaccio M, Bivona G, Bellia C (2008) Therapeutical approach to plasma homocysteine and cardiovascular risk reduction. Ther Clin Risk Manag 4(1):219–224
5. Karolczak K, Olas B (2009) Mechanism of action of homocysteine and its thiolactone in hemostasis system. Physiol Res 58(5):623–633
6. Kumar T, Sharma GS, Singh LR (2016) Homocystinuria: therapeutic approach. Clin Chim Acta 458:55–62
7. Boutayeb A, Boutayeb S (2005) The burden of non communicable diseases in developing countries. Int J Equity Health 4(1):2
8. Ganguly P, Alam SF (2015) Role of homocysteine in the development of cardiovascular disease. Nutr J 14:6
9. Nygård O, Nordrehaug JE, Refsum H, Ueland PM, Farstad M, Vollset SE (1997) Plasma homocysteine levels and mortality in patients with coronary artery disease. N Engl J Med 337(4):230–236
10. Graham MR, Grace FM, Boobier W, Hullin D, Kicman A, Cowan D, Davies B, Baker JS (2006) Homocysteine induced cardiovascular events: a consequence of long term anabolic-androgenic steroid (AAS) abuse. Br J Sports Med 40(7):644–648
11. Pushpakumar S, Kundu S, Sen U (2014) Endothelial dysfunction: the link between homocysteine and hydrogen sulfide. Curr Med Chem 21(32):3662–3672
12. McCully KS (1969) Vascular pathology of homocysteinemia: implications for the pathogenesis of arteriosclerosis. Am J Pathol 56(1):111–128
13. Zhang S, Yong-Yi B, Luo LM, Xiao WK, Wu HM, Ye P (2014) Association between serum homocysteine and arterial stiffness in elderly: a community-based study. J Geriatr Cardiol 11:32–38

14. Zorov DB, Juhaszova M, Sollott SJ (2014) Mitochondrial reactive oxygen species (ROS) and ROS-induced ROS release. Physiol Rev 94(3):909–950
15. Mohamed R, Sharma I, Ibrahim AS, Saleh H, Elsherbiny NM, Fulzele S, Elmasry K, Smith SB, Al-Shabrawey M, Tawfik A (2017) Hyperhomocysteinemia alters retinal endothelial cells barrier function and angiogenic potential via activation of oxidative stress. Sci Rep 7(1):11952
16. Wald DS, Law M, Morris JK (2002) Homocysteine and cardiovascular disease: evidence on causality from a meta-analysis. BMJ 325(7374):1202
17. Petras M, Tatarkova Z, Kovalska M, Mokra D, Dobrota D, Lehotsky J, Drgova A (2014) Hyperhomocysteinemia as a risk factor for the neuronal system disorders. J Physiol Pharmacol 65(1):15–23
18. Dan Dunn J, Alvarez LA, Zhang X, Soldati T (2015) Reactive oxygen species and mitochondria: a nexus of cellular homeostasis. Redox Biol 6:472–485
19. Murray TV, Dong X, Sawyer GJ, Caldwell A, Halket J, Sherwood R, Quaglia A, Dew T, Anilkumar N, Burr S, Mistry RK, Martin D, Schröder K, Brandes RP, Hughes RD, Shah AM, Brewer AC (2015) NADPH oxidase 4 regulates homocysteine metabolism and protects against acetaminophen-induced liver damage in mice. Free Radic Biol Med 89:918–930
20. Hwang SY, Siow YL, Au-Yeung KK, House J, O K (2011) Folic acid supplementation inhibits NADPH oxidase-mediated superoxide anion production in the kidney. Am J Physiol Ren Physiol 300(1):F189–F198
21. Huang A, Pinto JT, Froogh G, Kandhi S, Qin J, Wolin MS, Hintze TH, Sun D (2015) Role of homocysteinylation of ACE in endothelial dysfunction of arteries. Am J Physiol Heart Circ Physiol 308(2):H92–H100
22. Sipkens JA, Hahn N, van den Brand CS, Meischl C, Cillessen SA, Smith DE, Juffermans LJ, Musters RJ, Roos D, Jakobs C, Blom HJ, Smulders YM, Krijnen PA, Stehouwer CD, Rauwerda JA, van Hinsbergh VW, Niessen HW (2013) Homocysteine-induced apoptosis in endothelial cells coincides with nuclear NOX2 and peri-nuclear NOX4 activity. Cell Biochem Biophys 67(2):341–352
23. Sipkens JA, Krijnen PA, Hahn NE, Wassink M, Meischl C, Smith DE, Musters RJ, Stehouwer CD, Rauwerda JA, van Hinsbergh VW, Niessen HW (2011) Homocysteine-induced cardiomyocyte apoptosis and plasma membrane flip-flop are independent of S-adenosylhomocysteine: a crucial role for nuclear p47(phox). Mol Cell Biochem 358(1–2):229–239
24. Förstermann U, Sessa WC (2012) Nitric oxide synthases: regulation and function. Eur Heart J 33(7):829–837, 837a-837d
25. Angeline T, Aruna RM, Ramadevi K, Mohan G, Jeyaraj N (2005) Homocysteine status and acute myocardial infarction among Tamilians. Indian J Clin Biochem 20(1):18–20
26. Stamler JS, Osborne JA, Jaraki O, Rabbani LE, Mullins M, Singel D, Loscalzo J (1993) Adverse vascular effects of homocysteine are modulated by endothelium-derived relaxing factor and related oxides of nitrogen. J Clin Invest 91(1):308–318
27. Stühlinger MC, Tsao PS, Her JH, Kimoto M, Balint RF, Cooke JP (2001) Homocysteine impairs the nitric oxide synthase pathway: role of asymmetric dimethylarginine. Circulation 104(21):2569–2575
28. Bradic J, Dragojlovic Ruzicic R, Jeremic J, Petkovic A, Stojic I, Nikolic T, Zivkovic V, Srejovic I, Radovanovic D, Jakovljevic VL (2018) Comparison of training and detraining on redox state of rats: gender specific differences. Gen Physiol Biophys 37(3):285–297
29. Cheng Z, Yang X, Wang H (2009) Hyperhomocysteinemia and endothelial dysfunction. Curr Hypertens Rev 5(2):158–165
30. Cavalca V, Cighetti G, Bamonti F, Loaldi A, Bortone L, Novembrino C, De Franceschi M, Belardinelli R, Guazzi MD (2001) Oxidative stress and homocysteine in coronary artery disease. Clin Chem 47(5):887–892
31. Romerio SC, Linder L, Nyfeler J, Wenk M, Litynsky P, Asmis R, Haefeli WE (2004) Acute hyperhomocysteinemia decreases NO bioavailability in healthy adults. Atherosclerosis 176(2):337–344

32. Stühlinger MC, Tsao PS, Her JH, Kimoto M, Balint RF, Cook JP (2001) Homocysteine impairs the nitric oxide synthase pathway role of asymmetric dimethylarginine. Circulation 104:2569–2575
33. Sibal L, Agarwal SC, Home PD, Boger RH (2010) The role of Asymmetric Dimethylarginine (ADMA) in endothelial dysfunction and cardiovascular disease. Curr Cardiol Rev 6(2):82–90
34. Ekim M, Sekeroglu MR, Balahoroglu R, Ozkol H, Ekim H (2014) Roles of the oxidative stress and ADMA in the development of deep venous thrombosis. Biochem Res Int 2014:703128
35. Karolczak K, Olas B (2009) Mechanism of action of homocysteine and its thiolactone in hemostasis system. Physiol Res 58(5):623–633
36. Forde RC, Fitzerald DJ (1997) Reactive oxygen species and platelet activation in reperfusion injury. Circulation 95:787–789
37. Krotz F, Sonh HY, Pohl U (2004) Reactive oxygen species, players in the platelet game. Arterioscler Thromb Vasc Biol 24:1988–1996
38. Begonija AJ, Gambaryan S, Geiger J, Aktas B, Pozgajova M, Nieswandt B, Walter U (2005) Platelet NAD(P)H oxidase-generated ROS production regulates {alpha}IIb{beta}3 integrin activation independent of the NO/cGMP pathway. Blood 106:2757–2760
39. Miller DM, Grover TA, Nayini N, Aust SD (1993) Xantine-oxidase- and iron-dependent lipid peroxidation. Arch Biochem Biophys 301:1–7
40. Broeders MAW, Tangelder G-J, Slaaf DW, Reneman RS, Oude Egbrink MGA (1998) Endogenous nitric oxide protects against thromboembolism in venules but not in arterioles. Arterioscler Thromb Vasc Biol 18(1):139–145
41. Sauls DL, Arnold EK, Bell CW, Allen JC, Hoffman M (2007) Pro-thrombotic and prooxidant effects of diet-induced hyperhomocysteinemia. Thromb Res 120:117–126
42. Tyagi SC (1998) Homocysteine redox receptor and regulation of extracellular matrix components in vascular cells. Am J Phys 274(2 Pt 1):C396–C405
43. Stanger O, Weger M (2003) Interactions of homocysteine, nitric oxide, folate and radicals in the progressively damaged endothelium. Clin Chem Lab Med 41(11):1444–1454
44. Wilcken DE, Wang XL, Adachi T, Hara H, Duarte N, Green K, Wilcken B (2000) Relationship between homocysteine and superoxide dismutase in homocystinuria: possible relevance to cardiovascular risk. Arterioscler Thromb Vasc Biol 20(5):1199–1202
45. Drunat S, Moatti N, Paul JL, Cogny A, Benoit MO, Demuth K (2001) Homocysteine-induced decrease in endothelin-1 production is initiated at the extracellular level and involves oxidative products. Eur J Biochem 268(20):5287–5294
46. Zappacosta B, Mordente A, Persichilli S, Minucci A, Carlino P, Martorana GE, Giardina B, De Sole P (2001) Is homocysteine a pro-oxidant? Free Radic Res 35(5):499–505
47. Racek J, Rusnáková H, Trefil L, Siala KK (2005) The influence of folate and antioxidants on homocysteine levels and oxidative stress in patients with hyperlipidemia and hyperhomocysteinemia. Physiol Res 54(1):87–95
48. Clarke R (2000) Lowering blood homocysteine with folic acid-based supplements: meta-analysis of randomised trials. Indian Heart J 52(7 Suppl):S59–S64
49. Lonn E, Held C, Arnold JM, Probstfield J, McQueen M, Micks M, Pogue J, Sheridan P, Bosch J, Genest J, Yusuf S, HOPE-2 Investigators (2006) Rationale, design and baseline characteristics of a large, simple, randomized trial of combined folic acid and vitamins B6 and B12 in high-risk patients: the Heart Outcomes Prevention Evaluation (HOPE)-2 trial. Can J Cardiol 22(1):47–53
50. Toole JF, Malinow MR, Chambless LE, Spence JD, Pettigrew LC, Howard VJ, Sides EG, Wang CH, Stampfer M (2004) Lowering homocysteine in patients with ischemic stroke to prevent recurrent stroke, myocardial infarction, and death: the Vitamin Intervention for Stroke Prevention (VISP) randomized controlled trial. JAMA 291:565–575
51. Cheng Z, Yang X, Wang H (2009) Hyperhomocysteinemia and endothelial dysfunction. Curr Hypertens Rev 5(2):158–165

52. Armitage JM, Bowman L, Clarke RJ, Wallendszus K, Bulbulia R, Rahimi K, Haynes R, Parish S, Sleight P, Peto R, Collins R (2010) Effects of homocysteine-lowering with folic acid plus vitamin B12 vs placebo on mortality and major morbidity in myocardial infarction survivors: a randomized trial. Study of the Effectiveness of Additional Reductions in Cholesterol and Homocysteine (SEARCH) Collaborative Group. JAMA 303(24):2486–2494
53. Bunout D, Garrido A, Suazo M, Kauffman R, Venegas P, de la Maza P, Petermann M, Hirsch S (2000) Effects of supplementation with folic acid and antioxidant vitamins on homocysteine levels and LDL oxidation in coronary patients. Nutrition 16(2):107–110
54. Constans J, Blann AD, Resplandy F, Parrot F, Renard M, Seigneur M, Guérin V, Boisseau M, Conri C (1999) Three months supplementation of hyperhomocysteinaemic patients with folic acid and vitamin B6 improves biological markers of endothelial dysfunction. Br J Haematol 107(4):776–778

Nitrosative Stress and Cardiogenesis: Cardiac Remodelling Perturbs Embryonic Metabolome

15

Pavitra Kumar, Lakshmikirupa Sundaresan, and Suvro Chatterjee

Abstract

Nitrosative stress because of hyperactive redox milieu is thought to be associated with decrease in bioavailability of nitric oxide and the subsequent defective cardiogenesis. The role of nitrosative stress in pathophysiology of the heart in adults has been studied for several decades; however, very few studies link the structural deformities in the heart with nitrosative stress. In this article, we give a detailed discussion of evidence of the impact of nitrosative stress during cardiogenesis and also the effect of following cardiac remodelling on the metabolism of the embryo. We highlight specifically the reactive nitrogen species (RNS)-mediated structural changes in the cardiac looping and predicted its consequences on embryonic metabolism using transcriptome analysis. In the present study, we used thalidomide as RNS inducer, which increases peroxynitrite and superoxide levels in the developing heart. The transcriptome analysis of thalidomide-treated embryos showed that the treatment affected severely the protein and fatty acid metabolism that consequently might lead to thalidomide-mediated heart defects in the embryo. To summarize, our data suggest that fatty acid metabolism, which is a critical metabolic pathway during heart development, is perturbed under an oxidative and nitrosative environment due to thalidomide treatment.

Keywords

Nitrosative stress · Congenital heart diseases · Cardiogenesis · Thalidomide · Transcriptome · Metabolome

P. Kumar
AU-KBC Research Centre, Anna University, Chennai, Tamil Nadu, India

L. Sundaresan · S. Chatterjee (✉)
AU-KBC Research Centre, Anna University, Chennai, Tamil Nadu, India

Department of Biotechnology, Anna University, Chennai, Tamil Nadu, India

S. Chakraborti et al. (eds.), *Modulation of Oxidative Stress in Heart Disease*,
https://doi.org/10.1007/978-981-13-8946-7_15

15.1 Background

Structural deformities arising from abnormal formation of the heart or major blood vessels at the time of birth are commonly referred to as congenital heart defects, or diseases (CHDs) [1]. Congenital heart defects have frequency of 1 in 100 live births and are the most common birth defects [2–4]. CHDs alone account for approximately 10% of the present infant mortality [5]. In India, there is a large number of CHDs cases every year with a frequency of 3.7–17.5 per 1000 live births [6]. The primary aetiology of CHDs may be genetic, environmental or a combination of both. A few genetic factors linked to CHDs are known, whereas majority of them are still unknown. Therefore, it becomes a primary interest to understand the mechanisms regulating the heart formation which might be helpful to develop new strategies to prevent and treat CHDs. The major CHDs are tetralogy of Fallot (TOF), transposition of large vessels, hypoplastic right heart, acyanotic CHD, ventricular septal defect (VSD), patent ductus arteriosus (PDA), pulmonary stenosis and atrial septal defect (ASD). Aetiology of CHDs and their pattern of inheritance are multifactorial [7]. The risk factors of CHDs could be maternal ailments such as maternal smoking, gestational diabetes, urinary tract infections or foetal factors such as still birth, low birth weight and prematurity [8]. Nitric oxide and its products play a very important role during embryonic development, e.g. follicle development, germ cells differentiation and organogenesis. Supraphysiological concentration of NO is associated with various female reproductive irregularities including infertility [9], inhibition of implantation and inhibition of embryonic growth [10], whereas deficiency in NOS3 results in congenital heart defects [11]. In present study, our omics-based approach researches the link between nitrosative stress and anomaly of cardiac morphogenesis in the developing heart of chicken embryo.

15.2 Cardiac Morphogenesis

The heart is the first functional organ which begins to beat from 2nd week of gestation and is fully developed by 8 weeks of gestation [12]. Migration of epiblastic cells towards primitive streak differentiates the gastrula into three germ layers: ectoderm, mesoderm and endoderm [13]. Precardiac cells are multipotent and get differentiated into endothelial, smooth muscle cells and myocardiac cells [14]. Anterior mesodermal cells (precardiac cells) get differentiated into cardiac cells under the influence of endodermal signalling molecules such as BMPs [15]. Some of the highly conserved mesodermal origin transcription factors through the evolution are NKX2.5, GATA, Mef2, Tbx and Hand [16]. Myocardial cells differentiate into conduction cells and chamber-specific myocytes [17]. Structurally there are three intricately linked phases of heart development: looping, convergence and wedging (Fig. 15.1).

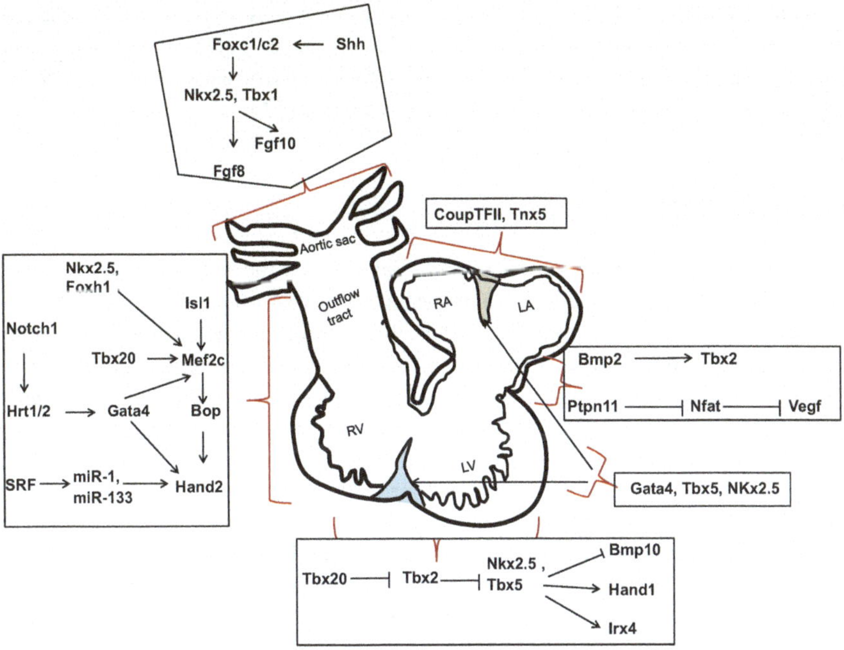

Fig. 15.1 Regulatory genes for chamber formation in the heart

15.2.1 Looping

The primitive linear heart tube folds into S-shaped structure towards the right and begin the right-left lateralization in the embryo [18]. Rotatory movement of cilia in primary node creates an extracellular flow current which bends the tube towards right. This step locates the future heart chambers into their relative spatial positions. Faulty looping results in the anomalies related to heart laterality such as heterotaxy syndrome and situs inversus [19].

15.2.2 Convergence

The convergence brings an outflow limb and the inflow limb together which permits the alignment of outflow tract with atrioventricular (AV), atrial and ventricular septa [18]. The atria and ventricles develop and segregate along anteroposterior and the right-left axes [20]. At this stage, the primitive heart develops into four transitional zones called sinus venosus, AV canal, primary fold and outflow tract endocardial cushions. The sinus venosus develops into atrial septation and contributes to the atrial conduction pathways [21]. Primary fold contributes to ventricular septation, ventricular conduction pathways and AV node [18]. The OT and the endocardial cushions of the AV canal delineate to the outlet segment and the inlet segment of the heart, respectively [22]. Defects at convergence stage lead to the retarded ventricular growth and the abnormal development of the right AV junction. The major congenital heart defects occurred as results of these malformations are ventricular hypoplasia, double-inlet ventricle and tricuspid atresia [23].

15.2.3 Wedging

Counterclockwise rotation in the myocardial wall of the OT establishes mitral-aortic continuity. Simultaneously, endocardial cushions of the OT are muscularized and fused to form conal septum [18, 24]. Conal septum together with the rearrangement of upper primitive ventricular septum establishes the mitroaortic fibrous continuity. Failure in the proper alignment of OT and ventricles inevitably results in ventricular septal defect [25]. TOF might result from the hampered OT looping which result in malalignment of the OT and the ventricles [26].

15.3 Molecular Candidates Involved in Chamber Specification

Shh-mediated Foxc1/c2 activation followed by involvement of Nkx2.5, Tbx5, Fgf8 and fgf10 regulates the formation of aortic sac during cardiogenesis [27]. Thalidomide is reported to bind with TBX5 at amino acids R81, R82 and K226, which are associated in DNA binding which further inhibits thalidomide's interaction with HAND [28]. However there is no report indicating the involvement of RNS in this interaction. CoupTFII and Tnx5 are responsible for the differentiation the left and right aorta. Tbx family, BMP10, NKX2.5 and HAND are associated with the formation of right and left ventricles [29]. GATA4, TBX5 and NKX2.5 mediate the formation of intra-atrial and intraventricular septa [30]. Nkx2.5, Notch,1, Isl1 and Hand2 pathways mediate the formation of outflow tract and right ventricle (Fig. 15.1) [31].

15.4 Nitrosative Stress During Embryogenesis

Excess production of reactive nitrogen species (RNS) overcomes the system's ability to neutralize and eliminate them and results in pathophysiological condition called nitrosative stress [32]. Generally, nitrosative stress and oxidative stress act collectively to damage the cells [33]. In the biological system, nitrosative stress is generated by nitric oxide-mediated modulation of biomolecules [34]. The molecular species responsible for nitrosative stress are referred as reactive nitrogen species (RNS). RNS family of molecules are formed by the reaction of superoxide ($O_2^{\bullet-}$)

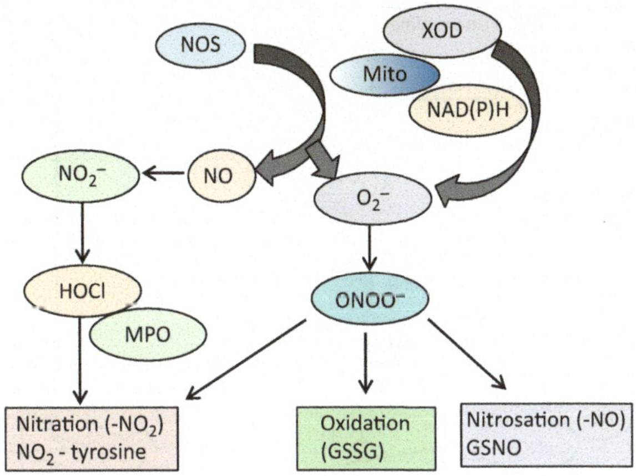

Fig. 15.2 Effects of reactive nitrogen species on biomolecules

with nitric oxide (•NO) [35]. The primary source of all RNS is nitric oxide. The major reactions of NO which lead to the oxidation of biomolecules are nitrosylation (NO), nitration (addition of NO_2) and nitrosation (addition of NO^+) (Fig. 15.2) [36].

In the cellular system, NO is produced by nitric oxide synthases (NOS) family using the semi-essential cationic amino acid L-arginine as substrate and mediates multiple cellular functions [37]. Emphasizing the fact that concentration of NO acts as an important factor in the production of RNS, the range of NO concentrations produced by NOS is 20 nM to 2 μM [38]. The best understood route of RNS production is the reaction of nitric oxide (•NO) with superoxide ($O_2^{\bullet-}$) to form peroxynitrite ($ONOO^-$) which might react with all the major classes of biomolecules and, therefore, has the potential to mediate cytotoxicity independent of NO or O_2^- (Fig. 15.3) [39, 40].

As stated previously, the cause of nitrosative stress could be maternal origin or foetal or both. In the case of maternal origin, pregnancy-induced hypertension is reported to be mediated by nitrosative stress [41]. However, during early pregnancy, the effect of nitrosative stress is not much studied. Particulate matter from air pollution is reported to increase placental nitrosative stress [42]. Several metabolic diseases during pregnancy such as diabetes mellitus are known to induce nitrosative stress during early embryonic development [43, 44]. Presence of nitrosative stress during early pregnancy majorly causes cardiac abnormalities as heart development occurs [45]. In our independent studies, we found that thalidomide causes structural deformities in developing heart. Supraphysiological dose of NO during early development causes situs inversus in the heart [72]. Thalidomide has been shown to increase RNS in developing heart [46]; therefore, there could be link between RNS during cardiogenesis and cardiac anomalies. Increased nitrosative stress also activates the nuclear enzyme poly(ADP-ribose) polymerase, which contributes to endothelial dysfunction and cardiac pathogenesis [47, 48].

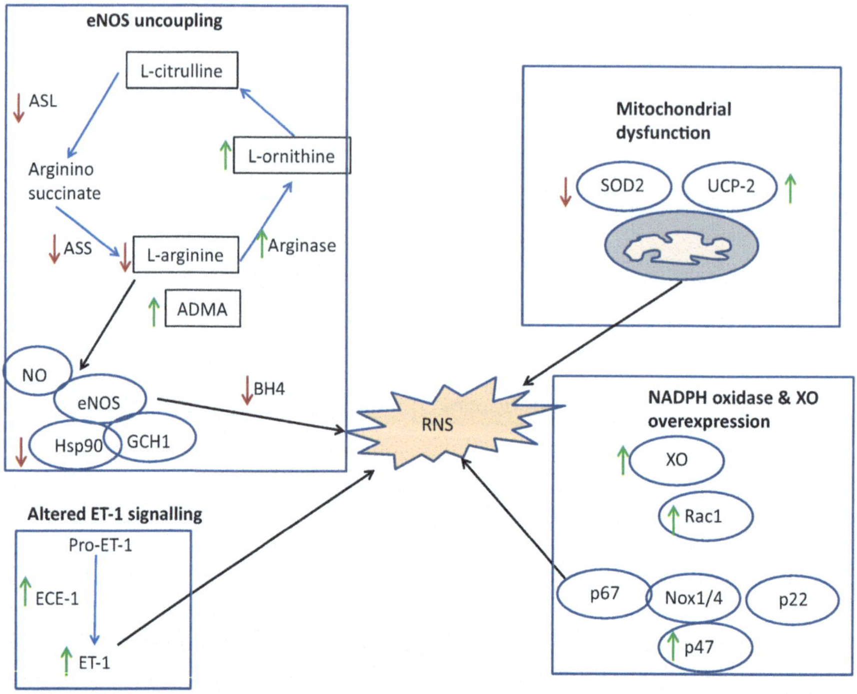

Fig. 15.3 Physiological sources of reactive nitrogen species

15.5 Thalidomide for Inducing Oxidative Stress in Embryo

On one hand, thalidomide is a well-known teratogen, but on the other hand, it shows a great potential against several pathologies such as Hansen's disease, multiple myeloma [49], Behcet's disease [50] and tuberculosis (TB) due to its anti-angiogenic and anti-inflammatory properties. In our previous study, we reported several structural deformities in developing heart using chick embryo model, predominantly a haematoma like structure named 'Suvro-Pavitra lump' rich in oxidative and nitrosative stress in heart muscles [46]. Taking the lead from the previous work, we planned to study the oxidative and nitrosative stress in early developing heart during cardiac looping under thalidomide treatment.

15.6 Materials and Methods

Thalidomide was obtained from TRC Canada (T058833). Nitro blue tetrazolium (NBT) was procured from Sigma-Aldrich®, India. DHR123 and DAF-FM were purchased from Thermo Fisher Scientific (India). Chick embryos were treated with thalidomide or vehicle as described elsewhere [46]. In brief, the embryonic stages were defined as described by Hamilton and Hamburger (HH) [51]. One small aperture was created in the egg shell over the air sac using sterile needle at HH/24 stage. A single dose of 20 μl of thalidomide solution of 1 mg/ml or vehicle was injected in the aperture of air sac. The aperture was sealed with Mediplast® tape, and eggs were incubated for further growth maximum of 6 days or as per experimental requirements.

Oxidative stress in the chick embryonic heart during cardiac looping at HH 15 was estimated. The heart was dissected out from embryos treated with vehicle or thalidomide (15 each group) and used to measure nitric oxide, peroxynitrite and superoxide using biochemical assays as described below.

15.6.1 Measurement of Nitric Oxide

Hearts from control or thalidomide-treated embryos were treated with 200 μl of 10 μM DAF-FM for 15 min at room temperature in the dark. The tissue was homogenized and centrifuged at 1000 g for 5 min, and then 100 μl of supernatant was collected and transferred into 96 well plate, and the fluorescence intensity was measured at excitation 488 nm/emission 519 nm using fluorescence spectrophotometer [52].

15.6.2 Measurement of Peroxynitrite

Hearts from control or thalidomide-treated embryos were treated with 200 μl of 10 μM of DHR123 probe at room temperature for 20 min in the dark. The tissue was homogenized and centrifuged at 1000 g for 5 min, and then 100 μl of supernatant was collected and transferred into 96 well plate, and fluorescence intensity was measured at excitation 500 nm/emission 536 nm using fluorescence spectrophotometer [46].

15.6.3 Measurement of Superoxide

Hearts from control or thalidomide-treated embryos were treated with 200 μl of 5 μM of NBT for 2 h at room temperature. The tissue was homogenized and centrifuged at 1000 g for 5 min, and then 100 μl of supernatant was collected and transferred into 96 well plate, and optical density was measured using adsorption spectroscopy at 560 nm using ELISA reader [46, 53].

15.6.4 Transcriptome Experiments and Analysis

Transcriptome data were acquired from our previously deposited database GEO (GSE69159). The data were obtained from mRNA isolated from HH29 (6th day) old chick embryos (from control or thalidomide-treated group). Cuffdiff (v2.2.0) programme was used to analyse the data for differentially expressed genes [52, 54]. The differentially expressed genes were used for the pathways enrichment analysis using Enrichr [55, 56].

15.6.5 Statistics

Each experiment was performed in triplicates (n = 3). The data in the results have been presented as mean ± standard error mean (SEM). Statistical analysis was done using student t-test or as specified. The significant differences among means were considered when $p \leq 0.05$.

15.7 Results and Discussion

Thalidomide hampers NO signalling in endothelial cells [57]; however, there was no significant difference observed in free NO while comparing the heart isolated from control group with thalidomide-treated (Fig. 15.4a). Peroxynitrite level was 1.76-fold higher in the heart isolated from thalidomide-treated embryos compared to the heart isolated from control group indicating an increased nitrosative stress (p=0.004) (Fig. 15.4b). Level of superoxide was 1.95-fold higher in the heart isolated from thalidomide-treated chick embryos compare to that of control group (p<0.001) (Fig. 15.4c). Altogether, these results showed that thalidomide increased nitrosative and oxidative stress in the heart during cardiac looping. Increased ROS and RNS act as cytotoxic agents in various cellular systems [58, 59]; therefore, thalidomide-mediated rise in RNS and ROS could be responsible for the structural and anatomical deformities in the heart.

15.8 Nitrosative Stress and CHDs

RNS, as described previously, can be transformed into other species such as nitroxyl anion (NO-), nitrosonium cation (NO+), and peroxynitrite ($ONOO^-$) [60]. There are several factors which might contribute to RNS formation during pregnancy including maternal metabolic pathology. Studies in mice showed that NOS3 expression in

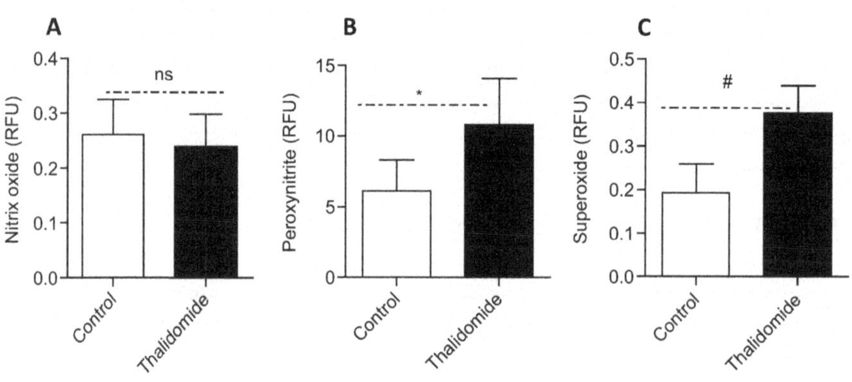

Fig. 15.4 Thalidomide increases ROS and RNS in developing heart

the heart starts at E9.5 stage, and it remains high until E13.5 stage. Thereafter, a low level of NOS3 is still detectable, and it remains to adulthood [61]. There is increased nitrosative stress in the leukocytes of maternal circulation 16th week of gestation onwards [62]. RNS modulate signalling of growth factors and transcription factors controlling gene expression associated with proliferation, differentiation and apoptosis [63]. In fact, ONOO- is a potent oxidizing agent that induces lipid oxidation and DNA fragmentation [60]. The cellular damage induced by ROS in the endothelium generates a reduced bioavailability of NO, leading to endothelial dysfunction [64]. Therefore, formation of ROS acts double-sided sword to the cellular system by reacting with NO which reduces bioavailability of NO and reducing the NOS activity [65]. Thalidomide is known to induce ROS and interfere with NO signalling pathways [57]. In our recent work, we have shown that thalidomide causes severe structural deformities in developing heart [46]. Thalidomide modulates the cardiac looping which changes the fluidics of the cardiovascular system. Biochemical assays shows that during cardiogenesis, there is increase in the levels of superoxide and peroxynitrite, whereas there is no change in the free NO as such (Fig. 15.4a–c). Therefore, there might be possible link between the defective looping mediated by ROS-RNS and altered fluidics.

15.9 Transcriptome Analysis

15.9.1 Cardiac Remodelling Alters Embryonic Metabolism

Thalidomide promotes an imbalance in the redox homeostasis resulting in the perturbation of various metabolic pathways. Remarkably, fatty acid metabolism and electron transport chain are the most significantly affected metabolic pathways (Fig. 15.5a and b). Enrichment of biological processes by genes significantly downregulated in thalidomide-treated embryos showed that key processes including metabolic pathways and superoxide biology are affected. Metabolic processes including mitochondrial electron transport chain, ATP synthesis, complex I biogenesis and assembly were inhibited. Cellular response to superoxide, removal of superoxide radicals and negative regulation of ROS metabolic process were significantly downregulated in thalidomide-treated embryos (Fig. 15.6a and b). Peptide biosynthetic process, α-amino acid biosynthetic process and transport were among the affected biological processes. Lipid transport and steroid biosynthetic process were also downregulated. Notably, there was no direct consequence of thalidomide treatment on the expression of genes associated with glycolysis/gluconeogenesis.

15.9.2 Altered Mitochondrial Redox Milieu in Thalidomide-Treated Heart

Majority of free radicals are generated in mitochondria. Mitochondria are also the most hit target of RNS/ROS as well. The global effect of nitrosation on metabolism

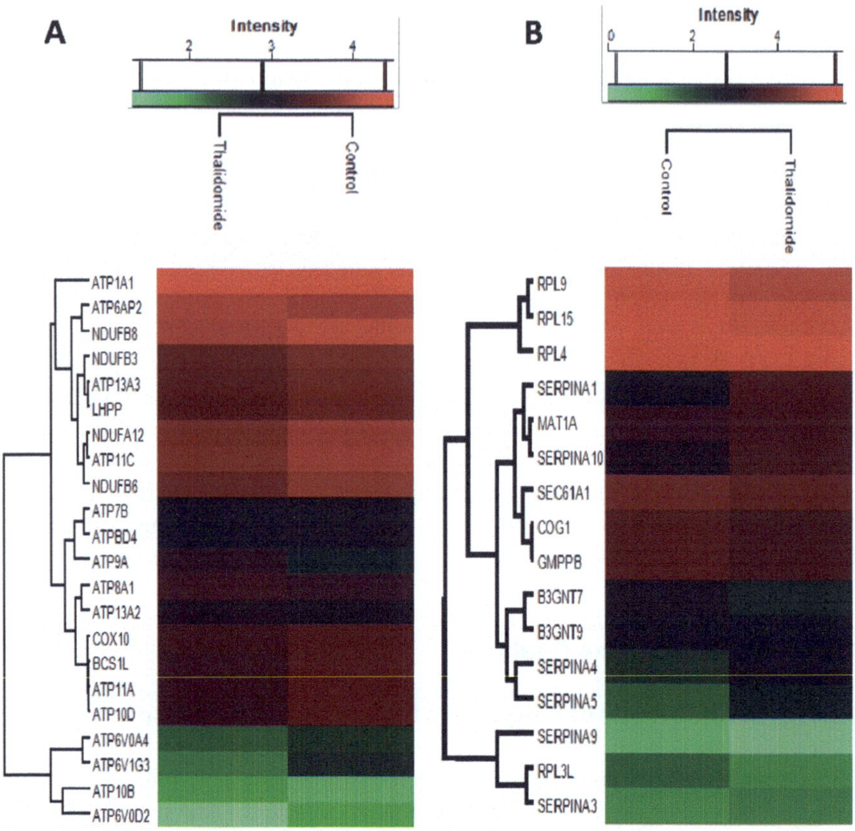

Fig. 15.5 Heatmaps representation of differentially regulated genes involved in (**a**) electron transport chain and (**b**) proteins metabolism

is substantial, and upon misregulation, many deleterious effects on metabolic enzymes have been reported. Nitrosative stress has been reported to inhibit the production of ATP in human spermatozoa by perturbing the processes of glycolysis as well as oxidative phosphorylation [66]. Oxidative and nitrosative stress have shown to cause deleterious effects on lipid-mediated signalling by directly interacting with lipids and affecting lipid membrane properties and palmitoylation [67]. Glucose 6 phosphate gene expression is upregulated under conditions of enhanced oxidative stress [68]. S-nitrosylation affects various enzymes which are key members of metabolic processes including glycolysis, gluconeogenesis, citric acid cycle, electron transport chain, amino acid and fatty acid metabolism [69]. A recent interesting study observed that S-nitrosation inhibits four particular metabolic enzymes, namely, 6PGD, ALDH41, COMT and PHGDH by interacting directly with the enzymes [70]. However, there are not many studies reporting the effect of nitrosative stress on gene expression of metabolic enzymes. We observed that majorly gene expression of enzymes involved in oxidative phosphorylation, fatty acid and amino

Fig. 15.6 Enriched metabolic pathways of differentially regulated genes of thalidomide-treated 6-day-old chick embryo. (**a**) Downregulated metabolic pathways. (**b**) Downregulated metabolic processes

acid metabolism were significantly downregulated under thalidomide treatment which induces nitrosative stress. During foetal heart development, mRNA transcripts of enzymes implicated in fatty acid metabolism specifically FABP4, FABP2 and NRAP have been observed to be highly expressed since 10 weeks of human heart development and increase as the heart continues to develop till 16–18 weeks of gestation [71]. Our data suggest that fatty acid metabolism which is a critical metabolic pathway during heart development is perturbed under an oxidative and nitrosative environment due to thalidomide treatment.

15.10 Summary

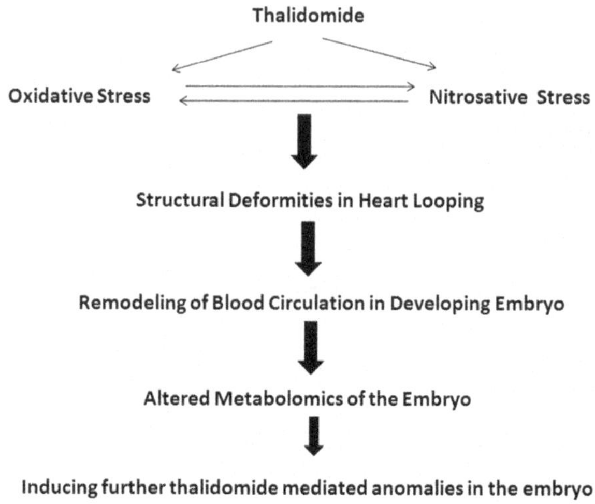

Taken together, the present study suggests that nitrosative stress during cardiogenesis is likely to cause structural deformities in the heart. The study further confirmed that thalidomide increased the level of peroxynitrite and superoxide in developing heart and caused structural deformities in the heart. The deformed heart might contribute to altered metabolome of the embryo. The transcriptome analysis of thalidomide-treated embryos showed that the genes involved in proteins and fatty acid metabolism were differentially regulated.

Acknowledgement This work was supported by a grant from University Grant Commission Faculty Recharge Programme (UGC-FRP) Government of India to SC. PK is thankful for the financial support from University Grant Commission-Senior Research Fellowship programme (UGC-SRF), Government of India. LS received financial support from the Department of Biotechnology-Senior Research Fellowship (DBT-SRF) programme, Government of India.

References

1. Gilboa SM, Salemi JL, Nembhard WN, Fixler DE, Correa A (2010) Mortality resulting from congenital heart disease among children and adults in the United States, 1999 to 2006. Circulation 122(22):2254–2263
2. Pei L, Kang Y, Zhao Y, Yan H (2017) Prevalence and risk factors of congenital heart defects among live births: a population-based cross-sectional survey in Shaanxi province, Northwestern China. BMC Pediatr 17(1):18
3. Leirgul E et al (2014) Birth prevalence of congenital heart defects in Norway 1994-2009—a nationwide study. Am Heart J 168(6):956–964
4. Øyen N, Poulsen G, Boyd HA, Wohlfahrt J, Jensen PKA, Melbye M (2009) National time trends in congenital heart defects, Denmark, 1977–2005. Am Heart J 157(3):467–473.e1
5. Nisale SH, Maske VG (2016) A study of prevalence and pattern of congenital heart disease and rheumatic heart disease among school children. Int J Adv Med 3(4):947–951

6. Bhardwaj R et al (2015) Epidemiology of congenital heart disease in India. Congenit Heart Dis 10(5):437–446
7. Nabulsi MM, Tamim H, Sabbagh M, Obeid MY, Yunis KA, Bitar FF (2003) Parental consanguinity and congenital heart malformations in a developing country. Am J Med Genet 116A(4):342–347
8. Tanner K, Sabrine N, Wren C (2005) Cardiovascular malformations among preterm infants. Pediatrics 116(6):e833–e838
9. Haney AF, Misukonist MA, Weinberg JB (1983) Macrophages and infertility: oviductal macrophages as potential mediators of infertility. Fertil Steril 39(3):310–315
10. Barroso YC, RP OC, Nagamani M (1998) Nitric oxide inhibits development of embryos and implantation in mice. – PubMed – NCBI. Mol Hum Reprod 5(4):503–507
11. Liu Y et al (2014) Nitric oxide synthase-3 deficiency results in hypoplastic coronary arteries and postnatal myocardial infarction. Eur Heart J 35(14):920–931
12. Valenti O et al (2011) Fetal cardiac function during the first trimester of pregnancy. J Prenat Med 5(3):59–62
13. Lawson A, Schoenwolf GC (2003) Epiblast and primitive-streak origins of the endoderm in the gastrulating chick embryo. Development 130(15):3491–3501
14. Rajala K, Pekkanen-Mattila M, Aalto-Setälä K (2011) Cardiac differentiation of pluripotent stem cells. Stem Cells Int 2011:383709
15. Brand T (2003) Heart development: molecular insights into cardiac specification and early morphogenesis. Dev Biol 258(1):1–19
16. Olson EN (2006) Gene regulatory networks in the evolution and development of the heart. Science 313(5795):1922–1927
17. Kane C, Terracciano CMN (2017) Concise review: criteria for chamber-specific categorization of human cardiac Myocytes derived from pluripotent stem cells. Stem Cell 35(8):1881–1897
18. Schleich J-M, Abdulla T, Summers R, Houyel L (2013) An overview of cardiac morphogenesis. Arch Cardiovasc Dis 106(11):612–623
19. Hirokawa N, Tanaka Y, Okada Y, Takeda S (2006) Nodal flow and the generation of left-right asymmetry. Cell 125(1):33–45
20. Christoffels VM et al (2000) Chamber formation and morphogenesis in the developing mammalian heart. Dev Biol 223(2):266–278
21. Moorman A, Webb S, Brown NA, Lamers W, Anderson RH (2003) Development of the heart: (1) formation of the cardiac chambers and arterial trunks. Heart 89(7):806–814
22. Lin C-J, Lin C-Y, Chen C-H, Zhou B, Chang C-P (2012) Partitioning the heart: mechanisms of cardiac septation and valve development. Development 139(18):3277–3299
23. Poelmann RE et al (2014) Evolution and development of ventricular Septation in the Amniote heart. PLoS One 9(9):e106569
24. Anderson RH, Webb S, Brown NA, Lamers W, Moorman A (2003) Development of the heart: (2) Septation of the atriums and ventricles. Heart 89(8):949–958
25. Dakkak W, Bhimji SS (2018) *Ventricular Septal Defect*. StatPearls Publishing, Treasure Island
26. Bailliard F, Anderson RH (2009) Tetralogy of Fallot. Orphanet J Rare Dis 4:2
27. Stennard FA, Harvey RP, Smoak I, Yamamura K, Meyers EN (2005) T-box transcription factors and their roles in regulatory hierarchies in the developing heart. Development 132(22):4897–4910
28. Khalil A et al (2017) A HAND to TBX5 explains the link between thalidomide and cardiac diseases. Sci Rep 7(1):1416
29. Takeuchi JK et al (2003) Tbx5 specifies the left/right ventricles and ventricular septum position during cardiogenesis. Development 130(24):5953–5964
30. Nadeau M et al (2010) An endocardial pathway involving Tbx5, Gata4, and Nos3 required for atrial septum formation. Proc Natl Acad Sci U S A 107(45):19356–19361
31. Zeisberg EM et al (2005) Morphogenesis of the right ventricle requires myocardial expression of Gata4. J Clin Invest 115(6):1522–1531

32. Ridnour LA et al (2004) The chemistry of nitrosative stress induced by nitric oxide and reactive nitrogen oxide species. Putting perspective on stressful biological situations. Biol Chem 385(1):1–10
33. Bhattacharyya A, Chattopadhyay R, Mitra S, Crowe SE (2014) Oxidative stress: an essential factor in the pathogenesis of gastrointestinal mucosal diseases. Physiol Rev 94(2):329–354
34. Uribe P, Boguen R, Treulen F, Sánchez R, Villegas JV (2015) Peroxynitrite-mediated nitrosative stress decreases motility and mitochondrial membrane potential in human spermatozoa. MHR Basic Sci Reprod Med 21(3):237–243
35. Iovine NM, Pursnani S, Voldman A, Wasserman G, Blaser MJ, Weinrauch Y (2008) Reactive nitrogen species contribute to innate host defense against Campylobacter jejuni. Infect Immun 76(3):986–993
36. Heinrich TA, da Silva RS, Miranda KM, Switzer CH, Wink DA, Fukuto JM (2013) Biological nitric oxide signalling: chemistry and terminology. Br J Pharmacol 169(7):1417–1429
37. Förstermann U, Sessa WC (2012) Nitric oxide synthases: regulation and function. Eur Heart J 33(7):829–837, 837a–837d
38. Wink DA et al (2011) Nitric oxide and redox mechanisms in the immune response. J Leukoc Biol 89(6):873–891
39. Beckman JS, Koppenol WH (1996) Nitric oxide, superoxide, and peroxynitrite: the good, the bad, and ugly. Am J Physiol Physiol 271(5):C1424–C1437
40. Radi R (2013) Peroxynitrite, a stealthy biological oxidant. J Biol Chem 288(37):26464–26472
41. Gaikwad KB, Joshi NG, Selkar SP (2017) Study of Nitrosative stress in 'Pregnancy Induced Hypertension. J Clin Diagn Res 11(3):BC06–BC08
42. Saenen ND et al (2016) Placental Nitrosative stress and exposure to ambient air pollution during gestation: a population study. Am J Epidemiol 184(6):442–449
43. Zhao Z, Reece EA (2013) New concepts in diabetic embryopathy. Clin Lab Med 33(2):207–233
44. Cao L, Tan C, Meng F, Liu P, Reece EA, Zhao Z (2016) Amelioration of intracellular stress and reduction of neural tube defects in embryos of diabetic mice by phytochemical quercetin. Sci Rep 6:21491
45. Zhao Z (2016) Reevaluation of Antioxidative strategies for birth defect prevention in diabetic pregnancies. J Biomol Res Ther 5(3):1–5
46. Kumar P et al (2018) Thalidomide remodels developing heart in chick embryo: discovery of a thalidomide mediated hematoma in heart muscle. Naunyn Schmiedeberg's Arch Pharmacol 391(10):1093–1105
47. Pacher P, Liaudet L, Soriano FG, Mabley JG, Szabó E, Szabó C (2002) The role of poly(ADP-ribose) polymerase activation in the development of myocardial and endothelial dysfunction in diabetes. Diabetes 51(2):514–521
48. Pacher P, Szabó C (2007) Role of poly(ADP-ribose) polymerase 1 (PARP-1) in cardiovascular diseases: the therapeutic potential of PARP inhibitors. Cardiovasc Drug Rev 25(3):235–260
49. Knobloch J, Rüther U (2008) Shedding light on an old mystery: thalidomide suppresses survival pathways to induce limb defects. Cell Cycle 7(9):1121–1127
50. Mujagić H, Chabner BA, Mujagić Z (2002) Mechanisms of action and potential therapeutic uses of thalidomide. Croat Med J 43(3):274–285
51. Hamburger V, Hamilton HL (1992) A series of normal stages in the development of the chick embryo. Dev Dyn 195(4):231–272
52. Kumar P et al (2016) Harvesting clues from genome wide transcriptome analysis for exploring thalidomide mediated anomalies in eye development of chick embryo: nitric oxide rectifies the thalidomide mediated anomalies by swinging back the system to normal transcriptome pattern. Biochimie 121:253–267
53. Dunand C, Crèvecoeur M, Penel C (2007) Distribution of superoxide and hydrogen peroxide in Arabidopsis root and their influence on root development: possible interaction with peroxidases. New Phytol 174(2):332–341
54. Veeriah V et al (2017) Transcriptomic analysis of thalidomide challenged Chick embryo suggests possible link between impaired vasculogenesis and defective organogenesis. Chem Res Toxicol 30(10):1883–1896

55. Chen EY et al (2013) Enrichr: interactive and collaborative HTML5 gene list enrichment analysis tool. BMC Bioinformatics 14(1):128
56. Kuleshov MV et al (2016) Enrichr: a comprehensive gene set enrichment analysis web server 2016 update. Nucleic Acids Res 44(W1):W90–W97
57. Majumder S et al (2009) Thalidomide attenuates nitric oxide-driven angiogenesis by interacting with soluble guanylyl cyclase. Br J Pharmacol 158(7):1720–1734
58. Kurutas EB (2015) The importance of antioxidants which play the role in cellular response against oxidative/nitrosative stress: current state. Nutr J 15(1):71
59. Di Meo S, Reed TT, Venditti P, Victor VM (2016) Role of ROS and RNS sources in physiological and pathological conditions. Oxidative Med Cell Longev 2016:1245049
60. Phaniendra A, Jestadi DB, Periyasamy L (2015) Free radicals: properties, sources, targets, and their implication in various diseases. Indian J Clin Biochem 30(1):11–26
61. Bloch W et al (1999) Nitric oxide synthase expression and role during cardiomyogenesis. Cardiovasc Res 43(3):675–684
62. Mihu D, Sabău L, Costin N, Ciortea R, Măluţan A, Mihu CM (2012) Implications of maternal systemic oxidative stress in normal pregnancy and in pregnancy complicated by preeclampsia. J Matern Neonatal Med 25(7):944–951
63. Kohlgrüber S, Upadhye A, Dyballa-Rukes N, McNamara CA, Altschmied J (May 2017) Regulation of transcription factors by reactive oxygen species and nitric oxide in vascular physiology and pathology. Antioxid Redox Signal 26(13):679–699
64. Widlansky ME, Gutterman DD (2011) Regulation of endothelial function by mitochondrial reactive oxygen species. Antioxid Redox Signal 15(6):1517–1530
65. Zhao J (2007) Interplay among nitric oxide and reactive oxygen species: a complex network determining cell survival or death. Plant Signal Behav 2(6):544–547
66. Uribe P, Treulen F, Boguen R, Sánchez R, Villegas JV (2017) Nitrosative stress by peroxynitrite impairs ATP production in human spermatozoa. Andrologia 49(3):e12615
67. Morris G, Walder K, Puri BK, Berk M, Maes M (2016) The deleterious effects of oxidative and nitrosative stress on Palmitoylation, membrane lipid rafts and lipid-based cellular signalling: new drug targets in neuroimmune disorders. Mol Neurobiol 53(7):4638–4658
68. Ursini MV, Parrella A, Rosa G, Salzano S, Martini G (1997) Enhanced expression of glucose-6-phosphate dehydrogenase in human cells sustaining oxidative stress. Biochem J 323(Pt 3):801–806
69. Doulias P-T, Tenopoulou M, Greene JL, Raju K, Ischiropoulos H (2013) Nitric oxide regulates mitochondrial fatty acid metabolism through reversible protein S-Nitrosylation. Sci Signal 6(256):rs1–rs1
70. Bruegger JJ, Smith BC, Wynia-Smith SL, Marletta MA (2018) Comparative and integrative metabolomics reveal that *S* -nitrosation inhibits physiologically relevant metabolic enzymes. J Biol Chem 293(17):6282–6296
71. Iruretagoyena JI et al (2014) Metabolic gene profile in early human fetal heart development. MHR Basic Sci Reprod Med 20(7):690–700
72. Siamwala J, Kumar P, Veeriah V, Muley A, Rajendran S, Konikkat S, Majumder S, Mani K, Chatterjee S (2019) Nitric oxide reverses the position of the heart during embryonic development. Int J Mol Sci 20(5):1157

Oxidative Stress in Pulmonary Artery Hypertension

16

Vinu Wilson and Subir Kumar Maulik

Abstract

Pulmonary artery hypertension (PAH) is a progressive disorder characterized by pulmonary vascular remodeling ultimately leading to right ventricular failure and death. The last few decades have seen considerable progress in PAH therapy based on drugs targeting three major mechanistic pathways, viz., prostacyclin, endothelin and nitric oxide pathways. A growing body of research has documented that "oxidative stress" is intricately associated with development of PAH. Experimental studies have shown that markers of oxidative tissue damage are present in different genetic and chemical models of PAH. Animal studies have also shown the preventive and therapeutic potential of endogenous antioxidants and/or drugs with antioxidant activity in experimental PAH. Though the evidence implicating oxidative stress in PAH has also been generated in human PAH studies, the clinical trials of antioxidants have not yet yielded encouraging results. Further studies are warranted to unravel the reason(s) underlying this paradox in order to develop potential curative drugs for this morbid disorder.

Keywords

Pulmonary hypertension · Oxidative stress · Reactive oxygen species · Antioxidants

V. Wilson
Department of Pharmacology, Govt. T. D. Medical College, Alappuzha, Alappuzha, Kerala, India

S. K. Maulik (✉)
Department of Pharmacology, All India Institute of Medical Sciences, New Delhi, India

S. Chakraborti et al. (eds.), *Modulation of Oxidative Stress in Heart Disease*,
https://doi.org/10.1007/978-981-13-8946-7_16

16.1 Introduction

Pulmonary arterial hypertension (PAH) can be idiopathic or associated with several heritable as well as acquired systemic disorders. PAH forms the first category of the current WHO clinical classification of pulmonary hypertension adopted in 2013 [1]. It is characterized by a resting mean pulmonary artery pressure ≥25 mm of Hg with elevated pulmonary vascular resistance (>3 Wood units) and a normal left atrial pressure (≤15 mm of Hg) [1]. PAH is a progressive disorder leading to right ventricular hypertrophy and failure reducing the median survival in affected patients to 2.8 years without treatment [2].

Despite the advent of several therapeutic agents such as prostaglandin analogues, phosphodiesterase 5 inhibitors, and endothelin receptor antagonists in the last few decades, PAH remains incurable, steadily progressive, and eventually fatal [3]. The symptoms of PAH are nonspecific making diagnosis difficult. Further, low awareness of PAH among primary caregivers as well as socioeconomic constraints of patients lead to a very low percentage of PAH patients actually being referred to the few tertiary centers equipped to perform definitive diagnoses [4].

16.2 Pathophysiological Mechanisms in PAH

The pathophysiology of PAH has been significantly unraveled in the past several decades as involving dynamic pulmonary artery vasoconstriction, thrombosis, and remodeling of small pulmonary arteries characterized by hypertrophy of pulmonary vascular smooth muscle cells and hyperplasia of endothelial and connective tissue cells resulting in plexiform lesions [5]. These pathologic processes are targeted with empirical treatment modalities such as oxygen therapy, oral anticoagulants, diuretics, digoxin, and vasodilators especially calcium channel blockers [6]. Calcium channel blockers are recommended only in patients showing a positive acute vasoreactivity to them during a right heart catheterization study. Unfortunately, none of these therapeutic measures have shown any long-term survival benefit in the limited number of uncontrolled studies done with them [6]. Over the past two decades, significant progress in therapy has been achieved by targeting three mechanistic pathways discovered to be dysregulated in PAH, viz., the prostacyclin pathway, the endothelin pathway, and the nitric oxide pathway [6].

16.2.1 Prostacyclin Pathway

Prostacyclin (PGI_2) and thromboxane A_2 (TxA_2) are the major derivatives of arachidonic acid metabolism in vascular cells. PGI_2 is a potent vasodilator, inhibits platelet activation, and has antiproliferative properties, while TxA_2 is a potent vasoconstrictor and platelet agonist. In PAH, the imbalance between these two molecules is found to be shifted toward TxA_2. In the urine of patients with pulmonary hypertension, the levels of 6-keto-prostacyclin $F_{2\alpha}$ (a metabolite of PGI_2) are decreased, whereas the levels of thromboxane B_2 (a metabolite of TxA_2) are increased [7]. Furthermore, the production of prostacyclin synthase is decreased in the small- and medium-sized pulmonary arteries of patients with pulmonary hypertension, particularly those with idiopathic PAH [8].

Based on these findings, intravenous epoprostenol (PGI_2 analogue) was first used in idiopathic PAH in the 1980s. Several randomized clinical trials have shown improvement in resting hemodynamics and clinical and functional status of NYHA class III and IV PAH patients given intravenous epoprostenol [6]. Epoprostenol is the only drug to have shown survival benefit in PAH patients. To obviate the need for cumbersome continuous intravenous administration of epoprostenol through central veins and associated complications, several longer acting prostacyclin analogues which could be given by intravenous (iloprost, treprostinil), subcutaneous (treprostinil), oral (beraprost, iloprost, treprostinil), or inhalational (iloprost, treprostinil) routes were developed and tested in clinical trials. An oral, non-prostanoid, selective prostacyclin receptor agonist, selexipag, was recently approved for PAH therapy [9]. Although these drugs reproduce the beneficial effects of prostacyclin, they are still far from being ideal treatments for PAH owing to their adverse effects, short half-lives necessitating frequent dosing, and high cost to the patients [6].

16.2.2 Endothelin Pathway

Endothelin-1 (ET-1), a potent vasoconstrictor chiefly produced by endothelial cells, stimulates the proliferation of pulmonary artery smooth-muscle cells. The plasma levels of ET-1 are found to be increased and inversely proportional to the magnitude of the pulmonary blood flow and cardiac output in PAH [5]. ET-1 can induce fibrosis and is a pro-inflammatory mediator by virtue of its capacity to enhance the expression of cellular adhesion molecules. The effects of ET-1 are mediated through the ET_A and ET_B endothelin receptors. Activation of ET_A receptors causes sustained vasoconstriction and proliferation of vascular smooth-muscle cells, whereas ET_B receptors mediate pulmonary endothelin clearance and induce the production of nitric oxide and PGI_2 by endothelial cells leading to vasodilatation [8]. Bosentan is an orally active dual (ET_A and ET_B) endothelin-receptor antagonist (ETRA) found to be beneficial in clinical trials of NYHA class III–IV PAH patients. Selective ET_A receptor antagonists (ambrisentan and sitaxsentan) have the theoretical advantage of sparing ET_B receptor mediated ET-1 clearance and vasodilatation and showed lesser perturbation of hepatic transaminase levels in clinical trials [6]. Macitentan, a tissue-targeting oral dual ET-1 receptor antagonist, was recently approved by the Food and Drug Administration (FDA) for PAH patients [10]. The use of ETRAs is, however, limited by their dose-limiting hepatotoxicity, teratogenic potential, and high cost [6].

16.2.3 Nitric Oxide Pathway

Nitric oxide (NO) is a potent endogenous, endothelium-derived vasodilator that directly relaxes the underlying vascular smooth muscle through stimulation of soluble guanylate cyclase (sGC) and increased production of intracellular cyclic guanosine monophosphate (cGMP). A number of experimental and clinical studies have documented that PAH is associated with a defect in NO availability and thereby decreased NO-induced vasodilatation [8]. Therapeutic trials showed that short-term NO administration improves pulmonary hemodynamics in PAH. However, long-term NO inhalation therapy is cumbersome to administer and associated with rebound deterioration in pulmonary hemodynamics on withdrawal [3].

An indirect strategy employed to increase the biological activity of endogenous NO in PAH is through inhibition of phosphodiesterase type 5 (PDE5), the predominant enzyme metabolizing cGMP in pulmonary vascular smooth muscle cells. PDE5 inhibitors (sildenafil, tadalafil) have shown improvement in pulmonary hemodynamics and functional status of patients when used as adjunctive treatments with prostacyclin analogues in New York Heart Association (NYHA) class III–IV PAH patients [6]. A direct sGC stimulator, Riociguat, which produces cGMP even in the absence of NO, is undergoing clinical trials in PAH [11]. However, all these drugs are expensive and associated with adverse effects including visual disturbances, dyspepsia, flushing, headache, and limb pain [3].

16.3 Oxidative Stress

As has already been discussed elsewhere in this book, oxidative stress is implicated in the pathophysiology of varied cardiovascular disorders. A considerable amount of literature generated over the last few decades supports its involvement in pulmonary vascular remodeling in PAH as well [12]. "Oxidative stress" is the abnormal cellular state of redox imbalance characterized by enhanced production of reactive oxygen species (ROS) and/or subdued antioxidant defenses. ROS contain at least one reactive oxygen atom and include relatively stable molecules such as NO and hydrogen peroxide (H_2O_2) as well as highly reactive ones such as superoxide ($O_2^{-\cdot}$) and hydroxyl ($OH^{\cdot}$) radicals. NO can react with superoxide to form the highly damaging peroxynitrite ($ONOO^-$) anion. While a low level of ROS is involved in cellular signaling, their excess production is shown to not only damage cellular macromolecules in a runaway "chain reaction" but also stimulate pathological cellular proliferation [12, 13]. Under physiological conditions, ROS overactivity is kept in check by endogenous enzymatic (catalase, superoxide dismutase, glutathione peroxidase) and nonenzymatic (glutathione, urate) antioxidant defenses. Pathological oxidative stress occurs when ROS production overwhelms the antioxidant defenses.

16.3.1 Sources of ROS in PAH

The multiple enzymatic and metabolic processes known to generate ROS within cells of the pulmonary vascular wall are similar to those found elsewhere in the body and are most abundant in the mitochondrion [14]. They include the nicotinamide adenine dinucleotide phosphate (NADPH) oxidases (Nox) [15], the mitochondrial electron transport chain complexes, xanthine oxidase (XO) [16], and uncoupled nitric oxide synthase (NOS) [17]. It is widely accepted that NADPH oxidases are not only the principal generator of ROS in the vasculature [18], but their activities regulate the activities of other ROS-generating oxidases such as XO [19]. Among the members of Nox enzyme family, Nox4 was selectively increased in the pulmonary vasculature and lungs of hypoxia-exposed mice and in pulmonary vascular tissue from patients with pulmonary arterial hypertension [20]. Hypoxia also upregulated Nox4 in pulmonary artery adventitial fibroblasts in vitro and in adventitial fibroblasts from patients with idiopathic pulmonary arterial hypertension [21]. Recently, Nox1, Nox2 (gp91phox), and Nox4 expression was found to be upregulated in monocrotaline (MCT)-induced model of PAH in rats which was shown to be attenuated by treatment with resveratrol [22].

16.3.2 Oxidative Stress in PAH

16.3.2.1 Experimental Studies

16.3.2.1.1 *Elevated ROS and/or Suppressed Antioxidant Defenses in Experimental PAH*

Monocrotaline (MCT)-induced PAH is one of the most commonly employed experimental models of PAH in rats. Oxidative stress has been documented in MCT-induced model in both lungs and the failing right ventricle. Elevated levels of lung malondialdehyde and inducible NOS (iNOS) expression and reduced levels of catalase, glutathione, and superoxide dismutase have been documented in MCT-treated rats [23]. In the right ventricle of MCT-treated rats, an initial rise and later decline in antioxidant enzyme (catalase, superoxide dismutase, and glutathione peroxidase) activity and increased lipid peroxidation have been shown [24].

In the mouse model of chronic hypoxia-induced pulmonary hypertension (CH-PH), intrapulmonary artery superoxide levels have been shown to be elevated [25]. Initially, it was thought that hypoxia would attenuate the generation of ROS due to the lack of molecular oxygen to generate superoxide radical. However, it was later recognized that hypoxia enhanced ROS generation in relative rather than absolute amounts [26].

The recently introduced caveolin-1 knockdown model of PAH also shows elevated ROS levels primarily derived from an uncoupled endothelial NOS (eNOS) [27]. Caveolin-1, a protein expressed in vascular smooth muscle caveolae, acts as a scaffold maintaining eNOS in an inactive form. Knockdown of caveolin-1 leads to widespread eNOS uncoupling and excess NO generation and resultant peroxynitrite anion formation. Experimental studies have shown that eNOS uncoupling also contributes to the persistent pulmonary hypertension of newborn [28].

16.3.2.1.2 *Genetic Loss/Gain of Function Studies*

It has been shown that Nox2 knockout mice fail to develop CH-PH which suggests a critical role for superoxide generated by Nox2 containing NADPH oxidases in this model [29]. Caveolin null mice have been shown to develop PAH due to elevated NO-mediated ROS production mediated by uncoupling of eNOS besides bone morphogenetic protein (BMP) receptor activation [30]. This observation was further strengthened by study which showed that rats with double knockout of caveolin and eNOS genes do not develop PAH due to lack of formation of peroxynitrite anion [27].

Intratracheal delivery of adenovirus transfected with gene for extracellular superoxide dismutase (EC-SOD) was shown to reverse pathological remodeling of pulmonary vascular cells as well as the right ventricle in MCT-treated rats [31]. Recombinant human SOD was shown to restore eNOS function, reduce oxidative stress, and reduce pulmonary vascular resistance while breathing 100% oxygen in a lamb model of persistent pulmonary hypertension of the newborn [32].

16.3.2.1.3 *Drug/Antioxidant Intervention Studies*

Several interventional studies employing drugs or herbal products with antioxidant properties have shown to attenuate the development of MCT-induced PAH and right ventricular hypertrophy. For instance, intratracheal delivery of adenovirus containing the gene for human extracellular SOD ameliorated development of MCT-PAH [31]. More recently, it was reported that the antioxidant resveratrol decreased pulmonary artery smooth muscle cell proliferation, NADPH oxidase-induced oxidative stress and prevented the development of MCT-PAH [22].

Our group has shown the preventive potential of the peroxisome proliferator-activated receptor α (PPARα) agonist, fenofibrate, and two herbal drugs, viz., *Ocimum sanctum* (Linn.) and *Terminalia arjuna* (Roxb.), against development of MCT-induced PAH in rats [33–35]. The antioxidant effect of these drugs is thought to be involved in their beneficial effect because all of them attenuated markers of oxidative stress and/or enhanced antioxidant defenses.

The pathological changes in experimental PAH associated with exposure to chronic hypoxia are abolished by administration of the antioxidant, N-acetylcysteine, or the XO inhibitor, allopurinol [36]. Excess iron has been implicated in accelerating the conversion of hydrogen peroxide to highly reactive superoxide and hydroxyl radicals by Fenton chemistry. Iron chelation therapy with deferoxamine has been shown to reverse chronic hypoxia-induced PAH in rats [37].

16.3.2.2 Clinical Studies

16.3.2.2.1 *Elevated ROS and/or Reduced Antioxidant Levels*

A large body of evidence attests to the involvement of oxidative stress in the lungs of patients with PAH. Oxidative stress has been shown to be associated with elevated pulmonary artery systolic pressure and with survival in PAH patients [38, 39].

Recently, it was shown that patients with idiopathic PAH have elevated XO activity compared to control patients and that XO-mediated oxidative stress could be reversed by treatment with XO inhibitors [40]. Lung biopsy samples of patients with idiopathic PAH have shown depletion of SOD and catalase and elevation of 3-nitrotyrosine, a widely used biomarker of oxidative protein damage caused by reaction of peroxynitrite with tyrosine residues [41]. 8-Hydroxyguanosine staining is present within the plexiform lesions from patients with PAH and is absent in the pulmonary vascular endothelium of control patients [42]. 8-Hydroxyguanosine is a biomarker of oxidative nuclear damage caused by reaction of superoxide with guanine. In the lungs of the same PAH patients, the amount and activity of SOD were lower, indicating decreased capacity to scavenge superoxide [42]. Genetic polymorphisms of antioxidant enzymes such as catalase and superoxide dismutase have been implicated in some cases of persistent primary hypertension of the newborn [43]. The valvular fibrosis caused by anti-obesity drugs such as fenfluramine and sibutramine has been shown to be due to excess serotonin-mediated monoamine oxidase-dependent superoxide generation [44]. The evidence from these studies suggests that the lungs of patients with PAH are under chronic oxidative stress.

16.3.2.2.2 *Effects of Drug/Antioxidant therapy*

The clinical trials of the currently approved drugs in PAH have shown beneficial effects in PAH patients by evaluating hemodynamic and functional endpoints [6]. However, studies exploring the effect of such drugs on markers of oxidative stress in PAH patients have been few and far between [12]. For instance, sildenafil has been shown to reduce serum 4-hydroxynonenal levels and improve heart rate variability in PAH patients [45]. Vardenafil administration in treatment-naïve PAH patients has been shown to reduce 8-iso-prostaglandin-F2α and 3-nitrotyrosine blood levels while significantly increasing NO levels [46]. Another study showed that the beneficial hemodynamic response to inhaled iloprost was attenuated in association with endothelial dysfunction and oxidative stress in PAH patients [47].

On the other hand, studies exploring the utility of antioxidants or of drugs with antioxidant properties in PAH patients have yielded disappointing results [6]. A variety of antioxidants showing beneficial effect in animal models of PAH failed to demonstrate similar effect in clinical studies. For instance, supplementation with coenzyme Q, a mitochondrial constituent, improved red blood cell redox status in PAH patients but not 6-min walk distance or BNP levels [48]. Similarly, in spite of promising experimental studies, neither atorvastatin nor simvastatin improved functional status of PAH patients in terms of the distance covered in the 6-min walking test [49, 50]. This was further endorsed by a recent meta-analysis of trials of statins in PAH patients [51].

16.4 Quasi-Cancerous Phenotype

A growing body of research has shown that PAH develops a quasi-cancerous phenotype over time characterized by pulmonary artery endothelial cell precursors and smooth muscle cells developing several hallmarks of cancerous cells [52]. These characteristics include self-sufficiency in several growth factors, resistance to apoptosis, and a metabolic switch to glycolysis instead of oxidative phosphorylation known as Warburg effect [52, 53]. Activation of several intracellular signaling pathways such as Rho kinase (ROCK) and mitogen-activated protein kinase (MAPK) have been implicated in conferring these properties [52]. Drugs targeting various mediators in these pathways such as imatinib (tyrosine kinase inhibitor), sorafenib (multikinase inhibitor), fasudil (Rho-kinase inhibitor), and dichloroacetate (mitochondrial pyruvate dehydrogenase inhibitor which inhibits glycolysis) have been tested in animal as well as clinical studies but have shown only modest benefit against risk of significant adverse effects [6].

16.5 Summary and Conclusion

The last few decades have seen considerable progress in the understanding of the molecular pathophysiology and drug therapy of PAH. Oxidative stress has been shown to be intricately involved in the underlying pathobiology of PAH. However, the exact pathways and mechanisms leading to dysregulated effects of oxidative stress signaling remain to be unraveled. In spite of the promising results shown by antioxidants and drugs with antioxidant properties in experimental studies, their clinical trials in PAH patients have yielded indeterminate results at best. What this points to is our incomplete knowledge of ROS kinetics, their subcellular compartmentalization, or the inability of drugs to reach appropriate subcellular targets. Whatever may be the reason(s), the answers to these questions will determine the fate of millions of patients suffering from this currently incurable disorder.

References

1. Simonneau G, Gatzoulis MA, Adatia I et al (2013) Updated clinical classification of pulmonary hypertension. J Am Coll Cardiol 62:D34–D41
2. D'Alonzo GE, Barst RJ, Ayres SM et al (1991) Survival in patients with primary pulmonary hypertension. Results from a national prospective registry. Ann Intern Med 115:343–349
3. Humbert M, Sitbon O, Simonneau G (2004) Treatment of pulmonary arterial hypertension. N Engl J Med 351:1425–1436
4. Menon S (2009) Pulmonary hypertension in the south east Asia region: an analysis of indexed publication profile. PVRI Rev 1:167
5. Humbert M, Morrell NW, Archer SL et al (2004) Cellular and molecular pathobiology of pulmonary arterial hypertension. J Am Coll Cardiol 43:13S–24S
6. Badlam JB, Bull TM (2017) Steps forward in the treatment of pulmonary arterial hypertension: latest developments and clinical opportunities. Ther Adv Chronic Dis 8:47–64
7. Christman BW, McPherson CD, Newman JH et al (1992) An imbalance between the excretion of thromboxane and prostacyclin metabolites in pulmonary hypertension. N Engl J Med 327:70–75

8. Farber HW, Loscalzo J (2004) Pulmonary arterial hypertension. N Engl J Med 351:1655–1665
9. Bruderer S, Hurst N, Remenova T, Dingemanse J (2017) Clinical pharmacology, efficacy, and safety of selexipag for the treatment of pulmonary arterial hypertension. Expert Opin Drug Saf 16:743–751
10. Sidharta PN, Treiber A, Dingemanse J (2015) Clinical pharmacokinetics and pharmacodynamics of the endothelin receptor antagonist macitentan. Clin Pharmacokinet 54:457–471
11. Ghofrani H-A, Galiè N, Grimminger F et al (2013) Riociguat for the treatment of pulmonary arterial hypertension. N Engl J Med 369:330–340
12. Wong C-M, Bansal G, Pavlickova L et al (2013) Reactive oxygen species and antioxidants in pulmonary hypertension. Antioxid Redox Signal 18:1789–1796
13. Pagel J-I, Deindl E (2012) Disease progression mediated by egr-1 associated signaling in response to oxidative stress. Int J Mol Sci 13:13104–13117
14. Wolin MS, Ahmad M, Gupte SA (2005) The sources of oxidative stress in the vessel wall. Kidney Int 67:1659–1661
15. Mohazzab KM, Wolin MS (1994) Sites of superoxide anion production detected by lucigenin in calf pulmonary artery smooth muscle. Am J Physiol 267:L815–L822
16. Terada LS, Guidot DM, Leff JA et al (1992) Hypoxia injures endothelial cells by increasing endogenous xanthine oxidase activity. Proc Natl Acad Sci USA 89:3362–3366
17. Grobe AC, Wells SM, Benavidez E et al (2006) Increased oxidative stress in lambs with increased pulmonary blood flow and pulmonary hypertension: role of NADPH oxidase and endothelial NO synthase. Am J Physiol Lung Cell Mol Physiol 290:L1069–L1077
18. Cai H, Griendling KK, Harrison DG (2003) The vascular NAD(P)H oxidases as therapeutic targets in cardiovascular diseases. Trends Pharmacol Sci 24:471–478
19. McNally JS, Davis ME, Giddens DP et al (2003) Role of xanthine oxidoreductase and NAD(P)H oxidase in endothelial superoxide production in response to oscillatory shear stress. Am J Physiol Heart Circ Physiol 285:H2290–H2297
20. Mittal M, Roth M, König P et al (2007) Hypoxia-dependent regulation of nonphagocytic NADPH oxidase subunit NOX4 in the pulmonary vasculature. Circ Res 101:258–267
21. Li S, Tabar SS, Malec V et al (2008) NOX4 regulates ROS levels under normoxic and hypoxic conditions, triggers proliferation, and inhibits apoptosis in pulmonary artery adventitial fibroblasts. Antioxid Redox Signal 10:1687–1698
22. Csiszar A, Labinskyy N, Olson S et al (2009) Resveratrol prevents monocrotaline-induced pulmonary hypertension in rats. Hypertension 54:668–675
23. Wang X, Yang Y, Yang D et al (2016) Tetrandrine prevents monocrotaline-induced pulmonary arterial hypertension in rats through regulation of the protein expression of inducible nitric oxide synthase and cyclic guanosine monophosphate-dependent protein kinase type 1. J Vasc Surg 64:1468–1477
24. Farahmand F, Hill MF, Singal PK (2004) Antioxidant and oxidative stress changes in experimental cor pulmonale. Mol Cell Biochem 260:21–29
25. Nisbet RE, Graves AS, Kleinhenz DJ et al (2009) The role of NADPH oxidase in chronic intermittent hypoxia-induced pulmonary hypertension in mice. Am J Respir Cell Mol Biol 40:601–609
26. Araneda OF, Tuesta M (2012) Lung oxidative damage by hypoxia. Oxid Med Cell Longev 2012:856918
27. Zhao Y-Y, Zhao YD, Mirza MK et al (2009) Persistent eNOS activation secondary to caveolin-1 deficiency induces pulmonary hypertension in mice and humans through PKG nitration. J Clin Invest 119:2009–2018
28. Konduri GG, Bakhutashvili I, Eis A, Pritchard K (2007) Oxidant stress from uncoupled nitric oxide synthase impairs vasodilation in fetal lambs with persistent pulmonary hypertension. Am J Physiol Heart Circ Physiol 292:H1812–H1820
29. Liu JQ, Zelko IN, Erbynn EM et al (2006) Hypoxic pulmonary hypertension: role of superoxide and NADPH oxidase (gp91phox). Am J Physiol Lung Cell Mol Physiol 290:L2–L10
30. Maniatis NA, Shinin V, Schraufnagel DE et al (2008) Increased pulmonary vascular resistance and defective pulmonary artery filling in caveolin-1-/- mice. Am J Physiol Lung Cell Mol Physiol 294:L865–L873

31. Kamezaki F, Tasaki H, Yamashita K et al (2008) Gene transfer of extracellular superoxide dismutase ameliorates pulmonary hypertension in rats. Am J Respir Crit Care Med 177:219–226
32. Farrow KN, Lakshminrusimha S, Reda WJ et al (2008) Superoxide dismutase restores eNOS expression and function in resistance pulmonary arteries from neonatal lambs with persistent pulmonary hypertension. Am J Physiol Lung Cell Mol Physiol 295:L979–L987
33. Galhotra P, Prabhakar P, Meghwani H et al (2018) Beneficial effects of fenofibrate in pulmonary hypertension in rats. Mol Cell Biochem. https://doi.org/10.1007/s11010-018-3355-3
34. Meghwani H, Prabhakar P, Mohammed SA et al (2018) Beneficial effect of ocimum sanctum (Linn) against monocrotaline-induced pulmonary hypertension in rats. Med Basel Switz 5. https://doi.org/10.3390/medicines5020034
35. Meghwani H, Prabhakar P, Mohammed SA et al (2017) Beneficial effects of aqueous extract of stem bark of Terminalia arjuna (Roxb.), An ayurvedic drug in experimental pulmonary hypertension. J Ethnopharmacol 197:184–194
36. Jankov RP, Kantores C, Pan J, Belik J (2008) Contribution of xanthine oxidase-derived superoxide to chronic hypoxic pulmonary hypertension in neonatal rats. Am J Physiol Lung Cell Mol Physiol 294:L233–L245
37. Wong C-M, Preston IR, Hill NS, Suzuki YJ (2012) Iron chelation inhibits the development of pulmonary vascular remodeling. Free Radic Biol Med 53:1738–1747
38. Ghasemzadeh N, Patel RS, Eapen DJ et al (2014) Oxidative stress is associated with increased pulmonary artery systolic pressure in humans. Hypertens Dallas Tex 63:1270–1275
39. Cracowski J-L, Degano B, Chabot F et al (2012) Independent association of urinary F2-isoprostanes with survival in pulmonary arterial hypertension. Chest 142:869–876
40. Spiekermann S, Schenk K, Hoeper MM (2009) Increased xanthine oxidase activity in idiopathic pulmonary arterial hypertension. Eur Respir J 34:276
41. Masri FA, Comhair SAA, Dostanic-Larson I et al (2008) Deficiency of lung antioxidants in idiopathic pulmonary arterial hypertension. Clin Transl Sci 1:99–106
42. Bowers R, Cool C, Murphy RC et al (2004) Oxidative stress in severe pulmonary hypertension. Am J Respir Crit Care Med 169:764–769
43. Dani C, Poggi C (2014) The role of genetic polymorphisms in antioxidant enzymes and potential antioxidant therapies in neonatal lung disease. Antioxid Redox Signal 21:1863–1880
44. Peña-Silva RA, Miller JD, Chu Y, Heistad DD (2009) Serotonin produces monoamine oxidase-dependent oxidative stress in human heart valves. Am J Physiol Heart Circ Physiol 297:H1354–H1360
45. Semen K, Yelisyeyeva O, Jarocka-Karpowicz I et al (2015) Sildenafil reduces signs of oxidative stress in pulmonary arterial hypertension: Evaluation by fatty acid composition, level of hydroxynonenal and heart rate variability. Redox Biol 7:48–57
46. Fan Y-F, Zhang R, Jiang X et al (2013) The phosphodiesterase-5 inhibitor vardenafil reduces oxidative stress while reversing pulmonary arterial hypertension. Cardiovasc Res 99:395–403
47. Gabrielli LA, Castro PF, Godoy I et al (2011) Systemic oxidative stress and endothelial dysfunction is associated with an attenuated acute vascular response to inhaled prostanoid in pulmonary artery hypertension patients. J Card Fail 17:1012–1017
48. Sharp J, Farha S, Park MM et al (2014) Coenzyme Q supplementation in pulmonary arterial hypertension. Redox Biol 2:884–891
49. Zeng W-J, Xiong C-M, Zhao L et al (2012) Atorvastatin in Pulmonary Arterial Hypertension (APATH) study. Eur Respir J 40:67–74
50. Kawut SM, Bagiella E, Lederer DJ et al (2011) Randomized clinical trial of aspirin and simvastatin for pulmonary arterial hypertension. Circulation 123:2985–2993
51. Rysz-Górzynska M, Gluba-Brzózka A, Sahebkar A et al (2016) Efficacy of statin therapy in pulmonary arterial hypertension: a systematic review and meta-analysis. Sci Rep 6:30060
52. Rai PR, Cool CD, King JAC et al (2008) The cancer paradigm of severe pulmonary arterial hypertension. Am J Respir Crit Care Med 178:558–564
53. Xu W, Koeck T, Lara AR et al (2007) Alterations of cellular bioenergetics in pulmonary artery endothelial cells. Proc Natl Acad Sci USA 104:1342–1347

Free Radicals and Reactive Oxygen Species in Cardiovascular Pathophysiology: An Overview

17

Shyamal K. Goswami

Abstract

Oxidative stress has long been attributed to the pathobiology of various degenerative diseases. However, despite its wide acceptance among the researchers and the clinicians, the mechanistic insight into the contribution of various oxidants to the aetiology of those disorders remained enigmatic for a long time. Also, the use of antioxidants as therapeutics had very limited success. In the past decade, a significant progress has been made in understanding the chemistry of various reactive oxygen and nitrogen species, their enzymatic mechanisms, their generation, their cellular locations and their targets of action. While some of the highly reactive species, viz. hydroxyl radical and peroxynitrite, are deleterious for the cell, others like hydrogen peroxide and superoxide often act as bona fide signalling molecules. Such knowledge has revealed that a close network of redox reactions mediated by these species intricately regulate cellular functions. Any perturbation in those circuitries affects the cell physiology, causing distress for the related tissue and the organ. This review summarizes the present-day knowledge of those redox processes in the context of certain cardiovascular disorders.

Keywords

Reactive oxygen species (ROS) · Free radicals · Cardiovascular disorders · Redox signalling · Apoptosis · Atherosclerosis

S. K. Goswami (✉)
School of Life Sciences, Jawaharlal Nehru University, New Delhi, India

S. Chakraborti et al. (eds.), *Modulation of Oxidative Stress in Heart Disease*,
https://doi.org/10.1007/978-981-13-8946-7_17

17.1 Introduction

17.1.1 Role of Free Radicals and Reactive Oxygen Species in Cellular Function: An Overview

Ageing is a natural process, but the ageing population are susceptible to various degenerating diseases like cancer, diabetes, Alzheimer's, heart failure and other cardiovascular disorders. Although the affected organs, pathobiology and the therapeutic options for all these diseases are very different, one commonality between them is oxidative stress. Nevertheless, despite the wide acceptance that oxidative stress is the causative agent for these diseases, the mechanistic views in this regard have undergone a paradigm shift in the past 20 years. While the early concept was undue generation of free radicals and reactive oxygen species (ROS) cause many of the age-related diseases, it is now established that low-intensity generation of ROS is an integral part of cell regulatory network and only their dysfunction and aberrant generation cause diseases.

17.2 The "Free Radical Theory of Ageing": A Historical Perspective

Biologists and clinicians have long observed the close relationship between ageing and degenerative diseases due to dysfunctional organs. A young gerontologist Denham Harman in 1955 first proposed that ageing is primarily caused by the damages to the organs and tissues by the free radicals [1]. Although his proposition was quite revolutionary, it was more of a hypothesis as the assays for free radicals and reactive species in biological samples were not available in those days. In the subsequent years, he validated his theory by studying the role of free radicals in ageing, cancer and atherosclerosis. Because of his phenomenal contribution, he is considered as the father of the free radical theory of ageing (https://en.wikipedia.org/wiki/Denham_Harman).

Among the major observations he had made:

1. Atherogenesis in the artery is caused due to the oxidative polymerization of serum lipoproteins, deposition of oxidized materials in the arterial wall and inflammation [2].
2. With the ageing, serum mercaptan level decreases. In this study he formally proposed the "Free radical theory of aging" [3].
3. Mutations induced by the free radicals cause cancer and ageing [4].
4. Antioxidants can prevent cancer and can extend the life span of mice [5].
5. Antioxidants boost the immunity in mice, thus connecting the age, immunity and the free radicals [6].

Superoxide dismutase (SOD) was the first antioxidant enzyme to be discovered in 1969. Since its biological relevance was not understood in those days, it was thought that since free radicals are deleterious for the cell, nature has created this enzyme to destroy them. Such interpretation thus gave credence to the "Free radical theory of aging" [7]. It was much later that cell biologists realized that the function of SOD is to convert superoxide, a strong oxidant, to hydrogen peroxide, a signalling molecule (details given in the following sections).

In a remarkable feat, almost quarter of a century after he proposed the "Free Radical Theory of Aging", Denham Harman rechristened it in 1984 as the "Free radical theory of diseases" [8]. He suggested that free radicals accumulate and cause damages to the body and the extent of damage increases with ageing. Although the environmental and genetic factors have a role to determine the magnitude of such damages which differ from one individual to another, aged population is generally susceptible for diseases because of those damages. He thus proposed that age-related diseases can be prevented by calorie restriction and taking the antioxidants that lowers the level of free radicals. Although the beneficial effects of calorie restriction have been proved over the years, that of antioxidant remains controversial [9–12]. Nevertheless, these groundbreaking and insightful studies by Denham Harman created the foundation of modern-day knowledge of the oxidative stress theory of diseases.

17.3 Emergence of the "Oxidative Stress Theory of Diseases"

The role of oxidation-reduction in cellular metabolism, e.g. Warburg effect, Krebs cycle, oxidative phosphorylation, etc., was established in the early part of the twentieth century. In the 1950s and 1960s, toxicologists, nutritionists and clinical biochemists realized that xenobiotic agents such as drugs, industrial products and environmental pollutants upon metabolism could become electrophiles, which damage the cellular DNA, proteins and lipids [13]. Some of these electrophiles were also associated with cancer. In 1985, the deleterious effects of these free radicals, especially those which damage the organs and tissues, were formally termed by H Sies, a pioneer in the field, as "Oxidative stress" [14].

According to his definition, oxidative stress is a disturbance in the balance between pro- and antioxidants in favour of the former. Further, early studies done on diseased tissues and organs from people suffering from cardiovascular diseases, atherosclerosis, Parkinson's and Alzheimer's showed higher incidences of oxidized lipids, DNA and proteins [15, 16]. However, in those days, most efforts were given to understand the cellular antioxidants defence and how to boost them rather than understanding the role of pro-oxidant system per se [17]. Genetic studies also corroborated this approach. It was shown that in *C. elegans*, mutations leading to the increased levels of superoxide dismutase and catalase, two major antioxidant enzymes, increases its life span [18]. It was also shown that small molecules that mimic these antioxidant enzymes could extend the life span of *C. elegans* [19]. Although these studies showed a close association between ageing and diseases, it also subscribed to the concept that oxidative stress is the cause of cancer, Parkinson's, Alzheimer's and cardiovascular disorders.

17.4 Limitations of the Oxidative Stress Theory of Diseases and Curing by Antioxidants

With the emergence of strong evidences associating oxidative stress and various degenerative diseases, use of antioxidants for their amelioration was also explored. In this context, it is necessary to remember that the term antioxidant is too generic as it includes a wide spectrum of biomolecules, which can neutralize the cellular ROS and thereby prevent their deleterious effects. Over the years, the commonly used antioxidants include natural products like vitamins C, E and A, carotenoids, polyphenols and synthetic compounds like N-acetylcysteine and allopurinol (an inhibitor of xanthine oxidase). However, numerous studies over the past quarter of a century and large-scale meta-analyses of those results led to the conclusion that the use of antioxidants does not necessarily prevent or cure those diseases [20]. Although such observations do not necessarily negate the role of ROS as a trigger for certain degenerative diseases, it definitely suggests that the function of ROS in cellular physiology is far more nuanced than perceived earlier.

17.5 Regulatory Role of ROS in Cell Physiology: A Paradigm Shift

During the 1990s, in a seminal discovery, it was found that when human carcinoma cells are treated with the EGF (epidermal growth factor), it generates H_2O_2. Further, this H_2O_2 inhibited protein tyrosine phosphatase by oxidizing certain cysteine residues and thereby sustaining the growth signalling [21]. Another major discovery was the existence of non-phagocytic NADPH oxidases in various tissues. NADPH oxidases are the membrane-associated multisubunit enzyme complexes that generate superoxide ($O_2^{\cdot -}$) from molecular oxygen and NADPH. It was first discovered in phagocytic cells, where the generation of superoxide and other reactive species derived from it is used as a strategy for combating invading pathogens. Therefore, with the discovery of the presence of other isoforms of NADPH oxidases, tissues other than phagocytic cells led to the proposition that they might generate superoxide (which is then converted into hydrogen peroxide) for physiological purposes [22].

Subsequent to these two original discoveries almost 25 years ago, it became more and more apparent that superoxide and hydrogen peroxide contribute to normal physiological functions including cell signalling and gene expression for physiological events like cell proliferation, differentiation, autophagy and apoptosis. Such role of ROS is also highly conserved in evolution as in plants; various stress responses are mediated through the ROS. Further study suggested that alike kinases, which reversibly and specifically phosphorylate target proteins, reactive oxygen species can selectively oxidize certain cysteine residues of various proteins, regulating their functions. Accordingly, the signalling by ROS is termed as redox signalling that has unique biochemical and cell biological characteristics [23].

One scepticism about the role of ROS in cell signalling was the tenet that any molecules mediating signal transduction must have target specificity and modifications induced by them on their targets ought to be reversible. Since physiological oxidants like superoxide ($O_2^{\cdot-}$) and hydrogen peroxide (H_2O_2) are highly diffusible, it was argued that they cannot have specific targets for the oxidation. However, soon it was proposed that since these molecules have moderate oxidation potential, they can selectively oxidize cysteine residues that have pKa values close to 6.4, which depends on the charge distribution of its adjacent amino acids in the 3D conformation, as superoxide and hydrogen peroxide can oxidize only the thiolate forms of cysteine [23]. Also, cellular antioxidants like thioredoxin and glutaredoxins can reverse such oxidations of cysteine in cellular environment [25]. These two observations thus established superoxide and hydrogen peroxide as the bona fide signalling agents that selectively and reversibly modify certain cysteine residues of their target proteins modulating their functions [26]. One such best-known example is the oxidative activation of the antioxidant transcription factor Nrf2 [27]. Emergence of redox proteomics has revealed the existence of numerous such modifications in various cellular and tissue contexts [28].

17.6 Pathological ROS Signals and the Onset of Diseases

In contrast to superoxide and hydrogen peroxide, other certain highly reactive species, viz. hydroxyl ($OH^{\cdot}$), hydroperoxyl ($HO_2^{\cdot}$) and peroxyl ($ROO^{\cdot}$) radicals, hypochlorous acid (HOCl), peroxynitrite ($ONOO^-$), etc., oxidize proteins and other biomolecules non-specifically [29]. In addition, the oxidative modifications induced by these highly reactive species are irreversible and thus are unable to have signalling functions. Besides cysteine, several other amino acids like methionine, tryptophan and tyrosine can also undergo such irreversible oxidation under highly oxidizing environment, resulting in the loss of functions and diseases [30–32].

17.7 Generation of ROS in the Cellular Milieu and Their Attenuation by Antioxidants

Different types of ROS are generated in the cellular milieu under various pathophysiological conditions. While the primary ROS like superoxide and hydrogen peroxide have distinct sources and subcellular localizations, several other subspecies are generated only transiently under specific conditions [33]. Superoxide anion ($O_2^{\cdot-}$) is generated by the one electron reduction of oxygen by the enzymes like NADPH oxidases, xanthine oxidase, lipoxygenase and cyclooxygenase as well as by the electrons leaking out of the electron transport system in the mitochondria [34, 35]. NADPH oxidase 2 (Nox 2) is a membrane-associated enzyme first characterized in the neutrophils where it produces superoxide in the phagosomes to kill the ingested bacteria and fungi. It is the prototype member of the family, and the presence of other Noxes was shown in various cell types thereafter [36]. In human, there are several isoforms of this enzyme, viz. NOX1, NOX3 and NOX4 and DUOX1 and DUOX2, which differ in their subunit compositions and tissue and subcellular distributions [37]. They are activated by various growth factors, cytokines, etc., and the superoxide generated by the Noxes in non-phagocytic cells is primarily involved in downstream signalling [38]. Being present in the plasma membrane or other intracellular locations, they can generate superoxide both extra- and intracellularly with distinctive functions. Aberrant generation of ROS by different Noxes has been associated with various diseases, and their pharmacological inhibition is being explored for therapeutic purposes [39].

Mitochondrial electron transport chain is another major source of superoxide generation. Extensive studies have established that complexes I and III are the two major sites where superoxide is generated. During normal respiration, electrons leak out from the chain at a low intensity. These electrons partially reduce the surrounding oxygen producing limited quantity of $O_2^{\cdot-}$ that acts as a signalling molecule [40]. However, under certain pathological conditions, the extent of generation of superoxide in the mitochondria increases which often leads to diseases [41].

During nucleotide metabolism, conversion of hypoxanthine to xanthine and xanthine to uric acid is catalysed by the enzyme xanthine dehydrogenase. In oxidizing environment, xanthine dehydrogenase is converted to another form of the enzyme called xanthine oxidase by the reversible oxidation of a cysteine residue to sulphenic acid (it can also be converted into xanthine oxidase by a proteolytic cleavage). Xanthine oxidase has the same catalytic activity like xanthine dehydrogenase except that the mechanisms of reactions are different. While xanthine dehydrogenase uses NAD^+ as a cofactor and reduces it to NADH, the cofactor for xanthine oxidase is molecular oxygen that is reduced to superoxide. Such conversion of xanthine dehydrogenase to xanthine oxidase is an example of ROS-induced activation of a ROS-producing enzyme. Superoxide derived from xanthine oxidase also has role in redox signalling under both physiological and pathological contexts [42, 43].

Since hydrogen peroxide does not carry any unpaired electron, it is not a free radical per se. However, when it comes in contact with Fe^{++}, it generates hydroxyl radical, a highly reactive species. Accordingly, hydrogen peroxide is a reactive oxygen species. As compared to superoxide, hydrogen peroxide is a mild oxidant that selectively oxidizes cysteine thiols, modulating protein functions. It is thus a signalling molecule [44]. It is formed by several enzymatic reactions of which the primary one is the catalytic dismutation of superoxide by the enzyme superoxide dismutases (SOD). In human, there are two different of superoxide dismutases: the Cu/Zn type that is cytosolic and the Mn types that are mitochondrial. Both the enzymes have major roles in regulating the superoxide and hydrogen peroxide levels under physiological and pathological conditions [45].

Both superoxide and hydrogen peroxide are mild oxidants, and when present at a lower concentration, they can act as signalling molecules, However, at an elevated level, they indiscriminately oxidize proteins, lipids and other macromolecules, causing cellular damage. Thus, their intracellular steady-state levels require being under tight regulation by the cellular antioxidant system [46]. Accordingly, apart from the ROS-generating enzymes, those that attenuate it also play a key role in the redox biology of diseases. Mammalian cells have several enzymatic and nonenzymatic antioxidants that maintain the redox homeostasis. While vitamins C and E are the primary nonenzymatic antioxidants, superoxide dismutase and catalase (degrades hydrogen peroxide) are the two main enzymatic antioxidants [47]. Two other antioxidant enzymes involved in the catalytic breakdown of peroxides are glutathione peroxidase (GPx) and thioredoxin peroxidase/peroxiredoxins (TPx/Prx). Mammalian cells have distinct genes encoding thioredoxin, thioredoxin reductase, and thioredoxin peroxidase/peroxiredoxin enzymes. Glutathione and thioredoxin peroxidases have been identified in multiple cellular locations. The localization, expression levels and activities of these antioxidant enzymes are the key determinants in maintaining ROS levels in cellular microdomains [48].

17.8 Role of ROS in Cardiovascular Functions

It is now well established that oxidative stress is an important contributor to the degenerative diseases like cancer, diabetes, Parkinson's, Alzheimer's and cardiovascular disorders. Since early studies were inconclusive about the precise role of the oxidants in these diseases of diverse aetiology, generalizations were made in terms of their prevention and cure by the antioxidants [49]. However, over the past 20 years, a large volume of information has emerged that clearly shows that the role of ROS in these diseases is very much context specific as they are an integral part of cellular function and dysfunctions leading to those diseases. As an example, in various types of cancers, ROS have now been found to be playing a delicate role in the sustenance of the cancer microenvironment and metastasis [50]. Therefore, the role of oxidants in these diseases is more specific than general and needs to be discussed separately.

17.9 The Role of Oxidative Stress in Cardiovascular Diseases

The cardiovascular system responds to a plethora of pathophysiological signals in a highly complex manner. Depending upon the origin, the complexity and the intensity of the signals, the changes in the biochemical and molecular biological parameters differ. It involves various means of cellular responses, viz. phosphorylation-dephosphorylation, intracellular protein trafficking, turnover of mRNA and protein, oxidative or reductive modifications of cysteine thiols, etc., that mediate the cognate responses. Any aberration in these signalling systems leads to conditions like coronary artery and other vascular diseases, atherosclerosis, obesity, diabetes, endothelial dysfunction, hypertension, ischaemia-reperfusion injury, heart failure, etc. Hyper-oxidation of cellular proteins, DNA and lipids has been commonly observed in these disorders. However, as discussed in previous sections, the consequences of the intracellular generation of ROS depend upon the factors like their intracellular locations, steady-state concentrations, duration of generation, presence or absence of antioxidant system, etc. [51]. Based upon these factors, the

ROS, mostly hydrogen peroxide and superoxide, either lead to redox signalling through the modification of cysteine thiols to sulphenic acids and disulphide bonds, by S-glutathionylation and S-nitrosylation, or global oxidative stress through the indiscreminate oxidation of cysteine to sulphonic acid [22–25, 51]. In the cardiovascular system, upon stimulation by various pathophysiological agonists, ROS is generated at multiple cellular locations that modify a plethora of targets like sarcolemmal Na^+/Ca^{++} exchangers, signalling kinases, ion channels, ryanodine receptor and gene regulatory proteins [52–57]. Despite the fact that various cardiovascular diseases are associated with an increase production of ROS and decreased antioxidant defence, their mechanism of action in inducing these diseases is likely to vary among the affected tissues and cell types. Certain environmental factors like tobacco smoking, pollution, etc. also contribute to oxidative stress promoting these diseases. In the following sections, I will be discussing the role of ROS in the context of two major cardiovascular disorders, i.e. atherosclerosis and heart failure.

17.10 Role of ROS in Atherosclerosis

The endothelium is the layer of cells that line the blood vessels maintaining the wall permeability. It also controls the vascular tone, proliferation of smooth muscle cells, platelet aggregation and inflammation. In atherosclerosis, plaques comprising of lipids and immune cells build up inside the blood vessels. It is a slow progressive disorder, and the process of developing such plaques is called atherogenesis. For many years, the presence of oxidized proteins and lipids have been found in those lesions, and the severity of the disease has been directly associated with the extent of oxidation of biomass present there [58]. Endothelial nitric oxide synthase (eNOS) is the key regulator of endothelial function. Nitric oxide produced by eNOS maintains vascular tone, regulates growth of smooth muscle cells and protects vessels from activated platelets and other circulatory cell types. Therefore, eNOS and its product NO play a key role in vascular homeostasis [59]. A major cause of atherosclerosis is the inadequacy of the eNOS function that results in a pathological condition called endothelial dysfunction [60]. The enzyme eNOS is a dimeric enzyme that has multiple cofactors including biopterin, and its substrate is NADPH and L-arginine. It oxidizes NADPH, and the released electron migrates through the electron carrier FAD and FMN to L-arginine and O_2 producing NO [61]. These reactions require the cofactor biopterin to be in its reduced form that is tetrahydrobiopterin (BH4). In pathophysiological conditions like diabetes, the activities of NADPH and xanthine oxidase increase resulting in an increased generation of superoxide which reacts with NO generated from eNOS to form peroxynitrite ($ONOO^-$), a potent oxidizing agent. Such oxidizing environment decreases the biopterin level (BH4 is oxidized to BH3 and BH2). BH2 competes with BH4 for binding to eNOS, disrupting its function, a process called uncoupling. The uncoupled eNOS then generates more superoxide with concomitant decrease in generation of NO [62]. It is therefore an example of ROS (superoxide generated from NADPH oxidases) induced generation of more ROS, i.e. further generation of superoxide by uncoupled eNOS. ROS generated from uncoupled eNOS has been associated with atherosclerosis in experimental mice and in human patients. It has also been seen in patients with hypertension, hypercholesterolemia and diabetes [63]. In experimentally induced atherosclerosis in mice and in human atherosclerotic plaque, increased activity of endothelial and circulatory xanthine oxidase has been demonstrated, suggesting the role of superoxide generated from xanthine oxidase in endothelial dysfunction. Inhibitor of xanthine oxidase like allopurinol reduces endothelial dysfunction in smokers and the development of atherosclerosis in experimental mice [64]. In mice, there are several isoforms of NADPH oxidases. While Nox2 and Nox4 are expressed in endothelial cells, Nox1 and Nox4 are expressed in vascular smooth muscle cells. Both Nox1 and Nox2 have been implicated in inducing atherogenesis. In apolipoprotein E-knockout mice (ApoE-KO; used for studying atherosclerosis), the deletion of *nox1* gene reduces atherogenesis induced by streptozotocin, an inducer of diabetes mellitus [65]. Interestingly, in contrast to the proatherogenic function of Nox1 and Nox2, Nox 4 has been shown to have a protective role in this process [66]. Nox4 produces hydrogen peroxide through the dismutation of superoxide. Therefore, upon activation of Nox4, peroxynitrite is not generated as in case of the activation of Nox1 and Nox2, rather hydrogen peroxide generated by it provides the protective signalling. This is thus an example of oxidative stress versus ROS signalling in the pathology of a disease. Apart from endothelial dysfunction, ROS generated in the endothelium from NADPH oxidases also increases the expression of adhesion molecules on endothelial cells and induces the proliferation of smooth muscle cells. These events lead to the infiltration of monocyte and macrophages. ROS generated from Nox enzymes in monocytes and macrophages are involved in the oxidation of low-density lipoprotein (LDL), a contributor to atherogenesis [67]. ROS generated from the mitochondria has also been shown to have roles in atherogenesis.

Based on the prevailing view that excessive generation of ROS leads to atherosclerosis and other cardiovascular diseases, various groups have tested the effects of boosting the antioxidant system, and that has led to interesting results. When kept in fatty diet, ApoE-knockout mice (used for atherosclerosis research) overexpressing SOD1 through transgene remains equally atherogenic as the control mice. However, combined expression of catalase and SOD1 or catalase alone reduces the atherogenic potential [68]. The possible explanation is that increased expression of SOD1 might shift the homoeostatic control resulting in the generation of other ROS like hydrogen peroxide and hydroxyl radical (generated by the reaction between hydrogen peroxide and metal ions). Antioxidant enzymes like glutathione peroxidases (Gpx) attenuate hydrogen and lipid peroxides. Deficiency of GPx1 in ApoE-KO mice increases the atherogenic potential by the oxidation of LDL, formation of foam cells and the proliferation of macrophages, while overexpression of Gpx4 reduces atherogenic potential [69]. Taken together, generation of atherosclerosis plaque involves decrease in eNOS activity, increased inflammation and expression of adhesion factors and advanced glycation and oxidation of low-density lipoprotein, which may or may not occur concurrently. Therefore, the nature of ROS involved and the pathways they trigger might vary from one experimental model to another. Thus, the pathobiology of atherosclerosis is far more complex than a simplistic correlation between increased and decreased level of ROS [70–73].

17.11 Redox Signalling in Cardiac Hypertrophy and Heart Failure

Adrenergic receptors (α and β) are the key modulator of cardiac output and its homoeostasis [74, 75]. Norepinephrine (NE), a catecholamine, is released from the sympathetic nervous system as an agonists for these receptors. Limited adrenergic stimulation is required for maintaining the contractile function of the heart. Under diseases like hypertension, valve defects, ischaemia-reperfusion injury, myocardial infarction, etc., cardiac output is reduced. To boost cardiac performance, sympathetic system increases the release of NE. Initially, under increased NE stimulation, myocytes are enlarged (myocytes are terminally differentiated, and they cannot proliferate), a process known as hypertrophy [76]. Although hypertrophic response under increased adrenergic stimulation is a compensatory process that recalibrates the adrenergic signalling and the downstream gene expression programmes, sustained adrenergic overdrive leads to the loss of cardiac myocytes by apoptosis, further compromising cardiac function. Such combination of hypertrophic and apoptotic responses eventually leads to substantial weakening of the heart, resulting in heart failure [77]. Earlier studies had shown that upon stimulation with lower doses of NE (~1–2 μM), cultured murine myocytes faithfully elicit hypertrophic responses, while at a higher dose (10 μM and above), they undergo apoptosis [78, 79]. This experimental system thus provided an excellent system for studying the role of oxidants in two distinct pathological responses culminating to a common disease, i.e. heart failure. The prevailing hypothesis in this regard was that while at a lower dose, NE elicits hypertrophic responses through oxidative signalling, increased adrenergic stimulation leads to oxidative stress followed by apoptosis [80–82]. We have been studying differential adrenergic signalling leading to two distinct responses, i.e. hypertrophy and apoptosis, as a model for understanding the role of ROS in cardiac pathobiology. We for the first time contested the concept that apoptosis is induced due to oxidative stress. We demonstrated that in H9c2 cardiac myoblast cells (rat), hypertrophic and apoptotic responses induced upon NE treatment are initiated by the low-intensity generation of ROS at comparable levels [83]. We then analysed the modulation of two redox-responsive transcription factors AP-1 and Nrf2 as the downstream targets of NE signalling. Both hypertrophic and apoptotic doses NE induced AP-1 and Nrf-2 activities, but the extent and the kinetics of induction of their DNA binding activities were not in direct correlation to the level of ROS generated by each treatment. Significantly, the AP-1 activities induced by the two doses of NE were functionally different. While 2 μM NE (hypertrophic dose) induced FosB: Jun dimer, at 100 μM (apoptotic dose), it activated Fra-1: Jun. We thus inferred that both the responses elicited by NE are characterized by distinct redox signalling and not an increase in the ROS level, i.e. oxidative stress [83]. We also demonstrated that the induction of both FosB and Fra-1 occurs at the level of transcription and were partially suppressed by catalase (converts hydrogen peroxide to water) and MnTMPyP (a mimetic of superoxide dismutase that converts superoxide to hydrogen peroxide), confirming the role of ROS in their induction. In order to decipher the mechanism of gene regulation by ROS, the promoter regions of *fosB* and *fra-1* genes were linked to the reporter gene luciferase and assayed for their expression under different treatment conditions. These assays showed that multiple binding sites for the transcription factors, viz. SP-1, CEBP and AP-1 in the *fosB* promoter, integrate the kinase and the ROS signalling. As an example, while SP-1 being a cysteine-zinc finger containing transcription factor is directly targeted by the ROS generated upon NE treatment, CEBP, a leucine zipper-containing transcription factor, is modulated by the upstream kinases which are in turn modulated by ROS further upstream [84]. In a parallel study, we observed that upon stimulation by the two different concentrations of NE, although multiple ROS are generated at comparable levels, they have distinctive kinetics. When treated with the hypertrophic dose of NE (2 μM), myoblast cells generate DCFH-DA positive ROS only for 2 h; but those treated with 100 μM NE (apoptotic dose) generated ROS of similar intensity as seen in the case of NE treatment at 2 μM dose but for 48 h. Noticeably, although certain highly reactive species of unknown nature were also detected under both treatment conditions, there were no major differences in their levels, further refuting the role of oxidative stress in triggering apoptosis [85]. Nevertheless, DNA damage commonly associated with apoptosis was only seen in cells treated with 100 μm NE [85]. Thus, from our study, it is evident that both the doses of NE elicit hypertrophic and apoptotic responses through characteristic redox signalling rather than a mere due to a quantitative difference in the level of ROS. In summary, our study therefore subscribes to hypothesis that a characteristic pattern of ROS generation modulates the cell signalling and gene expression programme leading to both hypertrophic and apoptotic responses.

To settle the ultimate question whether cardiac myocyte loss under adrenergic overdrive is due to oxidative stress or redox signalling, it is imperative to identify the sources of ROS and their targets in cells under specific stimuli. Among various sources of intracellular ROS, viz. mitochondrial electron transport system, enzymes involved in oxidation-reduction reactions (like nitric oxide synthase), NADPH oxidases (Noxes), etc., the Noxes are the important regulators of cardiovascular pathobiology [36, 39]. In a recent study, when we treated myoblast cells with 2 μM NE together with an inhibitor of Nox2, generation of ROS was inhibited. Expression of a number of marker genes of cardiac hypertrophy induced upon NE treatment was also prevented by the inhibition of Nox2. Also, organelle-specific GFP probe that specifically detects hydrogen peroxide (HyPerGFP) showed that upon NE treatment, ROS is primarily generated in the cytosol and not in the endoplasmic reticulum or the mitochondria [86]. Therefore, our study till date not only establishes the importance of ROS signalling over oxidative stress in mediating the pathological responses by NE; it also identified the possible source, nature and the downstream targets of the adrenergic signalling system.

17.12 Concluding Remarks

The exact role of free radicals and ROS in the pathobiology of degenerative diseases has been an enigma for about half a century. Despite substantial evidences of oxidizing environment in the diseased tissues and organs, the mechanistic insight into the source, nature and the targets of ROS is poorly understood till date. Accordingly, the therapeutic potential of antioxidants has been overrated and oversimplified. One among various reasons for such deficient outcome is the highly complex nature of redox reactions [87, 88]. While the specificity of most biochemical reactions is governed by the selective interaction between biomolecules, that of redox reactions is governed by the redox potentials of the reactants that are intrinsically relative in nature [35]. Further, the availability of appropriate reagents and probes for tracking redox reactions in the cellular milieu still remains a challenge. Nevertheless, the advent of stronger tools of cell biology and redox proteomics has tremendously helped in refining our knowledge about the redox biology of other neurodegenerative and cardiovascular diseases. Alike thc volume of information that has accumulated over the past quarter of a century on kinase signalling, it is expected that in the coming years, significant progresses will be made in understanding the redox regulation of cell function and dysfunction as well. Such advancement will not only help us expand our knowledge about this important cellular processes; it also will strengthen the process of development of better therapeutics for these crippling disorders.

References

1. Harman D (1956) Aging: a theory based on free radical and radiation chemistry. J Gerontol 11:298–300
2. Harman D (1957) Atherosclerosis: a hypothesis concerning the initiating steps on pathogenesis. J Gerontol 12:199–202
3. Harman D (1960) The free radical theory of aging: the effect of age on serum mercaptan levels. J Gerontol 15:38–40
4. Harman D (1961) Mutation, cancer, and ageing. Lancet 1:200–201
5. Harman D (1961) Prolongation of the normal lifespan and inhibition of spontaneous cancer by antioxidants. J Gerontol 16:247–254
6. Harman D, Heidrick ML, Eddy DE (1977) Free radical theory of aging: effect of free-radical-reaction inhibitors on the immune response. J Am Geriatr Soc 25:400–407
7. McCord JM, Fridovich I (1969) Superoxide dismutase. An enzymic function for erythrocuprein (hemocuprein). J Biol Chem 244:6049–6055
8. Harman D (1991) The aging process: major risk factor for disease and death. Proc Natl Acad Sci USA 88:5360–5363
9. Liang Y, Wang Z (2018) Which is the most reasonable anti-aging strategy: meta-analysis. Adv Exp Med Biol 1086:267–282
10. Gianfredi V, Vannini S, Moretti M, Villarini M, Bragazzi NL, Izzotti A, Nucci D (2017) Sulforaphane and epigallocatechin gallate restore estrogen receptor expression by modulating epigenetic events in the breast cancer cell line MDA-MB-231: a systematic review and meta-analysis. J Nutrigenet Nutrigenomics 10:126–135

11. Moser MA, Chun OK (2016) Vitamin C and heart health: a review based on findings from epidemiologic studies. Int J Mol Sci 17:1328
12. Buffenstein R, Edrey YH, Yang T, Mele J (2008) The oxidative stress theory of aging: embattled or invincible? Insights from non-traditional model organisms. Age (Dordr) 30:99–109
13. Guengerich FP, Liebler DC (1985) Enzymatic activation of chemicals to toxic metabolites. Crit Rev Toxicol 14:259–307
14. Sies H (1985) Oxidative stress: introductory remarks. In: Sies H (ed) Oxidative stress. Academic Press, London, pp 1–8
15. van Kuijk FJ, Dratz EA (1987) Detection of phospholipid peroxides in biological samples. Free Radic Biol Med 3:349–354
16. Hennig B, Chow CK (1988) Lipid peroxidation and endothelial cell injury: implications in atherosclerosis. Free Radic Biol Med 4:99–106
17. Halliwell B, Grootveld M (1987) The measurement of free radical reactions in humans. Some thoughts for future experimentation. FEBS Lett 213:9–14
18. Larsen PL (1993) Aging and resistance to oxidative damage in *Caenorhabditis elegans*. Proc Natl Acad Sci U S A 90:8905–8909
19. Melov S, Ravenscroft J, Malik S, Gill MS, Walker DW, Clayton PE, Wallace DC, Malfroy B, Doctrow SR, Lithgow GJ (2000) Extension of life-span with superoxide dismutase/catalase mimetics. Science 289:1567–1569
20. Ye Y, Li J, Yuan Z (2013) Effect of antioxidant vitamin supplementation on cardiovascular outcomes: a meta-analysis of randomized controlled trials. PLoS One 8(2):e56803
21. Bae YS, Kang SW, Seo MS, Baines IC, Tekle E, Chock PB, Rhee SG (1997) Role in EGF receptor-mediated tyrosine phosphorylation. J Biol Chem 272:217–221
22. Saran M (2003) To what end does nature produce superoxide? NADPH oxidase as an autocrine modifier of membrane phospholipids generating paracrine lipid messengers. Free Radic Res 37:1045–1059
23. Scialò F, Fernández-Ayala DJ, Sanz A (2017) Role of mitochondrial reverse electron transport in ROS signaling: potential roles in health and disease. Front Physiol 8:428
24. Mikkelsen RB, Wardman P (2003) Biological chemistry of reactive oxygen and nitrogen and radiation-induced signal transduction mechanisms. Oncogene 22:5734–5754
25. Jones DP, Go YM, Anderson CL, Ziegler TR, Kinkade JM Jr, Kirlin WG (2004) Cysteine/cysteine couple is a newly recognized node in the circuitry for biologic redox signaling and control. FASEB J 18:1246–1248
26. Di Marzo N, Chisci E, Giovannoni R (2018) The role of hydrogen peroxide in redox-dependent signaling: homeostatic and pathological responses in mammalian cells. Cell 7:156
27. Silva-Islas CA, Maldonado PD (2018) Canonical and non-canonical mechanisms of Nrf2 activation. Pharmacol Res 134:92–99
28. Foyer CH, Wilson MH, Wright MH (2018) Redox regulation of cell proliferation: bioinformatics and redox proteomics approaches to identify redox-sensitive cell cycle regulators. Free Radic Biol Med 122:137–149
29. Weidinger A, Kozlov AV (2015) Biological activities of reactive oxygen and nitrogen species: oxidative stress versus signal transduction. Biomol Ther 5:472–484
30. Yang J, Carroll KS, Liebler DC (2016) The expanding landscape of the thiol redox proteome. Mol Cell Proteomics 15:1–11
31. Devarie-Baez NO, Silva Lopez EI, Furdui CM (2016) Biological chemistry and functionality of protein sulfenic acids and related thiol modifications. Free Radic Res 50:172–194
32. Lushchak VI (2014) Free radicals, reactive oxygen species, oxidative stress and its classification. Chem Biol Interact 224:164–175
33. Zielonka J, Kalyanaraman B (2018) Small-molecule luminescent probes for the detection of cellular oxidizing and nitrating species. Free Radic Biol Med 128:3–22
34. Buettner GR (2011) Superoxide dismutase in redox biology: the roles of superoxide and hydrogen peroxide. Anti Cancer Agents Med Chem 11:341–346

35. Buettner GR, Wagner BA, Rodgers VG (2013) Quantitative redox biology: an approach to understand the role of reactive species in defining the cellular redox environment. Cell Biochem Biophys 67:477–483
36. Li Y, Pagano PJ (2017) Microvascular NADPH oxidase in health and disease. Free Radic Biol Med 109:33–47
37. Sirokmány G, Donkó Á, Geiszt M (2016) Nox/Duox family of NADPH oxidases: lessons from knockout mouse models. Trends Pharmacol Sci 37:318–327
38. Altenhöfer S, Radermacher KA, Kleikers PW, Wingler K, Schmidt HH (2015) Evolution of NADPH oxidase inhibitors: selectivity and mechanisms for target engagement. Antioxid Redox Signal 23(5):406–427
39. García-Redondo AB, Aguado A, Briones AM, Salaices M (2016) NADPH oxidases and vascular remodeling in cardiovascular diseases. Pharmacol Res 114:110–120
40. Shadel GS, Horvath TL (2015) Mitochondrial ROS signaling in organismal homeostasis. Cell 163:560–569
41. Cadenas S (2018) ROS and redox signaling in myocardial ischemia-reperfusion injury and cardioprotection. Free Radic Biol Med 117:76–89
42. Battelli MG, Bolognesi A, Polito L (1842) Pathophysiology of circulating xanthine oxidoreductase: new emerging roles for a multi-tasking enzyme. Biochim Biophys Acta 2014:1502–1517
43. Bredemeier M, Lopes LM, Eisenreich MA, Hickmann S, Bongiorno GK, d'Avila R, Morsch ALB, da Silva SF, Campos GGD (2018) Xanthine oxidase inhibitors for prevention of cardiovascular events: a systematic review and meta-analysis of randomized controlled trials. BMC Cardiovasc Disord 18(1):24
44. Banerjee R (2017) Introduction to the Thematic Minireview Series: redox metabolism and signaling. J Biol Chem 292:16802–16803
45. Wang Y, Branicky R, Noë A, Hekimi S (2018) Superoxide dismutases: dual roles in controlling ROS damage and regulating ROS signaling. J Cell Biol 217(6):1915–1928
46. Sies H, Berndt C, Jones DP (2017) Oxidative stress. Annu Rev Biochem 86:715–748
47. He L, He T, Farrar S, Ji L, Liu T, Ma X (2017) Antioxidants maintain cellular redox homeostasis by elimination of reactive oxygen species. Cell Physiol Biochem 44:532–553
48. Lu J, Holmgren A (2014) The thioredoxin antioxidant system. Free Radic Biol Med 66:75–87
49. Day BJ (2014) Antioxidant therapeutics: pandora's box. Free Radic Biol Med 66:58–64
50. Hawk MA, Schafer ZT (2018) Mechanisms of redox metabolism and cancer cell survival during extracellular matrix detachment. J Biol Chem 293:7531–7537
51. Ho E, Karimi Galougahi K, Liu CC, Bhindi R, Figtree GA (2013) Biological markers of oxidative stress: applications to cardiovascular research and practice. Redox Biol 1:483–491
52. Dhalla NS, Temsah RM, Netticadan T (2000) Role of oxidative stress in cardiovascular diseases. J Hypertens 18:655–673
53. Pimentel DR, Adachi T, Ido Y, Heibeck T, Jiang B, Lee Y et al (2006) Strain-stimulated hypertrophy in cardiac myocytes is mediated by reactive oxygen species-dependent Ras S-glutathiolation. J Mol Cell Cardiol 41:613–622
54. Sag CM, Wagner S, Maier LS (2013) Role of oxidants on calcium and sodium movement in healthy and diseased cardiac myocytes. Free Radic Biol Med 63:338–349
55. Donoso P, Sanchez G, Bull R, Hidalgo C (2011) Modulation of cardiac ryanodine receptor activity by ROS and RNS. Front Biosci (Landmark Edition) 16:553–567
56. Zhu H, Shan L, Schiller PW, Mai A, Peng T (2010) Histone deacetylase-3 activation promotes tumor necrosis factor-alpha (TNF-alpha) expression in cardiomyocytes during lipopolysaccharide stimulation. J Biol Chem 285:9429–9436
57. Takahashi M, Suzuki E, Takeda R, Oba S, Nishimatsu H, Kimura K et al (2008) Angiotensin II and tumor necrosis factor-alpha synergistically promote monocyte chemoattractant protein-1 expression: roles of NF-kappaB, p38, and reactive oxygen species. Am J Physiology Heart Circ Physiol 294:H2879–H2888
58. Förstermann U, Xia N, Li H (2017) Roles of vascular oxidative stress and nitric oxide in the pathogenesis of atherosclerosis. Circ Res 120:713–735

59. Tousoulis D, Kampoli AM, Tentolouris C, Papageorgiou N, Stefanadis C (2012) The role of nitric oxide on endothelial function. Curr Vasc Pharmacol 10:4–18
60. Gimbrone MA Jr, García-Cardeña G (2016) Endothelial cell dysfunction and the pathobiology of atherosclerosis. Circ Res 118:620–636
61. Poulos TL, Li H (2017) Nitric oxide synthase and structure-based inhibitor design. Nitric Oxide 63:68–77
62. Sun J, Druhan LJ, Zweier JL (2010) Reactive oxygen and nitrogen species regulate inducible nitric oxide synthase function shifting the balance of nitric oxide and superoxide production. Arch Biochem Biophys 494:130–137
63. Roe ND, Ren J (2012) Nitric oxide synthase uncoupling: a therapeutic target in cardiovascular diseases. Vasc Pharmacol 57:168–172
64. Alem MM (2018) Allopurinol and endothelial function: a systematic review with meta-analysis of randomized controlled trials. Cardiovasc Ther 36:e12432
65. Sobey CG, Judkins CP, Rivera J, Lewis CV, Diep H, Lee HW, Kemp-Harper BK, Broughton BR, Selemidis S, Gaspari TA, Samuel CS, Drummond GR (2015) NOX1 deficiency in apolipoprotein E-knockout mice is associated with elevated plasma lipids and enhanced atherosclerosis. Free Radic Res 49:186–198
66. Schürmann C, Rezende F, Kruse C, Yasar Y, Löwe O, Fork C, van de Sluis B, Bremer R, Weissmann N, Shah AM, Jo H, Brandes RP, Schröder K (2015) The NADPH oxidase Nox4 has anti-atherosclerotic functions. Eur Heart J 36:3447–3456
67. Manea A, Manea SA, Gan AM, Constantin A, Fenyo IM, Raicu M, Muresian H, Simionescu M (2015) Human monocytes and macrophages express NADPH oxidase 5; a potential source of reactive oxygen species in atherosclerosis. Biochem Biophys Res Commun 461:172–179
68. Yang H, Roberts LJ, Shi MJ, Zhou LC, Ballard BR, Richardson A, Guo ZM (2004) Retardation of atherosclerosis by overexpression of catalase or both Cu/Zn-superoxide dismutase and catalase in mice lacking apolipoprotein E. Circ Res 95:1075–1081
69. Tan SM, Sharma A, Yuen DY, Stefanovic N, Krippner G, Mugesh G, Chai Z, de Haan JB (2013) The modified selenenyl amide, M-hydroxy ebselen, attenuates diabetic nephropathy and diabetes-associated atherosclerosis in ApoE/GPx1 double knockout mice. PLoS One 8:e69193
70. Yuan T, Yang T, Chen H, Fu D, Hu Y, Wang J, Yuan Q, Yu H, Xu W, Xie X (2018) New insights into oxidative stress and inflammation during diabetes mellitus-accelerated atherosclerosis. Redox Biol 20:247–260
71. Vieceli Dalla Sega F, Aquila G, Fortini F, Vaccarezza M, Secchiero P, Rizzo P, Campo G (2017) Context-dependent function of ROS in the vascular endothelium: the role of the Notch pathway and shear stress. Biofactors 43:475–485
72. Byon CH, Heath JM, Chen Y (2016) Redox signaling in cardiovascular pathophysiology: a focus on hydrogen peroxide and vascular smooth muscle cells. Redox Biol 9:244–253
73. Murray TV, Ahmad A, Brewer AC (2014) Reactive oxygen at the heart of metabolism. Trends Cardiovasc Med 24:113–120
74. Ferrara N, Komici K, Corbi G, Pagano G, Furgi G, Rengo C, Femminella GD, Leosco D (2014) Bonaduce D β-Adrenergic receptor responsiveness in aging heart and clinical implications. Front Physiol 4:396
75. Ciccarelli M, Santulli G, Pascale V, Trimarco B, Iaccarino G (2013) Adrenergic receptors and metabolism: role in development of cardiovascular disease. Front Physiol 4:265
76. Rohini A, Agrawal N, Koyani CN, Singh R (2010) Molecular targets and regulators of cardiac hypertrophy. Pharmacol Res 61(4):269–280
77. Hunter JJ, Chien KR (1999) Signaling pathways for cardiac hypertrophy and failure. N Engl J Med 341(17):1276–1283
78. Singh K, Xiao L, Remondino A, Sawyer DB, Colucci WS (2001) Adrenergic regulation of cardiac myocyte apoptosis. J Cell Physiol 189(3):257–265
79. Xiao L, Pimental DR, Amin JK, Singh K, Sawyer DB, Colucci WS (2001) MEK1/2-ERK1/2 mediates alpha1-adrenergic receptor-stimulated hypertrophy in adult rat ventricular myocytes. J Mol Cell Cardiol 33(4):779–787

80. Siwik DA, Tzortzis JD, Pimental DR, Chang DL, Pagano PJ, Singh K, Sawyer DB, Colucci WS (1999) Inhibition of copper-zinc superoxide dismutase induces cell growth, hypertrophic phenotype, and apoptosis in neonatal rat cardiac myocytes in vitro. Circ Res 85(2):147–153
81. Clerk A (2003) The radical balance between life and death. J Mol Cell Cardiol 35:599–602
82. Fu YC, Chi CS, Yin SC, Hwang B, Chiu YT, Hsu SL (2004) Norepinephrine induces apoptosis in neonatal rat cardiomyocytes through a reactive oxygen species-TNF alpha-caspase signaling pathway. Cardiovasc Res 62:558–567
83. Gupta MK, Neelakantan TV, Sanghamitra M, Tyagi RK, Dinda A, Maulik S, Mukhopadhyay CK, Goswami SK (2006) An assessment of the role of reactive oxygen species and redox signaling in norepinephrine-induced apoptosis and hypertrophy of H9c2 cardiac myoblasts Antioxid. Redox Signal 8:1081–1093
84. Jindal E, Goswami SK (2011) In cardiac myoblasts, cellular redox regulates FosB and Fra-1 through multiple cis-regulatory modules. Free Radic Biol Med 51(8):1512–1521
85. Thakur A, Alam MJ, Ajayakumar MR, Ghaskadbi S, Sharma M, Goswami SK (2015) Norepinephrine-induced apoptotic and hypertrophic responses in H9c2 cardiac myoblasts are characterized by different repertoire of reactive oxygen species generation. Redox Biol 5:243–252
86. Saleem N, Goswami SK (2017) Activation of adrenergic receptor in H9c2 cardiac myoblasts co-stimulates Nox2 and the derived ROS mediate the downstream responses. Mol Cell Biochem 436(1–2):167–178
87. Kalyanaraman B, Cheng G, Hardy M, Ouari O, Bennett B, Zielonka J (2018) Teaching the basics of reactive oxygen species and their relevance to cancer biology: mitochondrial reactive oxygen species detection, redox signaling, and targeted therapies. Redox Biol 15:347–362
88. Kalyanaraman B (2017) Teaching the basics of cancer metabolism: developing antitumor strategies by exploiting the differences between normal and cancer cell metabolism. Redox Biol 12:833–842

Epigenetic Basis of Oxidative Stress in Diabetic Coronary Atherosclerosis: A Shift in Focus from Genetic Prerogative

18

Parimala Narne

Abstract

Hyperglycemia-induced excessive superoxide production is the single unifying link to development and progression of diabetes micro- and macrovascular complications. Oxidative stress and antioxidant defense systems, inter alia, are recognized as both antecedent and consequent factors in the development of major diabetic complications like diabetic coronary atherosclerosis. The attendant cellular sequelae of exacerbated oxidative stress in diabetic and coronary atherosclerotic milieu are entrenched in canonical epigenetic changes like DNA methylation and histone posttranslational modifications. They alter the chromatin accessibility to the transcriptional network and steer the transcriptional programs to invoke atherogenic and inflammatory phenotype in distinct cell types. They also act as portals for propagation of the effects of 'hyperglycemic or metabolic memory' or the 'legacy effect'. This chapter presents an update on the contribution of hyperglycemia and oxidative stress both singly and in connivance to accelerated development of coronary atherosclerosis through epigenetic modalities. Such a conceptual understanding would enable the identification of plausible therapeutic strategies for alleviating the burden of diabetic coronary atherosclerosis that is compounded by a formidable challenge posed by metabolic memory.

Keywords

Diabetes · Hyperglycemia · Oxidative stress · Coronary atherosclerosis · Genetic · Epigenetic · Metabolic memory

P. Narne (✉)
Department of Biotechnology and Bioinformatics, School of Life Sciences, University of Hyderabad, Hyderabad, India

S. Chakraborti et al. (eds.), *Modulation of Oxidative Stress in Heart Disease*,
https://doi.org/10.1007/978-981-13-8946-7_18

18.1 Introduction

Diabetes mellitus (DM) represents a syndrome characterized by disordered metabolism and chronic hyperglycemia due to deficiency of or resistance to insulin. Type 1 diabetes mellitus (T1DM) results from the lack of insulin production from pancreatic β cells due to an autoimmune antibody (insulin deficiency). Type 2 diabetes mellitus (T2DM) is a heterogeneous, multifactorial, polygenic condition typified by elevated blood glucose levels emanating from defective insulin secretion from pancreatic β cells and also its inaction [1]. Prolonged hyperglycemia is associated with microvascular complications subsuming diabetic nephropathy, retinopathy, and neuropathy and macrovascular complications like cardiovascular and cerebrovascular disease [2–4]. Atherosclerotic cardiovascular disease (CVD), manifested primarily as coronary artery disease (CAD), heightens the risk of premature mortality in T2DM individuals [3–6]. Metabolic syndrome, represented by a cluster of metabolic disorders subsuming abdominal obesity, hypertension, hyperlipidemia, hypercoagulability, and chronic inflammation, is an antecedent to T2DM and hence CVD [7, 8]. These comorbidities also act to precipitate CVD [9–14]. Nevertheless, T2DM is posited to be an independent risk factor for the development of ischemic heart disease, stroke, and death even in this co-occurrence of multiple risk factors [15]. Indeed, the American Diabetes Association (ADA) and American Heart Association (AHA) mandated diabetes be considered a CAD 'risk equivalent' rather than a risk factor [16, 17]. Alluding to this, CVD accounts for the greatest component of health care expenditure in people with diabetes [18, 19].

"Diabetes is a growing and massive silent epidemic that has the potential to cripple health services in all parts of the world," a quote from the deliberation by Dr. Robert Beaglehole, director of the Department of Chronic Diseases and Health Promotion, at the launch of Diabetes Action Now (a joint program of the World Health Organization and the International Diabetes Federation) in 2004. A strong degree of commiseration for the social and economic burden of any disease has been the triggering point for significant advancement of scientific research in the concerned area. In line with this, genomics and proteomics have been at the forefront in offering a concierge through high-throughput technologies, for disentangling the genetic basis of polygenic metabolic conditions like type 2 diabetes mellitus (T2DM) and associated CV complications [20–27]. In line with this, the quest for culpable genetic variants heralded the first wave of gene mapping studies comprising linkage and candidate gene association studies [20–27]. They were initiated with the primary objective of circumscribing genomic regions harboring candidate genes with intrinsic variation, which determine the susceptibility to diabetes and coronary atherosclerosis [28–39]. This was further extended across the entire genome through genome-wide association studies (GWAS) without a priori knowledge of functionality [40–45]. Nevertheless, each of these strategies had certain shortcomings.

18.2 Epigenetics: The Emerging Leitmotif

The first tranche of GWAS identified a large number of common genetic variants modestly associating with the disease with small effect sizes, which were postulated to coalesce in conferring disease susceptibility [20–45]. The resulting surfeit of genomic data showed up formidable gene-disease associations, notwithstanding the fact that many of them were implicated by 'winner's curse' and multiple hypothesis testing and were indeed conflicting due to lack of robust replication [46–48]. This was further compounded by the complex nature of T2DM and CVD that typically involves the interplay of a potpourri of genetic, metabolic, and environmental factors in determining disease susceptibility [20–45]. Amid this flood of data, a distinctive discipline termed 'epigenetics' forayed into the province of genomics making non-sequence-dependent mitotic and meiotic transmission of altered chromatin states and gene function, explicable from biological standpoint [49]. Accordingly, chromatin could be perceived as an intersection and storage point for various cellular signals and epigenetic modifications that interact in an intricate manner and elicit diverse functional upshots [50]. It is indeed averred that the bulk of heritability cannot be explained by the multitude of candidate loci identified through GWAS and that epigenetics is *de rigueur* explanation for 'missing heritability' of complex diseases [51, 52]. In keeping with this, only 6% of T2DM heritability has been explained by 18 loci identified using sibling recurrence risk as the heritability measure [53]. It also seems to address the paradoxical observations of monozygotic twin discordance for non-Mendelian disorders [54].

Conrad Waddington's doctrine of 'epigenetics' posited a deterministic link between genetic and epigenetic components of heredity. Over the years this definition has become more inclusive by accommodating several propositions. It could be unequivocally defined as "the structural adaptation of chromosomal regions so as to register, signal, or perpetuate altered activity states" [55]. Accordingly, it has been proposed to orchestrate the interactions between genetic and nongenetic entities with eventual realization of phenotype [56]. This led to the emergence of 'transgenerational epigenetic inheritance', which serves to explain as to how the altered transcriptional states and gene expression patterns are bequeathed to cell generations without the participation of an inherited genetic component [57, 58]. It also introduced a notable paradigm of 'integrative genetics' emerging from 'common disease genetic and epigenetic' (CDGE) hypothesis that, in part, refutes the linearity or derivative relationship of genotype and phenotype and espouses the concept of 'genome and epigenome' [58, 59].

Given the staggering complexity of gene regulation, chromatin could be perceived as the point of convergence of establishment signals comprising epigenetic states sculpted in response to environmental (both external and internal) and developmental contingencies [50, 55, 56]. This attribute stems from three distinct epigenetic processes, viz., post-replicative DNA methylation, posttranslational modifications of histones (PTMs), and ATP-dependent chromatin remodeling [60–63]. These are executed by a panoply of mediators catalogued under chromatin-modifying enzymes, viz., DNA methyltransferases (DNMTs), DNA demethylases, histone acetyl transferases (HATs), histone methyltransferases (HMTs) and histone deacetylases (HDACs), histone (lysine) demethylases (HDMs/KDMs), and the methyl-CpG binding protein (MeCP2) that contribute to chromatin fluidity. These enzymes often function in multiprotein modules encompassing chromatin modifiers and transcriptional coactivators and repressors that interact with the basal transcriptional machinery and regulate the promoter activity in a combinatorial manner. The assemblage of these trans-acting factors following the sequence cues at putative sites induces a "domino effect" resulting in self-perpetuation of altered chromatin states [55, 60–63].

A conglomerate of risk factors such as hyperglycemia, dyslipidemia, adiposity, vascular resistance, and smoking has been postulated to influence the risk of atherosclerosis [3–14]. The epigenetic modifications can accommodate internal and external environmental cues by ascertaining weights to them and dynamically altering the transcriptional programs. The reported association of epigenetic patterns with body mass index (BMI), blood lipids, insulin resistance (IR), and metabolism accords significance to this consideration [64–71]. This implicitly underscores the contribution of these factors in the development of late-onset phenotypes [58]. Therein, hyperglycemia-induced epigenetic changes do not function as standalone entities in predisposing to coronary atherosclerosis but function in concert with epigenetic modifications induced by intersecting pathways in the pathogenesis of T2DM as those of increased adiposity, hyperinsulinemia, elevated blood pressure, etc. Based on the accumulating evidence, a deliberation on the plausible mechanisms by which these risk factors influence epigenetic signaling points majorly towards oxidative stress among other factors as inflammation, endoplasmic reticulum stress, etc., as a common effector [71–75].

18.3 Atherosclerosis: The Primer for Acute Coronary Events in Diabetes Mellitus

Atherosclerosis, typified by significant narrowing of arterial walls, is central to the pathogenesis of macrovascular disease [3–5, 7, 11–13]. It involves accrual of modified low-density lipoproteins in arterial intima due to endothelial dysfunction and chronic inflammation. Following this is monocyte infiltration into the arterial wall and differentiation into macrophages, which morph into foam cells after accumulation of oxidized lipids. Thereafter, foam cells promote macrophage proliferation and attract T-lymphocytes. These, in turn, induce vascular smooth muscle cell (VSMC) proliferation and promote collagen accrual. The entire process of occlusive coronary thrombosis eventuates in a lipid-rich, deeply ulcerated atherosclerotic lesion with a fibrous cap, the rupture of which results in acute vascular infarction. In addition to atheroma formation, increased free radical formation in platelets, impaired nitric oxide (NO) generation, and altered calcium regulation are culpable in promoting platelet aggregation. As significant add-ons, hypercoagulability and impaired fibrinolysis perpetuate the risk of vascular occlusion and the attendant cardiovascular events in T2DM [4, 11–13].

18.4 Oxidative Stress: A '*Conditio Sine Qua Non*' of Diabetic Vascular Complications

Oxidative stress and the defects in antioxidant defense systems, inter alia, are recognized as both antecedent and consequent factors in the development of major diabetic complications like diabetic coronary atherosclerosis [72, 73, 76]. Oxidative stress is a cytopathic upshot of excessive production of reactive oxygen species (ROS), outstripping endogenous antioxidant defense systems. Indeed, oxidative stress, through a consolidated mechanism of superoxide anion ($O_2^{\cdot -}$) production, is the common denominator in the pathologies of insulin resistance, β-cell dysfunction, impaired glucose tolerance, and subsequent manifestation of T2DM that prognosticates diabetic atherosclerosis [72, 73, 77–79]. Under physiological conditions, defense against oxidant stress and maintenance of redox balance are achieved by the functioning of a plethora of cellular antioxidant systems. Cellular ROS are detoxified by an arsenal of antioxidative enzymatic systems including superoxide dismutases (SODs), catalase, and glutathione peroxidase (GPx) and the nonenzymatic system inclusive of alpha-tocopherol (vitamin E), ascorbic acid (vitamin C), glutathione (GSH), uric acid, etc.

18.4.1 Oxidative Stress, Hyperglycemia, and Atherosclerosis

Brownlee [77] has pioneered the concept that hyperglycemia-induced overproduction of $O_2^{\cdot -}$ is the single unifying link to diabetes complications [78]. This encompasses cellular activation of protein kinase C beta (PKCβ), hexosamine pathway, and polyol pathway which are the major pathways of hyperglycemic damage in endothelial cells (ECs) [79]. In addition, this mechanism has been insinuated in the pathologies of both macro and microvascular complications of T2DM thereby emphasizing upon the primacy of oxidative stress [72, 77–79]. This hypothesis is corroborated by the collective contribution of the above mentioned hyperglycemia-associated pathways to an increased free radical production [72, 77–79]. Augmented $O_2^{\cdot -}$production occurs following exposure of ECs to high glucose which avidly quenches NO, a potent endothelium-derived vasodilator that partakes in the general vascular homeostasis [80]. In the vascular system, ROS formation from ECs, VSMCs, and macrophages seems to be more pertinent in atherogenesis, in part, due to their reaction with NO [81]. NO is avidly scavenged by $O_2^{\cdot -}$ to yield cytotoxic peroxynitrite ($ONOO^-$), which can rearrange to generate cytotoxic NO_3^- and the highly reactive hydroxyl radical ($OH^\bullet$). Disrupted organ perfusion and systemic hypertension due to dampened endothelium-dependent dilation, cellular damage and inflammation, apoptosis induction, and deranged intracellular signaling processes are the significant pathophysiological sequelae of excessive ROS production in vascular system [81, 82].

VSMC proliferation is a prominent feature of atherosclerosis, and VSMC growth can be induced by ROS through augmented expression of fibroblast growth factor (FGF) and fibroblast growth factor receptor-1(FGFR-1), insulin-like growth factor-1 (IGF-1) and insulin-like growth factor-1 receptor (IGF-1R), and epidermal growth factor receptor (EGFR) [3, 4]. Compounding this, nicotinamide adenine dinucleotide phosphate (NADPH) oxidase-mediated $O_2^{\cdot -}$ generation plays a critical role in vasoconstriction, via angiotensin II (AngII)-induced VSMC proliferation and hypertrophy [82]. Hyperglycemia can potentially induce VSMC death through formation of cytotoxic hydrogen peroxide (H_2O_2) [83]. As regards antioxidant response, the ROS-scavenging heme oxygenase (HO) induced in the arterial wall attenuates monocyte adhesion [84].

18.4.2 Oxidative Stress, Hyperlipidemia, and Atherosclerosis

Oxidative stress prevails throughout the process of atherogenesis, initiating with endothelial dysfunction [80, 81]. With the progression of atherogenesis, ROS production is amplified through their release from inflammatory cells together with other atherosclerotic plaque component cells. This would have proatherosclerotic repercussions like LDL oxidation, EC dysfunction, VSMC growth, and monocyte migration [11–13]. The proatherosclerotic role of oxidized LDL (ox-LDL) becomes vivid from its effects on the components of the arterial wall. ox-LDL incites activation of the ECs resulting in the expression of several adhesion molecules like intercellular adhesion molecule 1 (ICAM1), vascular cell adhesion molecule 1 (VCAM1), monocyte chemoattractant protein (MCP-1), P-selectin, L-selectin, E-selectin, and platelet endothelial CAM-1 (PECAM-1) in vascular ECs that fosters monocyte or macrophage adhesion. Release of several growth factors from monocytes or macrophages is also instigated by ox-LDL, while VSMCs are subject to excess proliferation following exposure to ox-LDL [13]. Rupture of the soft plaque is also aided by ox-LDL through increased matrix metalloproteinase (MMP) generation in vascular ECs and fibroblasts. Additionally, in the initial stages of atherogenesis, ox-LDL fosters an increased expression of its endothelial receptor and other scavenger receptors on macrophages or monocytes that devour ox-LDL and promote foam cell formation [11–13].

18.5 Genomics: The Holy Grail

The susceptibility to diabetic coronary atherosclerosis is deeply entrenched in genetic antecedents as underscored by familial and genetic epidemiological studies together with those in animal models [6, 20–27, 85]. The genetic disposition of a multifactorial disease is often complex since it embodies multiple genes and environmental factors [6, 85]. Significant strides have been made in the field of genetic epidemiology of diabetes and atherosclerosis, in terms of deciphering the genetic underpinnings of disease susceptibility [26–46]. Early studies primarily egged on resolving the multigene contribution to genetic susceptibility to CAD [28–39, 86–101] (Fig. 18.1). The collaborative effort of numerous polymorphisms in the development and progression of atherosclerosis has been mapped to the genes involved

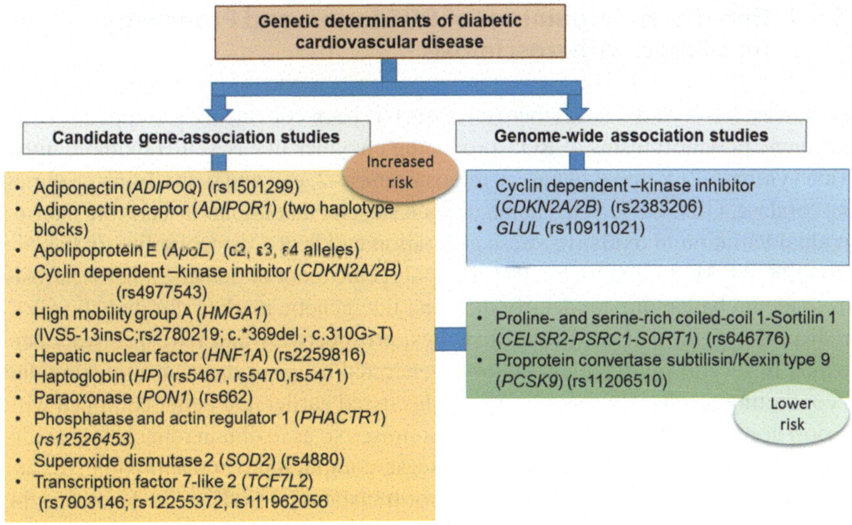

Fig. 18.1 Genetic variants associated with varied susceptibility to diabetic coronary atherosclerosis

in lipid metabolism, coagulation, endothelial function, and oxidative stress [20–46, 86–102].

Nonetheless, early candidate gene association studies were plagued by genotyping call inconsistencies, small sample sizes, ambiguities in phenotypic definitions, variations in the pattern of linkage disequilibrium structure, and lack of replicable associations in other independent populations due to population stratification and non-application of stringent 'p' value thresholds for the hypotheses tested (multiple hypothesis testing) [47, 48, 51–53, 103]. In a similar vein, precise dissection of the complex nature of diabetes and CVD has been counteracted by factors such as late onset of disease, genetic and phenotypic heterogeneity, phenocopies, variable penetrance, and the presence of confounding factors, technical artifacts, and problems generic to regional or global approach. Later refinements focused on gene-based or pathway analyses and detected genetic effects that replicated across independent cohorts [47, 48, 51–53, 103]. On a larger scale, GWAS has been relatively successful in identifying genetic loci for complex non-Mendelian traits such as diabetic coronary atherosclerosis [40–45, 95–102]. However, the requirement for an extremely stringent level of statistical significance to exclude false positives and the non-capture of information on non-single nucleotide polymorphism (SNP) gene variants such as insertions, deletions, and variations in gene copy numbers necessitated the replication in independent samples as for other genetic study designs.

18.5.1 Genetic Underpinnings of ROS Excess and Propensity for Diabetic Atherosclerosis

An increased risk of diabetic atherosclerosis has been convincingly mapped to heritable gene polymorphisms of ROS-producing enzymes like NADPH oxidase, nitric oxide synthase-2, 5-lipoxygenase, cyclooxygenase-2, etc. and detoxifying enzymes like catalase, GPx, GST, SOD, heme oxygenase, etc. that impart discernible interindividual variation in oxidative stress generation and hence susceptibility to diabetic CAD [28, 29, 31, 34, 38, 39, 88–94]. The polygenic tenor of diabetic atherosclerosis has made it challenging to identify the bona fide genetic markers for diabetic CVD predisposition. Several case-control studies, prospectively and retrospectively performed in several small patient cohorts, generated heaps of data for nitpicking due to conflicting results. Current GWAS, as discussed earlier, are better equipped with a priori approach and rigorous statistical premises so as to obtain reliable and ascertaining data. These could prioritize the disease-complicit pro and antioxidative candidate genes that could provide a beacon for translational studies and determine the robustness of pro/antioxidative system through information on their evolutionary conservation and predilection for disease like diabetic CVD. On a parallel plank, animal models for multifactorial disorders like diabetic CAD offer effective means for delineating the genetic contribution as with altered gene expression to metabolic disease development like diabetic atherosclerosis. Nevertheless, this arena is disadvantaged to some extent due to stipulations relating to biological microenvironment that is often described in the context of marked enzymatic deficiency or dysfunction. The tenacity of the model may further be influenced by caveats like environmental factors and therapeutic interventions that modulate the disease status and hence serve as an injunction against rigorous assessment of absolute effects of antioxidant enzyme deficiency.

18.6 Hyperglycemia and Metabolic Memory

The effects of persistent hyperglycemia and its contravention with intensive glycemic control are distinctively palpable as demonstrated in various clinical trials. Testing of "blood glucose hypothesis" through large randomized controlled clinical trials like Diabetes Complications Control Trial (DCCT) and Epidemiology of Diabetes Interventions and Complications (EDIC) has unequivocally established that transient hyperglycemia-induced damage persists for relatively longer periods even after achieving glycemic normalization through intensive glycemic control [104, 105]. Transient hyperglycemic excursions posit as an independent risk factor for diabetes complications in view of the absence of a correlation between HbA_{1c} and glycemic variability when adjusted for mean blood glucose. Empirically, time-averaged mean levels of glycemia measured in terms of hemoglobin A1c (HbA_{1C}) seemed to contribute trivially (<25%) towards variation in the risk of attendant diabetic complications. This is majorly ascribed to a phenomenon called 'metabolic memory' or 'legacy effect'. The profound effects of intensive glucose control in terms of a significantly delayed progression into the micro and macrovascular complications (stroke, nonfatal heart attack, death by CVD) in T1DM patients were evinced in DCCT and EDIC trials [104–106]. The prescribed benefits of intense glycemic control were also evident in (i) the United Kingdom Prospective Diabetes Study (UKPDS) where lower fasting glucose correlated with reduced CVD risk in T2DM patients [107–109], (ii) the Action in Diabetes and Vascular Disease: Preterax and Diamicron modified Release Controlled Evaluation (ADVANCE) trial where intensive glycemic control curtailed the progression of micro and macrovascular complications owing to a decline in nephropathy [110], and (iii) The Steno-2 study on T2DM patients where an intensive multifactorial therapy complementing glycemic control lessened the risk of CV events and death by CVD [111].

The long-term metabolic sequelae of persistent hyperglycemia are an upshot of metabolic memory of vascular dysfunction generated by acute hyperglycemic excursions or protracted hyperglycemia even after reinstating euglycemia with hypoglycemic agents [112–114]. This is best illustrated by the response of ECs and VSMCs to chronic hyperglycemia that sculpt the metabolic changes associated with CV complications in T2DM [112–118]. ECs cultured in high glucose or transiently exposed to high glucose sustained an incremental expression of fibronectin, collagen, extracellular matrix protein encoding genes, nuclear factor kappa-light-chain-enhancer of activated B cells (NF-κB)/p65 subunit, inflammatory genes, and oxidative stress even after glucose normalization [114–116]. The chronicity of hyperglycemia also impinges on oxidative stress that is sustained for up to 1 week even after euglycemia restoration, which could be partially counteracted with antioxidants or NADPH oxidase inhibitors [117]. A precocious phenotype and metabolic memory of precedent hyperglycemia seems to prevail in VSMCs and macrophages obtained from T2DM, insulin-resistant, or obese diabetic db/db mice that demonstrate a sustained increase in expression of NF-κB and cyclic AMP response element-binding protein (CREB) transcription factor (TF), inflammatory gene expression, monocyte migration, and oxidative stress [118]. These studies corroborate the role of oxidative stress in perpetuating the metabolic memory, essentially through DNA, lipid, and protein modifications [72, 77, 78, 119]. In relation with oxidative stress, the centrality of mitochondrial metabolism in inflicting far-flung effects of hyperglycemic spikes like endothelial dysfunction after subsequent normoglycemia lies in its ability to quantifiably produce ROS or $O_2^{\cdot -}$-induced oxoaldehydes. Hyperglycemia-induced mitochondrial ROS instigate four major pathways culpable in the development of CV complications in a diabetic setting as described earlier [72, 77–79]. Advanced glycation end products (AGEs) (generated by hyperglycemia and oxidative stress) through their interaction with receptor for AGEs (RAGE) in connivance with polyol pathway and putative downstream signaling events, beget existing oxidative stress and local inflammation through irreversible glycation of proteins and lipids [119, 120]. The dicarbonyl intermediate, methylglyoxal that is a key precursor of AGE, is complicit in development of diabetic vascular complications and can be disarmed with the enzyme glyoxylase-1 [119]. These deviant events contribute, inter alia, to hyperglycemic memory so as to precipitate long-term vascular and end organ damage [121].

18.7 Epigenetic Basis of Oxidative Stress in Hyperglycemic and Cardiac Milieu

A plethora of chromatin modifications either dynamic or in stable configuration induced by altered metabolic states and oxidative stress are prescribed to coordinate the transcriptional programs. They seemingly propel changes in redox regulation and synergize with altered signaling pathways to generate a coherent phenotype as atherosclerosis in a diabetic milieu [74, 75, 122]. These in combination with DNA methylation influence the interaction of transcriptional complexes with DNA and hence determine the directionality of transcriptional network. This eventually regulates the gene expression thereby enabling the cell to adapt to the alterations induced by exogenous and endogenous stimuli [68–71, 74, 75]. Though it might sound precocious for RO/NS and intermediary metabolism to fit into the providential hierarchy of 'epigenator-epigenetic initiator-epigenetic maintainer' proposed by Berger et al., 2007, the role essayed by these entities in modulating epigenetic events is increasingly becoming clear [49]. ROS act as signaling molecules and transduce the extracellular cues to the downstream signaling cascades thereby enabling cellular adaptation to bioenergetic source availability [122–124].

In continuity with the convention that hyperglycemia-induced ROS could perpetuate the legacy effect, efforts were made to disentangle the molecular underpinnings of the cellular memories. Based on the deluge of information from current research, mitochondrial ROS-mediated metabolic memory seems to be engendered in epigenetic changes induced by transient hyperglycemia and altered gene expression thereof, that persist even after restoration of euglycemia [117, 125–131]. Multivalent events such as histone PTMs and DNA methylation that could endure the effects of not only current glycemia but also the memory of precedent hyperglycemia are paramount in this consideration [112, 113, 116]. They integrate the short-term glucose exposure over time into stochastic or deviant transcriptional events in varied cell types. This form of metabolic memory could be visualized as a "palimpsest," with a superimposed and deleterious new epigenetic code written by prolonged or chronic hyperglycemia [125–131].

Empirical evidence in this direction was obtained from attenuation of glucose-induced NF-κB activation following overexpression of mitochondrial antioxidant manganese (MnSOD) and uncoupling protein 1 (UCP1) constructs in ECs [115]. This was explicitly linked with an increased localization of a HMT SET7 to chromatinized NF-κB/p65 template, which could further be precluded with a selective mitochondrial antioxidant idebenone [128–131]. In a similar vein, the complicity of PKC in furthering hyperglycemic memory could also be evaluated with the nonselective PKC inhibitor bisindolylmaleimide [132]. This abrogated glucose-induced accentuation of p65 gene expression through reduced methylation at histone 3 lysine 4 (H3K4me), a transcriptionally activating mark [128–131]. NADPH oxidase, the principal producer of ROS, can also plausibly be implicated in the perpetuation of hyperglycemic memory as it is complicit in NF-κB activation in diabetic context and its constituent p47phox subunit remains persistently elevated in ECs after glycemic reinstitution [117]. Taken together, the striking plasticity of precedent hyperglycemia in dynamically regulating the gene expression resonates with the idea that persistent hyperglycemia bequeaths the conserved epigenetic fates, conferred by spatial and temporal cues, to cell generations that translate into long-term implications in the diabetic heart.

18.7.1 DNA Methylation and Demethylation

Post-replicative methylation of DNA occurs at the 5-carbon ring of the cytosine base adjacent to guanine nucleotides (CpG) resulting in the generation of 5-methylcytosine (5mC) that summarily defines transcriptionally inactive DNA. DNMTs both denovo (DNMT3a, DNMT3b) and maintenance methylases (DNMT1) catalyze the addition of methyl marks to the CpG dyads and hence constitute the DNA "writers." Owing to the sequence symmetry of CpGs, 5mC is stably propagated through cell division [61]. DNA methylation is invariably associated with transcriptional silencing as it promotes the formation of transcriptionally repressive chromatin when occurring at or in the proximity of gene regulatory cis-acting elements. Being a thermodynamically stable epigenetic mark, 5mCs repress gene expression by acting as a structural perturbation per se, as they project into the major groove of DNA and impair the binding of trans-acting TFs like AP2, hypoxia-responsive hypoxia-inducible factor-1α (HIF-1α), and c-myc at the cognate sites [133]. Alternatively, a coterie of small molecules, viz., methylated CpG binding domain proteins, viz., MBD protein family (MBD1, MBD2, MBD4, and MeCP2), Kaiso and Kaiso-like proteins, and SRA domain proteins, preferentially bind methylated CpGs and recruit chromatin remodeler proteins [134–136]. These include epigenetic "erasers" like HDACs and "writers" like HMTs that cause a tight compaction of chromatin and preclude the assembly of transcriptional machinery. 5mC is also subject to removal by active demethylation catalyzed by ten-eleven translocases (TET1, TET2, TET3) through iterative oxidation, involving the generation of transient forms of oxidatively modified cytosine (oximC) like 5-hydroxymethylcytosine (5hmC), 5-formylcytosine (5fC), and 5-carboxylcytosine (5caC), followed by thymine DNA glycosylase-mediated base excision and DNA base excision repair (BER) eventually generating a demethylated cytosine [137, 138]. Further, a discernible variation in the effect of DNA methylation with respect to genome context and cell type contributes immensely to transcriptional plasticity. This alludes to the genome context and CpG content being considered crucial for conservation of DNA methylation signatures [133, 136]. DNA methylation in gene bodies and intergenic regions articulates with gene splicing, transcriptional elongation, and enhancer regulation [136].

Oxidative DNA damage that is featured inevitably among the pathological sequelae of hyperglycemia orchestrates an epigenetic phenomenon in collusion with DNA methylation. 8-oxo-2′-deoxyguanosine (8-oxodG), a common oxygen radical adduct of DNA and a molecular footprint of oxidative DNA damage, is significantly elevated in atherosclerotic vessels and in blood and urine of patients with atherosclerosis and cardiac failure [76, 139]. $OH^{\bullet}$, a ROS, can indirectly demethylate DNA by conversion of 5mC to 5hmC by abstraction of a hydrogen atom from the methyl group [140]. 5hmC then sterically hinders binding of DNMT1 thereby precluding the perpetuation of methylation signatures [130]. Mechanistically, in hemimethylated DNA, under conditions of escalating oxidative stress, the miscoding potential of oxidative DNA lesions encumbers the methylation of adjacent cytosines by interfering with the binding of MeCP2 to the oligonucleotide duplex thereby resulting in hypomethylation [141, 142]. 8-Oxoguanine DNA glycosylase (OGG1), instrumental in removing 8-oxodG, can also be recruited to the lesion sites, which in concert with TET1 can demethylate DNA [143] $O_2^{\cdot -}$ has also been postulated to nonenzymatically methylate cytosine by C5 deprotonation and direct transfer of a methyl group donated by S-adenosyl methionine (SAM), thereby circumventing DNMT utility [144]. The DNA-hypomethylating prowess of 8-oxodG could plausibly explain the transcriptional activation of NF-κB-dependent proinflammatory genes in response to tumor necrosis factor-α (TNF-α) and impaired binding of HIF-1α to VEGF promoter and other pro-angiogenic genes in ECs subjected to hypoxia [145, 146]. Structural perturbations of chromatin that impinge on epigenetic patterning can be induced by direct modification of histones by ROS. For instance, $ONOO^-$ can nitrosylate tyrosine residues in histones H1, H2B, and H3, while ROS can oxidatively modify lysine and arginine residues in H3 to form protein-bound carbonyl groups that could alter chromatin accessibility, transcriptional competence, and genome stability [147]. Chromatin openness can also be influenced by (i) lysine adducts on H2, H3 (H3K23, H3K27), and H4 formed by lipid peroxides like 4-oxo-2-nonenal in bacterial endotoxin lipopolysaccharide (LPS)-induced macrophages and (ii) redox-sensing S-glutathionylation of Cys^{110} residues in H3 [148, 149].

ROS can variably induce global and local DNA hypomethylation. In this regard, it can reduce DNMT1 activity by reducing methyl-donating SAM availability in a two-pronged manner, either by inhibition of SAM-synthesizing methionine adenosyl-transferase or that of methionine-producing methionine synthase. In line with this, LINE-1 hypomethylation is imminent following exposure to H_2O_2 due to diversion of methionine for cysteine synthesis so as to generate the antioxidant GSH, thereby severely impacting SAM generation [150]. Understandably, LINE-1 hypomethylation, a surrogate of global methylation, has been observed in patients with ischemic heart disease and stroke [151, 152]. Conversely, ROS can induce DNA hypermethylation in a context-dependent manner. This is exemplified by HIF-1α-mediated increase in activity of DNMT1, DNMT3A, and DNMT3B that invokes global hypermethylation and upregulated pro-fibrotic gene expression [153, 154]. Concurrently, locus-specific CpG hypermethylation could attenuate *SOD* gene expression in the event of ischemia or oxidative stress-induced injury which in turn can be ameliorated by DNMT inhibitors [123]. In a similar vein, hypermethylation of PKC-epsilon promoter is promoted by NADPH oxidase 1 (NOX1)-derived ROS in a norepinephrine-induced cardiac hypertrophy model that could also be effectively rescued by DNMT inhibition [155, 156]. ROS can also induce endothelial dysfunction by impinging on flow-mediated dilatation and triggering flow disturbances which purportedly associate with atherogenesis [157]. This involves hypermethylation of genes involved in mechanotransduction and that of a TF in particular, viz., KLF4, resulting in downregulated endothelial nitric oxide synthase gene *(eNOS)* expression. Alternatively, ROS can invoke hypermethylation, inter alia, by H_2O_2-induced enrichment of chromatin-silencing DNA-histone complexes comprising DNMT1, DNMT3B, SIRT1, and polycomb repressive complex 4 (PRC4) complexes at GC-rich, presumably CpG, islands, which in turn could become persistent [158]. This could conceivably relate with the observation of locus-specific aberrant hypermethylation in the context of global DNA hypomethylation [151, 152, 159–162].

Oxidative species like H_2O_2 and conditions like hypoxia significantly reduce TET expression and hence 5hmC with a concomitant rise in DNMT1, DNMT3A expression, and 5mC levels as observed in ECs [162–164]. ROS also impinge on the activity of oxygen, Fe(II), ascorbate, and 2-oxoglutarate (2-OG)-dependent epigenetic enzymes like TETs 1/2/3 and HIF prolylhydroxylases (PHDs), the latter of which are attenuated by hypoxia [165, 166]. Unequivocal tampering of TET and PHD activity by ROS is attributed to ROS-induced reduction in Fe(II) and ascorbate levels [165]. Further, PHDs exhibit an increased propensity to respond to fluctuating oxygen concentrations, while TETs with a K_M value of 30 μM for O_2 are inhibited only when it dips below 2% [167]. Preservation of TET activity over an unstipulated range of oxygen concentrations could be beneficial in that it could evoke a compensatory response by activating transcription of antioxidant defense genes [163]. Global reduction in 5hmC signatures with a partial enrichment of 5hmC at differentially methylated regions of the genes involved in oxidative stress and hypoxia pathways could align with this consideration [167]. In addition, Kreb's cycle substrates like succinate and fumarate potentiate mitochondrial ROS production in a pathological milieu like diabetic heart that could conceivably inhibit the activity of TETs and PHDs thereby significantly affecting the transcriptional cascades [167, 168].

TET2-mediated modulation of 5hmC occurrence is an exemplary instance of DNA demethylation-dependent modelling of contractile and dedifferentiated VSMC phenotypes [169]. While TET2 and 5hmC gain increased expression in VSMCs with contractile phenotype, a diminished expression of TET2 and 5hmC defines the demethylation status in experimental models of VSMC dedifferentiation and human atherosclerosis. Knockdown of TET2 represses the contractile phenotype while fostering the expression of synthetic genes like KLF4, KLF5, and OPN and proliferation of human coronary artery VSMCs (Fig. 18.2b). Conversely, TET2 overexpression confers contractile phenotype on VSMCs, in the absence of differentiation stimuli with a concomitant repression of dedifferentiation and proliferation [169]. The pivot of this phenotypic plasticity of VSMCs is the chromatin accessibility to contractile, synthetic, and proliferative genes arbitrated by TET2 and 5hmC, thus striking an effective balance in the molecular states [169] (Fig. 18.2b). A similar epigenetic modality can be conjectured in driving the erstwhile unknown trans-differentiation of VSMCs into macrophages in atherosclerosis that is a significant departure from the long-held view that lesioned macrophages are derived from circulating monocytes [170]. In this regard a therapeutic strategy

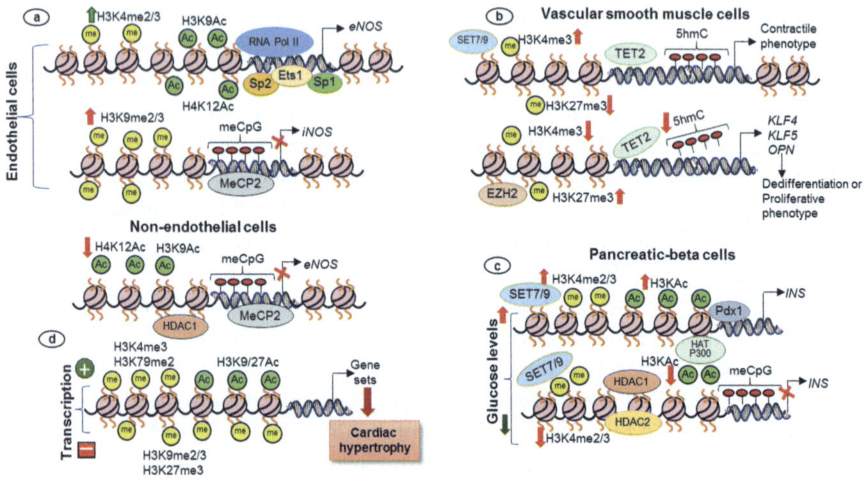

Fig. 18.2 Dynamic cooperation between DNA methylation and histone posttranslational modifications in different cell types relevant to diabetic coronary atherosclerosis. (**a**) Cell-type-specific *eNOS* and *iNOS* gene expression, (**b**) phenotypic plasticity in VSMCs, (**c**) glucose-dependent regulation of insulin (*INS*) gene expression, (**d**) histone code for transcriptional activation and repression of genes involved in cardiac hypertrophy (see text for details)

can be contemplated to modulate the paradigmatic transition in the phenotypes so as to avert the progression of VSMCs into the proatherosclerotic phenotype.

Atherosclerotic aortas exhibit a propensity for global hypomethylation characterized by reduced frequency of 5mC in the proliferating VSMCs in advanced human atherosclerotic plaques as compared with normal arteries and animal models like apolipoprotein E (ApoE) null mice [160]. A corollary to this could be altered target gene expression relating to HIF-1α, c-fos, p53, estrogen receptor, growth factors, eNOS, MMPs, and arachidonic acid metabolizing enzymes like 15-lipoxygense in the VSMCs and ECs, essential for increased cellular proliferation in atherosclerotic aortas [161, 170, 171]. DNA hypomethylation could also serve as an early predictor of atherosclerosis as observed in the leucocytes of ApoE knockout mice as early as 4 weeks of age much before the occurrence of histological atherosclerotic changes [172]. The coexistent conditions like inflammation and hyperhomocysteinemia could also provide cues for altered methylome in atherosclerosis [173, 174].

As regards glucose metabolism, insulin (*INS*) gene expression is epigenetically modulated as can be evinced from hypermethylation of *INS* gene promoter in mouse embryonic stem cells and its demethylation in pancreatic β cells [175]. Insulin production in pancreatic β cells is also affected by the promoter hypermethylation of pancreatic and duodenal homeobox 1 (*PDX1*) and peroxisome proliferator-activated receptor-γ coactivator 1α (*PGC-1α*) genes that encode TFs crucial for β-cell differentiation and *INS* gene expression [176–179]. Investigation of genome-wide methylation landscape in the pancreatic islets of T2DM individuals generated a dossier of differentially methylated regions (DMRs) covering loci with prescribed islet function, e.g., PDX1, TCF7L2, and ADCY5. The DMRs were enriched at TF binding sites, enhancers, and histone marks that precisely linked with impaired insulin secretion thereby linking differential methylation with islet cell dysfunction [180]. In an earlier study, the DMRs identified in the diabetic islets, affiliated to promoters of 254 genes which were annotated to pathways governing β-cell survival, cellular dysfunction, and stress adaptation, with a subgroup of them being concordant with transcriptional changes [181]. Further, changes in DNA methylome have been recently shown to be anticipatory of chronic conditions like diabetes, while those in transcriptome were pertinent to acute conditions [182]. This variation subsumes several differentially methylated regions that modulate the gene expression in an allele-specific manner. In a similar vein, a differentially methylated locus has been shown to be associated with triglycerides and HDL-C that sufficiently emphasize on the gene regulatory mechanisms linking serum lipid measures to CAD risk [183]. Intriguingly, this locus associates with cis-expression of reverse cholesterol transporter, viz., ABCG1, and purportedly links reverse cholesterol transport with incident CVD events. The magnitude of lipid level-associated CpGs is substantial, standing at 64% of 193 CpGs that constitute cis-methylation quantitative trait loci enriched with GWAS SNPs for lipid levels and CAD. For 17% of the genes, the methylation status seemed to affect the expression of adjacent genes. As regards T2DM, a systematic review of the studies inclusive of randomized control trials, cohorts, case-control, and cross-sectional studies in humans identified no consistent relationship between global DNA methylation and T2DM, glucose, insulin levels, and IR [159]. However, epigenetic variation associated with candidate genes in blood cells, adipose and muscle tissue, and placenta seemed to be pronounced, with no apparent overlap between the susceptible genes [159].

18.7.2 Posttranslational Modifications of Histones

Histones are one of the most conserved set of proteins, and the flexible tails of both canonical (H2A, H2B, H3, H4) and variant (H3.1, H3.3, and HTZ.1) histones are targets to an eclectic mix of PTMs, viz., acetylation, methylation, phosphorylation, poly(ADP)ribosylation, sumoylation, ubiquitination, biotinylation, hydroxylation, and citrullination [60]. PTMs of histones are readily reversible and principally affect the inter-nucleosomal interactions. They are vital for epigenetic regulation of gene expression as they regulate chromatin events such as histone eviction, nucleosome occupancy, positioning, and chromatin remodeling that define its transcriptional competence [49, 60, 62, 63]. They instigate nucleosome remodeling in response to intracellular signals that in turn modulates the binding of TFs together with gene regulatory proteins and other cognate factors. This majorly accounts for the dynamicity of DNA-templated functions or chromatin-associated processes such as DNA replication and repair, site-specific recombination, and gene expression [49, 60, 62, 63]. The preeminent histone PTMs are acetylation and methylation which exhibit a predilection for Lys and Arg residues in the unstructured N-terminal tails that extend from the histone octamer, while phosphorylation is directed at serine and threonine residues [49, 60, 62]. Distinct patterns of histone PTMs identified by genome-wide profiling can precisely delineate key regulatory regions including promoters, enhancers, gene bodies, and repetitive elements and define the transcriptional competence of chromatin.

18.7.2.1 Histone Acetylation and Deacetylation

Histone acetylation is considered a hallmark for transcriptionally active chromatin, as euchromatin is enriched in acetylation islands and heterochromatin is associated with global hypoacetylation [184]. It is an 'ephemeral' mark, as it could be reversed in response to cell signaling pathways. It constitutes a persistent activation signal that has the imminent effect of decondensing the chromatin and facilitating the binding of transcriptional effector complexes. It populates the active chromatin domains and contributes preeminently to the dynamic regulation of transcription by the recruitment of transcriptional coactivator complexes such as p33/CBP (E1A binding protein, p300/CREB binding protein), PCAF (p300/CBP-associated factor), and TAFII250 (transcription initiation factor IID 250-KDa subunit) [185]. Histone acetylation and deacetylation events are imperative for the maintenance of cellular homeostasis and involve mutually antagonizing activities of two classes of enzymes, viz., HATs and HDACs. The gene promoters flip from transcriptionally "off" to "on" state following displacement of bound HDACs by HATs. This dual-enzyme system predominantly regulates histone turnover and is instrumental in evoking homeostatic and adaptive transcriptional responses [184–186].

Perturbations in histone acetylation and deacetylation are associated with various coronary pathologies like atherosclerosis, systemic and pulmonary hypertension, coronary heart disease, cardiomyopathy, and heart failure [123, 124, 187–189]. In relation with this, ox-LDL-invoked chemokine (interleukin-8 [IL-8], MCP-1) gene expression requires H3K acetylation and phosphorylation with concomitant recruitment of HATs and NF-κB in cultured human ECs. This event is abrogated by pretreatment with statins which promoted binding of HDAC1/2 at these promoters [190, 191]. An altered acetylation of H3K9 and H3K27 occurs in SMCs and macrophages which is associated with cardiac hypertrophy, advanced atherosclerosis, and plaque severity [192, 193]. An unbridled production of ROS also increasingly associates with H3 and H4 acetylation in varied cell types owing to an increased activation of HATs [187, 194]. SOD overexpression tenably links with an increased MMP1 expression that is implicated in atherosclerotic plaque instability, through an increased recruitment of HAT p300/CBP to MMP1 gene promoter and H3 acetylation by H_2O_2 [195, 196]. Conceivably, SOD deficiency that augments $O_2^{\cdot -}$ levels decreases H3 acetylation at peroxiredoxin promoter, while GPx deficiency that elevates H_2O_2 levels reverses this event [197]. Insulin incites ROS production in hyperglycemic adipocytes that increases H3 acetylation, which can be abated only by catalase activity [198]. An acetylation event can also induce ROS production as exemplified by HAT p300/CBP-mediated H3K9 acetylation at NOX2 promoter [199]. Recruitment of HAT CBP/PCAF to the target inflammatory gene promoters in response to diabetic stimuli and RAGE ligand S100B so as to promote histone (H3K9/14) acetylation and an increased gene expression, is a decisive event in ECs and monocytes both in in vitro and in T1DM and T2DM individuals, which can be stymied by curcumin [128, 200–202]. This results in an increased expression of a panel of inflammatory molecules like COX-2, TNF-α, extracellular matrix components, and vasoactive factors and execution of oxidative stress-induced poly(ADP-ribose) polymerase-1 (PARP-1) and NF-κB signaling pathways that increasingly ally with atherogenesis [124, 193, 200–203].

Histone deacetylases class I (HDAC-1,2,3) and II (HDAC-4,5) are the perceived targets of oxidative stress in atherosclerotic disease [124, 187, 204]. ROS induce PTMs like S-glutathionylation, S-nitrosylation, acetylation, and phosphorylation of these enzymes that either directly dampen the activity of HDACs or indirectly impair their binding to DNA or recruitment to other regulatory complexes leading to an increased chromatin openness and an increased transcription [124, 204]. ROS like lipid peroxides or 4-hydroxynonenal can induce nitration of tyrosine residues or alkylation/carbonylation of HDAC1, HDAC2, and HDAC3 and also invoke casein kinase-induced phosphorylation of HDAC2 leading to its ubiquitination and proteasomal degradation [205–207]. Consequent of this, there is an increased H3 and H4 acetylation and upregulated expression of proinflammatory cytokines in macrophages and other proinflammatory cell types. Hypophosphorylation of HDAC2 corepressor complex entailing Mi2/mSin3A is also conceivable during oxidative stress [208]. HDACs can also contribute to the endophenotypes of cardiovascular risk like inflammation and insulin resistance [209]. In this regard, low-grade chronic inflammation and IR augment HDAC3 expression and activity that correlate positively with IL-6 and TNF-α and negatively with sirtuin 1 (SIRT1) expression [210]. DBC1, a HDAC3 repressor, is also decreased in the peripheral blood leucocytes of T2DM individuals [210]. Counteracting HDAC2 with pan-HDAC inhibitor like trichostatin A has been shown to elevate TNF-α, SRA, CD36, eNOS, and VCAM-1 expression [211]. It seems to be more pertinent to neointimal hyperplasia and maintenance of plaque stability as it precludes p21-mediated VSMC proliferation [212]. An upregulated eNOS expression by trichostatin is attributed to the loss of binding of HDAC2 to arginase gene promoter that promotes eNOS expression [213]. Yet another HDAC2 inhibitor, mocetinostat, was reportedly associated with ox-LDL-induced endothelial dysfunction [213]. Class II HDACs are less active than class I and are regulated by nucleus-cytoplasm shuttling that derepresses gene expression. Nuclear export of HDAC4 associates with cardiac hypertrophy as ROS potentiate oxidation of $Cys^{667/669}$ and $Cys^{274/276}$ of its coregulatory Dnajb5 [214]. NOX4 reportedly promotes nuclear exit of HDAC4 and NOX4 deficiency has been shown to protect against pressure overload-induced cardiac hypertrophy [188, 215]. ROS-induced increase in HDAC4 and 5 have been demonstrated in inflamed vessels in rats [216]. Nuclear accumulation of HDAC4 and 5 with a concomitant lowered expression of miR-424 and miR-503 can be purportedly linked with redox-associated pulmonary hypertension, as these miRs contribute towards maintenance of pulmonary vascular homeostasis [216, 217].

The NAD^+-dependent class III HDACs, viz., sirtuins (SIRTs), present an interesting premise for ROS-induced epigenetic regulation of metabolic programs [124, 187–189, 218]. The target motif for ROS in SIRTs is the highly conserved zinc tetrathiolate cluster in the deacetylase domain that is essential for deacetylase activity.

Thiol oxidation or S-nitrosylation of these clusters in SIRT1 and SIRT3 by ROS or NO donors results in accentuated deacetylation of target proteins like eNOS, thereby contributing to endothelial dysfunction [124, 189]. SIRT1 inhibition is associated with increased expression of p22phox subunit of NADPH oxidase with an exacerbated $O_2^{\cdot -}$ production resulting in increased DNA damage, apoptosis, and medial degeneration in human and animal models of atherosclerosis [219]. ROS-induced PTM of SIRTs can also stymie their activity resulting in varied repercussions like increased proteasomal degradation, altered binding to regulatory proteins (DBC1 and AROS), or cytoplasmic sequestration and localization to caveolae that associate with varied pathologies [124, 189, 204, 218–220]. Cellular insults like oxidative stress and hypoxia reportedly upregulate the expression and activity of SIRT1 in vascular cells and cardiac milieu so as to abrogate the expression of redox-sensitive genes *p53*, *FOXO3a*, *SOD2*, *GPX1*, *PGC-1 alpha*, or *NF-κB* [124, 189, 190, 204, 220, 221]. The expression of SIRTs can also be modulated by TFs like FOXO3A that potentiates SIRT1 expression, HIC1, E2F1, and miRs like miR-34a, miR-449a, and miR-199a in varied oxidative stress-related pathologies [218, 222, 223]. Another contender for NAD^+ utilization apart from SIRTs is PARP-1 which competes with SIRT1 and lowers its activity during oxidative stress and $ONOO^-$-induced nitrosative stress thereby significantly altering the target gene expression and hence transcriptional cascades in a pathological milieu like diabetes and atherosclerosis [203, 224–227]. The underlying epigenetic underpinnings and implications for a diabetic heart are yet to be delineated, which could constitute a therapeutic haven, given the promising nature of PARP-1 inhibitors and SIRT1 activators in alleviating various oxidative stress-related disease etiologies.

18.7.2.2 Histone Methylation and Demethylation

Histone lysine methylation is a nonobtrusive PTM in that it does not interfere with the tertiary structure of the protein and introduces slight hydrophobicity in protein-protein interactions [60, 62]. It differentially impacts gene expression, depending on the extent of methylation, the amino acid residue per se, and its location within the histone tail. Chromatinized templates that parallel transcriptional competence are distinguished by occurrence of H3K4me2 and H3K4me3 marks, with an enrichment of gene bodies with H3K36me3 and H3K79me1 marks. Trimethylated H3K9 (H3K9me3) and H3K27 (H3K27me3) are transcriptionally repressive as they populate constitutive and facultative heterochromatin, respectively [173, 174]. H3K9me3 initiates heterochromatin assembly and promotes the spatial spreading by recruiting adapters like heterochromatin protein 1 (HP1) dimers which provide the platform for the binding of other repressor proteins like DNMTs [49, 60, 62, 63]. The concurrent presence of activating and repressive histone marks, for instance, H3K4me3 and H3K27me3, respectively, causes the chromatin to assume a "poised" state, which can either promote or attenuate the gene expression. H3K4me1 gains high representation in promoters, while enhancers are enriched in H3K4me3 residues. The genome context-dependent variation in the activity of histone modifications could be exemplified by H3K36 and H3K9 methylation marks which are gene-activating when present in coding regions and gene-repressing when occurring in the promoter region [49, 60, 62, 63]. These methyl marks are written by HMTs, viz., (i) H3K4 HMTs, SET7/9, MLL, and Smyd3; (ii) K3K9 HMTs, Suv39H1 (suppressor of variegation 3–9), Suv39H2, ESET/SETDB1, and EHMT1; and (iii) H3K27 HMTs, EZH2 (enhancer of zeste-catalytic component or polycomb repressor complex [PRC2]) and G9a [75, 228].

ROS impinge on the activity of HMTs through methionine metabolism in a similar manner as DNMTs, as HMTs subsist on SAM for methyl groups [229, 230]. ROS decrease the activity of SET7/9 and MYND domain-containing protein 1 (SMYD1), the latter of which is a chromatin-binding protein and is seemingly restored by thioredoxin [231]. This could have therapeutic connotation as SYMD1 averts cardiac hypertrophy in a model of cardiac pressure overload [232, 233]. In a hyperglycemic milieu, ROS-induced activation of SET7 enables an increase in H3K4me1 marks in the promoter of Kelch-like ECH-associated protein (Keap-1) that attenuates Nrf2 activity and activates antioxidant gene expression [234]. This could have cardiac implications as Nrf2 is associated with increased ROS production in atherosclerotic setting [124, 235]. TNF-α-induced increased occupancy of SET7/9 that correlates with increased H3K9me occurs in a subset of inflammatory genes in monocytes [125]. As regards histone demethylation, ROS modulate Akt-dependent phosphorylation of H3K27 methyltransferase EZH2 that decreases its interaction with PRCs and decrease H3K27me3 marks, which in turn impinge on vascular function and atherosclerosis [124]. As regards HDMs/KDMs, an increased expression of LSD1 concomitant with reduced H3K4me1 mark was observed in hypertension and diabetes [236]. Incidentally H3K4me2, H3K9me2, and the corresponding eraser and writer enzymes, viz., LSD1 and SETDB1, have been implicated in adipogenesis [124, 125]. Intercepting LSD1 with natural polyphenols like resveratrol, quercetin, and also curcumin could thereby constitute a therapeutic allure in this direction. As LSD1 activity generates H_2O_2 that propels the formation of 8-oxodG, a pertinent link between ROS and LSD1 could be conjectured [237]. Further, jumonji domain-containing (jmjdc) KDMs (JMHDs) have an obligatory requirement of Fe(II), O_2, 2-OG, and ascorbate akin to TETs and PHDs and are affected by H_2O_2 and NO [166]. ROS, hypoxia, and NO have been shown to upregulate several KDMs [162, 238, 239]. Therein, a similar mode of regulation of JMHD activity by ROS and reactive nitrogen species like NO and ONOO- could be surmised in a pathological milieu like hyperglycemia in a diabetic heart. A coordinated action comprising of transcriptional activation by demethylation of H3K9me2 and H3K27me3 by KDM3A, KDM4B, and KDM4C and transcriptional repression by demethylation of H3K4me2/3 by KDM2B and KDM5B is invoked by HIF-1 alpha as a compensatory maneuver in response to ROS and hypoxia [124, 162, 165, 239, 240]. Also, two distinguished events of macrophage polarization and development of profibrotic phenotype are mediated by increased H3K27 demethylase KDM6/JMJD3 in a STAT6-dependent manner in response to H_2O_2-induced SOD overexpression [239]. A decrease in H3K9me2 owing to upregulation of KDM2A associates hypertrophic cardiomyopathy in mice and humans [241].

18.7.3 Molecular Tête-à-Tête: Cross Talk Between DNA Methylation and Histone Modifications

DNA methylation in combination with other PTMs of histones generates a remarkable array of possibilities and combinatorial diversity collectively referred to as "epigenetic indexing code" that provides a snapshot of homeostatic and pathological states. There exists a reciprocal relationship between DNA methylation and histone acetylation. However, there is no clear understanding on the preemptive role of either DNA methylation or histone modifications. Instead, they engage in mutual reinforcement as DNA methylation seems to provide a positive feedback for lysine modification. The interdependence of these two events is exemplified by the loss of H3K27me3 following the gain of 5hmC. DNA methylation can promulgate histone modifications in that it provides targets for HMTs and HDACs.

The endothelial-specific expression of eNOS serves as a prototype in this context, as it is governed by a preemptive epigenetic indexing code that is predicated on the associative and coordinated action of DNA methylation and histone PTMs together with cognate factors [242]. This is characterized by the occurrence of a DMR as detected by southern hybridization with methylation-sensitive isoschizomer mapping and nucleotide-resolution bisulfite genome sequencing strategies, in the proximal promoter of *eNOS* gene (−361/+3). This DMR undergoes robust symmetrical methylation in VSMCs and remains un- or less methylated in ECs [243]. Further, chromatin immunoprecipitation combined with quantitative real-time PCR uncovered the specific recruitment of Sp1, Sp3, and Ets1 TFs and RNA polymerase II to *eNOS* gene proximal promoter in ECs [244] (Fig. 18.2a). This was absent in VSMCs albeit the occurrence of similar global levels of these TFs as in ECs. With respect to histone PTMs, robust acetylation of H3 and H4, di- and tri-methylation of H3 (HK4me2, H3K4me3), and selective enrichment of H3K9 and H4K12 acetylation marks were detected in nucleosomes of *eNOS* proximal promoter and immediate downstream regions in ECs (Fig. 18.2a). Conversely, there was an increased localization of HDAC1 at *eNOS* proximal promoters in non-ECs [243, 245]. Binding of MeCP2 to methylated CpGs in *eNOS* proximal promoter is a vital cog in the epigenetic regulation of constitutively expressed *eNOS* gene in ECs [245]. MeCP2 recruits DNMTs, HDACs, and other requisite proteins that participate in chromatin remodeling. This process entails histone deacetylation, higher-order compaction of chromatin, and formation of transcriptionally repressive conformations [246]. This in turn hinders the access of transcriptional machinery to the regulatory DNA sequences, thereby resulting in *eNOS* gene silencing in non-ECs (Fig. 18.2a). At variation with *eNOS*, inducible nitric oxide synthase (*iNOS*) gene that is induced by cytokines is explicitly repressed in human vascular endothelium [242]. An augmented iNOS mRNA and protein expression has been documented in the neointima of atherosclerotic human blood vessels [247]. The hyporesponsiveness in *iNOS* gene is significantly attributed to the dense methylation of its proximal promoter, differential recruitment of MeCP2, and selective enrichment of H3K9me2 and me3 marks, all of which are refractory to cytokine stimulation in ECs (Fig. 18.2a). This is akin to the multimodal regulation of eNOS in non-ECs [248]. *iNOS* gene promoter is distinct with a canonical TATA box, lacking a CpG island and harbors multiple cis-elements essential for cytokine inducibility of the gene, that includes, inter alia, NF-κB, interferon regulatory factor 1 and STAT-3 [249]. This modality of *eNOS* gene expression is prototypical of the genes expressed in vascular endothelium like von Willebrand factor (vWF), vascular endothelial cadherin (VE cadherin), ICAM2, and VEGF receptors (FLT-1/VEGFR1) which are significantly implicated in diabetic atherosclerosis.

In a putative context, epigenome-wide analysis of aortic ECs stimulated with high glucose uncovered a dossier of epigenetic changes impinged on vascular chromatin by transient HG [250]. In this regard, hyperacetylated H3K9/K14 discernibly correlated with me-CpG content in an inverse relation. These marks were characterized by proximal and distal patterns of regionalization that explicitly associate with expression of genes and pathways subsuming endothelial dysfunction which were identified by ingenuity knowledge-based pathway and gene ontology analyses. This form of epigenetic cross talk also determines VSMC phenotypic plasticity, wherein a formidable influence of DNA demethylation represented by TET2/5hmC on transcriptional competence is exerted by its cross talk with repressive H3K27me3 and permissive H3K4me3, respectively, at the contractile and synthetic genes [169] (Fig. 18.2b). This subsequently affects the acquisition of the contractile or dedifferentiated/proliferative phenotype by VSMCs that could impose a proatherosclerotic phenotype. Another example is provided by HMT SET7/9 that regulates the stability of DNMT1 and demonstrates functional coupling of histone methylation with DNA methylation that enables it to effectively couple methylation of determinants involved in signal transduction with gene regulatory events [251]. Another striking observation is ROS-induced CpG hypomethylation of *p66shc* promoter with a concomitant rise in H3 acetylation which will be discussed in detail in the next section.

Another interesting facet is cross talk and coordination between histone acetylation and methylation. Most of these entities also participate in β-cell proliferation and regeneration [250]. It follows that islet cell-specific TF Pdx1 promotes *INS* gene expression by recruiting HAT-p300 and SET7/9 to *INS* promoter that relaxes the chromatin following acquisition of H3Kac and H3K4me2 marks and renders it transcriptionally active [250, 251] (Fig. 18.2c). Conversely, when glucose levels are limiting, insulin gene transcription is precluded by Pdx1-directed recruitment of HDAC-1/2 complexes [252]. Pdx also seemed to regulate beta cell-specific expression of SET7/9, regulating the genes involved in glucose-induced insulin secretion. TFs involved in islet differentiation are also regulated by acetylation events monitored by HATs/HDACs [250]. More importantly, SET7/9 has been identified to be a critical mediator of persistent gene-activating epigenetic changes such as histone methylation (H3Kme2) and acetylation in a hyperglycemic milieu [112, 115, 125–131]. Another pertinent example is the formulation of an epigenetic code, comprising transcription-activating H3K9ac, H3K27ac, H3K4me3, H3K79me2, and transcription-repressive H3K9me2, H3K9me3, and H3K27me3 histone marks, which regulates distinct gene sets, in lieu of cooperative action in promoter regulation in cardiac hypertrophy [192] (Fig. 18.2d).

18.7.4 Metabolic Memory: A "Palimpsest" of Epigenetic Marks

Epigenetic changes during transient hyperglycemic epoch can act as loose cannon that attenuate the effects of subsequent euglycemia restoration [112, 115]. They can steer the transcriptional programs in an unwarranted direction. Core components of chromatin and their variants can serve as insignia for persistence of epigenetic changes invoked by hyperglycemia-specifying signaling cascades thereby generating persistent adaptive or deviant transcriptional responses following stimulus reexposure [125–131]. This is best exemplified by incorporation of histone variants such as H2A.Z into nucleosomes that is emblematic of adaptive epigenetic memory associated with transcriptional activity. Robustness of ensuing transcriptional response following reexposure to environmental cues is in turn determined by future cell memories promulgated by chromatin structure-regulating transcriptional events [112, 115].

The epigenetic basis of "metabolic memory" or "legacy effect" can be formidably explained by a putative dialogue between histone and DNA methylation on the chromatinized template. This interaction sculpts the transcriptional output and consequence and is best illustrated by the gene regulatory response to hyperglycemic variability in a primary EC model. Changes in ambient glucose concentrations or hyperglycemic variability spur alterations in H3K4 and H3K9 methylation with an added dynamic cooperation between these marks to promulgate gene-activating events. Transient hyperglycemia invokes chromatinization of regulatory changes at the promoter of *RELA* gene encoding p65 subunit of NF-κB following exposure of aortic ECs to glucose [115, 116, 125–131]. The pivotal histone PTM in this scenario is H3K4me1 that protracts the transcriptional activation of NF-κB for 6 days following withdrawal of hyperglycemia and restoration of normoglycemia [115, 116, 125–131]. Two distinguished epigenomic events promulgate the effects of transient hyperglycemic excursions, viz., (i) sustained recruitment of SET7/9 to the endogenous promoter region of NF-κB-p65 (*RELA*) subunit and mono-methylation of H3K4 [116, 125–131] and (ii) concurrent reduction in the H3K9me1 and H3K4me2 marks by sustained recruitment of LSD1 on the NF-κB-p65 gene sequence (Fig. 18.3a). The exclusivity of this PTM occurrence is demonstrated by the absence of any change in the di- and trimethylation of this amino acid residue and the association of other HMTs like MLL1 with the NF-κB-p65 gene sequence in the hyperglycemic context. In essence, the above two events are indicative of a cooperative program that is emblematic of chromatinization of regulatory changes essential for perpetuation of hyperglycemic memory [115, 116, 125–131]. The ensuing heightened expression of proinflammatory or proatherogenic genes entails VCAM1 that fosters adhesion of monocytes to ECs and that of MCP-1, a chemokine that promotes macrophage infiltration (Fig. 18.3a). The same effects were recapitulated in macrophages with reduced H3K9me marks on the genes implicated in vascular inflammation. Summing up, an extracellular signal of remarkable plasticity like transient hyperglycemic gradient incites epigenetic transactions typified by an asymmetrical lysine methylation of H3K4 and H3K9 residues [115, 116, 125–131].

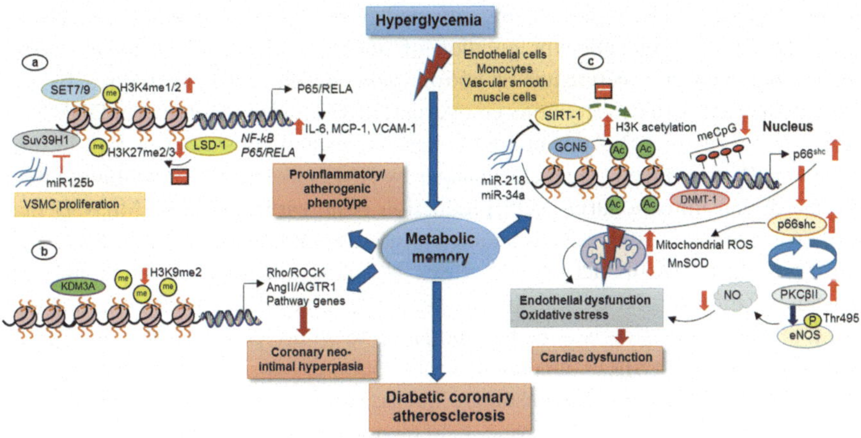

Fig. 18.3 Epigenetic mechanisms implicated in hyperglycemic memory in varied cell types relevant to diabetic coronary atherosclerosis (see text for details)

Sustenance of atherogenic and proinflammatory phenotype imposed by hyperglycemia in VSMCs subsumes histone methylation. A protracted expression of *IL-6* and *MCP-1* genes is observed in VSMCs from diabetic db/db mice for 8 weeks after removal of hyperglycemic stimulus and following culturing ex vivo in normal glucose [253]. Epigenetically, this event is arbitrated by a decrease in repressive H3K9me3 mark and its HMT Suv39H1 at the inflammatory gene promoters. Hyperglycemia-induced upregulation of miR-125b that targets Suv39h1 mRNA is the vital cog in enabling long-lasting effect of hyperglycemia on VSMC phenotype and extending hyperglycemic memory beyond ECs [254] (Fig. 18.3a). Another putative example is augmentation of AngII-induced transcription-permissive H3K4 and H3K36me3 marks that are contiguous with modulation of Lnc-Ang362 transcript which plays host to miR-221 and miR-222 that are crucial for VSMC proliferation [255]. Another epigenetic feature underlying metabolic memory in diabetic atherosclerosis is presented by an accelerated overexpression of KDM3A. This is accompanied by a sustained loss of H3K9me2 in diabetic rats that potentiates vascular neointimal hyperplasia through mediation of Rho/ROCK and AngII/AGTR1 pathways in VSMCs [256] (Fig. 18.3b). Alternatively, a preponderance of H3K9Ac in the promoters of NF-κB-related genes correlates with increased HbA_{1c} levels and presumably links with an increased propensity for microvascular complications as observed in the DCCT and EDIC cohorts [257].

A prototypical example of ROS-induced epigenetic changes that propel persistent vascular dysfunction and hence legacy effect is provided by p66shc that engages in a feed-forward mechanism of hyperglycemic memory. p66shc is a mitochondrial adaptor protein that can regulate intracellular redox state with an ability to foment endothelial dysfunction and vascular damage through increased mitochondrial ROS production and sustained activation of PKCβII. An augmented expression of p66shc occurs in a hyperglycemic milieu as demonstrated in wild-type diabetic mice that activates PKCβII in ECs and remains activated even after restitution of normoglycemia [258–261]. Epigenetic upregulation of p66shc is achieved by an increased GCN5-mediated H3 acetylation of *p66shc* gene which could be stymied by the coordinated action of deacetylating counterpart SIRT1 and upregulation of DNMT3b [262, 263] (Fig. 18.3c). Activated PKC begets elevated p66shc levels that in turn potentiate ROS-induced epigenetic changes by HMTs like SET7. Concomitant with p66shc-driven mitochondrial ROS production is downregulation of MnSOD that further exacerbates ROS accumulation in vascular endothelium [259]. Sustained p66shc-backed PKCβII activity decreases NO bioavailability owing to induced phosphorylation of eNOS ^{Thr495}. In endothelial-specific SIRT1 transgenic diabetic mice, p66shc expression is reined in compared to wild-type littermates that epigenetically aligns with SIRT1-mediated deacetylation of *p66shc* gene promoter in the aortas [264] (Fig. 18.3c). The ensuing sequelae include abrogation of PKCβII activation and the ensuing vicious cycle, thereby ameliorating endothelial dysfunction by restoration of NO levels [261]. Alteration in DNMT1/SIRT1 axis underlying p66shc expression in hyperglycemic milieu has recently been shown to be regulated by miR-218 and miR-34a that portends cardiovascular dysfunction [262] (Fig. 18.3c). A recent study demonstrated that intensification of glycemic control in T2DM subjects does not alleviate the detrimental effects of adverse epigenetic remodeling as evinced from the independent association of mean amplitude of glycemic excursion and postprandial incremental area under the curve (but not the HbA_{1c} levels) with altered epigenetic profile at *p66shc* gene promoter [263]. This model of ROS-mediated hyperglycemic memory powered by epigenetic cues provided by the opposing action of GCN5 and SIRT1 in the vascular endothelium can be dwelt upon to formidably explain as to how the legacy of transient hyperglycemic episodes is commuted to hyperglycemic memory that incites pathological sequelae.

Further, the downstream mechanisms of incident hyperglycemia such as ROS production are implicated in propagation of metabolic memory. Perhaps, ROS production need not necessarily correlate with existing glycemia, as precedent hyperglycemia could instigate the production of AGEs, continued glycation of mitochondrial proteins, and other reactive species which are indeed capable of sustaining altered gene expression of inflammatory molecules like HMOX1 *HMOX1*, MMP10, SLC7A11, MMP1, MCP-1, and ICAM1 even after normalization of glycemia [248]. This emphasizes on the need for an aggressive treatment of skewed/strayed glucose levels with a combined targeting of reduction of cellular reactive species and glycation of mitochondrial proteins.

18.8 Conclusions and Future Directions

The human genome project, candidate gene association studies, and GWAS have been instrumental in explicating the contribution of genetic variation to interindividual variation in disease susceptibility. They generated a potpourri of genetic variants that could be capitalized upon for generating drug targets and developing bespoke genotyping assays which heralded the era of personalized medicine. With the advent of epigenetics and initiation of epigenome-wide association studies, the aspect of missing heritability that seemed inexplicable from the genetic standpoint has begun to be understood. The pathologies of diabetes and associated macrovascular complications like diabetic coronary atherosclerosis are increasingly governed by epigenetic modifications. Prolonged hyperglycemia constitutes the persistent glycemic cue that instigates activated or repressed gene expression events that associate with epigenetic-context-dependent deviant signaling. It institutes adaptive molecular regimens that propagate adaptive or deviant transcriptional memory. A deeper understanding of such precise mechanisms in relation with oxidative stress could be translated towards developing necessary therapeutic regimens to alleviate the burden of diabetic coronary atherosclerosis. In line with this, currently available epigenetic therapy in the form of DNMT, HAT and HDAC inhibitors, and SIRT activators need to be harnessed for repurposing them to treat these ailments based on their suitability.

While gene polymorphisms in part offer an empirical explanation for inherent susceptibility to diabetes and attendant complications, the chasm of "missing heritability" in complex diseases like DM is seemingly filled by stochastic epigenetic changes. Interconnection of epigenetic information with articulate gene sets in a metabolic program necessitates robustness in chromatin cross talk and transcriptional coordination. Envisioning this as a splayed circuitry rather than as a linear on/off closed circuit switch and also as a process network in lieu of disconnected linear regulatory events would enable further exposition of the breadth of covalent epigenetic modifications. Employing advanced technologies inclusive of chromatin immunoprecipitation (ChIP) and CpG assays, accompanied by massive parallel sequencing (ChIP-seq and CpG-seq) that allows for parallel sequencing by synthesis of immunopurified content, and methods for analyzing chromatin interactions viz., Hi-C/3C-Seq/Capture-C, etc., can contribute immensely in this direction. Coupling this comprehension with the contribution of genetic variation to epigenome variation would institute an integrated approach in line with CDGE hypothesis that would allow the formulation of robust therapeutic modalities for taming the shrew as 'diabetic coronary atherosclerosis'.

References

1. Stumvoll M, Goldstein BJ, van Haeften TW (2005) Type 2 diabetes: principles of pathogenesis and therapy. Lancet 365(9467):1333–1346
2. Krolewski AS, Warram JH, Freire MB (1996) Epidemiology of late diabetic complications. A basis for the development and evaluation of preventive programs. Endocrinol Metab Clin N Am 2:217–242
3. Kannel WB, McGee DL (1979) Diabetes and cardiovascular disease. JAMA 241:2035–2038
4. Beckman JA, Creager MA, Libby P (2002) Diabetes and atherosclerosis. Epidemiology, pathophysiology and management. JAMA 287:2570–2581

5. Low Wang CC, Hess CN, Goldfine AB (2016) Clinical update: cardiovascular disease in diabetes mellitus: atherosclerotic cardiovascular disease and heart failure in type 2 diabetes mellitus – mechanisms, management, and clinical considerations. Circulation 133:124
6. Bowden DW, Cox AJ (2013) Diabetes: unravelling the enigma of type 2 DM and cardiovascular disease. Nat Rev Endocrinol 9:632–633
7. Hayden MR, Tyagi SC (2002) Intimal redox stress: accelerated atherosclerosis in metabolic syndrome and type 2 diabetes mellitus. Atheroscler Cardiovasc Diabetol 1:3
8. Grundy SM (2012) Pre-diabetes, metabolic syndrome, and cardiovascular risk. J Am Coll Cardiol 59:7
9. Hossain P, Kawar B, El Nahas M (2007) Obesity and diabetes in the developing world – a growing challenge. N Engl J Med 356:213–215
10. Sowers JR, Epstein M, Frohlich ED (2001) Diabetes, hypertension, and cardiovascular disease: an update. Hypertension 37:1053–1059
11. Wu L, Parhofer KG (2014) Diabetic dyslipidemia. Metabolism 63:1469–1479
12. Carr ME (2001) Diabetes mellitus: a hypercoagulable state. J Diabetes Complicat 15:44–54
13. Ross R (1999) Atherosclerosis – an inflammatory disease. N Engl J Med 340:115–126
14. Festa A, D'Agostino R, Howard G et al (2000) Chronic subclinical inflammation as part of the insulin resistance syndrome: the insulin resistance atherosclerosis study (IRAS). Circulation 102:42–47
15. Almdal T, Scharling H, Jensen JS, Vestergaard H (2004) The independent effect of type 2 diabetes mellitus on ischemic heart disease, stroke, and death: a population-based study of 13,000 men and women with 20 years of follow-up. Arch Intern Med 164:1422–1426
16. Buse JB, Ginsberg HN, Bakris GL et al (2007) Primary prevention of cardiovascular diseases in people with diabetes mellitus: a scientific statement from the American Heart Association and the American Diabetes Association. Diabetes Care 30:162–172
17. Benjamin EJ, Blaha MJ, Chiuve SE et al (2017) Heart disease and stroke statistics-2017 update: a report from the American Heart Association. Circulation 135:e146–e603
18. Guariguata L, Whiting DR, Hambleton I et al (2014) Global estimates of diabetes prevalence for 2013 and projections for 2035. Diabetes Res Clin Pract 103:137–149
19. American Diabetes Association (2017) Standards of medical care in diabetes-2017, classification and diagnosis of diabetes. Diabetes Care 40:S11–S24
20. Mohlke KL, Boehnke M, Abecasis GR (2008) Metabolic and cardiovascular traits: an abundance of recently identified common genetic variants. Hum Mol Genet 17(R2):R102–R108
21. O'Donnell CJ, Nabel EG (2011) Genomics of cardiovascular disease. N Engl J Med 365:2098–2109
22. Pfister R, Barnes D, Luben RN et al (2011) Individual and cumulative effect of type 2 diabetes genetic susceptibility variants on risk of coronary heart disease. Diabetologia 54:2283–2287
23. Kathiresan S, Srivastava D (2012) Genetics of human cardiovascular disease. Cell 148:1242–1257
24. Qi Q, Meigs JB, Rexrode KM et al (2013) Diabetes genetic predisposition score and cardiovascular complications among patients with type 2 diabetes. Diabetes Care 36:737–739
25. Cox AJ, Hsu FC, Ng MC et al (2014) Genetic risk score associations with cardiovascular disease and mortality in the diabetes heart study. Diabetes Care 37:1157–1164
26. Wang X, Strizivh G, Hu Y et al (2016) Genetic markers of type 2 diabetes: progress in genome-wide association studies and clinical application for risk prediction. J Diabetes 8:24–35
27. De Rosa S, Arcidiacono B, Chiefari E et al (2018) Type 2 diabetes mellitus and cardiovascular disease: genetic and epigenetic links. Front Endocrinol (Lausanne) 9:2
28. Pfohl M, Koch M, Enderle MD et al (1999) Paraoxonase 192 Gln/Arg gene polymorphism, coronary artery disease, and myocardial infarction in type 2 diabetes. Diabetes 48:623–627
29. Vendrell J, Fernandez-Real JM, Gutierrez C et al (2003) A polymorphism in the promoter of the tumor necrosis factor-alpha gene (-308) is associated with coronary heart diseases in type 2 diabetic patients. Atherosclerosis 167:257–264

30. Bacci S, Menzaghi C, Ercolino T et al (2004) The +276 G/T single nucleotide polymorphism of the adiponectin gene is associated with coronary artery disease in type 2 diabetic patients. Diabetes Care 27:2015–2020
31. Soccio T, Zhang YY, Bacci S et al (2006) Common haplotypes at the adiponectin receptor 1 (ADIPOR1) locus are associated with increased risk of coronary artery disease in type 2 diabetes. Diabetes 55:2770
32. Muendlein A, Saely CH, Geller-Rhomberg S et al (2011) Single nucleotide polymorphisms of TCF7L2 are linked to diabetic coronary atherosclerosis. PLoS One 3:e17978
33. Chiefari E, Tanyolaç S, Paonessa F et al (2011) Functional variants of the HMGA1 gene and type 2 diabetes mellitus. JAMA 305:309
34. Narne P, Ponnaluri KC, Singh S et al (2012) Relationship between angiotensin-converting enzyme gene insertion/deletion polymorphism, angiographically defined coronary artery disease and myocardial infarction in patients with type 2 diabetes mellitus. J Renin-Angiotensin-Aldosterone Syst 13:478–486
35. Qi L, Qi Q, Prudente S et al (2013) Association between a genetic variant related to glutamic acid metabolism and coronary heart disease in individuals with type 2 diabetes. JAMA 310:821–828
36. Beaney KE, Cooper JA, McLachlan S et al (2016) Variant rs10911021 that associates with coronary heart disease in type 2 diabetes, is associated with lower concentrations of circulating HDL cholesterol and large HDL particle but not with amino acids. Cardiovasc Diabetol 15:115
37. Narne P, Ponnaluri KC, Singh S et al (2013) Arg399Gln polymorphism of X-ray repair cross-complementing group 1 gene is associated with angiographically documented coronary artery disease in South Indian type 2 diabetic patients. Genet Test Mol Biomarkers 17(3):236–241
38. El-Lebedy D, Raslan HM, Mohammed AM (2016) Apolipoprotein E gene polymorphism and risk of type 2 diabetes and cardiovascular disease. Cardiovasc Diabetol 15:12
39. De Rosa S, Chiefari E, Salerno N et al (2017) HMGA1 is a novel candidate gene for myocardial infarction susceptibility. Int J Cardiol 227:331–334
40. Scott LJ, Mohlke KL, Bonnycastle LL et al (2007) A genome-wide association study of type 2 diabetes in Finns detects multiple susceptibility variants. Science 316:1341–1345
41. Zeggini E, Weedon MN, Lindgren CM et al (2007) Replication of genome-wide association signals in UK samples reveals risk loci for type 2 diabetes. Science 316:1336–1341
42. Sladek R, Rocheleau G, Rung J et al (2007) A genome-wide association study identifies novel risk loci for type 2 diabetes. Nature 445:881–885
43. Jarinova O, Stewart AF, Roberts R et al (2009) Functional analysis of the chromosome 9p21.3 coronary artery diseases risk locus. Arterioscler Thromb Vasc Biol 29:1671–1677
44. Cunnington MS, Santibanez Koref M, Mayosi BM et al (2010) Chromosome 9p21 SNPs associated with multiple disease phenotypes correlate with ANRIL expression. PLoS Genet 6:e1000899
45. Qi L, Parast L, Cai T et al (2011) Genetic susceptibility to coronary heart disease in type 2 diabetes: 3 independent studies. J Am Coll Cardiol 58:2675–2682
46. Welter D, MacArthur J, Morales J, Burdett T, Hall P, Junkins H et al (2014) The NHGRI GWAS catalog, a curated resource of SNP-trait associations. Nucleic Acids Res 42(Database issue):D1001–D1006
47. Xiao R, Boehnke M (2009) Quantifying and correcting for the winner's curse in genetic association studies. Genet Epidemiol 33(5):453–462
48. Zhong H, Prentice RL (2010) Correcting "winner's curse" in odds ratios from genomewide association findings for major complex human diseases. Genet Epidemiol 34(1):78–91
49. Berger SL (2007) The complex language of chromatin regulation during transcription. Nature 447(7143):407–412
50. Badeaux AI, Shi Y (2013) Emerging roles for chromatin as a signal integration and storage platform. Nat Rev Mol Cell Biol 14:211–224
51. Manolio TA, Collins FS, Cox NJ et al (2009) Finding the missing heritability of complex diseases. Nature 461:747–753

52. Eichler EE, Flint J, Gibson G et al (2010) Missing heritability and strategies for finding the underlying causes of complex disease. Nat Rev Genet 11:446–450
53. Zeggini E, Scott LJ, Saxena R et al (2008) Meta-analysis of genome-wide association data and large-scale replication identifies additional susceptibility loci for type 2 diabetes. Nat Genet 40:638–645
54. Kato T, Iwamoto K, Kakiuchi C et al (2005) Genetic or epigenetic difference causing discordance between monozygotic twins as a clue to molecular basis of mental disorders. Mol Psychiatry 10:622–630
55. Bird A (2007) Perceptions of epigenetics. Nature 447:396–398
56. Jaenisch R, Bird A (2003) Epigenetic regulation of gene expression: how the genome integrates intrinsic and environmental signals. Nat Genet 33:245–254
57. Jablonka E, Raz G (2009) Transgenerational epigenetic inheritance: prevalence, mechanisms, and implications for the study of heredity and evolution. Q Rev Biol 84:131–176
58. Bjornsson HT, Fallin MD, Feinberg AP (2004) An integrated epigenetic and genetic approach to common human disease. Trends Genet 20(8):350–358
59. McRae AF, Powell JE, Henders AK et al (2014) Contribution of genetic variation to transgenerational inheritance of DNA methylation. Genome Biol 15(5):R73
60. Kouzarides T (2007) Chromatin modifications and their function. Cell 128(4):693–705
61. Bird A (2002) DNA methylation patterns and epigenetic memory. Genes Dev 16:6–21
62. Tessarz P, Kouzarides T (2014) Histone core modifications regulating nucleosome structure and dynamics. Nat Rev Mol Cell Biol 15(11):703–708
63. Fenley AT, Anandakrishnan R, Kidane YH, Onufriev AV (2018) Modulation of nucleosomal DNA accessibility via charge-altering post-translational modifications in histone core. Epigenetics Chromatin 11(1):11
64. Dick KJ, Nelson CP, Tsaprouni L (2014) DNA methylation and body-mass index: a genome-wide analysis. Lancet 383(9933):1990–1998
65. Pfeiffer L, Wahl S, Pilling LC et al (2005) DNA methylation of lipid-related genes affects blood lipid levels. Circ Cardiovasc Genet 8:334–342
66. Irvin MR, Zhi D, Joehanes R et al (2014) Epigenome-wide association study of fasting blood lipids in the Genetics of Lipid-lowering Drugs and Diet Network study. Circulation 130:565–572
67. Arner P, Sahlqvist AS, Sinha I, Xu H, Yao X, Waterworth D et al (2016) The epigenetic signature of systemic insulin resistance in obese women. Diabetologia 59:2393–2405
68. Lu C, Thompson CB (2012) Metabolic regulation of epigenetics. Cell Metab 16:9–17
69. Gut P, Verdin E (2013) The nexus of chromatin regulation and intermediary metabolism. Nature 502:489–498. https://doi.org/10.1038/nature12752
70. Kaelin WG Jr, McKnight SL (2013) Influence of metabolism on epigenetics and disease. Cell 153:56–69. https://doi.org/10.1016/j.cell.2013.03.004
71. Fan J, Krautkramer KA, Feldman JL, Denu JM (2015) Metabolic regulation of histone post-translational modifications. ACS Chem Biol 10:95–108. https://doi.org/10.1021/cb500846u
72. Giacco F, Brownlee M (2010) Oxidative stress and diabetic complications. Circ Res 107:1058–1070
73. Otani H (2011) Oxidative stress as pathogenesis of cardiovascular risk associated with metabolic syndrome. Antioxid Redox Signal 15:1911–1926
74. Hitchler MJ, Domann FE (2007) An epigenetic perspective on the free radical theory of development. Free Radic Biol Med 43(7):1023–1036
75. Cyr AR, Domann FE (2011) The redox basis of epigenetic modifications: from mechanisms to functional consequences. Antioxid Redox Signal 15(2):551–589. https://doi.org/10.1089/ars.2010.3492
76. Nagayoshi Y, Kawano H, Hokamaki J et al (2009) Differences in oxidative stress markers based on the aetiology of heart failure: comparison of oxidative stress in patients with and without coronary artery disease. Free Radic Res 43:1159–1166
77. Brownlee M (2001) Biochemistry and molecular cell biology of diabetic complications. Nature 414:813–820

78. Brownlee M (2005) The pathobiology of diabetic complications: a unifying mechanism. Diabetes 54:1615–1625
79. Nishikawa TD, Eldestein D, Du XL, Yamagishi S, Matsumura T, Kaneda Y (2000) Normalizing mitochondrial superoxide production blocks three pathways of hyperglycemic damage. Nature 404:787–790
80. De Vriese AS, Verbeuren TJ, Van de Voorde J et al (2000) Endothelial dysfunction in diabetes. Br J Pharmacol 130:963–974
81. Davignon J, Ganz P (2004) Role of endothelial dysfunction in atherosclerosis. Circulation 109(23 Suppl 1):III27–III32
82. Rocic P, Seshiah P, Griendling KK (2003) Reactive oxygen species sensitivity of angiotensin II-dependent translation initiation in vascular smooth muscle cells. J Biol Chem 278:36973–36979
83. Peiro C, Lafuente N, Matesanz N et al (2001) High glucose induces cell death of cultured human aortic smooth muscle cells through the formation of hydrogen peroxide. Br J Pharmacol 133:967–974
84. Ishikawa K (2003) Heme oxygenase-1 against vascular insufficiency: roles of atherosclerotic disorders. Curr Pharm Des 9:2489–2497
85. Hauser ER, Pericak-Vance MA (2000) Genetic analysis for common complex disease. Am Heart J 140:36–44
86. Foti D, Chiefari E, Fedele M, Iuliano R, Brunetti L, Paonessa F et al (2005) Lack of the architectural factor HMGA1 causes insulin resistance and diabetes in human and mice. Nat Med 7:765–773
87. Grant SF, Thorleifsson G, Reynisdottir I, Benediktsson R, Manolescu A, Sainz J et al (2006) Variant of transcription factor 7-like 2 (TCF7L2) gene confers risk of type 2 diabetes. Nat Genet 38:320–323
88. Narne P, Ponnaluri KC, Singh S et al (2012) Relationship between NADPH oxidase p22phox C242T, PARP-1 Val762Ala polymorphisms, angiographically verified coronary artery disease and myocardial infarction in South Indian patients with type 2 diabetes mellitus. Thromb Res 130:e259–e265
89. Jones DA, Prior SL, Tang TS et al (2010) Association between the rs4880 superoxide dismutase 2(C>T) gene variant and coronary heart disease in diabetes mellitus. Diabetes Res Clin Pract 90:196–201
90. Narne P, Ponnaluri KC, Singh S et al (2013) Association of the genetic variants of endothelial nitric oxide synthase gene with angiographically defined coronary artery disease and myocardial infarction in South Indian patients with type 2 diabetes mellitus. J Diabetes Complicat 27(3):255–261
91. Adams JN, Cox AJ, Freedman BI et al (2013) Genetic analysis of haptoglobin polymorphisms with cardiovascular disease and type 2 diabetes in the diabetes heart study. Cardiovasc Diabetol 12:31
92. Ruiz J, Blanché H, James RW et al (1995) Gln-Arg192 polymorphism of paraoxonase and coronary heart disease in type 2 diabetes. Lancet 346:869–872
93. Blatter Garin MC, James P, Blanché H, Passa P et al (1997) Paraoxonase polymorphism Met-Leu54 is associated with modified serum concentration of the enzyme. A possible link between the paraoxonase gene and increased risk of cardiovascular disease. J Clin Invest 99:62–66
94. Asleh R, Marsch S, Shilkrut M et al (2003) Genetically determined heterogeneity in hemoglobin scavenging and susceptibility to diabetic cardiovascular disease. Circ Res 92:1193–1200
95. Wellcome Trust Case Control Consortium (2007) Genome-wide association study of 14,000 cases of seven common diseases and 3,000 shared controls. Nature 447:661–678
96. Helgadottir A, Thorleifsson G, Manolescu A et al (2007) A common variant on chromosome 9p21 affects the risk of myocardial infarction. Science 316:1491–1493
97. Diabetes Genetics Initiative of Broad Institute of Harvard and MIT, Lund University, and Novartis Institutes of BioMedical Research, Saxena R, Voight BF, Lyssenko V et al (2007)

Genome-wide association analysis identifies loci for type 2 diabetes and triglyceride levels. Science 316:1331–1336
98. Salonen JT, Uimari P, Aalto JM et al (2007) Type 2 diabetes whole-genome association study in four populations: the DiaGen consortium. Am J Hum Genet 81:338–345
99. Broadbent HM, Peden JF, Lorkowski S et al (2008) Susceptibility to coronary heart disease and diabetes is encoded by distinct, tightly linked SNPs in the ANRIL locus on chromosome 9p. Hum Mol Genet 17:806–814
100. Visel A, Zhu Y, May D et al (2010) Targeted deletion of the 9p21 non-coding coronary artery disease risk interval in mice. Nature 464:409–412
101. Harismendy O, Notani D, Song X et al (2011) 9p21 DNA variants associated with coronary artery disease impair interferon-γ signalling response. Nature 470:264–268
102. Stern MP (1995) Diabetes and cardiovascular disease. The "common soil" hypothesis. Diabetes 44:369–374
103. Palmer C, Pe'er I (2017) Statistical correction of the Winner's curse explains replication variability in quantitative trait genome-wide association studies. PLoS Genet 13(7):e1006916
104. The Diabetes Control and Complications Trial Research Group (1999) The effect of intensive treatment of diabetes on the development, and progression of long-term complications in insulin-dependent diabetes mellitus. N Engl J Med 329:977–986
105. Cleary PA, Orchard TJ, Genuth S et al (2006) The effect of intensive glycemic treatment on coronary artery calcification in type 1 diabetic participants of the Diabetes Control and Complications Trial/Epidemiology of Diabetes Interventions and Complications (DCCT/EDIC) study. Diabetes 55:3556–3565
106. EDIC (2003) Study sustained effect of intensive treatment of type 1 diabetes mellitus on development, and progression of diabetic nephropathy: the Epidemiology of Diabetes Interventions and Complications (EDIC) study. JAMA 290:2159–2167
107. UK Prospective Diabetes Study (UKPDS) Group (1998) Intensive blood-glucose control with sulphonylureas or insulin compared with conventional treatment and risk of complications in patients with type 2 diabetes (UKPDS 33). Lancet 352:837–853
108. Holman RR, Paul SK, Bethel MA et al (2008) 10-year follow-up of intensive glucose control in type 2 diabetes. N Engl J Med 359:1577–1589
109. Hayward RA, Reaven PD, Wiitala WL et al (2015) Investigators V follow-up of glycemic control and cardiovascular outcomes in type 2 diabetes. N Engl J Med 372:2197–2206
110. Patel A, MacMahon S, Chalmers J et al (2008) Intensive blood glucose control and vascular outcomes in patients with type 2 diabetes. N Engl J Med 358:2560–2572
111. Gaede P, Lund-Andersen H, Parving HH, Pedersen O (2008) Effect of a multifactorial intervention on mortality in type 2 diabetes. N Engl J Med 358:580–591
112. El-Osta A (2012) Glycemic memory. Curr Opin Lipidol 23:24–29
113. Miao F, Chen Z, Genuth S et al (2014) Evaluating the role of epigenetic histone modifications in the metabolic memory of type 1 diabetes. Diabetes 63(5):1748–1762
114. Roy S, Sala R, Cagliero E, Lorenzi M (1990) Overexpression of fibronectin induced by diabetes or high glucose: phenomenon with a memory. Proc Natl Acad Sci U S A 87:404–408
115. El-Osta A, Brasacchio D, Yao D et al (2008) Transient high glucose causes persistent epigenetic changes and altered gene expression during subsequent normoglycemia. J Exp Med 205:2409–2417
116. Brasacchio D, Okabe J, Tikellis C et al (2009) Hyperglycemia induces a dynamic cooperativity of histone methylase and demethylase enzymes associated with gene-activating epigenetic marks that coexist on the lysine tail. Diabetes 58:1229–1236
117. Ihnat MA, Thorpe JE, Kamat CD et al (2007) Reactive oxygen species mediate a cellular 'memory' of high glucose stress signalling. Diabetologia 50:1523–1531
118. Li SL, Reddy MA, Cai Q et al (2006) Enhanced proatherogenic responses in macrophages and vascular smooth muscle cells derived from diabetic db/db mice. Diabetes 55:2611–2619
119. Yao D, Brownlee M (2010) Hyperglycemia-induced reactive oxygen species increase expression of the receptor for advanced glycation end products (RAGE) and RAGE ligands. Diabetes 59:249–255

120. Meerwaldt R, Links T, Zeebregts C et al (2008) The clinical relevance of assessing advanced glycation endproducts accumulation in diabetes. Cardiovasc Diabetol 7:29
121. Calcutt NA, Cooper ME, Kern TS, Schmidt AM (2009) Therapies for hyperglycaemia-induced diabetic complications: from animal models to clinical trials. Nat Rev Drug Discov 8:417–429
122. Holmström KM, Finkel T (2014) Cellular mechanisms and physiological consequences of redox-dependent signalling. Nat Rev Mol Cell Biol 15:411–421
123. Kim GH, Ryan JJ, Archer SL (1920–1936) The role of redox signaling in epigenetics and cardiovascular disease. Antioxid Redox Signal 18:2013
124. Kietzmann T, Petry A, Shvetsova A et al (2017) The epigenetic landscape related to reactive oxygen species formation in the cardiovascular system. Br J Pharmacol 174:1533–1554
125. Cooper ME, El-Osta A (2010) Epigenetics: mechanisms and implications for diabetic complications. Circ Res 107:1403–1413
126. Siebel AL, Fernandez AZ, El-Osta A (2010) Glycemic memory associated epigenetic changes. Biochem Pharmacol 80:1853–1859
127. Okabe J, Orlowski C, Balcerczyk A et al (2012) Distinguishing hyperglycemic changes by Set7 in vascular endothelial cells. Circ Res 110:1067–1076
128. Reddy MA, Zhang E, Natarajan R (2015) Epigenetic mechanisms in diabetic complications and metabolic memory. Diabetologia 58(3):443–455
129. Li Y, Reddy MA, Miao F et al (2008) Role of the histone H3 lysine 4 methyltransferase, SET7/9, in the regulation of NF-kappaB-dependent inflammatory genes. Relevance to diabetes and inflammation. J Biol Chem 283:26771–26781
130. Paneni F, Costantino S, Battista R et al (2015) Adverse epigenetic signatures by histone methyltransferase Set7 contribute to vascular dysfunction in patients with type 2 diabetes mellitus. Circ Cardiovasc Genet 8:150–158
131. Keating ST, Plutzky J, El-Osta A (2016) Epigenetic changes in diabetes and cardiovascular risk. Circ Res 118:1706–1722
132. Geraldes P, King GL (2010) Activation of protein kinase C isoforms and its impact on diabetic complications. Circ Res 106:1319–1331
133. Baubec T, Schübeler D (2014) Genomic patterns and context specific interpretation of DNA methylation. Curr Opin Genet Dev 25:85–92
134. Klose RJ, Bird AP (2006) Genomic DNA methylation: the mark and its mediators. Trends Biochem Sci 31(2):89–97
135. Clouaire T, Stancheva I (2008) Methyl-CpG binding proteins: specialized transcriptional repressors or structural components of chromatin? Cell Mol Life Sci 65(10):1509–1522
136. Jones PA (2012) Functions of DNA methylation: islands, start sites, gene bodies and beyond. Nat Rev Genet 13(7):484–492
137. Branco MR, Ficz G, Reik W (2012) Uncovering the role of 5- hydroxymethylcytosine in the epigenome. Nat Rev Genet 13:7–13
138. Breiling A, Lyko F (2015) Epigenetic regulatory functions of DNA modifications: 5-methylcytosine and beyond. Epigenetics Chromatin 8:24
139. Di Minno A, Turnu L, Porro B et al (2016) 8-Hydroxy-2-deoxyguanosine levels and cardiovascular disease: a systematic review and meta-analysis of the literature. Antioxid Redox Signal 24:548–555
140. Madugundu GS, Cadet J, Wagner JR (2014) Hydroxyl-radical induced oxidation of 5-methylcytosine in isolated and cellular DNA. Nucleic Acids Res 42:7450–7460
141. Valinluck V, Tsai HH, Rogstad DK et al (2004) Oxidative damage to methyl-CpG sequences inhibits the binding of the methyl-CpG binding domain (MBD) of methyl-CpG binding protein 2 (MeCP2). Nucleic Acids Res 32:4100–41008
142. Fleming AM, Burrows CJ (2017) 8-Oxo-7,8-dihydroguanine, friend and foe: epigenetic-like regulator versus initiator of mutagenesis. DNA Repair (Amst) 56:75–83
143. Zhou X, Zhuang Z, Wang W et al (2016) OGG1 is essential in oxidative stress induced DNA demethylation. Cell Signal 28:1163–1171

144. Afanas'ev I (2014) New nucleophilic mechanisms of ros-dependent epigenetic modifications: comparison of aging and cancer. Aging Dis 5:52–62
145. Pastukh V, Roberts JT, Clark DW et al (2015) An oxidative DNA "damage" and repair mechanism localized in the VEGF promoter is important for hypoxia-induced VEGF mRNA expression. Am J Physiol Lung Cell Mol Physiol 309:L1367–L1375
146. Pan L, Zhu B, Hao W et al (2016) Oxidized guanine base lesions function in 8-oxoguanine DNA glycosylase-1-mediated epigenetic regulation of nuclear factor kappaB-driven gene expression. J Biol Chem 291:25553–25566
147. Khan MA, Alam K, Dixit K, Rizvi MM (2016) Role of peroxynitrite induced structural changes on H2B histone by physicochemical method. Int J Biol Macromol 82:31–38
148. Galligan JJ, Rose KL, Beavers WN et al (2014) Stable histone adduction by 4-oxo-2-nonenal: a potential link between oxidative stress and epigenetics. J Am Chem Soc 136:11864–11866
149. Garcia-Gimenez JL, Olaso G, Hake SB et al (2013) Histone h3 glutathionylation in proliferating mammalian cells destabilizes nucleosomal structure. Antioxid Redox Signal 19:1305–1320
150. Kloypan C, Srisa-art M, Mutirangura A, Boonla C (2015) LINE-1 hypomethylation induced by reactive oxygen species is mediated via depletion of S-adenosylmethionine. Cell Biochem Funct 33:375–385
151. Baccarelli A, Wright R, Bollati V et al (2010) Ischemic heart disease and stroke in relation to blood DNA methylation. Epidemiology 21:819–828
152. Narne P, Pandey V, Phanithi PB (2017) Interplay between mitochondrial metabolism and oxidative stress in ischemic stroke: an epigenetic connection. Mol Cell Neurosci 82:176–194
153. Watson CJ, Collier P, Tea I et al (2014) Hypoxia-induced epigenetic modifications are associated with cardiac tissue fibrosis and the development of a myofibroblast like phenotype. Hum Mol Genet 23:2176–2188
154. Bishop T, Ratcliffe PJ (2015) HIF hydroxylase pathways in cardiovascular physiology and medicine. Circ Res 117:65–79
155. Xiong F, Xiao D, Zhang L (2012) Norepinephrine causes epigenetic repression of PKCepsilon gene in rodent hearts by activating Nox1- dependent reactive oxygen species production. FASEB J 26:2753–2763
156. Xiao D, Dasgupta C, Chen M et al (2014) Inhibition of DNA methylation reverses norepinephrine-induced cardiac hypertrophy in rats. Cardiovasc Res 101:373–382
157. Dunn J, Simmons R, Thabet S, Jo H (2015) The role of epigenetics in the endothelial cell shear stress response and atherosclerosis. Int J Biochem Cell Biol 67:167–176
158. O'Hagan HM, Wang W, Sen S et al (2011) Oxidative damage targets complexes containing DNA methyltransferases, SIRT1, and polycomb members to promoter CpG Islands. Cancer Cell 20:606–619
159. Muka T, Nano J, Voortman T et al (2016) The role of global and regional DNA methylation and histone modifications in glycemic traits and type 2 diabetes: a systematic review. Nutr Metab Cardiovasc Dis 26(7):553–566
160. Laukkanen MO, Mannermaa S, Hiltunen MO et al (1999) Local hypomethylation in atherosclerosis found in rabbit ec-sod gene. Arterioscler Thromb Vasc Biol 19:2171–2178
161. Valencia-Morales Mdel P, Zaina S, Heyn H et al (2015) The DNA methylation drift of the atherosclerotic aorta increases with lesion progression. BMC Med Genet 8:7
162. Niu Y, Des Marais TL, Tong Z et al (2015) Oxidative stress alters global histone modification and DNA methylation. Free Radic Biol Med 82:22–28
163. Thienpont B, Steinbacher J, Zhao H et al (2016) Tumour hypoxia causes DNA hypermethylation by reducing TET activity. Nature 537:63–68
164. Kalani A, Kamat PK, Tyagi N (2015) Diabetic stroke severity: epigenetic remodeling and neuronal, glial, and vascular dysfunction. Diabetes 64:4260–4271
165. Salminen A, Kaarniranta K, Kauppinen A (2016) Hypoxia-inducible histone lysine demethylases: impact on the aging process and age related diseases. Aging Dis 7:180–200
166. Monfort A, Wutz A (2013) Breathing-in epigenetic change with vitamin C. EMBO Rep 14:337–346

167. Laukka T, Mariani CJ, Ihantola T, Cao JZ, Hokkanen J, Kaelin WG Jr et al (2016) Fumarate and succinate regulate expression of hypoxia-inducible genes via TET enzymes. J Biol Chem 291:4256–4265
168. Zhang J, Wang YT, Miller JH et al (2018) Accumulation of succinate in cardiac ischemia primarily occurs via canonical Krebs cycle activity. Cell Rep 23:2617–2628
169. Liu R, Jin Y, Tang WH et al (2013) Ten-eleven translocation-2 (TET2) is a master regulator of smooth muscle cell plasticity. Circulation 128(18):2047–2057
170. Feil S, Fehrenbacher B, Lukowski R et al (2014) Transdifferentiation of vascular smooth muscle cells to macrophage-like cells during atherogenesis. Circ Res 115:662–667
171. Hiltunen MO, Yla-Herttuala S (2003) DNA methylation, smooth muscle cells, and atherogenesis. Arterioscler Thromb Vasc Biol 23:1750–1753
172. Lund G, Andersson L, Lauria M et al (2004) DNA methylation polymorphisms precede any histological sign of atherosclerosis in mice lacking apolipoprotein E. J Biol Chem 279(28):29147–29154
173. Zaina S, Lindholm MW, Lund G (2005) Nutrition and aberrant DNA methylation patterns in atherosclerosis: more than just hyperhomocysteinemia? J Nutr 135:5–8
174. Stenvinkel P, Karimi M, Johansson S et al (2007) Impact of inflammation on epigenetic DNA methylation—a novel risk factor for cardiovascular disease? J Intern Med 261:488–499
175. Kuroda A, Rauch TA, Todorov I et al (2009) Insulin gene expression is regulated by DNA methylation. PLoS One 4(9):e6953
176. Yang BT, Dayeh TA, Kirkpatrick CL et al (2011) Insulin promoter DNA methylation correlates negatively with insulin gene expression and positively with HbA(1c) levels in human pancreatic islets. Diabetologia 54:360–367
177. Yang BT, Dayeh TA, Volkov PA et al (2012) Increased DNA methylation and decreased expression of PDX-1 in pancreatic islets from patients with type 2 diabetes. Mol Endocrinol 26(7):1203–1212
178. Ling C, Del Guerra S, Lupi R et al (2008) Epigenetic regulation of PPARGC1A in human type 2 diabetic islets and effect on insulin secretion. Diabetologia 51:615–622
179. Davegårdh C, García-Calzón S, Bacos K, Ling C (2018) DNA methylation in the pathogenesis of type 2 diabetes in humans. Mol Metab 14:12–25. Feb 7. pii: S2212-8778(17)31102-X
180. Volkov P, Bacos K, Ofori JK et al (2017) Whole-genome bisulfite sequencing of human pancreatic islets reveals novel differentially methylated regions in type 2 diabetes pathogenesis. Diabetes 66(4):1074–1085
181. Volkmar M, Dedeurwaerder S, Cunha DA et al (2012) DNA methylation profiling identifies epigenetic dysregulation in pancreatic islets from type 2 diabetic patients. EMBO J 31(6):1405–1426
182. Chen R, Xia L, Tu K, Duan M et al (2018) Longitudinal personal DNA methylome dynamics in a human with a chronic condition. Nat Med 24:1930–1939
183. Hedman ÅK, Mendelson MM, Marioni RE et al (2017) Epigenetic patterns in blood associated with lipid traits predict incident coronary heart disease events and are enriched for results from genome-wide association studies. Circ Cardiovasc Genet 10(1):e001487
184. Roh TY, Cuddapah S, Zhao K (2005) Active chromatin domains are defined by acetylation islands revealed by genome-wide mapping. Genes Dev 19:542–552
185. Margueron R, Reinberg D (2010) Chromatin structure and the inheritance of epigenetic information. Nat Rev Genet 11:285–296
186. Allis CD, Jenuwein T (2016) The molecular hallmarks of epigenetic control. Nat Rev Genet 17:487–500
187. Khyzha N, Alizada A, Wilson MD, Fish JE (2017) Epigenetics of atherosclerosis: emerging mechanisms and methods. Trends Mol Med 23:332–347
188. Matsushima S, Sadoshima J (2015) The role of sirtuins in cardiac disease. Am J Physiol Heart Circ Physiol 309:H1375–H1389
189. Santos L, Escande C, Denicola A (2016) Potential modulation of sirtuins by oxidative stress. Oxidative Med Cell Longev 2016:9831825

190. Bekkering S, Quintin J, Joosten LA et al (2014) Oxidized low-density lipoprotein induces long-term proinflammatory cytokine production and foam cell formation via epigenetic reprogramming of monocytes significance. Arterioscler Thromb Vasc Biol 34:1731–1738
191. Dje N'Guessan P, Riediger F, Vardarova K et al (2009) Statins control oxidized LDL-mediated histone modifications and gene expression in cultured human endothelial cells. Arterioscler Thromb Vasc Biol 29(3):380–386
192. Papait R, Cattaneo P, Kunderfranco P et al (2013) Genome-wide analysis of histone marks identifying an epigenetic signature of promoters and enhancers underlying cardiac hypertrophy. Proc Natl Acad Sci U S A 110(50):20164–20169
193. Greißel A, Culmes M, Burgkart R et al (2016) Histone acetylation and methylation significantly change with severity of atherosclerosis in human carotid plaques. Cardiovasc Pathol 25(2):79–86
194. Tomita K, Barnes PJ, Adcock IM (2003) The effect of oxidative stress on histone acetylation and IL-8 release. Biochem Biophys Res Commun 301:572–577
195. Bartling TR, Subbaram S, Clark RR et al (2014) Redox-sensitive gene-regulatory events controlling aberrant matrix metalloproteinase-1 expression. Free Radic Biol Med 74:99–107
196. Lehrke M, Greif M, Broedl UC et al (2009) MMP-1 serum levels predict coronary atherosclerosis in humans. Cardiovasc Diabetol 2009:8–50
197. Wang X, Vatamaniuk MZ, Roneker CA et al (2011a) Knockouts of SOD1 and GPX1 exert different impacts on murine islet function and pancreatic integrity. Antioxid Redox Signal 14:391–401
198. Gupta J, Tikoo K (2012) Involvement of insulin-induced reversible chromatin remodeling in altering the expression of oxidative stress-responsive genes under hyperglycemia in 3T3-L1 preadipocytes. Gene 504:181–191
199. Kikuchi H, Kuribayashi F, Kiwaki N, Takami Y, Nakayama T (2011) GCN5 regulates the superoxide-generating system in leukocytes via controlling gp91-phox gene expression. J Immunol 186:3015–3022
200. Miao F, Gonzalo IG, Lanting L, Natarajan R (2004) In vivo chromatin remodeling events leading to inflammatory gene transcription under diabetic conditions. J Biol Chem 279:18091–18097
201. Kaur H, Chen S, Xin X et al (2006) Diabetes-induced extracellular matrix protein expression is mediated by transcription coactivator p300. Diabetes 55:3104–3111
202. Xu B, Chiu J, Feng B, Chen S, Chakrabarti S (2008) PARP activation and the alteration of vasoactive factors and extracellular matrix protein in retina and kidney in diabetes. Diabetes Metab Res Rev 24:404–412
203. Pacher P, Szabó C (2007) Role of poly(ADP-ribose) polymerase 1 (PARP-1) in cardiovascular diseases: the therapeutic potential of PARP inhibitors. Cardiovasc Drug Rev 25(3):235–260
204. Pons D, de Vries FR, van den Elsen PJ et al (2009) Epigenetic histone acetylation modifiers in vascular remodelling: new targets for therapy in cardiovascular disease. Eur Heart J 30(3):266–277
205. Osoata GO, Yamamura S, Ito M, Vuppusetty C, Adcock IM, Barnes PJ et al (2009) Nitration of distinct tyrosine residues causes inactivation of histone deacetylase 2. Biochem Biophys Res Commun 384:366–371
206. Doyle K, Fitzpatrick FA (2010) Redox signaling, alkylation (carbonylation) of conserved cysteines inactivates class I histone deacetylases 1, 2, and 3 and antagonizes their transcriptional repressor function. J Biol Chem 285:17417–17424
207. Adenuga D, Yao H, March TH et al (2009) Histone deacetylase 2 is phosphorylated, ubiquitinated, and degraded by cigarette smoke. Am J Respir Cell Mol Biol 40:464–473
208. Rajendrasozhan S, Yang SR, Edirisinghe I et al (2008) Deacetylases and NF-kappaB in redox regulation of cigarette smoke-induced lung inflammation: epigenetics in pathogenesis of COPD. Antioxid Redox Signal 10:799–811
209. Sathishkumar C, Prabu P, Balakumar M et al (2016) Augmentation of histone deacetylase 3 (HDAC3) epigenetic signature at the interface of proinflammation and insulin resistance in patients with type 2 diabetes. Clin Epigenetics 8:125

210. Zhu H, Shan L, Schiller PW et al (2010) Histone deacetylase-3 activation promotes tumor necrosis factor-alpha (TNF-alpha) expression in cardiomyocytes during lipopolysaccharide stimulation. J Biol Chem 285:9429–9436
211. Choi JH, Nam KH, Kim J et al (2005) Trichostatin A exacerbates atherosclerosis in low density lipoprotein receptor-deficient mice. Arterioscler Thromb Vasc Biol 25:2404–2409
212. Okamoto H, Fujioka Y, Takahashi A et al (2006) Trichostatin A, an inhibitor of histone deacetylase, inhibits smooth muscle cell proliferation via induction of p21 (WAF1). J Atheroscler Thromb 13:183–191
213. Pandey D, Sikka G, Bergman Y et al (2014) Transcriptional regulation of endothelial arginase 2 by histone deacetylase 2. Arterioscler Thromb Vasc Biol 34:1556–1566
214. Ago T, Liu T, Zhai P, Chen W, Li H, Molkentin JD et al (2008) A redox-dependent pathway for regulating class II HDACs and cardiac hypertrophy. Cell 133:978–993
215. Zhang M, Brewer AC, Schroder K et al (2010) NADPH oxidase-4 mediates protection against chronic load-induced stress in mouse hearts by enhancing angiogenesis. Proc Natl Acad Sci U S A 107:18121–18126
216. Usui T, Okada M, Mizuno W et al (2012) HDAC4 mediates development of hypertension via vascular inflammation in spontaneous hypertensive rats. Am J Physiol Heart Circ Physiol 302:H1894–H1904
217. Kim J, Hwangbo C, Hu X et al (2015) Restoration of impaired endothelial myocyte enhancer factor 2 function rescues pulmonary arterial hypertension. Circulation 131:190–199
218. Rajendran R, Garva R, Krstic-Demonacos M, Demonacos C (2011) Sirtuins: molecular traffic lights in the crossroad of oxidative stress, chromatin remodeling, and transcription. J Biomed Biotechnol 2011:368276
219. Zarzuelo MJ, Lopez-Sepulveda R, Sanchez M et al (2013) SIRT1 inhibits NADPH oxidase activation and protects endothelial function in the rat aorta: implications for vascular aging. Biochem Pharmacol 85:1288–1296
220. de Kreutzenberg SV, Ceolotto G, Papparella I et al (2010) Downregulation of the longevity-associated protein sirtuin 1 in insulin resistance and metabolic syndrome: potential biochemical mechanisms. Diabetes 59:1006–1015
221. Yamamoto T, Tamaki K, Shirakawa K et al (2016) Cardiac Sirt1 mediates the cardioprotective effect of caloric restriction by suppressing local complement system activation after ischemia-reperfusion. Am J Physiol Heart Circ Physiol 310:H1003–H1014
222. Bonello S, Zahringer C, BelAiba RS, Djordjevic T, Hess J, Michiels C et al (2007) Reactive oxygen species activate the HIF-1alpha promoter via a functional NFkappaB site. Arterioscler Thromb Vasc Biol 27:755–761
223. Hwang JW, Yao H, Caito S, Sundar IK, Rahman I (2013) Redox regulation of SIRT1 in inflammation and cellular senescence. Free Radic Biol Med 61:95–110
224. Canto C, Sauve AA, Bai P (2013) Crosstalk between poly(ADP-ribose) polymerase and sirtuin enzymes. Mol Asp Med 34:1168–1201
225. Puthanveetil P, Zhang D, Wang Y et al (2012) Diabetes triggers a PARP1 mediated death pathway in the heart through participation of FoxO1. J Mol Cell Cardiol 53(5):677–686
226. Narne P, Pandey V, Phanithi PB (2018) Role of nitric oxide and hydrogen sulfide in ischemic stroke and the emergent epigenetic underpinnings. Mol Neurobiol. https://doi.org/10.1007/s12035-018-1141-6
227. Narne P, Pandey V, Simhadri PK, Phanithi PB (2017) Poly(ADP-ribose)polymerase-1 hyperactivation in neurodegenerative diseases: the death knell tolls for neurons. Semin Cell Dev Biol 63:154–166
228. Bannister AJ, Kouzarides T (2005) Reversing histone methylation. Nature 436:1103–1106
229. Chervona Y, Costa M (2012) The control of histone methylation and gene expression by oxidative stress, hypoxia, and metals. Free Radic Biol Med 53:1041–1047
230. Mentch SJ, Mehrmohamadi M, Huang L et al (2015) Histone methylation dynamics and gene regulation occur through the sensing of one-carbon metabolism. Cell Metab 22:861–873
231. Liu T, Wu C, Jain MR et al (1854) Master redox regulator Trx1 upregulates SMYD1 & modulates lysine methylation. Biochim Biophys Acta 2015:1816–1822

232. Franklin S, Kimball T, Rasmussen TL et al (2016) The chromatin-binding protein Smyd1 restricts adult mammalian heart growth. Am J Physiol Heart Circ Physiol 311:H1234–H1247
233. Zhang QJ, Liu ZP (2015) Histone methylations in heart development, congenital and adult heart diseases. Epigenomics 7:321–330
234. Rajasekar P, O'Neill CL, Eeles L et al (2015) Epigenetic changes in endothelial progenitors as a possible cellular basis for glycemic memory in diabetic vascular complications. J Diabetes Res 2015:436879
235. Al-Sawaf O, Clarner T, Fragoulis A et al (2015) Nrf2 in health and disease: current and future clinical implications. Clin Sci 129:989–999
236. Friso S, Carvajal CA, Fardella CE, Olivieri O (2015) Epigenetics and arterial hypertension: the challenge of emerging evidence. Transl Res 165:154–165
237. Li J, Braganza A, Sobol RW (2013) Base excision repair facilitates a functional relationship between guanine oxidation and histone demethylation. Antioxid Redox Signal 18:2429–2443
238. Hickok JR, Vasudevan D, Antholine WE, Thomas DD (2013) Nitric oxide modifies global histone methylation by inhibiting Jumonji C domain-containing demethylases. J Biol Chem 288:16004–16015
239. He C, Larson-Casey JL, Gu L, Ryan AJ, Murthy S, Carter AB (2016) Cu,Zn-SOD-mediated redox regulation of Jmjd3 modulates macrophage polarization and pulmonary fibrosis. Am J Respir Cell Mol Biol 55:58–71
240. Tausendschon M, Dehne N, Brune B (2011) Hypoxia causes epigenetic gene regulation in macrophages by attenuating Jumonji histone demethylase activity. Cytokine 53:256–262
241. Zhang QJ, Chen HZ, Wang L, Liu DP, Hill JA, Liu ZP (2011) The histone trimethyllysine demethylase JMJD2A promotes cardiac hypertrophy in response to hypertrophic stimuli in mice. J Clin Invest 121:2447–2456
242. Matouk CC, Marsden PA (2008) Epigenetic regulation of vascular endothelial gene expression. Circ Res 102:873–887
243. Gan Y, Shen YH, Wang J et al (2005) Role of histone deacetylation in cell-specific expression of endothelial nitric-oxide synthase. J Biol Chem 280:16467–16475
244. Fish JE, Matouk CC, Rachlis A et al (2005) The expression of endothelial nitric-oxide synthase is controlled by a cell-specific histone code. J Biol Chem 280:24824–24838
245. Wilcox JN, Subramanian RR, Sundell CL et al (1997) Expression of multiple isoforms of nitric oxide synthase in normal and atherosclerotic vessels. Arterioscler Thromb Vasc Biol 17:2479–2488
246. Chan GC, Fish JE, Mawji IA et al (2005) Epigenetic basis for the transcriptional hyporesponsiveness of the human inducible nitric oxide synthase gene in vascular endothelial cells. J Immunol 175:3846–3861
247. Chartrain NA, Geller DA, Koty PP et al (1994) Molecular cloning, structure, and chromosomal localization of the human inducible nitric oxide synthase gene. J Biol Chem 269:6765–6772
248. Pirola L, Balcerczyk A, Tothill RW et al (2011) Genome-wide analysis distinguishes hyperglycemia regulated epigenetic signatures of primary vascular cells. Genome Res 21:1601–1615
249. Estève PO, Chin HG, Benner J et al (2009) Regulation of DNMT1 stability through SET7-mediated lysine methylation in mammalian cells. Proc Natl Acad Sci U S A 106:5076–5081
250. Deering TG, Ogihara T, Trace AP et al (2009) Methyltransferase Set7/9 maintains transcription and euchromatin structure at islet-enriched genes. Diabetes 58:185–193
251. Maganti AV, Maier B, Tersey SA et al (2015) Transcriptional activity of the islet β cell factor Pdx1 is augmented by lysine methylation catalyzed by the methyltransferase Set7/9. J Biol Chem 290:9812–9822
252. Chakrabarti SK, Francis J, Ziesmann SM et al (2003) Covalent histone modifications underlie the developmental regulation of insulin gene transcription in pancreatic beta cells. J Biol Chem 278(26):23617–23623
253. Villeneuve LM, Reddy MA, Lanting LL et al (2008) Epigenetic histone H3 lysine 9 methylation in metabolic memory and inflammatory phenotype of vascular smooth muscle cells in diabetes. Proc Natl Acad Sci U S A 105(26):9047–9052

254. Villeneuve LM, Kato M, Reddy MA et al (2010) Enhanced levels of microRNA-125b in vascular smooth muscle cells of diabetic db/db mice lead to increased inflammatory gene expression by targeting the histone methyltransferase Suv39h1. Diabetes 59(11):2904–2915
255. Leung A, Trac C, Jin W, Lanting L et al (2013) Novel long noncoding RNAs are regulated by angiotensin II in vascular smooth muscle cells. Circ Res 113(3):266–278
256. Chen J, Zhang J, Yang J et al (2017) Histone demethylase KDM3a, a novel regulator of vascular smooth muscle cells, controls vascular neointimal hyperplasia in diabetic rats. Atherosclerosis 257:152–163
257. Chen Z, Miao F, Paterson AD et al (2016) Epigenomic profiling reveals an association between persistence of DNA methylation and metabolic memory in the DCCT/EDIC type 1 diabetes cohort. Proc Natl Acad Sci U S A 113(21):E3002–E3011
258. Paneni F, Mocharla P, Akhmedov A et al (2012) Gene silencing of the mitochondrial adaptor p66(Shc) suppresses vascular hyperglycemic memory in diabetes. Circ Res 111(3):278–289
259. Paneni F, Volpe M, Lüscher TF, Cosentino F (2013) SIRT1, p66(Shc), and Set7/9 in vascular hyperglycemic memory: bringing all the strands together. Diabetes 62(6):1800–1807
260. Keating ST, Ziemann M, Okabe J et al (2014) Deep sequencing reveals novel Set7 networks. Cell Mol Life Sci 71(22):4471–4486
261. Francia P, Cosentino F, Schiavoni M et al (2009) p66(Shc) protein, oxidative stress, and cardiovascular complications of diabetes: the missing link. J Mol Med (Berl) 87(9):885–891
262. Costantino S, Paneni F, Mitchell K et al (2018) Hyperglycaemia-induced epigenetic changes drive persistent cardiac dysfunction via the adaptor p66Shc. Int J Cardiol 268:179–186
263. Costantino S, Paneni F, Battista R et al (2017) Impact of glycemic variability on chromatin remodeling, oxidative stress, and endothelial dysfunction in patients with type 2 diabetes and with target HbA1c levels. Diabetes 66(9):2472–2482
264. Zhou S, Chen HZ, Wan YZ et al (2011) Repression of P66Shc expression by SIRT1 contributes to the prevention of hyperglycemia-induced endothelial dysfunction. Circ Res 109(6):639–648

Role of Oxidative Stress, Mitochondrial Dysfunction, and Autophagy in Cardiovascular Disease: Its Pathogenesis and Amelioration by Different Small Natural Molecules

19

Sharmistha Chatterjee, Uday Hossain, and Parames C. Sil

Abstract

The biggest cause of global mortality today is cardiovascular diseases. Not only old people, but even the younger generation gets afflicted now. This chapter will focus on the role of oxidative stress, mitochondrial dysfunction, and autophagy on the pathogenesis of the various forms of cardiovascular diseases including heart failure, atherosclerosis, hypertension, myocardial infarction, and ischemia-reperfusion injury. Various cell signaling pathways get modulated under external or internal stress stimuli to induce ROS which begets the oxidative stress condition. The antioxidant defense mechanisms by which the delicate balance between prooxidants and antioxidants in the cell is maintained in equilibrium get disrupted, and the structural and functional entities of the cell collapse. Mitochondrial dysfunction is directly implicated in the above process, as it is both the cause and outcome of oxidative stress. When dysfunctional mitochondria accumulate inside the cell, autophagy comes to the rescue. But excessive autophagy again is a cause of concern as it paves the way for a second type of programmed cell death, distinct from apoptosis. Antioxidants have mostly been proven highly effective against the plethora of cardiovascular diseases, as they have been successful in attenuating the oxidative stress in the vascular cells, as well as that in the myocardial cells, and have restored the physiological conditions close to the normal state. So they have been routed to be important drug leads for the development of effective therapeutics against cardiovascular diseases. With minimal or no toxicity, natural molecules have remained in the forefront to be tested and tried in this regard. So this field has and will continue to have importance in the research fraternity for decades to come.

S. Chatterjee · U. Hossain · P. C. Sil (✉)
Division of Molecular Medicine, Bose Institute, Kolkata, West Bengal, India
e-mail: parames@jcbose.ac.in

S. Chakraborti et al. (eds.), *Modulation of Oxidative Stress in Heart Disease*,
https://doi.org/10.1007/978-981-13-8946-7_19

Keywords
Cardiovascular diseases · Reactive oxygen species · Oxidative stress · Antioxidant · Mitochondrial dysfunction · Autophagy · Endothelial dysfunction · Atherosclerosis · Hypertension · Myocardial infarction · Ischemia-reperfusion · Amelioration · Small molecules

19.1 Introduction

In today's world, chronic diseases have come to be at par with acute ones and have even risen to become the dominant group contributing to the global burden of disease occurrence and mortality. Among them, cardiovascular diseases (CVDs) have been the largest contributor to world chronic disease epidemiology. This group, which includes atherosclerosis and cardiac ischemia, cardiomyopathy, coronary heart disease, heart failure, cardiac arrest, hypertension, and other related symptoms, claims around 17.9 million lives annually (Heart Disease and Stroke Statistics 2018 At-a-Glance, American Heart Association), making it the cause of 31% of all global deaths; hence it is officially acknowledged by WHO to be the number one cause of death globally. With the advancement in diagnosis and treatment and betterment of lifestyle and economy, death rates from cardiovascular diseases have declined significantly, but the situation in most low- and middle-income countries presents a stark contrast.

The total number of deaths from cardiovascular disease, mainly coronary heart disease, stroke, and rheumatic heart disease, had increased globally from 14.4 million in 1990 to 17.5 million by 2005, out of which 7.6 million resulted from coronary heart disease and 5.7 million were attributed to stroke. Not surprisingly, more than 80 percent of the deaths occurred in low- and middle-income countries [1]. The lowest age-adjusted mortality rates are in the advanced industrialized countries and parts of Latin America, whereas the highest rates today are found in Eastern Europe and a number of low- and middle-income countries, as reported by WHO in its 2008 report of World Health Statistics. A report by Beaglehole and Bonita in 2008 states that by 2030, noncommunicable diseases will account for more than three-quarters of deaths worldwide; and cardiovascular diseases alone will be responsible for more deaths in low-income countries than infectious diseases (including HIV/AIDS, tuberculosis, and malaria), maternal and perinatal conditions, and nutritional disorders combined, which numerically rounds up to nearly more than 23.6 million deaths across the world [2]. Ahead of every other major diagnostic group, cardiovascular diseases and stroke accounted for 14% of total health expenditures in 2013–2014, and it is projected that the total direct medical costs of CVD would increase to 749 billion dollars by 2035 (Heart Disease and Stroke Statistics 2018 At-a-Glance, American Heart Association). It could thus be extrapolated that this disease group will continue to dominate the global mortality trends in the future, unless interventions come at par in the developing countries with the developed ones, and more efficient and low-cost interventional therapeutics are developed, tested, and put to use.

In India, the condition is even worse. The Global Burden of Disease study estimated the age-standardized CVD death rate in India to be 272 per 100,000 population. This is higher than the global average of 235 per 100,000 population. In India, this cardiovascular disease epidemic is of particular concern, as here in this country, it shows accelerated buildup, has an early age of disease onset in the population, and also has a high case fatality rate. Premature mortality in terms of years of life lost resulting from cardiovascular diseases in India increased from 23.2 million in 1990 to 37 million in 2010, an increase of about 59%. In spite of the fact that the disease is widely heterogeneous in the prevalence of cardiovascular risk factors across different regions globally, cardiovascular diseases have emerged as the leading cause of death in all parts of India, across rural and urban areas as well as richer and poorer states without any distinction [3].

The main risk factors and causes of cardiovascular diseases, according to WHO as well as the American Heart Association, are smoking, physical inactivity, faulty nutrition, obesity, cholesterol, diabetes, and, last but not the least, high blood pressure.

Current therapeutic options encompass (a) interventional techniques ranging from cardiopulmonary resuscitation to heart transplant including angioplasty and stents, heart bypass surgery, valve disease treatment, cardioversion, EECP, pacemakers, implantable cardiovascular defibrillators (ICDs), lead extraction, and left ventricular assist device (LVAD); (b) medications like angiotensin-converting enzyme (ACE) inhibitors, angiotensin II receptor blockers, antiarrhythmics, antiplatelet drugs, aspirin therapy, beta-blocker therapy, calcium channel blocker drugs, clot buster drugs, digoxin, diuretics, nitrates, and warfarin and other blood thinners; and (c) therapeutic care involving plant-based diets for heart health, recovery after heart surgery, and finding strength during tough times.

This field of study would be of equal importance as has always been, if not more, owing to the fact that the younger age group is more afflicted with the disease, and the cumulative condition in developing countries is poor. Although people nowadays survive through one or more heart attacks with the wide variety of treatment options as discussed above, they are left behind with painful and dangerous heart failure with which they have to deal and live with, and hence, laboratories are obtaining more funds to understand how to repair a damaged heart efficiently to make it as healthy as possible.

This chapter will focus on the role of oxidative stress, mitochondrial dysfunction, and autophagy, in the pathogenesis of cardiovascular diseases, and how various drug molecules have been developed over the years to target the stressed pathways to ameliorate the disease conditions. We would focus on the natural molecules which have been tested and tried in this regard, to use them as efficient mediators of a healthier cardiovascular system.

19.2 Oxidative Stress, Mitochondrial Dysfunction, and Autophagy: Some Basic Ideas

Oxygen is one of the basic requirements of life for most living organisms, be it for respiration or photosynthesis [4]. The oxidative phosphorylation (OXPHOS) pathway takes place utilizing oxygen in the mitochondria to produce energy in the form of ATP. Oxygen has the tendency to generate transient but highly reactive free radicals like the superoxide radicals (O^{2-}), hydrogen peroxide (H_2O_2) and hydroxyl radical (OH^-) [5], nitric oxide (NO^-), peroxynitrite ($ONOO^-$), etc. [6] even in the steady state [7] and as natural by-products of normal oxygen metabolism. It can be released by the actions of lipoxygenase, reduced nicotinamide adenine dinucleotide phosphate oxidase or simply NAD(P)H oxidase, xanthine oxidase, or during the uncoupling of nitric oxide (NO) synthase inside the vascular cells, apart from the mitochondria [6] (Fig. 19.1). The electron transport chain (ETC) is the series of reactions which occur during the OXPHOS pathway, with electrons passing through a number of oxidation-reduction reactions where a number of protein complexes are reduced, ultimately reducing an oxygen atom to a water molecule and thus terminating the chain. But leakage of electrons from the ETC at a very low rate during the process sometimes causes the premature binding of electrons to the ultimate acceptor oxygen, thus producing superoxide radicals (O^{2-}). This is classified as the endogenous pathway of ROS production, and this basal amount of ROS is always

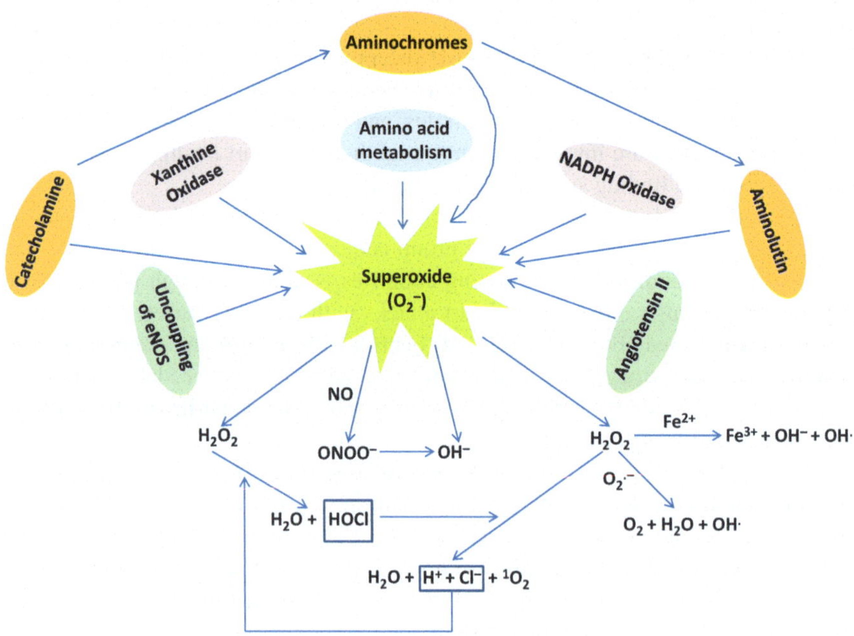

Fig. 19.1 Schematic diagram showing production of ROS in the body. Basic chemical reactions that lead to their formation have also been illustrated

present in our system and performs important functions as second messengers, mediates induction of host defense mechanisms, and controls mobilization of ions across transport systems. But a very large amount of ROS can wreak havoc in the cell system, disrupting cellular pathways and changing the nature of basic biomolecules including proteins [8], lipids (peroxidation), [9] and nucleic acids (strand breaks and DNA damage) [5, 10, 11] with severe oxidative damage [12, 13] (Fig. 19.2). This condition is generally referred to as oxidative stress, and in excess, it kills the cell by induction of apoptosis. In order to prevent such catastrophic events, all cells have a protective system to keep the level of ROS under check during normal physiological conditions, [14] which acts as an antioxidant defense system. This system includes some vital enzymes like superoxide dismutase (SOD), catalyzing the conversion of O^{2-} to molecular oxygen (O_2) and water; catalase (CAT), which scavenges H_2O_2 and forms similar products as that of SOD; glutathione reductase (GR), which converts oxidized GSSG (glutathione disulfide) to reduced GSH (glutathione); and also glutathione-S-transferase (GST) which helps in the radical scavenging of GSH [15]. There are also glutathione peroxidases (GPxs) which scavenge H_2O_2 along with GSH. ROS can also be induced by exogenous sources (exogenous ROS) by the action of environmental stress factors such as ionizing radiations [16], tobacco smoke [17], herbicides [18], heavy metals [19–24], drugs [25–32] and xenobiotics, and pollutants [12, 33–35]. Pathophysiological conditions like diabetes [27, 36–38] and cancer [39] can also induce ROS formation

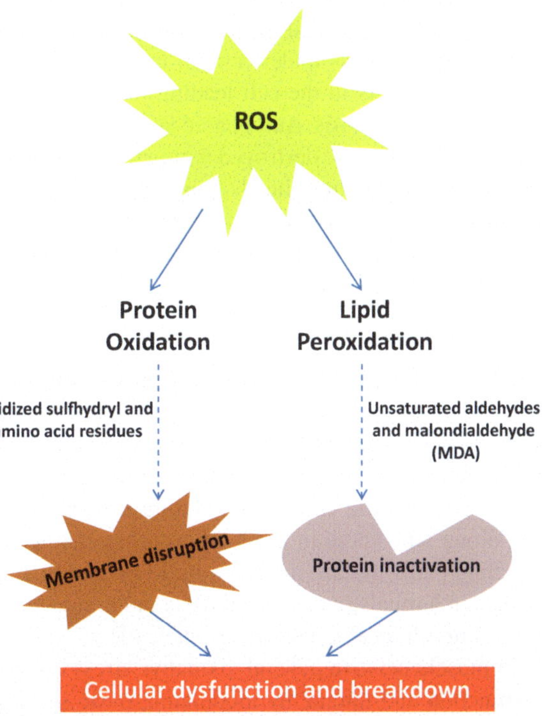

Fig. 19.2 Schematic illustration depicting the effect of ROS on body cells

and outburst. Oxidative stress develops when the delicate balance between the prooxidants and antioxidants gets disrupted and gets tilted in the favor of prooxidants, and then cell viability is compromised [4].

As the ETC takes place in the mitochondria, they are the primary sites of undergoing and detection of oxidative stress and damage. Generation of a huge amount of ROS under the influence of various prooxidants and development of a pathophysiological condition, due to which again ROS overproduction occurs, this vicious cycle of ROS production and amplification goes on and on [40]. Mitochondrial DNA (mtDNA) is affected with strand breaks, deletions, and mutations (in the absence of protective histone proteins) [41] and poor DNA damage repair mechanisms, [42] mitochondrial proteins are carbonylated, and translation is affected due to mtDNA mutations and transcriptional errors in producing tRNA and rRNA and posttranscriptional rearrangements [43]. The mitochondrial membrane potential drops and the mitochondrion starts to disintegrate. Before apoptosis, initially the cell drives an SOS response by trying to increase the number of mitochondria by biogenesis and hence mtDNA copy number increases with a concomitant increase in the expression of PGC-1α and its downstream components. These therefore are considered as markers for detecting mitochondrial dysfunction alongside mitochondrial membrane potential [40].

The cells' established SOS response is autophagy. This process is an evolutionarily conserved one which springs into action under external (like overcrowding and subsequent hypoxia, nutrient starvation, or high temperatures) or internal (like damaged or surplus organelle accumulation) stress stimuli, to repair, recycle, and sustain cellular life by controlled degradation within lysosomes [44–47]. Thus it is mainly a life-saving process under normal conditions. But uncontrolled and continuous stimulation results in the cell leading to programmed cell death type II, which is distinct from apoptosis. Autophagic vacuoles (autophagosomes) are formed and the internal organelles are destroyed within them, most noticeably the mitochondria and endoplasmic reticulum, thus leading to a total breakdown of all cellular structures and processes [48, 49]. Hence, this process, although poorly understood due to the scarcity of availability of biomarkers, is a death pathway which significantly affects organ physiology [50].

19.3 CVDs and the Role of Oxidative Stress

Atherogenesis and CVD is marked by vascular inflammation with the accumulation of monocytes in the vascular endothelium to subsequently transform into macrophages or dendritic cells accumulating lipoproteins (fat) and forming foam cells, [51] along with surge in levels of proinflammatory cytokines. These inflammatory signaling pathways are mediated by ROS, and this is supported by various disease models in animals (and clinical trials in humans) that indeed, ROS does happen to play a big role in atherosclerosis and CVD.

In muscle cells, in general, oxidative stress can wreak havoc causing contractile dysfunction and fatigue by modulating basic stress proteins [52] and calcium ion regulation [53]. In cardiac myocytes, similar deregulated signaling and desensitization might occur to cause arrhythmias and myocardial infarctions (Fig. 19.3).

ROS has been regarded as one of the main culprits of CVD, due to the accumulation of both direct and indirect evidences, including the fact that the level of ROS has been found to be elevated not only in the patients of CVD but also in people who are at high risk of developing CVD, like people with dyslipidemia, hypertension, diabetes, and smoking habits. Also the treatments with various antioxidants have not only found to decrease oxidative stress but also various cardiovascular disease clinical symptoms. When studied deeper, most of them revealed common mechanisms of molecular and cellular damage (Fig. 19.4). So, with greater understanding of these molecular mechanisms of action, interventions have been possible and are still under intense research.

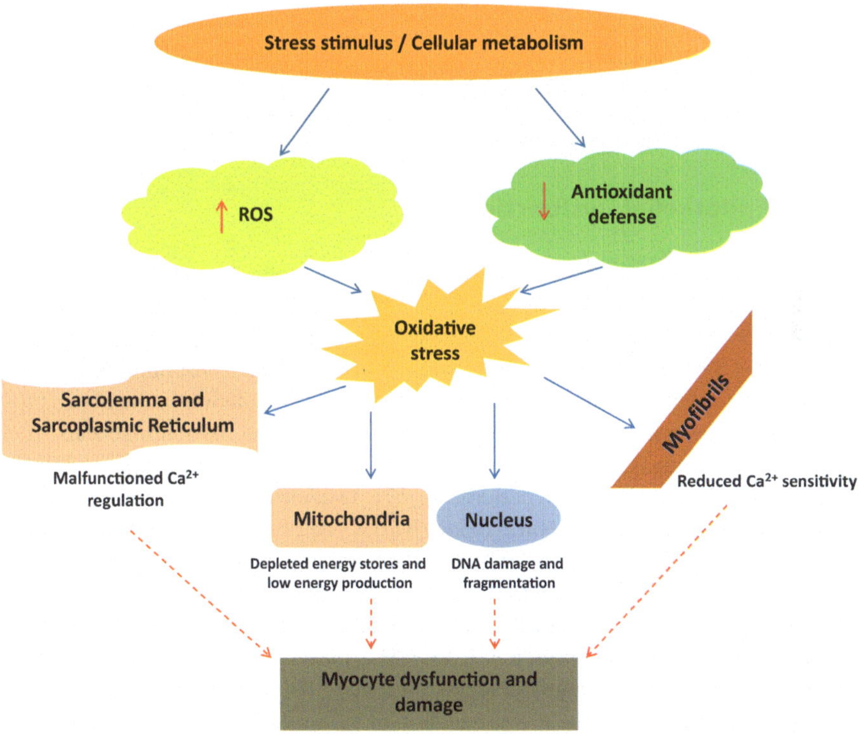

Fig. 19.3 Diagram depicting the effect of oxidative stress specifically on myocytes. Myocyte dysfunction is the basis of cardiovascular diseases like myocardial infarction and arrhythmias which could lead to ischemia and heart failure

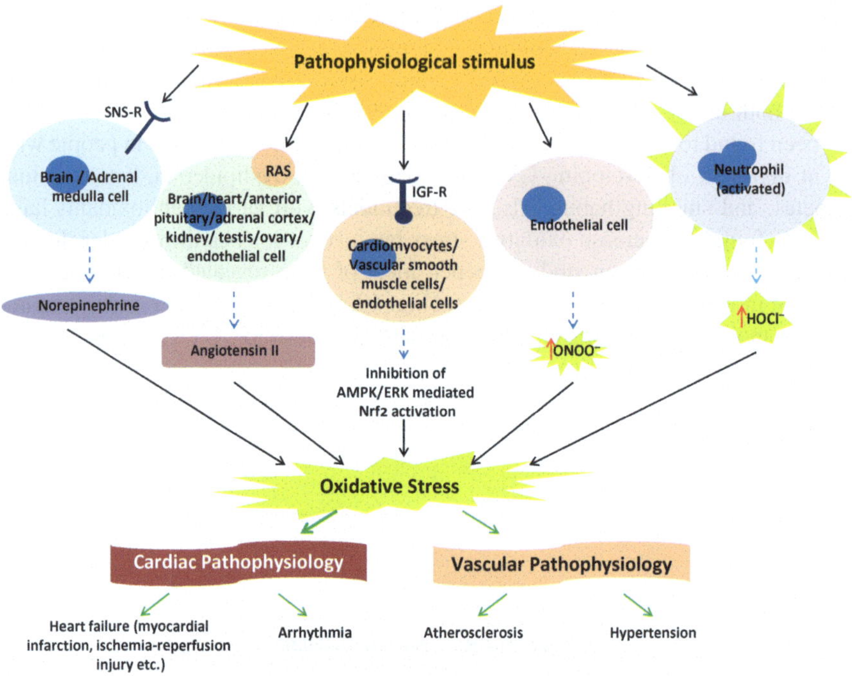

Fig. 19.4 Schematic diagram showing how the heart is affected under an oxidative outburst and what could be the main possible causes and outcomes

19.3.1 Endothelial Dysfunction

The major regulatory element of vascular homeostasis is the normal endothelium of vascular tissue. It is the site of metabolic processes, alongside mediating release of inflammatory cytokines and vasodilators and vasoconstrictors. So it has effects on the vascular tone, fibrinolysis, clot formation or thrombogenesis, and even proliferation and migration of smooth muscles [54]. Any dysregulation in the above processes leads to endothelial dysfunction, which underlines the pathogenesis of a plethora of cardiovascular disease pathophysiologies, including atherosclerosis [55], heart failure, [56] and hypertension [57]. Diabetes is also a major offset of endothelial dysfunction [58]. Nitric oxide (NO) is one of the known secondary messengers which aids in vasodilatation of vasculatures in animals; has antiplatelet, anti-inflammatory, antiproliferative, and permeability-decreasing properties; and also inhibits leukocyte rolling and adhesion to endothelial surface, besides inhibition of cytokine-induced expression of VCAM-1 (vascular cell adhesion protein-1) and MCP-1 (monocyte chemoattractant protein-1) [6]. The main culprit behind endothelial dysfunction is the low production or bioavailability of NO. ROS rapidly reacts with NO to produce peroxynitrite, itself a cellular toxin, while inactivating NO and reducing its bioavailability; this reaction occurs so fast that its rate exceeds the rate of dismutation of superoxide anions by superoxide dismutase. As has been discussed above, uncoupling of eNOS by oxidation of BH_4 inside the endothelial cells also releases ROS, thus amplifying the oxidative stress even further. The other sources of increase in ROS in vasculatures are NADPH and xanthine oxidases. Sometimes, even other sources might contribute synergistically although NADPH oxidase is the major contributor, [59] apart from mitochondrial ROS. With such high levels of ROS production inside the vasculatures and with the fact being established that NO mediates so many functions, it can be stated with disambiguation that ROS definitely is a major player in the pathogenesis of cardiovascular diseases. Indeed, ROS activates matrix metalloproteinases via NF-κB, leading to plaque instability and rupture, which has been experimentally shown to be reversed by administration of antioxidants [6]. ROS causes inflammation via NF-κB by stimulating the mitogen-activated protein kinase (MAP kinase) signaling (ERK1/2 and BMK1) pathway, tyrosine kinases (Src and Syk), and various protein kinase C (PKC) isoforms and also by stimulating the Rho-associated kinase or ROCK, which, along with its downstream components, have been found to be elevated in patients at high risk of developing CVD.

19.3.2 Hypertension

Oxidative stress has been recorded in genetic as well as experimental models of hypertension [54]. The $p22_{phox}$ promoter in humans with essential hypertension has been understood to be associated with endothelial dysfunction and an increased atherosclerosis [60] tendency. Several polymorphisms of this promoter have been found in spontaneously hypertensive rat (SHR) and stroke-prone spontaneously hypertensive rat (SP-SHR) models, [61] in which both the conduit and the resistance vessels show elevated NADPH oxidase activity, [62, 63] partly owing to increased levels of expression of component subunits [64]. The involvement of ROS is further evidenced from the fact that the hypertension pathophysiology and vascular superoxide production in these models could be ameliorated by the administration of antioxidant vitamins and SOD mimetics and eNOS cofactor BH_4 (tetrahydrobiopterin) [65]. Also, treatment with the NADPH oxidase inhibitor apocyanin and other free radical scavengers and antioxidant vitamins has attenuated hypertension in many models of hypertension, like mineralocorticoid-induced, angiotensin II-induced, obesity-induced, and hypertension in Dahl salt-sensitive rats, further evidencing the role of superoxides in hypertension development. Considering the fact that the renin-angiotensin system is an important pathway in various hypertension pathophysiological development and is the main pathway activating the NADPH oxidase, ACE inhibitors and angiotensin II receptor type 1 (AT1 receptor) antagonists have also been implicated in treating hypertension, and it has actually shown reduction in NADPH oxidase activity [66] and reduction in superoxide levels [67].

19.3.3 Atherosclerosis

The idea that oxidative stress is actually linked to hypertension and CVD as a whole had emerged from the fact that people having risk of developing CVD, for example, smokers [68], hypercholesterolemics, [69, 70] homocysteinemics, and diabetics, show elevated levels of the oxidative stress marker isoprostenes and also exhibit endothelial dysfunction. Numerous experiments conducted hold proof that ROS has to play a significant role in the development of atherosclerosis. All the components which comprise the atherosclerotic plaques have been shown to produce ROS at alarmingly high rates. NADPH oxidase activity, [71] uncoupling of eNOS, the renin-angiotensin system, [71, 72] and macrophages are the main sources identified which produce ROS in atherosclerotic disease models, apart from the classical mitochondrial ROS. In one experiment, cholesterol-fed rabbits showed superoxide production and low bioavailability of NO, [73] which was ameliorated by the administration of SOD mimetics [74] and L-arginine supplementation [75]. In another experiment, superoxide levels were found to be elevated in Watanabe heritable hyperlipidemic rabbits, [76] giving evidence for the involvement of ROS in hyperlipidemia as an upstream or downstream component. Also, ApoE (protein mediating cholesterol metabolism) and LDL receptor knockout mice showing atherosclerosis showed high levels of isoprostanes, which are considered as biomarkers for systemic oxidative stress [77, 78]. In yet another experiment, where CD-36 (oxidized LDL receptor)- and 12/15 lipoxygenase (enzyme effecting lipid peroxidation)-deficient mice were crossed with ApoE–/– mice, the progeny showed significantly smaller atherosclerotic lesions, providing insights into the role of macrophage-derived ROS in atherogenesis [79, 80]. The presence of fibrous tissue is a prominent feature of the atherosclerotic plaques, and ROS has been found to be crucial in modulating the matrix metalloproteinase (MMP) activity as discussed in the earlier section. The shoulder areas of atherosclerotic plaques show an overexpression of MMP-2 and MMP-9 [81, 82].

19.3.4 Heart Failure

Although the exact mechanisms which underlie heart failure are not completely understood yet, significant clinical and experimental evidences point to the fact that oxidative stress plays a major role in it. Oxidative stress has been found to have direct implications upon the conditions leading to heart failure such as cardiac hypertrophy and myocardial infarction.

19.3.4.1 Cardiac Hypertrophy

It is initially an adaptive process and minimizes wall stress, as an in vivo reaction to increased load initiated by a wide variety of stimuli, including mechanical stretch and hormones like angiotensin II, noradrenaline, and endothelin-I. However, continuous and maladapted responses have been shown to lead to heart failure. A part of the hypertrophic response in vascular smooth muscle cells (VSMCs) is due to

ROS, and the role of the $p22_{phox}$ subunit of NADPH oxidase has also been demonstrated. ROS levels were found to increase during the progression from hypertrophy to heart failure, which could be attenuated and rate of progression delayed with the treatment with vitamin E [83]. The role of ROS in cardiac hypertrophy has also been studied in in vitro models where hypertrophy induced by TNF-α and angiotensin II in cultured cardiac myocytes could be inhibited by the treatment with vitamin E and catalase [84]. In a study with experimental guinea pig model of pressure overload, the progression to heart failure was marked by left ventricular hypertrophy along with increase in levels of NADPH oxidase and increased expression of the $p22_{phox}$, $p47_{phox}$, $p67_{phox}$, and $gp91_{phox}$ subunits, besides diastolic dysfunction and inactivation of NO [85]. Also in an aldosterone infusion-induced cardiac hypertrophy model, the NADPH oxidase inhibitor apocyanin was able to block this response and caused reduced expression of $p22_{phox}$ [86].

It is also an established fact that excess interstitial fibrosis is a pathological symptom of hypertrophy, which has deleterious effects on cardiac function. It is interesting to note here that ROS is a major regulator of fibroblast collagen synthesis and causes increase in matrix metalloproteinase activity, thus causing distorted modeling unfavorable for the body [87]. That the superoxides have a role in fibrosis is also evidenced by the fact that fibrosis was absent in $gp91_{phox}$ knockout mice infused with angiotensin II, which implied a role of NADPH oxidase-derived superoxides in mediating the profibrotic effects of angiotensin II [88]. Increased NADPH oxidase has been directly reported in failing human hearts [89].

19.3.4.2 Myocardial Infarction and Ischemia-Reperfusion Injury

Myocardial infarction is the leading cause of heart failure, which causes either acute or chronic heart failure by a series of processes together known as cardiac remodeling. This is marked by a series of alterations in the cardiac structure, geometry, and volume which produce adverse effects on cardiac functions and output [54]. Large amount of ROS production is associated with reperfusion of the ischemic myocardium. There is also a decrease in antioxidant defense system activity with a concomitant increase in lipid peroxidation, [90] all leading to deleterious cardiac dysfunction. Malondialdehyde (MDA), which is a well-known marker of oxidative stress, shows marked increase along with decrease in catalase and SOD levels in hearts exposed to ischemia for 30 min [91]. Interestingly, separately conducted studies showed that in myocardial infarction models, the infarct size showed reduction and overall cardiac function improved upon treatment with SOD and catalase [92, 93]. Among the three forms of SOD (Mn, Zn, and Cu), Mn-SOD was shown to decrease ischemia-reperfusion injury in mice, [94] and when both Cu-SOD and Zn-SOD were disrupted, infarct sizes had increased and the recovery of contractile function after repeated ischemic periods was impaired [95]. Not surprisingly, when rats with myocardial infarction were treated with vitamin E, the infarct sizes reduced and levels of oxidative stress markers attenuated [96]. Oxidative stress not only leads to pathogenesis of myocardial infarction, but also affects the remodeling process post the infarction. Involvement of ROS in this process is clearly depicted by the results of studies where left ventricular dilatation and contractile impairment in a mouse model of myocardial infarction were attenuated by a superoxide radical scavenger dimethylthiourea (DMTU), [97] and an antioxidant probucol positively influenced the cardiac remodeling process in rats with myocardial infarction [98] by reducing cardiac fibrosis and preventing left ventricular dilatation thus improving left ventricular function, besides preventing wall thinning and increasing the thickness of the scar. Excess extracellular matrix deposition occurs in non-infarcted areas during remodeling. Nonetheless, ROS has effects on matrix metalloproteinases, [99] as has been discussed in earlier sections, so its role in infarction and reperfusion injury is further solidified.

But what is interesting to note here is that the signaling must be working both ways, i.e., it is self-regulatory. ROS has a role in the development of myocardial infarction, but at the same time, myocardial infarction and cardiac remodeling that follows produce greater amounts of ROS in a vicious cycle. This could be inferred from an experiment where cardiac inhibitor of metalloproteinase (CIMP) reduced NADPH oxidase activity besides decreasing matrix metalloproteinase activity and dilatation of left ventricles as expected [100].

19.4 Role of Mitochondrial Dysfunction and Autophagy in CVD

As already discussed above, mitochondrial dysfunction is the source of ROS and elevated ROS further causes mitochondrial damage, thus creating a vicious loop of mitochondrial destruction and breakdown of the cellular mechanisms as a whole. A damaged mitochondrion not only produces less ATP but also produces high amounts of ROS and thereby increases the propensity of the cell to undergo apoptosis [101]. All these phenomena are related to cardiac aging and subsequently cardiovascular disease development [102].

Continuous fusion and fission cycles regulate the function and morphology of mitochondrial networks. This is important as the basic functions of the cell which encompasses proper functioning of metabolic pathways, redistribution of proteins and metabolites, maintenance of mitochondrial DNA integrity, determination of organellar shape and integrity and transmittance of redox-sensitive signals, and quality control and cell death pathways are crucially regulated by these processes [103, 104]. The functional and morphological behavior of the mitochondrial networks thus largely determines the cellular, tissue, and organ bioenergetics [105] and, besides, imparts properties characteristic of complex systems, like redundancy of function, robustness, and plasticity, providing the cell the flexibility to adapt to the changing stresses and metabolic demands [106].

Mitochondria and autophagy form a quality control axis. When mitochondria are damaged, their functionality could be complemented and possibly even restored by their fusion with the neighboring intact mitochondria. But when they are severely damaged, they get separated from the mitochondrial network by fission and they are eventually eliminated by mitophagy [107]. The dysfunction of this quality control axis is regarded as a contributing factor to cardiovascular aging leading to cardiovascular diseases [102].

The fission-fusion balance in mitochondria is maintained by a set of complex dedicated machinery, among which some are widely known and well-characterized in mammals. These include the fusion proteins mitofusin 1 and mitofusin 2 (Mfn1 and Mfn2) [108, 109] and the fission proteins dynamin-related protein 1 (Drp1) [110] and fission protein 1 (Fis1) [111]. To elucidate the role of mitochondria on cardiovascular diseases, several studies have been conducted on vascular smooth muscle cells (VSMCs), neonatal cardiomyocytes, and primary vascular endothelial cells, [112] that is, some of the most rapidly metabolizing cells where the mitochondria undergo continuous fission-fusion processes while being arranged into a filamentous network within the cells. That mitochondrial dynamics might play a role in cardiac pathophysiology can be backed by the fact that excessive fission and/or reduced fusion has been found to have detrimental effects on ischemia, ischemic-reperfusion injury, [113] and heart failure, [114] besides diabetes [115] and hyperglycemia [116]. Several cardiovascular events rely on the disruption of this delicate balance and quality control checkpoints [117]. These include cardiac development and differentiation, [118] cardiomyocyte hypertrophy, [119] stem cell differentiation, [120] VSMC proliferation, [121] and myocardial ischemia-reperfusion injury [112, 113, 122].

In studies with ischemic-reperfusion injury models, it has been found that mitochondrial fission inhibition has cardioprotective effects [113]; in the same grounds it has also been found that the mitochondrial fusion proteins Mfn1, Mfn2, and OPA1 (mitochondrial dynamin-like GTPase) are not only required for proper health and functioning of mitochondria, but also their optimum expression aids in prevention of cardiac hypertrophy and heart failure [123–126]. Paradoxically, Mfn2, apart from mediating mitochondrial fusion, interacts with Bax and Bak to mediate apoptosis, [127, 128] is involved in mitophagy by acting as a substrate of Parkin which is a mitophagy-related protein, [129] and also acts as a tether between the mitochondria and endoplasmic reticulum [130] effecting the formation of the filamentous mitochondrial network. But contrary to its other functions, Mfn2 inhibits the proliferation of vascular smooth muscle cells under a variety of vascular proliferative conditions [131, 132] and induces oxidative stress-mediated apoptosis in these cells [133]. The understanding of the role that mitochondrial dynamics does play behind the pathogenesis of cardiovascular diseases is still in the preliminary stage. This area needs further research to develop this as a prospective target for treatment of cardiovascular diseases and related pathophysiological conditions [134].

As discussed above, autophagy is an evolutionarily conserved process which aids in the adjustment and survival of the cell under stress conditions. The process either could be selective, where protein aggregates and damaged or superfluous organelles are specifically targeted and removed from the cell, operating even in nutrient-rich conditions, or could be nonselective, which is primarily encountered in the classic definition of autophagy, where the process is activated in response to starvation and nutrient stress and serves to provide the cell with adequate nutrients necessary for survival by selective self-digestion. Each of the cargo-specific autophagic processes discovered in animals so far have been named after the specific organelle that is removed in selective autophagy. Like when peroxisomes are removed, it is termed as peroxophagy; when ribosomes are removed, it is ribophagy; when aggregates are removed, it is termed as aggrephagy; and it is ERphagy when the endoplasmic reticulum is degraded and removed or recycled [135]. The term for when damaged mitochondria are removed by this process is termed as mitophagy [136].

Mitophagy is an essentially important process as it serves as one part of the quality control measures taken by the cell to match the demand-supply ratio of cellular energy, along with removing the damaged mitochondria after they have undergone fission and segregation, thus also providing a cleanup quality check [103]. The trigger for mitophagy is provided by the loss of the $\Delta\psi_m$, that is, the mitochondrial membrane potential, although the opening of the mitochondrial permeability transition pores (mPTP) provides a major trigger for the initiation of selective removal of damaged mitochondria by autophagy [137]. The mPT pore opening dissipates the mitochondrial membrane potential [138] that leads to a chain of reactions mediated by the mitochondrial fission proteins, culminating in the fission of the damaged mitochondria and their segregation from the filamentous network of which they are a part and their subsequent take-up by the autophagolysosomes. Although very little is understood as of yet of the complete molecular regulation of mitophagy and mitochondrial biogenesis, the mTOR/AMPK pathway seems to be a major regulator of both the processes [139]. It stimulates the removal of damaged mitochondria by mitophagy as well as aids in mitochondrial biogenesis by enhancing the activity of sirtuin 1 (Sirt1) and its downstream target peroxisome proliferator-activated receptor coactivator 1 (PGC-1) [140].

Autophagy is a part of the vicious trio of pathogenesis of cardiovascular diseases along with ROS and mitochondrial dysfunction. If autophagy is experimentally inhibited, it results in the non-removal and subsequent accumulation of bioenergetically inefficient and damaged and dysfunctional mitochondria, leading to malfunction of cellular energetics unit and further ROS generation [101, 141, 142]. Autophagy is regulated by the Atg family of genes, and in a study where cardiac cell-specific knockdown of the Atg5 gene was done, it led to structural disorganization in sarcomere, reduced fractional shortening, left ventricular hypertrophy, and, most importantly, accumulation of damaged and dysfunctional mitochondria and resulting respiratory defects [143]. As discussed previously, it has been proposed that mitochondrial dysfunction could contribute to cardiomyocyte injury during ischemic-reperfusion by depletion of energy due to opening of mPT pores [144]. But among all these events, autophagy too is induced during ischemic-reperfusion to ensure cell survival by sourcing enough substrates to obtain energy from, during the time of reperfusion stress caused energy depletion [145]. This has been experimentally proven by inhibiting autophagy in models by pharmacological (using autophagy inhibitors wortmannin or 3-methyladenine) or genetic intervention (knockdown of beclin1 by RNAi or overexpression of dominant negative Atg5 gene) methods which resulted in cardiomyocyte death by apoptosis in ischemic-reperfusion simulated models. This condition could be reversed and cell survival is ensured by autophagy upregulation mediated by beclin1 overexpression or rapamycin treatment [146].

The role of both autophagy and diabetes in cardiovascular disease development can be reinforced from noting that the accumulation of dysfunctional mitochondria as a result of inefficient autophagy in the cardiac tissue has been proposed to be a major cause of development of diabetic cardiomyopathy. In diabetic mice, reduced AMPK activity and concomitant downregulation of cardiac autophagy have been recorded [147]. The ultrastructure of cardiac tissue in these models showed aberrant mitochondria with disrupted structure. The study was further extended to inquire whether AMPK downregulation was really the contributing factor in autophagy-mediated diabetic cardiomyopathy, and it was found that indeed, AMPK inhibition by a cardiac-specific dominant negative AMPK gene overexpression reduced autophagy even further, worsened the cardiac aberrations, exacerbated cardiac dysfunction, and caused mortality in diabetic mice. All of the above conditions were shown to be ameliorated upon metformin treatment, a widely known diabetes drug. Autophagy was enhanced while preserving the cardiac function and output. These results were not obtained from the diabetic rat group which overexpressed the dominant negative AMPK gene thus inhibiting autophagy. So, it was clear that the cardioprotective effect of metformin was achieved by the AMPK-mediated upregulation of autophagy.

While all this time we have discussed the positive effects of autophagy on the cardiac tissue ensuring its good health, there have been studies where negative impact of autophagy on cardiac tissue has also been recorded. It has been already discussed in the earlier sections of this chapter that excessive or deregulated autophagy led to cellular death. When diphtheria toxin was injected intramuscularly into a heart failure model comprising of transgenic mice overexpressing cardiac-specific diphtheria toxin receptor, the cardiomyocytes showed morphological signs of autophagic cell death and a degenerated heart phenotype [148]. Moreover, research done by another group of scientists also showed that upregulation of autophagy during reperfusion does more harm than good; it is maladaptive and increases the infarct size in cardiomyocytes [145]. So it could be concluded that depending upon the setting and conditions, autophagy can either be beneficial for cardiomyocyte survival or may sometimes cause their destruction.

With the growing age, mitochondrial dysregulation increases not only in the heart but also in the vascular endothelium as well as the vascular smooth muscle cells. With that, the mitochondrial ROS production increases owing to ETC disruption, elevated NADPH oxidase 4 activity, and declining antioxidant defense capacity, thus affirming the high probability of aged animals falling prey to cardiovascular diseases. Young animals are able to mount a powerful antioxidant defense driven by Nrf-2 (NF-E2-related factor 2)-mediated activation of antioxidant response elements or ARE transcription upregulation. This response is attenuated with growing age; thus older animals tend to have weaker antioxidant defense and concomitant oxidative stress resulting in a plethora of cardiovascular diseases [149]. But this trend is already changing. A significantly higher percentage of deaths resulting from cardiovascular diseases also happen to occur in the youth age group (35–64 years) in the developing countries as compared to developed ones. For example, in 2004, Leeder et al. reported that in the said age group, the proportion of cardiovascular disease deaths is 41% in South Africa, 35% in India, and 28% in Brazil, compared to only 12% in the United States and 9% in Portugal. So this area needs further research to find out the exact molecular mechanisms involved for developing efficient therapeutic strategies.

There also happens to be a causal relationship which could exist between insulin-like growth factor 1 (IGF-1) and the development of cardiovascular diseases via protection of mitochondria [150, 151]. It is interesting to note that in vitro treatment of cardiomyocytes as well as endothelial cells with IGF-1 decreased mitochondrial peroxide production [152]. Also the endothelial cells retained their mitochondrial membrane potential and inner cytochrome C and also reduced the exposure to caspase-3 upon H_2O_2 exposure [153]. It was even found that when IGF-1 was overexpressed in mice, they were protected from increase in ROS generation induced by feeding a high-fat diet, hence preventing mitochondrial damage [154]. Contrary to that, in Ames dwarf mice, low circulating levels of IGF-1 were associated with an increased production of mitochondrial ROS, both in the vasculature and the myocardium, [152] thus mimicking the aging phenotype (Fig. 19.5). There have also been studies which have demonstrated that angiotensin II in the cardiovascular system is critically mediated by mitochondrial ROS while mediating the cellular effects [155, 156]. As has been proposed, angiotensin II binds to angiotensin receptor 1

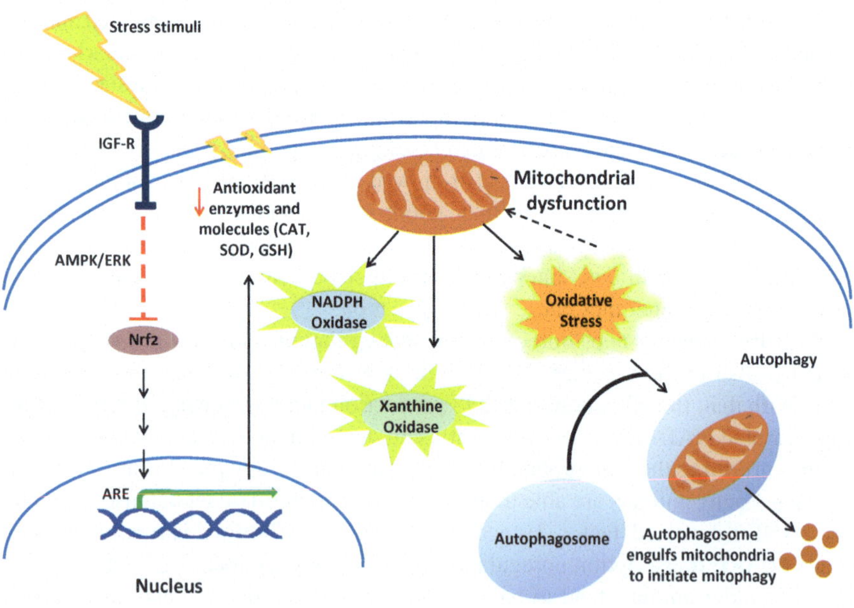

Fig. 19.5 Representative image of the vicious cycle that occurs inside a cell under stress conditions, mediated by oxidative stress and mitochondrial dysfunction and culminating in autophagy. Excessive autophagy leads to cell death

(ATR1), thus activating NADPH oxidases 2 and 4 (NOX2 and NOX4), further leading to an increase in the mitochondrial ROS production in both vascular endothelial cells and VSMCs as well as in the cardiac myocytes [136, 157, 158].

19.5 Amelioration of CVD by Small Natural Molecules

CVD had been treated symptomatically for a long time for angina and coronary blockages. Relief of symptoms and prevention of events which could lead to heart failure have been long standing for decades and are still in vogue. Even the early Romans used foxglove as treatment for heart failure, and leeches and letting out blood were also used historically. Until Sir William Harvey discovered circulation of blood in 1628, the understanding of the processes could not have been extant, and the discovery of X-rays by Röntgen and Einthoven's discovery of echocardiogram in the 1890s paved the way for investigation of heart diseases. In the early twentieth century, diuretics were developed, but the early mercurial agents conferred substantial toxicity. The later diuretics like thiazides were developed to combat that. Next came the angiotensin-converting enzyme inhibitors and the vasodilators [159]. With the advent of good surgical intervention techniques, stents and angioplasty and bypass surgeries became common. But as science progressed, the search for the root cause of cardiovascular diseases started and, as of today, has made quite a progress. Eliminating the root cause would probably be the best option for treatment against such a deadly disease with so many facets, and as toxicological studies report, molecules sourced from the nature itself would be the best option for executing this [160]. As has been discussed in the chapter, sources of oxidative stress, mitochondrial dysfunction, and autophagy mediated mainly by the AMPK pathway have remained the favorite targets of drug developers. The given table (Table 19.1) jots down some of the small natural molecules tested in this regard and their proposed mechanisms of action.

19.6 Conclusion

Cardiovascular diseases, being a deadly killer group, has needed and will continue to need serious attention for their treatment, increasing the options available while reducing the cost, and to successfully decrease the mortality rates. Deaths from CVD have no doubt declined since the last few years, but treatment costs and dearth in availability are still raging problems in developing countries, where age-adjusted mortality rates are quite high. There is a huge amount of scientific literature on this topic, but what we have really understood about the root causes of the diseases which fall in this category seems far from enough. The oxidative stress theory of aging [161] had given an idea about the role of free radicals, mainly superoxides, in cardiovascular aging. Since the diseased heart and vascular cells manifest signs of aging, it was hypothesized that oxidative stress might have some role to play in the development of CVD. When oxidative stress is an alleged cause, mitochondrial

Table 19.1 Amelioration of CVD by natural molecules

Bioactive molecule	Source	Mechanism	Restoration of normal physiology	Model
Cryptotanshinone (CTS) [162]	Danshen	1. Blocking LOX-1-mediated signaling pathway 2. Suppressing the ROS/NF-κB signaling pathway	1. Prevents formation of *atherosclerotic lesion* 2. Reduces ROS generation 3. Reduces serum level of IL-1β, IL-6, IL-17A, IFN-γ, and TNF-α 4. Decreased accumulation of macrophages 5. Reduce LOX-1 expression	*ApoE*−/− mice
Rutaecarpine [163, 164]	*Evodia rutaecarpa* (Chinese herb)	1. Inducing overexpression of ABCA1 and SR-BI/CLA-1 within RCT (reverse cholesterol transport) 2. ABCG1 overexpression	1. Promotes cholesterol efflux and inhibits lipid accumulation in vitro 2. *Atherosclerotic lesion* attenuation in aortic sinus 3. Reduces macrophage activity 4. Lowering TC, HDL-C, TG, and LDL-C 5. Decreases production of ROS and activation of NADPH oxidase	*ApoE*−/− mice
Palmitoleate (PAO) [165]	Breast milk, animal fats, vegetable oils, marine oils, macadamia oil, and sea buckthorn oil	1. Blocking of lipid-induced NLRP3 inflammasome formation and inflammation in macrophages 2. Remodeling of ER membranes 3. Decrease in high-sensitivity C-reactive protein (human)	1. Altered composition of plaque in lesions and prevents *atherosclerosis* development in mice 2. Prevented ER stress in both mice and human 3. Improved lipid levels in serum (human)	*ApoE*−/− mice and *human*
Isorhamnetin (Iso) [166] (flavonoid)	Chinese herb *Hippophae rhamnoides L.*	1. Activating the PI3K/AKT pathway and elevates HO-1 expression 2. Inhibiting MPO activity and elevating GSH-px activity 3. Inhibiting ox-LDL-induced ROS generation and reducing NOX activity in macrophages	1. Decreased caspase-3 expression and progression of macrophage-induced apoptosis in *atherosclerotic lesions* 2. Decreased intracellular lipid deposition in ox-LDL-induced macrophages 3. Prevented cell apoptosis in THP-1 macrophages	Male C57BL/6J mice

Catechin [160, 167, 168] (flavonoid)	Tea, apples, berries, and cocoa	1. Scavenging ROS and modulating ROS-generating enzymes (iNOS and XO) 2. Improving the ratio of ApoA-1/ApoB by reducing ApoB 3.Diminishing cholesterol absorption by formation of insoluble co-precipitates of cholesterol and reducing bile acid-mediated micellar solubility	1. Decreased *coronary heart disease* mortality 2. Decreased blood pressure in *cardiac stroke*-prone hypertensive rats 3. Reduced oxidative stress 4. Decreased blood cholesterol level and prevented its accumulation in heart and liver 5. Reduced high risk of CVD 6. Prevented development and progression of *atherosclerosis*	C57BL/6J and *ApoE−/−* mice, rats, and human
Quercetin [160, 169] (flavonoid)	Tea, onion, wines, apples	1. Scavenging free radicals and have chelating effect on transition metal ions 2. Inhibiting phospholipase A2 activity 3. Inhibition of amino acid metabolism pathways (LOX and COX) 4. Modulating NF-κB gene expression 5. Reducing the activity of HMG-CoA and elevating cholesterol acyl transferase activity in liver	1. Decreased platelet accumulation and superoxide production and elevated platelet-derived NO release 2. Prevented *ischemic-reperfusion injury* 3. Decrease plasma cholesterol level 4. Prevented *atherosclerotic plaque* formation 5. Reduced lipid peroxidation level	Rat
Naringenin [160, 170, 171] (flavonoid)	Grapefruit and orange	1. Increasing activity of nonenzymatic antioxidants and antioxidant enzymes 2. Irreversible inhibition of both LOX and COX pathways 3. Inhibiting HMG-CoA reductase activity 4. Suppressing mRNA expression of caspases	1. Diminished lipid peroxidation level and oxidative stress 2. Reduced ROS generation 3. Prevented *cardiomyopathy* 4. Reduced risk of ischemic stroke	Mice, rat, rabbit

(continued)

Table 19.1 (continued)

Bioactive molecule	Source	Mechanism	Restoration of normal physiology	Model
Anthocyanins [172–175] (flavonoid)	Berries, cherry, eggplant peel, ube, black rice, Okinawan sweet potato, Concord grape, red cabbage, etc.	1. Inhibiting NF-kB and TNF-α pathways 2. Sequestration of trace elements and inhibiting enzymes involved in oxidative stress 3.Activating AMPK ➔ inhibiting HMG-CoA reductase ➔ lowering cholesterol synthesis	1. Reduced inflammation 2. Suppressed ROS generation 3. Decreased cholesterol level and thus lowered risk of *atherosclerosis* 4. Decreased fatty acid synthesis and enhanced its oxidation	Mice, human
Omega-3 fatty acids (N-3 FA) [176]	Seafood (mackerel, salmon, tuna, herring and sardines) Seeds (flaxseed, chia seed, walnuts) Eggs, yoghurt, juices, milk, soy beverages	1. Decreasing production of proinflammatory mediators by substrate competition 2. Inducing fatty acid β-oxidation through activation of PPARα 3. Inhibiting de novo lipogenesis by reducing the expression of acetyl-CoA carboxylase (ACC) and fatty acid synthase (FAS)	1. Reduced inflammation 2. Lowered risk of CVD and mortality 3. Lowered *atherosclerotic plaque* inflammation and instability 4. Decreased lipid deposition and ROS generation 5.Reduced triacylglycerol (TAG) synthesis➔ decreased VLDL synthesis➔ hypotriglyceridemic effect	Human
Resveratrol [177, 178] (polyphenolic)	Peanuts, pistachios, grapes, red and white wine, blueberries, cranberries, cocoa and dark chocolate	1. Free radical scavenging ➔ blocking LDL oxidation 2. Upregulation of NF-κB in inflammatory cells 3. Activating eNOS, Nrf2, SIRT-1, and ARE and inhibiting TNF-α production 3. Inhibiting HMG-CoA reductase	1. Protected from *atherosclerosis, coronary heart disease, heart failure, cardiomyopathy* 2. Enhanced endothelial cell growth 3. Regulated circulation in *heart failure* 4. Prevented ROS generation 5. Decreased plasma level of TG, LDL-cholesterol level 6. Improved lipid profile	Rat, mice, and human
Myricitrin [179]	*Myrica rubra*	1. Inhibiting ox-LDL formation and downregulation of LOX-1 2. Activation of PI3K/Akt pathway ➔ increasing HO-1 expression via Nrf-2 activation	1. Reduced aortic lesion size and prevented *atherosclerotic plaque formation* 2. Diminished endothelial apoptosis and increased cell survival 3. Reduced occurrence of *cardiomyopathy*	*ApoE−/−* mice

Vitamin C and E [180]	Vitamin E: vegetable oils, nuts, sunflower seeds, spinach and broccoli, mango, kiwifruit, abalone, salmon, etc. Vitamin C: broccoli, brussels sprouts, cauliflower, spinach, cabbage, sweet and white potatoes, tomatoes	1. Suppressing protein kinase C and inhibiting IkB degradation ➔ inhibiting ox-LDL-induced NF-κB activation 2. Reducing expression of adhesion molecules 3. Reducing chemokine secretion and CD36 on macrophages and diminishing monocyte recruitment 4. Decreasing expression of intercellular adhesion molecule-1 gene 5. Prevention of apoptosis of vascular smooth muscle cell	1. Diminished inflammatory responses that cause CVD 2. Prevented foam cell formation and reduced *atherosclerotic plaque* formation 3. Prevented monocyte attachment to endothelium 4. Enhanced endothelial cell proliferation	
Ethanolic extracts [181]	*Citrus medica* "Otroj"	Unknown	1. Diminished serum levels of TC, TG, LDL, and VLDL 2. Significant decrease in MDA contents and increased the NP-SH and TP levels in heart muscle 3. Diminished serum level of cholesterol 4. Prevented ROS generation 5. Improved *cardiomyopathy* conditions	Male Wistar albino rat

The table above gives an idea of some of the small natural molecules tested for treatment of cardiovascular diseases and related complications. The mechanism(s) of action of these natural compounds which have been proposed for each one of these have also been given. *NP-SH* nonprotein sulfhydryl, *MDA* malondialdehyde, *TG* triglycerides, *MPO* myeloperoxidase, *GSH-px* glutathione peroxidase, *HO-1* heme oxygenase, *ApoA-1* apolipoprotein, *ApoB* apolipoprotein B, *HMG-CoA* 3-hydroxy-3-methylglutaryl-coenzyme A, *ox-LDL* oxidized low-density lipoprotein, *iNOS* inducible nitric oxide synthase, *eNOS* endothelial nitric oxide synthase, *CVD* cardiovascular disease dysfunction and autophagy cannot stay far behind. It has indeed been proven again and again from multiple experiments conducted by research groups across the world that oxidative stress, mitochondrial dysfunction, and autophagy form a vicious trio that mediate the development of various cardiac pathophysiologies, with the ERK1/2, BMK, and AMPK signaling cascades in work via suppression of antioxidant defenses.

History tells that cardiac problems have been occurring since the ancient ages. The Greeks, Indians, and Romans have taken attempts in treating them [159]. With subsequent discoveries of symptoms, diagnostic tools, interventions, molecular mechanisms, and biomarkers, treatment of cardiovascular diseases has come a long way, but still a longer path remains to be trodden. Nature has provided us with a plethora of chemical substances – magical molecules that act effectively without any significant toxicity to the healthy organs, to restore the normal physiology from a diseased state. So this field calls for further research in order to restrict this fatal epidemic within the human clutches.

References

1. Kelly BB, Fuster V (2010) Promoting cardiovascular health in the developing world: a critical challenge to achieve global health. National Academies Press, Washington, DC
2. Beaglehole R, Bonita R (2008) Global public health: a scorecard. Lancet 372(9654):1988–1996
3. Prabhakaran D, Jeemon P, Roy A (2016) Cardiovascular diseases in India: current epidemiology and future directions. Circulation 133(16):1605–1620
4. Farrugia G, Balzan R (2012) Oxidative stress and programmed cell death in yeast. Front Oncol 2:64
5. Matés JM, Segura JA, Alonso FJ, Márquez J (2012) Oxidative stress in apoptosis and cancer: an update. Arch Toxicol 86(11):1649–1665
6. Lakshmi SV, Padmaja G, Kuppusamy P, Kutala VK (2009) Oxidative stress in cardiovascular disease. Indian J Biochem Biophys 46(6):421–440
7. Sinha K, Das J, Pal PB, Sil PC (2013) Oxidative stress: the mitochondria-dependent and mitochondria-independent pathways of apoptosis. Arch Toxicol 87(7):1157–1180
8. Cabiscol Català E, Tamarit Sumalla J, Ros Salvador J (2000) Oxidative stress in bacteria and protein damage by reactive oxygen species. Int Microbiol 3(1):3–8
9. Biliński T, Litwińska J, Błszczyński M, Bajus A (1989) Superoxide dismutase deficiency and the toxicity of the products of autooxidation of polyunsaturated fatty acids in yeast. Biochim Biophys Acta 1001(1):102–106
10. Yakes FM, Van Houten B (1997) Mitochondrial DNA damage is more extensive and persists longer than nuclear DNA damage in human cells following oxidative stress. Proc Natl Acad Sci 94(2):514–519
11. Matés JM, Segura JA, Alonso FJ, Márquez J (2010) Roles of dioxins and heavy metals in cancer and neurological diseases using ROS-mediated mechanisms. Free Radic Biol Med 49(9):1328–1341
12. Halliwell B, Cross CE (1994) Oxygen-derived species: their relation to human disease and environmental stress. Environ Health Perspect 102(Suppl 10):5
13. Sies H (1991) Oxidative stress: from basic research to clinical application. Am J Med 91(3):S31–S38
14. Turrens JF, Boveris A (1980) Generation of superoxide anion by the NADH dehydrogenase of bovine heart mitochondria. Biochem J 191(2):421–427

15. Ueda S, Masutani H, Nakamura H, Tanaka T, Ueno M, Yodoi J (2002) Redox control of cell death. Antioxid Redox Signal 4(3):405–414
16. Bai H, Hai C, Xi M, Liang X, Liu R (2010) Protective effect of maize silks (Maydis stigma) ethanol extract on radiation-induced oxidative stress in mice. Plant Foods Hum Nutr 65(3):271–276
17. Altraja S, Mahlapuu R, Soomets U, Altraja A (2013) Cigarette smoke-induced differential regulation of glutathione metabolism in bronchial epithelial cells is balanced by an antioxidant tetrapeptide UPF1. Exp Toxicol Pathol 65(6):711–717
18. Ahmad I, Kumar A, Shukla S, Prasad Pandey H, Singh C (2008) The involvement of nitric oxide in maneb- and paraquat-induced oxidative stress in rat polymorphonuclear leukocytes. Free Radic Res 42(10):849–862
19. Morcillo P, Esteban MÁ, Cuesta A (2016) Heavy metals produce toxicity, oxidative stress and apoptosis in the marine teleost fish SAF-1 cell line. Chemosphere 144:225–233
20. Pal S, Pal PB, Das J, Sil PC (2011) Involvement of both intrinsic and extrinsic pathways in hepatoprotection of arjunolic acid against cadmium induced acute damage in vitro. Toxicology 283(2–3):129–139
21. Pal PB, Pal S, Das J, Sil PC (2012) Modulation of mercury-induced mitochondria-dependent apoptosis by glycine in hepatocytes. Amino Acids 42(5):1669–1683
22. Pal PB, Sinha K, Sil PC (2013) Mangiferin, a natural xanthone, protects murine liver in Pb (II) induced hepatic damage and cell death via MAP kinase, NF-κB and mitochondria dependent pathways. PloS one 8(2):e56894
23. Manna P, Roy A, Sil PC (2009) Prophylactic role of taurine on arsenic mediated oxidative renal dysfunction via MAPKs/NF-B and mitochondria dependent pathways. Free Radic Res 43(10):995–1007
24. Saha S, Rashid K, Sadhukhan P, Agarwal N, Sil PC (2016) Attenuative role of mangiferin in oxidative stress-mediated liver dysfunction in arsenic-intoxicated murines. Biofactors 42(5):515–532
25. Das J, Ghosh J, Manna P, Sil PC (2011) Taurine suppresses doxorubicin-triggered oxidative stress and cardiac apoptosis in rat via up-regulation of PI3-K/Akt and inhibition of p53, p38-JNK. Biochem Pharmacol 81(7):891–909
26. Das J, Ghosh J, Manna P, Sil PC (2012) Taurine protects rat testes against doxorubicin-induced oxidative stress as well as p53, Fas and caspase 12-mediated apoptosis. Amino Acids 42(5):1839–1855
27. Das J, Sil PC (2012) Taurine ameliorates alloxan-induced diabetic renal injury, oxidative stress-related signaling pathways and apoptosis in rats. Amino Acids 43(4):1509–1523
28. Das J, Ghosh J, Roy A, Sil PC (2012) Mangiferin exerts hepatoprotective activity against D-galactosamine induced acute toxicity and oxidative/nitrosative stress via Nrf2–NFκB pathways. Toxicol Appl Pharmacol 260(1):35–47
29. Banerjee S, Sinha K, Chowdhury S, Sil PC (2018) Unfolding the mechanism of cisplatin induced pathophysiology in spleen and its amelioration by carnosine. Chem Biol Interact 279:159–170
30. Das J, Ghosh J, Manna P, Sil PC (2010) Acetaminophen induced acute liver failure via oxidative stress and JNK activation: protective role of taurine by the suppression of cytochrome P450 2E1. Free Radic Res 44(3):340–355
31. Ghosh J, Das J, Manna P, Sil PC (2011) The protective role of arjunolic acid against doxorubicin induced intracellular ROS dependent JNK-p38 and p53-mediated cardiac apoptosis. Biomaterials 32(21):4857–4866
32. Bhattacharya S, Gachhui R, Sil PC (2011) Hepatoprotective properties of kombucha tea against TBHP-induced oxidative stress via suppression of mitochondria dependent apoptosis. Pathophysiology 18(3):221–234
33. Sarkar MK, Sil PC (2010) Prevention of tertiary butyl hydroperoxide induced oxidative impairment and cell death by a novel antioxidant protein molecule isolated from the herb, Phyllanthus niruri. Toxicol in Vitro 24(6):1711–1719
34. Gille G, Sigler K (1995) Oxidative stress and living cells. Folia Microbiol 40(2):131–152

35. Ghosh M, Das J, Sil PC (2012) D (+) galactosamine induced oxidative and nitrosative stress-mediated renal damage in rats via NF-κB and inducible nitric oxide synthase (iNOS) pathways is ameliorated by a polyphenol xanthone, mangiferin. Free Radic Res 46(2):116–132
36. Manna P, Sinha M, Sil PC (2009) Protective role of arjunolic acid in response to streptozotocin-induced type-I diabetes via the mitochondrial dependent and independent pathways. Toxicology 257(1–2):53–63
37. Manna P, Das J, Ghosh J, Sil PC (2010) Contribution of type 1 diabetes to rat liver dysfunction and cellular damage via activation of NOS, PARP, IκBα/NF-κB, MAPKs, and mitochondria-dependent pathways: prophylactic role of arjunolic acid. Free Radic Biol Med 48(11):1465–1484
38. Bhattacharya S, Manna P, Gachhui R, Sil PC (2013) D-Saccharic acid 1, 4-lactone protects diabetic rat kidney by ameliorating hyperglycemia-mediated oxidative stress and renal inflammatory cytokines via NF-κB and PKC signaling. Toxicol Appl Pharmacol 267(1):16–29
39. Reuter S, Gupta SC, Chaturvedi MM, Aggarwal BB (2010) Oxidative stress, inflammation, and cancer: how are they linked? Free Radic Biol Med 49(11):1603–1616
40. Dai D-F, Rabinovitch PS, Ungvari Z (2012) Mitochondria and cardiovascular aging. Circ Res 110(8):1109–1124
41. Kao S-H, Chao H-T, Wei Y-H (1998) Multiple deletions of mitochondrial DNA are associated with the decline of motility and fertility of human spermatozoa. Mol Hum Reprod 4(7):657–666
42. Clayton DA (1984) Transcription of the mammalian mitochondrial genome. Annu Rev Biochem 53(1):573–594
43. Ballinger SW (2005) Mitochondrial dysfunction in cardiovascular disease. Free Radic Biol Med 38(10):1278–1295
44. Klionsky DJ, Emr SD (2000) Autophagy as a regulated pathway of cellular degradation. Science 290(5497):1717–1721
45. Yoshimori T (2004) Autophagy: a regulated bulk degradation process inside cells. Biochem Biophys Res Commun 313(2):453–458
46. Levine B, Klionsky DJ (2004) Development by self-digestion: molecular mechanisms and biological functions of autophagy. Dev Cell 6(4):463–477
47. Yorimitsu T, Klionsky DJ (2005) Autophagy: molecular machinery for self-eating. Cell Death Differ 12(S2):1542
48. Debnath J, Baehrecke EH, Kroemer G (2005) Does autophagy contribute to cell death? Autophagy 1(2):66–74
49. Levine B, Yuan J (2005) Autophagy in cell death: an innocent convict? J Clin Invest 115(10):2679–2688
50. Edinger AL, Thompson CB (2004) Death by design: apoptosis, necrosis and autophagy. Curr Opin Cell Biol 16(6):663–669
51. Bobryshev YV (2006) Monocyte recruitment and foam cell formation in atherosclerosis. Micron 37(3):208–222
52. Powers SK, Ji LL, Kavazis AN, Jackson MJ (2011) Reactive oxygen species: impact on skeletal muscle. Compr Physiol 1(2):941–969
53. Espinosa A, Henriquez-Olguin C, Jaimovich E (2016) Reactive oxygen species and calcium signals in skeletal muscle: A crosstalk involved in both normal signaling and disease. Cell Calcium 60(3):172–179
54. Johar S, MacCarthy PA, Shah AM (2006) Oxidative stress and cardiovascular disease. In: Oxidative stress, disease and cancer. World Scientific, pp 519–535
55. Lüscher T, Barton M (1997) Biology of the endothelium. Clin Cardiol 20(11 Suppl 2):II-3
56. Landmesser U, Spiekermann S, Dikalov S, Tatge H, Wilke R, Kohler C, Harrison DG, Hornig B, Drexler H (2002) Vascular oxidative stress and endothelial dysfunction in patients with chronic heart failure: role of xanthine-oxidase and extracellular superoxide dismutase. Circulation 106(24):3073–3078
57. Park JB, Schiffrin EL (2001) Small artery remodeling is the most prevalent (earliest?) form of target organ damage in mild essential hypertension. J Hypertens 19(5):921–930

58. Rizzoni D, Porteri E, Guelfi D, Muiesan ML, Valentini U, Cimino A, Girelli A, Rodella L, Bianchi R, Sleiman I (2001) Structural alterations in subcutaneous small arteries of normotensive and hypertensive patients with non–insulin-dependent diabetes mellitus. Circulation 103(9):1238–1244
59. Li J-M, Shah AM (2004) Endothelial cell superoxide generation: regulation and relevance for cardiovascular pathophysiology. Am J Physiol Regul Integr Comp Physiol 287(5):R1014–R1030
60. Schächinger V, Britten M, Dimmeler S, Zeiher A (2001) NADH/NADPH oxidase p22 phox gene polymorphism is associated with improved coronary endothelial vasodilator function. Eur Heart J 22(1):96–101
61. Zalba G, José GS, Beaumont FJ, Fortuño MA, Fortuño A, Díez J (2001) Polymorphisms and promoter overactivity of the p22 phox gene in vascular smooth muscle cells from spontaneously hypertensive rats. Circ Res 88(2):217–222
62. Shokoji T, Nishiyama A, Fujisawa Y, Hitomi H, Kiyomoto H, Takahashi N, Kimura S, Kohno M, Abe Y (2003) Renal sympathetic nerve responses to tempol in spontaneously hypertensive rats. Hypertension 41(2):266–273
63. Park JB, Touyz RM, Chen X, Schiffrin EL (2002) Chronic treatment with a superoxide dismutase mimetic prevents vascular remodeling and progression of hypertension in salt-loaded stroke-prone spontaneously hypertensive rats. Am J Hypertens 15(1):78–84
64. Paravicini TM, Chrissobolis S, Drummond GR, Sobey CG (2004) Increased NADPH-oxidase activity and Nox4 expression during chronic hypertension is associated with enhanced cerebral vasodilatation to NADPH in vivo. Stroke 35(2):584–589
65. Touyz RM (2004) Reactive oxygen species, vascular oxidative stress, and redox signaling in hypertension: what is the clinical significance? Hypertension 44(3):248–252
66. Rueckschloss U, Quinn MT, Holtz J, Morawietz H (2002) Dose-dependent regulation of NAD (P) H oxidase expression by angiotensin II in human endothelial cells: protective effect of angiotensin II type 1 receptor blockade in patients with coronary artery disease. Arterioscler Thromb Vasc Biol 22(11):1845–1851
67. Berry C, Anderson N, Kirk AJ, Dominiczak AF, Mcmurray JJ (2001) Renin angiotensin system inhibition is associated with reduced free radical concentrations in arteries of patients with coronary heart disease. Heart 86(2):217–220
68. Morrow JD, Frei B, Longmire AW, Gaziano JM, Lynch SM, Shyr Y, Strauss WE, Oates JA, Roberts LJ (1995) Increase in circulating products of lipid peroxidation (F2-isoprostanes) in smokers—smoking as a cause of oxidative damage. N Engl J Med 332(18):1198–1203
69. Reilly MP, Pratico D, Delanty N, DiMinno G, Tremoli E, Rader D, Kapoor S, Rokach J, Lawson J, FitzGerald GA (1998) Increased formation of distinct F2 isoprostanes in hypercholesterolemia. Circulation 98(25):2822–2828
70. Yamaguchi Y, Kunitomo M, Haginaka J (2002) Assay methods of modified lipoproteins in plasma. J Chromatogr B 781(1–2):313–330
71. Warnholtz A, Nickenig G, Schulz E, Macharzina R, Brasen JH, Skatchkov M, Heitzer T, Stasch JP, Griendling KK, Harrison DG (1999) Increased NADH-oxidase–mediated superoxide production in the early stages of atherosclerosis: evidence for involvement of the renin-angiotensin system. Circulation 99(15):2027–2033
72. Weiss D, Kools JJ, Taylor WR (2001) Angiotensin II–induced hypertension accelerates the development of atherosclerosis in apoE-deficient mice. Circulation 103(3):448–454
73. Ohara Y, Peterson TE, Harrison DG (1993) Hypercholesterolemia increases endothelial superoxide anion production. J Clin Investig 91(6):2546–2551
74. Mügge A, Elwell JH, Peterson TE, Hofmeyer TG, Heistad DD, Harrison DG (1991) Chronic treatment with polyethylene-glycolated superoxide dismutase partially restores endothelium-dependent vascular relaxations in cholesterol-fed rabbits. Circ Res 69(5):1293–1300
75. Böger RH, Bode-Böger SM, Mügge A, Kienke S, Brandes R, Dwenger A, Frölich JC (1995) Supplementation of hypercholesterolaemic rabbits with L-arginine reduces the vascular release of superoxide anions and restores NO production. Atherosclerosis 117(2):273–284

76. Miller FJ, Gutterman DD, Rios CD, Heistad DD, Davidson BL (1998) Superoxide production in vascular smooth muscle contributes to oxidative stress and impaired relaxation in atherosclerosis. Circ Res 82(12):1298–1305
77. Praticò D, Tangirala RK, Rader DJ, Rokach J, FitzGerald GA (1998) Vitamin E suppresses isoprostane generation in vivo and reduces atherosclerosis in ApoE-deficient mice. Nat Med *4*(10):1189
78. Tangirala RK, Praticó D, FitzGerald GA, Chun S, Tsukamoto K, Maugeais C, Usher DC, Puré E, Rader DJ (2001) Reduction of isoprostanes and regression of advanced atherosclerosis by apolipoprotein E. J Biol Chem 276(1):261–266
79. Febbraio M, Podrez EA, Smith JD, Hajjar DP, Hazen SL, Hoff HF, Sharma K, Silverstein RL (2000) Targeted disruption of the class B scavenger receptor CD36 protects against atherosclerotic lesion development in mice. J Clin Investig 105(8):1049–1056
80. Cyrus T, Witztum JL, Rader DJ, Tangirala R, Fazio S, Linton MF, Funk CD (1999) Disruption of the 12/15-lipoxygenase gene diminishes atherosclerosis in apo E–deficient mice. J Clin Investig 103(11):1597–1604
81. Galis ZS, Sukhova GK, Lark MW, Libby P (1994) Increased expression of matrix metalloproteinases and matrix degrading activity in vulnerable regions of human atherosclerotic plaques. J Clin Investig 94(6):2493–2503
82. Faia KL, Davis WP, Marone AJ, Foxall TL (2002) Matrix metalloproteinases and tissue inhibitors of metalloproteinases in hamster aortic atherosclerosis: correlation with in-situ zymography. Atherosclerosis 160(2):325–337
83. Dhalla AK, Hill MF, Singal PK (1996) Role of oxidative stress in transition of hypertrophy to heart failure. J Am Coll Cardiol 28(2):506–514
84. Nakamura K, Fushimi K, Kouchi H, Mihara K, Miyazaki M, Ohe T, Namba M (1998) Inhibitory effects of antioxidants on neonatal rat cardiac myocyte hypertrophy induced by tumor necrosis factor-alpha and angiotensin II. Circulation 98(8):794–799
85. MacCarthy PA, Grieve DJ, Li JM, Dunster C, Kelly FJ, Shah AM (2001) Impaired endothelial regulation of ventricular relaxation in cardiac hypertrophy: role of reactive oxygen species and NADPH oxidase. Circulation 104(24):2967–2974
86. Park YM, Park MY, Suh YL, Park JB (2004) NAD(P)H oxidase inhibitor prevents blood pressure elevation and cardiovascular hypertrophy in aldosterone-infused rats. Biochem Biophys Res Commun 313(3):812–817
87. Siwik DA, Pagano PJ, Colucci WS (2001) Oxidative stress regulates collagen synthesis and matrix metalloproteinase activity in cardiac fibroblasts. Am J Physiol Cell Physiol 280(1):C53–C60
88. Bendall JK, Cave AC, Heymes C, Gall N, Shah AM (2002) Pivotal role of a gp91(phox)-containing NADPH oxidase in angiotensin II-induced cardiac hypertrophy in mice. Circulation 105(3):293–296
89. Heymes C, Bendall JK, Ratajczak P, Cave AC, Samuel JL, Hasenfuss G, Shah AM (2003) Increased myocardial NADPH oxidase activity in human heart failure. J Am Coll Cardiol 41(12):2164–2171
90. Palace V, Kumar D, Hill MF, Khaper N, Singal PK (1999) Regional differences in non-enzymatic antioxidants in the heart under control and oxidative stress conditions. J Mol Cell Cardiol 31(1):193–202
91. Prasad K, Lee P, Mantha SV, Kalra J, Prasad M, Gupta JB (1992) Detection of ischemia-reperfusion cardiac injury by cardiac muscle chemiluminescence. Mol Cell Biochem 115(1):49–58
92. Jolly SR, Kane WJ, Bailie MB, Abrams GD, Lucchesi BR (1984) Canine myocardial reperfusion injury. Its reduction by the combined administration of superoxide dismutase and catalase. Circ Res 54(3):277–285
93. Temsah RM, Netticadan T, Chapman D, Takeda S, Mochizuki S, Dhalla NS (1999) Alterations in sarcoplasmic reticulum function and gene expression in ischemic-reperfused rat heart. Am J Physiol 277(2 Pt 2):H584–H594

94. Chen EP, Bittner HB, Davis RD, Van Trigt P, Folz RJ (1998) Physiologic effects of extracellular superoxide dismutase transgene overexpression on myocardial function after ischemia and reperfusion injury. J Thoracic Cardiovasc Surg 115(2):450–458. Discussion 458–459
95. Yoshida T, Maulik N, Engelman RM, Ho YS, Das DK (2000) Targeted disruption of the mouse Sod I gene makes the hearts vulnerable to ischemic reperfusion injury. Circ Res 86(3):264–269
96. Sethi R, Takeda N, Nagano M, Dhalla NS (2000) Beneficial effects of vitamin E treatment in acute myocardial infarction. J Cardiovasc Pharmacol Ther 5(1):51–58
97. Kinugawa S, Tsutsui H, Hayashidani S, Ide T, Suematsu N, Satoh S, Utsumi H, Takeshita A (2000) Treatment with dimethylthiourea prevents left ventricular remodeling and failure after experimental myocardial infarction in mice: role of oxidative stress. Circ Res 87(5):392–398
98. Sia YT, Lapointe N, Parker TG, Tsoporis JN, Deschepper CF, Calderone A, Pourdjabbar A, Jasmin JF, Sarrazin JF, Liu P, Adam A, Butany J, Rouleau JL (2002) Beneficial effects of long-term use of the antioxidant probucol in heart failure in the rat. Circulation 105(21):2549–2555
99. Spinale FG (2002) Matrix metalloproteinases: regulation and dysregulation in the failing heart. Circ Res 90(5):520–530
100. Cox MJ, Hawkins UA, Hoit BD, Tyagi SC (2004) Attenuation of oxidative stress and remodeling by cardiac inhibitor of metalloproteinase protein transfer. Circulation 109(17):2123–2128
101. Marzetti E, Csiszar A, Dutta D, Balagopal G, Calvani R, Leeuwenburgh C (2013) Role of mitochondrial dysfunction and altered autophagy in cardiovascular aging and disease: from mechanisms to therapeutics. Am J Physiol Heart Circ Physiol 305(4):H459–H476
102. Dutta D, Calvani R, Bernabei R, Leeuwenburgh C, Marzetti E (2012) Contribution of impaired mitochondrial autophagy to cardiac aging: mechanisms and therapeutic opportunities. Circ Res 110(8):1125–1138
103. Calvani R, Joseph AM, Adhihetty PJ, Miccheli A, Bossola M, Leeuwenburgh C, Bernabei R, Marzetti E (2013) Mitochondrial pathways in sarcopenia of aging and disuse muscle atrophy. Biol Chem 394(3):393–414
104. Schafer A, Reichert AS (2009) Emerging roles of mitochondrial membrane dynamics in health and disease. Biol Chem 390(8):707–715
105. Benard G, Rossignol R (2008) Ultrastructure of the mitochondrion and its bearing on function and bioenergetics. Antioxid Redox Signal 10(8):1313–1342
106. Aon MA, Cortassa S (2012) Mitochondrial network energetics in the heart. Wiley Interdiscip Rev Syst Biol Med 4(6):599–613
107. Twig G, Hyde B, Shirihai OS (2008) Mitochondrial fusion, fission and autophagy as a quality control axis: the bioenergetic view. Biochim Biophys Acta 1777(9):1092–1097
108. Chen H, Detmer SA, Ewald AJ, Griffin EE, Fraser SE, Chan DC (2003) Mitofusins Mfn1 and Mfn2 coordinately regulate mitochondrial fusion and are essential for embryonic development. J Cell Biol 160(2):189–200
109. Cipolat S, Martins de Brito O, Dal Zilio B, Scorrano L (2004) OPA1 requires mitofusin 1 to promote mitochondrial fusion. Proc Natl Acad Sci USA 101(45):15927–15932
110. Smirnova E, Griparic L, Shurland DL, van der Bliek AM (2001) Dynamin-related protein Drp1 is required for mitochondrial division in mammalian cells. Mol Biol Cell 12(8):2245–2256
111. Stojanovski D, Koutsopoulos OS, Okamoto K, Ryan MT (2004) Levels of human Fis1 at the mitochondrial outer membrane regulate mitochondrial morphology. J Cell Sci 117(Pt 7):1201–1210
112. Ong SB, Hall AR, Hausenloy DJ (2013) Mitochondrial dynamics in cardiovascular health and disease. Antioxid Redox Signal 19(4):400–414
113. Ong SB, Subrayan S, Lim SY, Yellon DM, Davidson SM, Hausenloy DJ (2010) Inhibiting mitochondrial fission protects the heart against ischemia/reperfusion injury. Circulation 121(18):2012–2022
114. Chen L, Gong Q, Stice JP, Knowlton AA (2009) Mitochondrial OPA1, apoptosis, and heart failure. Cardiovasc Res 84(1):91–99

115. Makino A, Scott BT, Dillmann WH (2010) Mitochondrial fragmentation and superoxide anion production in coronary endothelial cells from a mouse model of type 1 diabetes. Diabetologia 53(8):1783–1794
116. Yu T, Sheu SS, Robotham JL, Yoon Y (2008) Mitochondrial fission mediates high glucose-induced cell death through elevated production of reactive oxygen species. Cardiovasc Res 79(2):341–351
117. Huss JM, Kelly DP (2005) Mitochondrial energy metabolism in heart failure: a question of balance. J Clin Invest 115(3):547–555
118. Hom JR, Quintanilla RA, Hoffman DL, de Mesy Bentley KL, Molkentin JD, Sheu SS, Porter GA Jr (2011) The permeability transition pore controls cardiac mitochondrial maturation and myocyte differentiation. Dev Cell 21(3):469–478
119. Pennanen C, Parra V, Lopez-Crisosto C, Morales PE, Del Campo A, Gutierrez T, Rivera-Mejias P, Kuzmicic J, Chiong M, Zorzano A, Rothermel BA, Lavandero S (2014) Mitochondrial fission is required for cardiomyocyte hypertrophy mediated by a Ca2+-calcineurin signaling pathway. J Cell Sci 127(Pt 12):2659–2671
120. Xu X, Duan S, Yi F, Ocampo A, Liu GH, Izpisua Belmonte JC (2013) Mitochondrial regulation in pluripotent stem cells. Cell Metab 18(3):325–332
121. Salabei JK, Hill BG (2013) Mitochondrial fission induced by platelet-derived growth factor regulates vascular smooth muscle cell bioenergetics and cell proliferation. Redox Biol 1:542–551
122. Sharp WW, Fang YH, Han M, Zhang HJ, Hong Z, Banathy A, Morrow E, Ryan JJ, Archer SL (2014) Dynamin-related protein 1 (Drp1)-mediated diastolic dysfunction in myocardial ischemia-reperfusion injury: therapeutic benefits of Drp1 inhibition to reduce mitochondrial fission. FASEB J 28(1):316–326
123. Chen Y, Liu Y, Dorn GW 2nd (2011) Mitochondrial fusion is essential for organelle function and cardiac homeostasis. Circ Res 109(12):1327–1331
124. Papanicolaou KN, Khairallah RJ, Ngoh GA, Chikando A, Luptak I, O'Shea KM, Riley DD, Lugus JJ, Colucci WS, Lederer WJ, Stanley WC, Walsh K (2011) Mitofusin-2 maintains mitochondrial structure and contributes to stress-induced permeability transition in cardiac myocytes. Mol Cell Biol 31(6):1309–1328
125. Papanicolaou KN, Ngoh GA, Dabkowski ER, O'Connell KA, Ribeiro RF Jr, Stanley WC, Walsh K (2012) Cardiomyocyte deletion of mitofusin-1 leads to mitochondrial fragmentation and improves tolerance to ROS-induced mitochondrial dysfunction and cell death. Am J Physiol Heart Circ Physiol 302(1):H167–H179
126. Piquereau J, Caffin F, Novotova M, Prola A, Garnier A, Mateo P, Fortin D, Huynhle H, Nicolas V, Alavi MV, Brenner C, Ventura-Clapier R, Veksler V, Joubert F (2012) Down-regulation of OPA1 alters mouse mitochondrial morphology, PTP function, and cardiac adaptation to pressure overload. Cardiovasc Res 94(3):408–417
127. Olichon A, Baricault L, Gas N, Guillou E, Valette A, Belenguer P, Lenaers G (2003) Loss of OPA1 perturbates the mitochondrial inner membrane structure and integrity, leading to cytochrome c release and apoptosis. J Biol Chem *278*(10):7743–7746
128. Ihara S, Maeda-Takekoshi F, Takekoshi M, Yokoyama M, Sakuma S, Watanabe Y (1991) Evidence that the tyrosine kinase domain of a small fraction of epidermal growth factor receptor molecules is exposed on the outer surface of A431 cells. Cell Struct Funct 16(3):217–223
129. Gegg ME, Cooper JM, Chau KY, Rojo M, Schapira AH, Taanman JW (2010) Mitofusin 1 and mitofusin 2 are ubiquitinated in a PINK1/parkin-dependent manner upon induction of mitophagy. Hum Mol Genet 19(24):4861–4870
130. de Brito OM, Scorrano L (2008) Mitofusin 2 tethers endoplasmic reticulum to mitochondria. Nature 456(7222):605–610
131. Chen KH, Guo X, Ma D, Guo Y, Li Q, Yang D, Li P, Qiu X, Wen S, Xiao RP, Tang J (2004) Dysregulation of HSG triggers vascular proliferative disorders. Nat Cell Biol 6(9):872–883
132. Guo YH, Chen K, Gao W, Li Q, Chen L, Wang GS, Tang J (2007) Overexpression of Mitofusin 2 inhibited oxidized low-density lipoprotein induced vascular smooth muscle cell

proliferation and reduced atherosclerotic lesion formation in rabbit. Biochem Biophys Res Commun 363(2):411–417
133. Guo X, Chen KH, Guo Y, Liao H, Tang J, Xiao RP (2007) Mitofusin 2 triggers vascular smooth muscle cell apoptosis via mitochondrial death pathway. Circ Res 101(11):1113–1122
134. Hall AR, Burke N, Dongworth RK, Hausenloy DJ (2014) Mitochondrial fusion and fission proteins: novel therapeutic targets for combating cardiovascular disease. Br J Pharmacol 171(8):1890–1906
135. Klionsky DJ, Baehrecke EH, Brumell JH, Chu CT, Codogno P, Cuervo AM, Debnath J, Deretic V, Elazar Z, Eskelinen EL, Finkbeiner S, Fueyo-Margareto J, Gewirtz D, Jaattela M, Kroemer G, Levine B, Melia TJ, Mizushima N, Rubinsztein DC, Simonsen A, Thorburn A, Thumm M, Toozc SA (2011) A comprehensive glossary of autophagy-related molecules and processes (2nd edition). Autophagy 7(11):1273–1294
136. Dai DF, Rabinovitch PS, Ungvari Z (2012) Mitochondria and cardiovascular aging. Circ Res 110(8):1109–1124
137. Twig G, Elorza A, Molina AJ, Mohamed H, Wikstrom JD, Walzer G, Stiles L, Haigh SE, Katz S, Las G, Alroy J, Wu M, Py BF, Yuan J, Deeney JT, Corkey BE, Shirihai OS (2008) Fission and selective fusion govern mitochondrial segregation and elimination by autophagy. EMBO J 27(2):433–446
138. Carreira RS, Lee Y, Ghochani M, Gustafsson AB, Gottlieb RA (2010) Cyclophilin D is required for mitochondrial removal by autophagy in cardiac cells. Autophagy 6(4):462–472
139. Hirota Y, Kang D, Kanki T (2012) The physiological role of mitophagy: new insights into phosphorylation events. Int J Cell Biol 2012:354914
140. Canto C, Gerhart-Hines Z, Feige JN, Lagouge M, Noriega L, Milne JC, Elliott PJ, Puigserver P, Auwerx J (2009) AMPK regulates energy expenditure by modulating NAD+ metabolism and SIRT1 activity. Nature 458(7241):1056–1060
141. Karbowski M, Kurono C, Wozniak M, Ostrowski M, Teranishi M, Nishizawa Y, Usukura J, Soji T, Wakabayashi T (1999) Free radical-induced megamitochondria formation and apoptosis. Free Radic Biol Med 26(3–4):396–409
142. Terman A, Kurz T, Navratil M, Arriaga EA, Brunk UT (2010) Mitochondrial turnover and aging of long-lived postmitotic cells: the mitochondrial-lysosomal axis theory of aging. Antioxid Redox Signal 12(4):503–535
143. Taneike M, Yamaguchi O, Nakai A, Hikoso S, Takeda T, Mizote I, Oka T, Tamai T, Oyabu J, Murakawa T, Nishida K, Shimizu T, Hori M, Komuro I, Takuji Shirasawa TS, Mizushima N, Otsu K (2010) Inhibition of autophagy in the heart induces age-related cardiomyopathy. Autophagy 6(5):600–606
144. Lesnefsky EJ, Moghaddas S, Tandler B, Kerner J, Hoppel CL (2001) Mitochondrial dysfunction in cardiac disease: ischemia–reperfusion, aging, and heart failure. J Mol Cell Cardiol 33(6):1065–1089
145. Matsui Y, Takagi H, Qu X, Abdellatif M, Sakoda H, Asano T, Levine B, Sadoshima J (2007) Distinct roles of autophagy in the heart during ischemia and reperfusion: roles of AMP-activated protein kinase and Beclin 1 in mediating autophagy. Circ Res 100(6):914–922
146. Hamacher-Brady A, Brady NR, Gottlieb RA (2006) Enhancing macroautophagy protects against ischemia/reperfusion injury in cardiac myocytes. J Biol Chem 281(40):29776–29787
147. Xie Z, Lau K, Eby B, Lozano P, He C, Pennington B, Li H, Rathi S, Dong Y, Tian R, Kem D, Zou MH (2011) Improvement of cardiac functions by chronic metformin treatment is associated with enhanced cardiac autophagy in diabetic OVE26 mice. Diabetes 60(6):1770–1778
148. Akazawa H, Komazaki S, Shimomura H, Terasaki F, Zou Y, Takano H, Nagai T, Komuro I (2004) Diphtheria toxin-induced autophagic cardiomyocyte death plays a pathogenic role in mouse model of heart failure. J Biol Chem 279(39):41095–41103
149. Ungvari Z, Bailey-Downs L, Gautam T, Sosnowska D, Wang M, Monticone RE, Telljohann R, Pinto JT, de Cabo R, Sonntag WE, Lakatta EG, Csiszar A (2011) Age-associated vascular oxidative stress, Nrf2 dysfunction, and NF-{kappa}B activation in the nonhuman primate Macaca mulatta. J Gerontol A Biol Sci Med Sci 66(8):866–875

150. Ungvari Z, Csiszar A (2012) The emerging role of IGF-1 deficiency in cardiovascular aging: recent advances. J Gerontol A Biol Sci Med Sci 67(6):599–610
151. Sonntag WE, Csiszar A, deCabo R, Ferrucci L, Ungvari Z (2012) Diverse roles of growth hormone and insulin-like growth factor-1 in mammalian aging: progress and controversies. J Gerontol A Biol Sci Med Sci 67(6):587–598
152. Csiszar A, Labinskyy N, Perez V, Recchia FA, Podlutsky A, Mukhopadhyay P, Losonczy G, Pacher P, Austad SN, Bartke A, Ungvari Z (2008) Endothelial function and vascular oxidative stress in long-lived GH/IGF-deficient Ames dwarf mice. Am J Physiol Heart Circ Physiol 295(5):H1882–H1894
153. Hao CN, Geng YJ, Li F, Yang T, Su DF, Duan JL, Li Y (2011) Insulin-like growth factor-1 receptor activation prevents hydrogen peroxide-induced oxidative stress, mitochondrial dysfunction and apoptosis. Apoptosis 16(11):1118–1127
154. Zhang D, Mott JL, Farrar P, Ryerse JS, Chang SW, Stevens M, Denniger G, Zassenhaus HP (2003) Mitochondrial DNA mutations activate the mitochondrial apoptotic pathway and cause dilated cardiomyopathy. Cardiovasc Res 57(1):147–157
155. Dai DF, Johnson SC, Villarin JJ, Chin MT, Nieves-Cintron M, Chen T, Marcinek DJ, Dorn GW 2nd, Kang YJ, Prolla TA, Santana LF, Rabinovitch PS (2011) Mitochondrial oxidative stress mediates angiotensin II-induced cardiac hypertrophy and Galphaq overexpression-induced heart failure. Circ Res 108(7):837–846
156. Pepe M, Mamdani M, Zentilin L, Csiszar A, Qanud K, Zacchigna S, Ungvari Z, Puligadda U, Moimas S, Xu X, Edwards JG, Hintze TH, Giacca M, Recchia FA (2010) Intramyocardial VEGF-B167 gene delivery delays the progression towards congestive failure in dogs with pacing-induced dilated cardiomyopathy. Circ Res 106(12):1893–1903
157. Doughan AK, Harrison DG, Dikalov SI (2008) Molecular mechanisms of angiotensin II-mediated mitochondrial dysfunction: linking mitochondrial oxidative damage and vascular endothelial dysfunction. Circ Res 102(4):488–496
158. Kimura S, Zhang GX, Nishiyama A, Shokoji T, Yao L, Fan YY, Rahman M, Abe Y (2005) Mitochondria-derived reactive oxygen species and vascular MAP kinases: comparison of angiotensin II and diazoxide. Hypertension 45(3):438–444
159. Davis RC, Hobbs FD, Lip GY (2000) ABC of heart failure. History and epidemiology. BMJ *320*(7226):39–42
160. Vasanthi HR, ShriShriMal N, Das DK (2012) Phytochemicals from plants to combat cardiovascular disease. Curr Med Chem 19(14):2242–2251
161. Kong Y, Trabucco SE, Zhang H (2014) Oxidative stress, mitochondrial dysfunction and the mitochondria theory of aging. Interdiscip Top Gerontol 39:86–107
162. Liu Z, Xu S, Huang X, Wang J, Gao S, Li H, Zhou C, Ye J, Chen S, Jin ZG, Liu P (2015) Cryptotanshinone, an orally bioactive herbal compound from Danshen, attenuates atherosclerosis in apolipoprotein E-deficient mice: role of lectin-like oxidized LDL receptor-1 (LOX-1). Br J Pharmacol 172(23):5661–5675
163. Xu Y, Liu Q, Xu Y, Liu C, Wang X, He X, Zhu N, Liu J, Wu Y, Li Y, Li N, Feng T, Lai F, Zhang M, Hong B, Jiang JD, Si S (2014) Rutaecarpine suppresses atherosclerosis in ApoE-/- mice through upregulating ABCA1 and SR-BI within RCT. J Lipid Res 55(8):1634–1647
164. Jia S, Hu C (2010) Pharmacological effects of rutaecarpine as a cardiovascular protective agent. Molecules 15(3):1873–1881
165. Cimen I, Kocaturk B, Koyuncu S, Tufanli O, Onat UI, Yildirim AD, Apaydin O, Demirsoy S, Aykut ZG, Nguyen UT, Watkins SM, Hotamisligil GS, Erbay E (2016) Prevention of atherosclerosis by bioactive palmitoleate through suppression of organelle stress and inflammasome activation. Sci Transl Med 8(358):358ra126
166. Luo Y, Sun G, Dong X, Wang M, Qin M, Yu Y, Sun X (2015) Isorhamnetin attenuates atherosclerosis by inhibiting macrophage apoptosis via PI3K/AKT activation and HO-1 induction. PLoS One 10(3):e0120259
167. Miura Y, Chiba T, Tomita I, Koizumi H, Miura S, Umegaki K, Hara Y, Ikeda M, Tomita T (2001) Tea catechins prevent the development of atherosclerosis in apoprotein E-deficient mice. J Nutr 131(1):27–32

168. Babu PV, Liu D (2008) Green tea catechins and cardiovascular health: an update. Curr Med Chem 15(18):1840–1850
169. Punithavathi VR, Stanely Mainzen Prince P (2011) The cardioprotective effects of a combination of quercetin and alpha-tocopherol on isoproterenol-induced myocardial infarcted rats. J Biochem Mol Toxicol 25(1):28–40
170. Joshipura KJ, Ascherio A, Manson JE, Stampfer MJ, Rimm EB, Speizer FE, Hennekens CH, Spiegelman D, Willett WC (1999) Fruit and vegetable intake in relation to risk of ischemic stroke. JAMA 282(13):1233–1239
171. Gorinstein S, Caspi A, Libman I, Lerner HT, Huang D, Leontowicz H, Leontowicz M, Tashma Z, Katrich E, Feng S, Trakhtenberg S (2006) Red grapefruit positively influences serum triglyceride level in patients suffering from coronary atherosclerosis: studies in vitro and in humans. J Agric Food Chem 54(5):1887–1892
172. Rangel-Huerta OD, Pastor-Villaescusa B, Aguilera CM, Gil A (2015) A systematic review of the efficacy of bioactive compounds in cardiovascular disease: phenolic compounds. Nutrients 7(7):5177–5216
173. Reis JF, Monteiro VV, de Souza Gomes R, do Carmo MM, da Costa GV, Ribera PC, Monteiro MC (2016) Action mechanism and cardiovascular effect of anthocyanins: a systematic review of animal and human studies. J Transl Med 14(1):315
174. Wallace TC (2011) Anthocyanins in Cardiovascular Disease. Adv Nutr 2(1):1–7
175. Wallace TC, Slavin M, Frankenfeld CL (2016) Systematic review of anthocyanins and markers of cardiovascular disease. Nutrients 8(1)
176. Scolaro B, Soo Jin Kim H, de Castro IA (2018) Bioactive compounds as an alternative for drug co-therapy: overcoming challenges in cardiovascular disease prevention. Crit Rev Food Sci Nutr 58(6):958–971
177. Bonnefont-Rousselot D (2016) Resveratrol and cardiovascular diseases. Nutrients 8(5)
178. Vidavalur R, Otani H, Singal PK, Maulik N (2006) Significance of wine and resveratrol in cardiovascular disease: French paradox revisited. Exp Clin Cardiol 11(3):217–225
179. Qin M, Luo Y, Meng XB, Wang M, Wang HW, Song SY, Ye JX, Pan RL, Yao F, Wu P, Sun GB, Sun XB (2015) Myricitrin attenuates endothelial cell apoptosis to prevent atherosclerosis: An insight into PI3K/Akt activation and STAT3 signaling pathways. Vasc Pharmacol 70:23–34
180. Honarbakhsh S, Schachter M (2009) Vitamins and cardiovascular disease. Br J Nutr 101(8):1113–1131
181. Al-Yahya MA, Mothana RA, Al-Said MS, El-Tahir KE, Al-Sohaibani M, Rafatullah S (2013) Citrus medica "Otroj": attenuates oxidative stress and cardiac dysrhythmia in isoproterenol-induced cardiomyopathy in rats. Nutrients 5(11):4269–4283

Parkin Protein: The Missing Link Between Cardiovascular and Parkinson's Disease

20

Angshuman Bagchi

Abstract
Parkinson's disease and cardiovascular diseases are two of the most frequently occurring disasters. Though these two diseases are quite common, molecular mechanisms of the onsets of these two diseases are still obscure. Recently, scientists have found an interrelation between these two diseases. One of the links between these two diseases is the Parkin protein. Mutations in Parkin lead to Parkinson's disease and also cardiovascular diseases. In this review, an attempt is made to describe the link between the Parkin mutations and the two diseases. This review would therefore be essential for the understanding of the molecular mechanism of the diseases.

Keywords
Cardiovascular diseases · Parkinson's disease · Parkin; Mutations

20.1 Introduction

Two of the most relevant and frequently occurring diseases are Parkinson 's disease (PD) and cardiovascular diseases. PD affects the motor functions, whereas cardiovascular diseases involve the heart. Interestingly, scientists recently found some links between PD and cardiovascular diseases. Recent studies have revealed that PD patients are nearly 50% more prone to getting cardiovascular diseases. It has also been proposed that damaged mitochondria have significant roles in the onset of both diseases [1–11]. Various studies revealed that PD patients who have a high or

A. Bagchi (✉)
Department of Biochemistry and Biophysics, University of Kalyani, Kalyani, West Bengal, India
e-mail: angshu@klyuniv.ac.in

S. Chakraborti et al. (eds.), *Modulation of Oxidative Stress in Heart Disease*,
https://doi.org/10.1007/978-981-13-8946-7_20

medium risk of cardiovascular diseases would have more problems with motor movements and memory. PD and cardiovascular diseases generally appear in older people. However, increasing latest evidences suggest the onset of both diseases in young populations as well. It has also been observed that patients with poorer cardiovascular health could have walking and memory issues, even in the early stages of PD [1–13].

20.2 A Few Words on Parkinson's Disease (PD)

Parkinson's disease (PD) is one of the most common age-related neurodegenerative disorders, with a prevalence rate of 1–2% [14–28]. Approximately 6.3 million people suffer from PD worldwide but the prevalence rate is higher in the developing countries. Akinesia, rigidity, postural instability, resting tremor, and bradykinesia are the clinical characteristics of PD [14–32]. PD is the second most common movement disorder caused by the progressive death of dopamine-producing neurons particularly located in the substantia nigra pars compacta region of the midbrain. The pathogenic aggregation of ubiquitin and α-synuclein-rich inclusion bodies called Lewy bodies is also produced. Such accumulation leads to the disease pathogenesis [33–40]. In 1817 James Parkinson first described the term Parkinson's disease on the paper called "the shaking palsy." In 1865 William Sanders coined the term "Parkinson's disease." Various environmental and genetic risk factors give rise to the sporadic, familial, and symptomatic form of PD [33–42]. Sporadic PD is the most common form of PD which appears due to a number of environmental and genetic influences [33–43]. Less than 10% familial PDs are caused by mutations in Parkin and PINK1 (PTEN-induced putative kinase protein 1 or PARK6) genes. Recent studies revealed that ubiquitylation and mitochondrial integrity are responsible for the disease pathogenesis. Defects in the electron transport chain during PD suggest that damaged mitochondria may play a vital role in the disease prognosis. Two recessive PD genes, PINK1 and Parkin (PARK2), contribute to the removal of damaged mitochondria by inducing mitophagy and autophagy [44–53].

20.3 A Few Words on Cardiovascular Disease

Diseases that involve the heart or blood vessels are called cardiovascular disease. Mitochondria are energy currencies of the cell that supply most of the ATP molecules required for cellular functions and integrity. Mammalian hearts also depend on mitochondrial oxidative phosphorylation process to fuel myocardial contraction and pump function. Damage, senescence, or diseases may disrupt the mitochondrial functionalities. The production of high levels of ATP can form superoxide and reactive oxygen species (ROS) radicals that can damage the mitochondria. Several clinical studies have reported that heart failure and increased risk of cardiovascular diseases are common in most of the PD patients [54, 55]. Although it is well characterized that Parkin is associated with the onset of PD, but now recent research deciphered its role in cardiovascular diseases as it plays as a central factor in mediating selective mitophagy of damaged mitochondria for mitochondrial quality control [44–53]. Not only Parkin but also other PD proteins, such as PINK1, DJ-1, α-synuclein, and LRRK2, may play crucial roles in cardiovascular diseases. The presence of Parkin protein in the heart, skeletal muscle, testis, brain, kidney, liver, and various other subcellular locations supports the notion that Parkin can exert its functions in these organs as well [44–57]. However, the biochemical mechanism of the relationship between Parkin protein and cardiovascular disease is still not clear. It has been observed that Ergot-derived dopamine receptor agonists are generally used to treat PD, which is associated with heart valve disease. Medication as well as other factors like concurrent comorbidities and cardiac sympathetic denervation may lead to cardiovascular disease [58–64].

20.4 Different Aspects of Parkin: The Cytosolic E3 Ubiquitin Ligase

Autosomal recessive Parkinson's disease (ARPD) is caused by mutations in Parkin, PINK1, or DJ-1 proteins. Parkin, also known as PARK2, is the most common gene responsible for early onset of PD. Parkin mutation is mostly associated with PD pathogenesis. Parkin protein, a RBR E3 ubiquitin-protein ligase which is coded by PARK2 gene, directly links to the ubiquitin-proteasome system and acts as a regulator of protein breakdown. PARK2 collaborates with PINK1 to involve in mitophagy. PARK2 and PARK6 (PTEN-induced kinase-1) gene products play major roles to control mitochondrial quality.

20.5 Structural Details of Parkin

Parkin (PARK2), a 52-kDa protein, is found in the heart, skeletal muscle, testis, brain, kidney, and liver and is associated with autosomal recessive PD. It is located on the human chromosome no 6. Parkin is a 465 amino acid residue long protein containing RBR type E3 ligase domain and it consists of N-terminal ubiquitin-like (Ubl) domain and four zinc-coordinating RING-like domains, viz., RING0, RING1, IBR, and RING2, at the C-terminal end of the protein. More than 120 pathogenic PD mutations have been reported to be spread throughout Parkin protein, which affect the structure and function of these domains. N-terminal ubiquitin-like (Ubl) domain is a 76 amino acid residue long part and is involved in substrate recognition, proteasome association, binding of SH3 and ubiquitin-interacting motif (UIM) domains, and regulation of cellular Parkin levels and activity. In eukaryotes the cysteine- and histidine-rich RING-IBR-RING (RBR) domain architecture is highly conserved. Parkin is stabilized by multiple hydrophobic interactions and forms a compact arrangement, resembling a coiled snake. Low-resolution structure of the full-length protein is available in the literature. Parkin possesses two RING finger

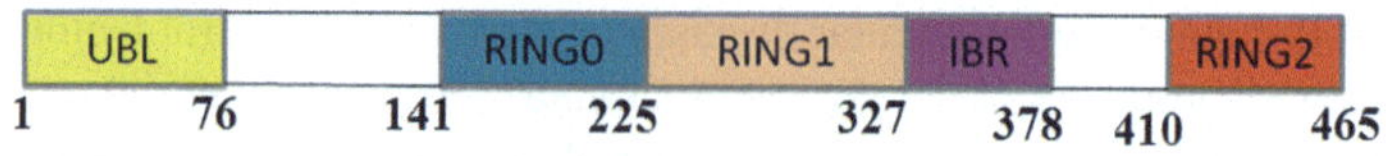

Fig. 20.1 Domain-wise representation of Parkin protein

domains and an in-between-RING (IBR) region. RING1 and RING2 are separated by an in-between-RING (IBR) domain as presented in Fig. 20.1.

The RING domains fold with distinct topologies. RING1 domain is the only domain with a classical cross-brace zinc-coordination topology, typical of other RING fingers. RING0 domain displays a hairpin topology, whereas the zinc-liganding residues are arranged in a sequential fashion in RING2 and IBR. RING0, RING1, IBR, and RING2 each coordinate two zinc ions. Primary amino acid sequences of RING0 and RING2 are similar but their structural topologies differ from a classical RING fold. The N-terminal Ubl domain binds to C-terminal RING1 domain and RING0 domain is tightly associated with the C-terminal catalytic domain. Its catalytic activity is repressed under normal cellular condition. Parkin activation requires phosphorylation of serine 65 in Ubl by serine/threonine kinase, PINK1 [65–69].

20.6 Mechanism of Ubiquitination

Proteasomal degradation, endocytosis, endosomal sorting, and DNA repair are the posttranslational modification events of a target protein conducted by ubiquitin signals. Majority of proteins are destined to be degraded by the ubiquitin-proteasome pathway. Ubiquitination is a posttranslational modification process in which proteins are marked by the covalent attachment of multiple ubiquitin molecules, and this ubiquitin provides a recognition signal for the 26S proteasome. Three enzymes – E1 ubiquitin-activating enzymes, E2 ubiquitin-conjugating enzymes, and E3 ubiquitin ligases – carry out the ubiquitination process. First, E1 activates ubiquitin; it is an ATP-dependent process to conjugate the C-terminal carboxylic acid group of ubiquitin to an active site cysteine. By forming a thioester linkage, an active site cysteine residue of E1 gets covalently linked to ubiquitin. Activated ubiquitin is then transferred by transacylation reaction to a thiol group of an active site cysteine residue of E2. Finally, the ubiquitin-charged E2 enzyme interacts with a specific E3 ubiquitin ligase and transfers the ubiquitin to the amino group of a substrate protein. E3 ubiquitin ligases are of three types – RING type (including U-box ligases), HECT type, and RING-HECT hybrid type – based on their chemistry and structure. Parkin is a member of RBR E3 ubiquitin ligases, which combine the chemistry of HECT-type ligases with structural similarity to RING-type ligases. In other words, Parkin belongs to the RING/HECT hybrid ligase-type protein family. Parkin contains a RING domain that binds the E2 enzyme and a catalytic cysteine that transfers ubiquitin to the substrate. Parkin activity is repressed in normal individuals although various conditions can activate this protein such as depolarization of mitochondria or epidermal growth factor signalling. Activation of Parkin protein is tightly regulated. It can perform the ligation of ubiquitin to various cytosolic and outer mitochondrial membrane proteins when mitochondrial depolarization occurs. K63, K48, K11, and K6 ubiquitin chains are formed by Parkin. Accumulations of these polyubiquitin chains on mitochondria act as the signals for recruitment of the autophagosome and proteasome machinery to initiate mitophagy [70–75].

20.7 Parkin Mutations: The Causative Agents of Diseases

Frameshift mutations, missense mutations, exon rearrangements, and point mutations are the different classes of mutations that are spread over the different parts of the Parkin protein. Generally, frameshift mutations and nonsense mutations are considered as pathogenic ones which destabilize the protein structure by destroying the structural integrity of the protein. On the other hand, amino acid substitutions and nonsense mutation required more detailed analysis to determine pathogenicity. However, these mutations also affect the binding affinities of the protein with its partners. PDmutDB, Cosmic, and some other databases were developed to store the information about mutations, pathogenicity, and structure of the mutant Parkin protein. Mutations in Ubl and RING1 domain of Parkin lead to unfolding of the protein structure, thereby making changes to the binding affinities of Parkin protein. Mutations in zinc finger domain can prevent binding of this protein to the DNA. Heterogeneous and homogeneous mutations in Parkin protein are responsible for the onset of PD. Even a single mutation in the conserved domain of Parkin may lead to loss of functionality of the protein. In Ubl domain E28K mutation disrupts the domain architecture. The two other mutations, A379V and P437L, could destroy the RING2 domain architecture. Auto-inhibition is a general feature of RBR ligases. Parkin shows similar mechanism of auto-inhibition like other RBR ligases. Mutations in Parkin protein could change its function in the heart [76–80]. Mutations also result in the early activation of Parkin protein and also unfold the protein which results in the excessive removal of mitochondria. Removals of excess mitochondria are not good for cellular processes needed for the proper maintenance of the heart. Under basal conditions ubiquitin ligase activity of Parkin is inhibited by these mutations. Under various stress conditions, PINK1 recruits Parkin from the cytoplasm to mitochondria and then PINK1-dependent Parkin ubiquitination occurs. This results in mitochondrial fragmentation, degradation, and mitophagy. The abundance and activity of PINK1 are maintained at very low levels in healthy mitochondria with normal membrane potential. However, in damaged mitochondria the depolarization of the membrane potential or activation of the unfolded protein allows PINK1 to stabilize the outer mitochondrial membrane. After accumulation of PINK1, it shows the kinase activity on the mitochondria, which helps to recruit and activate Parkin. PINK1 phosphorylates polyubiquitin and ubiquitin-like (Ubl) domain of Parkin protein [81–85].

20.8 Mitochondrial Quality Control

The presence and function of Parkin protein in the brain is well understood. However, it is highly expressed in the heart as well. Interestingly, the functional role of Parkin in the heart is not yet fully explored. Parkin plays an important role in the removal of dysfunctional mitochondria via autophagy in neurons. Autophagy is also very important in the heart; it is associated with a wide variety of cardiovascular pathologies. Damaged mitochondria produce reactive oxygen species that can cause further damage to nearby mitochondria and result in the release of pro-apoptotic proteins. Pro-apoptotic proteins disrupt the normal cell death pathway. Activation of this type of protein may cause the death of healthy mitochondria, which may lead to the depletion of ATP. So, damaged mitochondria are quickly removed by the cell in different mechanisms. In acute and chronic myocardial ischemia, heart failure, and dilated cardiomyopathy conditions, autophagic mechanism is increased to remove the damaged mitochondria and it acts as a protective response by the cell. Deletion of Atg5 in the adult heart is known to develop cardiac dysfunction by accumulating damaged mitochondria. Some studies suggest that Parkin plays an essential role in adapting stress in the myocardium by enhancing autophagy. The gene expression of PARK2 is upregulated transcriptionally under stress conditions. In Parkin-deficient mice, loss of Parkin protein results in formation of disorganized mitochondria. Parkin adapts to the stress condition by activating myocardium mitophagy. In cardiac and skeletal muscle, Parkin routinely removes dysfunctional mitochondria by autophagic mechanism. A disruption of this process accumulates damaged mitochondria which leads to heart failure or other multiple complexities. In *Drosophila melanogaster*, Parkin regulates mitochondrial function. The role of PINK1/Parkin pathway is important for marking and removal of dysfunctional mitochondria. Latest studies suggest that in mammalian cells Parkin does not contribute to mitochondrial function normally, but it comes into play under various stress conditions, such as infection (chronic systemic inflammation induced by lipopolysaccharide administration). It has also been observed that after translocation and ubiquitination, Parkin could activate several mitochondrial proteins. Parkin deficiency leads to loss of ubiquitination of mitochondrial proteins for their activations. This evidence suggests that Parkin plays crucial roles in the regulation of mitochondrial functionalities. Parkin also interacts with autophagy-promoting protein Ambra1 which activates the PI3K complex to form a new autophagosome. Ambra1-mediated Parkin-dependent autophagy could reveal that Parkin may play important roles to maintain normal mitochondrial quality in the heart. In healthy mitochondria with functional membrane potential, PINK1 is cleaved by the protease PARL. Dysfunctional mitochondria upon changing membrane potential would lead to accumulation of PINK1 in the outer mitochondrial membrane. The serine/threonine-protein kinase PINK1 then drags the cytosolic Parkin protein into the damaged mitochondria. After recruitment, Parkin activates several outer mitochondrial membrane proteins which then generate signals for autophagic clearance. Recent findings have revealed that Parkin serves as an important mediator to regulate mitochondrial degradation in the heart and cardiac myocytes. Although Parkin mediates mitochondrial autophagy in response to myocardial infarction, mechanisms of Parkin recruitment and activation of myocytes are not well understood. Parkin mediates selective mitochondrial clearance by autophagy in cells and also plays important roles in the mitochondrial dysfunction and disruption of autophagosomal clearance, associated with autophagy-related marker LC3 I to LC3 II in cardiomyocytes. In PD patients, oxidative damage of complex I subunits of mitochondria could disrupt the electron transport chain (ETC). In addition, in the case of PD patients, mitochondrial DNA is deleted in the dopaminergic neurons. It is reported that in Drosophila fruit fly, Parkin and PINK1 deficiency give rise to mitochondrial abnormalities in flight muscles and increased male sterility. PINK1-Mfn2-Parkin pathway is responsible for clearance of damaged mitochondria. Abnormalities in Mnf2 receptor give rise to nonspecific mitophagy. In mouse hearts Nix/Bnip3L and Bnip3 death proteins appoint autophagosome into the mitochondria in Parkin-dependent and Parkin-independent manners. The involvements of these proteins prove that mitochondrial quality is controlled in the heart by the direct involvements of Parkin [47, 72, 86–102].

20.9 Concluding Remarks

Recent studies have revealed that mutations in the Parkin protein are responsible for autosomal recessive inheritance of Parkinson's disease. In this review we tried to analyze the link between PD and cardiovascular disease through the involvements of Parkin protein. Recent studies suggest that patients suffering from Parkinson's disease are prone to heart failure. However, it is not well described how Parkinson's and heart disease are linked. We are trying to provide an overview of how Parkin protein controls the mitochondrial quality. We are also trying to decipher the effects of mutations on Parkin in PD pathogenesis and how these mutations are responsible for the onset of cardiovascular disease. In *Drosophila melanogaster*, Parkin plays important roles to remove the dysfunctional mitochondria. In mammalian system the mechanism of Parkin is not clear. Ageing is the common factor that gives rise to Parkinson's and cardiovascular disease. Recent data suggest that mitochondrial Parkin protein not only regulates the brain mitochondrial function but also is effective for cardioprotection. Due to the beneficial roles in the mitochondrial function, Parkin and its partners can be used as therapeutic agents for cardiac disease as well.

Acknowledgments The author is thankful to UGC-SAP-DSR-II program, Government of India, DST-PURSE2, and University of Kalyani for their support. The infrastructural supports provided by DBT-funded BIF Center and ICMR (BIC/12(02)/2014) are duly acknowledged.

References

1. Chen Y, Dorn GW (2013) PINK1-phosphorylated mitofusin 2 is a Parkin receptor for culling damaged mitochondria. Science 340(6131):471–475. https://doi.org/10.1126/science.1231031
2. Dorn GW II (2016) Central Parkin: the evolving role of Parkin in the heart. Biochim Biophys Acta (BBA) Bioenerg 1857(8):1307–1312
3. Swallow DMA, Lawton MA, Grosset KA et al (2016) Statins are underused in recent-onset Parkinson's disease with increased vascular risk: findings from the UK Tracking Parkinson's and Oxford Parkinson's Disease Centre (OPDC) discovery cohorts. J Neurol Neurosurg Psychiatry 87:1183–1190
4. Günaydın ZY, Özer FF, Karagöz A, Bektaş O, Karataş MB, Vural A, Bayramoğlu A, Çelik A, Yaman M (2016) Evaluation of cardiovascular risk in patients with Parkinson disease under levodopa treatment. J Geriatr Cardiol 13(1):75–80. https://doi.org/10.11909/j.issn.1671-5411.2016.01.003
5. Jones JD, Tanner JJ, Okun M, Price CC, Bowers D (2017) Are Parkinson's patients more vulnerable to the effects of cardiovascular risk: a neuroimaging and neuropsychological study. J Int Neuropsychol Soc. Author manuscript; available in PMC 2017 Dec 14. Published in final edited form as: J Int Neuropsychol Soc. 2017 Apr; 23(4): 322–331. Published online 2017 Feb 6. doi: 10.1017/S1355617717000017
6. Montastruc F, Moulis F, Araujo M, Chebane L, Rascol O, Montastruc JL (2017) Ergot and non-ergot dopamine agonists and heart failure in patients with Parkinson's disease. Eur J Clin Pharmacol 73(1):99–103. https://doi.org/10.1007/s00228-016-2142-x. Epub 2016 Oct 27
7. Solla P, Cadeddu C, Cannas A, Deidda M, Mura N, Mercuro G, Marrosu F (2015) Heart rate variability shows different cardiovascular modulation in Parkinson's disease patients with tremor dominant subtype compared to those with akinetic rigid dominant subtype. J Neural Transm (Vienna) 122(10):1441–1446. https://doi.org/10.1007/s00702-015-1393-5. Epub 2015 Mar 24
8. Sommerauer M, Imbach LL, Jarallah M, Baumann CR, Valko PO (2015) Diminished event-related cortical arousals and altered heart rate response in Parkinson's disease. Mov Disord 30(6):866–870. https://doi.org/10.1002/mds.26165. Epub 2015 Mar 4
9. Scorza FA, Scorza CA, Finsterer J (2017a) The heart in Parkinson's disease: opening Pandora's box. Auton Neurosci 17:pii: S1566-0702(17)30186-8. https://doi.org/10.1016/j.autneu.2017.10.004. [Epub ahead of print]
10. Scorza FA, Tufik S, Scorza CA, Andersen ML, Cavalheiro EA (2017b) Sudden unexpected death in Parkinson's disease (SUDPAR): sleep apnea increases risk of heart attack. Sleep Breath 21(4):965–966. https://doi.org/10.1007/s11325-017-1511-8. Epub 2017 May 23
11. Piqueras-Flores J, López-García A, Moreno-Reig Á, González-Martínez A, Hernández-González A, Vaamonde-Gamo J, Jurado-Román A (2018) Structural and functional alterations of the heart in Parkinson's disease. Neurol Res 40(1):53–61. https://doi.org/10.1080/01616412.2017.1390933. Epub 2017 Oct 23
12. http://parkinsonhope.org/link-between-parkinsons-and-heart-disease-discovered/
13. http://www.heart.org
14. Pajarillo E, Rizor A, Lee J, Aschner M, Lee E (2018) The role of posttranslational modifications of α-synuclein and LRRK2 in Parkinson's disease: potential contributions of environmental factors. Biochim Biophys Acta Mol basis Dis. pii: S0925-4439(18)30478-2. https://doi.org/10.1016/j.bbadis.2018.11.017
15. Sveinbjornsdottir S (2016) The clinical symptoms of Parkinson's disease. J Neurochem 139(Suppl 1):318–324. https://doi.org/10.1111/jnc.13691. Epub 2016 Jul 11
16. Schneider RB, Iourinets J, Richard IH (2017) Parkinson's disease psychosis: presentation, diagnosis and management. Neurodegener Dis Manag 7(6):365–376. https://doi.org/10.2217/nmt-2017-0028. Epub 2017 Nov 21

17. Fredericks D, Norton JC, Atchison C, Schoenhaus R, Pill MW (2017) Parkinson's disease and Parkinson's disease psychosis: a perspective on the challenges, treatments, and economic burden. Am J Manag Care 23(5 Suppl):S83–S92
18. Opara J, Małecki A, Małecka E, Socha T (2017) Motor assessment in Parkinson's disease. Ann Agric Environ Med 24(3):411–415. https://doi.org/10.5604/12321966.1232774. Epub 2017 May 11
19. Gregory R, Miller S (2015) Parkinson's disease and the skin. Pract Neurol 15(4):246–249. https://doi.org/10.1136/practneurol-2015-001107. Epub 2015 Apr 10
20. Suttrup I, Warnecke T (2016) Dysphagia in Parkinson's disease. Dysphagia 31(1):24–32. https://doi.org/10.1007/s00455-015-9671-9. Epub 2015 Nov 21
21. George S, Brundin P (2015) Immunotherapy in Parkinson's disease: micromanaging alpha-synuclein aggregation. J Park Dis 5(3):413–424. https://doi.org/10.3233/JPD-150630
22. Simons JA (2017) Swallowing dysfunctions in Parkinson's disease. Int Rev Neurobiol 134:1207–1238. https://doi.org/10.1016/bs.irn.2017.05.026. Epub 2017 Jul 13
23. Goldman SM (2014) Environmental toxins and Parkinson's disease. Annu Rev Pharmacol Toxicol 54:141–164. https://doi.org/10.1146/annurev-pharmtox-011613-135937. Epub 2013 Sep 16
24. Csoti I, Jost WH, Reichmann H (2016) Parkinson's disease between internal medicine and neurology. J Neural Transm (Vienna) 123(1):3–17. https://doi.org/10.1007/s00702-015-1443-z. Epub 2015 Aug 23
25. O'Callaghan C, Lewis SJG (2017) Cognition in Parkinson's disease. Int Rev Neurobiol 133:557–583. https://doi.org/10.1016/bs.irn.2017.05.002. Epub 2017 Jun 16
26. Pagonabarraga J, Kulisevsky J (2017) Apathy in Parkinson's disease. Int Rev Neurobiol 133:657–678. https://doi.org/10.1016/bs.irn.2017.05.025. Epub 2017 Jul 10
27. Cosgrove J, Alty JE, Jamieson S (2015) Cognitive impairment in Parkinson's disease. Postgrad Med J 91(1074):212–220. https://doi.org/10.1136/postgradmedj-2015-133247. Epub 2015 Mar 26
28. Antony PM, Diederich NJ, Krüger R, Balling R (2013) The hallmarks of Parkinson's disease. FEBS J 280(23):5981–5993. https://doi.org/10.1111/febs.12335. Epub 2013 Jun 10
29. Vivekanantham S, Shah S, Dewji R, Dewji A, Khatri C, Ologunde R (2015) Neuroinflammation in Parkinson's disease: role in neurodegeneration and tissue repair. Int J Neurosci 125(10):717–725. https://doi.org/10.3109/00207454.2014.982795. Epub 2015 May 22
30. Delgado-Alvarado M, Gago B, Navalpotro-Gomez I, Jiménez-Urbieta H, Rodriguez-Oroz MC (2016) Biomarkers for dementia and mild cognitive impairment in Parkinson's disease. Mov Disord 31(6):861–881. https://doi.org/10.1002/mds.26662. Epub 2016 May 19
31. Reichmann H (2017) Premotor diagnosis of Parkinson's disease. Neurosci Bull 33(5):526–534. https://doi.org/10.1007/s12264-017-0159-5. Epub 2017 Aug 3
32. Cebrián C, Loike JD, Sulzer D (2015) Neuroinflammation in Parkinson's disease animal models: a cell stress response or a step in neurodegeneration? Curr Top Behav Neurosci 22:237–270. https://doi.org/10.1007/7854_2014_356
33. Poewe W, Seppi K, Tanner CM, Halliday GM, Brundin P, Volkmann J, Schrag AE, Lang AE (2017) Parkinson disease. Nat Rev Dis Primers 3:17013. https://doi.org/10.1038/nrdp.2017.13
34. Beitz JM (2014) Parkinson's disease: a review. Front Biosci (Schol Ed) 6:65–74
35. Oertel W, Schulz JB (2016) Current and experimental treatments of Parkinson disease: a guide for neuroscientists. J Neurochem 139(Suppl 1):325–337. https://doi.org/10.1111/jnc.13750. Epub 2016 Aug 30
36. Miraglia F, Betti L, Palego L, Giannaccini G (2015) Parkinson's disease and alpha-synucleinopathies: from arising pathways to therapeutic challenge. Cent Nerv Syst Agents Med Chem 15(2):109–116
37. Uemura N, Yagi H, Uemura MT, Hatanaka Y, Yamakado H, Takahashi R (2018) Inoculation of α-synuclein preformed fibrils into the mouse gastrointestinal tract induces Lewy body-like aggregates in the brainstem via the vagus nerve. Mol Neurodegener 13(1):21. https://doi.org/10.1186/s13024-018-0257-5

38. Wakabayashi K, Tanji K, Mori F, Takahashi H (2007) The Lewy body in Parkinson's disease: molecules implicated in the formation and degradation of alpha-synuclein aggregates. Neuropathology 27(5):494–506
39. Lasser-Katz E, Simchovitz A, Chiu WH, Oertel WH, Sharon R, Soreq H, Roeper J, Goldberg JA (2017) Mutant α-synuclein overexpression induces stressless pacemaking in vagal motoneurons at risk in Parkinson's disease. J Neurosci 37(1):47–57. https://doi.org/10.1523/JNEUROSCI.1079-16.2016
40. Ip CW, Klaus LC, Karikari AA, Visanji NP, Brotchie JM, Lang AE, Volkmann J, Koprich JB (2017) AAV1/2-induced overexpression of A53T-α-synuclein in the substantia nigra results in degeneration of the nigrostriatal system with Lewy-like pathology and motor impairment: a new mouse model for Parkinson's disease. Acta Neuropathol Commun 5(1):11. https://doi.org/10.1186/s40478-017-0416-x
41. Blanz J, Saftig P (2016) Parkinson's disease: acid-glucocerebrosidase activity and alpha-synuclein clearance. J Neurochem 139(Suppl 1):198–215. https://doi.org/10.1111/jnc.13517. Epub 2016 Feb 10
42. Leak RK (2018) Conditioning against the pathology of Parkinson's disease. Cond Med 1(3):143–162. Epub 2018 Apr 28
43. Weilnau JN, Carcella MA, Miner KM, Bhatia TN, Hutchison DF, Pant DB, Nouraei N, Leak RK (2018) Evidence for cross-hemispheric preconditioning in experimental Parkinson's disease. Brain Struct Funct 223(3):1255–1273. https://doi.org/10.1007/s00429-017-1552-6. Epub 2017 Nov 4
44. Rüb C, Wilkening A, Voos W (2017) Mitochondrial quality control by the Pink1/Parkin system. Cell Tissue Res 367(1):111–123. https://doi.org/10.1007/s00441-016-2485-8. Epub 2016 Sep 2
45. Bingol B, Sheng M (2016) Mechanisms of mitophagy: PINK1, Parkin, USP30 and beyond. Free Radic Biol Med 100:210–222. https://doi.org/10.1016/j.freeradbiomed.2016.04.015. Epub 2016 Apr 16
46. Nguyen TN, Padman BS, Lazarou M (2016) Deciphering the molecular signals of PINK1/Parkin mitophagy. Trends Cell Biol 26(10):733–744. https://doi.org/10.1016/j.tcb.2016.05.008. Epub 2016 Jun 10
47. Truban D, Hou X, Caulfield TR, Fiesel FC, Springer W (2017) PINK1, Parkin, and mitochondrial quality control: what can we learn about Parkinson's disease pathobiology? J Park Dis 7(1):13–29. https://doi.org/10.3233/JPD-160989
48. Kazlauskaite A, Muqit MM (2015) PINK1 and Parkin – mitochondrial interplay between phosphorylation and ubiquitylation in Parkinson's disease. FEBS J 282(2):215–223. https://doi.org/10.1111/febs.13127. Epub 2014 Nov 20
49. Barodia SK, Creed RB, Goldberg MS (2017) Parkin and PINK1 functions in oxidative stress and neurodegeneration. Brain Res Bull 133:51–59. https://doi.org/10.1016/j.brainresbull.2016.12.004. Epub 2016 Dec 23
50. Pickrell AM, Youle RJ (2015) The roles of PINK1, parkin, and mitochondrial fidelity in Parkinson's disease. Neuron 85(2):257–273. https://doi.org/10.1016/j.neuron.2014.12.007
51. van der Merwe C, Jalali Sefid Dashti Z, Christoffels A, Loos B, Bardien S (2015) Evidence for a common biological pathway linking three Parkinson's disease-causing genes: parkin, PINK1 and DJ-1. Eur J Neurosci 41(9):1113–1125. https://doi.org/10.1111/ejn.12872. Epub 2015 Mar 11
52. Whitworth AJ, Pallanck LJ (2017) PINK1/Parkin mitophagy and neurodegeneration-what do we really know in vivo? Curr Opin Genet Dev 44:47–53. https://doi.org/10.1016/j.gde.2017.01.016. Epub 2017 Feb 16
53. Mukherjee UA, Ong SB, Ong SG, Hausenloy DJ (2015) Parkinson's disease proteins: novel mitochondrial targets for cardio-protection. Pharmacol Ther 156:34–43
54. Zhang SX, Zhuang LL, Liu J, Jing YY, Sun J, Gong L, Liu XY (2018a) The role of Parkin protein in cardiac function and ventricular remodelling in myocardial infarction rats. Eur Rev Med Pharmacol Sci 22(15):5004–5013

55. Zhang SX, Zhuang LL, Liu J, Jing YY, Sun J, Gong L, Liu XY (2018b) The role of Parkin protein in cardiac function and ventricular remodeling in myocardial infarction rats. Eur Rev Med Pharmacol Sci 22(15):5004–5013. https://doi.org/10.26355/eurrev_201808_15641
56. Wu L, Maimaitirexiati X, Jiang Y, Liu L (2016) Parkin regulates mitochondrial autophagy after myocardial infarction in rats. Med Sci Monit 22:1553–1559
57. Kim T, Vemuganti R (2017) Mechanisms of Parkinson's disease-related proteins in mediating secondary brain damage after cerebral ischemia. J Cereb Blood Flow Metab 37(6):1910–1926. https://doi.org/10.1177/0271678X17694186. Epub 2017 Jan 1
58. Borovac JA (2016) Side effects of a dopamine agonist therapy for Parkinson's disease: a mini-review of clinical pharmacology. Yale J Biol Med 89(1):37–47. eCollection 2016 Mar
59. Ceravolo R, Rossi C, Del Prete E, Bonuccelli U (2016) A review of adverse events linked to dopamine agonists in the treatment of Parkinson's disease. Expert Opin Drug Saf 15(2):181–198. https://doi.org/10.1517/14740338.2016.1130128. Epub 2016 Jan 20
60. Stocchi F, Torti M, Fossati C (2016) Advances in dopamine receptor agonists for the treatment of Parkinson's disease. Expert Opin Pharmacother 17(14):1889–1902. https://doi.org/10.1080/14656566.2016.1219337. Epub 2016 Aug 29
61. Lockett K, DeBacker D, Cauthon KA (2015) The link between non-ergot-derived dopamine agonists and heart failure: how strong is it? Consult Pharm 30(3):136–140. https://doi.org/10.4140/TCP.n.2015.136
62. Adachi N, Yoshimura A, Chiba S, Ogawa S, Kunugi H (2018) Rotigotine, a dopamine receptor agonist, increased BDNF protein levels in the rat cortex and hippocampus. Neurosci Lett 662:44–50. https://doi.org/10.1016/j.neulet.2017.10.006. Epub 2017 Oct 6
63. Zhang J, Tan LC (2016) Revisiting the medical management of Parkinson's disease: levodopa versus dopamine agonist. Curr Neuropharmacol 14(4):356–363
64. Elshoff JP, Cawello W, Andreas JO, Mathy FX, Braun M (2015) An update on pharmacological, pharmacokinetic properties and drug-drug interactions of rotigotine transdermal system in Parkinson's disease and restless legs syndrome. Drugs 75(5):487–501. https://doi.org/10.1007/s40265-015-0377-y
65. Byrd RA, Weissman AM (2013) Compact Parkin only: insights into the structure of an autoinhibited ubiquitin ligase. EMBO J 32(15):2087–2089. https://doi.org/10.1038/emboj.2013.158. Epub 2013 Jul 12
66. Caulfield TR, Fiesel FC, Springer W (2015) Activation of the E3 ubiquitin ligase Parkin. Biochem Soc Trans 43(2):269–274. https://doi.org/10.1042/BST20140321
67. Dove KK, Klevit RE (2017) RING-between-RING E3 ligases: emerging themes amid the variations. J Mol Biol 429(22):3363–3375. https://doi.org/10.1016/j.jmb.2017.08.008. Epub 2017 Aug 19
68. Marín I, Lucas JI, Gradilla AC, Ferrús A (2004) Parkin and relatives: the RBR family of ubiquitin ligases. Physiol Genomics 17(3):253–263
69. Smit JJ, Sixma TK (2014) RBR E3-ligases at work. EMBO Rep 15(2):142–154. https://doi.org/10.1002/embr.201338166. Epub 2014 Jan 27
70. Wang L, Cao C, Wang F, Zhao J, Li W (2017) H2B ubiquitination: conserved molecular mechanism, diverse physiologic functions of the E3 ligase during meiosis. Nucleus 8(5):461–468. https://doi.org/10.1080/19491034.2017.1330237. Epub 2017 Jun 19
71. Swiader A, Nahapetyan H, Faccini J, D'Angelo R, Mucher E, Elbaz M, Boya P, Vindis C (2016) Mitophagy acts as a safeguard mechanism against human vascular smooth muscle cell apoptosis induced by atherogenic lipids. Oncotarget 7(20):28821–28835. https://doi.org/10.18632/oncotarget.8936
72. Eiyama A, Okamoto K (2015) PINK1/Parkin-mediated mitophagy in mammalian cells. Curr Opin Cell Biol 33:95–101. https://doi.org/10.1016/j.ceb.2015.01.002. Epub 2015 Feb 17
73. Trempe JF, Sauvé V, Grenier K, Seirafi M, Tang MY, Ménade M, Al-Abdul-Wahid S, Krett J, Wong K, Kozlov G, Nagar B, Fon EA, Gehring K (2013) Structure of parkin reveals mechanisms for ubiquitin ligase activation. Science 340(6139):1451–1455. https://doi.org/10.1126/science.1237908. Epub 2013 May 9

74. Kumar A, Chaugule VK, Condos TEC, Barber KR, Johnson C, Toth R, Sundaramoorthy R, Knebel A, Shaw GS, Walden H (2017) Parkin-phosphoubiquitin complex reveals cryptic ubiquitin-binding site required for RBR ligase activity. Nat Struct Mol Biol 24(5):475–483. https://doi.org/10.1038/nsmb.3400. Epub 2017 Apr 17
75. Spratt DE, Walden H, Shaw GS (2014) RBR E3 ubiquitin ligases: new structures, new insights, new questions. Biochem J 458(3):421–437. https://doi.org/10.1042/BJ20140006
76. Ren Y, Jiang H, Hu Z, Fan K, Wang J, Janoschka S, Wang X, Ge S, Feng J (2015) Parkin mutations reduce the complexity of neuronal processes in iPSC-derived human neurons. Stem Cells 33(1):68–78. https://doi.org/10.1002/stem.1854
77. Biswas R, Bagchi A (2017) A comprehensive computational study on pathogenic mis-sense mutations spanning the RING2 and REP domains of Parkin protein. Gene 610:49–58. https://doi.org/10.1016/j.gene.2017.02.008. Epub 2017 Feb 9
78. Levin L, Srour S, Gartner J, Kapitansky O, Qutob N, Dror S, Golan T, Dayan R, Brener R, Ziv T, Khaled M, Schueler-Furman O, Samuels Y, Levy C (2016) Parkin somatic mutations link melanoma and Parkinson's disease. J Genet Genomics 43(6):369–379. https://doi.org/10.1016/j.jgg.2016.05.005. Epub 2016 May 13
79. Zhong P, Hu Z, Jiang H, Yan Z, Feng J (2017) Dopamine induces oscillatory activities in human midbrain neurons with Parkin mutations. Cell Rep 19(5):1033–1044. https://doi.org/10.1016/j.celrep.2017.04.023
80. Fiesel FC, Caulfield TR, Moussaud-Lamodière EL, Ogaki K, Dourado DF, Flores SC, Ross OA, Springer W (2015) Structural and functional impact of Parkinson disease-associated mutations in the E3 ubiquitin ligase Parkin. Hum Mutat 36(8):774–786. https://doi.org/10.1002/humu.22808. Epub 2015 Jun 3
81. Puschmann A (2017) New genes causing hereditary Parkinson's disease or Parkinsonism. Curr Neurol Neurosci Rep 17(9):66. https://doi.org/10.1007/s11910-017-0780-8
82. Hernandez DG, Reed X, Singleton AB (2016) Genetics in Parkinson disease: mendelian versus non-Mendelian inheritance. J Neurochem 139(Suppl 1):59–74. https://doi.org/10.1111/jnc.13593. Epub 2016 Apr 18
83. Sassone J, Serratto G, Valtorta F, Silani V, Passafaro M, Ciammola A (2017) The synaptic function of parkin. Brain 140(9):2265–2272. https://doi.org/10.1093/brain/awx006
84. Hang L, Thundyil J, Lim KL (2015) Mitochondrial dysfunction and Parkinson disease: a Parkin-AMPK alliance in neuroprotection. Ann N Y Acad Sci 1350:37–47. https://doi.org/10.1111/nyas.12820. Epub 2015 Jun 29
85. Dias V, Junn E, Mouradian MM (2013) The role of oxidative stress in Parkinson's disease. J Park Dis 3(4):461–491. https://doi.org/10.3233/JPD-130230
86. Bose A, Beal MF (2016) Mitochondrial dysfunction in Parkinson's disease. J Neurochem 139(Suppl 1):216–231. https://doi.org/10.1111/jnc.13731. Epub 2016 Aug 21
87. Connolly BS, Lang AE (2014) Pharmacological treatment of Parkinson disease: a review. JAMA 311(16):1670–1683
88. Moyzis AG, Sadoshima J, Gustafsson ÅB (2014) Mending a broken heart: the role of mitophagy in cardio-protection. Am J Phys Heart Circ Phys 308(3):H183–H192
89. Cieri D, Brini M, Calì T (2017) Emerging (and converging) pathways in Parkinson's disease: keeping mitochondrial wellness. Biochem Biophys Res Commun 483(4):1020–1030
90. Bhandari P, Song M, Chen Y, Burelle Y, Dorn GW (2014) Mitochondrial contagion induced by Parkin deficiency in drosophila hearts and its containment by suppressing mitofusin novelty and significance. Circ Res 114(2):257–265
91. Narendra D, Walker JE, Youle R (2012) Mitochondrial quality control mediated by PINK1 and Parkin: links to parkinsonism. Cold Spring Harb Perspect Biol 4(11):a011338
92. Kubli DA, Zhang X, Lee Y, Hanna RA, Quinsay MN, Nguyen CK, Gustafsson ÅB (2013) Parkin protein deficiency exacerbates cardiac injury and reduces survival following myocardial infarction. J Biol Chem 288(2):915–926
93. Matsuda N, Sato S, Shiba K, Okatsu K, Saisho K, Gautier CA et al (2010) PINK1 stabilized by mitochondrial depolarization recruits Parkin to damaged mitochondria and activates latent Parkin for mitophagy. J Cell Biol 189(2):211–221

94. Kubli DA, Cortez MQ, Moyzis AG, Najor RH, Lee Y, Gustafsson ÅB (2015) PINK1 is dispensable for mitochondrial recruitment of parkin and activation of mitophagy in cardiac myocytes. PLoS One 10(6):e0130707
95. Dorn GW (2016) Parkin-dependent mitophagy in the heart. J Mol Cell Cardiol 95:42–49
96. Qiao H, Ren H, Du H, Zhang M, Xiong X, Lv R (2018) Liraglutide repairs the infarcted heart: the role of the SIRT1/Parkin/mitophagy pathway. Mol Med Rep 17(3):3722–3734
97. Han K, Hassanzadeh S, Singh K, Menazza S, Nguyen TT, Stevens MV, Nguyen A, San H, Anderson SA, Lin Y, Zou J, Murphy E, Sack MN (2017) Parkin regulation of CHOP modulates susceptibility to cardiac endoplasmic reticulum stress. Sci Rep 7(1):2093
98. Vásquez-Trincado C, García-Carvajal I, Pennanen C, Parra V, Hill JA, Rothermel BA, Lavandero S (2016) Mitochondrial dynamics, mitophagy and cardiovascular disease. J Physiol 594(3):509–525
99. Van Humbeeck C, Cornelissen T, Hofkens H, Mandemakers W, Gevaert K, De Strooper B, Vandenberghe W (2011) Parkin interacts with Ambra1 to induce mitophagy. J Neurosci 31(28):10249–10261
100. Hailey DW, Rambold AS, Satpute-Krishnan P, Mitra K, Sougrat R, Kim PK, Lippincott-Schwartz J (2010) Mitochondria supply membranes for autophagosome biogenesis during starvation. Cell 141(4):656–667
101. Fimia GM, Stoykova A, Romagnoli A, Giunta L, Di Bartolomeo S, Nardacci R, Corazzari M, Fuoco C, Ucar A, Schwartz P, Gruss P, Piacentini M, Chowdhury K, Cecconi F (2007) Ambra1 regulates autophagy and development of the nervous system. Nature 447(7148):1121
102. Nakai A, Yamaguchi O, Takeda T, Higuchi Y, Hikoso S, Taniike M, Omiya S, Mizote I, Matsumura Y, Asahi M, Nishida K, Hori M, Mizushima N, Otsu K (2007) The role of autophagy in cardiomyocytes in the basal state and in response to hemodynamic stress. Nat Med 13(5):619

Part III

Prevention and Therapeutics

New Technologies in Drug Development Provide New Hope in Targeting of Dysregulated Redox Signalling in Cardiovascular Disease

21

Soloman Saleh, Kristen Bubb, and Gemma A. Figtree

Abstract

Ever-accumulating evidence supports the pivotal role of dysregulated redox signalling in a broad spectrum of cardiovascular disease and degenerative ageing. Until now, therapeutic strategies have involved non-specific dietary antioxidants which have failed to demonstrate clinical benefits. Indirect success has been seen in the context of effective receptor-based pharmacotherapies such as antagonists of angiotensin or β1-adrenergic signalling. A major challenge has been to successfully target key subcellular compartments, each with separate redox microenvironments, but communicating with each other through a network of signalling pathways and cascades. Caveolar, mitochondrial, inflammasome, and transcriptional regulation have all proven to have redox-sensitive elements. The expanding 'tool box' available in the modern drug development field has opened the door for new approaches to treating or even reversing dysregulated redox signalling in these microdomains. Small molecules, novel genetic vectors, and biologics combined with nanoparticle delivery mechanisms are all emerging approaches to tackle shortcomings of our existing pharmacological toolset. This

S. Saleh
Cardiothoracic and Vascular Health, Kolling Institute, University of Sydney, St Leonards, NSW, Australia

School of Medical Sciences, Faculty of Medicine, University of New South Wales, Sydney, NSW, Australia

K. Bubb
Cardiothoracic and Vascular Health, Kolling Institute, University of Sydney, St Leonards, NSW, Australia

G. A. Figtree (✉)
Cardiothoracic and Vascular Health, Kolling Institute, University of Sydney, St Leonards, NSW, Australia

Department of Cardiology, Royal North Shore Hospital, Sydney, NSW, Australia
e-mail: gemma.figtree@sydney.edu.au

S. Chakraborti et al. (eds.), *Modulation of Oxidative Stress in Heart Disease*,
https://doi.org/10.1007/978-981-13-8946-7_21

chapter reviews recent advances in molecular targets of cardiovascular therapy, emerging technologies for their delivery, and approaches in subcellular targeting of pharmaceuticals.

Keywords
Redox signaling · Reactive oxygen species · Caveolae · Drug development · Nitroso-redox balance · Mitochondria · Mitochondrial targeting · Inflammasome · Nanoparticles · microRNA · Gene therapy · Micropeptide therapy

21.1 Introduction: Heart Disease and the Shortcomings of Current 'Antioxidant' Therapies

Cardiovascular disease remains one of the leading causes of both death and morbidity globally [77]. Despite widespread improvements in pharmacotherapy, health literacy, and minimally invasive intervention, the burden of disease remains massive in the face of a globally ageing population and in the context of an epidemic of obesity and metabolic abnormalities [30]. As such, accumulation of irreversible myocardial and vascular damage inevitably converges on the all-too-common failing cardiac phenotype of normal ageing and chronic disease.

A common mechanism implicated in CVD is dysregulated redox signalling. This occurs when there is excessive generation of oxidative free radicals. Pathological redox signalling is complex and variable within the heart and vasculature. This makes pharmacological intervention aimed at targeting dysregulated signalling domains difficult and no clear guidelines for specific antioxidant therapies exist for clinicians. Previous attempts to utilise antioxidants have been limited by their indiscriminate tissue distribution, non-specific activity, and unpredictable pharmacokinetic profiles [85]. The initial excitement of promising small-scale trials of antioxidant therapies to target cardiovascular disease was met with disappointment in larger clinical studies, failing to meet clinically relevant endpoints despite solid preclinical grounds [37]. Investigations have been conducted into the reasons for the failures of antioxidant clinical trials, such as the highly anticipated HOPE Vitamin E trial. It was revealed that systemic and untargeted administration of antioxidants may not only fail to reach the biological target, but may in fact disturb otherwise intact redox signalling domains [59].

However, recent advances in our understanding of the redox state in cardiovascular disease, alongside new developments in the applications of biochemical engineering, offer hope of a paradigm shift in the delivery of agents capable of modifying dysregulated oxidative signalling domains in the cardiovascular system. The advent of highly targeted drug delivery platforms offers the opportunity to exploit our growing knowledge of redox compartmentalisation to modulate these pathways with precision and specificity. In this chapter we explore promising subcellular redox targets in cardiac disease, in the context of recent successes with the traditional small-molecule approach. We then review the changing landscape of technologies with the potential to expand the scope of biological targets available for pharmacological manipulation. Finally, the third subchapter discusses how these technologies may be used to precisely deliver therapeutic agents to specific subcellular redox microdomains.

21.2 The Redox Subcellular Microdomains: Promising Molecular Targets and Small-Molecule Approaches

Dysregulated production of, or protection against, reactive oxygen species (ROS) – superoxide ($O_2^{\cdot-}$), hydrogen peroxide (H_2O_2), reactive nitrogen species (RNS), and nitric oxide (NO) – has been implicated in the origins of key pathological motifs common across cardiovascular disease [62]. Cardiometabolic, endothelial, post-translational, and immunological function have all been shown to have a redox component to their progressive decline in cardiac impairment. To date, current pharmacological formulations to address these imbalances can be broadly divided into two categories: small-molecule drugs and biologics. Small molecules – such as receptor-based therapies and neurohormonal modulators – make up the overwhelming majority of established therapies in cardiovascular disease, including β-blockers, statins, and renin-angiotensin system (RAS) antagonists which act, at least in part, through redox-protective means [36, 67]. Conversely, larger but more complex biologics – including antibodies, proteins, and genetic material – make up an emerging field showing great promise in manipulating redox signalling pathways. However, these often suffer from poor stability and trafficking through membranes and to the site of disease [41]. Here we discuss promising targets and small-molecule therapies to directly modulate these pathways implicated in the progression of cardiac ageing and chronic disease. We address key redox compartments within the cell: the caveolae, the mitochondria, the inflammasome, and finally the nucleus.

21.2.1 Neurohormonal Pathways Regulate Nitroso-Redox Imbalance in the Caveolae

The cardiovascular system receives substantial neurohormonal input controlling heart rate, inotropy, and cellular growth [36]. Invariably, this signal is conducted via membrane hormone receptors anchored to flask-shaped plasma membrane invaginations known as caveolae [19]. The capacity of these caveolae as a signalling domain is underpinned by tightly controlled ROS generation interacting with downstream effector proteins, kinases, and transporters in a highly targeted post-translational fashion [119]. Within relevant CV cells, including endothelial cells, vascular smooth muscle cells, and cardiomyocytes, the bulk of this originates from the NADPH oxidase (Nox) superfamily. This is a group of 'primary ROS generator' enzymes with the express function of localising synthesis of $O_2^{\cdot-}$ or H_2O_2 for the purpose of adaptive signalling and reversible protein modifications [4]. More specifically, the Nox family contains seven members, four of which have been detected in cardiovascular tissue: Nox-1, Nox-2, Nox-4, and Nox-5. Of these, all but Nox-4 have been shown to produce $O_2^{\cdot-}$ in an inducible neurohormonal-sensitive manner, whereas Nox-4 constitutively generates H_2O_2 and appears to act as an intracellular oxygen sensor [39]. Caveolar $O_2^{·-}$ levels in turn facilitate S-glutathionylation, a post-translational thiol modification process acting as a functional 'switch', reversibly modulating the activity of target proteins within the caveolae, such as the Na^+/K^+/ATPase pump, ion channels, and signalling kinases [11]. While important for adaptive physiological signalling, this leads to indiscriminate ROS production and inappropriate post-translational redox modification of caveolar contents in the setting of the chronic neurohormonal overstimulation associated with decompensating cardiac disease.

Importantly, the position of Nox activity downstream of existing neurohormonal therapies, such as RAS antagonism, suggests that direct Nox inhibition may have synergistic effects with these established therapies. Pharmacological inhibition of Nox may be achieved through two main mechanisms: by targeting the membrane-bound catalytic subunits of Nox itself or by preventing the translocation and docking of regulatory subunits from the cytosol to the caveolae [21]. While the latter of these is the topic of peptide therapy technologies (see 21.3.5), the former has been employed in small-molecule form. GKT137831 is a direct Nox inhibitor with specificity to Nox-1 and Nox-4 isoforms. While the mechanism of action has not been formally elucidated, GKT137831 was designed through a rational structural analysis of Nox itself, implying a direct inhibitory binding to the catalytic subunit [92]. Furthermore, treatment of apolipoprotein E knockout (atherosclerosis-prone) diabetic mice with GKT reduced atherosclerotic plaque size, reduced macrophage accumulation, and attenuated inflammatory response as well as a similar renal protective effect [38]. The association of all of these improvements with a reduced ROS in vascular and renal tissues leaves specific Nox inhibitors as a promising avenue for targeting caveolar ROS in a wide range of diseases, either alone or used synergistically with upstream neurohormonal and downstream effector therapies.

Nitric oxide (NO) is another key player in the caveolae, and its downstream signalling more broadly influences cellular redox state. NO levels are closely coupled to Nox activity through multiple mechanisms and are intimately involved in controlling cardiac contractility, vascular tone, and remodelling [115]. At the heart of this circuit is endothelial NO synthase (eNOS), a caveolae-resident enzyme that co-localises with Nox and is the chief producer of NO within the compartment [35]. eNOS exists in one of two distinct states: 'coupled' to its substrate L-arginine and cofactor BH_4, or in an uncoupled state in which $O_2^{·-}$ is generated in the place of NO. Several stimuli can trigger eNOS uncoupling, including substrate depletion, but also oxidative modification of either BH_4 or glutathionylation of eNOS itself [63]. NO and ROS may have synergistic or antagonistic effects depending on their relative concentrations. At low/physiological levels, $O_2^{·-}$ potentiates NO's ability to act in a cardioprotective manner via S-nitrosylation, a post-translational thiol modification analogous to glutathionylation and mediated by NO [45]. At higher concentrations however, $O_2^{·-}$ 'quenches' NO to form peroxynitrite which reduces NO bioavailability and subsequently compromises vasodilation, triggers inflammatory pathways, and promotes hypercoagulability [31]. This gives rise to the concept of the nitroso-redox balance – the ratio between NO and $O_2^{·-}$ levels – as an important determinant of the post-translational status of caveolar proteins and, by extension, general cardiovascular health. Crucially, the position of eNOS both upstream and downstream of ROS generation, alongside its proximity to Nox within the caveolae, allows for a 'spark' of ROS to destabilise nitroso-redox balance. This can trigger an exacerbating cycle of eNOS glutathionylation, uncoupling, and spiralling $O_2^{\cdot-}$ excess. This interplay between neurohormonal input, Nox activity, and eNOS state has been shown experimentally where angiotensin II treatment resulted in increased eNOS glutathionylation in a Nox-dependent manner in human endothelial cultures [34]. However, while this led to decreased bioavailability of NO and increased $O_2^{\cdot-}$ levels, transfection with 'non-glutathionylatable' eNOS restored this imbalance, reinforcing the synergistic therapeutic potential of targeting redox dysfunction both at the level of Nox and in upstream neurohormonal pathways (Fig. 21.1).

It then comes as little surprise that neurohormonal therapies show promise in regulating eNOS state and restoring nitroso-redox balance within the caveolae. β_3-adrenergic stimulation in particular has been shown to have an inhibitory impact on glutathionylation, both of eNOS and other key proteins. Notably, excessive glutathionylation of the Na^+/K^+/ATPase pump is implicated in heart failure and vascular dysfunction [110, 16]. Infusion of a β_3-agonist, CL316243, abolished elevated $O_2^{\cdot-}$ levels and restored deficient NO levels in a hyperglycaemic rabbit model, alongside

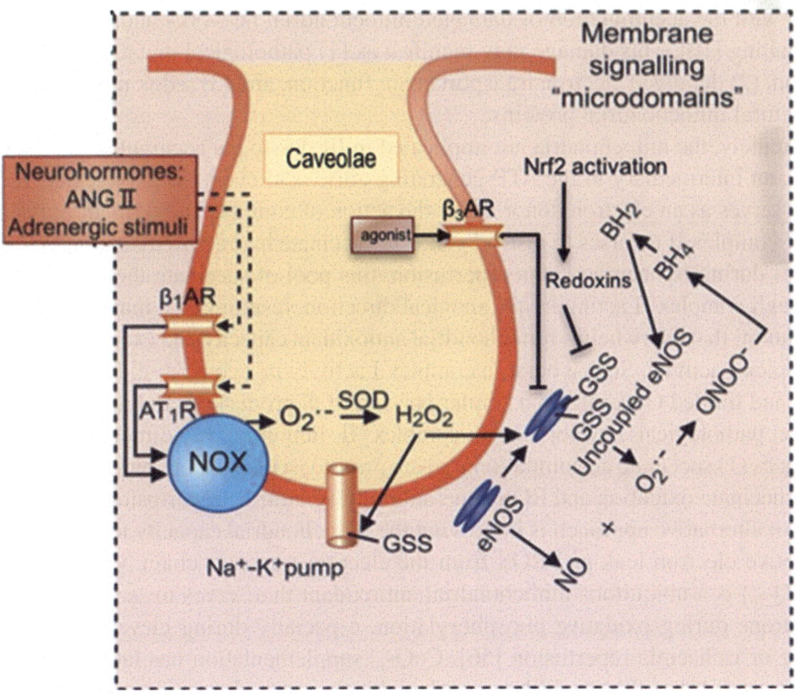

Fig. 21.1 eNOS uncoupling/Nox, downstream effects, post-trans mods. (Adapted with permission from [13])

a normalisation of eNOS glutathionylation in the treated group [51]. This was accompanied by a restored endothelial relaxation capacity. The impact on heart failure was investigated in a clinical study, the phase II BEAT-HF trial where the β_3-agonist mirabegron was given to a heart failure population. While there was no significant effect in major cardiovascular outcomes between the mirabegron and placebo-treated groups, there was a significant benefit in a subgroup with severe systolic dysfunction and an LV ejection fraction <40% [15]. These encouraging results suggest that rebalancing of NO and ROS in a caveolae-targeted manner may be a promising prospect.

21.2.2 Targeting Metabolic, Enzymatic, and Structural Components of the Mitochondria

The heart exhausts and replenishes its ATP reserves approximately once every 10 s [111]. As such, matching metabolic supply against the highly variable demand of daily life becomes a delicate balancing act between metabolic excess and ischaemic injury. Cellular machinery must be exquisitely 'aware' of the nutritional state of the tissue, and as the greatest source of intracellular ROS, the mitochondria are naturally situated to convey this information to the cell proper via redox signals. Conversely, normal ageing, metabolic disarray, and chronic diseases are all associated with the accumulation of damaged mitochondrial networks and dysfunctional signalling [10]. This damage may manifest as (1) pathological metabolite accumulation, (2) deranged electron transport chain function, and (3) redox modification of structural mitochondrial proteins.

Acutely, the mitochondria are implicated in IR due to the accumulation of succinate, an intermediary in the ATP-generating citric acid cycle [86]. Typically, succinate serves as an electron donor under the action of complex II, but in an ischaemic state complex II reverses to instead generate succinate in order to maintain ATP synthesis during ischaemia. Upon reperfusion, this pool of succinate then passes back through complex II acting in its canonical direction, resulting in a massive 'electron donation' that overwhelms mitochondrial antioxidant capacity and excessively drives complex I activity. Such a burst in complex I activity in a context of depleted cofactors and limited O_2 results in a similar burst in ROS production. In line with both of these pathological developments, complex II inhibition by dimethyl malonate reduces (1) succinate accumulation if given prior to ischaemia, but more importantly, (2) succinate oxidation and ROS generation if given during reperfusion [20].

An alternative approach is improving the mitochondrial capacity to survive maladaptive electron leak and ROS from the electron transport chain. Coenzyme Q_{10} (CoQ-$_{10}$) is a ubiquitous mitochondrial antioxidant that serves to 'soak up' excess electrons during oxidative phosphorylation, especially during elevated metabolic drive or ischaemia-reperfusion [56]. CoQ-$_{10}$ supplementation has largely failed to prove useful, but this could be contributed to by poor absorption and delivery to mitochondria and low-powered studies. More promising recent results show that CoQ-$_{10}$ can reduce major cardiac events compared to placebo, alongside reductions in both cardiovascular and all-cause mortality [74]. Building on this is MitoQ, a CoQ-10 formulation tagged with TPP^+, an ion that improves drug accumulation within the mitochondria (see 21.4.3). In a 6-week randomised controlled trial, MitoQ reduced aortic stiffness, improved brachial flow, and reduced oxidised LDL in an adult human cohort with impaired endothelial function [96].

More chronically, oxidative ageing of the mitochondria affects cell survival and inflammation through disruption of the structural protein cardiolipin. Cardiolipin is a phospholipid embedded in the inner mitochondrial membrane with a key role in enzyme integrity and mitochondrial turnover [100]. The best-characterised function of cardiolipin is as a structural component of the electron transport chain, where it stabilises oxidative phosphorylation enzyme super-complexes and promotes anchoring of mitochondrial constituents to the inner membrane. Cardiolipin is also a potent regulator of the removal of damaged mitochondria through the organelle recycling process known as mitophagy, wherein cardiolipin localises to the outer mitochondrial membrane as a signal to trigger fusion with lysosomes and subsequent degradation [84]. However, its physical proximity to ROS generation puts cardiolipin at considerable risk of maladaptive oxidative modification. When disrupted, oxidised cardiolipin loses the ability to retain mitochondrial subunits to the inner membrane, compromising metabolic capacity and releasing the pro-apoptotic factor cytochrome C (Fig. 21.2).

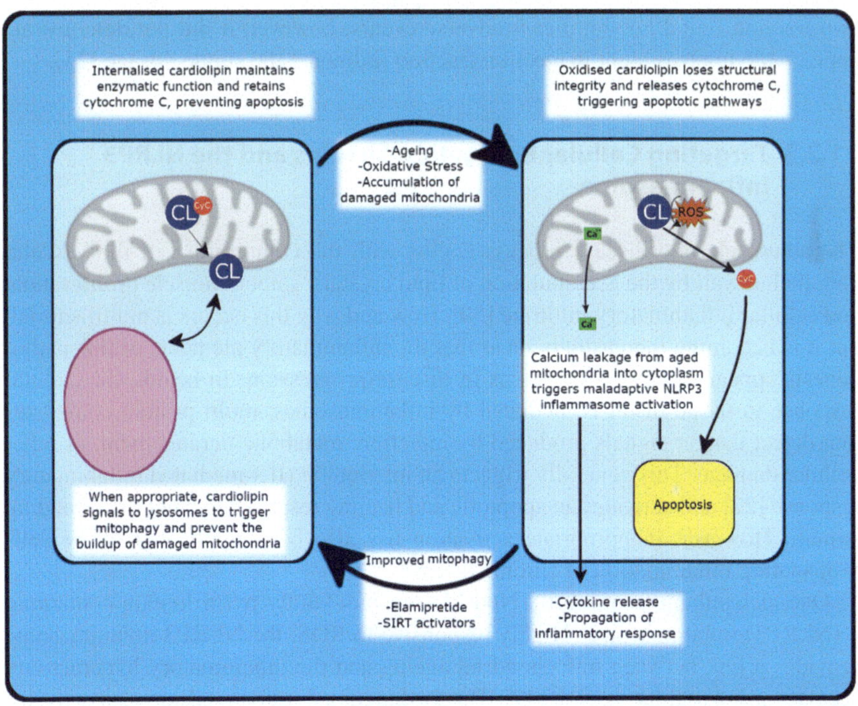

Fig. 21.2 Mitochondrial integrity in the healthy and aged heart. *CL* cardiolipin, *CyC* cytochrome C, *ROS* reactive oxygen species

Accordingly, cardiolipin levels are known to decrease in the ageing or ischaemic heart – associated with increased apoptosis, decreased mitophagy, and accumulation of dysfunctional mitochondria [27].

Cardiolipin-based therapies have recently been met with success in both human and animal trials. Elamipretide is a tetrapeptide that localises to the mitochondrial inner membrane and stabilises cardiolipin, supressing its oxidative modification by binding to its hydrophobic domains and maintaining membrane cristae structure [8]. Cardiolipin, as well as cytochrome C and active caspase-3 (other associated regulators of mitochondrial kinetics), improved mitochondrial and cardiac functional parameters in a preclinical heart failure model [101]. Improvements included normalisation of mitochondrial membrane potential, ATP synthesis, and ATP/ADP ratios which were accompanied by increases in expression of both eNOS and SERCA-2a activity – both associated with the heart failure phenotype. Cardiolipin (elamipretide) has also proven successful in a preclinical model of atherosclerosis [128]. Importantly, reduced plaque burden was not only associated with improved ATP synthesis but also the attenuation of a systemic inflammatory response seen in atherosclerosis, with reduced IL-6 and LOX-1 receptor downregulation reducing macrophage lipid influx and foam cell formation. This suggests that elamipretide restores a healthy cardiac metabolic phenotype through mitochondrial stabilisation and subsequently leads to a reduction in ROS-related immune dysfunction. Finally, human phase I trials of single-dose elamipretide for heart failure with reduced ejection fraction found no significant adverse events; however, it did not demonstrate appreciable improvement in ejection fraction following the single dosage [25].

21.2.3 Targeting Cellular Energetics: Sirtuins and the NLRP3 Inflammasome

The atherosclerotic disease process begins with the compromise of the vascular endothelial wall by the accumulation of lipid crystals, smooth muscle proliferation, and cellular inflammatory infiltrate [50]. How and why this occurs is multifactorial, but it is becoming increasingly clear that the inflammatory elements of this pathogenesis appear to be driving forces in disease progression. In health, the cellular response to tissue insult is mediated by inflammasomes, multi-protein complexes that detect danger signals produced by infection, metabolic derangement, or other cellular damage. This generally triggers an interleukin (IL)-mediated inflammatory response [22] and activates an apoptotic and healing response to clear the offending stimuli. However, inappropriate activation has also been implicated in the self-propagating inflammation of chronic disease.

One such inflammasome is the Nod-like receptor family, pyrin domain-containing 3 (NLRP3) inflammasome. Highly responsive to ROS, the NLRP3 inflammasome provides a link between mitochondrial ageing and the inflammatory hyperactivity seen in cardiovascular disease [49]. The dysfunctional mitochondrion also activates the NLRP3 inflammasome through several indirect pathways. Ca^{2+} leakage from ageing mitochondria to the cytoplasm, mitochondrial $NAD/NADH^{+}$ imbalance, and cardiolipin translocation to the outer mitochondrial membrane can all activate NLRP3 cascades (Fig. 21.2) [42]. As the inflammatory response mounts, attempted phagocytosis of cholesterol crystals is followed by lysosomal rupture and release of pro-inflammatory enzymes that further exacerbate NLRP3 activation. The net effect is that as damaged mitochondria accumulate due to insufficient mitophagy, NLRP3 is inappropriately activated and leads to the release of inflammatory cytokine IL-1β in a magnifying spiral of apoptosis and growing inflammation.

One endogenous target for the rectification of maladaptive NLRP3 inflammasome activity is the sirtuin (SIRT) family. The sirtuin family of enzymes are responsible for cellular responses to metabolic fluctuation by catalysing de-acetylation reactions, a post-translational modification targeting proteins and DNA histones [120]. The endogenous substrates of sirtuins are broad – indeed, interactions with NLRP3 may be only one of many mechanisms by which sirtuins normalise cardiovascular tissue. In particular, SIRT1 and SIRT3 have both shown protective effects against mitochondrial overactivation of the NLRP3 inflammasome [24]. SIRT1 resides primarily in the nucleus, where its targets include a regulator of mitochondrial biogenesis, peroxisome proliferator-activated receptor gamma coactivator 1-alpha (PGC-1a). Alternatively, SIRT3 – localised to the mitochondria – improves mitophagy, upregulates mitochondrial antioxidant manganese superoxide dismutase (MnSOD), and reduces IL-1β in the human aortic endothelium [58]. Conversely, ROS may directly inhibit sirtuins by reacting with and exhausting NAD^+, an essential cofactor.

Unsurprisingly, sirtuin activators are a class of molecules with putative cardioprotective effects through immunomodulatory, anti-remodelling, and endothelial-protective functions. One such molecule is SRT1720, a SIRT1 activator which has previously improved lifespan in obese mice and is suggested to have SIRT3 reactivity [70]. More recently, SRT1720 has been evaluated in a model of left ventricle pressure overload and successfully attenuated the left ventricular remodelling response and restored ejection fraction [14]. In light of the declining sirtuin levels found in heart failure, sirtuin activators stand as a promising therapy for rectifying mitochondrial and redox imbalances common to many cardiovascular diseases [32].

21.2.4 Targets in the Nucleus: Total Cellular Redox Protection by Nrf2-Dependent Transcriptional Regulation

Keeping these processes in check is the intracellular antioxidant defence network, which must be finely sensitive to ROS production as to prevent oxidative damage without perturbing physiological redox signalling. To this end, many endogenous protective enzymes rely on antioxidant response elements (AREs) for their transcription [23]. The transcription factor nuclear factor erythroid-2-related factor 2 (Nrf2) is a master regulator of antioxidant defences, translocating to AREs to promote the synthesis of key antioxidants [2]. However, at its basal state Nrf2 is bound to its inhibitor Keap1. It is only under the influence of excess ROS that Keap1 is modified and releases Nrf2. This is a remarkable mechanism that allows the intracellular antioxidant machinery fine temporal and spatial.

Dihydro-CDDO-trifluoroethyl amide (DH404) is a synthetic small molecule with the capacity to stimulate endogenous antioxidant defences through the Nrf2 pathway [102]. DH404 disrupts the ubiquitination and subsequent degradation of Nrf2, increasing its translocation to antioxidant response elements (AREs) and upregulating antioxidant synthesis. In line with this, DH404 admission in a rat MI model markedly decreased infarct size and ameliorated ventricular remodelling post-infarct [12]. These changes were associated with a reduction of eNOS glutathionylation. Given DH404's interactions with Nrf2 and the antioxidant system, this implies a causative relationship through a stabilisation of eNOS in its coupled state, restoring caveolar health indirectly through therapeutic targeting of the nucleus.

21.3 Emerging Technologies in Drug Discovery and Moving Beyond Small Molecules

Our ever-increasing characterisation of the redox network in cardiac and vascular tissues leaves no shortage of molecular targets for novel therapeutics. Less sophisticated, however, has been our ability to reach these molecules and pathways in a controlled and precise manner. To this end, cardiovascular pharmacotherapy has gravitated towards more intricate development strategies in recent years, pursuing both improvements in small-molecule discovery and novel alternatives capable of more complex pharmacological interactions [72]. As we move toward more powerful drug discovery techniques, biochemically engineered microstructures, and larger molecule therapies, we expand our ability to target individual cell populations, diseased tissues, or even subcellular compartments with precision and specificity. In this section, we give an overview of advances in small-molecule discovery platforms and move on to the growing role of nanoparticle formulations in cardiovascular redox therapy, before finally examining the various classes of biologics as they enter the preclinical and clinical spheres.

21.3.1 Small-Molecule Libraries and Fragment-Based Drug Discovery

Modern-day small-molecule discovery is typically library-based – that is, a diverse 'library' of molecules is screened for those with the desired pharmacological action to use as starting points for drug formulations [29]. There are several techniques used in the discovery of small molecules, most prominently high-throughput screening and more recently fragment-based lead discovery (FBLD). Both techniques follow a similar workflow; a library is generated, filtered for elements with a binding potential to the target, and finally analysed structurally to tweak into a final optimised drug [75]. The difference between the two techniques lies in the size of each molecule in the library – where compounds screened in high-throughput screening are restricted to 500 Da, each 'fragment' analysed in FBLD is generally kept below 200 Da. Initially, these smaller fragments make for much lower-affinity compounds; however, FBLD brings with it a number of advantages that ultimately allow for the development of a more effective drug. More specifically, multiple promising fragments may be combined into the one drug; thus even a small library makes for an exponentially larger number of potential permutations of several fragments [54]. Indeed, despite the use of libraries several orders of magnitude smaller than HTS libraries, FBLD more effectively mines its 'chemical space' – that is, it tests a greater proportion of all possible drugs for efficacy. Further, FBLD naturally favours the development of hydrophilic drugs, leaving greater potential for optimising affinity through the addition of lipophilic adducts without resulting in an excessively lipophilic drug [97].

In turn, the chief challenge of FBLD methods is the need to detect binding activity at these initially low binding affinities. X-ray crystallography, nuclear magnetic resonance, and surface plasmon resonance (SPR) are all high-sensitivity methods that typically outperform biochemical assays at low binding affinities [122]. SPR in particular benefits from a low amount of target protein required and a resistance to drug aggregation; in this method, a metal surface is conjugated with the target protein and coated in a drug-containing solution. A light source is reflected off the metal surface, while ligand formation between drug and protein causes minute shifts in the metal plate and thus the oscillations of the reflected light [87]. SPR is sensitive over a wide range of affinities (min-max) and resistant to error from drug aggregation which limits detection of false positives. This is doubly important in the high drug concentrations needed to detect fragment binding. These recent advances in library-based methods and ligand detection pave the way for small-molecule formulations to have a very real place in highly targeted redox therapy.

21.3.2 Nanoparticles for Targeted Drug Delivery

'Nanoparticle' is a catch-all term for a wide range of structures produced on a nanometre scale, often loaded with drugs or other small molecules, with designer pharmacokinetic profiles for use in treatment or diagnostics [48]. Functionally, nanoparticles are characterised by physical and chemical properties determined not only by their material but also by their structural arrangement. This allows for an overwhelming level of fine-tuning in the biological properties of these compounds. Nanoparticles may serve several functions in pharmacology, but are most commonly used as carriers for other drugs. They are robust against degradation and exhibit other attractive qualities such as potential for tissue targeting and reduced toxicity [6]. A simplistic and familiar example is found in liposomes, which may be 'loaded' with a hydrophilic target that then inherits the membrane permeability of its lipid shell. Indeed, such liposomes are often used as non-viral vectors in genetic transfection to assist membrane traversal and shield genetic material from degradation [65].

Lipid structures such as liposomes are just one class of nanoparticles. In addition, metallic, polymer, and non-metal lattices all have a solid proof of concept for use in targeted cancer therapy and are gaining traction in cardiovascular therapy [68]. Increasingly complex or novel structures may be manufactured from an organic or artificial scaffold, loaded with a bioactive payload, and finally functionalised by stabilising or targeting moieties to improve half-life and tissue specificity (Table 21.1).

An example of this is the small-molecule pitavastatin. In vitro, statins are known to protect against cardiovascular disease by lipid-lowering activity, as well as by restoring redox balance. This leads to improved remodelling and decreased inflammation at high doses [5]. Despite this theoretical backing however, statins have failed to demonstrate any clinically observable cardioprotective effect if administered during acute infarct or ischaemia-reperfusion (ref). A series of preclinical studies have shown that combining pitavastatin with a polymeric nanoparticle increases atherosclerotic plaque stability and decreases vascular monocyte recruitment. This strategy has also proven advantageous in preclinical work when administered during reperfusion of the ischaemic heart [47, 52]. These effects – at least in part due to the nanoparticle's preferential uptake by monocytes – were not detected in the pitavastatin only group. Pitavastatin nanoparticles have undergone phase I trials with no significant adverse outcomes and are currently undergoing phase II trials in critical limb ischaemia and pulmonary hypertension [78].

21.3.3 miRNA Therapies May Offer Cardiac Regenerative Potential

MicroRNAs (miRNA or miR) are short non-coding nucleotide arrays with potent translational regulation of coding messenger RNAs [98]. miRNAs inhibit protein translation through both direct mRNA degradation and formation of an 'RNA-induced silencing complex', a protein complex capable of cleaving the miRNA targets. Both mechanisms allow a given miRNA up to hundreds of inhibitory targets, and roughly 40 miRNAs have been implicated in the progression or attenuation of cardiovascular disease through putative effects on the redox network [55].

AntagomiRs are synthetic oligonucleotides with an anti-sense sequence matching their target miRNA, binding to and thus inhibiting their action [61]. Over the past 15 years, several side-chain modifications have been developed to protect against endogenous nucleases and encourage binding to the target miRNA. However even with these modifications, antagomiRs face pharmacokinetic challenges and naturally accumulate in the liver and kidneys, posing a hepatotoxic or nephrotoxic threat [57]. This poor targeting is doubly concerning – not only do lower concentrations reach the target site, but delivery of potentially regeneration-inducing agents to unintended tissues may pose an oncogenic risk. Table 21.1 includes several nanoparticle formulations to circumvent these concerns with targeted antagomiR delivery to cardiovascular tissue and further reducing exposure to nuclease activity.

Table 21.1 Classes of nanoparticles and examples of cardiovascular applications

Indication	Class	Contents	Functional groups	Outcomes
Heart failure	Polyketal polymer	S100A1	N-acetylglucosamine	Improved Ca^{2+} handling in cardiomyocytes [69]
	Calcium phosphate	R7W-MP	–	Inhaled nanoparticles localise to myocardium and improve fractional shortening in diabetic cardiomyopathy [71]
Atherosclerosis	PLGA polymer	Simvastatin	High-density lipoprotein (improves endothelial uptake)	Preferential uptake by vessel wall, attenuated inflammation, marked decrease in atherosclerotic burden [28]
	PEG/PEI polymer	miR-146a & miR-181b	E-selectin (improves endothelial uptake)	Reduced vascular inflammatory markers, chemokinetic response, plaque size [64]
	PLGA polymer	Pioglitazone	–	Increased plaque stability, normalised macrophage differentiation [80]
Myocardial infarction/ ischaemia-reperfusion	PLGA polymer	Pitavastatin	–	Reduced left ventricle remodelling post-infarct, reduced infarct size – not seen in standard pitavastatin arm [47]
	Liposome	BH_4	–	Improved eNOS coupling during ischaemia and reperfusion, increased NO, reduced ROS [123]
	PLGA polymer	VEGF gene	–	Increased vascular density, reduced infarct size, improved ejection fraction [83]

(continued)

Table 21.1 (continued)

Indication	Class	Contents	Functional groups	Outcomes
Cardiac regeneration	PLGA polymer	miR-132	Cyclic RGD (improves clathrin-mediated uptake)	Improved angiogenesis in vitro, followed by greater survival on transplantation [26]
	Carbon nanotube	–	–	Improved conductivity in tissue engineering scaffolds [3]

Despite these challenges, miRNA therapies have yielded promising results *in vivo*. miR-34a is a tumour suppressor with transcriptional inhibition of the cell cycle that controls protein B-cell lymphoma 2 (Bcl2), cyclin D1, and SIRT1 [126]. Both Bcl2 and cyclin D1 appear to play a role in the short-lived regenerative capacity of the neonatal heart, while SIRT1 has anti-apoptotic and redox-protective effects. Administration of a miR-34a inhibitor displayed a remarkable partial rescue of both function and tissue viability in infarcted myocardium in adult mice. It remains questionable as to whether miR-34a simply reduces post-infarct remodelling or truly reactivates myocardial regenerative machinery, and this warrants future investigation.

21.3.4 Gene Therapy

Over the past 15 years a plethora of genetic targets, from cancer to congenital mutations, have been brought to clinical trial – almost all of which have stubbornly remained in the sphere of research [121]. In fact, it was not until 2015 with Glybera, a treatment for lipoprotein lipase deficiency, that the first gene therapy entered clinical use in the western world [73]. Each gene therapy consists of two parts – a genetic payload and a delivery vector. Of these, the latter takes chiefly one of three overarching approaches: (1) naked plasmid, (2) viral vector, or (3) nanoparticle carrier, each with its own profile of advantages and challenges. Viral vectors are the most commonly employed and best studied for their shielding of fragile genetic material and relative simplicity compared to nanoparticles [94]. More recently, however, on the platform of success in cancer therapy, nanoparticles have crossed over to cardiovascular gene therapy. They offer real promise for overcoming some of the roadblocks that have held gene therapy back from clinical use.

The sarcoplasmic-endoplasmic reticulum Ca^{2+}/ATPase (SERCA) is a critical regulatory pump controlling myocardial calcium homeostasis and contractility implicated in the progression of cardiac failure [44]. Built on the back of promising preclinical data, the recent CUPID-2 phase II trial delivered SERCA-2 gene therapy via adenoviral vector to a cardiac failure and cardiomyopathy population [40]. However, despite high expectations, CUPID-2 failed to demonstrate any significant change in left ventricular ejection fraction, seemingly due to insufficient gene delivery by the viral vector. While disappointing, the results of CUPID-2 do not preclude a future for SERCA gene therapy, especially given the recent advances in non-viral, nanoparticle-based DNA delivery covered below. A genetic target with more clinical success is vascular endothelial growth factor (VEGF). VEGF stimulates all of the major pro-angiogenesis pathways. It promotes redox balance by reducing Nox activity, promoting the coupling of eNOS, and improving Nrf-2-mediated antioxidant expression [81]. Accordingly, the phase I/II KAT301 study administering percutaneous intramyocardial injections of adenoviral VEGF in coronary artery disease demonstrated not only safety but improved myocardial perfusion at 1 year on PET scan [43].

Despite their complexity, nanoparticles have several advantages over adenoviruses. Adenoviruses suffer from endogenous antibody resistance which can limit the drug viability and efficacy in an unpredictable manner and development of tolerance making subsequent dosing less effective [46]. Meanwhile nanoparticles offer a more controllable pharmacokinetic profile, targeting, and controlled release. Building on this is the delivery of a VEGF plasmid using a redox-sensitive heparin-loaded polymeric nanoparticle. The abundance of intracellular redox agents favours the reduction of the heparin elements in the nanoparticle, freeing them and displacing the relatively less electronegative plasmid DNA in a self-accelerating feedforward loop [82]. This results in maximised drug unloading, but only after it has entered the intracellular space, where the particle may be exposed to a reducing agent and thus 'unlocked'. A similar strategy has been employed in the delivery of the eNOS gene, using a redox-sensitive polymer nanoparticle [116]. However, in this case, the nanoparticle carrier was targeted to α_2-adrenergic receptors, thus preferentially targeting endothelial cells. Moreover, α_2-receptors have been shown to be upregulated in atherosclerotic lesions, further improving precision towards diseased tissue. In vitro, this corresponded to increased endothelial NO production and a reduced inflammatory response and reduced plaque burden in an atherosclerotic mouse model. The observation of both a more selective uptake by diseased endothelium and the apparent improvements in intracellular delivery make redox-sensitive nanoparticles a promising candidate for delivery of genetic materials.

21.3.5 Peptides and Micropeptides Regulate Cellular Functions

Over the past decade, progress in the fields of bioinformatics and genome-wide analysis have led to the discovery of previously unrecognised coding regions, misannotated as non-coding due to their small size [66]. These encode for short amino acid sequences known as micropeptides, a group of molecules with diverse regulatory function influencing cellular proliferation, organogenesis, and metabolic function. While relatively novel, micropeptides have garnered interest in the field of cardiovascular therapy as regulators of cellular and organelle function, including caveolar signalling.

The Nox enzymes do not function in isolation – rather they require the translocation and docking of a number of subunits to facilitate their action [92]. This intrinsic assembled regulation system makes Nox a well-suited target for peptide therapies. The specific subunits vary from isoform to isoform, but each Nox complex may be largely considered as membrane-bound subunits, organisers that assist in enzyme complex assembly, and activators that promote catalytic activity. Several of these are shared across isoforms and may be used as broad inhibitors of Nox activity, such as the membrane-bound subunit $p22^{phox}$ required for the activity of Nox-1–4 [21]. The binding of $p22^{phox}$ with the Nox-2 activator subunit $p47^{phox}$ may be blocked with the peptide PR39, while structural similarities between various Nox-activating subunits give cross-activity to both Nox and non-Nox targets. More specific inhibition may be achieved by targeting subunits involved in the activation of individual isoforms, for example, Nox-1 activator (NoxA1) is a cytosolic subunit that translocates from the cytosol to activate Nox-1 and has not been observed to act on any other isoforms [89]. Based on this, NoxA1 docking sequence (NoxA1ds) is a peptide competitively binding to the Nox-1 site at which NoxA1 docks, resulting in a potent inhibition of $O_2^{\cdot-}$ production specifically at Nox-1 without impacting other Nox isoforms in vitro. This approach appears to offer greater isoform control than similar small-molecule inhibitors, and application of these peptides *in vivo* will shed light on their effects in a more complex disease system. This is especially important in the context of the role of Nox-2 in neutrophils and the prevention of bacterial infections [90].

The small peptide N-acetyl-Ser-Asp-Lys-Pro (Ac-SDKP) is a potent angiogenic and anti-fibrotic agent, acting in part as an angiotensin-converting enzyme inhibitor [95]. Chronic treatment in a preclinical cardiomyopathy model resulted in higher cardiomyocyte density, reduced fibrotic change and macrophage infiltration, and improved ventricular contractility [104]. There was synergistic benefit of combined Ac-SDKP and the stem cell chemotactic stromal-derived factor 1 (SDF-1) used in loaded hydrogels applied in an MI model [109], largely by further improvement of similar endpoints.

21.4 Overcoming Barriers to Reaching the Subcellular Redox Microdomains

The ability to target dysregulated oxidative signalling in specific subcellular compartments without disturbing homeostatic mechanisms remains the elusive 'Holy Grail' of redox therapeutics of high relevance to cardiovascular disease. The promise of greater efficacy and reduced systemic toxicity of this approach has not yet reached its potential, due to a series of bioengineering challenges. Here we discuss each of these challenges in turn – first, the drug must enter the cell, then it must localise to the appropriate subcellular location, and finally, the drug must cross structural subcellular barriers or organelle membranes to reach its ultimate site of action.

21.4.1 Penetrating the Cell Membrane and Escaping Lysosomal Degradation

Moving from the circulation and entering the cell is the first step to accessing the predominantly intracellular network of redox pathways. Where small-molecule drugs are typically small, uncharged, and hydrophobic enough to passively diffuse across the membrane, larger nanoparticles, peptides, and genetic material require active transport mechanisms [9]. Attempts to cross the cell membrane and reach the cytoplasm can broadly be considered under three strategies: clathrin-dependent endocytosis, caveolar-dependent endocytosis, and receptor-independent endocytosis.

Clathrin is a structural protein that mediates the formation of membrane invaginations involved in endocytosis and the primary route of entry into the cell [125]. During endocytosis, clathrin rapidly accumulates at the cell membrane, pocketing and pinching off a part of the membrane as an endosome; this endosome is destined for lysosomal fusion, exposing its contents to high acidity and enzymatic degradation. Thus, utilising clathrin dependency requires overcoming two chief obstacles. First, the drug must interact with the membrane and trigger endocytosis. Following this, the drug must be able to enter the cytoplasm and avoid lysosomal degradation – a feat known as endosomal escape [60]. Neutral or cationic nanoparticles are ideally suited to interact with the membrane due to the negative charge of the cell, while pH-sensitive agents may be designed to undergo a conformational change as acidity rises in the endosome, triggering a permeability through the endosome. Lastly, membrane-destabilising cations or peptides have also been employed to disrupt endosomal membranes, typically through conjugation with a nanoparticle carrier [105].

While we now know the caveolae as a potent signalling domain, their first described function in the cell was as a site of entry for key nutrients including folic acid and albumin [9]. Caveolin facilitates the pinching of the caveolar space to form the vesicular 'caveosome' in response to the binding of an extracellular receptor on the surface of the caveolae. Notably, caveolae-dependent endocytosis at least partially bypasses lysosomal degradation, and the relatively neutral pH of the caveolae allows for significantly improved drug survival. Several strategies have been described to facilitate caveolar drug uptake. Most simply, drugs may be functionalised with a canonical substrate of caveolae such as folic acid, ferritin, or albumin [93]. Alternatively, nanoparticles constructed of synthetic polyelectrolytes may be 'physiochemically tuned' to have a customisable affinity for caveolae by modifying the polyelectrolyte backbone and charge, without the need for ligand targeting [117]. Whatever the strategy employed, targeting caveolar uptake has the convenient side effect of affecting tissues proportional to the density of caveolae in their membranes – notably higher in cardiovascular tissues such as endothelial cells [18].

Lastly, cell entry may be independent of both clathrin and caveolae. Cell-penetrating peptides (CPP) are short amino acid sequences that, when conjugated with a target molecule, improve its ability to cross biological membranes [79]. This is thought to be through macropinocytosis – a receptor-independent endocytic pathway for the absorption of fluids and non-specific solutes into an endosome similar to the clathrin pathway. The exact pathway – or pathways – employed by the various CPPs is still the subject of debate, and it is quite possible that at least part of this membrane permeability comes from caveolar-mediated mechanisms [99]. Regardless of mechanism, CPPs are notable in that they may be directly conjugated onto proteins, small molecules, or even some nucleotide structures with or without the use of a nanoparticle carrier.

21.4.2 Physical and Chemical Barriers to Caveolar Therapy: Plvap Tagging, Flexible Nanoparticles, and Caveolin Scaffolding Domain Peptides

Caveolae were first described in 1953, and their signalling capacity elucidated in 1994 [127]. Despite this, no pharmacotherapy directly targeting and localising to the caveolae has been brought into clinical use during the intervening 25 years. This seeming lack of progress may be explained by the difficulty of reaching the caveolae pharmacologically – not only do caveolae form and dissipate dynamically in response to myriad stimuli, but also possess a restrictive geometry, blocking larger particles from entering [91]. With this in mind, three strategies have been employed with success in delivering pharmaceuticals to the caveolae, tagging with the caveolar-specific marker plasmalemmal vesicle-associated protein (Plvap), employing flexible nanoparticles, and modulating caveolin action with caveolin scaffolding domain (CSD) mimics.

Plvap is a caveolae-specific marker that has demonstrated preclinical potential [106]. Alternatively, platelet endothelial cell adhesion molecule 1 (PECAM-1) is expressed extensively across the endothelial membrane. To assess the relative efficacy of Plvap over PECAM-1, human endothelial cells were stimulated at toll-like receptor 4 (TLR4) with lipopolysaccharide to trigger a caveolae-centred influx of $O_2^{\cdot-}$ as a model or ROS excess [108]. In this model, treatment with MnSOD-loaded antibodies conjugated to either Plvap or PECAM-1 both significantly reduced the inflammatory response, with Plvap-associated SOD having a greater effect than those conjugated with PECAM-1. Interestingly, this was in the face of an overall reduction in binding to endothelial cells and greater effectiveness at lower doses, implying a reduction in non-specific endothelial uptake and a greater proportion of SOD reaching the caveolae.

While Plvap targeting has proven effective in concentrating small molecules such as MnSOD in caveolae, the 'bottleneck' of the flask-shaped caveolae restricts the size of drug formations to ~50 nm, sterically blocking any larger molecules – including most nanoparticles [107]. This may be addressed through the use of flexible nanoparticles, capable of deforming to fit through the caveolar opening and interact with Plvap. In particular, a deformable lysozyme-dextran nanogel has allowed for the transport of drugs to the caveolar membrane, while analogous polystyrene nanoparticles failed to significantly enter endothelium [76].

Finally, caveolar signalling can be targeted in a specific manner by directly modulating interactions of caveolar constituents, chiefly of caveolin proteins. Beyond their role in mediating caveolar endocytosis, caveolin proteins serve as functional regulators of the caveolar redox signalling network [17]. Caveolin-1 (Cav-1) in particular has been identified to interact through its CSD in an inhibitory manner, in turn reducing endogenous NO production [53]. CavNOxin is a synthetic peptide mimicking the CSD with key amino acid substitutions making for a non-inhibitory mutant and conjugated to a CPP allowing for greater membrane permeability. The net result is that CavNOxin competitively disrupts caveolin-dependent inhibition of eNOS, increases NO, and markedly reduces atherosclerotic burden in an eNOS-dependent fashion alongside reduced leukocyte recruitment and oxidative damage in vivo [103].

21.4.3 Delivery of Drugs, Antioxidants, and Genetic Material to the Mitochondria

Since supposedly symbiosing with eukaryotic life billions of years ago, mitochondria have become essential not only in metabolic function but also in mediating inflammation, apoptosis, and autocrine signalling [118]. Despite this, in many ways the mitochondria still function as an independent entity from the cell. With highly selective organelle membranes, self-maintained genomes, and autonomous turnover kinetics, these mitochondrial properties provide both great challenge and opportunity for targeted pharmaceutical therapy.

Pursuit of this has led to the discovery of 'mitochondriotropics', compounds that may be adducted to pharmaceutical agents to improve targeting of the mitochondria [112]. The most widely used mitochondriotropics are delocalised lipophilic cations. In a similar principle to crossing the cell's own membrane, delocalised lipophilic cations such as triphenylphosphonium (TPP^+) take advantage of the mitochondrial transmembrane potential – the negative charge within the mitochondria – allowing for drug accumulation within the inner mitochondrial membrane. TPP^+ has been conjugated extensively with various antioxidant formulations, for example, MitoQ, discussed earlier. A mitochondriotropic which has had particular success in delivering DNA to the mitochondria is dequalinium (DQA) [7]. DQA may be built into liposome preparations known as DQAsomes, which are able to selectively unload their genetic payload only after crossing the mitochondrial membrane rather than the cellular membrane due to its relatively larger electrochemical gradient.

Even so, mitochondriotropics have a limitation; while cations such as TPP^+ and DQA rely on the mitochondrial membrane potential for their function, this polarity is often diminished in disease, reducing efficacy and favouring accumulation only in healthy mitochondria [10]. MITO-Porter is a multilayered nanoparticle system designed for the delivery of drugs or genetic material to the mitochondria independent of the size or properties of the payload [124]. The CPP octaarginine is expressed on the surface of a liposome facilitating endocytosis, followed by activation of a pH-sensitive fusogenic peptide as the endosome acidifies, to trigger endosomal escape. Finally, octaarginine interacts with the mitochondrial membrane in a similar fashion to the cellular membrane, offloading its payload. This system has been tested in a number of preclinical studies with encouraging results, including small molecules, RNA, and antioxidants [1, 33].

Finally, mitochondrial peptides utilise their own intrinsic mechanisms to localise to the mitochondria and may act to improve their function through endogenous means; an example already familiar to the reader is the mitochondrial tetrapeptide elamipretide. Humanin – a micropeptide encoded within the mitochondrial genome – has been shown to reduce cellular damage from the mitochondria during ischaemia-reperfusion injury [88]. Endogenously, humanin is secreted into the circulation in response to certain stress stimuli and taken up by target tissues, including the myocardium. The mechanics of humanin trafficking from its cytosolic uptake to the mitochondria is still emerging, but it is known that humanin supresses complex I activity and reduces mitochondrial ROS generation [113]. Additionally, in a cardiac ischaemia-reperfusion injury rat model, humanin supplementation reduced mitochondrial swelling and restored membrane potential, alongside reductions in infarct size, arrhythmia, and apoptosis [114]. Given the endogenous route of humanin in the vascular circulation – and presumably subsequent subcellular targeting – humanin and similar mitochondrial peptides are particularly attractive as mitochondrial therapies.

21.5 Concluding Remarks: The Future of Targeted Cardiovascular Therapy

Disease processes that in the past have been defined as singular, blanket syndromic phenomena – cardiac failure, atherosclerosis, diabetes, and infarction – are being increasingly characterised as highly heterogenous states with phenotypic variability from patient to patient that has previously gone otherwise unappreciated. Until recently, considering individual patients in the context of these differences has seemed extraneous due to our inability to act on this information, and treatment has remained 'one-size-fits-all' across a range of disease patterns at the subcellular level. However, the targets and technologies discussed here represent an emerging toolbox available for correcting specific imbalances within diseased tissue and across a wider spectrum of systemic cardiovascular disease.

The next logical step then is taking on a personalised approach to antioxidant-based cardiovascular therapy. This demands a number of prerequisites: (1) an understanding and pharmacological isolation of the key motifs in cardiovascular disease, (2) 'high-resolution' diagnostic capacity to provide a snapshot of an individual's redox state at a compartment level, and (3) drug design and delivery platforms capable of not only reducing but also rectifying imbalance and damage. In this chapter, we have seen that the first of these is an ongoing challenge, yet progress is steadily increasing. Similarly, the third is more quickly progressing due to rapid technological advances. As for the second point, advancements such as diagnostic and theranostic nanoparticles, microRNA signatures, and circulating redox biomarker panels may hold the key.

Taken in combination, this paints a promising picture for the future direction of cardiovascular therapy, as the gap between our knowledge and our ability to generate change draws ever closer.

References

1. Abe J, Yamada Y, Takeda A, Harashima H (2017) P2537Cardiac progenitor cells with resveratrol delivered into mitochondria can ameliorate survival of doxorubicin-induced cardiomyopathy mouse via mitochondrial activation of damaged myocardium. Eur Heart J 38(suppl_1):ehx502.P2537–ehx2502.P2537. https://doi.org/10.1093/eurheartj/ehx502.P2537
2. Abrahao TB, Griendling KK (2015) Nuclear factor (erythroid-derived 2)-like 2, the brake in oxidative stress that nicotinamide adenine dinucleotide phosphate-oxidase-4 needs to protect the heart. Hypertension 65(3):499–501. https://doi.org/10.1161/HYPERTENSIONAHA.114.04347
3. Ahadian S, Davenport Huyer L, Estili M, Yee B, Smith N, Xu Z et al (2017) Moldable elastomeric polyester-carbon nanotube scaffolds for cardiac tissue engineering. Acta Biomater 52:81–91. https://doi.org/10.1016/j.actbio.2016.12.009
4. Altenhöfer S, Radermacher KA, Kleikers PW, Wingler K, Schmidt HH (2015) Evolution of NADPH oxidase inhibitors: selectivity and mechanisms for target engagement. Antioxid Redox Signal 23(5):406–427
5. Athyros VG, Katsiki N, Karagiannis A, Mikhailidis DP (2015) Short-, mid-, and long-term benefits of peri-procedural high-intensity statin administration in patients undergoing percutaneous coronary intervention. Current Medical Research and Opinion 31(2):191–5
6. Azevedo C, Macedo MH, Sarmento B (2018) Strategies for the enhanced intracellular delivery of nanomaterials. Drug Discov Today 23(5):944–959. https://doi.org/10.1016/j.drudis.2017.08.011
7. Bae Y, Jung MK, Song SJ, Green ES, Lee S, Park H-S et al (2017) Functional nanosome for enhanced mitochondria-targeted gene delivery and expression. Mitochondrion 37:27–40. https://doi.org/10.1016/j.mito.2017.06.005
8. Birk AV, Liu S, Soong Y, Mills W, Singh P, Warren JD et al (2013) The mitochondrial-targeted compound SS-31 re-energizes ischemic mitochondria by interacting with cardiolipin. J Am Soc Nephrol 24(8):1250–1261. https://doi.org/10.1681/ASN.2012121216
9. Blanco E, Shen H, Ferrari M (2015) Principles of nanoparticle design for overcoming biological barriers to drug delivery. Nat Biotechnol 33:941. https://doi.org/10.1038/nbt.3330
10. Bonora M, Wieckowski MR, Sinclair DA, Kroemer G, Pinton P, Galluzzi L (2018) Targeting mitochondria for cardiovascular disorders: therapeutic potential and obstacles. Nat Rev Cardiol. https://doi.org/10.1038/s41569-018-0074-0
11. Bubb KJ, Birgisdottir AB, Tang O, Hansen T, Figtree GA (2017a) Redox modification of caveolar proteins in the cardiovascular system- role in cellular signalling and disease. Free Radic Biol Med 109:61–74. https://doi.org/10.1016/j.freeradbiomed.2017.02.012
12. Bubb KJ, Kok C, Tang O, Rasko NB, Birgisdottir AB, Hansen T et al (2017b) The NRF2 activator DH404 attenuates adverse ventricular remodeling post-myocardial infarction by modifying redox signalling. Free Radic Biol Med 108:585–594. https://doi.org/10.1016/j.freeradbiomed.2017.04.027
13. Bubb KJ, Drummond GR, Figtree GA (2019) New opportunities for targeting redox dysregulation in cardiovascular disease. Cardiovasc Res cvz183. https://doi.org/10.1093/cvr/cvz183
14. Bugyei-Twum A, Ford C, Civitarese R, Seegobin J, Advani SL, Desjardins J-F et al (2018) Sirtuin 1 activation attenuates cardiac fibrosis in a rodent pressure overload model by modifying Smad2/3 transactivation. Cardiovasc Res 114(12):1629–1641. https://doi.org/10.1093/cvr/cvy131

15. Bundgaard H, Axelsson A, Hartvig Thomsen J, Sørgaard M, Kofoed KF, Hasselbalch R et al (2017) The first-in-man randomized trial of a beta3 adrenoceptor agonist in chronic heart failure: the BEAT-HF trial. Eur J Heart Fail 19(4):566–575. https://doi.org/10.1002/ejhf.714
16. Cannavo A, Koch WJ (2017) Targeting β3-adrenergic receptors in the heart: selective agonism and β-blockade. J Cardiovasc Pharmacol 69(2):71
17. Chen Z, Oliveira SDS, Zimnicka AM, Jiang Y, Sharma T, Chen S et al (2018) Reciprocal regulation of eNOS and caveolin-1 functions in endothelial cells. Mol Biol Cell 29(10):1190–1202. https://doi.org/10.1091/mbc.E17-01-0049
18. Cheng JP, Mendoza-Topaz C, Howard G, Chadwick J, Shvets E, Cowburn AS et al (2015) Caveolae protect endothelial cells from membrane rupture during increased cardiac output. J Cell Biol 211(1):53–61
19. Cheng JP, Nichols BJ (2016) Caveolae: one function or many? Trends Cell Biol 26(3):177–189
20. Chouchani ET, Pell VR, Gaude E, Aksentijević D, Sundier SY, Robb EL et al (2014) Ischaemic accumulation of succinate controls reperfusion injury through mitochondrial ROS. Nature 515:431. https://doi.org/10.1038/nature13909. https://www.nature.com/articles/nature13909#supplementary-information
21. Cifuentes-Pagano ME, Meijles DN, Pagano PJ (2015) Nox inhibitors & therapies: rational design of peptidic and small molecule inhibitors. Curr Pharm Des 21(41):6023–6035
22. Coll RC, Robertson AAB, Chae JJ, Higgins SC, Muñoz-Planillo R, Inserra MC et al (2015) A small-molecule inhibitor of the NLRP3 inflammasome for the treatment of inflammatory diseases. Nat Med 21:248. https://doi.org/10.1038/nm.3806. https://www.nature.com/articles/nm.3806#supplementary-information
23. Cominacini L, Mozzini C, Garbin U, Pasini A, Stranieri C, Solani E et al (2015) Endoplasmic reticulum stress and Nrf2 signaling in cardiovascular diseases. Free Radic Biol Med 88:233–242. https://doi.org/10.1016/j.freeradbiomed.2015.05.027
24. Dai H, Sinclair DA, Ellis JL, Steegborn C (2018) Sirtuin activators and inhibitors: promises, achievements, and challenges. Pharmacol Ther 188:140–154. https://doi.org/10.1016/j.pharmthera.2018.03.004
25. Daubert MA, Yow E, Dunn G, Barnhart H, Douglas P, Udelson J et al (2016) Effects of a novel tetrapeptide in heart failure with reduced ejection fraction (HFREF): a phase I randomized, placebo-controlled trial of elamipretide. J Am Coll Cardiol 67(13 Supplement):1283
26. Devalliere J, Chang WG, Andrejecsk JW, Abrahimi P, Cheng CJ, Jane-wit D et al (2014) Sustained delivery of proangiogenic microRNA-132 by nanoparticle transfection improves endothelial cell transplantation. FASEB J 28(2):908–922. https://doi.org/10.1096/fj.13-238527
27. Dudek J, Hartmann M, Rehling P (2018) The role of mitochondrial cardiolipin in heart function and its implication in cardiac disease. Biochim Biophys Acta (BBA) – Mol Basis Dis. https://doi.org/10.1016/j.bbadis.2018.08.025
28. Duivenvoorden R, Tang J, Cormode DP, Mieszawska AJ, Izquierdo-Garcia D, Ozcan C et al (2014) A statin-loaded reconstituted high-density lipoprotein nanoparticle inhibits atherosclerotic plaque inflammation. Nat Commun 5:3065. https://doi.org/10.1038/ncomms4065. https://www.nature.com/articles/ncomms4065#supplementary-information
29. Erlanson DA, Fesik SW, Hubbard RE, Jahnke W, Jhoti H (2016) Twenty years on: the impact of fragments on drug discovery. Nat Rev Drug Discov 15(9):605
30. Ezzati M, Obermeyer Z, Tzoulaki I, Mayosi BM, Elliott P, Leon DA (2015) Contributions of risk factors and medical care to cardiovascular mortality trends. Nat Rev Cardiol 12(9):508
31. Förstermann U, Xia N, Li H (2017) Roles of vascular oxidative stress and nitric oxide in the pathogenesis of atherosclerosis. Circ Res 120(4):713–735
32. Fu Y, Kinter M, Hudson J, Humphries KM, Lane RS, White JR et al (2016) Aging promotes sirtuin 3–dependent cartilage superoxide dismutase 2 acetylation and osteoarthritis. Arthritis Rheumatol 68(8):1887–1898
33. Furukawa R, Yamada Y, Kawamura E, Harashima H (2015) Mitochondrial delivery of antisense RNA by MITO-Porter results in mitochondrial RNA knockdown, and has a

functional impact on mitochondria. Biomaterials 57:107–115. https://doi.org/10.1016/j.biomaterials.2015.04.022
34. Galougahi KK, Liu CC, Gentile C, Kok C, Nunez A, Garcia A et al (2014) Glutathionylation mediates angiotensin II–Induced eNOS uncoupling, amplifying NADPH oxidase-dependent endothelial dysfunction. J Am Heart Assoc 3(2):e000731
35. Garcia V, Sessa WC (2019) Endothelial nitric oxide synthase (eNOS): perspective and recent developments. Br J Pharmacol 176(2):189–96
36. Goldsmith SR, Bart BA, Pin IL (2018) Neurohormonal imbalance: a neglected problem—and potential therapeutic target—in acute heart failure. Curr Probl Cardiol 43(7):294–304
37. Goszcz K, Deakin SJ, Duthie GG, Stewart D, Leslie SJ, Megson IL (2015) Antioxidants in cardiovascular therapy: panacea or false hope? Front Cardiovasc Med 2(29). https://doi.org/10.3389/fcvm.2015.00029
38. Gray SP, Jha JC, Kennedy K, Van Bommel E, Chew P, Szyndralewiez C et al (2017) Combined NOX1/4 inhibition with GKT137831 in mice provides dose-dependent reno-and atheroprotection even in established micro-and macrovascular disease. Diabetologia 60(5):927–937
39. Gray SP, Marco ED, Kennedy K, Chew P, Okabe J, El-Osta A et al (2016) Reactive oxygen species can provide atheroprotection via NOX4-dependent inhibition of inflammation and vascular remodeling. Arterioscler Thromb Vasc Biol 36(2):295–307. https://doi.org/10.1161/ATVBAHA.115.307012
40. Greenberg B, Butler J, Felker GM, Ponikowski P, Voors AA, Desai AS et al (2016) Calcium upregulation by percutaneous administration of gene therapy in patients with cardiac disease (CUPID 2): a randomised, multinational, double-blind, placebo-controlled, phase 2b trial. Lancet 387(10024):1178–1186. https://doi.org/10.1016/S0140-6736(16)00082-9
41. Guillard S, Minter RR, Jackson RH (2015) Engineering therapeutic proteins for cell entry: the natural approach. Trends Biotechnol 33(3):163–171. https://doi.org/10.1016/j.tibtech.2014.12.004
42. Gurung P, Lukens JR, Kanneganti T-D (2015) Mitochondria: diversity in the regulation of the NLRP3 inflammasome. Trends Mol Med 21(3):193–201
43. Hartikainen J, Hassinen I, Hedman A, Kivelä A, Saraste A, Knuuti J et al (2017) Adenoviral intramyocardial VEGF-DΔNΔC gene transfer increases myocardial perfusion reserve in refractory angina patients: a phase I/IIa study with 1-year follow-up. Eur Heart J 38(33):2547–2555
44. Hayward C, Banner NR, Morley-Smith A, Lyon AR, Harding SE (2015) The current and future landscape of SERCA gene therapy for heart failure: a clinical perspective. Hum Gene Ther 26(5):293–304. https://doi.org/10.1089/hum.2015.018
45. Htet Hlaing K, Clément M-V (2014) Formation of protein S-nitrosylation by reactive oxygen species. Free Radic Res 48(9):996–1010
46. Hulot J-S, Ishikawa K, Hajjar RJ (2016) Gene therapy for the treatment of heart failure: promise postponed. Eur Heart J 37(21):1651–1658. https://doi.org/10.1093/eurheartj/ehw019
47. Ichimura K, Matoba T, Nakano K, Tokutome M, Honda K, Koga J-i, Egashira K (2016) A translational study of a new therapeutic approach for acute myocardial infarction: nanoparticle-mediated delivery of pitavastatin into reperfused myocardium reduces ischemia-reperfusion injury in a preclinical porcine model. PLoS One 11(9):e0162425. https://doi.org/10.1371/journal.pone.0162425
48. Jiang W, Rutherford D, Vuong T, Liu H (2017) Nanomaterials for treating cardiovascular diseases: a review. Bioact Mater 2(4):185–198. https://doi.org/10.1016/j.bioactmat.2017.11.002
49. Jo E-K, Kim JK, Shin D-M, Sasakawa C (2016) Molecular mechanisms regulating NLRP3 inflammasome activation. Cell Mol Immunol 13(2):148
50. Karasawa T, Takahashi M (2017) Role of NLRP3 inflammasomes in atherosclerosis. J Atheroscler Thromb 24(5):443–451. https://doi.org/10.5551/jat.RV17001
51. Karimi Galougahi K, Liu CC, Garcia A, Gentile C, Fry NA, Hamilton EJ et al (2016) β3 Adrenergic stimulation restores nitric oxide/redox balance and enhances endothelial function in hyperglycemia. J Am Heart Assoc 5(2):e002824

52. Katsuki S, Matoba T, Nakashiro S, Sato K, Koga J-i, Nakano K et al (2014) Nanoparticle-mediated delivery of pitavastatin inhibits atherosclerotic plaque destabilization/rupture in mice by regulating the recruitment of inflammatory monocytes. Circulation 129(8):896–906. https://doi.org/10.1161/CIRCULATIONAHA.113.002870
53. Kraehling JR, Sessa WC (2017) Contemporary approaches to modulating the nitric oxide-cGMP pathway in cardiovascular disease. Circ Res 120(7):1174–1182. https://doi.org/10.1161/CIRCRESAHA.117.303776
54. Lamoree B, Hubbard RE (2017) Current perspectives in fragment-based lead discovery (FBLD). Essays Biochem 61(5):453–464. https://doi.org/10.1042/EBC20170028
55. Leisegang MS, Schröder K, Brandes RP (2017) Redox regulation and noncoding RNAs. Antioxid Redox Signal 29(9):793–812. https://doi.org/10.1089/ars.2017.7276
56. Lesnefsky EJ, Chen Q, Hoppel CL (2016) Mitochondrial Metabolism in Aging Heart. Circ Res 118(10):1593–1611. https://doi.org/10.1161/CIRCRESAHA.116.307505
57. Li Z, Rana TM (2014) Therapeutic targeting of microRNAs: current status and future challenges. Nat Rev Drug Discov 13:622. https://doi.org/10.1038/nrd4359
58. Liu P, Huang G, Wei T, Gao J, Huang C, Sun M et al (2018) Sirtuin 3-induced macrophage autophagy in regulating NLRP3 inflammasome activation. Biochim Biophys Acta (BBA)-Mol Basis Dis 1864(3):764–777
59. Lonn E, Bosch J, Yusuf S, Sheridan P, Pogue J, Arnold J et al (2005) Effects of long-term vitamin E supplementation on cardiovascular events and cancer: a randomized controlled trial. JAMA 293(11):1338–1347
60. Lönn P, Kacsinta AD, Cui X-S, Hamil AS, Kaulich M, Gogoi K, Dowdy SF (2016) Enhancing endosomal escape for intracellular delivery of macromolecular biologic therapeutics. Sci Rep 6:32301. https://doi.org/10.1038/srep32301. https://www.nature.com/articles/srep32301#supplementary-information
61. Lucas T, Dimmeler S (2016) RNA therapeutics for treatment of cardiovascular diseases. Circ Res 119(7):794–797. https://doi.org/10.1161/CIRCRESAHA.116.308730
62. Lundberg JO, Gladwin MT, Weitzberg E (2015) Strategies to increase nitric oxide signalling in cardiovascular disease. Nat Rev Drug Discov 14(9):623
63. Luo S, Lei H, Qin H, Xia Y (2014) Molecular mechanisms of endothelial NO synthase uncoupling. Curr Pharm Des 20(22):3548–3553
64. Ma S, Tian XY, Zhang Y, Mu C, Shen H, Bismuth J et al (2016a) E-selectin-targeting delivery of microRNAs by microparticles ameliorates endothelial inflammation and atherosclerosis. Sci Rep 6:22910. https://doi.org/10.1038/srep22910. https://www.nature.com/articles/srep22910#supplementary-information
65. Ma X, Gong N, Zhong L, Sun J, Liang X-J (2016b) Future of nanotherapeutics: targeting the cellular sub-organelles. Biomaterials 97:10–21. https://doi.org/10.1016/j.biomaterials.2016.04.026
66. Makarewich CA, Olson EN (2017) Mining for micropeptides. Trends Cell Biol 27(9):685–696. https://doi.org/10.1016/j.tcb.2017.04.006
67. Malinowski JT, St Jean DJ Jr (2018) Next-generation small molecule therapies for heart failure: 2015 and beyond. Bioorg Med Chem Lett 28(9):1429–1435. https://doi.org/10.1016/j.bmcl.2018.03.064
68. Matoba T, Koga J-I, Nakano K, Egashira K, Tsutsui H (2017) Nanoparticle-mediated drug delivery system for atherosclerotic cardiovascular disease. J Cardiol 70(3):206–211. https://doi.org/10.1016/j.jjcc.2017.03.005
69. Maxwell JT, Somasuntharam I, Gray WD, Shen M, Singer JM, Wang B et al (2015) Bioactive nanoparticles improve calcium handling in failing cardiac myocytes. Nanomedicine 10(22):3343–3357. https://doi.org/10.2217/nnm.15.126
70. Minor RK, Baur JA, Gomes AP, Ward TM, Csiszar A, Mercken EM et al (2011) SRT1720 improves survival and healthspan of obese mice. Sci Rep 1:70–70. https://doi.org/10.1038/srep00070

71. Miragoli M, Ceriotti P, Iafisco M, Vacchiano M, Salvarani N, Alogna A et al (2018) Inhalation of peptide-loaded nanoparticles improves heart failure. Sci Transl Med 10(424). https://doi.org/10.1126/scitranslmed.aan6205
72. Mitragotri S, Burke PA, Langer R (2014) Overcoming the challenges in administering biopharmaceuticals: formulation and delivery strategies. Nat Rev Drug Discov 13:655. https://doi.org/10.1038/nrd4363. https://www.nature.com/articles/nrd4363#supplementary-information
73. Morrison C (2015) $1-million price tag set for Glybera gene therapy. Nat Biotechnol 33:217. https://doi.org/10.1038/nbt0315-217
74. Mortensen SA, Rosenfeldt F, Kumar A, Dolliner P, Filipiak KJ, Pella D et al (2014) The effect of coenzyme Q10 on morbidity and mortality in chronic heart failure. results from Q-SYMBIO: a randomized double-blind trial. JACC Heart Fail 2(6):641–649. https://doi.org/10.1016/j.jchf.2014.06.008
75. Murray CW, Rees DC (2016) Opportunity knocks: organic chemistry for Fragment-Based Drug Discovery (FBDD). Angew Chem Int Ed 55(2):488–492. https://doi.org/10.1002/anie.201506783
76. Myerson JW, Braender B, McPherson O, Glassman PM, Kiseleva RY, Shuvaev VV et al (2018) Flexible nanoparticles reach sterically obscured endothelial targets inaccessible to rigid nanoparticles. Adv Mater 30(32):1802373. https://doi.org/10.1002/adma.201802373
77. Naghavi M, Abajobir AA, Abbafati C, Abbas KM, Abd-Allah F, Abera SF et al (2017) Global, regional, and national age-sex specific mortality for 264 causes of death, 1980–2016: a systematic analysis for the Global Burden of Disease Study 2016. Lancet 390(10100):1151–1210
78. Nakano K, Matoba T, Koga J-i, Kashihara Y, Fukae M, Ieiri I et al (2018) Safety, tolerability, and pharmacokinetics of NK-104-NP. Int Heart J 59(5):1015–1025
79. Nakase I, Noguchi K, Aoki A, Takatani-Nakase T, Fujii I, Futaki S (2017) Arginine-rich cell-penetrating peptide-modified extracellular vesicles for active macropinocytosis induction and efficient intracellular delivery. Sci Rep 7(1):1991. https://doi.org/10.1038/s41598-017-02014-6
80. Nakashiro S, Matoba T, Umezu R, Koga J-I, Tokutome M, Katsuki S et al (2016) Pioglitazone-incorporated nanoparticles prevent plaque destabilization and rupture by regulating monocyte/macrophage differentiation in ApoE−/− mice. Arterioscler Thromb Vasc Biol 36(3):491–500. https://doi.org/10.1161/ATVBAHA.115.307057
81. Neves KB, Rios FJ, Mey L, Alves-Lopes R, Cameron AC, Volpe M et al (2018) VEGFR (Vascular Endothelial Growth Factor Receptor) inhibition induces cardiovascular damage via redox-sensitive processes. Hypertension 71(4):638–647. https://doi.org/10.1161/HYPERTENSIONAHA.117.10490
82. Nie J-J, Qiao B, Duan S, Xu C, Chen B, Hao W et al (2018) Unlockable nanocomplexes with self-accelerating nucleic acid release for effective staged gene therapy of cardiovascular diseases. Adv Mater 30(31):1801570. https://doi.org/10.1002/adma.201801570
83. Oduk Y, Zhu W, Kannappan R, Zhao M, Borovjagin AV, Oparil S, Zhang J (2017) VEGF nanoparticles repair the heart after myocardial infarction. Am J Phys Heart Circ Phys 314(2):H278–H284. https://doi.org/10.1152/ajpheart.00471.2017
84. Orogo AM, Gustafsson ÅB (2015) Therapeutic targeting of autophagy. Circ Res 116(3):489–503. https://doi.org/10.1161/CIRCRESAHA.116.303791
85. Palacio C, Mooradian AD (2016) Clinical trials and antioxidant outcomes. In: Oxidative stress and antioxidant protection: the science of free radical biology and disease, pp 493–506
86. Pell VR, Chouchani ET, Frezza C, Murphy MP, Krieg T (2016) Succinate metabolism: a new therapeutic target for myocardial reperfusion injury. Cardiovasc Res 111(2):134–141. https://doi.org/10.1093/cvr/cvw100
87. Pfaff SJ, Chimenti MS, Kelly MJS, Arkin MR (2015) Biophysical methods for identifying fragment-based inhibitors of protein-protein interactions. In: Meyerkord CL, Fu H (eds) Protein-protein interactions: methods and applications. Springer New York, New York, NY, pp 587–613

88. Qin Q, Mehta H, Yen K, Navarrete G, Brandhorst S, Wan J et al (2018) Chronic treatment with the mitochondrial peptide humanin prevents age-related myocardial fibrosis in mice. Am J Phys Heart Circ Phys 315(5):H1127–H1136. https://doi.org/10.1152/ajpheart.00685.2017
89. Ranayhossaini DJ, Rodriguez AI, Sahoo S, Chen BB, Mallampalli RK, Kelley EE et al (2013) Selective recapitulation of conserved and nonconserved regions of putative NOXA1 protein activation domain confers isoform-specific inhibition of Nox1 oxidase and attenuation of endothelial cell migration. J Biol Chem 288(51):36437–36450. https://doi.org/10.1074/jbc.M113.521344
90. Rane D, Patil T, More V, Patra SS, Bodhale N, Dandapat J, Sarkar A (2018) Neutrophils: interplay between host defense, cellular metabolism and intracellular infection. Cytokine. https://doi.org/10.1016/j.cyto.2018.07.009
91. Rappaport J, Papademetriou I, Muro S (2015) Endocytosis and the endolysosomal route in drug delivery. In: Drug delivery across physiological barriers
92. Rastogi R, Geng X, Li F, Ding Y (2017) NOX activation by subunit interaction and underlying mechanisms in disease. Front Cell Neurosci 10(301). https://doi.org/10.3389/fncel.2016.00301
93. Rewatkar PV, Parton RG, Parekh HS, Parat M-O (2015) Are caveolae a cellular entry route for non-viral therapeutic delivery systems? Adv Drug Deliv Rev 91:92–108. https://doi.org/10.1016/j.addr.2015.01.003
94. Rincon MY, Vanden Driessche T, Chuah MK (2015) Gene therapy for cardiovascular disease: advances in vector development, targeting, and delivery for clinical translation. Cardiovasc Res 108(1):4–20. https://doi.org/10.1093/cvr/cvv205
95. Romero CA, Kumar N, Nakagawa P, Worou ME, Liao T-D, Peterson EL, Carretero OA (2018) Renal release of Ac-SDKP is part of an antifibrotic peptidergic system in the kidney. Am J Physiol Renal Physiol 316(1):195–203
96. Rossman MJ, Santos-Parker JR, Steward CAC, Bispham NZ, Cuevas LM, Rosenberg HL et al (2018) Chronic supplementation with a mitochondrial antioxidant (MitoQ) improves vascular function in healthy older adults. Hypertension 71(6):1056–1063. https://doi.org/10.1161/HYPERTENSIONAHA.117.10787
97. Rühmann E, Betz M, Heine A, Klebe G (2015) Fragment binding can be either more enthalpy-driven or entropy-driven: crystal structures and residual hydration patterns suggest why. J Med Chem 58(17):6960–6971
98. Rupaimoole R, Slack FJ (2017) MicroRNA therapeutics: towards a new era for the management of cancer and other diseases. Nat Rev Drug Discov 16:203. https://doi.org/10.1038/nrd.2016.246. https://www.nature.com/articles/nrd.2016.246#supplementary-information
99. Choi YS, David AE (2014) Cell penetrating peptides and the mechanisms for intracellular entry. Curr Pharm Biotechnol 15(3):192–199
100. Sabbah HN (2016) Targeting mitochondrial dysfunction in the treatment of heart failure. Expert Rev Cardiovasc Ther 14(12):1305–1313. https://doi.org/10.1080/14779072.2016.1249466
101. Sabbah HN, Gupta RC, Kohli S, Wang M, Hachem S, Zhang K (2016) Chronic therapy with elamipretide (MTP-131), a novel mitochondria-targeting peptide, improves left ventricular and mitochondrial function in dogs with advanced heart failure. Circ Heart Fail 9(2):e002206–e002206. https://doi.org/10.1161/CIRCHEARTFAILURE.115.002206
102. Sharma A, Rizky L, Stefanovic N, Tate M, Ritchie RH, Ward KW, de Haan JB (2017) The nuclear factor (erythroid-derived 2)-like 2 (Nrf2) activator dh404 protects against diabetes-induced endothelial dysfunction. Cardiovasc Diabetol 16(1):33
103. Sharma A, Sellers S, Stefanovic N, Leung C, Tan SM, Huet O et al (2015) Direct endothelial nitric oxide synthase activation provides atheroprotection in diabetes-accelerated atherosclerosis. Diabetes 64(11):3937–50
104. Sharma UC, Sonkawade SD, Spernyak JA, Sexton S, Nguyen J, Dahal S et al (2018) A small peptide Ac-SDKP inhibits radiation-induced cardiomyopathy. Circ Heart Fail 11(8):e004867. https://doi.org/10.1161/CIRCHEARTFAILURE.117.004867

105. Shete HK, Prabhu RH, Patravale VB (2014) Endosomal escape: a bottleneck in intracellular delivery. J Nanosci Nanotechnol 14(1):460–474
106. Shuvaev VV, Brenner JS, Muzykantov VR (2015) Targeted endothelial nanomedicine for common acute pathological conditions. J Control Release 219:576–595. https://doi.org/10.1016/j.jconrel.2015.09.055
107. Shuvaev VV, Khoshnejad M, Pulsipher KW, Kiseleva RY, Arguiri E, Cheung-Lau JC et al (2018a) Spatially controlled assembly of affinity ligand and enzyme cargo enables targeting ferritin nanocarriers to caveolae. Biomaterials 185:348–359. https://doi.org/10.1016/j.biomaterials.2018.09.015
108. Shuvaev VV, Kiseleva RY, Arguiri E, Villa CH, Muro S, Christofidou-Solomidou M et al (2018b) Targeting superoxide dismutase to endothelial caveolae profoundly alleviates inflammation caused by endotoxin. J Control Release 272:1–8. https://doi.org/10.1016/j.jconrel.2017.12.025
109. Song M, Jang H, Lee J, Kim JH, Kim SH, Sun K, Park Y (2014) Regeneration of chronic myocardial infarction by injectable hydrogels containing stem cell homing factor SDF-1 and angiogenic peptide Ac-SDKP. Biomaterials 35(8):2436–2445. https://doi.org/10.1016/j.biomaterials.2013.12.011
110. Srikanthan K, Shapiro J, Sodhi K (2016) The role of Na/K-ATPase signaling in oxidative stress related to obesity and cardiovascular disease. Molecules 21(9):1172
111. Suzuki T (2018) Biomarkers for heart failure, an issue of heart failure clinics. Elsevier-Health Science
112. Teixeira J, Deus CM, Borges F, Oliveira PJ (2018) Mitochondria: targeting mitochondrial reactive oxygen species with mitochondriotropic polyphenolic-based antioxidants. Int J Biochem Cell Biol 97:98–103. https://doi.org/10.1016/j.biocel.2018.02.007
113. Thummasorn S, Apaijai N, Kerdphoo S, Shinlapawittayatorn K, Chattipakorn SC, Chattipakorn N (2016) Humanin exerts cardioprotection against cardiac ischemia/reperfusion injury through attenuation of mitochondrial dysfunction. Cardiovasc Ther 34(6):404–414. https://doi.org/10.1111/1755-5922.12210
114. Thummasorn S, Shinlapawittayatorn K, Khamseekaew J, Jaiwongkam T, Chattipakorn SC, Chattipakorn N (2018) Humanin directly protects cardiac mitochondria against dysfunction initiated by oxidative stress by decreasing complex I activity. Mitochondrion 38:31–40. https://doi.org/10.1016/j.mito.2017.08.001
115. Treuer AV, Gonzalez DR (2015) Nitric oxide synthases, S-nitrosylation and cardiovascular health: from molecular mechanisms to therapeutic opportunities. Mol Med Rep 11(3):1555–1565
116. Ul Ain Q, Chung H, Chung JY, Choi J-H, Kim Y-H (2017) Amelioration of atherosclerotic inflammation and plaques via endothelial adrenoceptor-targeted eNOS gene delivery using redox-sensitive polymer bearing l-arginine. J Control Release 262:72–86. https://doi.org/10.1016/j.jconrel.2017.07.019
117. Voigt J, Christensen J, Shastri VP (2014) Differential uptake of nanoparticles by endothelial cells through polyelectrolytes with affinity for caveolae. Proc Natl Acad Sci 111(8):2942–2947. https://doi.org/10.1073/pnas.1322356111
118. Wang Z, Wu M (2015) An integrated phylogenomic approach toward pinpointing the origin of mitochondria. Sci Rep 5:7949. https://doi.org/10.1038/srep07949. https://www.nature.com/articles/srep07949#supplementary-information
119. Ward JPT (2017) From physiological redox signalling to oxidant stress. Adv Exp Med Biol 967:335–342. https://doi.org/10.1007/978-3-319-63245-2_21
120. Winnik S, Auwerx J, Sinclair DA, Matter CM (2015) Protective effects of sirtuins in cardiovascular diseases: from bench to bedside. Eur Heart J 36(48):3404–3412. https://doi.org/10.1093/eurheartj/ehv290
121. Wong JKL, Mohseni R, Hamidieh AA, MacLaren RE, Habib N, Seifalian AM (2017) Will nanotechnology bring new hope for gene delivery? Trends Biotechnol 35(5):434–451. https://doi.org/10.1016/j.tibtech.2016.12.009

122. Woods LA, Dolezal O, Ren B, Ryan JH, Peat TS, Poulsen S-A (2016) Native state mass spectrometry, surface plasmon resonance, and X-ray crystallography correlate strongly as a fragment screening combination. J Med Chem 59(5):2192–2204
123. Xie L, Talukder MAH, Sun J, Varadharaj S, Zweier JL (2015) Liposomal tetrahydrobiopterin preserves eNOS coupling in the post-ischemic heart conferring in vivo cardioprotection. J Mol Cell Cardiol 86:14–22. https://doi.org/10.1016/j.yjmcc.2015.06.015
124. Yamada Y, Harashima H (2018) Targeting the mitochondrial genome through a nanocarrier and the regulation of mitochondrial gene expression. In: Oliveira PJ (ed) Mitochondrial biology and experimental therapeutics. Springer International Publishing, Cham, pp 491–498
125. Yang NJ, Hinner MJ (2015) Getting across the cell membrane: an overview for small molecules, peptides, and proteins. Methods Mol Biol (Clifton, NJ) 1266:29–53. https://doi.org/10.1007/978-1-4939-2272-7_3
126. Yang Y, Cheng H-W, Qiu Y, Dupee D, Noonan M, Lin Y-D et al (2015) MicroRNA-34a plays a key role in cardiac repair and regeneration following myocardial infarction. Circ Res 117(5):450–459. https://doi.org/10.1161/CIRCRESAHA.117.305962
127. Yang Y, Ma Z, Hu W, Wang D, Jiang S, Fan C et al (2016) Caveolin-1/-3: therapeutic targets for myocardial ischemia/reperfusion injury. Basic Res Cardiol 111(4):45. https://doi.org/10.1007/s00395-016-0561-6
128. Zhang M, Zhao H, Cai J, Li H, Wu Q, Qiao T, Li K (2017) Chronic administration of mitochondrion-targeted peptide SS-31 prevents atherosclerotic development in ApoE knockout mice fed Western diet. PLoS One 12(9):e0185688

22 An Exercise Mimetic That Targets Nitroso-Redox Balance as a Therapeutic for Heart Disease

Vikram Shettigar and Mark T. Ziolo

Abstract

Numerous investigations performed over the last few decades clearly indicate that exercise leads to adaptations of the heart resulting in improved function and protection against various cardiovascular diseases (CVD) and is a powerful rehabilitation tool for cardiovascular patients. However, attaining the required intensity of exercise for these cardiac adaptations may not be possible due to numerous constraints such as physical inability, mental unwillingness, stress, etc. Thus, an exercise mimetic drug (i.e., a drug that can mimic the effects of exercise at the cellular level) can be a novel therapeutic approach for cardiovascular patients. An increase in reactive oxygen species (ROS), known as oxidative stress, in the cardiomyocyte and vasculature initiates adverse signaling mechanisms causing numerous detrimental effects. This oxidative stress is a key contributor to the development of CVD. Unfortunately, antioxidant therapy has shown little clinical benefit. In addition to the increased ROS, there is also a concurrent reduction in nitric oxide (NO) levels in CVD. Studies have shown that NO signaling has a protective effect in the cardiomyocyte and vasculature. Contrary to CVD, exercise decreases ROS levels (i.e., antioxidant effect) and increases NO production. Recent studies have shown that ROS and NO do not function independently but work in tandem, known as the nitroso-redox balance, and help clarify the failure of antioxidant therapies. The nitroso-redox balance is critical to maintain a healthy cardiac state. In CVD, there is a detrimental shift in the nitroso-redox balance, while exercise results in a positive shift of this balance. A possible exercise mimetic may be a drug that restores this nitroso-redox balance. We have developed such a drug (EMEPO) that can simultaneously decrease ROS levels

V. Shettigar · M. T. Ziolo (✉)
Department of Physiology and Cell Biology, Davis Heart and Lung Research Institute, The Ohio State University, Columbus, OH, USA
e-mail: ziolo.1@osu.edu; vikram.shettigar@osumc.edu

S. Chakraborti et al. (eds.), *Modulation of Oxidative Stress in Heart Disease*,
https://doi.org/10.1007/978-981-13-8946-7_22

and increase NO levels. Studies at the cellular level and pre-clinical models of CVD have demonstrated that EMEPO, by restoring the nitroso-redox balance, improved cardiomyocyte and heart function. Thus, mimicking exercise effects on the heart via EMEPO may be a paradigm-shifting therapeutic strategy.

Keywords
Exercise · Nitric oxide · Reactive oxygen species · Excitation-contraction coupling

22.1 Introduction

Heart failure (HF) is the culmination of a wide spectrum of disorders that affect cardiovascular function. The clinical condition of HF is generally defined as the inability of the heart to support the metabolic demands of the peripheral tissues and organs [1]. This inability leads to conditions such as fatigue or dyspnea at rest [1]. HF has a significant worldwide impact on mortality and morbidity in the adult population. In the USA alone, HF is the largest cause of death responsible for as many as one in four fatalities and a significant proportion of all hospital admissions [2]. Unfortunately, even with our improved treatment strategies such as early diagnosis, prevention, and management of acute cardiovascular events, the prevalence of HF is gradually increasing [2]. Although newer pharmacological strategies were successful in pre-clinical models of HF, these treatments provided little to no meaningful respite or improvement in the patient's condition in the clinical setting [3]. With the increase of the aging population and the obesity pandemic, HF is only going to continue to grow [2, 3]. Thus, it is imperative to design novel treatment strategies that will be clinically useful. This approach may entail exercise or drugs that mimic exercise physiology [4, 5]. There exists a strong correlation between cardiac health and the ability to perform exercise (known as exercise tolerance) [6]. Hence, a diseased heart's function may be enhanced with the use of an exercise mimetic [7].

22.2 Exercise in Health

Several research investigations and clinical studies have been done over the last five decades to understand the effects of physical exercise on cardiovascular health [8, 9]. Physical activity/fitness clearly decreases the risk of cardiovascular diseases, has favorable effects on risk factors, and improves overall health culminating in lower mortality [8, 9]. Conversely, a sedentary lifestyle has risen to become a top risk factor for the development of cardiovascular disease, especially in the developed countries [10]. Less than half of the population is involved in regular aerobic activity (including workplace physical activity) leading to an increased prevalence of diabetes and coronary heart disease [11]. Thus, there is an ever-increasing need to determine the basic minimum exercise regimen tailored for an individual based on age, gender, race, and genetic predisposition [9]. There is also a need to determine the physiological and molecular mechanisms of exercise to develop the next-generation therapeutics that can mimic the effects of exercise and help prevent cardiovascular disease.

22.3 Exercise: Benefits and Risks

Clinically, exercise is determined as any activity which requires more energy than in the resting stage. There are different types of physical activities or exercises which can be categorized as aerobic, muscle-strengthening, bone-strengthening, and/or stretching. In general, it is the aerobic exercises that have the most significant impact on the cardiovascular system [12]. Aerobic activities can vary from mild (such as walking or gardening) to intense (such as running, mountain-climbing, etc.). Generally, to derive the positive benefits of physical activity, the Centers for Disease Control and Prevention (CDC) and the American College for Sports Medicine (ACSM) recommend 30 min of continuous or accumulated moderate activity daily [12, 13]. Moderate activity is when the heart rate increases 40–50% above resting heart rate as achieved by walking briskly at 3–4 miles per hour, swimming, dancing, etc.

Benefits of regular physical activity or exercise are hard to disregard. There is a direct connection between the effects of exercise and heart adaptations [14, 15]. Exercise increases stroke volume in trained subjects due to structural and myocyte adaptations (discussed further below) resulting in an increased cardiac reserve and VO2max [16–18]. VO2max is the maximal amount of O2 consumed during peak intensity of physical activity and is dependent on the ability of the cardiorespiratory system to deliver O2 to the exercising muscle. Cardiovascular benefits of physical activity begin to occur when the activity is performed at 50% VO2max. Studies have shown that the greatest benefit of exercise to the heart itself is observed at ~75–90% of VO_{2max}. Regular exercise also dilates blood vessels leading to improved circulation and decreased blood pressure. Additionally, it has been shown that VO2max mostly depends on cardiac stroke volume [19]. Thus, since the VO2max level correlates strongly to cardiac fitness, it used as a predictor of cardiovascular mortality in clinics.

Additionally, exercise increases the cardiac reserve of the myocardium. Cardiac reserve is defined as maximum output the heart can provide over the resting stage. This increased cardiac reserve with exercise is a major contributing factor to the beneficial effects of exercise. This is especially true in cardiac patients as impaired cardiac reserve (leading to exercise intolerance) is one of the first indices that is depressed at the onset of cardiac disease [20, 21].

Some of the other commonly known effects of exercise are improvement in the strength of skeletomuscular system, maintenance of healthy weight, improvement in the proportion of high-density lipoproteins to low-density lipoproteins, and improvement in insulin sensitivity [9]. Apart from this, exercise also has positive effects on cognitive function, autonomic nervous system, and immune function which can indirectly effect the heart. Thus, overall, exercise has an enormous impact on all the physiological processes in the body to improve the quality of life.

22.4 Risks of Exercise

Since heart adaptations are directly linked to the intensity of exercise, it should be mentioned that performing a strenuous exercise regimen can have certain serious complications. The effects of such exercises range from mild arrhythmia, fibrosis, myocardial infarction, and sudden cardiac death [22–24]. These effects of extreme exercise regimen have been recapitulated in a rodent study [25]. However, these cases are extremely rare and occur only in 1 in 800,000 of exercise hours. Also such outcomes generally occur in people with existing cardiac condition or in people who have congenital heart defects, myocarditis, or hypertrophic cardiomyopathy.

22.5 Exercise in Disease

In terms of exercise recommendations, the general population can be divided into four major categories: (i) people with no instance of CVD or any elevation in risk factors, (ii) people with elevated levels of risk factors but no clinical presentation of CVD, (iii) people with clinically evident CVD but without any major adverse event such as acute myocardial infarctions, and (iv) people with congestive HF. For people with no evident CVD and with or without elevated risk factors (groups i and ii), there is enormous amount of literature affirming that regular exercise has a huge effect on cardiovascular health and overall constitution (as discussed above). Healthy individuals with a regular exercise regimen display better health indices and are the group with the least risk of getting CVD or associated chronic diseases such as obesity and diabetes [26]. Furthermore, regular exercise reverses many risk factors like hypertension, unhealthy weight gain, glucose intolerance, insulin resistance, and inflammation. Interestingly, the biggest modification in health indices after exercise is seen in people who move from a sedentary lifestyle to implementing regular exercise.

Even in people with mild to moderate cardiac dysfunction (group iii), exercise has shown to improve cardiac contractility and quality of life and reduce hospitalization and mortality. For people with advanced HF (group iv), cardiac function is severely impaired leading to trouble performing daily activities. This leads to further inactivity which worsens the cardiac disease. Thus, their physical workout is generally supervised and is of minimal intensity owing to frequent events of shortness of breath, dizziness, chest pain, and arrhythmias. Nonetheless, patients who undergo supervised physical workout or cardiac rehabilitation programs demonstrate increased VO2max and left ventricular ejection fraction (EF) without any serious cardiac complications [27]. Thus exercise has clear benefits across different stages of cardiac disease and other chronic diseases.

22.6 Molecular Effects of Exercise

Exercise has profound effects on the cardiac muscle, vasculature, skeletal muscle, lung function, blood composition, and other internal organs function such as the kidney, liver, and GI tract [13, 28–30]. Owing to the advances in the field of molecular biology, it has been possible to understand numerous molecular events occurring in these different tissues [31–33]. Numerous studies have deciphered different aspects of cellular changes in response to stress and strain caused by exercise [34, 35]. From an exercise physiology perspective, a comprehensive understanding of the molecular signaling and gene expression network will allow us to design therapies which could not only reduce the impact of chronic diseases but also potentially mimic the effects of exercise and treat cardiovascular diseases.

22.6.1 Vascular Function

Exercise induces shear stress on the blood vessels [36]. Shear stress is the increase in the pressure induced on the blood vessel due to the increased flow of blood. An increase in shear stress has an enormous effect on the overall architecture of the vasculature [29, 37]. Studies have reported formation of new blood vessels after exercise [38]. The formation of new blood vessels or angiogenesis is mediated by a complex interplay of growth factors and other enzymes. Growth factors such as vascular endothelial growth factor (VEGF), fibroblast growth factor (FGF), and angiopoietins along with matrix metalloproteases (MMPs) and tissue plasminogen activators orchestrate the formation of new blood vessels based on exercise intensity and training duration [39–41]. Additionally, there is a simultaneous change in vessel diameter of large and small arteries [42]. This phenomenon known as arteriogenesis is mediated by stimulation by VEGF, vascular cell adhesion molecules (VCAM-1), intracellular adhesion molecules (ICAMs), and integrins [41, 43]. Numerous studies in different disease models ranging from rodents to humans have highlighted the importance of arteriogenesis as an exercise adaptation of vascular function.

In the vasculature, the vascular smooth muscle cells are less affected by shear stress compared to the vascular endothelial cells. Endothelial cells display a variety of gene expression changes and protein posttranslational modifications. One of the important proteins to increase expression with exercise is endothelial nitric oxide synthase (eNOS or NOS3) [44]. eNOS oxidizes L-arginine to L-citrulline and releases nitric oxide (NO) [45]. Under normal circumstances eNOS undergoes dimerization with the help of the cofactor 5,6,7,8-tetrahydro-L-biopterin (BH4) [45]. NO released from dimerized eNOS diffuses to smooth muscle cells to stimulate guanylyl cyclase to increase cellular cyclic guanosine monophosphate (cGMP) resulting in vessel relaxation [46]. This dilates the blood vessels and lowers the resistance allowing better perfusion of tissues [47, 48]. Additionally, NO also has anti-apoptotic and antioxidant properties, hence improving the health and flexibility of blood vessels. This is why exercise-induced NO bioavailability and activity in the endothelial cells are considered central to observe the beneficial effects of exercise on the whole body.

eNOS gene expression and enzyme activity are regulated by numerous factors. The expression of eNOS gene is modulated by factors such as shear stress and cytokines [44, 49, 50]. Additionally, eNOS enzyme activity is subject to various post-translational modifications, regulation of cellular localization, and activity of other proteins which change the efficiency of eNOS function [51]. One of the common modification leading to eNOS activation as a result of exercise is the phosphorylation of eNOS at residue serine 1177 [51]. This phosphorylation is modulated in response to shear stress by AMP-activated protein kinase (AMPK), protein kinase A (PKA), and Akt. Besides, eNOS undergoes acetylation, glutathionylation, O-glycosylation, S-nitrosylation, and acylation at different residues which affects eNOS subcellular localization and activity [51].

Contrarily, there is sufficient evidence that the absence of physical activity or a sedentary lifestyle has the reverse effect on eNOS expression and function [52]. Young healthy mice when forced into physical inactivity (5 weeks) showed drastic reduction in vascular eNOS expression and development of vascular dysfunction [29]. Additionally, the decrease in eNOS and NO is generally associated with an increase in cellular reactive oxygen species (ROS) levels (i.e., oxidative stress) [46]. ROS is a group of short-lived highly reactive compounds which include free radicals, oxygen ions, and peroxides. These are formed during oxygen metabolism and determine key aspects of cell survival and death. Numerous studies have shown that the shift from a healthy vascular tone to a diseased state in endothelial cells is caused by oxidative stress [53]. In endothelial cells, ROS is mainly produced by nicotinamide adenine dinucleotide phosphate (NADPH) oxidases such as Nox2 and Nox4, mitochondria, and xanthine oxidase [54]. Nox4 produces low levels of ROS which is essential for maintaining cell function [54]. Nox2 however, under conditions such as hypertension, generates large amounts of ROS and drives the endothelial cell into a pathological state [54]. Antioxidant enzymes such as superoxide dismutase (SOD) and catalase are not capable of countering the increased levels of ROS and alleviating oxidative stress. As seen in the myocardium (see below), ROS is pro-apoptotic and brings about a variety of detrimental gene expression and protein modification and degradation in the endothelial cell [55].

Additionally, ROS also reacts with NO to produce peroxynitrite. Peroxynitrite oxidizes BH4 and makes it less available for eNOS dimerization. Absence of BH4 and reduced dimerization of eNOS change the enzyme activity of eNOS to produce ROS instead of NO. Thus, "uncoupled" eNOS is a critical aspect in the development of vascular oxidative stress [56]. eNOS uncoupling can also occur due to shortage of L-arginine and heat-shock protein 90. Such changes in vascular function or endothelial dysfunction are observed as a consequence of smoking, aging, physical inactivity, hypertension, diabetes, hypercholesterolemia, and genetic predisposition. Thus, the impaired nitric oxide and redox levels (also known as nitroso-redox imbalance) is the most critical factor of overall vascular health which directly correlates with the cardiovascular health and the outcomes in diseases [57, 58].

22.6.2 Myocardium

The myocardium is highly sensitive to activities of the human body and can acutely alter function to adjust to the required activity. For example, movement such as standing up from a chair will almost instantaneously change heart rate and preload, major determinants of cardiac output [59]. During exercise, the heart rate can increase up to threefold from the resting level. To constantly readjust as per the demand, the heart is equipped with an array of mechanisms which allows it to deliver the necessary work and power to meet the body's requirements. The heart also has the ability for chronic adaptations. For example, with hypertension, the heart will initially undergo adaptive hypertrophy for the increased workload [60]. Additionally, the heart also has direct adaptions with exercise [14]. This is observed intrinsically at the level of the ventricular myocyte. The three major changes to the myocyte are (1) greater contraction and accelerated relaxation [61]; (2) increased size [62]; and (3) greater antioxidant properties [63].

22.6.3 Excitation-Contraction Coupling

Myocyte contraction is regulated by a process termed excitation-contraction coupling (ECC) [64]. The greater contraction and accelerated relaxation with exercise occurs via changes in the ECC machinery [65]. Under normal circumstances in the myocyte, depolarization via an action potential triggers influx of Ca^{2+} via the L-type $Ca2^{+}$ channel (LTCC) [66]. This $Ca2^{+}$ stimulates the ryanodine receptor (RyR2) to open, which releases an enormous amount of $Ca2^{+}$ from the intracellular organelle, the sarcoplasmic reticulum. This causes a sharp tenfold increase in the intracellular $Ca2^{+}$ level which becomes available for the sarcomere assembly to power contraction and generate the required force to eject blood out of the chamber. At the end of contraction, $Ca2^{+}$ is rapidly removed out of the cytosol allowing the sarcomeric contractile apparatus to relax [67]. The sequestration of free Ca^{2+} from the cytosol into the sarcoplasmic reticulum is conducted by the sarcoplasmic reticulum Ca^{2+} ATPase (SERCA2a) [67]. SERCA2a is one of the most critical proteins in the whole ECC machinery of the myocardium. Its expression and activity are tightly correlated with overall heart function [68]. SERCA2a activity is regulated by various posttranslational modifications; however, its allosteric regulation by phospholamban (PLB) is the most critical [69]. PLB is subject to critical phosphorylation which modifies its association with SERCA2a and modulates its Ca^{2+} uptake activity. Dephosphorylated PLB binds tightly to SERCA2a reducing its Ca^{2+} uptake rate, whereas upon phosphorylation, it dissociates from SERCA2a to enhance Ca^{2+} uptake into the sarcoplasmic reticulum [70]. This SERCA2a/PLB interaction forms a critical aspect in modifying heart function based on various demands and conditions such as exercise and disease [70]. To maintain homeostasis, the Ca^{2+} that entered via LTCC must be extruded. The most important pathway to extrude Ca^{2+} from the myocyte is via the $Na^{+}/Ca2^{+}$ exchanger (NCX) [71].

22.6.4 β-AR Signaling

Since every myocyte contracts with every heartbeat; the heart has designed signaling pathways to alter its function to meet the metabolic demands of the body (i.e., moving from a resting stage to a state of physical activity/exercise). The most important pathway is activation of the sympathetic nervous system [72, 73]. The catecholamines norepinephrine and epinephrine or β-agonists (e.g., dobutamine) bind to the adrenergic receptors (AR) on the heart (mostly β1-AR and β2-AR in 4:1 ratio) which are G-protein-coupled receptors that initiate the cascade of converting ATP into cyclic AMP (cAMP) 7 [73]. Within atria, activation of the sympathetic nervous system rapidly increases the heart rate, with the magnitude tightly correlated with the intensity of the activity and the circulatory demand of the body [74]. In the ventricular myocyte, increased intracellular levels of cAMP via stimulation of β-AR receptors activate cAMP-dependent protein kinase (PKA) which phosphorylates numerous critical proteins in the entire excitation-contraction machinery [75]. It phosphorylates and stimulates LTCC and RyR2 to release higher amounts of Ca^{2+} from the sarcoplasmic reticulum. PKA phosphorylates ser^{16} PLB to disassociate it and SERCA2a to accelerate Ca^{2+} cycling rates [72]. The larger Ca^{2+} levels result in greater contraction (inotropy). Additionally, PKA also modifies key proteins of the sarcomeric assembly such as sites $ser^{23/24}$ on troponin I (TnI) and ser^{270}, ser^{282}, and ser^{302} of myosin-binding protein C (MyBP-C), along with the faster Ca^{2+} cycling to accelerate relaxation (lusitropy) [76–78]. These inotropic and lusitropic modifications are necessary to achieve the force generating and hemodynamic change required to meet the increased systemic circulatory demand [79, 80].

22.6.5 Nitric Oxide

Nitric oxide (NO) is also an important signaling molecule involved in the regulation of ECC and is produced on a beat per beat basis within the heart [81, 82]. In cardiac myocytes, NO is produced by nitric oxide synthases (NOS) from the substrate L-arginine via the constitutively expressed isozymes nNOS and eNOS [45].

nNOS produces low amounts of NO in phase with the $Ca2^{+}$ transient due to its regulation by Ca-calmodulin [83]. In nNOS knockout mice and with acute nNOS inhibition, myocyte contraction is blunted via a depressed $Ca2^{+}$ transient and a slowed $[Ca]_i$ decline [84–87]. There is also a reduced functional response to β-AR stimulation. The NO produced from nNOS, which is localized to the SR by binding to RyR, targets PLB and RyR directly via protein S-nitrosylation and through regulation of phosphorylation levels via the modulation of PKA and phosphatase activity.

As with nNOS, eNOS produces low amounts of NO in phase with the Ca^{2+} transient via regulation by Ca-calmodulin. eNOS knockout mice (vs wild type) had an increased functional response to β-AR stimulation; while transgenic mice with cardiac myocyte overexpression of eNOS showed a decreased response to β-AR stimulation [86, 88, 89]. In addition, there does not appear to be an effect of eNOS in regulating basal contractility or lusitropy (opposite of nNOS). Thus, it appears the main effect of eNOS is to reduce the β-AR response, possibly to protect against Ca^{2+} overload and the generation of arrhythmias. Indeed, this appears to be the case as eNOS knockout myocytes have increased early and delayed afterdepolarizations [90].

22.6.6 Heart Disease

ECC is altered in heart disease resulting in decreased contraction and slowed relaxation [91]. Changes in the ECC machinery are often observed such as downregulation of SERCA2a [68]. In fact, reduction in expression of SERCA2a in clinical and experimental heart failure models is associated with loss of contractile function, impaired $Ca2^{+}$ handling, and survival [92]. Therapeutic strategies have been designed to restore SERCA2a function (either directly via increasing SERCA2a expression levels or changing PLB function), in both animal experiments and clinical trials, and have shown to improve contractile function, electrical remodeling, and energetics [93–96]. Other changes in ECC machinery are also observed such as upregulation of NCX and phosphorylation changes in PLB and TnI [97, 98]. With these changes, the myocardium is incapable of generating sufficient force and output to meet the systemic circulatory demand. To compensate for the changes in ECC, the body resorts to mechanisms to maintain the necessary cardiac output to sustain survival. One of the primary modes of compensatory adaptation is increased sympathetic stimulation [79, 99]. As described above, enhanced sympathetic stimulation increases contractility and heart rate to maintain the necessary cardiac output [79]. However, chronic stimulation or β-adrenergic hyperactivity overwhelms the heart and leads to numerous detrimental modifications [100]. Sustained sympathetic stimulation leads to internalization and desensitization of surface β1-AR [79]. Downregulation of β1-AR necessitates further elevation of catecholamine secretion to maintain the required output [79]. These elevated levels of catecholamines initiate a complex set of signaling pathways (e.g., Ca-CaM-dependent protein kinase and calcineurin). These pathways activate numerous detrimental modifications that lead to myocardial fibrosis, inflammation, and apoptosis [101]. It also activates fetal gene reprogramming and drives the myocardium toward pathological hypertrophy (discussed further below).

Chronic catecholaminic stimulation is also strongly associated with increase in ROS production in the myocytes [102]. In the myocardium ROS is produced by mitochondrial NADH dehydrogenase, NADPH oxidase, xanthine oxidase, and a few other enzymes. In a healthy state, there are strong antioxidant mechanisms in the cell that keep the ROS level under check and negate its adverse effects [103–105]. These antioxidant mechanisms include superoxide dismutase (cytosolic Cu/SOD, mitochondrial MnSOD, and extracellular Cu/ZnSOD and ecSOD), catalase, glutathione peroxidases, and thioredoxin system [106]. There are also various compounds (e.g., vitamin E, ascorbic acid) and peptides (glutathione) that are also antioxidants [107, 108]. In disease states, however, the levels of ROS exceed the buffering capacity of the antioxidant mechanisms and cause oxidative stress [104,

105]. ROS activates a series of signal transduction mechanisms that have immense effects on the myocardium [109]. It activates various redox-sensitive transcription factors (such as NFkB, AP-1, and Ets) which express genes that cause inflammation and apoptosis [110, 111]. It also activates the matrix metalloproteinases that remodel the myocardium and enhance fibrosis [112]. It stimulates Akt, Src, and MAPK to enhance hypertrophy and cellular senescence [113]. Additionally, it also activates a variety of phosphatases that decrease phosphorylation levels of critical proteins such as LTCC, RyR, and PLB to decrease contraction and slow relaxation [16]. Thus ROS has an exceedingly enormous damaging effect on the myocardium and hastens the development of heart failure [103]. While there are β-AR overdrive and oxidative stress, there is actually a decrease in NO bioavailability in heart disease. Within the myocytes, there are a downregulation of eNOS and a translocation of nNOS from the SR to the caveolae [114, 115]. The net result is less production and different subcellular localization and signaling. This contributes to the contractile dysfunction observed in heart disease [58].

22.6.7 Exercise

The myocytes from trained animals exhibit greater SR Ca^{2+} cycling, stronger contraction, and faster relaxation kinetics. Contrary to heart disease, exercise, at the cellular level, modifies the ECC machinery in a beneficial way [80]. Studies have found that exercise results in increased expression of SERCA2a [116, 117]. As explained above, SERCA2a is a critical ECC protein and, thus, a major contributor to the greater inotropy and lusitropy of the trained myocyte. However, a study in SERCA2a knockout mice observed that there were still cardiac adaptations with exercise, suggesting that there are additional pathways besides SERCA2a [118]. We have found that increased expression of nNOS (neuronal NOS or NOS1) within ventricular myocytes is necessary for the exercise-induced inotropy and lusitropy [119]. Similar to eNOS (as explained above), nNOS and its cofactors produce NO. The increased expression of nNOS and the increased NO levels deliver the beneficial effects of exercise by directly enhancing contraction and accelerating relaxation by increasing Ca^{2+} transient amplitude and cycling rates [119]. This is consistent with the role nNOS plays in regulating SR $Ca2^{+}$ handling. Interestingly, it has been shown in nNOS knockout mice that most of the exercise adaptations are ablated and the heart does not show improvement in function. In addition to changes in protein expression, it has been demonstrated that exercise shifts the balance toward greater protein phosphorylation of ECC proteins such as RyR2, PLB, and sarcomeric proteins. In fact, we have shown that exercise does result in a shift in the kinase/phosphatase balance within cardiac myocytes resulting in greater protein phosphorylation levels. We have previously shown that nNOS signaling can also regulate PKA activity. That is, nNOS (via the formation of peroxynitrite) can directly activate PKA, even in the absence of cAMP [120]. Conversely, nNOS signaling has also been found to inhibit phosphatase activity. Thus, the greater protein phosphorylation observed with exercise is nNOS-mediated. Taken together, the nNOS-mediated shift in kinase/phosphatase balance resulting in enhanced SR Ca^{2+} cycling and myocyte contraction will increase stroke volume and thus VO2max. The net effect of increasing nNOS with exercise is a decreased sympathetic tone at rest and reduced resting heart rate as seen in the athletic heart.

On the contrary, a sedentary lifestyle leads to the opposite of the beneficial adaptation of exercise on nNOS, that is, an altered localization of nNOS and decreased NO production. This directly affects the functioning of SERCA2a via PLB leading to lower systolic levels of Ca^{2+}, reduced rates of Ca^{2+} uptake from the cytosol into the SR by SERCA2a, and overall reduced force generation. This will gradually result in myocyte apoptosis, fibrosis, arrhythmias, and quicker progression toward heart failure. These changes are similar to the effects seen in aging, after elevation of risk factors and/or after an acute cardiovascular event.

The detrimental changes in ECC and Ca^{2+} handling that have been observed in heart disease are the complete opposite of what is observed with exercise on ECC and Ca^{2+} handling. This can explain why exercise not only slows the progression of heart disease but is the only treatment that actually reverses the disease phenotype [121].

22.6.8 Athlete's Heart

Long-term aerobic exercise is known to change the myocardium to a characteristic "athletic heart" [122]. Just as skeletal muscle, exercise can increase the size of heart (i.e., hypertrophy) resulting in bigger cardiac myocytes [123]. The athletic heart is very different from a normal or diseased heart in not only function but also structure. An athletic heart undergoes ventricular remodeling resulting in greater end-diastolic volumes [124]. Also based on the different types of exercises, the myocardium can hypertrophy in different ways. Endurance training will cause eccentric left ventricular hypertrophy, and resistance training will cause concentric left ventricular hypertrophy, while a mixture of both training causes eccentric-concentric left ventricular hypertrophy [122, 125, 126]. With the exercise-induced increase in ECC to increase contractility and the physiological hypertrophy resulting in larger end-diastolic volume (i.e., preload), the net result is an increased stroke volume [124]. As a consequence, the resting heart rate drops to between 40 and 60 beats per minute. Molecular studies are being performed to identify the pathways involved in forming the athletic heart. To date, key mediators identified have been the insulin-like growth factor 1 (IGF1), phosphoinositide 3-kinase (PI3K)-Akt signaling axis, and downregulation of C/EBPβ [62, 127, 128]. Similar to the effects on ECC, we have found that nNOS is also required for the observed physiological hypertrophy [119]. When nNOS knockout mice were trained, there was no physiological hypertrophy observed. Furthermore, when inducible, myocyte-specific nNOS transgenic mice were studied, the mice had bigger hearts [16, 119].

Contrary to being beneficial, pathological hypertrophy reduces the chamber size and increases the wall stress on the myocardium [129, 130]. This wall stress eventually results in chamber dilation. Pathological hypertrophy also decreases cardiac contractility and the inotropic reserve and increases the risk for the development of heart failure [131]. Interestingly, the signaling pathways for pathological hypertrophy are distinct from physiological hypertrophy [132]. Studies have shown that a key pathway is calcineurin/NFAT [130].

Hence, the molecular mechanisms of exercise result in activation of different signaling pathways resulting in an athlete's heart. Along with the restored ECC (discussed above) and reduced heart rate, exercise negates the need for the compensatory sympathetic overdrive to reverse the adverse remodeling and reverse pathological hypertrophy [129], once more highlighting the importance of exercise and nNOS as a therapeutic approach for heart disease [133].

22.7 Antioxidant Properties

During exercise, the increased energetic demand results in the mitochondria producing more ROS [134]. As described above, increased ROS levels are detrimental to the heart and other organs [135]. Thus, the body enhances its antioxidant capabilities. In the heart, exercise results in increased expression of mitochondrial SOD [135, 136]. This has been touted as the major pathway responsible for the beneficial effects of exercise on cardiac patients. Similar to the role of nNOS in the exercise-mediated effects on ECC and physiological hypertrophy, our study also found that nNOS is required for the enhanced myocyte antioxidant properties after exercise [16]. We observed that our trained myocytes had less ROS levels compared to myocytes from sedentary mice. However, when we isolated myocytes from our trained nNOS knockout mice, this effect of exercise was absent. In fact, these myocytes had increased ROS levels, consistent with the enhanced ROS production of exercise [16]. Remarkably, in studies performed on our conditional nNOS overexpressing transgenic mouse (that were not trained), these myocytes also had decreased ROS levels. In cardiac disease, not only is ROS production increased (via NOX, XO, mitochondria, MAO); there is also a decrease in the ROS scavenging properties of the heart resulting in oxidative stress [137]. Hence, the effects of exercise are opposite of that cardiac disease in terms of ROS levels in the heart, with nNOS playing a key role.

22.8 Nitroso-Redox Balance

The vast data demonstrates that exercise will prevent and possibly cure heart disease [26]. Unfortunately, many people are mentally unwilling or physically unable to reach the intensities needed (especially for the adaptions to the myocyte). So, we need to design a therapy that can mimic the effects of exercise on the heart. As discussed above, our data illustrate that the beneficial effects of exercise on the myocyte (ECC, physiological hypertrophy, and antioxidant) are due to the upregulation of nNOS. So what makes nNOS so important? While nNOS directly modulates protein function via S-nitrosylation and formation of cGMP [138], we believe the importance of nNOS is it being the central hub to regulate the nitroso-redox balance, which is the balance between ROS and NO levels [105].

As it is apparent from the descriptions above, the dominance of either ROS or NO clearly tilts the balance toward either disease (ROS) or fitness (NO). Over the last decade, it has become increasingly clear that, irrespective of the risk factors, accumulation of ROS (or oxidative stress) is the predominant mechanism by which the heart progresses toward failure [103, 139]. At basal levels, ROS is essential for maintaining normal functioning and myocyte physiology. However, with age and other cardiac diseases, the antioxidant capacity gradually goes down and is overwhelmed by the increased levels of ROS leading to oxidative stress and disease.

One may believe that reducing ROS may be an elixir to all the cardiac problems. However, clinical trials using potent antioxidants clearly disapprove this idea. Clinical trials (such as GISSI and HOPE) failed miserably and actually hastened death in the treatment group [140]. This could be because NO is the molecular opposite of ROS and offsets the detrimental effects. However, NO bioavailability is decreased in cardiac disease, and its levels will not be restored via antioxidant treatment. ROS and NO activate almost complementary sets of pathways which determine if the cell moves toward apoptosis and dysfunction (ROS) or survival and healthy state (NO) [58]. The "proof in the pudding" would be to mimic the effects of exercise wherein the antioxidant mechanisms get upregulated to scavenge ROS while increasing the levels of NO [4]. Thus it is absolutely critical that newer and novel pharmaceutical therapies be developed along the lines of reduction of ROS levels and simultaneously increase in the NO levels to observe beneficial effects.

22.9 EMEPO

Along these lines, a novel strategy was adopted for the development of an exercise-mimetic therapeutic or an "exercise pill" [5]. The basic premise of this approach would be to reduce the level of cellular ROS and increase the cellular levels of NO. Nitrone spin traps, used in EPR measurements, have the biological potential of delivering these desired effects. Nitrone spin traps react with the superoxide anion ($O_2^{\cdot-}$) to "scavenge" it. Concomitantly, the nitrone spin trap will release NO as a chemical by-product [4, 141]. Such a pharmacological strategy would be ideal in various cardiovascular disease conditions. Nitrone spin traps such as 5,5-dimethylpyrroline N-oxide (DMPO) have been shown to be effective in ischemia reperfusion injury of the heart and brain [142–144]. We made a novel modification to DMPO by adding ester groups to enhance membrane-permeating abilities allowing for cell entry. We termed this new compound EMEPO, 2-(2-ethoxy-2-oxoethyl)-2-(ethoxycarbonyl)-3,4-dihydro-2H-pyrrole 1-oxide(4). Our data has shown that EMEPO is able to enter the myocyte and restore the nitroso-redox balance(4). Using a genetic model of nitroso-redox imbalance (nNOS knockout mice), EMEPO and antioxidants were able to decrease myocyte ROS levels. However, unlike antioxidants, only EMEPO was able to increase myocyte NO levels. Further characterization of EMEPO revealed that like exercise, EMEPO was able to enhance myocyte contraction via improved Ca^{2+} handling. Interestingly, the effect of EMEPO on myocyte function was much greater than antioxidants. We believe that this is due to EMEPO restoring the nitroso-redox balance, unlike antioxidants which will only fix one side of this balance. Furthermore, we did not observe any effect of DMPO on myocyte function highlighting the importance of cellular entry. The molecular mechanisms of EMEPO are also similar to exercise by increasing PLB phosphorylation(4). We further demonstrated that, like exercise, EMEPO also shifts the kinase/phosphatase balance. Thus, EMEPO's effects on ECC, its antioxidant properties, and means to increase NO levels are able recapitulate the effects of exercise in the myocyte. These novel characteristics of EMEPO should improve in vivo heart function and relieve the need for compensatory catecholaminic stimulation observed in disease. Thus, theoretically, it should have profound effects on remodeling, fibrosis, inflammation, and apoptosis. These desirable benefits, even if provided to a moderate extent, will go a long way in improving the health and outcome of numerous patients of CVD across the globe.

22.10 Conclusion

Exercise clearly has beneficial effects which cannot be ignored. Going by the old adage, "Prevention is the best medicine," it is easy to say that a preventative exercise regimen will be undoubtedly desirable in reducing the impact of CVD worldwide. However, exercise during heart disease may not be possible (or reaching the intensities required) owing to the complexities associated with different cardiovascular diseases such as physical inability, anxiety, emotional stability, etc. Thus, we must also be better equipped with resources that will allow us to handle risks entailed with CVD particularly given that developed countries have a growing aging population and the rise of obesity. The major molecular mechanism to the observed effects of exercise on the heart is upregulation of nNOS. The foremost influence of nNOS in the ventricular myocyte is to control the nitroso-redox balance to modulate ECC, antioxidant properties, and growth.

Thus, any drug that could bring a positive shift to the nitroso-redox balance should be able to provide the benefits derived from exercise alone. To highlight this point, HF patients were treated with isosorbide dinitrate (NO donor) and hydralazine (vasodilator) in clinical trials (e.g., HeFT and A-HeFT). The HF patients treated with these drugs exhibited improved survival, greater ejection fraction, and enhanced quality of life. While the mechanisms of the beneficial effects of isosorbide dinitrate and hydralazine [145] are not completely understood, we postulate that this treatment is partially mimicking exercise by correcting the nitroso-redox balance. However, these trials used a combination of drugs whose primary purpose is not to fix the nitroso-redox imbalance. We believe a superior approach will be the design of a single compound whose primary purpose is to target the nitroso-redox imbalance (i.e., EMEPO). Hence, we believe that mimicking exercise effects on the heart via EMEPO will be a paradigm-shifting therapeutic approach.

References

1. Inamdar AA, Inamdar AC (2016) Heart failure: diagnosis, management and utilization. J Clin Med 5(7):62. https://doi.org/10.3390/jcm5070062
2. Benjamin EJ et al (2017) Heart disease and stroke statistics-2017 update: a report from the American Heart Association. Circulation 135:e146–e603
3. Udelson JE, Stevenson LW (2016) The future of heart failure diagnosis, therapy, and management. Circulation 133:2671–2686
4. Traynham CJ et al (2012) Diesterified nitrone rescues nitroso-redox levels and increases myocyte contraction via increased SR Ca(2+) handling. PLoS One 7:e52005
5. Li S, Laher I (2015) Exercise pills: at the starting line. Trends Pharmacol Sci 36:906–917
6. Morris CK, Froelicher VF (1993) Cardiovascular benefits of improved exercise capacity. Sports Med 16:225–236
7. Guerrieri D, Moon HY, van Praag H (2017) Exercise in a pill: the latest on exercise-mimetics. Brain Plast 2:153–169
8. Choi HE, Kim C, Sohn Y (2017) Cardiac rehabilitation exercise training for high-risk cardiac patients. Ann Rehabil Med 41:650–658
9. Warburton DE, Nicol CW, Bredin SS (2006) Health benefits of physical activity: the evidence. CMAJ 174:801–809
10. Booth FW, Roberts CK, Laye MJ (2012) Lack of exercise is a major cause of chronic diseases. Compr Physiol 2:1143–1211
11. Booth FW, Laye MJ, Roberts MD (2011) Lifetime sedentary living accelerates some aspects of secondary aging. J Appl Physiol (1985) 111:1497–1504
12. Agarwal SK (2012) Cardiovascular benefits of exercise. Int J Gen Med 5:541–545
13. Schuler G, Adams V, Goto Y (2013) Role of exercise in the prevention of cardiovascular disease: results, mechanisms, and new perspectives. Eur Heart J 34:1790–1799
14. Evans DL (1985) Cardiovascular adaptations to exercise and training. Vet Clin North Am Equine Pract 1:513–531
15. Hellsten Y, Nyberg M (2015) Cardiovascular adaptations to exercise training. Compr Physiol 6:1–32
16. Roof SR et al (2015) Obligatory role of neuronal nitric oxide synthase in the heart's antioxidant adaptation with exercise. J Mol Cell Cardiol 81:54–61
17. Kemi OJ et al (2005) Moderate vs. high exercise intensity: differential effects on aerobic fitness, cardiomyocyte contractility, and endothelial function. Cardiovasc Res 67:161–172
18. Le TT et al (2017) Assessing exercise cardiac reserve using real-time cardiovascular magnetic resonance. J Cardiovasc Magn Reson 19:7
19. Bassett DR, Howley ET (2000) Limiting factors for maximum oxygen uptake and determinants of endurance performance. Med Sci Sports Exerc 32:70–84
20. Strait JB, Lakatta EG (2012) Aging-associated cardiovascular changes and their relationship to heart failure. Heart Fail Clin 8:143–164
21. Norman HS et al (2011) Decreased cardiac functional reserve in heart failure with preserved systolic function. J Card Fail 17:301–308
22. La Gerche A, Heidbuchel H (2014) Can intensive exercise harm the heart? You can get too much of a good thing. Circulation 130:992–1002
23. Heidbüchel H et al (2003) High prevalence of right ventricular involvement in endurance athletes with ventricular arrhythmias. Role of an electrophysiologic study in risk stratification. Eur Heart J 24:1473–1480
24. La Gerche A et al (2012) Exercise-induced right ventricular dysfunction and structural remodelling in endurance athletes. Eur Heart J 33:998–1006
25. Benito B et al (2011) Cardiac arrhythmogenic remodeling in a rat model of long-term intensive exercise training. Circulation 123:13–22
26. Myers J (2003) Cardiology patient pages. Exercise and cardiovascular health. Circulation 107:e2–e5

27. Wisløff U et al (2007) Superior cardiovascular effect of aerobic interval training versus moderate continuous training in heart failure patients: a randomized study. Circulation 115:3086–3094
28. Ergen N et al (2005) The effects of aerobic exercise on skeletal muscle metabolism, morphology and in situ endurance in diabetic rats. J Sports Sci Med 4:472–481
29. Kojda G, Hambrecht R (2005) Molecular mechanisms of vascular adaptations to exercise. Physical activity as an effective antioxidant therapy? Cardiovasc Res 67:187–197
30. Ritland S (1988) Exercise and liver disease. Sports Med 6:121–126
31. Booth FW (1989) Application of molecular biology in exercise physiology. Exerc Sport Sci Rev 17:1–27
32. Golbidi S, Laher I (2012) Exercise and the cardiovascular system. Cardiol Res Pract 2012:210852
33. Gielen S, Schuler G, Adams V (2010) Cardiovascular effects of exercise training: molecular mechanisms. Circulation 122:1221–1238
34. Booher MA, Smith BW (2003) Physiological effects of exercise on the cardiopulmonary system. Clin Sports Med 22:1–21
35. Stults-Kolehmainen MA, Sinha R (2014) The effects of stress on physical activity and exercise. Sports Med 44:81–121
36. Padilla J et al (2011) Vascular effects of exercise: endothelial adaptations beyond active muscle beds. Physiology (Bethesda) 26:132–145
37. Niebauer J, Cooke JP (1996) Cardiovascular effects of exercise: role of endothelial shear stress. J Am Coll Cardiol 28:1652–1660
38. Huber-Abel FA, Gerber M, Hoppeler H, Baum O (2012) Exercise-induced angiogenesis correlates with the up-regulated expression of neuronal nitric oxide synthase (nNOS) in human skeletal muscle. Eur J Appl Physiol 112:155–162
39. Gustafsson T, Kraus WE (2001) Exercise-induced angiogenesis-related growth and transcription factors in skeletal muscle, and their modification in muscle pathology. Front Biosci 6:D75–D89
40. Bellafiore M et al (2013) The involvement of MMP-2 and MMP-9 in heart exercise-related angiogenesis. J Transl Med 11:283
41. Nagy JA, Dvorak AM, Dvorak HF (2003) VEGF-A(164/165) and PlGF: roles in angiogenesis and arteriogenesis. Trends Cardiovasc Med 13:169–175
42. Schirmer SH et al (2015) Exercise promotes collateral artery growth mediated by monocytic nitric oxide. Arterioscler Thromb Vasc Biol 35:1862–1871
43. Cai WJ et al (2009) Activation of the integrins alpha 5beta 1 and alpha v beta 3 and focal adhesion kinase (FAK) during arteriogenesis. Mol Cell Biochem 322:161–169
44. Ramírez-Vélez R, Bustamante J, Czerniczyniec A, Aguilar de Plata AC, Lores-Arnaiz S (2013) Effect of exercise training on eNOS expression, NO production and oxygen metabolism in human placenta. PLoS One 8:e80225
45. Förstermann U, Sessa WC (2012) Nitric oxide synthases: regulation and function. Eur Heart J 33:829–837, 837a–837d
46. Fleissner F, Thum T (2011) Critical role of the nitric oxide/reactive oxygen species balance in endothelial progenitor dysfunction. Antioxid Redox Signal 15:933–948
47. Maiorana A, O'Driscoll G, Taylor R, Green D (2003) Exercise and the nitric oxide vasodilator system. Sports Med 33:1013–1035
48. Wang J, Wolin MS, Hintze TH (1993) Chronic exercise enhances endothelium-mediated dilation of epicardial coronary artery in conscious dogs. Circ Res 73:829–838
49. Seidel M, Billert H, Kurpisz M (2006) Regulation of eNOS expression in HCAEC cell line treated with opioids and proinflammatory cytokines. Kardiol Pol 64:153–158; discussion 159–160
50. Chen CA et al (2010) S-glutathionylation uncouples eNOS and regulates its cellular and vascular function. Nature 468:1115–1118
51. Qian J, Fulton D (2013) Post-translational regulation of endothelial nitric oxide synthase in vascular endothelium. Front Physiol 4:347

52. Suvorava T, Lauer N, Kojda G (2004) Physical inactivity causes endothelial dysfunction in healthy young mice. J Am Coll Cardiol 44:1320–1327
53. Li H, Horke S, Förstermann U (2014) Vascular oxidative stress, nitric oxide and atherosclerosis. Atherosclerosis 237:208–219
54. Fan LM et al (2014) Endothelial cell-specific reactive oxygen species production increases susceptibility to aortic dissection. Circulation 129:2661–2672
55. Higashi Y, Noma K, Yoshizumi M, Kihara Y (2009) Endothelial function and oxidative stress in cardiovascular diseases. Circ J 73:411–418
56. Kuzkaya N, Weissmann N, Harrison DG, Dikalov S (2003) Interactions of peroxynitrite, tetrahydrobiopterin, ascorbic acid, and thiols: implications for uncoupling endothelial nitric-oxide synthase. J Biol Chem 278:22546–22554
57. Menazza S et al (2015) Molecular signature of nitroso-redox balance in idiopathic dilated cardiomyopathies. J Am Heart Assoc 4:e002251
58. Hare JM (2004) Nitroso-redox balance in the cardiovascular system. N Engl J Med 351:2112–2114
59. Sipinková I, Hahn G, Meyer M, Tadlánek M, Hájek J (1997) Effect of respiration and posture on heart rate variability. Physiol Res 46:173–179
60. Lavie CJ, Messerli FH (1986) Cardiovascular adaptation to obesity and hypertension. Chest 90:275–279
61. Kemi OJ, Ellingsen O, Smith GL, Wisloff U (2008) Exercise-induced changes in calcium handling in left ventricular cardiomyocytes. Front Biosci 13:356–368
62. Boström P et al (2010) C/EBPβ controls exercise-induced cardiac growth and protects against pathological cardiac remodeling. Cell 143:1072–1083
63. Powers SK, Sollanek KJ, Wiggs MP, Demirel HA, Smuder AJ (2014) Exercise-induced improvements in myocardial antioxidant capacity: the antioxidant players and cardioprotection. Free Radic Res 48:43–51
64. Bers DM (2002) Cardiac excitation-contraction coupling. Nature 415:198–205
65. Rissoli RZ, Vasconcelos EDS, Rantin FT, Kalinin AL (2017) Effects of exercise training on excitation-contraction coupling, calcium dynamics and protein expression in the heart of the Neotropical fish Brycon amazonicus. Comp Biochem Physiol A Mol Integr Physiol 214:85–93
66. Bouchard RA, Clark RB, Giles WR (1995) Effects of action potential duration on excitation-contraction coupling in rat ventricular myocytes. Action potential voltage-clamp measurements. Circ Res 76:790–801
67. Koivumäki JT, Korhonen T, Takalo J, Weckström M, Tavi P (2009) Regulation of excitation-contraction coupling in mouse cardiac myocytes: integrative analysis with mathematical modelling. BMC Physiol 9:16
68. Lipskaia L et al (2014) Expression of sarco (endo) plasmic reticulum calcium ATPase (SERCA) system in normal mouse cardiovascular tissues, heart failure and atherosclerosis. Biochim Biophys Acta 1843:2705–2718
69. Mattiazzi A, Mundiña-Weilenmann C, Vittone L, Said M, Kranias EG (2006) The importance of the Thr17 residue of phospholamban as a phosphorylation site under physiological and pathological conditions. Braz J Med Biol Res 39:563–572
70. Gustavsson M et al (2013) Allosteric regulation of SERCA by phosphorylation-mediated conformational shift of phospholamban. Proc Natl Acad Sci U S A 110:17338–17343
71. Ottolia M, Torres N, Bridge JH, Philipson KD, Goldhaber JI (2013) Na/Ca exchange and contraction of the heart. J Mol Cell Cardiol 61:28–33
72. Gordan R, Gwathmey JK, Xie LH (2015) Autonomic and endocrine control of cardiovascular function. World J Cardiol 7:204–214
73. Madamanchi A (2007) Beta-adrenergic receptor signaling in cardiac function and heart failure. Mcgill J Med 10:99–104
74. White DW, Raven PB (2014) Autonomic neural control of heart rate during dynamic exercise: revisited. J Physiol 592:2491–2500

75. Bers DM, Ziolo MT (2001) When is cAMP not cAMP? Effects of compartmentalization. Circ Res 89:373–375
76. Rao V et al (2014) PKA phosphorylation of cardiac troponin I modulates activation and relaxation kinetics of ventricular myofibrils. Biophys J 107:1196–1204
77. Yang Q et al (2001) PKA-dependent phosphorylation of cardiac myosin binding protein C in transgenic mice. Cardiovasc Res 51:80–88
78. Salhi HE et al (2016) Myofilament calcium sensitivity: mechanistic insight into TnI Ser-23/24 and Ser-150 phosphorylation integration. Front Physiol 7:567
79. de Lucia C, Eguchi A, Koch WJ (2018) New insights in cardiac β-Adrenergic signaling during heart failure and aging. Front Pharmacol 9:904
80. Leosco D et al (2013) Effects of exercise training on cardiovascular adrenergic system. Front Physiol 4:348
81. Hare JM (2003) Nitric oxide and excitation-contraction coupling. J Mol Cell Cardiol 35:719–729
82. Ziolo MT, Kohr MJ, Wang H (2008) Nitric oxide signaling and the regulation of myocardial function. J Mol Cell Cardiol 45:625–632
83. Zhang YH, Jin CZ, Jang JH, Wang Y (2014) Molecular mechanisms of neuronal nitric oxide synthase in cardiac function and pathophysiology. J Physiol 592:3189–3200
84. Wang H et al (2010) Regulation of myocyte contraction via neuronal nitric oxide synthase: role of ryanodine receptor S-nitrosylation. J Physiol 588:2905–2917
85. Wang H et al (2008) Neuronal nitric oxide synthase signaling within cardiac myocytes targets phospholamban. Am J Physiol Cell Physiol 294:C1566–C1575
86. Khan SA et al (2003) Nitric oxide regulation of myocardial contractility and calcium cycling: independent impact of neuronal and endothelial nitric oxide synthases. Circ Res 92:1322–1329
87. Roof SR, Biesiadecki BJ, Davis JP, Janssen PM, Ziolo MT (2012) Effects of increased systolic Ca(2+) and β-adrenergic stimulation on Ca(2+) transient decline in NOS1 knockout cardiac myocytes. Nitric Oxide 27:242–247
88. Wang H, Kohr MJ, Traynham CJ, Ziolo MT (2009) Phosphodiesterase 5 restricts NOS3/Soluble guanylate cyclase signaling to L-type Ca2+ current in cardiac myocytes. J Mol Cell Cardiol 47:304–314
89. Wang H, Kohr MJ, Wheeler DG, Ziolo MT (2008) Endothelial nitric oxide synthase decreases beta-adrenergic responsiveness via inhibition of the L-type Ca2+ current. Am J Physiol Heart Circ Physiol 294:H1473–H1480
90. Rakhit A et al (2001) In vivo electrophysiologic studies in endothelial nitric oxide synthase (eNOS)-deficient mice. J Cardiovasc Electrophysiol 12:1295–1301
91. Zima AV et al (2014) Ca handling during excitation-contraction coupling in heart failure. Pflugers Arch 466:1129–1137
92. Lipskaia L, Chemaly ER, Hadri L, Lompre AM, Hajjar RJ (2010) Sarcoplasmic reticulum Ca(2+) ATPase as a therapeutic target for heart failure. Expert Opin Biol Ther 10:29–41
93. Shareef MA, Anwer LA, Poizat C (2014) Cardiac SERCA2A/B: therapeutic targets for heart failure. Eur J Pharmacol 724:1–8
94. Hayward C, Banner NR, Morley-Smith A, Lyon AR, Harding SE (2015) The current and future landscape of SERCA gene therapy for heart failure: a clinical perspective. Hum Gene Ther 26:293–304
95. Hulot JS, Ishikawa K, Hajjar RJ (2016) Gene therapy for the treatment of heart failure: promise postponed. Eur Heart J 37:1651–1658
96. Ziolo MT, Martin JL, Bossuyt J, Bers DM, Pogwizd SM (2005) Adenoviral gene transfer of mutant phospholamban rescues contractile dysfunction in failing rabbit myocytes with relatively preserved SERCA function. Circ Res 96:815–817
97. Lee A, Oh JG, Gorski PA, Hajjar RJ, Kho C (2016) Post-translational modifications in heart failure: small changes, big impact. Heart Lung Circ 25:319–324

98. Pogwizd SM, Qi M, Yuan W, Samarel AM, Bers DM (1999) Upregulation of Na(+)/Ca(2+) exchanger expression and function in an arrhythmogenic rabbit model of heart failure. Circ Res 85:1009–1019
99. Lohse MJ, Engelhardt S, Eschenhagen T (2003) What is the role of beta-adrenergic signaling in heart failure? Circ Res 93:896–906
100. Triposkiadis F et al (2009) The sympathetic nervous system in heart failure physiology, pathophysiology, and clinical implications. J Am Coll Cardiol 54:1747–1762
101. Koch WJ, Lefkowitz RJ, Rockman HA (2000) Functional consequences of altering myocardial adrenergic receptor signaling. Annu Rev Physiol 62:237–260
102. Costa VM et al (2011) Contribution of catecholamine reactive intermediates and oxidative stress to the pathologic features of heart diseases. Curr Med Chem 18:2272–2314
103. Moris D et al (2017) The role of reactive oxygen species in myocardial redox signaling and regulation. Ann Transl Med 5:324
104. Tsutsui H, Kinugawa S, Matsushima S (2011) Oxidative stress and heart failure. Am J Physiol Heart Circ Physiol 301:H2181–H2190
105. Ziolo MT, Houser SR (2014) Abnormal Ca(2+) cycling in failing ventricular myocytes: role of NOS1-mediated nitroso-redox balance. Antioxid Redox Signal 21:2044–2059
106. Pellegrino D (2016) Antioxidants and cardiovascular risk factors. Diseases 4(1):11. https://doi.org/10.3390/diseases4010011
107. Panth N, Paudel KR, Parajuli K (2016) Reactive oxygen species: a key hallmark of cardiovascular disease. Adv Med 2016:9152732
108. Mangge H, Becker K, Fuchs D, Gostner JM (2014) Antioxidants, inflammation and cardiovascular disease. World J Cardiol 6:462–477
109. Hafstad AD, Nabeebaccus AA, Shah AM (2013) Novel aspects of ROS signalling in heart failure. Basic Res Cardiol 108:359
110. Hamid T et al (2011) Cardiomyocyte NF-κB p65 promotes adverse remodelling, apoptosis, and endoplasmic reticulum stress in heart failure. Cardiovasc Res 89:129–138
111. von Harsdorf R, Li PF, Dietz R (1999) Signaling pathways in reactive oxygen species-induced cardiomyocyte apoptosis. Circulation 99:2934–2941
112. Kong P, Christia P, Frangogiannis NG (2014) The pathogenesis of cardiac fibrosis. Cell Mol Life Sci 71:549–574
113. Takano H et al (2003) Oxidative stress-induced signal transduction pathways in cardiac myocytes: involvement of ROS in heart diseases. Antioxid Redox Signal 5:789–794
114. Damy T et al (2004) Increased neuronal nitric oxide synthase-derived NO production in the failing human heart. Lancet 363:1365–1367
115. Damy T et al (2003) Up-regulation of cardiac nitric oxide synthase 1-derived nitric oxide after myocardial infarction in senescent rats. FASEB J 17:1934–1936
116. Pshennikova MG, Khaspekov GL, Tatarenko AO, Malyshev IIu, Bibilashvili RSh (1999) [Adaptation to physical exertion increases expression of Ca-ATPase genes in heart muscle sarcoplasmatic reticulum]. Biull Eksp Biol Med 128:24–28
117. Kemi OJ, Ceci M, Condorelli G, Smith GL, Wisloff U (2008) Myocardial sarcoplasmic reticulum Ca2+ ATPase function is increased by aerobic interval training. Eur J Cardiovasc Prev Rehabil 15:145–148
118. Ericsson M et al (2010) Exercise training before cardiac-specific Serca2 disruption attenuates the decline in cardiac function in mice. J Appl Physiol (1985) 109:1749–1755
119. Roof SR et al (2013) Neuronal nitric oxide synthase is indispensable for the cardiac adaptive effects of exercise. Basic Res Cardiol 108:332
120. Kohr MJ, Traynham CJ, Roof SR, Davis JP, Ziolo MT (2010) cAMP-independent activation of protein kinase A by the peroxynitrite generator SIN-1 elicits positive inotropic effects in cardiomyocytes. J Mol Cell Cardiol 48:645–648
121. Howden EJ et al (2018) Reversing the cardiac effects of sedentary aging in middle age-A randomized controlled trial: implications for heart failure prevention. Circulation 137:1549–1560

122. Pluim BM, Zwinderman AH, van der Laarse A, van der Wall EE (2000) The athlete's heart. A meta-analysis of cardiac structure and function. Circulation 101:336–344
123. Hart G (2003) Exercise-induced cardiac hypertrophy: a substrate for sudden death in athletes? Exp Physiol 88:639–644
124. Lee BA, Oh DJ (2016) The effects of long-term aerobic exercise on cardiac structure, stroke volume of the left ventricle, and cardiac output. J Exerc Rehabil 12:37–41
125. Vinereanu D et al (2002) Left ventricular long-axis diastolic function is augmented in the hearts of endurance-trained compared with strength-trained athletes. Clin Sci (Lond) 103:249–257
126. Baggish AL et al (2010) Differences in cardiac parameters among elite rowers and subelite rowers. Med Sci Sports Exerc 42:1215–1220
127. Roof SR et al (2016) Insulin-like growth factor 1 prevents diastolic and systolic dysfunction associated with cardiomyopathy and preserves adrenergic sensitivity. Acta Physiol (Oxf) 216:421–434
128. Weeks KL, Bernardo BC, Ooi JYY, Patterson NL, McMullen JR (2017) The IGF1-PI3K-Akt signaling pathway in mediating exercise-induced cardiac hypertrophy and protection. Adv Exp Med Biol 1000:187–210
129. Shimizu I, Minamino T (2016) Physiological and pathological cardiac hypertrophy. J Mol Cell Cardiol 97:245–262
130. Samak M et al (2016) Cardiac hypertrophy: an introduction to molecular and cellular basis. Med Sci Monit Basic Res 22:75–79
131. Frey N, Olson EN (2003) Cardiac hypertrophy: the good, the bad, and the ugly. Annu Rev Physiol 65:45–79
132. Diwan A, Dorn GW (2007) Decompensation of cardiac hypertrophy: cellular mechanisms and novel therapeutic targets. Physiology (Bethesda) 22:56–64
133. Tang L, Wang H, Ziolo MT (2014) Targeting NOS as a therapeutic approach for heart failure. Pharmacol Ther 142:306–315
134. He F et al (2016) Redox mechanism of reactive oxygen species in exercise. Front Physiol 7:486
135. Sallam N, Laher I (2016) Exercise modulates oxidative stress and inflammation in aging and cardiovascular diseases. Oxidative Med Cell Longev 2016:7239639
136. Powers SK et al (1993) Rigorous exercise training increases superoxide dismutase activity in ventricular myocardium. Am J Phys 265:H2094–H2098
137. Khaper N, Kaur K, Li T, Farahmand F, Singal PK (2003) Antioxidant enzyme gene expression in congestive heart failure following myocardial infarction. Mol Cell Biochem 251:9–15
138. Ziolo MT (2008) The fork in the nitric oxide road: cyclic GMP or nitrosylation? Nitric Oxide 18:153–156
139. Sugamura K, Keaney JF (2011) Reactive oxygen species in cardiovascular disease. Free Radic Biol Med 51:978–992
140. Marchioli R, Schweiger C, Levantesi G, Tavazzi L, Valagussa F (2001) Antioxidant vitamins and prevention of cardiovascular disease: epidemiological and clinical trial data. Lipids 36(Suppl):S53–S63
141. Locigno EJ, Zweier JL, Villamena FA (2005) Nitric oxide release from the unimolecular decomposition of the superoxide radical anion adduct of cyclic nitrones in aqueous medium. Org Biomol Chem 3:3220–3227
142. Tosaki A, Braquet P (1990) DMPO and reperfusion injury: arrhythmia, heart function, electron spin resonance, and nuclear magnetic resonance studies in isolated working guinea pig hearts. Am Heart J 120:819–830

143. Dogan S, Ozlem Elpek G, Kirimlioglu Konuk E, Demir N, Aslan M (2012) Measurement of intracellular biomolecular oxidation in liver ischemia-reperfusion injury via immuno-spin trapping. Free Radic Biol Med 53:406–414
144. Zuo L et al (2009) The radical trap 5,5-dimethyl-1-pyrroline N-oxide exerts dose-dependent protection against myocardial ischemia-reperfusion injury through preservation of mitochondrial electron transport. J Pharmacol Exp Ther 329:515–523
145. Taylor AL et al (2004) Combination of isosorbide dinitrate and hydralazine in blacks with heart failure. N Engl J Med 351:2049–2057

Myocardial Injury Secondary to Intestinal Ischemia/Reperfusion or Microbiota Disturbance: Preventive and Therapeutic Concerns

23

Emmanuel E. Douzinas and Aikaterini Apeiranthitis

Abstract

The cardiovascular disease of atherosclerosis (myocardial ischemia and cerebrovascular disease) is the principal cause of death globally. Remote organ ischemic events or remote organ underlying disturbances and pathology may either injure the heart acutely or contribute to cardiovascular disease chronically. The most typical remote organ system that may provoke such effects to the heart is the intestine, either due to (a) thrombotic disease of its vasculature, producing acutely intestinal ischemia, or (b) gut-originated disturbance, deteriorating atherosclerosis. In the first case, at reperfusion, the heart acutely becomes a victim, causing circulatory shock due to myocardial ischemic changes. In the second case, when the intestinal microbial community has altered from symbiotic to dysbiotic, metabolites are composed in the intestine, resulting to chronic advance of atherosclerosis and CVD worsening. Specifically, food phosphatidylcholine is metabolized to TMA (trimethylamine) in the intestine, which is transformed in the liver to TMAO (TMA-N-oxide). Interest about TMAO is rapidly growing. This substance is being studied thoroughly, in an effort to understand and explain its adverse cardiovascular effects, the so-called cardio-intestinal axis. It appears that there exist many sites for possible therapeutic interventions to limit its effect.

Keywords

Ischemia/reperfusion injury · Intestinal postischemic shock · Cardiovascular disease · Microbiota · Phosphatidylcholine · Lecithin · Trimethylamine · Trimethylamine-N-oxide

E. E. Douzinas (✉)
3rd Department of Critical Care, Evgenidio Hospital, Athens University School of Medicine, Athens, Greece

A. Apeiranthitis
2nd Department of Internal Medicine, Evangelismos Hospital, Athens, Greece

S. Chakraborti et al. (eds.), *Modulation of Oxidative Stress in Heart Disease*,
https://doi.org/10.1007/978-981-13-8946-7_23

Abbreviations

CVD	Cardiovascular disease
FMOs	Flavin-containing monooxygenases
I/R	Ischemia/reperfusion
LPS	Lipopolysaccharide
TMA	Trimethylamine
TMAO	Trimethylamine-N-oxide

23.1 Introduction

In the vast majority of medical literature, atherosclerosis and heart and stroke ischemic diseases are defined as cardiovascular disease (CVD), while myocardial ischemia and necrosis appear to constitute the main cause of heart failure. Reactive oxygen species may largely contribute to the beginning or evolution of CVD [1, 2] and the progression of atherosclerosis [3]. It seems that ROS-induced oxidation of LDL in the vessel wall contributes as pathogenetic factor to atherosclerosis [3]. Moreover, vessel plaque fissure, the link of thrombosis to atherosclerosis, may well correlate with the activation of matrix metalloproteinases (MMPs) that is ROS mediated [4].

However, a heart without any apparent previous disease may become a victim of a distally occurring damage or disturbance. The former (damage) may be represented by a distally occurring oxidative stress, e.g., secondary to intestinal ischemia, producing an injury that principally affects the myocardium, leading to acute cardiogenic shock. The latter (disturbance) is the result of an altered intestinal microbial community that has been associated with acceleration of atherosclerosis and CVD.

23.2 Cardiac Injury Secondary to Intestinal I/R

Intestinal ischemia-reperfusion is a well-described experimental model that consistently leads to circulatory shock and multiple organ failure [5, 6]. It replicates most of the features of the multiple, progressive systemic failure that follows acute intestinal ischemia in humans, which is characterized by a high mortality rate [5]. In the 1980s intravascular volume depletion was considered to be the main mechanism for the circulatory derangement, but, at the time, intravascular volume had not been assessed by measurements of left heart filling pressures [7, 8]. The association between intestinal reperfusion and myocardial dysfunction has been experimentally delineated by the observation that cardiac contractile depression and increased lipid peroxidation of cardiac membranes occurred in rats after untreated intestinal reperfusion [9]. The fact that allopurinol pretreatment prevented ischemia-reperfusion-mediated deficits in cardiac contraction and relaxation [10] further confirmed the aforementioned association. Additionally, the cardiovascular consequences were significantly attenuated and the survival rate improved after acute intestinal I/R, when animals were pretreated with anti-TNF-alpha [8]. These data suggested that myocardial contractility might be impaired, implying a cardiogenic component in the intestinal postischemic shock.

In an attempt to clarify the pathophysiology of intestinal postischemic shock, an experimental study [11] was carried out by our group, testing the effect of normoxemic (control) over hypoxemic reperfusion (HR) after ischemia. Two pilot studies were comprised before the performance of the main experimental protocol. The first one included eight animals (four in each group), with reperfusion of 120 min that followed a 120-min intestinal ischemia introduced by superior mesenteric artery clamping. Crystalloid fluids were given before the onset of reperfusion to reach a pulmonary artery occlusion pressure (PAOP) of 10 mm Hg, as prevention against oncoming shock. However, severe irreversible shock followed and was the cause of death in three out of four control animals, which occurred within 15 min of reperfusion, and in one out of four HR animals, which occurred at 90 min of reperfusion. The myocardial histology showed severe ischemic lesions in all succumbed animals. A second pilot study that included another eight animals (all control) showed that prompt use of epinephrine substantially decreased mortality (two out of five) in contrast to its use when shock had been manifested (two out of three). These two pilot studies dictated that fluid load had no, or even negative, circulatory effect in contrast with early applied inotropes that exhibited a favorable effect.

In the main protocol, a complete hemodynamic study was realized [11]. Five of the 13 animals of the control group died in intractable shock; no animal of 8 HR group died ($p = .11$). The decrease in the mean arterial pressure during reperfusion is shown in Fig. 23.1, lower panel; it was more pronounced in the control group ($p < .008$) despite the larger doses of epinephrine administered, compared with the HR group ($p < .02$, Fig. 23.1, upper panel). During reperfusion, both groups exhibited a decrease in cardiac index (CI) (Fig.23.2, lower panel); this was more pronounced in the control group ($p = .0007$). Pulmonary artery occlusion pressure (PAOP) (Fig. 23.2, upper panel) increased during reperfusion in both groups and was more pronounced in the control group ($p = .04$ at 60 min). Although mixed venous blood oxygen saturation of the control animals was higher at 30 min of reperfusion ($p = .005$), it declined after 60 min and became lower than that of HR animals at the end of reperfusion ($p < .02$). Representative pictures of myocardial injury appear in Fig. 23.3, and the myocardial histopathologic injury score was higher in the control group (2.0 ± 0.69 and 3.4 ± 0.89 for the HR and control groups, respectively; $p < .03$). The concentrations of intestinal mucosa malondialdehyde were significantly higher in the control group at 60 min of reperfusion ($p < .03$).

It becomes evident that tissue-damaging effects of I/R are not limited exclusively to the tissue undergoing the initial ischemic insult [12]. In fact, a frequent occurrence of injury to other organ systems follows reperfusion of localized organ ischemia, the so-called distant or remote organ injury (ROI). Remote organs may suffer oxidative injury because their vascular bed becomes exposed to metabolites

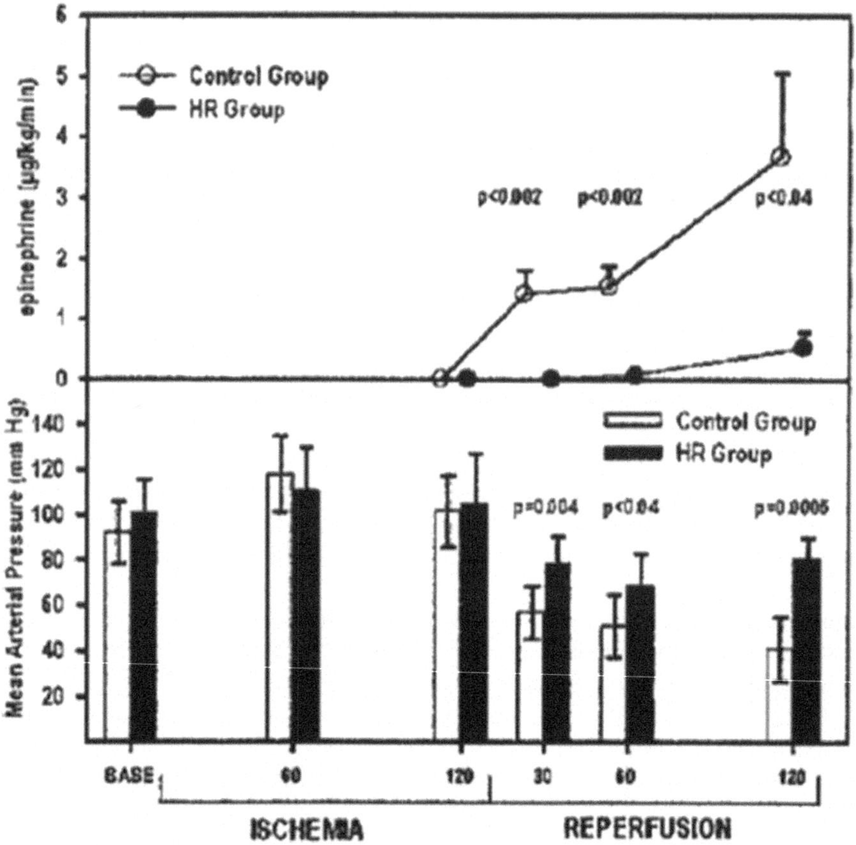

Fig. 23.1 The variation of mean arterial pressure (lower panel) together with the need for epinephrine infusion (upper panel) to maintain a mean arterial pressure of >60 mm Hg. HR, hypoxemic reperfusion. Values are mean ± SE, from Ref. [11] with permission

contained in the postischemic blood, liberated during reperfusion of the primary ischemic organ. In most cases, the organs whose I/R may result in ROI are the gut [13–15], the aorta cross-clamping and reperfusion during graft replacement surgery [13], the lung [16], the liver [17], and the heart [18]. The cardinal expression of ROI is the multiple organ failure syndrome (MOFS). However, the most affected organ seems to be the lung, since respiratory dysfunction is one of the first symptoms preceding MOFS [14, 15].

23.3 Gut Hypothesis of Heart Failure

The human body contains ~10^{14} bacteria belonging to more than 2000 species, their larger part being in the gut [19]. Their weight is approximately 2 kg and their genetic material is 100 times more than the human, and importantly their population is 10

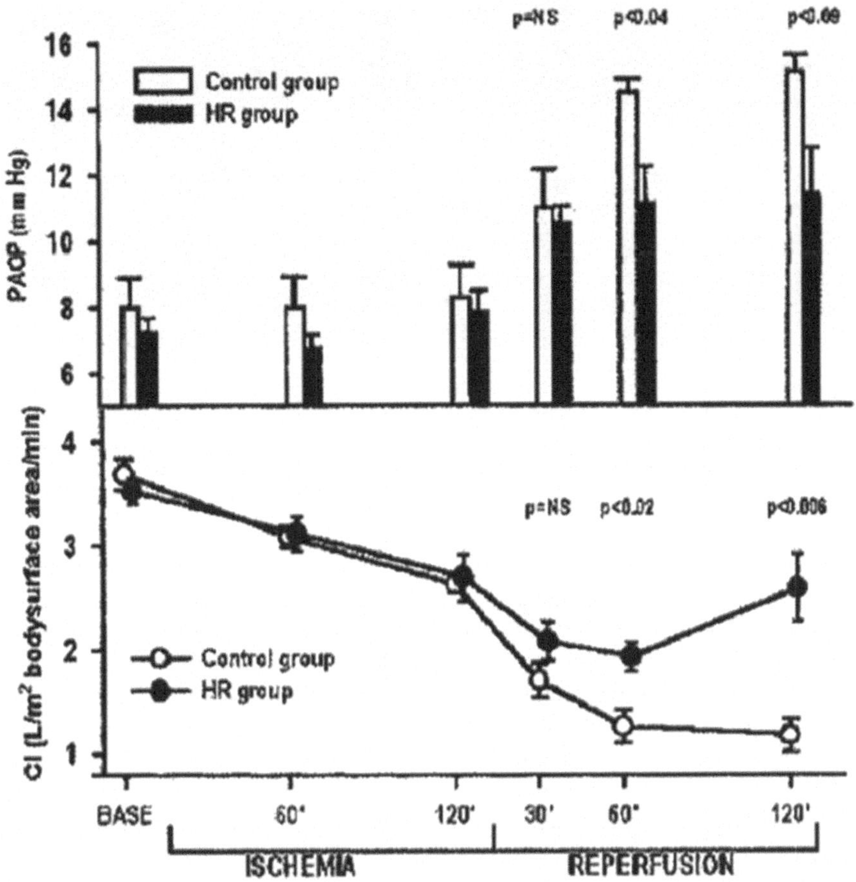

Fig. 23.2 The variation of cardiac index (CI; lower panel) and pulmonary artery occlusion pressure (PAOP; upper panel) in the various stages of the experiment. HR, hypoxemic reperfusion. Values are mean ± SE, from Ref. [11] with permission

times more than the human cells. This population is self-renewed every 3 days, the same as the metabolically active human organs [20–22]. This diverse dynamic community of microorganisms is called microbiota ("gut flora"). The coevolution for years within the human intestinal environment has led to a reciprocity in coexistence that exerts an effect to human's essential life processes from digestion and absorption to the maintenance of physiology of the host [20], called symbiosis.

Changes of the composition of intestinal microbial community result in dysbiosis. In that case, a production of metabolites that may promote atherosclerosis and cardiovascular disease (CVD) may occur [23]. On the other hand, a mechanistic link is provided between therapeutic modifications of dysbiotic intestinal microbiota that may result in a reduction of myocardial infarction in rats [24]. These observations have only recently been suggested, and therefore, the intestine, its microbiota, and their production of metabolites associated with the western diet have attracted

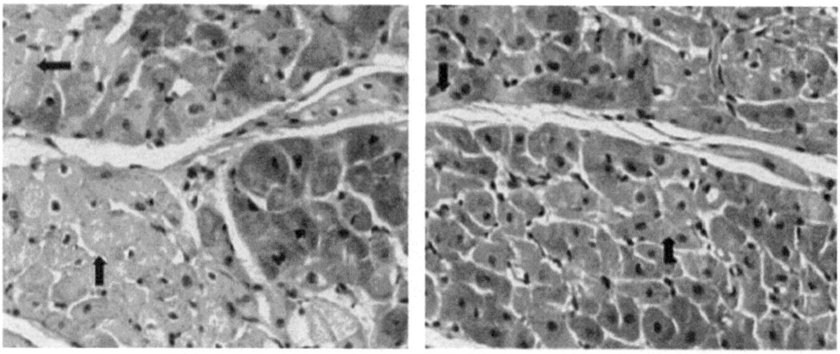

Fig. 23.3 Left, large areas of ischemic pale myocytes (arrows) from a heart of a control animal graded as grade IV, hematoxylin and eosin staining. Right, scattered ischemic myocytes (arrows) from a heart of a hypoxemic reperfusion animal graded as II, hematoxylin and eosin staining. (From Ref. [11] with permission)

enormous interest in the field of CVD and atherosclerosis including leading causes of death such as myocardial infarction and stroke. Furthermore, gut hypothesis seems to represent a probable pathophysiological background to systemic disease processes, including susceptibility for obesity [25] and ease in developing resistance to insulin and fatty liver disease of nonalcoholic origin [26].

The relationship between blood lipid levels and risk of CVD is well recognized. Interestingly, the role of phospholipids, the third class of lipids (the other two are triglycerides and sterols), in the pathogenesis of atherosclerosis and CVD had until recently escaped wide consideration. Particularly, in the gut, L-carnitine or phosphatidylcholine (lecithin), which is the principal source of choline in the food, may be metabolized by microbiota to trimethylamine (TMA) [23]. This substance, after its absorption in the gut, is converted to TMA-N-oxide (TMAO) in the presence of flavin- monooxygenases (FMOs) in the liver. Large quantities of phosphatidylcholine and choline are contained in foods, namely, eggs and meat, and have been related with higher levels of TMAO and betaine. Betaine comes from oxidation of choline in the kidney and the liver and also is converted to TMA by microbiota and to TMAO, consequently, in the liver. These metabolites have been correlated with a greater risk of severe cardiovascular insults in individuals who suffer from coronary heart disease [23, 27].

Similarly, an association was found between fasting plasma TMAO levels and incidence of major cardiovascular events in a 3-year follow-up of 4007 patients who underwent elective coronary angiography [27]. In this study, TMAO levels after the administration of antibiotics were significantly reduced but recovered after antibiotics' withdrawal. These data strongly support the concept of association between the gut microbiota and TMAO production. The additional clinical significance, introduced by this study relative to the fasting TMAO plasma levels, was the potential of risk prediction of major cardiovascular events, independently of risk factors regarded as traditional (age, sex, low-density and high-density lipoprotein cholesterol levels and triglyceride levels, smoking status, blood pressure, diabetes). The potential of TMAO levels for prediction includes the subgroup of participants without angiographic stenosis in main coronary vessels or those with levels of apolipoprotein and lipids of low risk. Another study [28] showed that plasma TMAO levels predict both near- and long-term (30 days to 6 months and 7 years, respectively) risks of incident cardiovascular events, among patients presenting with chest pain, signifying an eventually important biochemical tool of clinical utility in risk assessment for coronary syndromes.

Choline given in excess to mice increases platelet responsiveness, an effect regarded as pro-thrombotic. This effect was not observed if choline preceded a period of orally given antimicrobial agents or was given to germ-free animals [29]. The identification of more than 15 taxa of bacteria was accompanied with amplified risk of thrombosis in mice. Also, it has been shown that germ-free animals transplanted with microbiota carry a higher threat of thrombosis, indicating that this threat is a transferable feature in mice. Unfortunately, in eukaryotes a receptor for TMAO has not yet been recognized.

The unfavorable effect of intestinal metabolites on atherosclerosis and CVD may also act inversely, namely, the failing heart may provoke risks to the intestine, leading to gut failure. For instance, congestion due to splanchnic stagnation because of heart failure may lead to transmural bowel edema, loss of barrier function, and enhanced permeability of intestinal wall, resulting in bacterial translocation. Bacterial and toxins, like LPS detected by Limulus test [30], travelling through the portal blood to the liver, stimulate Kupffer macrophages that produce mediators, thus increasing the body inflammatory state. These in turn are thought to further contribute to heart failure and atherosclerosis progression [31]. In particular, the case of impaired function of intestinal barrier, as it happens in sepsis, cirrhosis, postischemic intestinal reperfusion, severe acute pancreatitis, and burns, frequently results or aggravates sepsis and organ failure [32–36].

23.4 Therapeutic Targets

23.4.1 Restriction of Phosphatidylcholine and Choline-Rich Diet

The concept of TMAO reduction by dietic changes to reduce choline intake is welcome, up to a certain degree. Severe reduction in choline availability may be involved to several critical functions of the cell, such as synthesis of phospholipids which are essential parts of membrane structure and synthesis of the neurotransmitter acetylcholine and methyl group metabolism. Safer therapeutic approaches than diet that decrease systemic TMAO levels are under investigation. In order to conclude whether decreasing dietary choline would be harmful or not, the enrollment of clinical studies would be helpful. However, the limited consumption of red meat, eggs, and cheese should be advised.

23.4.2 The Modulation of Microbiota

There are few studies for the modulation of microbiota. Nevertheless, the composition and the abundance of the intestinal microbial community may be modified in the following ways: (a) diet [37], where strict vegetarian diet alters the gut microbiota and reduces intestinal inflammation, and (b) antibiotics [27], which, in short term, reduce the TMAO levels. However, recolonization of the gut occurs after the cessation of antibiotics and TMAO reappears. No favorable effect has been shown in studies after the administration of antibiotics for secondary prophylaxis of CVD insults [38–40].

However, rifaximin given in patients with alcoholic cirrhosis increased natriuresis and glomerular filtration rate and reduced plasma endotoxin, IL-6, and TNF-a levels [41]. These findings do not imply that the favorable effect exerted via the modulation of intestinal microbial population could dispose a chronic duration. Unpublished data of this author, however, show that there is a more protracted, sometimes permanent, favorable effect of this nonabsorbable oral antibiotic at least on the normalization of the bowel movements in the ambulatory patient. In any case, antibiotics should be given cautiously, since the bothersome complication of intestinal flora change or the severe one, selection of antibiotic-resistant bacteria, occurs often.

An important observation of structural modulation of microbiota comes from a study testing the effect of different diets in composition of fat or voluntary exercise in mice with or without calorie restriction. It was shown that calorie restriction in lifelong-fed mice changes significantly the structure of the gut microbiota enriching with genus of intestinal flora that is associated with longer life expectancy such as *Lactobacillus*. This effect was observed in both high-fat or low-fat diet, but not in mice with voluntary exercise and free feeding [42]. The modification of microbiota was associated with significant reduction of LPS-binding protein in the serum implying that calorie restriction may lessen the antigen load and prevent the intestinal mucosa integrity, reducing the inflammatory stress and exerting a health benefit to the host. Although calorie restriction is regarded as almost the only experimental approach in the elongation of life, the mechanism of its effect remains to be clarified.

23.4.3 Blocking FMOs to Prevent TMAO Production

Levels of FMO3 have been shown to be amplified in both insulin-resistant obese humans and male mice [43]. Mice FMO3 knockdown obviated to develop hyperglycemia, hyperlipidemia, and atherosclerosis [43]. Since FMO3 seems to be a central controller for the hepatic metabolism of cholesterol [44], therapeutic targeting of FMO3 eventually represents an effective alternative to diet for the limitation of atherosclerosis in the treatment of cardiovascular diseases.

23.5 Conclusion

The unfavorable effects exerted by the intestine as remote organ to the heart, either acute, as mesenteric vessel insult, or chronic, due to metabolite production resulting from microbiota change, contribute to cardiovascular disease. The common denominator of these unfavorable effects is atherosclerosis. For this reason, during the last decade, relevant research is focused in identifying methods to reduce the incidence of atherosclerosis.

References

1. Khatri JJ, Johnson C, Magid R et al (2004) Vascular oxidant stress enhances progression and angiogenesis of experimental atheroma. Circulation 109:520–525
2. Giordano FJ (2005) Oxygen, oxidative stress, hypoxia, and heart failure. J Clin Invest 115(3):500–508
3. Witztum JL, Steinberg D (1991) Role of oxidized low-density lipoprotein in atherogenesis. J Clin Invest 88:1785–1792
4. Rajagopalan S, Meng XP, Ramasamy S et al (1996) Reactive oxygen species produced by macrophage-derived foam cells regulate the activity of vascular matrix metalloproteinases in vitro. Implications for atherosclerotic plaque stability. J Clin Invest 98:2572–2579
5. Harward TR, Brooks DL, Flynn TC et al (1993) Multiple organ dysfunction after mesenteric artery revascularization. J Vasc Surg 18:459–469
6. Haglund U, Lundgren O (1978) Intestinal ischemia and shock factors. Fed Proc 37:2729–2733
7. Marston A (1989) Vascular occlusion. In: Marston A, Bulkley GR, Fiddian-Green RG et al (eds) Splanchnic ischemia and multiple organ failure. Edward Arnold, London, pp 51–71
8. Yao YM, Bahrami S, Redl H et al (1996) Monoclonal antibody to tumor necrosis factor-α attenuates hemodynamic dysfunction secondary to intestinal ischemia/reperfusion in rats. Crit Care Med 24:1547–1553
9. Horton JW, White DJ (1993) Lipid peroxidation contributes to cardiac deficits after ischemia and reperfusion of the small bowel. Am J Phys 264:H1686–H1692
10. Horton JW, White DJ (1991) Cardiac contractile injury after intestinal ischemia-reperfusion. Am J Phys 261:1164–1170
11. Douzinas EE, Pitaridis MT, Patsouris E et al (2003) Myocardial ischemia in intestinal post-ischemic shock: the effect of hypoxemic reperfusion. Crit Care Med 31:2183–2189
12. Kalogeris T, Baines CP, Krenz M et al (2012) Cell biology of ischemia/reperfusion injury. Int Rev Cell Mol Biol 298:229–317. Review
13. Carden DL, Granger DN (2000) Pathophysiology of ischemia-reperfusion injury. J Pathol 190:255–266
14. He GZ, Dong LG, Chen XF et al (2011) Lymph duct ligation during ischemia/reperfusion prevents pulmonary dysfunction in a rat model with omega-3 polyunsaturated fatty acid and glutamine. Nutrition 27:604–614
15. Santora RJ, Lie ML, Grigoryev DN et al (2010) Therapeutic distant organ effects of regional hypothermia during mesenteric ischemia reperfusion injury. J Vasc Surg 52:1003–1014
16. Esme H, Fidan H, Koken T et al (2006) Effect of lung ischemia-reperfusion on oxidative stress parameters of remote tissues. Eur J Cardiothorac Surg 29:294–298
17. Hirsch J, Niemann CU, Hansen KC et al (2008) Alterations in the proteome of pulmonary alveolar type II cells in the rat after hepatic ishchemia-reperfusion. Crit Care Med 36:1846–1854
18. Barry MC, Wang JH, Kelly CJ et al (1997) Plasma factors augment neutrophil and endothelial cell activation during aortic surgery. Eur J Vasc Endovasc Surg 13:381–387
19. Neish AS (2009) Microbes in gastrointestinal health and disease. Gastroenterology 136:65–80
20. Zununi Vahed S, Barzegari A, Zuluaga M et al (2018) Myocardial infarction and gut microbiota: an incidental connection. Pharmacol Res 129:308–317
21. Griffin JL, Wang X, Stanley E (2015) Does our gut microbiome predict cardiovascular risk? A review of the evidence from metabolomics. Circ Cardiovasc Genet 8:187–191
22. Mazidi M, Rezaie P, Kengne AP et al (2016) Gut microbiome and metabolic syndrome. Diabetes Metab Syndr 10(2 Suppl 1):S150–S157
23. Wang Z, Klipfell E, Bennett BJ et al (2011) Gut flora metabolism of phosphatidylcholine promotes cardiovascular disease. Nature 472:57–63
24. Lam V, Su J, Koprowski S, Hsu A et al (2012) Intestinal microbiota determine severity of myocardial infarction in rats. FASEB J 26(4):1727–1735
25. Turnbaugh PJ, Ley RE, Mahowald MA et al (2006) An obesity-associated gut microbiome with increased capacity for energy harvest. Nature 444:1027–1131

26. Dumas ME, Barton RH, Toye A et al (2006) Metabolic profiling reveals a contribution of gut microbiota to fatty liver phenotype in insulin-resistant mice. Proc Natl Acad Sci U S A 103:12511–12516
27. Tang WH, Wang Z, Levison BS et al (2013) Intestinal microbial metabolism of phosphatidylcholine and cardiovascular risk. N Engl J Med 368(17):1575–1584
28. Li XS, Obeid S, Klingenberg R et al (2017) Gut microbiota-dependent trimethylamine N-oxide in acute coronary syndromes: a prognostic marker for incident cardiovascular events beyond traditional risk factors. Eur Heart J 38(11):814–824
29. Zhu W, Gregory JC, Org E et al (2016) Gut microbial metabolite TMAO enhances platelet hyperreactivity and thrombosis risk. Cell 165:111–124
30. Douzinas EE, Kollias S, Tiniakos D et al (2004) Hypoxemic reperfusion after 120 mins of intestinal ischemia attenuates the histopathologic and inflammatory response. Crit Care Med 32:2279–2283
31. Tang WH, Kitai T, Hazen SL (2017) Gut microbiota in cardiovascular health and disease. Circ Res 120(7):1183–1196
32. Doig CJ, Sutherland LR, Sandham JD et al (1998) Increased intestinal permeability is associated with the development of multiple organ dysfunction syndrome in critically ill ICU patients. Am J Respir Crit Care Med 158:444–451
33. Cirera I, Bauer TM, Navasa M et al (2001) Bacterial translocation of enteric organisms in patients with cirrhosis. J Hepatol 34:32–37
34. Deitch EA (2002) Bacterial translocation or lymphatic drainage of toxic products from the gut: what is important in human beings? Surgery 131:241–244
35. Riddington DW, Venkatesh B, Boivin CM et al (1996) Intestinal permeability, gastric intramucosal pH, and systemic endotoxemia in patients undergoing cardiopulmonary bypass. JAMA 275:1007–1012
36. Krack A, Sharma R, Figulla HR et al (2005) The importance of the gastrointestinal system in the pathogenesis of heart failure. Eur Heart J 26:2368–2374
37. Kim MS, Hwang SS, Park EJ et al (2013) Strict vegetarian diet improves the risk factors associated with metabolic diseases by modulating gut microbiota and reducing intestinal inflammation. Environ Microbiol Rep 5:765–775
38. Nagatomo Y, Tang WH (2015) Intersections between microbiome and heart failure: revisiting the gut hypothesis. J Card Fail 21(12):973–980. Review
39. Grayston JT, Kronmal RA, Jackson LA et al (2005) Azithromycin for the secondary prevention of coronary events. N Engl J Med 352:1637–1645
40. Cannon CP, Braunwald E, McCabe CH et al (2005) Antibiotic treatment of Chlamydia pneumoniae after acute coronary syndrome. N Engl J Med 352:1646–1654
41. Kalambokis GN, Mouzaki A, Rodi M et al (2012) Rifaximin improves systemic hemodynamics and renal function in patients with alcohol-related cirrhosis and ascites. Clin Gastroenterol Hepatol 10:815–818
42. Zhang C, Li S, Yang L et al (2013) Structural modulation of gut microbiota in life-long calorie-restricted mice. Nat Commun 4:2163–2173
43. Miao J, Ling AV, Manthena PV et al (2015) Flavin-containing monooxygenase 3 as a potential player in diabetes-associated atherosclerosis. Nat Commun 6:6498
44. Worrier M, Shih DM, Burrows AC et al (2015) The TMAO-generating enzyme flavin monooxygenase 3 is a central regulator of cholesterol balance. Cell Rep 10(3):S2211–S1247

Multi-target Approach for Oxidative Stress Modulation by Aspirin, Salicylates and Other NSAIDs: Clinical Implications in Atherosclerosis

24

Eugenia Yiannakopoulou

Abstract

Atherosclerosis is a multifactorial process with oxidative stress being implicated in its pathophysiology. Aspirin and salicylates have pleiotropic effects. Among these pleiotropic effects, modulation of stress response by salicylates is quite interesting. Salicylates modulate stress response in prokaryotic organisms as well as in eukaryotic cells. Modulation of stress response by salicylates is due to the effect of salicylates on cell signalling pathways as well as to the pro-oxidant–antioxidant effects of salicylates. Aspirin and salicylates target oxidative stress in atherosclerosis through multiple antiplatelet-independent mechanisms of action, including scavenging of reactive oxygen species, enhancement of nitrous oxide release, inhibition of superoxide anion release, induction of GSH-dependent antioxidant mechanisms and epigenetic regulation of antioxidant enzymes. Thus, aspirin and salicylates are promising multi-target agents against oxidative stress implicated in atherosclerosis. Based on this evidence, the role of aspirin in the primary prevention of atherosclerosis should be revisited.

Keywords

Atherosclerosis · Oxidative stress · Aspirin · Salicylates · Pharmacological modulation

E. Yiannakopoulou (✉)
Department of Biomedical Sciences, Faculty of Health Sciences, University of West Attica, Athens, Greece
e-mail: nyiannak@teiath.gr; nyiannak@uniwa.gr

S. Chakraborti et al. (eds.), *Modulation of Oxidative Stress in Heart Disease*,
https://doi.org/10.1007/978-981-13-8946-7_24

24.1 Introduction

Atherosclerosis is a chronic inflammatory disease of the arteries that is characterized by the abnormal accumulation of lipids, inflammatory cells, matrix deposits and smooth muscle cell proliferation in the wall of medium- and large-calibre arteries. Lipid deposition is the first step in the pathophysiology of atherosclerosis followed by the development of a chronic inflammatory reaction [1]. Thus, lipid deposition leads to the proliferation of certain cell types within the arterial wall which gradually impinge on the vessel lumen and impede blood flow. Atherosclerosis starts early in life, even in the second decade of life, and is the main cause of cardiovascular disease and fatal cardiovascular events. Atherosclerosis has increased mortality and morbidity in the western countries. Thus, modulation of atherosclerosis would have a beneficial effect in the population health. Modulation of atherosclerosis could result in reversal of atherosclerosis or in deceleration of its development. Modulation of atherosclerosis could be achieved through targeting the multiple factors implicated in its pathophysiology.

Pathophysiology of atherosclerosis is quite challenging. Atherosclerosis is a multifactorial process with oxidative/nitrosative stress, imbalance of vasoconstrictor–vasodilator production, platelet aggregation and modification of LDL cholesterol being implicated in its pathophysiology. Among the above-cited factors, oxidative stress plays a critical role in the pathophysiology of atherosclerosis.

Since decades, research efforts have been focused on the primary and secondary prevention of atherosclerosis. Aspirin has an established role in the secondary prevention of atherosclerosis, while its role in the primary prevention of atherosclerosis is still under investigation [2–6]. Although the antiplatelet effect of aspirin via cyclooxygenase inhibition is well established [7], there are a number of cyclooxygenase-independent mechanisms of action of aspirin, including inhibition of transcription factors, regulation of epigenetic targets, modulation of lymphangiogenesis and modulation of oxidative stress, that are currently under investigation [8–15]. Aspirin and the other salicylates, especially in high doses, are well-recognized inhibitors of I(kappa)B kinase-beta, an enzyme complex that is involved in the propagation of the cellular response to inflammation and also part of the upstream NF-κB signal transduction cascade [15]. Given that aspirin and the other salicylates act via multiple mechanisms, it is expected that these agents are able to modulate a biological process through a multifactorial approach. This chapter aims to review evidence on modulation of oxidative stress in atherosclerosis through a multi-target approach via aspirin, salicylates and the other nonsteroidal anti-inflammatory agents.

24.2 Biology of Oxidative Stress

Cellular oxidative stress results from an imbalance between pro-oxidant and antioxidant mechanisms due to either increased production of free oxygen radicals or deficiency of antioxidant mechanisms. Under physiological conditions, in a normal cell, there is continuous production of free oxygen radicals. It is well known that free radicals are products of normal cellular metabolism and have a dual role being either beneficial or harmful for the cell [8–10].

Low levels of free oxygen radicals have beneficial properties including a crucial role in:

- Cell signalling
- Cellular stress response
- Cellular differentiation
- Gene transcription
- Cellular proliferation
- Apoptosis

On the other hand, high levels of free oxygen radicals lead to oxidative damage to cellular constituents as DNA, lipids, proteins and sugars.

The redox homeostasis that is quite crucial for the cellular physiology is defined as the balance between pro-oxidant and antioxidant substances. This homeostasis is kept through the antioxidant mechanisms.

24.2.1 Reactive Oxygen Species

Reactive oxygen species are produced in the mitochondria, in peroxisomes, in the cytoplasm as well as in the cell membrane. Molecular oxygen is mildly reactive, due to its chemical structure, being a biradical that includes two electrons in the orbits. ROS formation is derived from molecular oxygen through electron transfer or energy absorption processes. Reactive oxygen species include oxygen-free radicals as well as non-radical compounds [8–10].

Oxygen-free radicals are:

- The hydroxyl radical (HO)
- Superoxide anion (O_2•−)
- Peroxyl (ROO•)
- Alkoxyl (RO•)
- Nitric oxide (NO)

In chemistry, superoxide anion is derived from the addition of one electron on molecular oxygen. The production of superoxide anion consists the first step in the chain of reactive oxygen species production. In vivo, superoxide anion is produced either enzymatically or non-enzymatically. Mitochondria constitute the main cellular source of superoxide anion. Enzymatic sources of superoxide anion include NADPH oxidases that are situated in the cellular membrane of polymorphonuclear cells, macrophages and endothelial cells as well as cytochrome P450-dependent oxygenases. Another enzymatic source of superoxide anion is the proteolytic conversion of xanthine oxidoreductase to xanthine oxidase.

The hydroxyl radical (HO) with an extremely short in vivo half-life is the most highly reactive and therefore toxic form of oxygen. Hydroxyl radical reacts in the vicinity of its site of production.

Non-radical compounds are oxygen, hydrogen peroxide and transition metals such as copper.

Although hydrogen peroxide does not contain unpaired electrons and chemically it is not a radical, it is included in the reactive oxygen species due to its high reactivity. In addition to reactive oxygen species, reactive nitrogen species have similar effects on the cells being able to induce oxidative stress.

24.2.2 Antioxidant Mechanisms

Antioxidant mechanisms are divided into two broad categories. The first includes antioxidant enzymes and the second includes the non-enzymatic antioxidants [8–10].

24.2.2.1 Antioxidant Enzymes

- Superoxide dismutase

Superoxide dismutase and catalase are the more effective enzymatic antioxidants in vivo. Human superoxide dismutase can be classified into cytosolic CuZnSOD, mitochondrial MnSOD and extracellular SOD. Cytosolic SOD is an enzyme with a molecular weight of 32 kDa and is composed of two identical subunits. CuZnSOD contains copper and zinc in its active centre and has an enzymatic activity that is not dependent on pH, if the pH ranges from 5 to 9.5. MnSOD consists of four subunits and has a molecular weight of 95 kDa. Extracellular SOD, which is the major SOD in the vascular extracellular space, is highly expressed in the vessels, the heart, the kidney and the placenta. Extracellular SOD is a glycoprotein with four subunits. Extracellular SOD plays a vital role in endothelial function as it protects against inactivation of endothelium-released NO by the superoxide anion.

Superoxide dismutase catalyses the conversion of superoxide anion into either hydrogen peroxide or molecular oxygen. Dismutation of superoxide yields oxygen and hydrogen peroxide ($20_2' - + 2H+ \sim 0_2 + H_20_2$). Thus, protection by SOD is incompletely achieved if H_20_2 is not subsequently degraded. The elimination of H_20_2 can be achieved by catalase.

- Catalase

Catalase converts hydrogen peroxide to water and molecular oxygen in two steps ($2H_20_2 \sim 2H_20 + 0_2$). Catalase is a highly active enzyme that converts one million molecules of hydrogen peroxide per minute.

- Glutathione peroxidases

Glutathione peroxidase family contains three groups. The first group contains human GPX1 and GPX2. The second group contains human GPX3, GPX5 and GPX6. The third group, named the phospholipid hydroperoxide GPX (PHGPX) group, includes human GPX4 as well as GPXs from a number of organisms including plants. Glutathione peroxidases catalyse the reduction of hydrogen peroxide or of lipid peroxides using glutathione as the reducing substance ($2GSH + H_20_2 = GSSG + 2H_20$).

- Glutathione reductase

Glutathione reductase is an enzyme, in humans encoded by the GSR gene, that maintains glutathione in its reduced form. Enzymatic (by glutathione peroxidases) as well as non-enzymatic neutralization of reactive oxygen species by reduced glutathione (GSH) leads to the production of oxidized glutathione (GSSG).

GSSG is exported from the cell resulting in decrease of total intracellular glutathione. In order to ensure a high level of antioxidant action of glutathione, it is essential to keep a high intracellular GSH/GSSG ratio. This is achieved through the action of glutathione reductase.

24.2.2.2 Non-enzymatic Antioxidants

- Non-enzymatic antioxidants, i.e. hydrophilic antioxidants including glutathione, ascorbate, and flavonoids and lipophilic antioxidants including tocopherol, carotenoid and ubiquinol.

24.3 Oxidative Stress and Atherosclerosis

Oxidative stress is implicated in the pathogenesis of atherosclerosis, arterial hypertension, heart failure, diabetes mellitus, pancreatitis and carcinogenesis [8, 9].

In particular, in atherosclerosis, oxidative modification of low-density lipoprotein (LDL) cholesterol, which is initiated by a free radical-driven lipid peroxidation process, is considered to play a key role in the initiation as well as in the progression of atherosclerosis. Cardiovascular risk factors, such as smoking, hypercholesterolaemia and hyperglycaemia associated with atherosclerosis, are at the same time associated with oxidative stress. In the case of myocardial infarction, ROS-induced ischaemia/reperfusion injury plays a critical role in myocardial damage. However, although relative research evidence supports the impact of oxidative stress in the pathophysiology of atherosclerosis, a direct causative role of LDL oxidation for atherosclerosis has not been established.

Data derived from animal models of atherosclerosis suggest that reactive oxygen species released from nicotinamide adenine dinucleotide phosphate oxidases, xanthine oxidases and lipooxygenases as well as enhanced ROS production from dysfunctional mitochondrial respiratory chain indeed have a causative role in atherosclerosis. Furthermore, impairment of vascular function and enhanced atherogenesis have been observed in animal models that have deficiencies in antioxidant enzymes. Investigation in humans supports the oxidative stress hypothesis of atherosclerosis.

Undoubtedly, since atherosclerosis is one of the leading underlying causes of death worldwide, there is need for novel approaches on the modulation of atherosclerosis. In that aspect, effective targeting of oxidative stress is a promising approach that is anticipated to have a clinical effect on atherosclerosis. More importantly, since atherosclerosis is a multifactorial process, a multi-target approach on the modulation of atherosclerosis would be anticipated to result in reduction of cardiovascular events. This multi-target approach could be achieved through lifestyle modification, i.e. weight reduction, low-fat diet, smoking cessation and exercise, as well as through pharmacological approaches [16, 17]. These pharmacological approaches should target the factors implicated in the pathophysiology of atherosclerosis. In that aspect, there is ongoing research interest on the modulation of oxidative stress [18].

24.4 Modulation of Oxidative Stress in Atherosclerosis

Experimental evidence suggests that pharmacological modulation of oxidative stress response is feasible and might alter the natural history of relevant pathological states [19–21]. Thus, initial efforts in the modulation of oxidative stress of atherosclerosis have been attempted through treatment with antioxidants. Based on preclinical data, antioxidants were anticipated to inhibit initiation, progression and development of atherosclerosis via a number of mechanisms including inhibition of LDL oxidation, inhibition of leucocyte adhesion to the endothelium as well as inhibition of vascular endothelial dysfunction.

Indeed, data derived from relevant epidemiological studies have suggested an inverse relationship between antioxidant vitamin consumption and cardiovascular disease. The National Health and Nutrition Examination Survey epidemiological follow-up study included 11,348 participants, aged 25–74 years. Based on the data, the authors expressed the conclusion that individuals who received a high dose of vitamin C (>50 mg/d) had lower overall total mortality rate after 10 years, and in particular, with mortality rate from cardiovascular diseases being lower [22]. In accordance with the previous study, data from the Health Professionals Follow-up Study showed that the participants whose vitamin C intake exceeded 50 mg/d tended to have a lower rate of death from all cardiovascular diseases [23].

Furthermore, in a number of prospective cohort studies of thousands of healthy men and women free of cardiovascular disease, cancer and diabetes followed for a long period of time ranging from 8 to 20 years, the authors have found that the consumption of fruits and vegetables, particularly green leafy vegetables and vitamin C-rich fruits and vegetables, appeared to have a protective effect against coronary heart disease (i.e. nonfatal acute myocardial infarction or fatal coronary heart disease) [24, 25].

However, despite the favourable results of epidemiological studies, the beneficial effect of antioxidant supplementation has failed to be reproduced by randomized clinical studies. Thus, data derived from clinical trials have been giving a more confused picture than expected, with results ranging from a significant protective action to the absence of any effect.

Thus, a double-blind, randomized, placebo-controlled cardiovascular and cancer prevention trial of a combination of antioxidants (120 mg vitamin C, 30 mg vitamin E, 6 mg b-carotene, 100 mg selenium and 20 mg zinc) showed that long-term daily low-dose supplementation of antioxidant vitamins and minerals had no beneficial effects on carotid atherosclerosis and arterial stiffness.

Another clinical trial, the Physicians' Health Study, a randomized, prospective, double-blind placebo-controlled study of 14,641 male physicians, investigated the effect of oral supplementation of 400 IU of vitamin E every other day and 500 mg of vitamin C daily. The results of the study did not show any significant difference on the clinical implications of atherosclerosis between the treated and control groups in respect to the risk of major cardiovascular events [26].

Therefore, there is a discordance between epidemiological studies and clinical studies in relation to the effect of exogenous antioxidants in atherosclerosis. This discordance could be attributed to the lack of knowledge on the effect of interaction of exogenous antioxidant supplementation with one another or on the effect of interaction of exogenously administered antioxidant vitamins with endogenous ones.

In that context, a relevant systematic review has investigated the possible synergistic, additive or antagonistic effect of exogenous antioxidants in atherosclerosis [27]. This systematic review evidenced that the co-administration of exogenous antioxidants results in synergistic or additive antioxidant effect, while there was no evidence of antagonistic effect in the case of co-administration of exogenous antioxidants. The main message of this systematic review is that targeting oxidative stress via multiple mechanisms might be the ideal approach [27]. This approach could be achieved either by targeting the oxidative stress in atherosclerosis with multiple antioxidants or by administering one multi-target agent. In fact, aspirin and salicylates provide the ideal agent for the multi-target approach of oxidative stress in atherosclerosis.

24.5 Aspirin and Salicylates: Multi-target Agents

Acetylsalicylic acid or aspirin is a prototype of nonsteroidal anti-inflammatory agents. In addition, acetylsalicylic acid belongs to the family of salicylates, having in common with the other members of the family the active agent salicylic acid. Salicylic acid is a benzene ring with two radicals, one carboxyl and one hydroxyl radical. In aspirin, the hydroxyl group of salicylates has been transformed into an acetyl group by esterification. In aspirin, both salicylate and acetyl groups are active and act at different sites independently of each other. Thus, aspirin shares common biological actions with salicylates, and in addition aspirin has biological effects that are modulated by the acetyl group.

Before 1970, knowledge on the mechanism of action of aspirin was quite restricted. At that time, it was known that aspirin has anti-inflammatory action and that this action is qualitatively and quantitatively different from the action of anti-inflammatory steroids as well as analgesic action that differs from the action of opioid analgesics. In 1971, anti-inflammatory and analgesic actions of salicylates and other nonsteroidal anti-inflammatory agents were attributed to the inhibition of prostaglandin biosynthesis.

Nowadays, it is known that aspirin selectively acetylates the hydroxyl radical of the residue Ser-530 situated 70 amino acids before the carboxy-terminal of cyclo-oxygenase. Acetylation leads to irreversible inhibition of cyclooxygenase, and thus, prostaglandin biosynthesis is feasible only if cyclooxygenase is newly synthesized. In low concentrations, aspirin acetylates the enzyme within minutes and acetylation is selective. Under high concentrations of aspirin and when exposure to aspirin is of long duration, aspirin acetylates non-selectively a large number of proteins and nucleic acids.

Although some decades have passed since the discovery of the mechanism of action of aspirin, there is still ongoing research interest on the mechanism of action of aspirin and salicylates. Experimental studies showing weak inhibition of prostaglandin biosynthesis by salicylates created doubts on the dogma that the anti-inflammatory action of salicylates was due to the inhibition of prostaglandin biosynthesis. Nowadays, it is well known that aspirin and salicylates have pleiotropic effects. Among these pleiotropic effects, modulation of stress response by salicylates is quite interesting. Salicylates modulate stress response in prokaryotic organisms as well as in eukaryotic cells [8, 9, 20]. Modulation of stress response by salicylates is due to the effect of salicylates on cell signalling pathways as well as to the pro-oxidant–antioxidant effects of salicylates [8, 9, 20].

24.6 Targeting Oxidative Stress by Aspirin, Salicylates and Other Non-steroidal Anti-inflammatory Agents

Salicylates, including aspirin and the other nonsteroidal anti-inflammatory agents (NSAIDs), apart from inhibiting cyclooxygenase enzymes are known to target multiple pathways. Experimental evidence suggests that aspirin and salicylates modulate oxidative stress.

Evidence suggests that salicylates and the other nonsteroidal anti-inflammatory agents have both pro-oxidant and antioxidant actions. The antioxidant actions of salicylates have been linked with the beneficial effects of these agents, while their pro-oxidant actions have been associated with the adverse reactions of these drugs. However, this approach is superficial and does not take into account the beneficial effect of low-level pro-oxidant action. Low levels of pro-oxidant modulate cell signalling pathways. Salicylates are known to modulate cell signalling pathways, possibly through their pro-oxidant action.

The mechanisms of antioxidant action of salicylates are not well delineated. Sodium salicylate acts as a chemical trap against hydrogen peroxide radicals, the most detrimental reactive oxygen species, thus reducing ischaemia/reperfusion injury [28, 29]. In addition, nonsteroidal anti-inflammatory agents including indomethacin and sulindac have been reported to scavenge free oxygen radicals, thus exerting a protective effect against cellular oxidative stress. Although salicylates trap free oxygen radicals, this action is weak and does not seem to interpret the antioxidant action of salicylates. Evidence suggests that aspirin and salicylates enhance the activity of cellular protective antioxidant mechanisms. In particular, aspirin elicits nitric oxide release by a direct activation of the endothelial NO synthase. In addition, aspirin and salicylates downregulate superoxide production and enhance GSH-dependent antioxidant mechanisms [30, 31]. Sulindac protects normal cell by oxidative stress by initiating a preconditioning response. Furthermore, salicylic acid functions as a signalling molecule in plants involved in the expression of a number of genes [32].

The most direct evidence on the protective effect of salicylates against oxidative stress is based on experimental data from *S. cerevisiae* [8, 9, 20]. Post-logarithmic cell cultures of *S. cerevisiae* were exposed to hydrogen peroxide for 1 h. The authors investigated the effect of long-term pretreatment, the effect of short-term pretreatment with salicylates as well as the effect of exposure to salicylates during the oxidative stress. For chronic pretreatment, cells were exposed to salicylates for 22 h prior to oxidative stress. For the short-term treatment, cells were exposed to salicylates 1 h prior to the oxidative stress.

Experimental data from *Saccharomyces cerevisiae* have shown protective effect of salicylates against hydrogen peroxide stress in yeast. Importantly, it has been shown that treatment with low-dose aspirin confers long-term resistance against hydrogen peroxide-induced oxidative stress in yeast. In addition, other non-inflammatory agents have shown to protect against hydrogen peroxide stress in *Saccharomyces cerevisiae*. In an attempt to investigate the mechanisms of this protection, the effects of antioxidants glutathione and N-acetylcysteine have been investigated in the same experimental model. NAC has been administered in the dose of 30 mM. Glutathione was administered in the form of oxidized glutathione. Both NAC and glutathione had protective action against hydrogen peroxide stress when administered concurrently with hydrogen peroxide. In addition, both NAC and GSSG exerted their protective pro-oxidant action when administered 1 h prior to oxidative stress. Given that NAC and GSSG exerted the same pattern of protection with sodium salicylate, it was suggested that the mechanisms of protection of sodium salicylate against oxidative stress included the induction of cellular antioxidant mechanisms [8, 9, 20].

Sodium salicylate has been reported to inhibit the hydrogen peroxide-induced stress in HeLa cells and the effect was dose dependent. In addition, Oliveira et al. in a recently published paper have reported protective effect of acetylsalicylic acid against mitomycin C-induced carcinogenicity in *Drosophila melanogaster* through the antioxidant action of aspirin [33]. Wrobel et al. have investigated the effect of intraperitoneal injection of acetylsalicylic acid against mouse liver. The results of the study suggested that acetylsalicylic acid stimulates the GSH-dependent antioxidant system, thus protecting liver cells from oxidative stress [31]. Furthermore, other investigators have shown that aspirin and salicylates exert their antioxidant actions by the stimulation of cellular antioxidant mechanisms, i.e. Cu/Zn superoxide dismutase (SOD1) induction has been implicated in the antioxidative activity of aspirin in HCV-expressing cells [30]. In human melanocytes, it has been reported that aspirin protects against hydrogen peroxide oxidative stress through the induction of Nrf2-mediated transcriptional activation of haem oxygenase-1 [34]. Furthermore, antioxidant action has been reported for the willow bark extract that contains salicin, prodrug of salicylates [35–37].

In haemodialysis patients, aspirin has been shown to reduce inflammation through modulation of oxidative stress [38]. In addition, continued aspirin treatment until surgery in patients undergoing coronary artery bypass grafting has been shown to reduce surgery-associated oxidative stress [39]. On the other hand, preoperative withdrawal of aspirin has been shown to increase oxidative stress markers in patients undergoing coronary artery bypass grafting [40].

24.7 Aspirin and Atherosclerosis

The beneficial effects of aspirin in cardiovascular disease have well been established. Currently, evidence supports the role of aspirin in the secondary prevention of cardiovascular disease. However, the role of aspirin for primary prevention remains controversial [2–6]. Recently published European guidelines on cardiovascular disease prevention did not support prophylactic use of aspirin in individuals without established cardiovascular disease because the risk of major bleeding outweighs the minor decrease in rate of major adverse cardiac events. On the other hand, European guidelines on the management of arterial hypertension have suggested consideration of aspirin use for primary prevention in patient with high cardiovascular risk or reduced kidney function based on a more balanced risk-benefit profile in these categories of patients.

Recently, the US Preventive Services Task Force has published updated recommendations on the use of aspirin for the primary prevention of cardiovascular disease and colorectal cancer. According to these recommendations, low-dose aspirin is now supported in men and women aged 50–59 years who have a predicted risk for myocardial infarction or stroke of at least 10% over 10 years, with no elevated bleeding risk, and are willing to take aspirin within 10 years or longer. In patients aged 60–69 years, a decision has to be made on an individual basis based on personal history and comorbidities No recommendations have been included for other age groups due to the lack of relevant evidence [6].

On the other hand, UK recommendations for aspirin use for primary prevention of cardiovascular disease have adopted similar principles. Aspirin has been recommended in hypertensive patients over 50 years with a high cardiovascular risk, defined as 10-year risk of greater than 20%, or reduced renal function (e.g. estimated glomerular filtration rate less than 45 mL/min/1.73 m^2). There is also controversy on the use of aspirin in the primary prevention of cardiovascular risk in diabetic patients.

Aspirin targets atherosclerosis through multiple mechanisms including antiplatelet action, endothelial cell modulation and modulation of oxidative stress [41–54].

A number of studies have shown that aspirin exerts an inhibitory effect on the generation of reactive oxygen species. Experimental data have shown that aspirin reduced ox-LDL-mediated LOX-1 expression, MMP-1 expression and activity, p38MAPK activation and superoxide anion generation in human coronary artery endothelial cells with the effect being time dependent and dose dependent. In addition, treatment of human coronary artery endothelial cells with salicylate has resulted in effects similar to those of aspirin.

Grosser and Schröder have showed that pretreatment of endothelial cells with aspirin, but not salicylate or indomethacin, protected them from hydrogen peroxide-mediated toxicity and increased their viability. The effect was concentration dependent. In the same experimental model, pretreatment with salicylate or indomethacin failed to reproduce the protective effect of aspirin. This effect was abrogated in the presence of a NO scavenger and arginine analogs. Under the same experimental conditions, aspirin enhanced activity and intracellular cyclic GMP accumulation in endothelial cells [41, 42]. In addition, aspirin has also been shown to inhibit hydrogen peroxide-induced caspase-3, caspase-9 and NF-κB activation through inhibition of phosphorylation and degradation of IκB2 and IκBβ, with the effect being dose dependent. Heme oxygenase-1 induction via NO-dependent pathways has been suggested as another mechanism by which aspirin prevents cellular injury in cardiovascular disease [41].

In addition, the antioxidant action of aspirin has been shown in hypertensive rats. Aspirin has been shown to provide cerebrovascular protection from oxidant damage in salt-loaded, stroke-prone rats. Pretreatment of rats with aspirin and zinc complex has been shown to protect after the onset of myocardial injury through upregulation of antioxidant enzymes.

In humans it is not straightforward to prove the antioxidant action of any substance as the measurement of oxidative stress has not been standardized. In the majority of the papers, investigators measure a limited number of redox parameters that are inadequate for the valid estimation of the redox state. Despite that, the antioxidant effect of aspirin has been evidenced in human cell cultures.

24.8 Conclusion

The role of oxidative stress is well established in the pathophysiology of atherosclerosis. Aspirin and salicylates target oxidative stress in atherosclerosis through multiple antiplatelet-independent mechanisms of action, including scavenging of reactive oxygen species, enhancement of nitrous oxide release, inhibition of superoxide anion release and induction of GSH-dependent antioxidant mechanisms, epigenetic regulation of antioxidant enzymes. Thus, aspirin and salicylates are promising multi-target agents against oxidative stress implicated in atherosclerosis. Based on this evidence, the role of aspirin in the primary prevention of atherosclerosis should be revisited.

References

1. Cervadoro A, Palomba R, Vergaro G et al (2018) Targeting inflammation with nanosized drug delivery platforms in cardiovascular diseases: immune cell modulation in atherosclerosis. Front Bioeng Biotechnol 6:177. https://doi.org/10.3389/fbioe.2018.00177. eCollection 2018
2. Uchiyama S (2017) Aspirin for primary stroke prevention in elderly patients with vascular risk factors. J Gen Fam Med 18:331–335
3. Lee SK, Khambhati J, Varghese T et al (2017) Comprehensive primary prevention of cardiovascular disease in women. Clin Cardiol 40:832–838
4. Saito Y, Okada S, Ogawa H et al (2017) Low-dose aspirin for primary prevention of cardiovascular events in patients with type 2 diabetes mellitus: 10-year follow-up of a randomized controlled trial. Circulation 135:659–670
5. Ogawa H, Nakayama M, Morimoto T et al (2008) Low-dose aspirin for primary prevention of atherosclerotic events in patients with type 2 diabetes: a randomized controlled trial. JAMA 300:2134–2141
6. Wolff T1, Miller T, Ko S (2009) Aspirin for the primary prevention of cardiovascular events: an update of the evidence for the U.S. preventive services task force. Ann Intern Med Mar 150:405–410
7. Yiannakopoulou E (2011) Hemoptysis under diclofenac and antiplatelet doses of aspirin. Pharmacology 87:1–4
8. Yiannakopoulou E (2005) Cellular mechanisms of adaptation in oxidative stress and in heat shock in Saccharomyces cerevisiae: the effect of antioxidants. Msc thesis for the Msc Medical Biology Medical School and Department of Biology, University of Athens
9. Yiannakopoulou E (2007) Investigation of the role of salicylates in preconditioning. Final report for postdoctoral research (IKY Scholarship 15109/07.10.2005)
10. Yiannakopoulou EC, Delitheos A, Tiligada E (2005) Dose-dependent effect of non-steroidal anti-inflammatory agents on the cellular stress response. Epitheor Klin Farmacol Farmakokinet 23:39–41
11. Yiannakopoulou EC, Tiligada E (2007) Pharmacological preconditioning in the oxidative stress response of eukaryotic cells: in process method validation. Epitheor Klin Farmacol Farmakokinet 25:30–32
12. Yiannakopoulou E (2009) Oxidative stress-antioxidant mechanisms-clinical implications. Arch Greek Med 26:23–35
13. Yiannakopoulou E (2014) Targeting epigenetic mechanisms and microRNAs by aspirin and other non steroidal anti-inflammatory agents--implications for cancer treatment and chemoprevention. Cell Oncol (Dordr) 37:167–178
14. Yiannakopoulou E (2012) Modulation of lymphangiogenesis: a new target for aspirin and other nonsteroidal anti-inflammatory agents? A systematic review. J Clin Pharmacol 52:1749–1754
15. Yin MJ, Yamamoto Y, Gaynor RB (1999) The anti-inflammatory agents aspirin and salicylate inhibit the activity of I(kappa)B kinase-beta. Nature 396:77–80
16. Yiannakopoulou E (2003) Quality of diagnosis and treatment of arterial hypertension in Greece. PhD thesis, Medical School University of Athens, Greece
17. Yiannakopoulou E, Papadopulos IS, Cokkinos DV et al (2005) Adherence to antihypertensive treatment: a critical factor for blood pressure control. Eur J Cardiovasc Prev Rehabil 2005(12):243–249
18. Lu L, Qin Y, Chen C et al (2018) Beneficial effects exerted by paeonol in the management of atherosclerosis. Oxidative Med Cell Longev 2018:1098617
19. Yiannakopoulou E, Nikiteas N, Perrea D et al (2012) Pharmacological preconditioning of oxidative stress response: systematic review. Surg Laparosc Endosc Percutan Tech 22:200–204
20. Yiannakopoulou E, Tiligada E (2009) Preconditioning effect of salicylates against oxidative stress in yeast. J Appl Microbiol 106:903–908
21. Yiannakopoulou EC, Tiligada E (2006) Acetaminophen modulates the oxidative stress response in eucaryotic cells. Rev Clin Pharmacol Pharmacokinet 20:125–127

22. Radimer K, Bindewald B, Hughes J et al Dietary supplement use by US adults: data from the national health and nutrition examination survey, 1999–2000. Am J Epidemiol 160:339–349
23. Kim HJ, Giovannucci E, Rosner B et al (2014) Longitudinal and secular trends in dietary supplement use: nurses' health study and health professionals follow-up study, 1986–2006. J Acad Nutr Diet 114:436–443
24. Joshipura KJ, Hu FB, Manson JE et al (2001) The effect of fruit and vegetable intake on risk for coronary heart disease. Ann Intern Med 134:1106–1114
25. Miedema MD, Petrone A, Shikany JM et al (2015) Association of fruit and vegetable consumption during early adulthood with the prevalence of coronary artery calcium after 20 years of follow-up: the coronary artery risk development in young adults (CARDIA) study. Circulation 132:1990–1998
26. Christen WG, Gaziano JM, Hennekens CH (2000) Design of physicians' health study II—a randomized trial of beta-carotene, vitamins E and C, and multivitamins, in prevention of cancer, cardiovascular disease, and eye disease, and review of results of completed trials. Ann Epidemiol 10:125–134
27. Yiannakopoulou E (2012) Does pharmacodynamic interaction of non enzymatic antioxidants modify response to anti-oxidant therapy in the process of atherosclerosis? J Cardiovasc Pharmacol 17:366–372
28. Mohanakumar KP, Muralikrishnan D, Thomas B (2000) Neuroprotection by sodium salicylate against 1-methyl-4-phenyl-1,2,3, 6-tetrahydropyridine-induced neurotoxicity. Brain Res 864:281–290
29. Bektaşoğlu B, Ozyürek M, Güçlü K et al (2008) Hydroxyl radical detection with a salicylate probe using modified CUPRAC spectrophotometry and HPLC. Talanta 77:90–97
30. Rivas-Estilla AM, Bryan-Marrugo OL, Trujillo-Murillo K et al (2012) Cu/Zn superoxide dismutase (SOD1) induction is implicated in the antioxidative and antiviral activity of acetylsalicylic acid in HCV-expressing cells. Am J Physiol Gastrointest Liver Physiol 302:G1264–G7327
31. Wróbel M, Góralska J, Jurkowska H et al (2017) Similar effect of sodium nitroprusside and acetylsalicylic acid on antioxidant system improvement in mouse liver but not in the brain. Biochimie 135:181–185
32. Gondor OK, Pál M, Darkó É et al (2016) Salicylic acid and sodium salicylate alleviate cadmium toxicity to different extents in maize (Zea mays L.). PLoS One 11:e0160157
33. Oliveira VC, Constante SAR, Polloni L et al (2018) Protective effect of aspirin against mitomycin C-induced carcinogenicity, assessed by the test for detection of epithelial tumor clones (warts) in Drosophila melanogaster. Drug Chem Toxicol 41:330–33733
34. Jian Z, Tang L, Yi X et al (2016) Aspirin induces Nrf2-mediated transcriptional activation of haem oxygenase-1 in protection of human melanocytes from H2 O2 -induced oxidative stress. J Cell Mol Med 20:1307–1318
35. Bonaterra GA, Heinrich E, Kelber O et al (2010a) Anti-inflammatory effects of the willow bark extract STW 33-I (Proaktiv®) in LPS-activated human monocytes and differentiated macrophages. Phytomedicine 17:1106–1113
36. Enayat S, Banerjee S (2009) Comparative antioxidant activity of extracts from leaves, bark and catkins of Salix aegyptiaca sp. Food Chem 116:23–28
37. Ishikado A, Sono Y, Matsumoto M et al (2013) Willow bark extract increases antioxidant enzymes and reduces oxidative stress through activation of Nrf2 in vascular endothelial cells and Caenorhabditis elegans. Free Radic Biol Med 65:1506–1515
38. Maniqlia FP, Costa JA (2016) Effects of acetylsalicylic acid usage on inflammatory and oxidative stress markers in hemodialysis patients. Inflammation 9:243–24734
39. Berg K, Langaas M, Ericsson M et al (2013) Acetylsalicylic acid treatment until surgery reduces oxidative stress and inflammation in patients undergoing coronary artery bypass grafting. Eur J Cardiothorac Surg 43:1154–1163
40. Berg K, Haaverstad R, Astudillo R et al (2006) Oxidative stress during coronary artery bypass operations: importance of surgical trauma and drug treatment. Scand Cardiovasc J 40:291–297
41. Grosser N, Abate A, Oberle S et al (2003) Heme oxygenase-1 induction may explain the antioxidant profile of aspirin. Biochem Biophys Res Commun 308:956–960

42. Grosser N, Schroder H (2003) Aspirin protects endothelial cells from oxidant damage via the nitric oxide-cGMP pathway. Arterioscler Thromb Vasc Biol 23:1345–1351
43. He B, Zhao S, Zhang W et al (2010) Salicylate prevents hepatic oxidative stress activation caused by short-term elevation of free fatty acids in vivo. Diabetes Res Clin Pract 89:150–156
44. He B1, Zhao S, Zhang W et al (2010) Effect of sodium salicylate on oxidative stress and insulin resistance induced by free fatty acids. Hepatobiliary Pancreat Dis Int 9:49–53
45. Wu R, Lamontagne D, de Champlain J (2002) Antioxidative properties of acetylsalicylic acid on vascular tissues from normotensive and spontaneously hypertensive rats. Circulation 105:387–9235
46. El MA, Wu R, de Champlain J (2002) Prevention of hypertension, hyperglycemia and vascular oxidative stress by aspirin treatment in chronically glucose-fed rats. J Hypertens 20:1407–1412
47. Wu R, Laplante MA, De Champlain J (2004) Prevention of angiotensin II-induced hypertension, cardiovascular hypertrophy and oxidative stress by acetylsalicylic acid in rats. J Hypertens 22:793–801
48. Tauseef M, Sharma KK, Fahim M (2007) Aspirin restores normal baroreflex function in hypercholesterolemic rats by its antioxidative action. Eur J Pharmacol 556:136–143
49. Mei L, Daud MK, Ullah N et al (2015) Pretreatment with salicylic acid and ascorbic acid significantly mitigate oxidative stress induced by copper in cotton genotypes. Environ Sci Pollut Res Int 22:9922–9931
50. Tauseef M, Shahid M, Sharma KK et al (2008) Antioxidative action of aspirin on endothelial function in hypercholesterolaemic rats. Basic Clin Pharmacol Toxicol 103:314–321
51. Podhaisky H-P, Abate A, Polte T et al (1997) Aspirin protects endothelial cells from oxidative stress–possible synergism with vitamin E. FEBS Lett 417:349–351
52. Korkmaz-Icöz S, Atmanli A, Radovits T, Li S et al (2016) Administration of zinc complex of acetylsalicylic acid after the onset of myocardial injury protects the heart by upregulation of antioxidant enzymes. J Physiol Sci 66:113–125
53. Korkmaz S, Atmanli A, Li S, Radovits T et al (2015) Superiority of zinc complex of acetylsalicylic acid to acetylsalicylic acid in preventing postischemic myocardial dysfunction. Exp Biol Med (Maywood) 240:1247–1255
54. Amel Zabihi N, Mahmoudabady M, Soukhtanloo M et al (2018) Salix alba attenuated oxidative stress in the heart and kidney of hypercholesterolemic rabbits. Avicenna J Phytomed 8:63–72

The Evolving Concept of Mitochondrial Dynamics in Heart: Interventional Opportunities

25

Ashok Sivasailam, Mahalaxmi Ganjoo, Varghese T. Panicker, Vivek V. Pillai, and Srinivas Gopala

Abstract

The cardiac tissue with its enormous task of continuous pumping relies heavily on the mitochondria. The different subpopulations of the mitochondria support the cardiac contractile function in various ways. These organelles are established as a continuous network in the cardiac tissue, i.e. in a highly dynamic state undergoing biogenesis, fusion, fission and degradation. This dynamic nature of the organelle helps in maintaining a healthy mitochondrial circuit which in turn is necessary for optimal cardiac functioning. There are increasing empirical evidences suggesting that the cardiovascular diseases are primarily associated with the decrease in the mitochondrial capacity of ATP synthesis, ROS handling and calcium homeostasis. This implies that the quantity and quality of mitochondria is crucial for its optimal fucntion, particularly during energy challenges faced by heart. Available data suggest that for prevention and therapy for most of the cardiovascular diseases, mitochondria could be an ideal target. There are various therapies that have focused on improving the mitochondrial efficiency through multifarious means, ranging from repairing the ROS-mediated damage to inducing the mitochondrial biogenesis and degradation, thus ensuring a newer and adept network of mitochondria. The mechanisms behind the compounds hitherto believed to be beneficial for the heart are also examined. This chapter summa-

All the Authors are equally contributed to this Chapter

A. Sivasailam · M. Ganjoo · S. Gopala (✉)
Department of Biochemistry, Sree Chitra Tirunal Institute for Medical Sciences and Technology, Trivandrum, Kerala, India
e-mail: srinivasg@sctimst.ac.in

V. T. Panicker · V. V. Pillai
Department of Cardiovascular and Thoracic Surgery, Sree Chitra Tirunal Institute for Medical Sciences and Technology, Trivandrum, Kerala, India

S. Chakraborti et al. (eds.), *Modulation of Oxidative Stress in Heart Disease*,
https://doi.org/10.1007/978-981-13-8946-7_25

rizes the importance of mitochondria and its quality control in the cardiac tissue and some of the therapeutic interventions targeting the same.

Keywords
Mitocondria · ROS · Mitochondrial dynamics · Mitochondrial biogenesis · Mitochondrial fission and fusion · Mitophagy · Mitochondrial therapies

25.1 Introduction

The onerous task of oxygenating the entire organ system in a living being proceeds through the incessant, rhythmic contraction and relaxation of a muscular organ, the heart. This fist-sized organ is no less than a wonder by way of its incredible work load, of pumping ~7500 l of blood a day in a healthy adult human being, and its amazingly resilient mechanical nature, together with the other components of the circulatory system, helping in maintaining the haemodynamic stability of the organism. The history of the discovery of the heart and its involvement in circulation dates back to ~2000 years, but more comprehensive knowledge on cardiovascular circulation and the functioning of the heart was deciphered by William Harvey, in the seventeenth century AD [1, 2].

The heart consists predominantly of muscle fibres, formed from a functional syncytium of cardiomyocytes widely connected via gap junctions, and these muscle fibres are intensely interspersed with blood vessels for optimal nutrient/oxygen supply. The functioning of the heart is highly energy demanding, and the reserve of energy currencies like ATP, creatine phosphate, etc. within the myocardial cells is nominal. This necessitates a tight coordination of metabolic production of ATP and the energy requiring process of contraction and relaxation [3]. The energy needs are met mostly by the oxidative metabolism of the fuels in specialized organelles called mitochondria.

Mitochondria are double membrane-bound symbiotic organelles residing in the cytoplasm of almost all eukaryotic cells, which acts as the epicentre of energy production, by disposing organic substrates into electrochemical gradient. The complex architecture of the ion-permeable inner membrane and the more porous outer membrane (permeable to particles $^<$5 kDa) helps in devising the primary function, oxidative phosphorylation and other cellular events ranging from cellular differentiation to cell death via integration of diverse signalling pathways [4] (Fig. 25.1). Mitochondria, often illustrated as small oval structures, are labyrinthine in nature with regard to composition, structure and function. In normal physiology, these organelles are dynamic, constantly fusing and dividing under the influence of cellular stress. The contractile units of the heart, cardiomyocytes, are densely populated by mitochondria (~1/3 the cell volume) which ensure the preservation of metabolic need of the organ [5]. Thickly populated mitochondrial reticulum within the contractile apparatus of the myotubes appears to be a continuous scaffold, but irrespective of their propinquity, they continue to maintain their identity as single

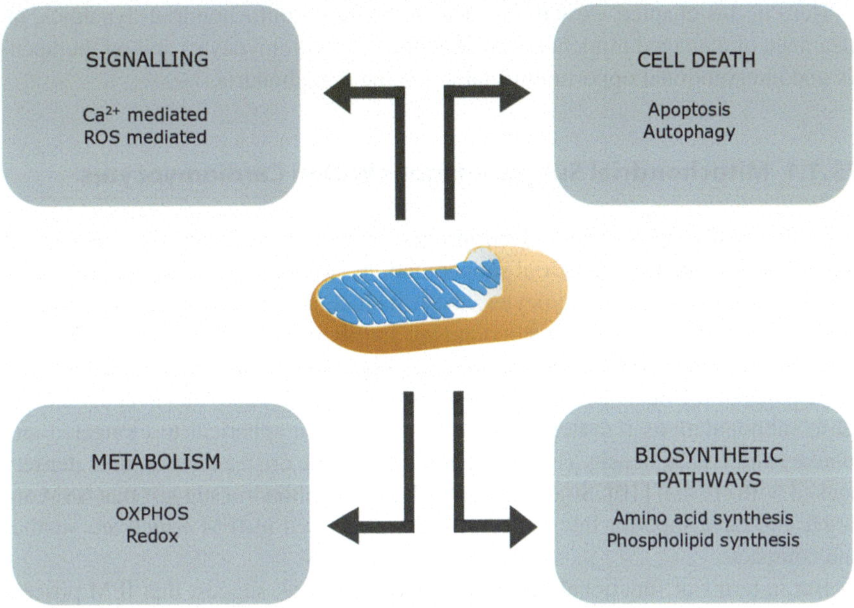

Fig. 25.1 Mitochondrial signalling: Mitochondria are involved in various cellular events like energy generation, maintaining redox homeostasis, playing a role in various signalling events mainly calcium- and ROS-mediated signalling and synthesis of various metabolites like phospholipids, amino acids, etc. and programmed cell death event

entities connected by specific inter-mitochondrial junctions [6]. This segmentation of mitochondrial entities within the reticulum helps in rapid separation of dysfunctional mitochondrial units even when they are highly interconnected and coupled [6, 7]. Even though the adult myocardium is competent in using all categories of energy substrates (which counts carbohydrates, lipids, amino acids and ketone bodies) for ATP generation, ~70% of energy is derived from fatty acid substrates. Therefore under normal physiology, most of the reducing equivalents, NADH and FADH, that feed the mitochondrial complexes will be generated from fatty acid oxidation. But being a metabolic omnivore by nature, the myocardial metabolic network would be flexible enough to utilize the abundant fuel and supply energy for the process of contraction-relaxation cycle. For example, the use of lactate by myocardial tissue is enhanced when lactate production from skeletal muscle increases by way of exercise; while ketone bodies released by way of fasting can result in its utilization, and this metabolic adaptability of myocardium is substantiated by experimental evidences from isolated heart [8, 9].

Alterations in substrate selection occur at the level of mitochondria, and this depends on the physiological condition, nutritional status of the cell and the presence or absence of any diseases; altogether determining the nature of the mitochondrial network that wires the cell. Thus, it is not surprising that heart diseases are associated with mitochondrial changes.

Here in this chapter, we will be focusing more on mitochondrial dynamics, on clearance of damaged mitochondrial structures in cardiomyocytes and on therapeutic and interventional opportunities targeting the mitochondria.

25.1.1 Mitochondrial Subpopulations Within Cardiomyocytes

Ultrastructural analyses using electron microscopy have revealed the existence of spatially estranged mitochondrial subpopulations, with distinct structural and functional characters in cardiomyocytes; these include subsarcolemmal mitochondria (SSM) that resides below the sarcolemma, interfibrillar mitochondria (IFM) that dwell between the myofibrils and, a specific population of this organelle that resides at the poles of the nucleus, the perinuclear mitochondria. Also, the diversity among these subpopulations is reflected in size, shape (from spherical to elongated and tubular) and cristae density (from structures with little cristae to structures densely packed with cristae) [10]. Studies using isolated mitochondria suggest that SSM are larger and have superior internal complexity compared to IFM which are smaller and compact.

But in terms of functional performance, Palmer et al. suggest that IFM possess higher respiration rates and enzyme activities (specifically succinate dehydrogenase and citrate synthase activity) compared with SSM. All lipid and nonlipid substrates are oxidized ~1.5 times faster by the IFM than the SSM [11].

IFM span the entire length of the sarcomeric unit from one Z-disk to another and are bordered by the sarcoplasmic reticulum. This positional arrangement of IFM is thought to favour calcium-mediated interaction with sarcoplasmic reticulum relating calcium signalling and mitochondrial function. Thus, IFM is hypothesized to effectively power contraction [12]. However, SSM are thought to be involved in active transport of electrolytes and metabolites across the sarcolemmal membrane. C. Crochemore et al. reported distinct superoxide production profiles of SSM and IFM, SSM producing more ROS on incubation with complex 1 substrates compared to IFM, and this difference in ROS production profile is attributed to their functional differences, IFM with higher respiratory activity than SSM [13]. Also, these distinct populations of mitochondria have been reported to have different effects on cardiac-related pathologies like hypoxia, pressure overload, diabetes and heart failure [14] (Table 25.1).

Although the perinuclear mitochondria are well observed as a different microscopic subpopulation, their biochemical and functional properties are not well described. Nevertheless, it has been suggested to power mitochondrial metabolism close to the nucleus. The structural and functional differences between the members of different mitochondrial subpopulations can be attributed to their localization [11, 15]. The molecular mechanisms underlying ultrastructural remodelling of mitochondrial population is a hot topic of research.

Table 25.1 Different subtype of mitochondria found in myocyte

Type of mitochondria	Location	Size and shape	ATP produced for
Perinuclear	Poles of the nucleus	Spherical in shape with diameter ranging from 0.8 to 1.4 μm	Powers mitochondrial metabolism close to nucleus
		Well-developed cristae with very little matrix	
Interfibrillar	Longitudinally arranged in rows between the myofibrils	Elongated in shape with usually one mitochondrion existing per sarcomere	Powers muscular contraction
		~1.5–2.0 μm in length	
		Curved cristae structure	
Subsarcolemmal	Beneath the plasma membrane	Observed in variety of shapes like oval, spherical, polygonal and horse-shoe patterns	Powers transport of solutes and metabolites
		Sizes range from 0.4 to 3.0 μm	

25.2 Mitochondrial Dynamics: Structural and Functional Remodelling of Mitochondria

Mitochondrial network, a vital component of cardiac cell architecture, forms a highly dynamic and intricate structure, adopting diverse distribution patterns, morphologies and functional roles in accordance with the cellular signals, both internal and external. This structural and functional plasticity of the organelle is proficiently maintained by orchestrating a proper balance between the counteracting forces like fission and fusion processes as well as the biosynthesis of new mitochondrial masses and the programmed elimination of the mitochondrial fragments, which depends heavily on the metabolic cues and changes in cellular stresses. The term mitochondrial dynamics encompass the structural and functional remodelling of the organelle network by way of fission and fusion, the subcellular mitochondrial mobility across the cell cytoplasm (to reach out to areas based on energy needs) mitochondrial biogenesis that helps increase mitochondrial mass and the programmed clearance of mitochondria, either damaged or functional. Together these processes form a quality control mechanism which ensures a healthy mitochondrial circuit [16, 17].

25.2.1 Mitochondrial Biogenesis

As discussed earlier, the cytoplasmic space of terminally differentiated cardiomyocytes in adult myocardium is densely packed with high-capacity mitochondrial networks. And this assembly of complex mitochondrial structures starts with a surge in the process of mitochondrial biogenesis at birth, followed by dynamic restructuring

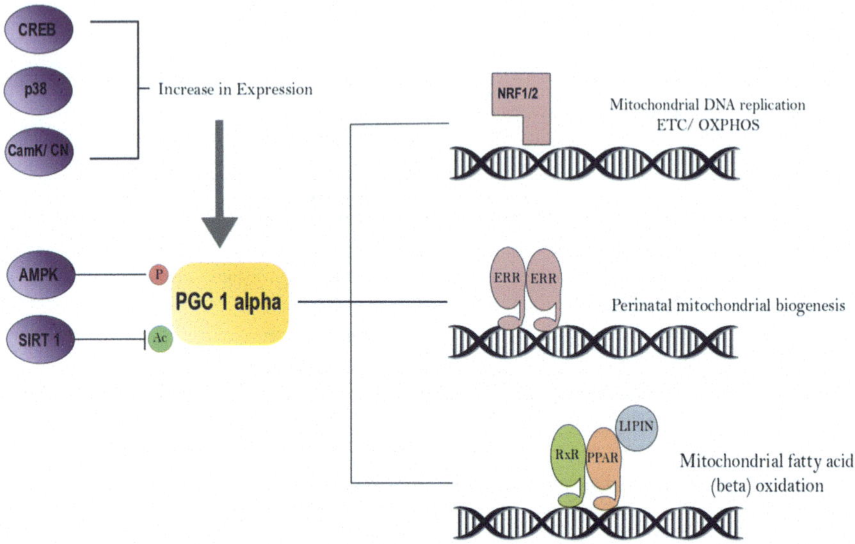

Fig. 25.2 The upstream inputs and downstream targets of PGC-1 α transcriptional coactivators: the PGC-1 α expression and activity is dependent on various physiological and metabolic stimuli which ultimately regulate mitochondrial biogenesis and function. PGC-1 α interacts with various transcriptional factors and controls mitochondrial dynamics and protein levels

and distribution of matured mitochondria between sarcomeres (IFM), around the nucleus (perinuclear), and in the subsarcolemmal (SSM) regions.

Mammalian mitochondria contain ~1200 proteins, and these may vary significantly according to the cellular environment and cell or tissue types [18]. Mitochondrial DNA (mtDNA) encodes 13 of these proteins, and the rest are encoded by nuclear genome, thus making it obvious that mitochondrial homeostasis is under strict nuclear control. The presence of a self-replicating genome for the organelle necessitates a tight coordination between the transcriptional and replication machinery of nuclear and mitochondrial genome for the process of biogenesis. The complex transcriptional network that choreographs these coordinated processes must not only be active during development but also be responsive to the physiological indications of changes in energy demands and substrate availability. Peroxisome proliferator-activated receptor-γ (PPARγ) coactivator 1α (PGC-1α), a transcriptional co-regulator, has emerged as a metabolic node and the master regulator of mitochondrial function and biogenesis [19] ever since its discovery in brown adipose tissue as a key regulator of adipogenesis [20]. PGC-1α belongs to PGC-1 family of transcriptional coactivators alongside the closest homolog, PGC-1β, and a more distant, PGC-1-related coactivator (PRC). Knockdown and overexpression studies of these factors have revealed their critical role in driving mitochondrial biogenesis in the mitochondria-rich tissues like cardiac muscle, skeletal muscle, adipose tissue, etc. (Fig. 25.2).

Table 25.2 Transcription factors regulated by PGC1 coactivators

Transcription factors	Functions
NRF-1 and NRF-2 [27–31]	Regulation of expression of protein complexes of ETC
	Regulation of replication and transcription of mitochondrial genome via factors like TFAM and TFB2M
ERRα, ERRβ and ERRγ [32–36]	Regulation of the transcription of enzymes involved in fatty acid oxidation, TCA cycle and OXPHOS
	Plays important role in metabolic maturation of postnatal myocardium
	Shown to regulate the expression of genes that control contractile function of myocytes
PPARα, PPARβ and PPARγ [37–43]	Plays decisive role in the genesis of high-capacity mitochondrial network specialized for fatty acid oxidation

Cardiac-specific gene manipulation of PGC-1 coactivators in perinatal and postnatal mouse models has helped in demonstrating their decisive and overlapping roles in cardiac mitochondrial biogenesis, maturation and maintenance of proper dynamics (fission, fusion), which helps in development as well as proper functioning of the organ especially under stress like pressure overload [21–24]. In contrast to these observations, the induced knockdown of PGC-1 coactivators in adult myocardium shows no deleterious effects but exhibits a lower respiratory capacity which can be attributed to the reduced expression of components of metabolic pathways including fatty acid oxidation, TCA cycle, electron transport chain, etc. Also observed was a subset of mitochondrial population having an abnormal cristae structure by way of disrupted phospholipid (cardiolipin) synthesis. These observations along with the studies on mitochondrial fusion [25] suggest a lower rate of mitochondrial turnover in adult heart compared to developing myocardium. PGC-1 coactivators coordinate their function by interacting with transcription factors of nuclear receptor superfamily, and most of the studies were focused on PGC-1 α, which interacts with the target transcription factors by means of specific LXXLL recognition domains. This helps in recruiting molecules that mediate chromatin remodelling via histone acetylation as well as recruitment of RNA polymerase II, via interaction with TRAP/DRIP complex [26]. The effector transcription factors that interact with PGC-1 coactivators include the members of the PPAR (PPARα, PPARβ and PPARγ), oestrogen-related receptor (ERRα, ERRβ and ERRγ) and nuclear respiratory factor (NRF-1 and NRF-2) transcription factor families (Table 25.2).

PGC-1 α is best known to be regulated by exercise and cold exposure, the effect mediated through the stimulation of β-adrenergic receptors and the cAMP/CREB (cyclic AMP response element binding protein) [44, 45]. Akimoto et al. have shown to induce PGC-1 α expression in skeletal muscle even with a single exercise session controlled through the ATF2 transcription factor mediated by the p38 MAPK signalling; the AMPK activation additionally regulates [46] the PGC-1 α in the skeletal muscle [47]. In the cardiac tissue, the proviral integration site for Moloney murine leukaemia virus (Pim) kinases is a prominent regulator of PGC-1 α as the studies on site-specific deletions of Pim1, Pim2 and Pim3 kinases result in expedited ageing response in the heart concomitantly showing mitochondrial defects and depleted ATP levels [48]. Post-translational modifications of phosphorylation and acetylation govern the fine-tuned response of PGC-1 α especially through the phosphorylation by AMPK. Deactivated PGC-1 α through the acetylation of its multiple lysine residues form a substrate for NAD+ -dependent deacetylase sirtuin 1 (SIRT1) [49–51]. With the upstream regulators constituting the AMPK, NAD+ and SIRT1 of PGC-1 α, the mitochondrial biogenesis can be appropriately regulated in response to the energy requirement and the redox status of the cell [49], helping the mitochondria adapt to the cues and determine the most apt course of action as to whether to undergo biogenesis or to dynamically remodel the mitochondrial network.

25.2.2 Mitochondrial Fusion and Fission

Mitochondrial morphology is intricately connected to most of its functions, which include oxidative metabolism, maintenance of calcium homeostasis, ROS generation and programmed cell death. And the changes in mitochondrial morphology are choreographed by mitochondrial fission and fusion processes; that are continuous processes essential for cell survival, cell development, cell division and cellular adaptation to different stress and also play a decisive role in the rewiring of mitochondrial network in physiology as well as pathology. Research outputs in this field evolving from pioneering studies in simple eukaryotes, yeast, prove that these processes are mediated by a set of evolutionarily conserved proteins, most of which belongs to GTPase superfamily.

25.2.2.1 Machinery of Fusion

Mitochondrial fusion process brings together different mitochondrial units by way of membrane fusion and matrix integration. This process is a well-organized reaction towards cellular signals than just a simple mixing of mitochondrial contents; making a single large network of mitochondria, for efficient production of energy currencies (both as ATP and GTP) rather than relying on individual isolated factories.

Being a fundamental process, the mitochondrial fusion and its associated mechanism is evolutionarily conserved from yeast to mammals. The core machinery of mitochondrial fusion includes three proteins of dynamin family GTPases: Mitofusin1 and Mitofusin2 which act on outer mitochondrial membrane and optic atrophy1 (OPA1) that mediates inner mitochondrial membrane fusion (Fig. 25.3) (Table 25.3).

25.2.2.1.1 Mitofusins/Outer Membrane Fusion Proteins

Mfn1 and Mfn2, the mammalian orthologs of the protein components that coordinate mitochondrial outer membrane fusion, are carved up in a similar molecular construction. Both the proteins possess an N-terminal catalytic domain with GTPase activity; two heptad repeat (HR) regions, HR1 and HR2; and two transmembrane

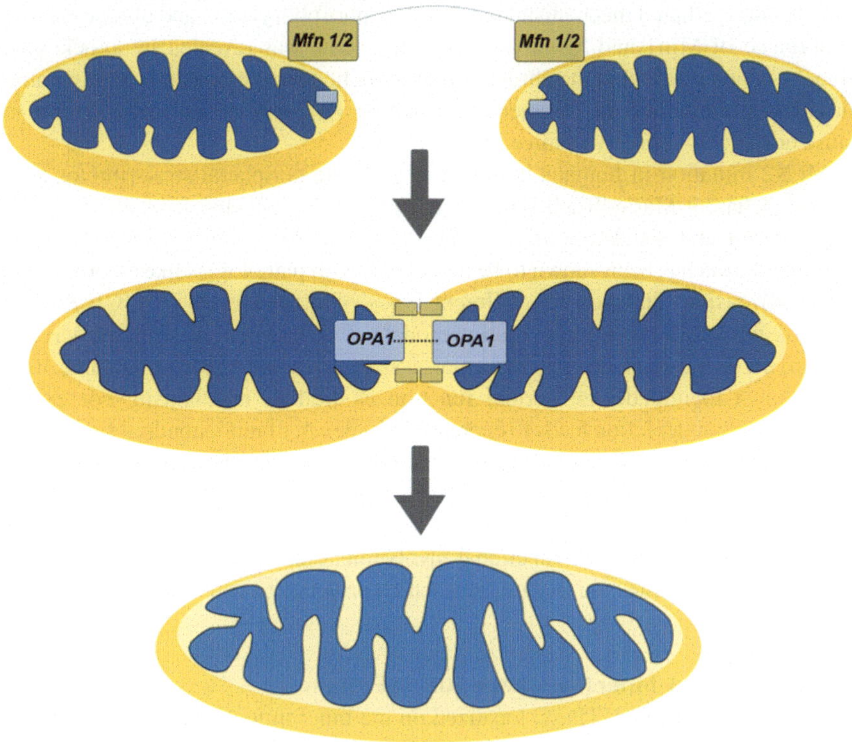

Fig. 25.3 Mitochondrial fusion: Mitochondrial fusion process mediated by mitofusins (Mfn1/2-regulating outer membrane fusion) and OPA1 (regulating inner membrane fusion)

Table 25.3 Protein components, dynamin family GTPases, important in the mitochondrial fusion process

Protein		Location	Function
Mitofusin1/ Mfn1		Outer mitochondrial membrane	Outer membrane fusion
Mitofusin2/ Mfn2		Outer mitochondrial membrane	Outer membrane fusion
OPA1	L-OPA1	Inner mitochondrial membrane	Inner membrane fusion
	S-OPA1	Inner mitochondrial membrane & inter membrane space	

(TM) domains, sandwiched between the HR regions, which keep the mitofusins to be anchored to outer mitochondrial membrane (OMM). The heptad repeats HR1 and HR2 project into the cytosol, where HR2 is shown to act as a go-between their counterparts in adjoining mitochondrial membranes.

Mitofusin expression and activity are regulated via transcriptional and post-transcriptional routes, and clearance of the protein proceeds through proteasome-mediated mechanism, regulated by phosphorylation and ubiquitylation. Importance of Mfn1 and Mfn2 protein components was revealed by knockdown models, which resulted in the abrogation of mitochondrial fusion and consequential unopposed mitochondrial fragmentation that in vivo studies have also shown to cause embryonic lethality in mice [52].

MFN2 mutation in humans is linked to a classic peripheral sensorimotor neuropathy, Charcot-Marie-Tooth type 2A (CMT2A) disease, and it involves axonal degeneration and distal muscular atrophy [53, 54]. Also, defective MFN2 or its downregulation has been shown to be associated with pathologies like atherosclerosis, hypertension, diabetes and different types of malignancies [55–62]. Studies using tissue-specific gene manipulation models of mitofusin proteins in the nervous system, cardiovascular system, muscular system, or other major organs like liver have shown to impede the normal function as well as metabolism. For example, knockdown/depletion leads to decreased efficiency of mitochondrial fuel oxidation while overexpression leads to hyperactivation of mitochondria via increased expression of mitochondrial complexes as well as other proteins involved in energy generation. [25, 53, 54, 63–70].

Contrasting evidences of functional involvement and idleness of these proteins have been reported in some adult tissues which projects it as an interesting research endeavour.

25.2.2.1.2 OPA1/Inner Membrane Fusion Protein

OPA1, a dynamin-like GTPase, localized on the inner mitochondrial membrane as well as intermembrane space, mediates the process of inner mitochondrial membrane (IMM) fusion. OPA1 is named after optic atrophy, an autosomal dominant optic neuropathy caused by mutations in the gene. The protein has also been shown to mediate several functions of mitochondria other than the fusion processes, which include cristae morphology and mitochondrial DNA stability [71–75].

OPA1 is originally expressed with an N-terminal mitochondrial targeting sequence, a transmembrane domain, a heptad repeat region, GTP binding domain and a C-terminal GTPase effector domain [76]. Depending on the tissue of expression, eight different OPA1 splice variants are generated by alternative splicing of the mRNA [77]. These long forms of OPA1 (L-OPA1) are susceptible for proteolytic cleavage by peptidases like PARL (presenilin-associated rhomboid-like protease), i-AAA metalloprotease Ymel, m-AAA metalloprotease paraplegin, metallopeptidase which localizes in the IMM OMA-1 and zinc metalloprotease [78–85]. Proteolytic modification occurs in the membrane-spanning region of long form which results in the generation of a soluble form, S-OPA1. Thus, the unprocessed long form of OPA1(L-OPA1) remains anchored to the IMM by its TM domain, while S-OPA1 which lacks the TM domain is targeted to the intermembrane space via its interaction with the IMM-anchored L-OPA1. Mammalian cells lacking OPA1 processing enzymes have shown to exhibit balanced fission and fusion processes, and these studies claim the dispensable nature of OPA1 processing for maintaining structural and functional integrity of mitochondrial network.

Mitochondrial membrane potential disruption results in the accumulation of inactive OPA1 isoforms, by induced proteolytic cleavage, thereby inhibiting fusion and targeting the mitochondrial fragments for clearance by mitophagy. Also, pro-apoptotic signals have shown to activate OMA1-mediated cleavage of OPA1, ultimately resulting in excessive fragmentation of mitochondria, suggesting its role in programmed cell death. Knockdown studies in mouse embryonic fibroblasts have revealed the existence of hyperfused mitochondrial network even under stress, thus exposing their role as a stress sensor essential to maintain mitochondrial homeostasis [81, 86, 87].

L-OPA1 cleavage serves as an important regulatory role in balancing mitochondrial fusion and fission processes, preserving mitochondrial architecture. OPA1 expression and associated changes in mitochondrial morphology depend heavily on the metabolic alterations and the type of substrates available to the cell; and is transcriptionally regulated by factors like NF-kB which serves as a master controller of metabolic reprogramming, cell survival, etc. [88].

Gene knockout models of Opa1 have fragmented mitochondria due to impaired fusion, and re-expression of L-OPA1 lacking the cleavage sites for the peptidases has shown to rescue it, highlighting its importance for maintaining a balance between fusion and fission processes. In mice models, complete knockout of Opa1 is lethal, but heterozygous expression makes them viable except showing 50% reduction in steady-state levels making them susceptible for tissue-specific pathologies like slow-onset retinal degeneration and decline in vision, as well as cardiac hypertrophy and associated dysfunction [89, 90]. In the same line, overexpression studies with functionally active OPA1 have been shown to confer cytoprotective advantages in a range of pathophysiological conditions, emphasizing the significance of OPA1 function and its regulation. However, it remains elusive as to what function of OPA1 confers the cytoprotective effect, whether it is the fusion activity or other functions that are yet to be discovered.

Current understanding of the process of mitochondrial fusion proposes it as an outcome of highly coordinated interaction between the outer membrane and inner membrane fusion proteins. The GTP powered fusion process occurs in various steps: tethering of different mitochondrial units via the interaction between the outer membrane proteins (Mfn1 and Mfn2), outer membrane fusion followed by inner membrane fusion mediated by OPA1. After the fusion of the outer and the inner mitochondrial membranes, the two separate mitochondrial entities fuse to either become a larger mitochondrion or the discrete mitochondria become part of the already formed network. The fusion process, though coalesces the mitochondrial units, the mitochondria can nonetheless maintain some discontinuity; that safeguards the entire network from depolarizing when just a part of the network becomes damaged; thus making possible the severing of the damaged part of the network alone while maintaining the integrity of the network.

25.2.2.2 Machinery of Fission

Mitochondrial fission involves the fragmentation of existing mitochondrial entity into separate units, which can either be part of a programmed series of events like embryogenesis, cell differentiation and development, apoptosis, etc. or can be an induced process to combat cellular stress resulting from mitochondrial dysfunction. Mitochondrial fission has been shown to be a part of numerous events, although the physiological significance of the process still remains elusive. In mammalian cells, the process of mitochondrial fission is regulated by dynamin-related protein (Drp) 1, mitochondrial fission factor (Mff) and fission 1 (Fis1).

25.2.2.2.1 Dynamin-Related Protein (Drp1)

Dynamin-related protein, Drp1, is an 80 kDa protein that belongs to the GTPase family and is found in the cytosol of mammalian cells. Mitochondrial fission stimulus induces the localization of this large protein to the outer mitochondrial membrane eventually resulting in the fragmentation of the constricted organelle into separate entities [91, 92].

Drp1 consists of an N-terminal GTPase domain, followed by a middle domain, a variable domain/insert B and a C-terminal GTPase effector domain. Here, the insert B helps in interacting with the target membrane via membrane proteins, while GTPase effector domain helps in self-association to form oligomers. In its soluble form, Drp1 may occur as monomer, dimer or tetramer, but once recruited to OMM via interaction with other players like fission protein 1, mitochondrial fission factor and mitochondrial dynamics proteins of 49 and 51 kDa, they can oligomerize forming higher-order structures [91, 93, 94].

25.2.2.2.2 Mammalian Fission Protein 1 (Fis1)

Fis1, a 17 kDa single pass transmembrane protein with C-terminal end anchored to outer mitochondrial membrane and an N-terminal tetratrico-peptide repeat motif facing cytosol, was the only protein thought to mediate the process of mitochondrial fission in mammals. The tetratrico-peptide repeat motif facing the cytosol facilitates protein-protein interaction and protein transport.

Recent research, like those with knockout models of Fis1, exposed the limited role of the protein in fission process [94, 95], thus suggesting that though Fis1 participates in the process of fission, it acts as one among the array of adaptor proteins that helps in recruitment of Drp1 and its interaction with mitochondrial membrane and has a dispensable role in the process.

Alternative receptors for Drp1 include Mff, ganglioside-induced differentiation-associated protein 1 (GDAP1), Mid49 and Mid51/Mief1; knockdown studies of these factors suggest a dispensable association of these proteins as well with Drp1 in driving the process of mitochondrial fission [94, 96] (Table 25.4).

Current concept of mitochondrial fission process involves the recruitment of Drp1 to the OMM via interaction with various adaptor proteins on the membrane where they oligomerize to form spiral structures followed by GTPase-driven constriction of the structure forcing the severing of both inner and outer mitochondrial membrane, giving two separate mitochondrial units (Fig. 25.4).

As mentioned earlier, the turnover rate of the high-capacity mitochondrial network in terminally differentiated cardiomyocytes is low. Irrespective of this, the protein components of fusion and fission machinery are highly expressed and are inevitable for proper functioning of adult heart. Emerging evidences suggest their imperative role in the mitochondrial quality control.

Table 25.4 Protein components important in the mitochondrial fission process

Protein	Location	Function
Dynamin-related protein 1 (Drp1)	Cytosol and outer mitochondrial membrane	Severs both outer and inner mitochondrial membranes; forms oligomeric structures that constrict the organelle in a GTP-dependent manner
Fission protein 1 (Fis1)	Outer mitochondrial membrane	Helps in Drp1 recruitment to outer membrane fission
Mitochondrial fission factor (Mff)	Outer mitochondrial membrane	Helps in Drp1 recruitment to outer membrane fission
Mitochondrial dynamics proteins of 49/51 kDa (Mid49/51)	Outer mitochondrial membrane	Helps in Drp1 recruitment to outer membrane fission
Ganglioside-induced differentiation-associated protein 1; GDAP1	Outer mitochondrial membrane	Helps in Drp1 recruitment to outer membrane fission

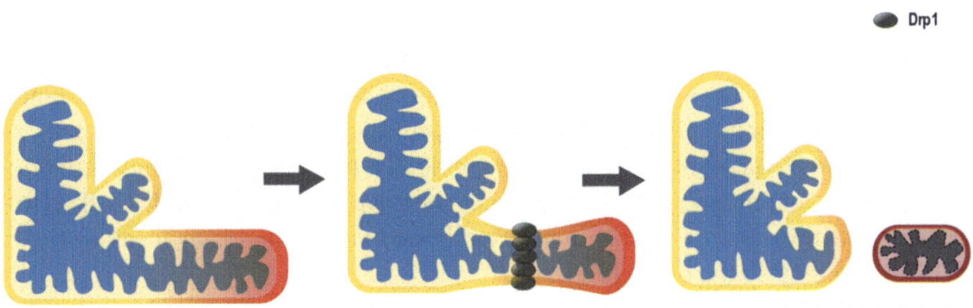

Fig. 25.4 Mitochondrial fission process is mediated by Drp1 and associated proteins like mitochondrial fission factor (MFF), mitochondrial dynamics proteins of 49 kDa and 51 kDa (MiD 49/51) and fission 1 protein (Fis1). The process requires the localization of Drp1 from the cytosol to the outer membrane of damaged/depolarized mitochondria which is mediated by the associated proteins, causing a constriction in the damaged part of the mitochondria and eventual budding off of the damaged mitochondria

25.2.3 Mitophagy: Pathway for the Clearance of Dysfunctional Mitochondria

Mitochondria are involved not only in the energy production but also in the signal transduction pathways ranging from calcium signalling in the cell to the cell death mechanisms through apoptosis. The damaged mitochondria can be a source of free radicals that cause damage to the cellular constituents; it can also be a source of calcium leaching into the cytoplasm and release of the caspases leading to the cell death cascade. To maintain the health of cell, it thus becomes imperative to clear off damaged mitochondria. Indeed, the stable number of mitochondria that is observed in the terminally differentiated cell types, like the cardiomyocytes, is rather in a dynamic flux with a balanced mitochondrial biogenesis and degradation of the damaged or dysfunctional mitochondria, thus maintaining mitochondrial homeostasis [97, 98]. A disruption of this balance can contribute to the pathologies of various disease conditions which ultimately could result in heart failure. The cellular clearance of dysfunctional mitochondrial structures is achieved by the process of autophagy, more specifically mitophagy, a term suggested in the last decade by Lemasters et al., for the degradation and recycling of the mitochondria [99].

25.2.3.1 Autophagy in Myocardium

Autophagy is a lysosome-dependent recycling mechanism that works at cellular level. It helps to maintain cellular homeostasis by way of recycling long-lived proteins and damaged organelles by sequestrating them in double membrane vesicles called 'autophagosomes', which ultimately fuses with lysosome to degrade the cargo. So far, ~35 autophagy-related (ATG) genes have been identified in yeast along with their mammalian counterparts; and three different types of autophagy have been classified: macroautophagy, microautophagy and chaperone-mediated autophagy [100]. Microautophagy is the process by which small cellular fractions are degraded by indentation in the lysosomal membrane and engulfment of the adjoining cytoplasm into the lumen of the lysosome. In chaperone-mediated autophagy, the macromolecules with the amino acid motif KFERQ (lysine-phenylalanine-glutamic acid-arginine-glutamine) associate with the Hsc70 which translocates to the LAMP molecule on the lysosomal membrane, while macroautophagy (hitherto referred to as autophagy) is the random and selective degradation of the bulk cytoplasmic constituents in a double membrane-bound vesicle that later fuses with the lysosome to recycle the cargo [100].

On induction of autophagy, many Atg proteins, including the lone kinase in the mammalian Atg family, Unc-51-like autophagy activating kinase 1 (ULK1), help in formation of autophagosomes by sourcing membranes from various cytosolic membranous structures, primarily endoplasmic reticulum [101, 102], endosomes, trans-Golgi network [103, 104] and even in constrained conditions, nuclear envelope [105]. The isolation membrane, known as omegasome, is sourced from the ER network, and autophagosomal structure gets assembled with the support of various protein factors. Ultimately the matured autophagosome with the cargo targeted for degradation fuses with the lysosome where degradation takes place. While bulk autophagy is a nonselective process, various types of selective autophagy have also been characterized, including mitophagy, pexophagy, chlorophagy and xenophagy. Of these, mitophagy or mitochondrial-specific autophagy is the most extensively investigated process, and some mitophagy-specific regulators have also been identified. Most of the molecular machinery of autophagosome formation are same between the nonselective form of autophagy and selective degradation process like mitophagy, but it also possesses some unique molecular mechanisms, like the PTEN-induced putative protein kinase 1 (PINK1)-Parkin pathway and some mitophagy receptor proteins that function as adaptors.

Under physiological conditions, basal level of autophagy helps in recycling long-lived proteins and dysfunctional organelles, having a significant role especially in an organ like the heart where terminally differentiated cardiomyocytes are the functional players. In normal physiology, autophagy levels are increased in myocytes in

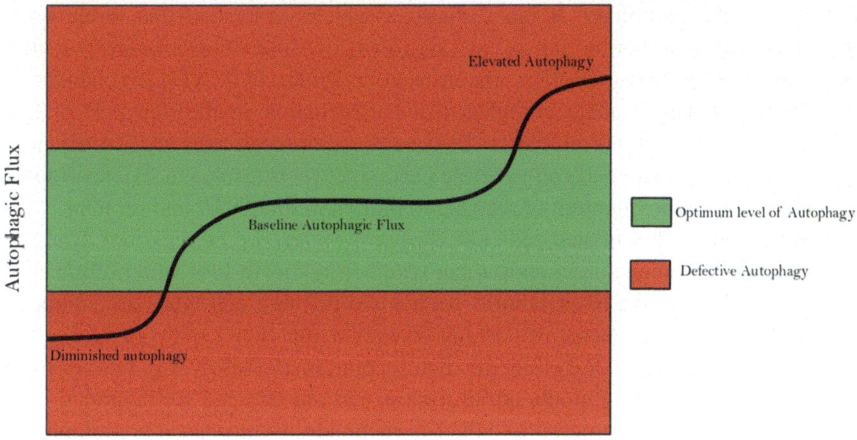

Fig. 25.5 Autophagic flux: All tissues maintain a basal healthy level of autophagy that helps in keeping the essential resources in dynamic homeostasis. When the autophagy becomes defective, either more than or less than what is required for that tissue type in the given physiological setting, the proper functioning of the tissue gets hampered

response to nutritional stress to cope with energy demands, providing adequate levels of ATP to maintain myocyte contractile force. Here, the nutritional stress can either be starvation induced or exercise induced, both acting via adenosine monophosphate-activated protein kinase (AMPK). In addition, autophagy levels are also altered in pathological conditions like myocardial infarction and heart failure models like cardiac pressure overload. Also, chronic hyperactivation of autophagy has shown to damage the myocardium (Fig. 25.5).

Conditional knockdown of autophagy-related genes in myocardium results in rapid cardiac abnormalities demonstrating the importance of baseline autophagy for myocyte survival [106, 107]. Cardiac-specific knockdown of proteins like *Atg5*, LAMP-2 and anti-apoptotic protein BCL-2 has been shown to result in dysregulated autophagy and ultimately heart failure [107–109]. Most of the cardiovascular diseases are associated with risk factors like diabetes, obesity, hypertension and hyperlipidaemia. And these risk factors have been experimentally proven to alter autophagic status of the myocardium which ultimately results in the accumulation of dysfunctional organelles and proteins or chronic hyperactivation and loss of cardiac health especially in conditions like reperfusion after an ischaemic insult [110–112]. Although autophagy was once thought to be a bulk, nonselective process, emerging evidences give more insights into the selective nature of autophagy in removing protein aggregates and dysfunctional organelles like mitochondria.

Mitochondria, with its vast variety of roles in processes ranging from cell survival to cell death, operate as a vital regulator of cellular homeostasis. Being the hub of oxidative metabolism in eukaryotic cells, mitochondria have to face the inevitable challenge of managing enormous amount of oxygen and its reactive radicals while having its own as well as the cell's redox status in check. Thus, if dysfunctional, this organelle can be a major contributor for the increased cellular ROS levels and thus to the pathology of many diseases which varies with the tissue type. Mitochondrial dysfunction can occur due to (a) the loss of membrane potential essential for its primary function, chemi-osmotic synthesis of ATP, (b) changes in the electron transport chain complexes or (c) reduction in the transport of key metabolites to mitochondrial matrix. Dysfunctional mitochondria are less efficient in terms of metabolism and can generate excessive reactive oxygen species (ROS), which can add to the quantum of damage to mitochondrial DNA and proteins.

In terminally differentiated cells like cardiomyocytes, accumulation of dysfunctional or less efficient mitochondria can result in myocyte loss which ultimately leads to conditions like heart failure. To tackle this, these cell types are equipped with a quality check mechanism which ensures a functional network of healthy mitochondria. Damaged or dysfunctional mitochondria are selectively severed from the network, by way of mitochondrial fission, and are targeted for degradation in autophagosomes. This process is called mitochondrial autophagy or mitophagy. Evidence from research targeting the process of mitophagy confirms its indispensable role in maintenance of a healthy network of mitochondria in myocardium [113–117].

25.2.3.2 Molecular Mechanisms of Mitophagy

Mitochondrial autophagy or mitophagy depends on various molecular mediators like PTEN-induced putative protein kinase 1 (PINK1), Parkin, Bcl2/adenovirus E1B 19 kDa protein-interacting protein 3 (Bnip3), Nip3-like protein X (NIX, also known as Bnip3L), Fun14 domain containing 1 (FUNDC1) and cardiolipin. And the mediators involved in the process of mitophagy often vary with the kind of stress that serves as induction: PINK1/Parkin-mediated mitophagy targeting depolarized/damaged mitochondria and other identified receptors like Bnip3, NIX and FUNDC1 involved mainly in the hypoxic induced clearance of mitochondria (Fig. 25.6).

25.2.3.2.1 PINK1/Parkin-Dependent Mitophagy

Mitophagy is intensively investigated by induction of mitochondrial damage/depolarization; perhaps for this reason, Parkin-mediated mitophagy remains one of the most thoroughly studied forms of mitochondrial clearance [118, 119].

PTEN-induced putative protein kinase 1, PINK1, is a mitochondria-targeted serine/threonine kinase; in healthy mitochondria, it is imported to the matrix via translocase of outer membrane/TOM complex, processed by mitochondrial processing peptidase followed by digestion by presenilin-associated rhomboid-like (PARL) protease, marking it for degradation [118, 120]. However, in compromised mitochondria, with abated membrane potential and the electrochemical gradient or an inefficient antioxidant system, PINK1 accumulates in the OMM evading degradation and phosphorylating several OMM proteins. Mfn2, the fusion protein, is a key substrate of the PINK1 phosphorylation acting at both Thr^{111} and Ser^{442}, lending it to function as a receptor for Parkin, phosphoubiquitin-dependent cytosolic E3-ubiquitin ligase, recruiting it to the mitochondrial membrane [121]. Mfn2 phosphorylation by PINK1 has been shown to have a dispensable role in recruitment of Parkin and hence mitophagy in MEF cells. Phosphorylation of Parkin at Ser^{65} and

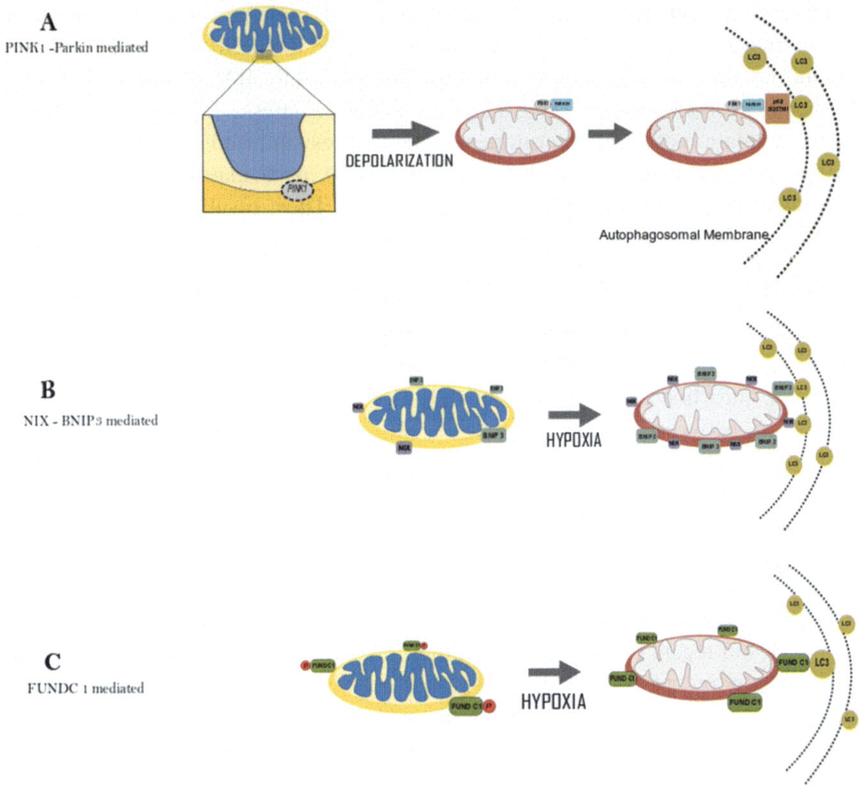

Fig. 25.6 A graphical illustration of different mitophagic pathways. (**a**) In a healthy mitochondria, PINK1 is processed by the mitochondrial peptidases, while in a depolarized mitochondria, PINK1 accumulates on the outer mitochondrial membrane and gets activated promoting Parkin translocation to the damaged mitochondria. Parkin, an E3 ubiquitin ligase, ubiquitinates different outer mitochondrial membrane proteins helping it to interact with the autophagic adaptor proteins like p62, thus targeting it for autophagic degradation. (**b**, **c**) BNIP3/NIX and FUNDC1 are alternate pathways where they serve as receptors that tether mitochondria to LC3 in the autophagosomal membrane

ubiquitin at Ser^{65} by PINK1 activates these proteins recruiting them to OMM. This is followed by ubiquitination of OMM proteins like VDAC1, mitofusin and Miro, thus initiating the autophagic clearance of mitochondria [122–125]. Structural studies of the proteins reveal that Ser^{65} phosphorylation of ubiquitin interacts with Parkin via a conserved phosphate pocket triggering a conformational change in Parkin that causes the release of a ubiquitin-like domain from its core. This results in activation of Parkin through phosphorylation by PINK1 [126] (Fig. 25.6a).

Recent findings have shown that the increased expression of Parkin compensates for PINK1 deficiency in cardiomyocytes suggesting the existence of multiple routes of Parkin translocation and the replaceable nature of PINK1 involvement, despite the crucial role it has shown to play in Parkin-dependent mitochondrial clearance [127, 128].

Increasing evidences unveil the involvement of more upstream regulators of Parkin translocation to OMM. RNAi studies revealed that HSPA1L (HSP70 family member) positively regulates Parkin translocation while BAG4 inhibits it [129]. Further, heat shock protein (HSP) 72, a mitochondrial stress sensor, is shown to act via the regulation of Parkin recruitment and thus stress-induced mitophagy. Also, in cardiomyocytes, BAG3, a co-chaperone of HSP70, participates in mitophagy by co-translocating to OMM along with Parkin [130]. These results point towards the complexity of mitophagy signalling and its importance in maintaining cellular homeostasis.

25.2.3.2.2 Parkin-Independent Mitophagy

Emerging evidences rope in the idea of existence of Parkin-independent mechanisms for mitochondrial clearance [131–134]. And these routes rely mainly on the interaction between OMM components, which serve as mitophagy receptors/autophagy adaptors, and LC3 on autophagosomes, via LC3-interacting motifs. The mitophagy receptors on OMM include Bcl2/adenovirus E1B 19 kDa protein-interacting protein 3 (Bnip3), Nip3-like protein X (NIX, also called Bnip3L), Fun14 domain containing 1 (FUNDC1), Bcl-2-like protein 13 (Bcl2-L-13), activating molecule in Beclin1-regulated autophagy (AMBRA1) and cardiolipin.

BNIP3 and NIX, these Bcl-2 family proteins are critical players in cell death pathways, but recent evidences suggest their involvement in cell survival by acting as mitophagy receptors [133, 134]. Both the proteins localize on the OMM where they act as receptors which directly interact with LC3/γ-aminobutyric acid receptor-associated protein (GABARAP) on the autophagosomal membrane. This eliminates the need for other adaptor proteins like P62 and can tether mitochondria to autophagosomal membrane. Bnip3 and NIX are upregulated by hypoxia-inducing factor-1 (HIF-1). Under hypoxic conditions, both these proteins act in the induction of mitophagy and help the cells in regulating ROS. Knockdown studies of Bnip3 and NIX reveal redundancy in their roles in mediating mitophagy, since both of them can compensate for each other's loss while allowing for the progression of mitophagy (Fig. 25.6b).

Recent evidences hint the existence of crosstalks between different mitophagic pathways. Bnip3 is found to interact with PINK1 promoting its accumulation on the OMM. Also, Bnip3 inactivation enhanced the proteolytic degradation of PINK1, thereby suppressing PINK1/Parkin-dependent autophagy. Thus, activated BNIP3 not only causes mitophagy through the BNIP3-mediated route but also enhances the mitochondrial clearance by assisting the PINK1/Parkin-dependent autophagy.

FUNDC1, another mitophagy receptor protein, mediates hypoxia-induced mitophagy in mammalian cell [132]. It is an OMM protein and acts by interacting with LC3 through its LC3 interaction motif. Phosphoglycerate mutase family member 5 phosphatase (PGAM5), a mitochondrially localized phosphatase, activates FUNDC1 during hypoxia or under elevated ROS levels by dephosphorylating the protein at Ser^{13}. Wider role of this protein in modulation of mitophagy is evidenced by its interaction with mitochondrial fission and fusion proteins, Drp1 and OPA1, respectively. The mitochondria in a network better evade the autophagic degradation as the growing phagophore; though a means of macrodegradation of organelles; has spatial constraints for its cargo, thus making the role of Drp1 critical as it severs the mitochondria from the network and makes it accessible for the autophagic engulfment. Hence, the interaction of FUNDC1 with Opa1, which keeps the mitochondrial membranes in a fused state and thus maintains the network, results in the suppression of mitochondrial fission inevitably suppressing mitophagy as well. But, when activated by PGAM5 through its dephosphorylation, FUNDC1 dissociates from Opa1 and associates with Drp1, thereby promoting mitochondrial fission and, ultimately, mitophagy (Fig. 25.6c).

Cardiolipin, a phospholipid dimer, exclusively located in the inner mitochondrial membrane (IMM) is a critical component of functional mitochondria, and its importance in mitochondrial bioenergetics is evidenced by the intimate association it has with the energy-transducing membranes. Damage to mitochondrial membrane integrity results in the redistribution of cardiolipin to the outer mitochondrial membrane resulting in the commencement of a cascade of events culminating in apoptosis, thus cardiolipin was initially thought to be associated only with cell death pathways. However, a recent finding demonstrated the capacity of externalized cardiolipin to induce mitophagy by interacting with LC3 on autophagosomes [131]. The peroxidation status of externalized cardiolipin regulates both autophagy and apoptosis. Nonperoxidized cardiolipin on OMM can interact with LC3, thus protecting cells from cell death pathways, and this phospholipid is peroxidized in the absence of autophagy, giving way for apoptosis.

AMBRA1, an upstream autophagy regulator, was first identified as one among the Beclin1-interacting components that mediates Ulk1-Beclin interaction. It is found to be localized both in the cytosol and on mitochondria; and a recent study suggests that AMBRA1 plays a role in mitophagy through its interaction with Parkin in HEK293 cells, SH-SY5Y cells and adult mouse brain. Also, the strength of the interaction of endogenous Parkin and Ambra1 increased during prolonged mitochondrial depolarization [135, 136]. These evidences suggest the Parkin dependence of mitochondrial clearance involving AMBRA1. More recently, Strappazzon et al. showed that OMM-localized AMBRA1 can stimulate a substantial amount of Parkin and p62-independent but LC3-dependent mitophagy.

Bcl2-L-13 is an OMM protein, assumed to be a functional homolog of yeast protein Atg32, which is a key player in orchestrating mitophagy in yeast, by localizing on mitochondria and interacting with mitophagic adaptors. A recent report has shown that the overexpression of Bcl2-L-13 protein induces mitochondrial fragmentation in neonatal cardiomyocytes and HEK293 cells, derived from human embryonic kidney cells [137, 138]. Involvement of this protein in the process of mitophagy in metazoans remains understudied, making this an alluring target for research.

Advances in research in the field of mitophagy helped in deciphering different routes as well as components involved in the clearance of mitochondria: in physiology and pathology. Many of these pathways are shown to have a crosstalk establishing their cumulative action and interdependence for an efficient clearance of the organelle.

Mitophagy serves as the quality control mechanism in terminally differentiated myocytes, and if impaired, it results in the accumulation of dysfunctional mitochondria, ultimately leading to aberrant cardiac function and associated pathologies. Mouse models deficient in key players of mitophagy like Parkin, Pink1, Bnip3 and Nix have shown to develop cardiac hypertrophy and ultimately heart failure with progression of age, demonstrating the imperative role played by the process of mitophagy in maintaining cardiac homeostasis [218]. Also, removal of dysfunctional mitochondria via autophagy is shown to decline with age, and overexpression of Parkin has shown to revert the age-associated effect [219]. These studies suggest a plausible target for therapies against various pathologies as well as age-associated cardiac dysfunction. A thorough understanding about the long-term as well as short-term modulation of autophagy on mitochondrial homeostasis, cardiac output and overall health of the organ system can help to develop a better intervention.

25.3 Therapies Targeting Mitochondrial Dynamics

The discussions in the previous sections suggest the importance of uninterrupted mitochondrial dynamics in maintaining proper functioning of myocardium. This highlights the tantalizing possibility of targeting these processes for therapeutic gain.

25.3.1 Mitochondrial Biogenesis as a Therapy

There are evidences emerging that nicotinamide riboside helps in mitochondrial biogenesis [139, 140]. Nicotinamide phosphoribosyltransferase (NPT) enzyme is under the control of PPAR, the family of transcription factors that are involved in a myriad of energy and metabolic signalling in the cell. PPAR as previously described is also involved in transcription of the transporters associated with shuttling of substrates into the mitochondria. PPAR is associated with the major mitochondrial biogenesis switch in the cell, the PGC-1 α. Thus, if the nicotinamide phosphoribosyltransferase is acted upon by the transcription factor PPAR which also influences the PGC-1 α, through which it regulates the formation of new mitochondria in the cell, it seems plausible that NPT is influenced by the PGC-1 α and could be influencing it in return, in a feedback fashion. This, when looked at from the overall perspective of the cell, would make a beneficial alternative for the cell metabolism because the switch that causes mitochondrial biogenesis must in effect have a role in formation of the substrates in turn assuring the proper functioning of the newly generated mitochondria. Though still being empirically validated, a highly probable mechanism of the observed beneficial effects of dietary supplementation of NR could be mitochondrial biogenesis and providing the newly formed mitochondria with the major substrate required for electron shuttle in the ETC, along with handling the ROS generated in the process; thus reducing the damage to the new mitochondria and maintaining the overall mitochondrial health.

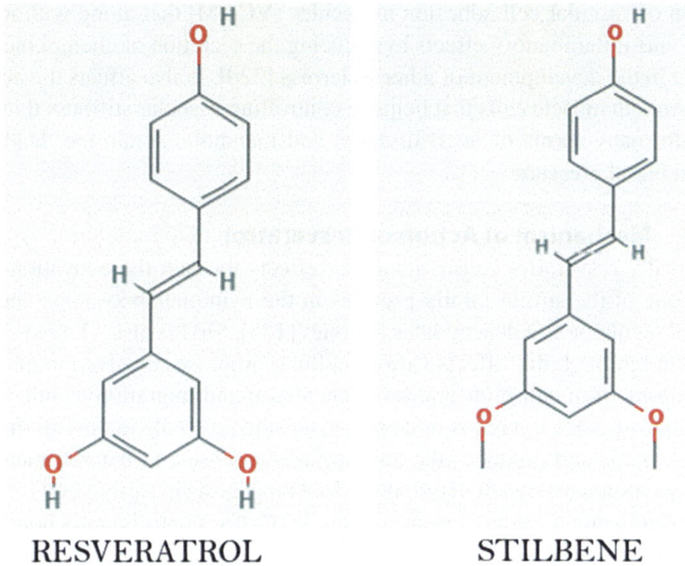

Fig. 25.7 Structure of resveratrol and stilbene

25.3.1.1 Resveratrol/Pterostilbene

Pterostilbene (trans-3, 5-dimethoxy-4-hydroxystilbene) is a type of stilbene, an active compound found in many plants especially blueberries, grapes, hardwood of the red sandalwood tree, etc. [141]. It is secreted in response to environmental insults to the plant like bacterial infection and excessive ultraviolet light exposure. The stilbenes are thus compounds having anti-inflammatory and antioxidant effects. Resveratrol is a type of stilbene compound secreted from many grapevines and popularized recently as a beneficial agent for its anti-ageing and heart health-promoting properties. Pterostilbenes have slightly different structures than resveratrol because of the two extra methoxy groups in its structure which makes it more bioavailable than resveratrol. The pterostilbene can be absorbed nearly 80% through oral supplementation as opposed to just 20% absorption of resveratrol; also it has longer half-life that makes it an interesting compound for therapeutic purposes [142] (Figs. 25.7).

Resveratrol (3, 4′, 5-trihydroxystilbene) is a type of stilbene and chemically a phytoalexin that is a common nutritional additive [143]. It was first discovered in the 1940s and isolated from the plant white hellebore (*Veratrum grandiflorum*) [144]. It is not just present in the fresh plant material like fruits, leaves, bark, etc. but is also found in the processed plant products like tea leaves, red wine, etc.

Resveratrol has been shown to affect many metabolic signalling events and has an effect on the mitochondrial function. It has also been shown to modulate the expression of vascular cell adhesion molecules (VCAM) that along with its ability to induce anti-inflammatory effects by reducing the secretion of chemokines causes a decrease in the development of atherosclerosis [220]. It also affects the activity of vascular smooth muscle cells that helps in controlling vascular stiffness that is associated with many forms of heart diseases and metabolic syndrome, leading to a controlled blood pressure.

25.3.1.2 Mechanism of Action of Resveratrol

Biochemically resveratrol causes its active effects through the activation of Sirt1 which is one of the sirtuin family proteins in the mammalian systems and a transcriptional regulator as it deacetylates histones [145]. Sirt1 is also a known regulator of NFκB and through this affects various cellular processes involved in diverse cellular functions from cell proliferation, metabolism and migration to inflammatory responses [146]. Sirt1 is a redox and energy-sensing molecule in the cell that senses the energy status and produces the appropriate responses. In the cell metabolism, Sirt1 has been shown to result in mitochondrial biogenesis in various cell lines [221]. The major mitochondrial biogenesis switch, PGC-1 α, controls mitochondrial biogenesis to meet the energy demands of the cell, thus establishing a fine-tuned mechanism of ensuring balanced energy demand and supply. This implies that PGC-1 α needs to constantly monitor the energy status of the cell. The means it adopts to sense the energy status is through AMPK and Sirt1. The reversible phosphorylation of PGC-1 α by AMPK and deacetylation by Sirt1 determine its activity status. The acetylated form of PGC-1 α deactivates the transcription factor, while the deacetylation by activated Sirt1 causes its activation and eventual mitochondrial biogenesis along with many other metabolic effects. Thus resveratrol and other stilbene compounds like pterostilbene cause the beneficial effects in the energy utilization and oxidative free radical scavenging through the generation of healthy mitochondria.

25.3.1.3 Why Is Dark Chocolate and Red Wine Good for the Heart?

The famous French paradox had baffled the scientific community for long. The paradox is the reduced risk of heart diseases and development of cardiovascular illness among the French population, especially the southern French people as compared to the average healthy European, even when their consumption of saturated fats is very high, way higher than found in the diets of the average European person [147]. The consumption of the cheese, dairy and meat products being high in the so-called unhealthy fats (trans-fats, saturated fats, etc.) among this group should have resulted in the increased incidence of atherosclerosis and metabolic syndrome. But they are surprisingly found to have healthy hearts. The most prominent correlation in this group is the consumption of red wine. This results in the natural intake of essential stilbenes like resveratrol that would result in their mitochondria being better adept to handling the fatty acid oxidation and scavenging oxidative radicals.

Like red wine, the cocoa compounds are also high in the active resveratrol [148]. The cocoa trees are found in the tropical regions of the world which receive intense sunlight and thus intense ultraviolet exposure. To avoid any UV-related damage that the plant at its cellular and genome level would inevitable endure, it produces resveratrol. The cocoa and grape plant and their products are thus very high in their resveratrol content that helps in protecting the plant from oxidative damage caused by the high UV exposure. This resveratrol is not washed out or leached during the processing steps in making chocolate and wine, giving dark chocolate (which does not contain excessive sugar or dairy compounds that may have their own harmful effects for the heart and overall health) and red wine a natural capacity to promote mitochondrial health. As any defect in the mitochondria has a direct bearing on the organs that depend heavily on the continuous energy supply, like the heart and the nervous tissue; these compounds dark chocolate, red wine, etc. are heart friendly.

There are a number of studies validating the aforementioned association between heart health and these natural compounds [149–151].

25.3.1.4 Can PGC-1 α Be a Therapeutic Target?

If the PGC-1 α can have these many beneficial effects on the heart health, should it be a treatment modality for cardiovascular diseases? The issue with giving PGC-1 α or its precursors as supplements is that it would result in uncontrolled mitochondrial growth and other metabolic changes associated with the activation of this transcription factor. The well-being of a cell or a tissue is dependent on the appropriate number of functional mitochondria; if in excess, it would lead to too much energy generation and depletion of the substrates and intermediates of the biochemical pathways that could be important signalling molecules for the maintenance of homeostasis. The unwanted increase in the mitochondrial content could lead to biogenesis-induced cardiomyopathy. By inducing the activation of PGC-1 α through its upstream regulators, it is subject to a stricter control and can be maintained without adverse side effects.

Also, the mitochondria in an individual or in a single cell are not always uniform. It has its own genome that is subject to its specific mutations. The mutated mitochondria are intermingled among the population of healthy mitochondria, thus showing mitochondrial heteroplasmy. It has been observed that the mitochondrial DNA mutations are present at the rate of 1 in 200 individuals which is many times higher than the rate of occurrence of mitochondrial diseases (also believed to be caused by defective and mutated mitochondria) [152] postulated to be around 1 in 5000 individuals in UK, which is a greater than 20-fold increase [140]. This implies that there is a critical threshold that needs to be reached for the damaged and mutated mitochondria to show any detrimental effects. Thus, if the mitochondrial biogenesis is enhanced, without directing the degradation of the damaged mitochondria, the mutated mitochondria will also divide causing there to be an increased population of the same and increasing the susceptibility of development of related disorders. It hence follows that a linear relationship between mitochondrial number and efficiency in the cell is dependent on the level of heteroplasmy and the number of mitochondria that are healthy.

25.3.2 Enhancing Mitochondrial Efficiency by Clearance of Damaged Mitochondria

There are various physiological events that lead to enhanced mitophagy and thus to an efficacious mitochondrial network in the cells. The various tissues and organ systems are affected simultaneously with the initiation of the autophagic process, and the benefits are not confined to a specific tissue alone. With respect to the general cardiovascular salubrity and heart efficiency, it is also benefited as the mitochondrial population in the tissue remains efficient. The most common mechanism of enhancing mitophagy (and autophagy in general) is nutrient deprivation.

25.3.2.1 Calorie Restriction

Starvation is one of the most well-known and widely studied phenomena that induces autophagy. But it cannot be utilized on a regular basis for healthy, normal weight adults. Prolonged starvation can cause more harm than good becoming counterproductive for health promotion [153]. To assimilate the beneficial aspects of starvation without the detrimental side effects, the best method is restricting the food intake or calorie restriction (CR). This means to restrict calories in a particular meal or throughout the meals had during the day such that the total calories consumed is slightly less than the total caloric requirement for weight maintenance of the individual, calculated as per the age, sex, physical activity performed on a regular basis, etc. These periods of CR are interspersed with the periods of normal caloric intake (as many calories as required for that individual) so that an absolute deficiency of nutrients is not created in the body. It is observed that calorie restriction induces the clearance of pre-existing mitochondria through the process of mitophagy induction [154], and when the body is released off the stress of starvation/CR, the mitochondrial biogenesis is enhanced [155]. This CR not only has proven to be helpful in the maintenance of the potency of the heart in normal weight healthy adults but has also shown to increase the cardiac efficiency in the diseased condition. CR has been claimed to be the means to halt the heart ageing and promote cardiac performance [156, 157].

The means by which CR leads to the mitophagy is through the blocking of the mechanistic target of rapamycin (mTOR) complex1 [158]. mOTR is a serine/threonine kinase that functions as a major regulator of cellular growth, proliferation, etc. It is activated by the presence of growth factors, insulin and insulin-like growth factors, which initiate a cascade of signalling through the PI3K–Akt–mTOR axis. mTOR is also activated by the 5' AMP-activated protein kinase (AMPK), thus combining the growth factor cues with the cellular energy availability. mTOR negatively influences the autophagic machinery, so that the growth signals do not coincide with the cell recycling processes. With the reduction in the energy reserves in the cell, the AMP to ATP ratio gets increased, and this activates the AMPK, which in turn negatively signals the mTOR complex1 [159]. The mTOR complex1 is an inhibitor of first of the autophagic proteins that get activated and are prerequisites for the initiation of the process of autophagy, the ULK1/ULK2–ATG13–FIP200 complex that associates and forms the autophagic initiation membrane [160] (Figure 25.8). Thus,

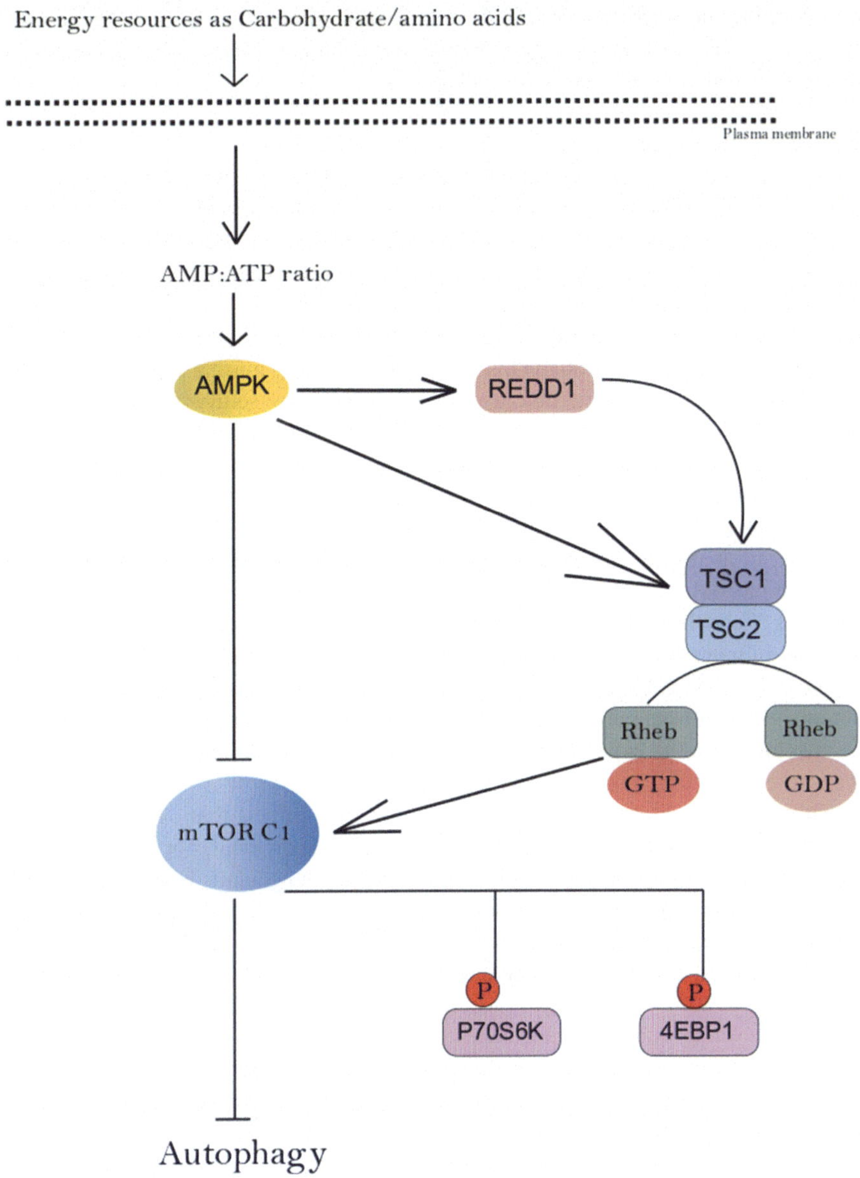

Fig. 25.8 Reduction in cellular levels of ATP stimulates AMP-activated protein kinase (AMPK) which induces the expression of REDD1 inhibiting TSC1-TSC2 complex. This subsequently inhibits mTOR complex 1 activity. Also AMPK phosphorylates TSC2 activating its GTPase activity converting Rheb GTP to Rheb-GDP and thus inhibiting mTOR complex 1. Similarly, AMPK can phosphorylate mTORC1 (in its Raptor domain on Ser722 and Ser792) inhibiting mTORC1 activity. mTOR C1 negatively regulates autophagy and thus, the ATP depletion in the cell causes autophagic upregulation

by inhibiting the activation of mTOR, the AMPK initiates the autophagy and also mitophagy in the cell, as the general increase of autophagy in the cell also causes an increase in selective autophagy (mitophagy).

25.3.2.2 Exercise

Exercise is a known source of mitochondrial renewal by the elimination of old existing mitochondria and the generation of new mitochondria during the periods of rest amidst the periods of intense exercise [161]. The exercising muscle is at a stress of intense energy requirement provided through the oxidative phosphorylation that in turn generates reactive oxygen species. With the depletion of the energy reserves of the cell and limited availability of the adequate oxygen necessary to sustain the fast burst of ATP generation through OXPHOS, the exercising muscle shifts from oxidative phosphorylation to anaerobic glycolysis as the prime source of energy, in turn generating lactic acid. This quick burst of ATP generation from the mitochondria results in the membrane depolarization of the mitochondria. The depolarized mitochondria, with the ROS generated during exercise, furthermore mark the mitochondria for the autophagic degradation [162]. The mitochondrial membrane depolarization results in the stability of the membrane molecule PINK1 which would in physiological basal conditions be degraded by the membrane proteosomal enzymes. This PINK1 recruits the Parkin assembly to the depolarized mitochondria, and this initiates a cascade of events with the signalling molecules like ubiquitin ligase and p62/SQSTM1 being recruited and activated that later associate with the autophagic membrane protein LC3, thus marking the depolarized mitochondria to be enveloped by the expanding autophagic membrane to be later merged with the lysosome, thus completing the degradation process [163].

There are numerous studies showing the positive effects that exercise, mild to moderate, can have on the overall well-being of the individual, not to mention the benefits on the cardiovascular health of not only the young adults but also the ageing population or the patients with cardio vascular disease [230]. Exercise thus provides the anti-ageing benefits that are observed in the numerous clinical trials through its direct regulation of the clearance of damaged mitochondria by initiating the process of autophagy.

25.3.2.3 Synthetic Means of Enhancing Mitophagy

Exercise and calorie restriction have been found to be of immense help to not only the cardiovascular health but the overall well-being of the individual, for both normal adults and ageing or diseased groups. Though difficult, it has been a lucrative task for the pharmacological industry to find an exercise and CR mimetic. The aim of having a mimetic is that it can provide the benefits in the population that cannot be on mild to moderate exercise and calorie restriction regimes, like the physically challenged individuals, stroke patients with lasting comorbidities, chronic hypoglycaemic patients, underweight patients and patients with various forms of eating disorders. The best known method of achieving this is to target autophagy with external compounds. As the mTOR complex is a negative regulator of the autophagic and the mitophagic processes, the compounds that inhibit mTOR have been promising. The best known is the immunosuppressant and antifungal drug rapamycin. Rapamycin blocks the mTOR complex1 and thus activates autophagy; as the activation of the general process of autophagy also in turn activates the process of mitophagy, it can be considered as a pharmacological mitophagic inducer [158]. As this drug is an FDA-approved treatment modality for transplant rejection and antifungal properties and is now also being administered as an antitumour agent, the use of rapamycin instead of the development of novel targets to induce autophagy becomes easier for the medical community. The pharmacological intervention to find another potent mitophagy inducer is thus not a sound drug development strategy and has hence had unsubstantial enthusiasm from the pharmaceutical sector. Other compounds that have also shown to affect the mTOR pathway and that in turn would induce mitophagy are far less utilized, but are nonetheless beginning to gain eminence in the scientific and the pharmacological communities as they realize the necessity of an alternative before the use of rapamycin becomes so rampant that the recipient becomes tolerant, like the case with the traditional antibiotics.

25.4 Mitochondria and Reactive Oxygen Species

ROS is generated from the mitochondria during respiration (oxidative phosphorylation) by the flow of electrons through complexes I and II, through the NADH and succinate, respectively, and later by ubiquinone and complex III to complex IV ultimately reducing molecular oxygen to water. This is accompanied with the flow of protons through complex I, II and III generating an electromotive force across the mitochondrial membrane later utilized by complex V (ATP synthase) to produce ATP. Not all the oxygen consumed by the mitochondria is used for its reduction, and some gets partially reduced generating oxygen radical, O_2^-. The ROS produced, although contributes as an intermediate in intracellular signals by activating MEKK1 and NFκB [164], is highly reactive and can induce damage to lipids, DNA, proteins, etc. Mitochondria are at an increased susceptibility of oxidative stress-mediated damage as opposed to the other subcellular organelles: first by the virtue of the proximity as mitochondrion is the chief ROS producer and second, the mitochondrial DNA lacks the protective histones and associated DNA coiling along with lacking many DNA damage repair enzymes.

The ROS is generated at either the NADH/NAD+ isopotential group or the ubiquinone/ubiquinol isopotential group. The ROS generation capacity is variable in these two systems based on the enzymes involved and the tissue localization. For, e.g. in the NADH/NAD+ isopotential group, the enzymes α-ketoglutarate dehydrogenase and pyruvate dehydrogenase are high ROS contributors in the muscle and liver but not in the cardiac tissue, while complex I produces high ROS in the cardiac tissue but not in the muscle and liver [165, 166]. Within the ubiquinone/ubiquinol isopotential group, the complex II, complex III and glycerol 3-phosphate dehydrogenase have high ROS generation capacity in the muscle, liver and heart [165, 167].

The oxygen free radical is reduced to H_2O_2 by mitochondrial superoxide dismutase isozymes present in the mitochondria, the Mn-superoxide dismutase

(Mn-SOD) that is matrix bound, and the Cu/Zn-SOD predominant in the intermembrane space; H_2O_2 later being converted to water by catalase and glutathione peroxidase. The major antioxidant systems in the mitochondria are the glutathione and the thioredoxin systems. Two glutathione molecules in the presence of H_2O_2 convert to glutathione disulphide (GSSG) through the enzyme glutathione peroxidase (GPX). The two GPX isozymes, GPX1 and GPX4, in the mitochondria have high Km values and thus high efficiency of H_2O_2 quenching (approx. Km of GPX1 = 6×10^7 M^-s^-; approx. Km of GPX4 = 3×10^6 M^-s^-) [168]. The restoration of the glutathione from GSSG is catalysed by glutathione reductase with NADPH as a coenzyme. The mechanism of H_2O_2 sequestration of the thioredoxin system is through the peroxiredoxin (PRX), having two isozymes in the mitochondria PRX3 and PRX5. The peroxidiatic cysteine present in the active site of peroxiredoxin gets oxidized in the presence of H_2O_2 and reacts with a neighbouring cysteine moiety forming an intramolecular disulphide bridge (approx. Km of PRX3 = 2×10^7 M^-s^-; while approx. Km of PRX5 = 3×10^5 M^-s^-) [168]. Thioredoxin 2 helps in reactivating the PRX3 and PRX5 through a disulphide exchange reaction, permitting another round of H_2O_2 neutralization through PRX. In the mitochondria the thioredoxin 2 is reactivated by the thioredoxin reductase, with NADPH as a coenzyme. Thus, the antioxidant buffering capacity of the glutathione and the thioredoxin systems is dependent on the availability of NADPH. The cellular NADPH is maintained by the pentose phosphate pathway, but predominantly by the mitochondrial inner membrane-bound nicotinamide nucleotide transhydrogenase (NNT) [169]. The NNT catalyses the production of NADPH by utilizing the electromotive force generated by the proton gradient across the inner mitochondrial membrane [170]. The loss of this proton gradient consequently reduces the antioxidant buffering capacity of the mitochondria, increasing the overall ROS levels and ROS-mediated damage.

Mitochondria undergo ROS-mediated damage when either the ROS production exceeds the normal levels or the intrinsic ROS scavenging system becomes overwhelmed and incompetent. Thus, the ROS generated in the mitochondria through oxidative respiration of energy substrates as a normal part of oxidative energy metabolism, if not regulated and appropriately quenched, would result in mitochondrial DNA damage and subsequent mutations in the mitochondrial genome. The DNA damage repair mechanisms in the mitochondria are not as well established as their nuclear counterparts and their role and mechanism of action has only recently emerged [171]. The mitochondrial division occurs through mitosis, resulting in the distribution of mutated mitochondrial DNA to the daughter mitochondria upon division. The mutated copies of the mitochondria can co-exist with the non-mutated copies of the healthy mitochondria having no adverse effect to the cell until the mutated mitochondria reach a critical threshold beyond which the damaged mitochondria become the predominant population in the cell and prove detrimental. This tipping point is influenced by the nature of the insult to mitochondrial DNA, the severity of damage to the mitochondrial membrane and associated respiratory complexes and the efficiency or lack thereof of the damage repair mechanisms.

25.4.1 ROS Damage in Failing and Ageing Heart

Reactive species have been implicated in a lot of disease conditions, and heart disease is no exception. It is observed that the ROS levels are increased in patients with heart failure, both heart failure with reduced and preserved ejection fraction. Whether it is the cause of heart failure or the effect thereof is yet to be determined conclusively. But in preclinical models of heart failure, cardiovascular disease and metabolic syndrome, the ROS levels have been shown to be higher than that in controls [222]. This implies that the ROS generation is a critical initial step preceding heart damage through ischaemia or infarct. Increased ROS levels have also been noted in ageing heart, implying that either the ROS generation exceeds critical (levels to be taken care of by the antioxidant system) or the endogenous scavenging by the antioxidant system is compromised in preclinical models of heart failure, entailing that efficient antioxidant defence systems are imperative for the maintenance of mitochondrial and, in effect, heart vigour.

25.4.2 Can Scavenging ROS Be the Solution?

If the ROS overproduction and/or the reduced capacity of the ROS handling systems of the cell contribute to the cardiovascular disease progression, it seems logical to believe that the exogenously given ROS scavengers could remedy the problem. Thus, there were clinical trials and research endeavours taken up to determine if the antioxidants like vitamin C or vitamin E supplementation could result in improvement of symptoms and vital assessor measures in the cardiovascular patients [223]. It was observed that the antioxidant therapy could prove beneficial when given at low doses and played a more prominent role in the prevention of cardiovascular diseases rather than as a therapy. Although positive outcomes are observed in low doses, the higher doses instead had a negative effect in the treated versus the control group [224]. The antioxidant intervention also showed no beneficial effects when it was given as a treatment modality rather than a preventive strategy, especially so when the cardiovascular damage had progressed to a disease or to a myocardial infarct.

25.4.2.1 ROS as a Signalling Molecule

Physiologically, the ROS is generated in mitochondria as a by-product of the oxidative metabolism of substrates and has shown to mediate pathways that are critical for organismal homeostasis, stress response, health and longevity, proving detrimental when in excess.

With a host of studies in the field, it has now become accepted the ROS in the form of H_2O_2 and superoxide form important signalling molecules [172]. With earlier studies on growth factor-mediated ROS burst, it was demonstrated that the activation of receptor tyrosine kinase directs NADPH oxidase (NOX)-mediated ROS increase that bring about the downstream effects of growth factor stimulation. The NOX-mediated increase in H_2O_2 causes the inactivation of the tyrosine phosphatase proteins, thereby affecting its downstream signalling, mediated through substrate dephosphorylation. This is brought about by the inactivation of certain thiol groups in the tyrosine phosphatases like cysteine residues. These cysteine moieties can undergo sequential redox-dependent oxidation reactions like sulphenic–sulphinic–sulphonic acid, etc., thus modifying these regulatory proteins [173]. As with other post-translational modifications, these modifications alter the cellular response to external and intrinsic cues.

The intracellular ROS-mediated activities of some important intermediates like JNK are shown to be influenced by the cellular redox state mediated by inactivation of tyrosine phosphatases. ROS-induced damage in conditions like diabetes mellitus has been shown to be governed by its association of transcription factor NFκB [174].

Being essential in a multitude of cellular signalling events, the elimination of ROS through synthetic scavengers is not always apt especially if the ROS levels start depleting beyond the basal essential levels. Thus, the use of dietary antioxidants for the benefit of subjects with cardiovascular disease has been disputable. Although the excess ROS generation is detrimental, the basal levels are indispensable for the normal physiologic functioning of the cell.

25.4.3 Reducing ROS-Induced Damage Through Metabolic Shift

The oxidation of fatty acids results in generation of more ATP molecules than that produced by the oxidation of per molecule of glucose. The number of ATP produced depends upon the length of the fatty acid chain, wherein every two carbon moieties are converted to a molecule of acetyl co-A, through beta oxidation, that enters the TCA cycle of the mitochondria. For example, a molecule of palmitate (C_{16}) generates a net of 129 molecules of ATP (taking into account the two ATP equivalents utilized during the conversion of palmitate to palmitoyl-CoA). While a molecule of glucose generates a net of 36 ATP molecules. This makes fatty acid oxidation a preferred source of energy production when the energy demands of the organ are high. The heart amounts to about 0.5% of the body weight but consumes about 8% of the total energy in the body. As the continuous beating of the heart is an immensely energy-intensive process, evolutionarily the heart depends on beta oxidation of fatty acids for its energy requirements. In a stark contrast to the metabolism in the heart, though highly energy consuming, the brain does not use fats as fuels and solely relies on glucose for meeting its requirements. This is to avoid the ROS-mediated damage to the non-dividing neurons that do not have a good antioxidant pool to cope with the inadvertent ROS generation caused by beta oxidation. On account of the more potent antioxidant system, the cardiomyocytes relay on fatty acid oxidation.

25.4.4 The Reasons for the Metabolic Shift Under Diseased Condition

The scenario is different in failing heart. It has been shown that in the cardiovascular diseases that precede heart failure (ischaemia and/or congestive heart failure) like aortic insufficiency, aortic stenosis, coronary artery disease, dilated cardiomyopathy, hypertensive heart disease, ventricular hypertrophies, etc., there is a reduction in the utilization of fatty acids as the main source of fuel. As much as the beta oxidation of fatty acids would have been the preferred source of fuel for the energy-intense contractile activity of the heart, the resultant high ROS generation makes it more detrimental under the environment of reduced efficiency of ROS scavengers. The failing and diseased heart has inefficient mitochondria with reduced electrochemical gradient across the membrane, thus reducing its antioxidant capacity. With reduced ROS scavenging, the homeostatic mechanisms of the heart make arrangements in preventing excessive ROS generation. As discussed previously, the cardiac disease or damage is accompanied with the abnormal clearance of damaged mitochondria, resulting in the build-up of inefficient mitochondria within the cardiomyocytes. With plummeted mitophagy and subsequent increase in the number of damaged mitochondria, there is more mitochondrial ROS generation, creating a vicious cycle. This cycle could prove detrimental to the mitochondrial network resulting in their total damage and release of the cleaved caspases and other pro-apoptotic factors which could culminate into cardiomyocyte loss. To prevent this doom, the inefficient mitochondria shift their metabolic preference to glucose from fatty acids. Heart diseases and heart failure can hence be considered to be an aftereffect of metabolic disorder.

25.4.5 The Innate Precautions Taken by the Failing Heart

Not only does the failing cardiac cells show altered preference for the primary substrate fuel (glucose instead of fatty acids); the cells actively take steps to avoid the utilization of any fatty acids through beta oxidation. The fatty acid oxidation starts with the transport of the same into the mitochondria through various transporters. The peroxisome proliferator-activated receptor-α (PPARα) is a transcription factor that aids the transport of fatty acids into the mitochondria and is abundantly found in the healthy heart, but is decreased in the diseased heart [225]. This reduction in its expression partly drives the altered substrate selectivity.

Carnitine o-palmitoyltransferase 1 (CPT1) is a rate-limiting enzyme in the uptake of fatty acids in the mitochondria, which later shunts these fatty acid moieties into the beta oxidation taking place inside the mitochondria, and its deficiency has been reported in various cardiovascular pathologies [175, 176]. CPT2 is also recently implicated in beta oxidation in heart [177]. The loss of cardiac-specific CPT2 reduces the oxidation of long-chain fatty acids in the cardiac homogenates by 93% also demonstrating that the CPT2 functions to flux the fatty acid oxidation and sensitized the heart cells to insulin-mediated signalling. As the significance of

CPT2-mediated flux through beta oxidation is recently elucidated, the upstream mechanisms are not as well established as that for CPT1. Malonyl-coenzyme A negatively influences the activity of CPT1. As malonyl CoA levels increase, the inhibition of CPT1 blocks the transfer of fatty acids into the mitochondria [178]. The levels of malonyl-CoA increase by either the overexpression of its synthesizer acetyl-CoA carboxylase or the inhibition of its degrading enzyme, malonyl-CoA-decarboxylase. These have been tried experimentally in animal models through targeted gene overexpression and deletion, respectively. The clinical trials with the acetyl-CoA carboxylase overexpression face the intrinsic problems with any gene delivery treatment modality, whereas the inhibition of malonyl Co-A has not reached the clinical trials as the site-directed cellular inhibition is of paramount importance and the cross-reactions with its inhibitions in other organs could pose serious problems. Thus, more work on the translational sections need to be taken up by the pharmacological industry to remedy this problem.

25.5 Interventions for Enhancing Mitochondrial Efficiency

The damaged mitochondria with the 'leaky' membranes are a reservoir for the generation of reactive oxygen and nitrogen free radicals. During mitochondrial biogenesis, both the nuclear and mitochondrial genes are transcribed and the optimal electron transport chain enzymes assembly is ensured along with the new and functional antioxidant enzymes produced. With the generation of new mitochondria, the ROS scavenging capacity of the organelle is enhanced and the cell can sustain high levels of metabolism without undergoing oxidative damage in the process. The master regulator of the mitochondrial biogenesis is the transcription factor peroxisome proliferator-activated receptor (PPAR)-γ coactivator 1 α (PGC-1 α). It has been observed in animal studies that the level of this transcription factor is decreased in the heart failure models [179, 180]. It is yet to be conclusively established if this is the cause or the effect of the diseased condition; nevertheless it is proven beyond doubt that the lack of generation of new mitochondria has an association with the cardiovascular diseases [181]. It thus follows that the methods that enhance mitochondrial biogenesis and work upon the PGC-1 α would have beneficial effects on high risk population or patients with cardiovascular disorders.

25.5.1 Mitochondrial Permeability Transition Pores

Mitochondrial permeability transition pores (MPTP) are a group of pores associated with the mitochondrial membrane that open in response to calcium overload in the mitochondria. These pores have as yet unidentified components, but recent research points towards the complex V of the electron transport chain (ATP synthase) as being a part of it; and the way it is dependent on the mitochondrial matrix calcium content is through its association with cyclophilin D [182]. These were previously thought to be associated only with the pathological conditions related to the heart disorders, but contemporary reports have confirmed their role in the normal physiological response by showing that the MPTPs open in normal heart as well [226]. This has resulted in the hypothesis that these act as calcium vaults that open in response to calcium build-up within the mitochondrial matrix and not only help in maintaining calcium homeostasis but also contribute to the overall stability of the mitochondria by keeping the calcium levels within optimum ranges [183]. Previous work that focused on the MPTPs as therapeutic targets saw them in a negative light, and therapies were based on blocking these pores. The MPTPs are found to be open in both acute and chronic heart diseases that led to confirmation of the belief that these have a pathological outcome. The use of cyclosporin, NIM 811, and TRO 40303 as some of the potential drugs which were used in the early trials to assess their beneficial action on heart diseases [184, 185] showed mixed results [186, 187].

But the recent work showing its importance in the normal physiological processes of handling calcium overload in the mitochondria has brought some concerns in the use of MPTP blockers as therapy. The clinical studies undertaken have focused on the reduction of ischaemia-reperfusion injury with the delivery of cyclosporin. There were studies with mild efficiency with this treatment modality [187, 188], but the results from other studies have shown contradicting outcomes with no benefit in the ischaemia-reperfusion [186, 189]. Also, there have not been studies on the overall heart function or left ventricular efficiency in patients with heart diseases. Furthermore, cyclosporin administration cannot be continued as a chronic treatment modality as its chronic delivery has shown association with immunosuppression and renal problems [190, 191]. A more comprehensive analysis of the potential of MPTP as a therapeutic target is possible only with more fundamental research into its components, functions and mechanism of action. Until then other mitochondrial targets must be the focus for drug development.

25.5.2 Coenzyme Q10

The coenzyme Q (ubiquinone/ubiquinone CoQ) is composed of an enzyme found in the electron transport chain that is involved in the redox shuttle. It is a part of the ETC and undergoes the two-electron reduction from the substrates of complex I (NADH-CoQ reductase) and complex II (succinate-CoQ reductase) of the ETC while getting oxidized as it transfers electrons to the complex III (CoQH2-cytochrome c reductase). This positions the CoQ as a redox cycler where it can either accept or donate the electrons to the respective complexes based on its own redox potential [192]. When the cellular concentration of CoQ reduces below a critical threshold, instead of shuttling electrons from complexes I and II to complex III, it can pump these electrons back to complex I causing the backflow of electrons that is a potent source of ROS generation in the mitochondria [193]. In some clinical studies the blood levels of CoQ were observed in patients with heart failure and preclinical heart failure, and the levels of CoQ were found to be much lower for the patient population as compared to the control group [194, 195]. It was thus believed that heart diseases could have CoQ deficiency as an aetiology and the replacement of the deficient CoQ became an attractive endeavour.

Idebenone and EPI-743, synthetic analogues of CoQ10, are short-chain quinones that have been recently found use as dietary supplements as a treatment strategy for a number of diseases of not just cardiovascular nature. These CoQ10 analogues can not only function as shuttler for electrons in the ETC much like the function of native CoQ10 but also as scavenger for free radicals generated at the mitochondria. It has received much attention as a potential therapy as it can be easily delivered non-invasively and has the potential to cross the blood–brain barrier, thus making its use more widespread. It has been used not only in mitochondrial dysfunctional diseases and heart diseases but also in many neurological illnesses like Friedreich's ataxia and Alzheimer's disease [201–203].

Rosenfeldt et al. [196] published a systemic review on the effects of delivery of CoQ10 and its synthetic derivatives like idebenone. With their own randomized control trial and their meta-analysis of the available literature on trials, they came to the conclusion that the CoQ10 treatment resulted in significantly improved exercise times in the patients as compared to the non-treated group. It also suggested that though not statistically significant, there was a trend in the patients with heart failure towards improved ejection fraction and reduced mortality. In their own randomized placebo controlled trial of 3 months with 35 subjects on the effects on oral coenzyme Q10 in patients with heart failure, they observed significant improvements in symptoms in the treatment but not in the control group. When the studies evaluating the mean blood pressure were assessed, it was observed that the participants had a better outcome with their hypertension and their blood pressure decreased, the mean systolic decrease was around 16 mm Hg, while the mean diastolic decrease was around 10 mm Hg. Its mechanism of action is twofold, both by bypassing defective respiratory chain enzymes and in helping scavenge the oxygen free radicals.

25.5.3 Cardiolipin

The mitochondria produce ATP after the energy substrates are broken down in the TCA cycle and the electrons are shuttled to the electron carriers NADH and FADH2 that later transport the electrons to the complexes of the electron transport chain in the inner mitochondrial membrane ultimately reducing molecular oxygen to water. Thus, the core of oxidative phosphorylation in the mitochondria is through the complexes of the electron transport chain. The individual complexes of the ETC are composed of many subunits synthesized as a coordinated orchestrated event of expression of both nuclear and mitochondrial genes and then their correct assembly into supramolecular complexes that render them functional to carry out the process of electron shuttle. If the correct assembly of these molecular complexes to form the functional complexes of the mitochondrial ETC is disturbed, the mitochondrial section becomes less efficient at ATP generation and more prone to having reduced electrochemical gradient, increased ROS generation and reduced membrane stability and thus leaky nature of the mitochondrial membrane [197]. The molecule cardiolipin is shown to assist the correct assembly of these membrane complexes into their right supra-complex structures. Cardiolipin is found in the inner mitochondrial membrane and has four acyl tails instead of the normal two for most phospholipids. It's named such after it was shown to be associated with the heart diseases of both neonatal origin [198] and adult onset [199, 200]. The loss or damage of cardiolipin causes its dissociation from the complex III cytochrome C oxidase, making the cytochrome C act as a peroxidase instead of its regular role of an electron carrier; resulting in loss of ETC efficiency. The presence of excessive ROS/RNS causes damage to the mitochondrial proteins, lipids and DNA. The damage to extremely ROS-sensitive cardiolipin is more dangerous as it causes lowered efficiency of respiratory complex assembly, leading to leaky mitochondrial membrane and more ROS generation, creating a vicious cycle. More research into this molecule has established its role in a host of other important mitochondrial functions like ROS production, managing the activity of membrane transporters of the mitochondria, mitochondrial ion homeostasis, etc.

After the deficiency of cardiolipin had been shown in many pathological conditions associated with energy deficiency and myocardial diseases, it gained a prominence as a drug target. A cardiolipin-associating compound was serendipitously discovered by Szeto and Schiller that was named SS-31. SS-31 and its analogue called MTP-131 have been used in many studies assessing cardiovascular and associated functions. It was observed that the delivery of SS-31 and MTP-131 during reperfusion greatly reduces reperfusion-based necrosis in the cardiomyocytes. During the ischaemic attack, the mitochondria are damaged and the respirosomes (the respiratory complexes of the ETC) become disassembled. This makes the mitochondria inefficient and creates ATP decline, ROS generation, etc. associated with the ischaemic episode. With reperfusion, the substrate and oxygen again become available to the mitochondria, but as the respiratory complexes are disassembled and the membrane lacks the electeochemical gradient, mitochondria are unable to produce proportional levels of ATP and much of the oxygen is turned towards production of ROS. But, when the cardiolipin associating SS-31, MTP-131 was given during reperfusion, as in the case of many trials [201–203], it was found to be beneficial in ischaemia-reperfusion injury. The cardiolipin restores the damaged ETC supercomplex assemblies and reduces the membrane permeability and enhances the shuttle of electrons through the respiratory complexes, thus generating ATP efficiently while reducing ROS production and associated cardiomyocyte damage.

The decrease in the functional capacity of the failing heart before it is afflicted with an infarct is associated with the decreased capacity of various energy-intensive organs like renal insufficiency and skeletal muscle inefficiency that results in heart disease-associated exercise intolerance. The trials that studied the effect of SS-31 and MTP-131 also have shown improvement in patients with renal insufficiency and enhanced the exercise tolerance capacity of the subjects [204–206]. Cardiolipin is thus an attractive therapeutic target as it restores not only the cardiac function but alleviates associated deficiencies of the skeletal muscle and renal tissue.

25.5.4 Phospholipid Replacement Therapy

The major damage to the mitochondria during the heart disease is to the integrity of their membranes. The mechanisms that are dealt with in the preceding sections talk about saving mitochondria from the additional burden of ROS and RNS so that their proteins, phospholipids and DNA do not undergo additional damage. The antioxidant therapy has been widely used in the treatment of cardiovascular and other diseases sharing the aetiology of oxidative stress; but that does not take care of the damage that the membrane phospholipids already underwent [207, 208]. The only means of correcting that damage is through the clearance of the damaged mitochondria through mitophagy; as already mentioned this process is reduced in the diseased and failing myocardium. The replacement of the damaged membrane phospholipids provides a reasonable means to compensate for the damaged mitochondrial membranes.

There are dietary means through which the membrane phospholipids of the mitochondria can be replaced. These have mainly been used in the management of mitochondrial illnesses resulting in chronic fatigue syndrome, with promising results. In a trial by Ellithorpe et al. [209], the subjects with chronic fatigue and fibromyalgia experienced a 40.5% decrease in fatigue with 8 weeks of therapy. Other trials with Propax and NT factor have also observed benefit in mitochondrial capacity. Though these trials focused on mitochondrial disorders, the alleviation of mitochondrial inefficiency shows great promise in trying these combinations of drugs along with antioxidant therapies for the treatment of heart diseases and failing myocardial capacity.

25.5.5 Vitamin B3 Precursor

One of the features of the inefficient mitochondria is that it has inefficient ATP production for the particular cell or tissue type, thus making it energy deficient; and it performs its function at a less than optimal level. One of the ways to enhance the efficiency of mitochondria is to increase the substrate/cofactor concentration. The ultimate mechanism of energy production in the mitochondria is sequential passage of electrons through the electron transport chain that generates an electrochemical gradient across the mitochondrial matrix and the inner mitochondrial membrane, which causes the controlled return of the electrons back through the complex V (ATP synthase) driving the phosphorylation of ADP to ATP with the reduction of molecular oxygen to water. The electron carrier from the TCA cycle to the electron transport chain is through the NAD+/NADH (nicotinamide adenine dinucleotide) and FAD+/FADH2. The increase in the cellular NAD+ levels has shown to be beneficial to the mitochondria and the overall well-being of the organism by reduction in the oxidative damage and an increase in the ATP-generating capacity of the cell and tissue [139, 210]. Dietary NAD+ is unstable; thus, a precursor of NAD+ has been widely used to elucidate the effects of NAD+ on mitochondria. The NAD+ is produced in the cell through the salvage pathway with the vitamin B3 (nicotinic acid) as the prime precursor molecule. A variant of vitamin B3 is employed that has better stability and tolerability in humans, nicotinamide riboside (NR).

In human failing heart, it is observed that the levels of nicotinamide phosphoribosyltransferase, key enzyme required for the recycling of the NAD+, are reduced [211]. It is accompanied by the rise in the levels of another enzyme, nicotinamide riboside kinase 2, the enzyme that phosphorylates and thus makes unavailable the precursors for NAD+. This implies that in the failing heart, the levels of NAD+ are reduced and the generation of new NAD+ by the intrinsic mechanisms of the salvage pathway gets restrained. It was thus considered a worthwhile endeavour to observe the effects of supplementation of NAD+ precursors through diet. Many studies attempted it in rodent models of heart failure and cardiovascular disease. The vitamin B3 (nicotinic acid) is a precursor of the NAD+, but a more stable dietary supplement is nicotinamide riboside (NR). NR supplementation enhances the levels of NAD+ in the tissues and blood of the recipients and is observed to enhance the mitochondrial efficiency [212, 213]. The ATP generation of the mitochondria increases with augmentation of the overall efficiency of the tissue. NAD+ is also a redox-sensitive molecule and can help reduce the overall oxidative damage. It thus has a dual role in helping in shuttling more electrons through the ETC for higher ATP generation while reducing the ROS load that might result from the process.

Diguet et al. [214] created the murine models of heart failure by producing dilated cardiomyopathy through Serum Response Factor transcription factor depletion (SRF^{HKO} mice). A 30% reduction in the levels of NAD+ was observed in failing heart with dilated cardiomyopathy and a decline, but to a lesser extent, was observed in aortic constriction. The dietary supplementation of NR was delivered and the observed symptoms were alleviated in cardiomyocytes by increase in NAD+ levels. Thus, NAD+ precursor intervention forms an attractive and easy model for the mitochondria-targeted cardiovascular therapy.

25.6 Mitochondrial Transplantation: A Magic or Menace?

The most recent development in the area of mitochondria-targeted therapy is mitochondrial transplantation, transplantation of healthy functional mitochondria from autologous sources to the diseased tissue. As we previously discussed, cardiomyocytes are densely populated by mitochondrial structures (~1/3 of the volume), and many of the heart-associated diseases are coupled with mitochondrial dysfunction, especially myocardial ischaemia-reperfusion injury mainly related to surgical interventions. McCully's group devised a new therapeutic approach towards the mitochondria-related problems which has been recently taken to clinical trials in paediatric patients with myocardial ischaemia [215]. This transplantation approach was first validated in animal models where the authors reported decreased ROS generation and cell death and a reduction in infarct size [216, 217]. But still, there remain a plethora of unanswered questions with regard to the uptake of perfused/injected mitochondria into the functional cells where they have to power the contractile filaments and other essential cell survival processes. Also, the high calcium content in the blood raises questions about the viability of the perfused mitochondrial units. This array of unresolved puzzles raises questions on the rapid rush towards translation of the procedure.

25.7 Conclusion

The cardiac tissue, with its onus task of rhythmic contraction for a lifetime resting a mere 0.8 s in between, is dependent upon continuous and sustained levels of ATP, thereby being dependent upon its mitochondria. The labyrinthine and dynamic mitochondrial network in cardiomyocytes is maintained by the appropriate and counterbalanced biogenesis, dynamics and degradation, working at all times to respond to the energy and physiological demands. It thus follows that the regulation of the mitochondria is imperative to maintain the adeptness of the heart failing which, as is evidenced by many studies, is the damage to the mitochondria finally culminate in CVD progression (Fig. 25.9a, b).

Thereupon, many emerging therapies for CVD are aimed at the mitochondrial efficiency and quality control. The mitochondria-based interventions range from *(i)* bypass agents that bypass the defective ETC, like coenzyme Q10; *(ii)* replacement supplementation that replaces the defective components in the mitochondrial membrane like the cardiolipin and membrane phospholipid replacement; *(iii)* antioxidant treatments; *(iv)* precursor supplementation, for, e.g. resveratrol and nicotinamide; *(v)* or the more natural and advantageous combination therapies that target various components in the mitochondria and its dynamics together, like exercise and calorie restriction. With more pinpointed therapies towards maintaining the efficient

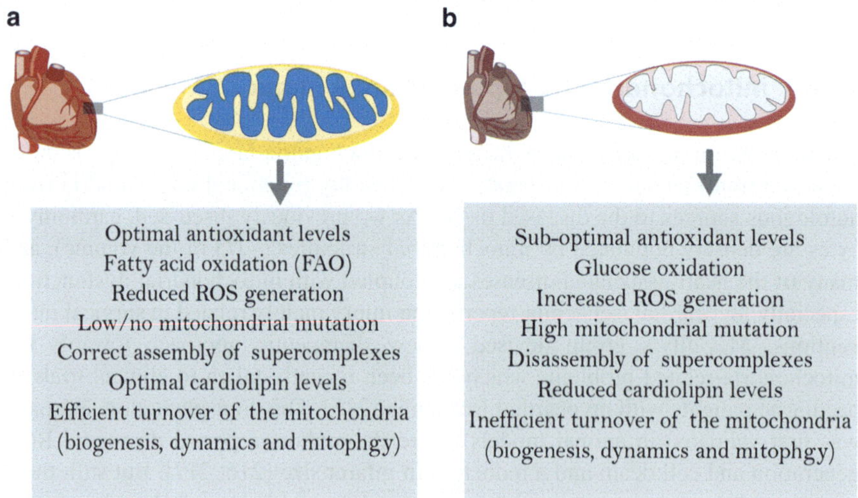

Fig. 25.9 Healthy vs damaged heart and their mitochondrial states

mitochondria in cardiomyocytes, it may be possible to alleviate the failing myocardium and rescue the cardiomyocytes without progression towards heart failure.

25.8 Future Perspectives

Cardiovascular disorders have a strong underlying metabolic dysregulation, hence it would be less surprising that the therapies to lower the global CVD burden will focus more on the mitochondria. The biggest problem that the mitochondrial-targeted therapies face is the specificity of the intervention. With the advancement in the tissue-specific delivery of pharmacological agents and gene delivery systems that employ an alternative to the viral delivery method, the core of the problem, i.e. organelle-specific therapy, can be accomplished. As the mitochondrial DNA damage and mitochondrial depolarization cause the mitochondria to release calcium and caspases, targeting mitochondrial DNA repair through the gene delivery systems becomes attractive. With devising a means to degrade the damaged mitochondria alone while retaining and inducing genesis of their healthy counterparts, the efficacious mitochondrial network can be re-established. It can be hoped that the CVD epidemic can be targeted to the very root of the problem; at the damaging mitochondrial network, commencing the cascade of signals resulting in reduced cardiac efficiency and eventually to heart failure; and the burden can be nipped at the bud.

References

1. Friedland G (2009) Discovery of the function of the heart and circulation of blood. Cardiovasc J Afr 20:160
2. Loe MJ, Edwards WD (2004) A light-hearted look at a lion-hearted organ (or, a perspective from three standard deviations beyond the norm) part 1 (of two parts). Cardiovasc Pathol 13:282–292
3. Kolwicz SC, Purohit S, Tian R (2013) Cardiac metabolism and its interactions with contraction, growth, and survival of cardiomyocytes. Circ Res 113:603–616
4. Balaban RS (1990) Regulation of oxidative phosphorylation in the mammalian cell. Am J Phys Cell Phys 258:C377–C389
5. Hom J, Sheu S-S (2009) Morphological dynamics of mitochondria – a special emphasis on cardiac muscle cells. J Mol Cell Cardiol 46:811–820
6. Glancy B, Hartnell LM, Combs CA et al (2017) Power grid protection of the muscle mitochondrial reticulum. Cell Rep 19:487–496
7. Hoppel CL, Tandler B, Fujioka H et al (2009) Dynamic organization of mitochondria in human heart and in myocardial disease. Int J Biochem Cell Biol 41:1949–1956
8. Jeffrey FMH, Diczku V, Sherry AD et al (1995) Substrate selection in the isolated working rat heart: effects of reperfusion, afterload, and concentration. Basic Res Cardiol 90:388–396
9. Schönekess BO (1997) Competition between lactate and fatty acids as sources of ATP in the isolated working rat heart. J Mol Cell Cardiol 29:2725–2733
10. Woods DC (2017) Mitochondrial heterogeneity: evaluating mitochondrial subpopulation dynamics in stem cells. Stem Cells Int 2017:1–7. https://www.hindawi.com/journals/sci/2017/7068567/
11. Palmer JW, Tandler B, Hoppel CL (1977) Biochemical properties of subsarcolemmal and interfibrillar mitochondria isolated from rat cardiac muscle. J Biol Chem 252:8731–8739

12. Schwarzer M, Schrepper A, Amorim PA et al (2013) Pressure overload differentially affects respiratory capacity in interfibrillar and subsarcolemmal mitochondria. Am J Physiol Heart Circ Physiol 304:H529–H537
13. Crochemore C, Mekki M, Corbière C et al (2015) Subsarcolemmal and interfibrillar mitochondria display distinct superoxide production profiles. Free Radic Res 49:331–337
14. Hollander JM, Thapa D, Shepherd DL (2014) Physiological and structural differences in spatially distinct subpopulations of cardiac mitochondria: influence of cardiac pathologies. Am J Phys 307:H1–H14
15. Shimada T, Horita K, Murakami M et al (1984) Morphological studies of different mitochondrial populations in monkey myocardial cells. Cell Tissue Res 238:577–582
16. Bereiter-Hahn J (1990) Behavior of mitochondria in the living cell. Int Rev Cytol 122:1–63
17. Soubannier V, McBride HM (2009) Positioning mitochondrial plasticity within cellular signaling cascades. Biochim Biophys Acta (BBA) – Mol Cell Res 1793:154–170
18. Stefely JA, Kwiecien NW, Freiberger EC et al (2016) Mitochondrial protein functions elucidated by multi-omic mass spectrometry profiling. Nat Biotechnol 34:1191–1197
19. Fernandez-Marcos PJ, Auwerx J (2011) Regulation of PGC-1α, a nodal regulator of mitochondrial biogenesis. Am J Clin Nutr 93:884S–890S
20. Puigserver P, Wu Z, Park CW et al (1998) A cold-inducible coactivator of nuclear receptors linked to adaptive thermogenesis. Cell 92:829–839
21. Arany Z, Novikov M, Chin S et al (2006) Transverse aortic constriction leads to accelerated heart failure in mice lacking PPAR-γ coactivator 1α. Proc Natl Acad Sci U S A 103:10086–10091
22. He X, Sun C, Wang F et al (2012) Peri-implantation lethality in mice lacking the PGC-1-related coactivator protein. Dev Dyn 241:975–983
23. Lai L, Leone TC, Zechner C et al (2008) Transcriptional coactivators PGC-1α and PGC-1β control overlapping programs required for perinatal maturation of the heart. Genes Dev 22:1948–1961
24. Martin OJ, Ling L, Soundarapandian MM et al (2014) A role for peroxisome proliferator-activated receptor γ coactivator-1 in the control of mitochondrial dynamics during postnatal cardiac growth. Circ Res 114:626–636
25. Yun C, Yingqiu L, Dorn GW (2011) Mitochondrial fusion is essential for organelle function and cardiac homeostasis. Circ Res 109:1327–1331
26. Ge K, Guermah M, Yuan C-X et al (2002) Transcription coactivator TRAP220 is required for PPARγ2-stimulated adipogenesis. Nature 417:563–567
27. Andersson U, Scarpulla RC (2001) PGC-1-related coactivator, a novel, serum-inducible coactivator of nuclear respiratory factor 1-dependent transcription in mammalian cells. Mol Cell Biol 21:3738–3749
28. Gleyzer N, Vercauteren K, Scarpulla RC (2005) Control of mitochondrial transcription specificity factors (TFB1M and TFB2M) by nuclear respiratory factors (NRF-1 and NRF-2) and PGC-1 family coactivators. Mol Cell Biol 25:1354–1366
29. Huo L, Scarpulla RC (2001) Mitochondrial DNA instability and peri-implantation lethality associated with targeted disruption of nuclear respiratory factor 1 in mice. Mol Cell Biol 21:644–654
30. Ristevski S, O'Leary DA, Thornell AP et al (2004) The ETS transcription factor GABPα is essential for early embryogenesis. Mol Cell Biol 24:5844–5849
31. Scarpulla RC (2008) Nuclear control of respiratory chain expression by nuclear respiratory factors and PGC-1-related coactivator. Ann N Y Acad Sci 1147:321–334
32. Dufour CR, Wilson BJ, Huss JM et al (2007) Genome-wide orchestration of cardiac functions by the orphan nuclear receptors ERRα and γ. Cell Metab 5:345–356
33. Huss JM, Torra IP, Staels B et al (2004) Estrogen-related receptor α directs peroxisome proliferator-activated receptor α signaling in the transcriptional control of energy metabolism in cardiac and skeletal muscle. Mol Cell Biol 24:9079–9091
34. Huss JM, Kopp RP, Kelly DP (2002) Peroxisome proliferator-activated receptor coactivator-1α (PGC-1α) coactivates the cardiac-enriched nuclear receptors estrogen-related receptor-α

and -γ identification of novel leucine-rich interaction motif within PGC-1α. J Biol Chem 277:40265–40274
35. Sladek R, Bader JA, Giguère V (1997) The orphan nuclear receptor estrogen-related receptor alpha is a transcriptional regulator of the human medium-chain acyl coenzyme A dehydrogenase gene. Mol Cell Biol 17:5400–5409
36. Vega RB, Kelly DP (1997) A role for estrogen-related receptor α in the control of mitochondrial fatty acid β-oxidation during brown adipocyte differentiation. J Biol Chem 272:31693–31699
37. Issemann I, Green S (1990) Activation of a member of the steroid hormone receptor superfamily by peroxisome proliferators. Nature 347:645
38. Kersten S (2014) Integrated physiology and systems biology of PPARα. Mol Metab 3:354–371
39. Mascaró C, Acosta E, Ortiz JA et al (1998) Control of human muscle-type carnitine palmitoyltransferase I gene transcription by peroxisome proliferator-activated receptor. J Biol Chem 273:8560–8563
40. McMullen PD, Bhattacharya S, Woods CG et al (2014) A map of the PPARα transcription regulatory network for primary human hepatocytes. Chem Biol Interact 209:14–24
41. Prosdocimo DA, John JE, Zhang L et al (2015) KLF15 and PPARα cooperate to regulate cardiomyocyte lipid gene expression and oxidation. PPAR Res 2015:201625. https://www.hindawi.com/journals/ppar/2015/201625/
42. Prosdocimo DA, Anand P, Liao X et al (2014) Kruppel-like factor 15 is a critical regulator of cardiac lipid metabolism. J Biol Chem 289:5914–5924
43. van der Meer DLM, Degenhardt T, Väisänen S et al (2010) Profiling of promoter occupancy by PPARα in human hepatoma cells via ChIP-chip analysis. Nucleic Acids Res 38:2839–2850
44. Baar K, Wende AR, Jones TE et al (2002) Adaptations of skeletal muscle to exercise: rapid increase in the transcriptional coactivator PGC-1. FASEB J 16:1879–1886
45. Wu Z, Puigserver P, Andersson U et al (1999) Mechanisms controlling mitochondrial biogenesis and respiration through the thermogenic coactivator PGC-1. Cell 98:115–124
46. Akimoto T, Pohnert SC, Li P et al (2005) Exercise stimulates Pgc-1α transcription in skeletal muscle through activation of the p38 MAPK pathway. J Biol Chem 280:19587–19593
47. Little JP, Safdar A, Cermak N et al (2010) Acute endurance exercise increases the nuclear abundance of PGC-1α in trained human skeletal muscle. Am J Phys Regul Integr Comp Phys 298:R912–R917
48. Shabana D, Konstandin MH, Bevan J et al (2014) Metabolic dysfunction consistent with premature aging results from deletion of Pim kinases. Circ Res 115:376–387
49. Cantó C, Gerhart-Hines Z, Feige JN et al (2009) AMPK regulates energy expenditure by modulating NAD^+ metabolism and SIRT1 activity. Nature 458:1056–1060
50. Coste A, Louet J-F, Lagouge M et al (2008) The genetic ablation of SRC-3 protects against obesity and improves insulin sensitivity by reducing the acetylation of PGC-1{alpha}. Proc Natl Acad Sci U S A 105:17187–17192
51. Gerhart-Hines Z, Rodgers JT, Bare O et al (2007) Metabolic control of muscle mitochondrial function and fatty acid oxidation through SIRT1/PGC-1α. EMBO J 26:1913–1923
52. Chen H, Detmer SA, Ewald AJ et al (2003) Mitofusins Mfn1 and Mfn2 coordinately regulate mitochondrial fusion and are essential for embryonic development. J Cell Biol 160:189–200
53. Misko A, Sasaki Y, Tuck E et al (2012) Mitofusin2 mutations disrupt axonal mitochondrial positioning and promote axon degeneration. J Neurosci 32:4145–4155
54. Züchner S, Mersiyanova IV, Muglia M et al (2004) Mutations in the mitochondrial GTPase mitofusin 2 cause charcot-marie-tooth neuropathy type 2A. Nat Genet 36:449–451
55. Chen K-H, Guo X, Ma D et al (2004) Dysregulation of HSG triggers vascular proliferative disorders. Nat Cell Biol 6:872–883
56. Cheng X, Zhou D, Wei J et al (2013) Cell-cycle arrest at G2/M and proliferation inhibition by adenovirus-expressed mitofusin-2 gene in human colorectal cancer cell lines. Neoplasma 60:620–626

57. Gao Q, Wang X-M, Ye H-W et al (2012) Changes in the expression of cardiac mitofusin-2 in different stages of diabetes in rats. Mol Med Rep 6:811–814
58. Guo X, Chen K-H, Guo Y et al (2007) Mitofusin 2 triggers vascular smooth muscle cell apoptosis via mitochondrial death pathway. Circ Res 101:1113–1122
59. Rehman J, Zhang HJ, Toth PT et al (2012) Inhibition of mitochondrial fission prevents cell cycle progression in lung cancer. FASEB J 26:2175–2186
60. Shen T, Zheng M, Cao C et al (2007) Mitofusin-2 is a major determinant of oxidative stress-mediated heart muscle cell apoptosis. J Biol Chem 282:23354–23361
61. Wang W, Zhu F, Wang S et al (2010) HSG provides antitumor efficacy on hepatocellular carcinoma both in vitro and in vivo. Oncol Rep 24:183–188
62. Zhang G-E, Jin H-L, Lin X-K et al (2013) Anti-tumor effects of Mfn2 in gastric cancer. Int J Mol Sci 14:13005–13021
63. Bach D, Pich S, Soriano FX et al (2003) Mitofusin-2 determines mitochondrial network architecture and mitochondrial metabolism a novel regulatory mechanism altered in obesity. J Biol Chem 278:17190–17197
64. Chen H, McCaffery JM, Chan DC (2007) Mitochondrial fusion protects against neurodegeneration in the cerebellum. Cell 130:548–562
65. Chen H, Chomyn A, Chan DC (2005) Disruption of fusion results in mitochondrial heterogeneity and dysfunction. J Biol Chem 280:26185–26192
66. Eura Y, Ishihara N, Yokota S et al (2003) Two mitofusin proteins, mammalian homologues of FZO, with distinct functions are both required for mitochondrial fusion. J Biochem (Tokyo) 134:333–344
67. Liu R, Jin P, LiqunYu, et al (2014) Impaired mitochondrial dynamics and bioenergetics in diabetic skeletal muscle. PLoS One 9:e92810
68. Papanicolaou KN, Ryosuke K, Ngoh GA et al (2012) Mitofusins 1 and 2 are essential for postnatal metabolic remodeling in heart. Circ Res 111:1012–1026
69. Pich S, Bach D, Briones P et al (2005) The charcot–marie–tooth type 2A gene product, Mfn2, up-regulates fuel oxidation through expression of OXPHOS system. Hum Mol Genet 14:1405–1415
70. Sebastián D, Hernández-Alvarez MI, Segalés J et al (2012) Mitofusin 2 (Mfn2) links mitochondrial and endoplasmic reticulum function with insulin signaling and is essential for normal glucose homeostasis. PNAS 109:5523–5528
71. Elachouri G, Vidoni S, Zanna C et al (2011) OPA1 links human mitochondrial genome maintenance to mtDNA replication and distribution. Genome Res 21:12–20
72. Frezza C, Cipolat S, de Brito OM et al (2006) OPA1 controls apoptotic cristae remodeling independently from mitochondrial fusion. Cell 126:177–189
73. Meeusen S, DeVay R, Block J et al (2006) Mitochondrial inner-membrane fusion and crista maintenance requires the dynamin-related GTPase Mgm1. Cell 127:383–395
74. Tsutsui H, Kinugawa S, Matsushima S (2008) Oxidative stress and mitochondrial DNA damage in heart failure. Circ J 72:A31–A37
75. Varanita T, Soriano ME, Romanello V et al (2015) The Opa1-dependent mitochondrial cristae remodeling pathway controls atrophic, apoptotic, and ischemic tissue damage. Cell Metab 21:834–844
76. Delettre C, Lenaers G, Griffoin J-M et al (2000) Nuclear gene *OPA1*, encoding a mitochondrial dynamin-related protein, is mutated in dominant optic atrophy. Nat Genet 26:207–210
77. Delettre C, Griffoin JM, Kaplan J et al (2001) Mutation spectrum and splicing variants in the OPA1 gene. Hum Genet 109:584–591
78. Baker MJ, Tatsuta T, Langer T (2011) Quality control of mitochondrial proteostasis. Cold Spring Harb Perspect Biol 3
79. Ehses S, Raschke I, Mancuso G et al (2009) Regulation of OPA1 processing and mitochondrial fusion by m-AAA protease isoenzymes and OMA1. J Cell Biol 187:1023–1036
80. Griparic L, Kanazawa T, van der Bliek AM (2007) Regulation of the mitochondrial dynamin-like protein Opa1 by proteolytic cleavage. J Cell Biol 178:757–764

81. Head B, Griparic L, Amiri M et al (2009) Inducible proteolytic inactivation of OPA1 mediated by the OMA1 protease in mammalian cells. J Cell Biol 187:959–966
82. Ishihara N, Otera H, Oka T et al (2012) Regulation and physiologic functions of GTPases in mitochondrial fusion and fission in mammals. Antioxid Redox Signal 19:389–399
83. Ishihara N, Fujita Y, Oka T et al (2006) Regulation of mitochondrial morphology through proteolytic cleavage of OPA1. EMBO J 25:2966–2977
84. Pellegrini L, Passer BJ, Canelles M et al (2001) PAMP and PARL, two novel putative metalloproteases interacting with the COOH-terminus of Presenilin-1 and -2. J Alzheimers Dis 3:181–190
85. Song Z, Chen H, Fiket M et al (2007) OPA1 processing controls mitochondrial fusion and is regulated by mRNA splicing, membrane potential, and Yme1L. J Cell Biol 178:749–755
86. Jiang X, Jiang H, Shen Z et al (2014) Activation of mitochondrial protease OMA1 by Bax and Bak promotes cytochrome c release during apoptosis. Proc Natl Acad Sci U S A 111:14782–14787
87. Quirós PM, Ramsay AJ, Sala D et al (2012) Loss of mitochondrial protease OMA1 alters processing of the GTPase OPA1 and causes obesity and defective thermogenesis in mice. EMBO J 31:2117–2133
88. Laforge M, Rodrigues V, Silvestre R et al (2016) NF-κB pathway controls mitochondrial dynamics. Cell Death Differ 23:89–98
89. Piquereau J, Caffin F, Novotova M et al (2012) Down-regulation of OPA1 alters mouse mitochondrial morphology, PTP function, and cardiac adaptation to pressure overload. Cardiovasc Res 94:408–417
90. White KE, Davies VJ, Hogan VE et al (2009) OPA1 deficiency associated with increased autophagy in retinal ganglion cells in a murine model of dominant optic atrophy. Invest Ophthalmol Vis Sci 50:2567–2571
91. Smirnova E, Shurland DL, Ryazantsev SN et al (1998) A human dynamin-related protein controls the distribution of mitochondria. J Cell Biol 143:351–358
92. Yoon Y, Pitts KR, Dahan S et al (1998) A novel dynamin-like protein associates with cytoplasmic vesicles and tubules of the endoplasmic reticulum in mammalian cells. J Cell Biol 140:779–793
93. Liesa M, Palacín M, Zorzano A (2009) Mitochondrial dynamics in mammalian health and disease. Physiol Rev 89:799–845
94. Losón OC, Song Z, Chen H et al (2013) Fis1, Mff, MiD49, and MiD51 mediate Drp1 recruitment in mitochondrial fission. Mol Biol Cell 24:659–667
95. Otera H, Wang C, Cleland MM et al (2010) Mff is an essential factor for mitochondrial recruitment of Drp1 during mitochondrial fission in mammalian cells. J Cell Biol 191:1141–1158
96. Niemann A, Ruegg M, La Padula V et al (2005) Ganglioside-induced differentiation associated protein 1 is a regulator of the mitochondrial network. J Cell Biol 170:1067–1078
97. Andres AM, Stotland A, Queliconi BB et al (2015) A time to reap, a time to sow: Mitophagy and biogenesis in cardiac pathophysiology. J Mol Cell Cardiol 78:62–72
98. Palikaras K, Lionaki E, Tavernarakis N (2015) Balancing mitochondrial biogenesis and mitophagy to maintain energy metabolism homeostasis. Cell Death Differ 22:1399–1401
99. Lemasters JJ (2005) Selective mitochondrial autophagy, or mitophagy, as a targeted defense against oxidative stress, mitochondrial dysfunction, and aging. Rejuvenation Res 8:3–5
100. Glick D, Barth S, Macleod KF (2010) Autophagy: cellular and molecular mechanisms. J Pathol 221:3–12
101. Hayashi-Nishino M, Fujita N, Noda T et al (2009) A subdomain of the endoplasmic reticulum forms a cradle for autophagosome formation. Nat Cell Biol 11:1433–1437
102. Ylä-Anttila P, Vihinen H, Jokitalo E et al (2009) 3D tomography reveals connections between the phagophore and endoplasmic reticulum. Autophagy 5:1180–1185
103. Axe EL, Walker SA, Manifava M et al (2008) Autophagosome formation from membrane compartments enriched in phosphatidylinositol 3-phosphate and dynamically connected to the endoplasmic reticulum. J Cell Biol 182:685–701

104. Mizushima N, Klionsky DJ (2007) Protein turnover via autophagy: implications for metabolism. Annu Rev Nutr 27:19–40
105. English L, Chemali M, Duron J et al (2009) Autophagy enhances the presentation of endogenous viral antigens on MHC class I molecules during HSV-1 infection. Nat Immunol 10:480–487
106. Nakai A, Yamaguchi O, Takeda T et al (2007) The role of autophagy in cardiomyocytes in the basal state and in response to hemodynamic stress. Nat Med 13:619–624
107. Nishino I, Fu J, Tanji K et al (2000) Primary LAMP-2 deficiency causes X-linked vacuolar cardiomyopathy and myopathy (Danon disease). Nature 406:906–910
108. Tanaka Y, Guhde G, Suter A et al (2000) Accumulation of autophagic vacuoles and cardiomyopathy in LAMP-2-deficient mice. Nature 406:902–906
109. Thomas RL, Roberts DJ, Kubli DA et al (2013) Loss of MCL-1 leads to impaired autophagy and rapid development of heart failure. Genes Dev 27:1365–1377
110. Sciarretta S, Hariharan N, Monden Y et al (2011) Is autophagy in response to ischemia and reperfusion protective or detrimental for the heart? Pediatr Cardiol 32:275–281
111. Xu X, Kobayashi S, Chen K et al (2013) Diminished autophagy limits cardiac injury in mouse models of type 1 diabetes. J Biol Chem 288:18077–18092
112. Xu X, Hua Y, Sreejayan N et al (2013) Akt2 knockout preserves cardiac function in high-fat diet-induced obesity by rescuing cardiac autophagosome maturation. J Mol Cell Biol 5:61–63
113. Dorn GW (2016) Parkin-dependent mitophagy in the heart. J Mol Cell Cardiol 95:42–49
114. Gong G, Song M, Csordas G et al (2015) Parkin-mediated mitophagy directs perinatal cardiac metabolic maturation in mice. Science:350–aad2459
115. Moyzis AG, Sadoshima J, Gustafsson ÅB (2014) Mending a broken heart: the role of mitophagy in cardioprotection. Am J Phys Heart Circ Phys 308:H183–H192
116. Shires SE, Gustafsson ÅB (2015) Mitophagy and heart failure. J Mol Med 93:253–262
117. Thomas RL, Gustafsson AB (2013) Mitochondrial autophagy–an essential quality control mechanism for myocardial homeostasis. Circ J 77:2449–2454
118. Matsuda N, Sato S, Shiba K et al (2010) PINK1 stabilized by mitochondrial depolarization recruits Parkin to damaged mitochondria and activates latent Parkin for mitophagy. J Cell Biol 189:211–221
119. Narendra D, Tanaka A, Suen D-F et al (2008) Parkin is recruited selectively to impaired mitochondria and promotes their autophagy. J Cell Biol 183:795–803
120. Greene AW, Grenier K, Aguileta MA et al (2012) Mitochondrial processing peptidase regulates PINK1 processing, import and Parkin recruitment. EMBO Rep 13:378–385
121. Chen Y, Dorn GW (2013) PINK1-phosphorylated mitofusin 2 is a parkin receptor for culling damaged mitochondria. Science 340:471–475
122. Gegg ME, Cooper JM, Chau K-Y et al (2010) Mitofusin 1 and mitofusin 2 are ubiquitinated in a PINK1/parkin-dependent manner upon induction of mitophagy. Hum Mol Genet 19:4861–4870
123. Geisler S, Holmström KM, Skujat D et al (2010) PINK1/Parkin-mediated mitophagy is dependent on VDAC1 and p62/SQSTM1. Nat Cell Biol 12:119–131
124. Poole AC, Thomas RE, Yu S et al (2010) The mitochondrial fusion-promoting factor mitofusin is a substrate of the PINK1/parkin pathway. PLoS One 5:e10054
125. Wang X, Winter D, Ashrafi G et al (2011) PINK1 and Parkin target Miro for phosphorylation and degradation to arrest mitochondrial motility. Cell 147:893–906
126. Gladkova C, Maslen SL, Skehel JM et al (2018) Mechanism of parkin activation by PINK1. Nature 559:410
127. Deas E, Piipari K, Machhada A et al (2014) PINK1 deficiency in β-cells increases basal insulin secretion and improves glucose tolerance in mice. Open Biol 4
128. Kubli DA, Cortez MQ, Moyzis AG et al (2015) PINK1 is dispensable for mitochondrial recruitment of parkin and activation of mitophagy in cardiac myocytes. PLoS One 10:e0130707

129. Hasson SA, Kane LA, Yamano K et al (2013) High-content genome-wide RNAi screens identify regulators of parkin upstream of mitophagy. Nature 504:291–295
130. Tahrir FG, Knezevic T, Gupta MK et al (2017) Evidence for the role of BAG3 in mitochondrial quality control in cardiomyocytes. J Cell Physiol 232:797–805
131. Chu CT, Bayır H, Kagan VE (2014) LC3 binds externalized cardiolipin on injured mitochondria to signal mitophagy in neurons: implications for Parkinson disease. Autophagy 10:376–378
132. Liu L, Feng D, Chen G et al (2012) Mitochondrial outer-membrane protein FUNDC1 mediates hypoxia-induced mitophagy in mammalian cells. Nat Cell Biol 14:177–185
133. Quinsay MN, Thomas RL, Lee Y et al (2010) Bnip3-mediated mitochondrial autophagy is independent of the mitochondrial permeability transition pore. Autophagy 6:855–862
134. Sandoval H, Thiagarajan P, Dasgupta SK et al (2008) Essential role for nix in autophagic maturation of erythroid cells. Nature 454:232–235
135. Strappazzon F, Vietri-Rudan M, Campello S et al (2011) Mitochondrial BCL-2 inhibits AMBRA1-induced autophagy. EMBO J 30:1195–1208
136. Van Humbeeck C, Cornelissen T, Hofkens H et al (2011) Parkin interacts with Ambra1 to induce mitophagy. J Neurosci 31:10249–10261
137. Murakawa T, Yamaguchi O, Okamoto K et al (2015) The novel mitophagic receptor protein, Bcl2-like protein 13: new insights for the molecular mechanisms of the pathogenesis of heart disease. J Card Fail 21:S147
138. Murakawa T, Yamaguchi O, Hashimoto A et al (2015) Bcl-2-like protein 13 is a mammalian Atg32 homologue that mediates mitophagy and mitochondrial fragmentation. Nat Commun 6:7527
139. Cantó C, Houtkooper RH, Pirinen E et al (2012) The NAD(+) precursor nicotinamide riboside enhances oxidative metabolism and protects against high-fat diet-induced obesity. Cell Metab 15:838–847
140. Nicotinamide Riboside and Mitochondrial Biogenesis – Full Text View – ClinicalTrials.gov, https://clinicaltrials.gov/ct2/show/NCT03432871
141. Chen R-J, Lee Y-H, Yeh Y-L et al (2017) Autophagy-inducing effect of pterostilbene: a prospective therapeutic/preventive option for skin diseases. J Food Drug Anal 25:125–133
142. McCormack D, McFadden D (2012) Pterostilbene and cancer: current review. J Surg Res 173:e53–e61
143. Berman AY, Motechin RA, Wiesenfeld MY et al (2017) The therapeutic potential of resveratrol: a review of clinical trials. NPJ Precis Oncol 1
144. Lightowlers RN, Chrzanowska-Lightowlers ZM (2014) Salvaging hope: is increasing NAD+ a key to treating mitochondrial myopathy? EMBO Mol Med 6:705–707
145. Chung S, Yao H, Caito S et al (2010) Regulation of SIRT1 in cellular functions: role of polyphenols. Arch Biochem Biophys 501:79–90
146. Kundu JK, Shin YK, Kim SH et al (2006) Resveratrol inhibits phorbol ester-induced expression of COX-2 and activation of NF-kappaB in mouse skin by blocking IkappaB kinase activity. Carcinogenesis 27:1465–1474
147. Ferrières J (2004) The French paradox: lessons for other countries. Heart 90:107–111
148. Hurst WJ, Glinski JA, Miller KB et al (2008) Survey of the trans-resveratrol and trans-piceid content of cocoa-containing and chocolate products. J Agric Food Chem 56:8374–8378
149. Galleano M, Oteiza PI, Fraga CG (2009) Cocoa, chocolate, and cardiovascular disease. J Cardiovasc Pharmacol 54:483–490
150. (2009) Is cocoa good for the heart? Eur Heart J 30:2951–2952
151. Saleem TSM, Basha SD (2010) Red wine: a drink to your heart. J Cardiovasc Dis Res 1:171–176
152. Bioenergetics – 3rd Edition. https://www.elsevier.com/books/bioenergetics/nicholls/978-0-12-518121-1
153. Lanza IR, Zabielski P, Klaus KA et al (2012) Chronic caloric restriction preserves mitochondrial function in senescence without increasing mitochondrial biogenesis. Cell Metab 16:777–788

154. Ruetenik A, Barrientos A (2015) Dietary restriction, mitochondrial function and aging: from yeast to humans. Biochim Biophys Acta 1847:1434–1447
155. López-Lluch G, Hunt N, Jones B et al (2006) Calorie restriction induces mitochondrial biogenesis and bioenergetic efficiency. Proc Natl Acad Sci U S A 103:1768–1773
156. Cruzen C, Colman RJ (2009) Effects of caloric restriction on cardiovascular aging in non-human primates and humans. Clin Geriatr Med 25:733–743
157. Han X, Ren J (2010) Caloric restriction and heart function: is there a sensible link? Acta Pharmacol Sin 31:1111–1117
158. Bartolomé A, García-Aguilar A, Asahara S-I et al (2017) MTORC1 regulates both general autophagy and mitophagy induction after oxidative phosphorylation uncoupling. Mol Cell Biol 37:e00441-17
159. Laplante M, Sabatini DM (2012) mTOR signaling in growth control and disease. Cell 149:274–293
160. Jung CH, Ro S-H, Cao J et al (2010) mTOR regulation of autophagy. FEBS Lett 584:1287–1295
161. Hood DA (2009) Mechanisms of exercise-induced mitochondrial biogenesis in skeletal muscle. Appl Physiol Nutr Metab 34:465–472
162. Adhihetty PJ, Ljubicic V, Hood DA (2007) Effect of chronic contractile activity on SS and IMF mitochondrial apoptotic susceptibility in skeletal muscle. Am J Physiol Endocrinol Metab 292:E748–E755
163. Huang C, Andres AM, Ratliff EP et al (2011) Preconditioning involves selective mitophagy mediated by Parkin and p62/SQSTM1. PLoS One 6:e20975
164. Zhang J, Wang X, Vikash V et al (2016) ROS and ROS-mediated cellular signaling. Oxidative Med Cell Longev 2016:4350965. https://www.hindawi.com/journals/omcl/2016/4350965/
165. Chalker J, Gardiner D, Kuksal N et al (2017) Characterization of the impact of glutaredoxin-2 (GRX2) deficiency on superoxide/hydrogen peroxide release from cardiac and liver mitochondria. Redox Biol 15:216–227
166. Slade L, Chalker J, Kuksal N et al (2017) Examination of the superoxide/hydrogen peroxide forming and quenching potential of mouse liver mitochondria. Biochim Biophys Acta Gen Subj 1861:1960–1969
167. Kuksal N, Gardiner D, Qi D et al (2018) Partial loss of complex I due to NDUFS4 deficiency augments myocardial reperfusion damage by increasing mitochondrial superoxide/hydrogen peroxide production. Biochem Biophys Res Commun 498:214–220
168. Chance B, Sies H, Boveris A (1979) Hydroperoxide metabolism in mammalian organs. Physiol Rev 59:527–605
169. Zhang Q, Padayatti PS, Leung JH (2017) Proton-translocating nicotinamide nucleotide transhydrogenase: a structural perspective. Front Physiol 8
170. Nickel AG, von Hardenberg A, Hohl M et al (2015) Reversal of mitochondrial transhydrogenase causes oxidative stress in heart failure. Cell Metab 22:472–484
171. Zinovkina LA (2018) Mechanisms of mitochondrial DNA repair in mammals. Biochem Mosc 83:233–249
172. Finkel T (2012) Signal transduction by mitochondrial oxidants. J Biol Chem 287:4434–4440
173. Go Y-M, Chandler JD, Jones DP (2015) The cysteine proteome. Free Radic Biol Med 84:227–245
174. Gloire G, Piette J (2009) Redox regulation of nuclear post-translational modifications during NF-κB activation. Antioxid Redox Signal 11:2209–2222
175. Martín MA, Gómez MA, Guillén F et al (2000) Myocardial carnitine and carnitine palmitoyltransferase deficiencies in patients with severe heart failure. Biochim Biophys Acta 1502:330–336
176. Pereyra AS, Hasek LY, Harris KL et al (2017) Loss of cardiac carnitine palmitoyltransferase 2 results in rapamycin-resistant, acetylation-independent hypertrophy. J Biol Chem 292:18443–18456
177. Ellis JM, Hasek LY, Yurovich EJ et al (2016) Mouse carnitine palmitoyltransferase 2 (CPT2) is required to sustain cardiac function. FASEB J 30:684.8–684.8

178. Foster DW (2012) Malonyl-CoA: the regulator of fatty acid synthesis and oxidation. J Clin Invest 122:1958–1959
179. Sack MN, Rader TA, Park S et al (1996) Fatty acid oxidation enzyme gene expression is downregulated in the failing heart. Circulation 94:2837–2842
180. Sihag S, Cresci S, Li AY et al (2009) PGC-1alpha and ERRalpha target gene downregulation is a signature of the failing human heart. J Mol Cell Cardiol 46:201–212
181. Nisoli E, Clementi E, Carruba MO et al (2007) Defective mitochondrial biogenesis: a hallmark of the high cardiovascular risk in the metabolic syndrome? Circ Res 100:795–806
182. Halestrap AP, Clarke SJ, Javadov SA (2004) Mitochondrial permeability transition pore opening during myocardial reperfusion–a target for cardioprotection. Cardiovasc Res 61:372–385
183. Bernardi P, von Stockum S (2012) The permeability transition pore as a Ca(2+) release channel: new answers to an old question. Cell Calcium 52:22–27
184. Sharov VG, Todor A, Khanal S et al (2007) Cyclosporine A attenuates mitochondrial permeability transition and improves mitochondrial respiratory function in cardiomyocytes isolated from dogs with heart failure. J Mol Cell Cardiol 42:150–158
185. Sharov VG, Todor AV, Imai M et al (2005) Inhibition of mitochondrial permeability transition pores by cyclosporine A improves cytochrome C oxidase function and increases rate of ATP synthesis in failing cardiomyocytes. Heart Fail Rev 10:305–310
186. Ghaffari S, Kazemi B, Toluey M et al (2013) The effect of prethrombolytic cyclosporine-A injection on clinical outcome of acute anterior ST-elevation myocardial infarction. Cardiovasc Ther 31:e34–e39
187. Mewton N, Croisille P, Gahide G et al (2010) Effect of cyclosporine on left ventricular remodeling after reperfused myocardial infarction. J Am Coll Cardiol 55:1200–1205
188. Piot C, Croisille P, Staat P et al (2008) Effect of cyclosporine on reperfusion injury in acute myocardial infarction. N Engl J Med 359:473–481
189. MITOCARE Study Group (2012) Rationale and design of the "MITOCARE" study: a phase II, multicenter, randomized, double-blind, placebo-controlled study to assess the safety and efficacy of TRO40303 for the reduction of reperfusion injury in patients undergoing percutaneous coronary intervention for acute myocardial infarction. Cardiology 123:201–207
190. Naesens M, Kuypers DRJ, Sarwal M (2009) Calcineurin inhibitor nephrotoxicity. Clin J Am Soc Nephrol 4:481–508
191. Tábara LC, Poveda J, Martin-Cleary C et al (2014) Mitochondria-targeted therapies for acute kidney injury. Expert Rev Mol Med 16:e13
192. Guidelines for assignment to e-books I International ISBN Agency. https://www.isbn-international.org/content/guidelines-assignment-e-books
193. Chouchani ET, Pell VR, Gaude E et al (2014) Ischaemic accumulation of succinate controls reperfusion injury through mitochondrial ROS. Nature 515:431–435
194. Molyneux SL, Florkowski CM, George PM et al (2008) Coenzyme Q10: an independent predictor of mortality in chronic heart failure. J Am Coll Cardiol 52:1435–1441
195. Okonko DO, Shah AM (2015) Heart failure: mitochondrial dysfunction and oxidative stress in CHF. Nat Rev Cardiol 12:6–8
196. Rosenfeldt F, Hilton D, Pepe S et al (2003) Systematic review of effect of coenzyme Q10 in physical exercise, hypertension and heart failure. Biofactors 18:91–100
197. Maranzana E, Barbero G, Falasca AI et al (2013) Mitochondrial respiratory supercomplex association limits production of reactive oxygen species from complex I. Antioxid Redox Signal 19:1469–1480
198. Chatfield KC, Sparagna GC, Sucharov CC et al (2014) Dysregulation of cardiolipin biosynthesis in pediatric heart failure. J Mol Cell Cardiol 74:251–259
199. Saini-Chohan HK, Holmes MG, Chicco AJ et al (2009) Cardiolipin biosynthesis and remodeling enzymes are altered during development of heart failure. J Lipid Res 50:1600–1608
200. Sparagna GC, Lesnefsky EJ (2009) Cardiolipin remodeling in the heart. J Cardiovasc Pharmacol 53:290–301

201. Frasier CR, Moukdar F, Patel HD et al (2013) Redox-dependent increases in glutathione reductase and exercise preconditioning: role of NADPH oxidase and mitochondria. Cardiovasc Res 98:47–55
202. Kloner RA, Hale SL, Dai W et al (2012) Reduction of ischemia/reperfusion injury with bendavia, a mitochondria-targeting cytoprotective peptide. J Am Heart Assoc 1:e001644
203. Sloan RC, Moukdar F, Frasier CR et al (2012) Mitochondrial permeability transition in the diabetic heart: contributions of thiol redox state and mitochondrial calcium to augmented reperfusion injury. J Mol Cell Cardiol 52:1009–1018
204. Eirin A, Ebrahimi B, Zhang X et al (2014) Mitochondrial protection restores renal function in swine atherosclerotic renovascular disease. Cardiovasc Res 103:461–472
205. Siegel MP, Kruse SE, Percival JM et al (2013) Mitochondrial-targeted peptide rapidly improves mitochondrial energetics and skeletal muscle performance in aged mice. Aging Cell 12:763–771
206. Szeto HH, Liu S, Soong Y et al (2011) Mitochondria-targeted peptide accelerates ATP recovery and reduces ischemic kidney injury. J Am Soc Nephrol 22:1041–1052
207. Agadjanyan M, Vasilevko V, Ghochikyan A et al (2003) Nutritional supplement (NT Factor™) restores mitochondrial function and reduces moderately severe fatigue in aged subjects. J Chronic Fatigue Syndr 11:23–36
208. Nicolson GL (2010) Lipid replacement therapy: a nutraceutical approach for reducing cancer-associated fatigue and the adverse effects of cancer therapy while restoring mitochondrial function. Cancer Metastasis Rev 29:543–552
209. Ellithorpe RR, Settineri RA, Nicolson GL et al (2003) Pilot study: reduction of fatigue by use of a dietary supplement containing glycophospholipids. J Am Nutraceut Assoc 6(1):23-8
210. Cantó C, Menzies KJ, Auwerx J (2015) NAD(+) metabolism and the control of energy homeostasis: a balancing act between mitochondria and the nucleus. Cell Metab 22:31–53
211. Hsu C-P, Yamamoto T, Oka S et al (2014) The function of nicotinamide phosphoribosyltransferase in the heart. DNA Repair (Amst) 23:64–68
212. Khan NA, Auranen M, Paetau I et al (2014) Effective treatment of mitochondrial myopathy by nicotinamide riboside, a vitamin B3. EMBO Mol Med 6:721–731
213. Martens CR, Denman BA, Mazzo MR et al (2018) Chronic nicotinamide riboside supplementation is well-tolerated and elevates NAD + in healthy middle-aged and older adults. Nat Commun 9:1286
214. Diguet N, Trammell SAJ, Tannous C et al (2018) Nicotinamide riboside preserves cardiac function in a mouse model of dilated cardiomyopathy. Circulation 137:2256–2273
215. Emani SM, Piekarski BL, Harrild D et al (2017) Autologous mitochondrial transplantation for dysfunction after ischemia-reperfusion injury. J Thorac Cardiovasc Surg 154:286–289
216. Kaza AK, Wamala I, Friehs I et al (2017) Myocardial rescue with autologous mitochondrial transplantation in a porcine model of ischemia/reperfusion. J Thorac Cardiovasc Surg 153:934–943
217. Masuzawa A, Black KM, Pacak CA et al (2013) Transplantation of autologously derived mitochondria protects the heart from ischemia-reperfusion injury. Am J Physiol Heart Circ Physiol 304:H966–H982
218. Bravo-San Pedro JM, Kroemer G, Galluzzi L (2017) Autophagy and Mitophagy in cardiovascular disease. Circ Res 120(11):1812–1824
219. Rana A, Rera M, Walker DW (2013) Parkin overexpression during aging reduces proteotoxicity, alters mitochondrial dynamics, and extends lifespan. Proc Natl Acad Sci 110(21):8638–8643
220. Ferrero ME, Bertelli AE, Fulgenzi A, Pellegatta F, Corsi MM, Bonfrate M, Ferrara F, De Caterina R, Giovannini L, Bertelli A (1998) Activity in vitro of resveratrol on granulocyte and monocyte adhesion to endothelium. Am J Clin Nutr 68(6):1208–1214
221. Tang BL (2016) Sirt1 and the mitochondria. Mol Cells 39(2):87–95
222. Bonomini F, Rodella LF, Rezzani R (2015) Metabolic syndrome, aging and involvement of oxidative stress. Aging Dis 6(2):109

223. Tribble DL (1999) Antioxidant consumption and risk of coronary heart disease: emphasis on vitamin C, vitamin E, and β-carotene. Circulation 99(4):591–595
224. Bjelakovic G, Nikolova D, Gluud LL, Simonetti RG, Gluud C (2015) Antioxidant supplements for prevention of mortality in healthy participants and patients with various diseases. Sao Paulo Med J 133(2):164–165
225. Kaimoto S, Hoshino A, Ariyoshi M, Okawa Y, Tateishi S, Ono K, Uchihashi M, Fukai K, Iwai-Kanai E, Matoba S (2017) Activation of PPAR-α in the early stage of heart failure maintained myocardial function and energetics in pressure-overload heart failure. Am J Phys Heart Circ Phys 312(2):H305–H313
226. Pérez MJ, Quintanilla RA (2017) Development or disease: duality of the mitochondrial permeability transition pore. Dev Biol 426(1):1–7
227. Cooper JM, Schapira AHV (2007) Friedreich's ataxia: coenzyme Q10 and vitamin E therapy. Mitochondrion 7:S127–S135
228. Koh S-H, Choi H, Park H-H, Lee K-Y, Lee YJ, Kim SH (2010) Neuroprotective effects of coenzyme Q10 against beta-amyloid–induced neural stem cell death. Alzheimers Dement 6(4):S209
229. Dumont M, Kipiani K, Yu F, Wille E, Katz M, Calingasan NY, Gouras GK, Lin MT, Beal MF (2011) Coenzyme Q10 decreases amyloid pathology and improves behavior in a transgenic mouse model of Alzheimer's disease. J Alzheimers Dis 27(1):211–223
230. Lanza IR, Short DK, Short KR, Raghavakaimal S, Basu R, Joyner MJ, McConnell JP, Nair KS (2008) Endurance exercise as a countermeasure for aging. Diabetes 57(11):2933–2942

Pharmacogenetic Implications of Statin Therapy on Oxidative Stress in Coronary Artery Disease

26

Nivas Shyamala and Surekha Rani Hanumanth

Abstract

Coronary artery disease (CAD) remains the leading global public health burden in cardiovascular diseases. Atherosclerosis is a primary mechanism to cause CAD with the contribution of epidemiological, traditional, genetic, and epigenetic risk factors. Statins, prescribed drugs for lowering of cholesterol levels, also have pleiotropic effect on oxidative stress, inflammation, apoptosis, etc. Reactive oxygen species (ROS)-induced oxidative stress associates with risk factors and participates in initiation and progression of disease. ROS molecules generated as superoxides ($O_2^{\bullet-}$), singlet/triplet oxygen, peroxides (H_2O_2, $ONOO^-$), and hydroxyl radicals ($HO^\bullet$) via reactions catalyzed by endothelial nitric oxide synthase, myeloperoxidase, NADPH oxidase, and xanthine oxidase enzyme are encoded by eNOS, MPO, NOX, and XO genes, respectively. Polymorphisms in eNOS, MPO, NOX, and XO genes influence the expression and attributes to interindividual variation in response to statin drugs. Differential response to statin drug insights into emerging of pharmacogenetic studies to understand the genetic makeup and treat the patient with suitable drug and dose. In clinical practice, pharmacogenetic approach toward oxidative stress is a future emerging trend in personalized medicine development.

Keywords

Coronary artery disease · Oxidative stress · Reactive oxygen species · Statins · Pharmacogenetics

N. Shyamala · S. R. Hanumanth (✉)
Department of Genetics & Biotechnology, Osmania University, Hyderabad, Telangana, India

S. Chakraborti et al. (eds.), *Modulation of Oxidative Stress in Heart Disease*,
https://doi.org/10.1007/978-981-13-8946-7_26

26.1 Introduction

Coronary artery disease (CAD) is the foremost leading cause of cardiovascular diseases (CVD), and it is estimated that the CVD annual deaths may rise from 17.5 million to 22.2 million from 2012 to 2030 [1]. In India, CAD is the second rising burden among the noncommunicable diseases, and the occurrence of ischemic heart disease is increased to 8.7% from 3.7% since 1990 to 2016 [2].

Coronary atherosclerosis is the chief underlying mechanism of the coronary artery disease. Atherosclerosis is preceded by fatty streak formation, accumulation of lipids and lymphocytes, inflammation, and thrombosis. Atherosclerotic plaque narrows the lumen of coronary artery and diminishes blood flow to the myocardium [3, 4]. CAD is a multifactorial disease influenced by epidemiological, traditional, and novel risk factors for its initiation and development [5–7]. Recent studies also implicate the importance of genetic and epigenetic factors in the pathophysiology of coronary artery disease. Evidences suggest oxidative stress (OS) is a key contributor to the initiation and exacerbation of atherosclerosis [8, 9] (Fig. 26.1).

Reactive oxygen species (ROS) are generated endogenously by mitochondria, peroxisomes, endoplasmic reticulum, and phagocytes and exogenously by cigarette smoking, ultraviolet rays, radiation, pesticides, alcohol, and metals as superoxides ($O_2^{\bullet-}$), singlet/triplet oxygen, peroxides (H_2O_2, $ONOO^-$), hydroxyl radicals ($HO^\bullet$), etc. [10]. Increased levels of ROS have various effects including endothelial dysfunction by loss of nitric oxide (NO) activity, increased lipid peroxidation by regulation of oxidized low-density lipoprotein (oxLDL) production, inflammation by NF-kβ activation, and thrombosis by vascular smooth muscle cell apoptosis [8].

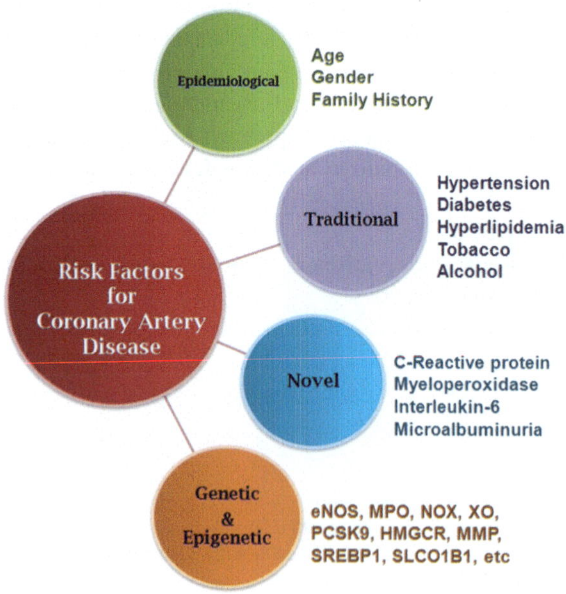

Fig. 26.1 Risk factors for coronary artery disease

Regulation of ROS production is a potential mechanism to control CAD initiation and progression.

Statins (hydroxymethylglutaryl coenzyme A reductase inhibitors) are common drugs used for the treatment of coronary artery disease [11]. These drugs significantly reduce the cholesterol levels by competitively inhibiting hydroxymethylglutaryl coenzyme A reductase (HMGCR) enzyme in hepatic cholesterol biosynthetic (mevalonate) pathway [12]. In clinical practice, statins show primarily cholesterol-dependent and additionally cholesterol-independent (pleiotropic) beneficial effects in CAD patients [13]. Cholesterol-independent beneficial effects include antioxidant, anti-inflammatory, anti-angiogenic, and anti-apoptotic activities [14, 15].

However, pharmacogenetic studies revealed that there is a variability in clinical response to statin treatment in CAD patients depending upon their genetic variations and expression of genes involved in absorption, transportation, and metabolic pathways. Genetic variations in CYP, ABC, Apo, IL family genes, HMGCR, PCSK9, LDLR, SLCO1B1, ACE, CETP, SREBP1, MMP, eNOS, NOX, XO, MPO, etc. genes are significantly affecting pharmacokinetics and dynamic properties of statins [14, 15]. Pharmacogenetic investigation insights into response to statin drug and doses and novel treatment strategies in CAD patients based on the genetic makeup of an individual. The present chapter is focused to discuss the impact of oxidative stress-associated candidate gene polymorphisms and their relative expression on efficacy of statin drugs in the treatment of coronary artery disease.

26.2 Oxidative Stress in Atherosclerosis

Oxidative stress is a form of imbalance between oxidants (ROS) and antioxidants of cells. Oxygen (O_2) is a major molecule for all the metabolic processes and generates as free radical by reduction. Enzymatic and non-enzymatic reactions, auto-oxidation, electron transport chain, etc. are the major sources for superoxide generation by transferring an electron to molecular oxygen [16].

Enzymatic and non-enzymatic reaction

$$O_2 + e^- \rightarrow O_2^{\bullet -}\,(\text{superoxide})$$

Auto-oxidation

$$O_2 + Fe^{2+} \rightarrow Fe^{3+} + O_2^{\bullet -}\,(\text{superoxide})$$

Accumulating evidence suggests that various metabolic pathways including enzymes like endothelial nitric oxide synthase (eNOS), myeloperoxidase (MPO),

NOX family enzymes (NOXs), xanthine oxidase (XO), etc are involved in the ROS production and imbalance between oxidants and antioxidants resulting in oxidative stress [10, 17–20].

Increased ROS has a vital role in initiation and progression of lesions at coronary arteries, for example, superoxide radical reacts with $NO^{\bullet}$ forming peroxynitrite ($ONOO^{-}$) which consequently reduces the bioavailability of nitric oxide (NO). In addition to superoxides, $NO^{\bullet}$ reacts with hydroxyl ($HO^{\bullet}$) and lipid radicals ($LO^{\bullet}$ and $LOO^{\bullet}$) forming OLNO and LOONO, respectively [10]. Peroxynitrite inactivates metal-centric eNOS enzymes, mitochondrial enzymes, and creatinine kinase and activates MMPs, NF-kβ, PARP, etc. by cysteine oxidation attributing to the pathology of CAD [21].

Initially, ROS modifies phospholipids by lipid peroxidation and results in the formation of oxidized LDL (oxLDL). Further OxLDL activates immune cells such as T cells, dendritic cells, monocytes, and macrophages and evokes the synthesis of inflammatory cytokines like IL-1, 6, TNFα, etc. These OxLDL molecules are taken up by macrophage receptors CD36, scavenger receptor class A, and lectin-like oxLDL receptor-1 and develop into foam cells and further trigger the formation of thrombus in the arterial layers as plaque [22, 23]. The plaque fibrous cap made up of VSMCs, collagen, proteoglycans, and elastin. Apoptosis of VSMCs and macrophages ruptures the fibrous cap and releases thrombosis into the blood stream and obstructs the blood flow to the myocardium [3, 8, 24].

26.3 Statins (Hydroxymethylglutaryl Coenzyme A (HMGCoA) Reductase Inhibitors)

Statin drugs are commercially approved in 1987 by the Food and Drug Administration, USA; these drugs act as HMGCoA analogues to inhibit the HMGCoA reductase enzyme at mevalonate pathway and regulate the cholesterol biosynthesis in hepatocytes. As per the 2013 ACC/AHA Guidelines, statin therapy is the most predominant treatment to patients with increased CAD risk [25]. Lovastatin is the first commercialized statin in the market. Based on the synthesis, statins are synthetic and semisynthetic statins. Synthetic statins include fluvastatin, atorvastatin, rosuvastatin, and pitavastatin, whereas semisynthetic statins include mevastatin, lovastatin, simvastatin, and pravastatin (Fig. 26.2) [26]. Among these, atorvastatin and rosuvastatin are worldwide chief drugs to treat CAD patients to reduce cholesterol levels.

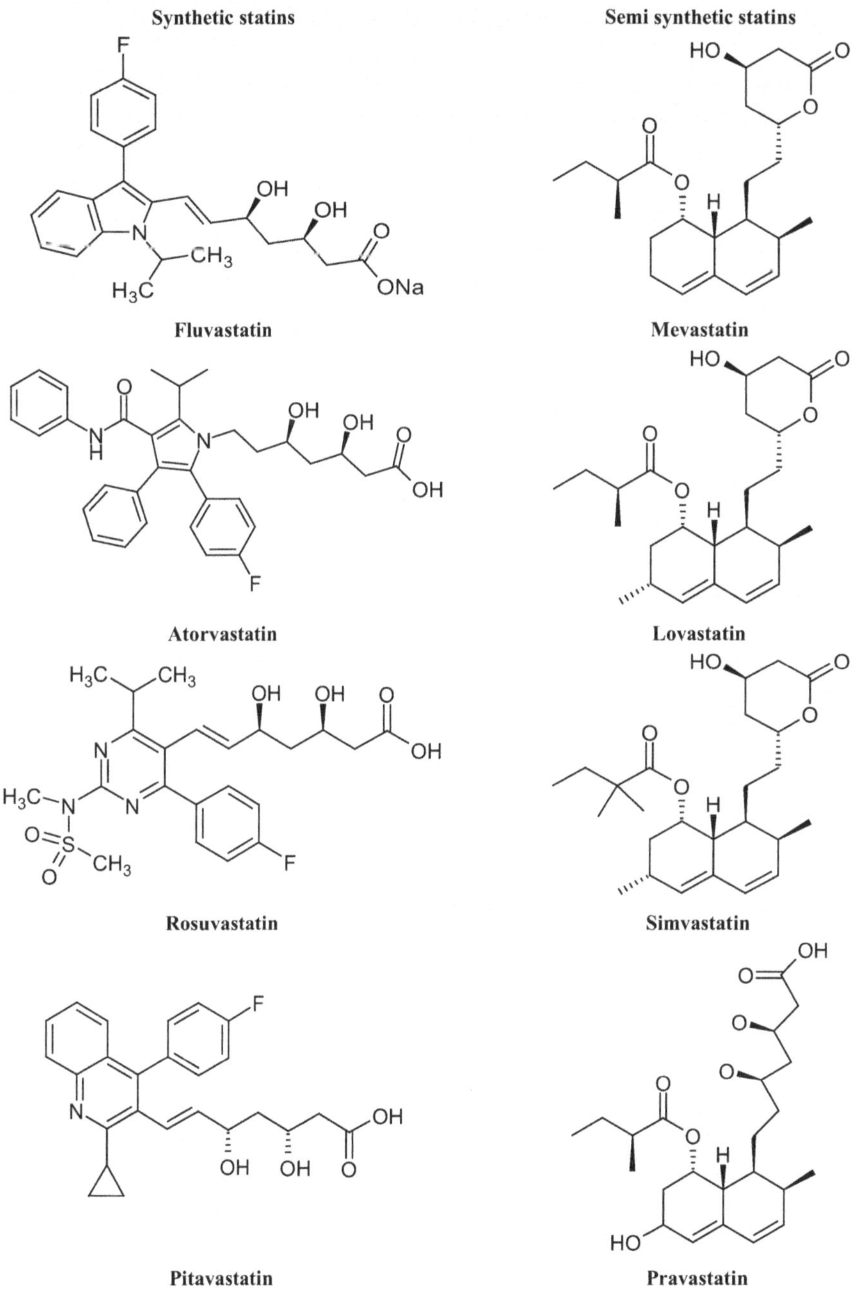

Fig. 26.2 Chemical structures of synthetic and semisynthetic statins

26.3.1 Cholesterol Biosynthesis and Its Inhibition by Statins

Cholesterol biosynthesis by mevalonate pathway includes mevalonate, isopentyl phosphate, squalene, and lanosterol synthetic reactions. Mevalonate pathway converts acetyl coenzyme A to sterol (squalene) and non-sterol (farnesylated pyrophosphate and geranylgeranyl pyrophosphate) isoprenoids. Sterol isoprenoids participate in cholesterol synthesis while non-sterol in Rho, Ras, Rab, and nuclear laminin synthesis [27]. HMGCoA to mevalonate reduction is a rate-limiting step, catalyzed by HMGCR enzyme (Fig. 26.3). Statins are class of drugs designed to bind active site of HMGCoA reductase (HMGCR) and inhibit the enzyme activity in cholesterol biosynthetic pathway. Three decades of research and clinical studies established that statins have also antioxidant, anti-inflammatory, anti-angiogenic, and anti-apoptotic activities as pleiotropic effects [28].

26.4 Statins and Oxidative Stress

Statins apart from lowering the LDL also have other pleiotropic effects like regulation of genes involved in ROS production and their expression by inhibiting various pathways [28–30]. Endothelial nitric oxide synthase (eNOS), myeloperoxidase (MPO), nicotinamide adenine dinucleotide phosphate (NADPH) oxidase (NOX), and xanthine oxidase (XO) genes are associated with reactive oxygen intermediate production. Studies show that genetic variations in these genes and their expression attribute to the interindividual differences in the efficacy of statins [31, 32]. The pharmacogenetic implications of statins on regulation of genes involved in oxidative stress are summarized as below:

26.4.1 Endothelial Nitric Oxide Synthase (eNOS) Gene

Endothelial nitric oxide synthase (NOS3/eNOS) gene located on chromosome 7q36.1 with 28 exons encodes endothelial nitric oxide synthase enzyme. eNOS enzyme couples with cofactors tetrahydrobiopterin (BH4) and oxygen to produce nitric oxide (NO) by oxidizing L-arginine to L-citrulline (Fig. 26.4). Coupled eNOS inhibits endothelial leukocyte adhesion, platelet aggregation, and VSMC migration and proliferation to prevent atherogenesis [33, 34]. Previous reports suggested that uncoupled eNOS generates superoxides ($O_2^{\bullet-}$) which react with NO and form peroxynitrite ($ONOO^-$) and inactivates NO [35, 36]. Endothelial dysfunction is also due to downregulation of eNOS expression in endothelial cells [37].

Studies evidenced that the statins attribute to upregulate the expression of endothelium nitric oxide synthase gene by extending half-life of mRNA [38], inhibiting mevalonate pathway and Rho kinase activity [39–41]. In addition, statins activate phosphatidylinositol 3-kinase signal (PI3K)-Akt pathway to enhance the bioavailability of nitric oxide [28].

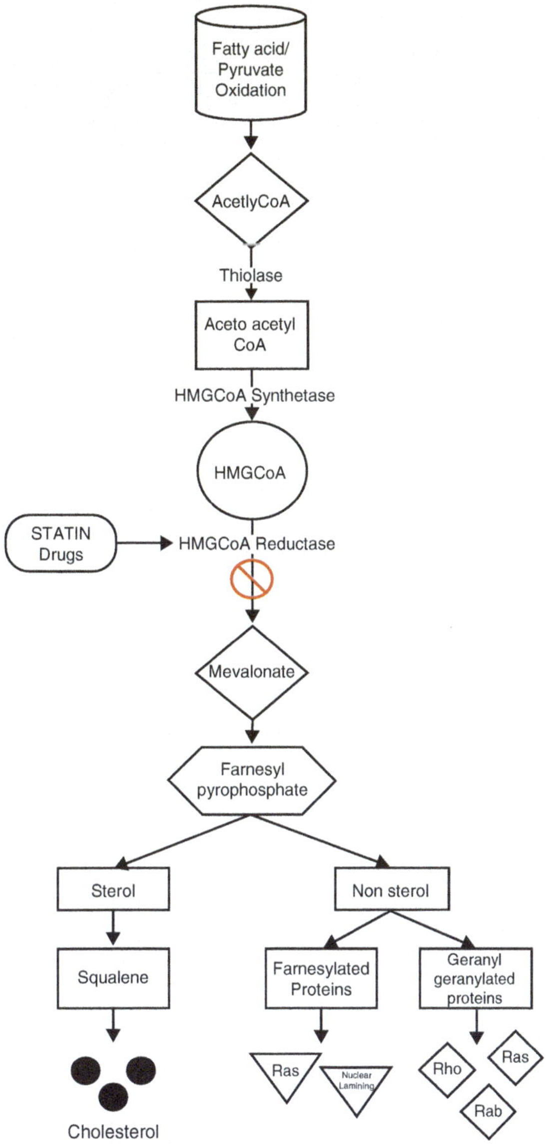

Fig. 26.3 Statins inhibiting the sterol (cholesterol) and non-sterol (Rho, Ras, Rab) synthesis in mevalonate pathway

In our earlier study, we have reported significantly higher levels of nitric oxide and malondialdehyde (MDA) levels in CAD patients [37, 42]. Further when CAD patients were treated with ATV 40 mg/day for 6 months, there was a significant reduction in NOx and MDA levels in both men and women (unpublished data). Another study by Kureishi et al. suggested that simvastatin and pravastatin increase Akt serine 473 phosphorylation in endothelial cells to produce NO, which leads to

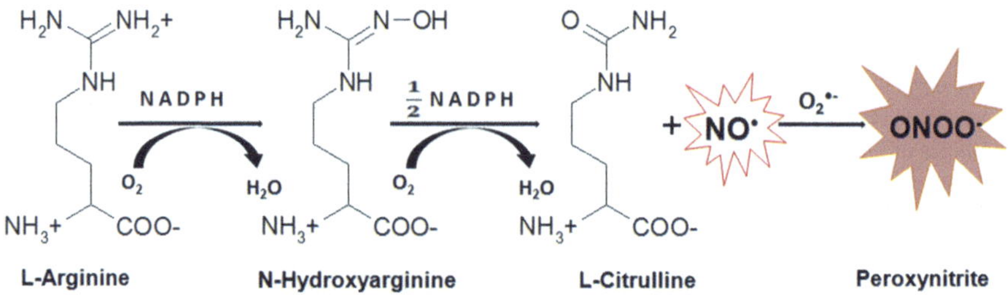

Fig. 26.4 Generation of NO• radical and peroxynitrite

the improvement of endothelium function [43]. Besides cholesterol biosynthesis inhibition, statins also inhibit GTP binding proteins Rho/Rho kinase, Ras, and Rac synthesis in mevalonate pathway. Inhibition of these proteins decrease VSMC contraction and oxidative stress and increases NO bioavailability, which are favorable factors for the efficacy of statins in treatment [44].

Pharmacogenetic studies suggested that fluvastatin and atorvastatin are significantly increasing eNOS gene expression in endothelial cells by regulating transcriptional activity and mRNA stability. It has been reported that RPA1 binds to the promoter of eNOS to repress the expression and this activity of RPA1 is regulated by statin drugs [45]. Studies reporting functional implications of eNOS gene promoter -786T>C polymorphism have been found that the individual with CC genotype has lower NO levels compared to TT genotype [29, 45].

Abe et al. treated human umbilical vein endothelial cells (HUVECs) with fluvastatin and observed that the cells with eNOS -786CC genotype have improved eNOS mRNA levels [31]. Nagassaki et al. treated eNOS -786TT and -786CC genotype subjects with 10 mg/day atorvastatin and placebo for 14 days. Interestingly they found that individuals with CC genotype have significantly reduced nitrite levels compared to TT genotype in subjects treated with ATV. Consequently nitrite level reduction in subjects with CC genotype implies the importance of genotype in modulating the response to drug [32]. These in vitro and clinical studies reported fluvastatin and atorvastatin to be associated with reduction of elevated levels of plasma nitrite concentrations in CC genotype individuals. These results indicate statins have capacity to restore diminished nitric oxide production in those carrying CC genotype of -786T>C polymorphism and are good responders for statin drug treatment [31, 32].

26.4.2 Myeloperoxidase (MPO) Gene

Myeloperoxidase (MPO) gene localized at 17q22 with 12 exons translates as myeloperoxidase enzyme. It is synthesized as translational product with 80 kDa, subsequently converts into Apopro MPO (90 kDa) and proMPO (90 kDa), and undergoes proteolytic processing to produce homodimeric matured MPO (74 kDa) [46]. MPO enzyme is present in neutrophils, monocytes, macrophages, etc. and a key contributor for inflammation in cardiovascular diseases. MPO catalyzes various reactions in

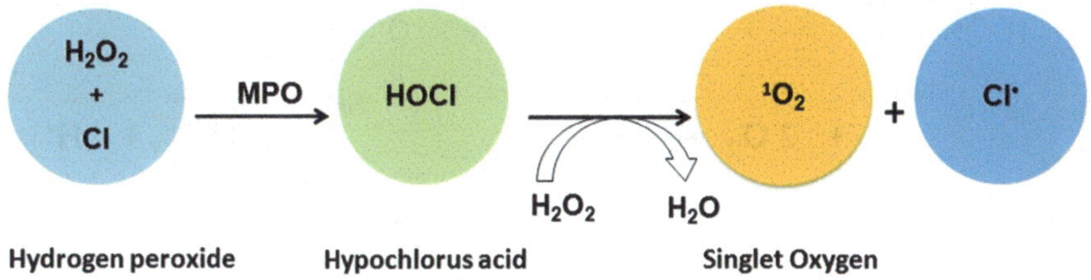

Fig. 26.5 Generation of ROS by myeloperoxidase

biological system and generates reactive oxygen species, cytotoxic hypochlorous acid, tyrosyl radical (Fig. 26.5) [6, 47, 48].

Studies on MPO gene polymorphisms have shown association with the risk of coronary artery disease. MPO promoter polymorphic variants potentially influence transcription factors binding and MPO levels. Yan Wang et al., in their meta-analysis study, have observed that the MPO -463G/A and -129G/A polymorphisms regulate the gene expression and A allele of -463G/A and A allele of -129G/A polymorphisms are associated with the lower levels of MPO [49].

Evidences suggest that the different concentrations of lovastatin, simvastatin, atorvastatin, and pravastatin are significantly downregulating the expression of MPO mRNA. Kumar et al. reported that 50 μM of lovastatin and simvastatin are showing greatest effect with 194 ± 8-fold and 45 ± 5-fold reduction in MPO mRNA expression, respectively, in peripheral blood monocytes [47]. Ndrepepa et al. reported that the statins are significantly ($p < 0.005$) reducing the MPO levels by regulating expression of MPO gene in acute coronary syndrome patients [50]. Sygitowicz et al. treated acute myocardial infarction (MI) patients with ATV 40 mg/40 days and found significantly decreased MPO gene expression in 60.5% of MI patients. The differences in the efficacy of ATV might be due to the promoter polymorphism of MPO gene [51].

26.4.3 NADH/NADPH Oxidase (NOX) Gene

Nicotinamide adenine dinucleotide phosphate (NADPH) oxidase (NOX 1) gene, located at Xq22.1 with 14 exons, encodes NADPH family of enzymes. NOX enzyme is involved in the production of reactive oxygen species, i.e., superoxide, in the vascular system (Fig. 26.6).

NOX isoforms and component subunits are shown in Table 26.1. Among NOX isoforms, NOX1, 2, 4, and 5 isoforms catalyze to release superoxide/hydrogen peroxide influencing proliferation, differentiation, endothelial impairment, and vascular structure in coronary atherosclerosis [52, 53].

NOX enzyme has complex, membrane-bound subunits gp91phox and p22phox; cytosolic subunits p40phox, p47phox, and p67phox; and small GTP binding protein Rac to form complexes and transfer electrons in biological system as represented in

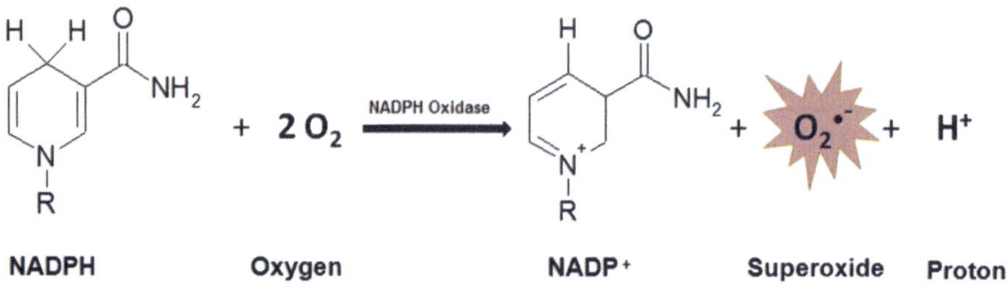

Fig. 26.6 Superoxide generation by NADPH oxidase

Table 26.1 Isoforms of NADPH oxidase enzyme

NOX isoforms	Component subunits
NOX1	Rac1, NOXA1, p22phox, NOXO1, p47phox
NOX2	Rac1 and 2, p40phox, p67phox, p22phox, p47phox
NOX3	NOXA1, p22phox, NOXO1
NOX4	p22phox, POLDIP2
NOX5	4 EF hands

NOX nicotinamide adenine dinucleotide phosphate oxidase

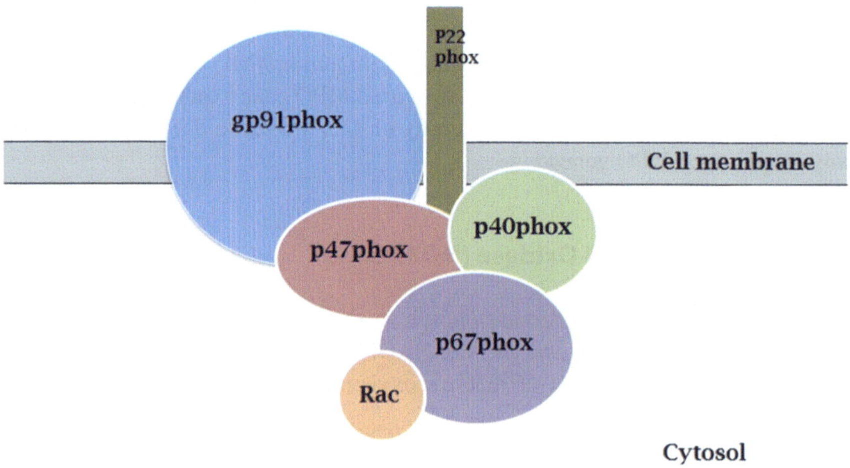

Fig. 26.7 Structure of NADPH oxidase

Fig. 26.7 [54]. NOX1, 2, and 5 are expressed in endothelial cells, VSMCs, and NOX4 in vascular cell walls [20, 55].

Guzik et al. measured the NOX-produced superoxide in blood vessels, which reacts with nitric oxide and forms peroxynitrite, and found a proportionately deficit NO bioavailability leading to endothelial impairment in atherosclerosis [56]. Zhang

Table 26.2 Genes encoding NOX enzyme subunits

NOX subunit	Encoding genes
p22phox	Cytochrome b-245 alpha chain (CYBA)
p40phox	Neutrophil cytosolic factor (NCF) 4
p47phox	Neutrophil cytosolic factor (NCF) 1
p67phox	Neutrophil cytosolic factor (NCF) 2
gp91PHOX	Cytochrome b-245 beta chain (CYBB)
Rac	Rac family small GTPase

et al. evidenced that the mRNA expression of NOX subunits was significantly higher in endothelial progenitor cells in CAD. Out of all subunits, p47phox and p22phox regulate the activity of NADPH for production of superoxide radicals and hydrogen peroxide. Activation of p47phox occurs when it is translocated from cytosol to plasma membrane of endothelial cells, and it was observed that the activation rate is enhanced in CAD patients ($p < 0.05$) [53, 57]. The genes encoding NOX enzyme subunits are shown in Table 26.2.

Genetic variations in genes encoding NOX subunits influence the activity of enzyme and generation of reactive oxygen species. One of the chief components of NOX is p22phox, encoded by CYBA/p22phox gene located at 16q24.2 with seven exons. Cahilly et al. suggested that the T-allele of C242T polymorphism in p22phox gene is significantly associated with 3- to 5-fold loss in minimum lumen diameter and disease progression [58]. Ito et al. observed a high frequency of T allele of C242T polymorphism in CVD patients than the controls in Japanese population [59].

Meta-analysis conducted by Xu et al. included functional studies which suggested the association of p22phox 640G allele with mRNA stability and processing in CAD patients and also found significant decrease in ROS formation. Further it has been suggested that the individuals with 640G allele might show protection against CAD [60, 61]. Antioxidant capacity of statins includes the regulation of ROS production in cells participating in coronary atherosclerotic process. A number of studies evidenced that the statins are reducing the ROS production by inhibiting the NOX enzyme and Rac. Hamilton et al. evidenced 10/20 mg/day atorvastatin (ATV) reduces the Rac GTPases on membranes of platelet in hyperlipidemia patients, which may reduce the activity of NOX [62].

Antoniades et al. treated preoperative coronary artery bypass-grafted patients with 40 mg/day atorvastatin for 3 days to find the redox rate in vein graft and found significant reduction in basal and vascular NOX stimulating $O_2^{\bullet-}$ and Rac1 activation in vein grafts. ATV treatment has no impact on NOX1/2/4 protein levels but significantly reduced Rac1 and p67phox of NOX [63]. Studies have indicated that atorvastatin and simvastatin were involved in downregulating the expression of Rac1 gene [30]. Furthermore, evidences by Inoue et al. have shown that HUVECs treated with different concentrations of fluvastatin, simvastatin, pravastatin, and cerivastatin showed a significantly downregulated expression of p22phox mRNA and decreased p47phox protein levels in response to fluvastatin and simvastatin [64].

26.4.4 Xanthine Oxidase (XO) Gene

Xanthine oxidase (XO)/xanthine dehydrogenase (XDH) gene located at 2p23.1 with 37 exons, encodes xanthine oxidase enzyme. It exists as a homodimer with approximately 290 kDa molecular mass [65]. Xanthine oxidase catalyzes the oxidation of hypoxanthine to xanthine, followed by xanthine to uric acid in purine metabolism (Fig. 26.8). In the process of oxidation, XO reduces molecular oxygen (O_2) to superoxide radical ($O_2^{\bullet-}$) and peroxides (H_2O_2). Chung et al. reported that XO is highly expressed in endothelial, epithelial, and polymorphonuclear cells [66]. Previous studies evidenced that superoxides and peroxides were involved in a variety of clinicopathological conditions including endothelial dysfunction, elevated uric acid levels, and chemoattractant for neutrophils in coronary artery disease [66, 67]. Landmesser et al. evidenced an enhanced expression of XO protein and subsequent XO-dependent endothelial superoxide production in response to the stimulus of angiotensin II hormone in bovine aortic endothelial cells [68].

Kudo et al. functionally characterized various polymorphisms in XO gene and observed the loss of enzyme activity for subjects with 445C>T (Arg149Cys) and 2729C>A (Thr910Lys) variations and decreased enzyme activity for 1663C>T (Pro555Ser), 1820G>A (Arg607Gln), 1868C>T (Thr623Ile), 2727C>A (Asn909Lys), 3449C>G (Pro1150Arg), and 3953G>A (Cys1318Tyr) [65].

Recent study on rs2073316 (g.31583C>T), rs1054889 (g.85304C>T) and rs1042039 (g.84306A>G) polymorphisms of XDH gene revealed an association with hypertension. Frequency of C allele for rs1042039 is higher, while C allele of rs1054889 and A allele of rs2073316 are significantly lower in hypertensives compared to controls. These polymorphisms may regulate the expression of XDH gene and might be associated with hypertension in Chinese population [69]. CAD patients had higher levels of XO protein and activity [68]; several studies evidenced that XO inhibition improved the endothelial function and decreased the free radical and uric acid production levels [70].

Greig et al. reported that 4 weeks of atorvastatin 20 mg/day treatment independently decreased the levels of MDA, uric acid and flow-dependent endothelial-mediated vasodilation in heart failure patients. Possibly statins might have decreased the expression of endothelial XO by inhibiting Rac1 or NOX and transcription of

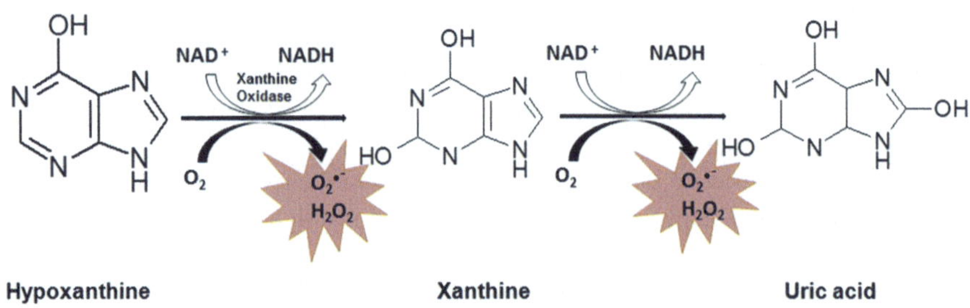

Fig. 26.8 Superoxide generation by xanthine oxidase

XO gene [70]. In addition, simvastatin prevented 50% superoxide anion production by angiotensin II-dependent ROS production in rats, which plays a pivotal role in XO activity and endothelial dysfunction [71]. The above reports suggest that increased expression of XO and angiotensin II genes might be key factors for the stimulation of enhanced ROS production to initiate the atherosclerotic plaque and inhibition of these genes may be additional therapeutic targets of statins.

26.5 Conclusion and Future Directions

Coronary artery disease is a devastating disease, and oxidative stress plays a crucial role in initiation and progression of disease. Statins, the prescribed drugs for lowering of cholesterol levels, have also other pleiotropic effects on oxidative stress, inflammation, apoptosis, etc. The generation of oxidative stress is influenced by the genetic variations in eNOS, MPO, XO, NOX, etc. Differential response to statin drug insights into emerging of pharmacogenetic studies to understand the genetic makeup and treat the patient with suitable drug and dose. In clinical practice, pharmacogenetic approach toward oxidative stress is a future emerging trend in personalized medicine development.

Acknowledgments This work was supported by UGC, New Delhi, India: MRP-2013 (F.No.42-52/2013[SR]), BSR-Fellowship, CPEPA, and OU-DST-PURSE-II Programme (C-DST-PURSE-II/23/2017), ICMR-SRF (ID No. 2019-4183).

References

1. World Health Organization (2016) Technical package for cardiovascular disease management in primary health care. Hearts 76
2. MoHFW (2017) India: health of the nation's states. Government report, pp 70–75
3. Hansson GK (2005) Inflammation, atherosclerosis, and coronary artery disease. N Engl J Med 352:1685–1695
4. Hansson GK, Libby P, Tabas I (2015) Inflammation and plaque vulnerability. J Intern Med 278:483–493
5. Mack M, Gopal A (2016) Epidemiology, traditional and novel risk factors in coronary artery disease. Heart Fail Clin 12:1–10
6. Teng N, Maghzal GJ, Talib J, Rashid I, Lau AK, Stocker R (2017) The roles of myeloperoxidase in coronary artery disease and its potential implication in plaque rupture. Redox Rep 22:51–73
7. Libby P, Rocha VZ (2018) All roads lead to IL-6: a central hub of cardiometabolic signaling. Int J Cardiol 259:213–215
8. Forrester SJ, Kikuchi DS, Hernandes MS, Xu Q, Griendling KK (2018) Reactive oxygen species in metabolic and inflammatory signaling. Circ Res 122:877–902
9. Vichova T, Motovska Z (2013) Oxidative stress: predictive marker for coronary artery disease. Exp Clin Cardiol 18:88–91
10. Cai H, Harrison DG (2000) The role of oxidant stress. Circ Res 87:840–844
11. Reiner, De Backer G, Fras Z, Kotseva K, Tokgözoglu L, Wood D, De Bacquer D (2016) Lipid lowering drug therapy in patients with coronary heart disease from 24 European countries – findings from the EUROASPIRE IV survey. Atherosclerosis 246:243–250

12. Catapano AL, Graham I, De Backer G et al (2016) 2016 ESC/EAS guidelines for the management of dyslipidaemias. Eur Heart J 37:2999–3058l
13. Oesterle A, Laufs U, Liao JK (2017) Pleiotropic effects of statins on the cardiovascular system. Circ Res 120:229–243
14. Kitzmiller JP, Binkley PF, Pandey SR, Suhy AM, Baldassarre D, Hartmann K (2013) Statin pharmacogenomics: pursuing biomarkers for predicting clinical outcomes. Discov Med 16:45–51
15. Mangravite LM, Thorn CF, Krauss RM (2006) Clinical implications of pharmacogenomics of statin treatment. Pharm J 6:360–374
16. Phaniendra A, Jestadi DB, Periyasamy L (2015) Free radicals: properties, sources, targets, and their implication in various diseases. Indian J Clin Biochem 30:11–26
17. Panth N, Paudel KR, Parajuli K (2016) Reactive oxygen species: a key hallmark of cardiovascular disease. Adv Med 2016:1–12
18. Madamanchi NR, Vendrov A, Runge MS (2005) Oxidative stress and vascular disease. Arterioscler Thromb Vasc Biol 25:29–38
19. Ushio-fukai M, Zafari AM, Fukui T, Ishizaka N, Griendling KK (1996) p22 phox is a critical component of the superoxide-generating NADH/NADPH oxidase system and regulates angiotensin II- induced hypertrophy in vascular smooth muscle cells ∗. J Biol Chem 271:23317–23321
20. Sorescu D, Weiss D, Lassègue B et al (2002) Superoxide production and expression of Nox family proteins in human atherosclerosis. Circulation 105:1429–1435
21. Pacher P, Beckman JS, Liaudet L (2007) Nitric oxide and peroxynitrite in health and disease. Physiol Rev 87:315–424
22. Chistiakov DA, Melnichenko AA, Myasoedova VA, Grechko AV, Orekhov AN (2017) Mechanisms of foam cell formation in atherosclerosis. J Mol Med 95:1153–1165
23. Yu XH, Fu YC, Zhang DW, Yin K, Tang CK (2013) Foam cells in atherosclerosis. Clin Chim Acta 424:245–252
24. Mishra TK, Mishra C, Das B (2013) An approach to the classification, diagnosis and management of vulnerable plaque. J Indian Coll Cardiol 3:57–66
25. Stone NJ, Robinson JG, Lichtenstein AH et al (2014) 2013 ACC/AHA guideline on the treatment of blood cholesterol to reduce atherosclerotic cardiovascular risk in adults: a report of the American college of cardiology/American heart association task force on practice guidelines. J Am Coll Cardiol 63:2889–2934
26. Endo A (2010) A historical perspective on the discovery of statins. Proc Jpn Acad Ser B 86:484–493
27. Singh P, Saxena R, Srinivas G, Pande G, Chattopadhyay A (2013) Cholesterol biosynthesis and homeostasis in regulation of the cell cycle. PLoS One 8(3):e58833. https://doi.org/10.1371/journal.pone.0058833
28. Liao JK, Laufs U (2005) Pleiotropic effects of statins. Annu Rev Pharmacol Toxicol 45:89–118
29. Oliveira-Paula GH, Lacchini R, Tanus-Santos JE (2017) Clinical and pharmacogenetic impact of endothelial nitric oxide synthase polymorphisms on cardiovascular diseases. Nitric Oxide Biol Chem 63:39–51
30. Laufs U, Kilter H, Konkol C, Wassmann S, Bohm M, Nickenig G (2002) Impact of HMG CoA reductase inhibition on small GTPases in the heart. Cardiovasc Res 53:911–920
31. Abe K, Nakayama M, Yoshimura M et al (2005) Increase in the transcriptional activity of the endothelial nitric oxide synthase gene with fluvastatin: a relation with the -786T>C polymorphism. Pharmacogenet Genomics 15:329–336
32. Nagassaki S, Sertório JTC, Metzger IF, Bem AF, Rocha JBT, Tanus-Santos JE (2006) eNOS gene T-786C polymorphism modulates atorvastatin-induced increase in blood nitrite. Free Radic Biol Med 41:1044–1049
33. Gimbrone MA, García-Cardeña G (2016) Endothelial cell dysfunction and the pathobiology of atherosclerosis. Circ Res 118:620–636
34. Verhaar MC (2004) Free radical production by dysfunctional eNOS. Heart 90:494–495

35. Sullivan JC, Pollock JS (2006) Coupled and uncoupled NOS: separate but equal? Uncoupled NOS in endothelial cells is a critical pathway for intracellular signaling. Circ Res 98:717–719
36. Matsubara K, Higaki T, Matsubara Y, Nawa A (2015) Nitric oxide and reactive oxygen species in the pathogenesis of preeclampsia. Int J Mol Sci 16:4600–4614
37. Kumar GR, Spurthi KM, Kumar GK et al (2016) Genetic polymorphisms of eNOS (-786T/C, Intron 4b/4a & 894G/T) and its association with asymptomatic first degree relatives of coronary heart disease patients. Nitric Oxide Biol Chem 60:40–49
38. Laufs U, Liao JK (1998) Post-transcriptional regulation of endothelial nitric oxide synthase mRNA stability by Rho GTPase. J Biol Chem 273:24266–24271
39. Laufs U, La Fata V, Plutzky J, Liao JK (1998) Upregulation of endothelial nitric oxide synthase by HMG CoA reductase inhibitors. Circulation 97:1129–1135
40. Ruan F, Zheng Q, Wang J (2012) Mechanisms of bone anabolism regulated by statins. Biosci Rep 32:511–519
41. Wolfrum S, Dendorfer A, Rikitake Y, Stalker TJ, Scalia R, Dominiak P, Liao JK (2009) Inhibition of rho-kinase leads to rapid activation of phosphatidylinositol 3-kinase/protein kinase Akt and cardiovascular protection. Arterioscler Thromb Vasc Biol 24:1842–1847. NIH Public Access
42. Gundapaneni KK, Galimudi RK, Kondapalli MS, Gantala SR, Mudigonda S, Padala C, Shyamala N, Sahu SK, Hanumanth SR (2017) Oxidative stress markers in coronary artery disease patients with diabetes mellitus. Int J Diab Dev Ctries 37:190–194
43. Kureishi Y, Luo Z, Shiojima I, Bialik A, Fulton D, David J, Sessa WC, Walsh K (2010) The HMGCoA reductase inhibitor simvastatin activates the protein kinase Akt and promotes angiogenesis in normocholesterolemic animals. Nat Med 6:1004–1010. NIH Public Access
44. Shimokawa H, Takeshita A (2005) Rho-kinase is an important therapeutic target in cardiovascular medicine. Arterioscler Thromb Vasc Biol 25:1767–1775
45. Silva PS, Lacchini R, Gomes Vde A, Tanus-Santos JE (2011) Pharmacogenetic implications of the eNOS polymorphisms for cardiovascular action drugs. Arq Bras Cardiol 96:e27–e34
46. Grishkovskaya XI, Paumann-Page M, Tscheliessnig R et al (2017) Structure of human promyeloperoxidase (proMPO) and the role of the propeptide in processing and maturation. J Biol Chem 292:8244–8261
47. Kumar AP, Reynolds WF (2005) Statins downregulate myeloperoxidase gene expression in macrophages. Biochem Biophys Res Commun 331:442–451
48. Zhang R, Brennan ML, Shen Z, MacPherson JC, Schmitt D, Molenda CE, Hazen SL (2002) Myeloperoxidase functions as a major enzymatic catalyst for initiation of lipid peroxidation at sites of inflammation. J Biol Chem 277:46116–46122
49. Wang Y, Chen X-Y, Wang K, Li S, Zhang X-Y (2017) Myeloperoxidase polymorphism and coronary artery disease risk: a meta-analysis. Medicine (Baltimore) 96:e7280
50. Ndrepepa G, Braun S, Schömig A, Kastrati A (2011) Impact of therapy with statins, beta-blockers and angiotensin-converting enzyme inhibitors on plasma myeloperoxidase in patients with coronary artery disease. Clin Res Cardiol 100:327–333
51. Sygitowicz G, Maciejak A, Piniewska-Juraszek J, Pawlak M, Gora M, Burzynska B, Dluzniewski M, Opolski G, Sitkiewicz D (2015) Interindividual variability of atorvastatin treatment influence on the MPO gene expression in patients after acute myocardial infarction. Acta Biochim Pol 63(1):89–95. https://doi.org/10.18388/abp.2015_1014
52. Touyz RM, Briones AM, Sedeek M, Burger D, Montezano AC (2011) NOX isoforms and reactive oxygen species in vascular health. Mol Interv 11:27–35
53. Kwok JMF, Ma CCH, Ma S (2013) Recent development in the effects of statins on cardiovascular disease through Rac1 and NADPH oxidase. Vasc Pharmacol 58:21–30
54. Maack C, Kartes T, Kilter H, Schäfers HJ, Nickenig G, Böhm M, Laufs U (2003) Oxygen free radical, release in human failing myocardium is associated with increased activity of Rac1-GTPase and represents a target for statin treatment. Circulation 108:1567–1574
55. Antonopoulos AS, Margaritis M, Shirodaria C, Antoniades C (2012) Translating the effects of statins: from redox regulation to suppression of vascular wall inflammation. Thromb Haemost 108:840–848

56. Guzik TJ, West NEJ, Black E, McDonald D, Ratnatunga C, Pillai R, Channon KM (2000) Vascular superoxide production by NAD(P)H oxidase: association with endothelial dysfunction and clinical risk factors. Circ Res 86:e85–e90
57. Zhang J, Wang M, Li Z, Bi X, Song J, Weng S, Fu G (2016) NADPH oxidase activation played a critical role in the oxidative stress process in stable coronary artery disease. Am J Transl Res 8:5199–5210
58. Cahilly C, Ballantyne CM, Lim DS, Gotto A, Marian AJ (2000) A variant of p22(phox), involved in generation of reactive oxygen species in the vessel wall, is associated with progression of coronary atherosclerosis. Circ Res 86:391–395
59. Ito D, Murata M, Watanabe K, Yoshida T (2000) C242T polymorphism of NADPH oxidase p22 PHOX gene and ischemic cerebrovascular disease in the Japanese population. Stroke 31(4):936–940
60. Xu Q, Yuan F, Shen X, Wen H, Li W, Cheng B, Wu J (2014) Polymorphisms of C242T and A640G in CYBA gene and the risk of coronary artery disease: a meta-analysis. PLoS One 9(1):e84251. https://doi.org/10.1371/journal.pone.0084251
61. Gardemann A, Mages P, Katz N, Tillmanns H, Haberbosch W (1999) The p22 phox A640G gene polymorphism but not the C242T gene variation is associated with coronary heart disease in younger individuals. Atherosclerosis 145:315–323
62. Hamilton PK, Hughes SMT, Plumb RD, Devine A, Leahey W, Lyons KS, Johnston D, McVeigh GE (2010) Statins have beneficial effects on platelet free radical activity and intracellular distribution of GTPases in hyperlipidaemia. Clin Sci 118:359–366
63. Antoniades C, Bakogiannis C, Tousoulis D et al (2010) Preoperative atorvastatin treatment in CABG patients rapidly improves vein graft redox state by inhibition of Rac1 and NADPH-oxidase activity. Circulation 122(11 Suppl):S66–S73. https://doi.org/10.1161/CIRCULATIONAHA.109.927376
64. Inoue I, Goto S, Mizotani K, Awata T, Mastunaga T, Kawai S, Nakajima T, Hokari S, Komoda T, Katayama S (2000) Lipophilic HMG-CoA reductase inhibitor has an anti-inflammatory effect: reduction of MRNA levels for interleukin-1beta, interleukin-6, cyclooxygenase-2, and p22phox by regulation of peroxisome proliferator-activated receptor alpha (PPARalpha) in primary e. Life Sci 67:863–876
65. Kudo M, Moteki T, Sasaki T, Konno Y, Ujiie S, Onose A, Mizugaki M, Ishikawa M, Hiratsuka M (2008) Functional characterization of human xanthine oxidase allelic variants. Pharmacogenet Genomics 18:243–251
66. Chung HY, Baek BS, Song SH, Kim MS, Huh JI, Shim KH, Kim KW, Lee KH (1997) Oxidation xanthine oxidase I [(reversible XOD) Proteolys ~ Xant ~ ine]. Age (Omaha) 20:127–140
67. Zdrenghea M, Sitar-Tăut A, Cismaru G, Zdrenghea D, Pop D (2017) Xanthine oxidase inhibitors in ischaemic heart disease. Cardiovasc J Afr 28:201–204
68. Landmesser U, Spiekermann S, Preuss C, Sorrentino S, Fischer D, Manes C, Mueller M, Drexler H (2007) Angiotensin II induces endothelial xanthine oxidase activation: role for endothelial dysfunction in patients with coronary disease. Arterioscler Thromb Vasc Biol 27:943–948
69. Wu B, Hao Y, Shi J et al (2015) Association between xanthine dehydrogenase tag single nucleotide polymorphisms and essential hypertension. Mol Med Rep 12:5685–5690
70. Greig D, Alcaino H, Castro PF et al (2011) Xanthine-oxidase inhibitors and statins in chronic heart failure: effects on vascular and functional parameters. J Heart Lung Transplant 30:408–413
71. Delbosc S, Cristol JP, Descomps B, Mimran A, Jover B (2002) Simvastatin prevents angiotensin II-induced cardiac alteration and oxidative stress. Hypertension 40:142–147

Basic Mechanisms of Ischemia/Reperfusion Injury Leading to Cellular and Tissue Damage: Therapeutic Implications

27

Emmanuel E. Douzinas and Aikaterini Apeiranthitis

Abstract

Produced free radicals exert their physiological function and thereafter become neutralized by antioxidants. In contrast, if they are produced in abundance, e.g., during ischemia/reperfusion (I/R), after they exhaust antioxidant reserves, they exert their harmful effect on cellular structures. The major significant reactive oxygen species (ROS) are the anion of superoxide ($O_2^{\bullet -}$), the hydrogen peroxide (H_2O_2), and the hydroxyl radical ($OH^{\bullet}$). The collective term reactive nitrogen species (RNS) mainly includes the radical of nitric oxide ($NO^{\bullet}$), the peroxynitrite potent oxidant ($ONOO^-$), the radical nitrogen dioxide ($NO_2^{\bullet}$), and other nitrogen oxides. During ischemia, the tissue cells begin to suffer, when the oxygen delivery (DO_2) to tissues decreases beyond the critical DO_2 (cDO_2) level, namely, when the metabolism becomes anaerobic. The cell suffering maximizes, when beta-oxidation of fatty acids is the last fuel that still feeds oxidative phosphorylation. Further drop of DO_2 leads to severe ischemia with intracellular conversion of xanthine dehydrogenase to xanthine oxidase and increased concentration of xanthine and hypoxanthine. Upon reperfusion and abundant O_2 reentry, free radical burst follows, with membrane destruction and massive cellular damage, mainly coming from the peroxidation of lipid bilayer arrangement. Suggested methods of I/R injury prophylaxis are the use of antioxidants, scavengers, and preconditioning techniques. The new approaches that seem to be promising focus on the progressive reentry of O_2 to the thirsty for O_2 ischemic tissues: initially in low concentrations of O_2 to meet the low potentials of biochemical pathways to use O_2 and thereafter in gradually increasing concentrations toward normal. Large, open, double blind, multicenter trials are still lacking.

E. E. Douzinas (✉)
3rd Department of Critical Care, Evgenidio Hospital, Athens University School of Medicine, Athens, Greece

A. Apeiranthitis
4th Department of Internal Medicine, Evangelismos Hospital, Athens, Greece

S. Chakraborti et al. (eds.), *Modulation of Oxidative Stress in Heart Disease*,
https://doi.org/10.1007/978-981-13-8946-7_27

Keywords
I/R injury · Adenosine triphosphate · Oxidative stress · Reactive oxygen species · Reactive nitrogen species · Xanthine oxidase · Xanthine dehydrogenase

Abbreviations

ATP	adenosine triphosphate
cNOS	constitutive NO synthase
DO_2	oxygen delivery
GR	glutathione reductase
GSH	glutathione reduced
GSSG	glutathione disulfide
iNOS	inducible NO synthase
LTF	lactoferrin
MBPs	metal-binding proteins
MPO	myeloperoxidase
NAD^+/NADH	oxidized/reduced adenine dinucleotide
$NADP^+$/NADPH	oxidized/reduced adenine dinucleotide phosphate
O_2ER	oxygen extraction ratio
RNS	reactive nitrogen species
ROS	reactive oxygen species
VO_2	oxygen consumption
XDH	xanthine dehydrogenase
XO	xanthine oxidase
XOR	xanthine oxidoreductase

27.1 Introduction

Ischemia/reperfusion (I/R)-induced injury has been described as one of the main factors that contribute to the observed morbidity and mortality in a variety of clinical entities. In these entities, tissue hypoxia due to ischemia is the common denominator of either regional distribution such as myocardial infarction and mesenteric embolism or of systemic involvement such as cardiac arrest and hemorrhagic shock, the latter representing the equivalent of whole body ischemia [1].

Cells utilize oxygen to produce energy by the mitochondria in the form of ATP molecules, and in this process, a number of free radicals are normally generated. These very active products may be involved in chain reactions, joining with different radicals to produce other more harming species, a process called "oxidative stress" observed in various pathological clinical conditions [2, 3]. When free radicals are produced in normal amounts, besides exerting their physiological functions

[4], they become neutralized by the inherent cellular defenses that are collectively termed as antioxidants. In contrast, if they are produced in abundance, e.g., during ischemic states, after they exhaust antioxidant reserves, they exert their harmful effect on cellular structures. Different types of oxygen and nitrogen reactive species (ions, molecules, and atoms) that possess unpaired electrons, react readily with DNA, proteins, and lipids, producing harmful products such as lipid peroxides and cause cellular damage. These rapidly evolving reactions characterize the so-called I/R injury.

The point of this chapter is to signify and indicate our present comprehension of the multifactorial systems that add to the I/R damage development, looking thoroughly toward possible therapeutic approaches that target to the root of pathologic processes in order to increase resistance of cellular death and/or eliminate or attenuate injury.

27.2 Energy Production

The essential energy substrate for aerobic eukaryotic cells is adenosine triphosphate (ATP). Its hydrolysis produces 30.5 kJ per mole (Fig. 27.1), useful for the energetic needs of the cell. In the presence of oxygen, ATP production in these cells is particularly effective when the glucose and fatty acid degradation are coupled with the oxidative phosphorylation through a hyper-molecular complex of mitochondrial inner surface [5].

Oxidative phosphorylation, without oxygen stops and the mitochondrial effective synthesis of ATP, is interrupted. During ischemia, coming up either by no tissue blood flow or by protracted low tissue blood flow, the aerobic glycolysis stops, and the main source of ATP production becomes the anaerobic glycolysis to metabolize the existent cytosolic glucose. This way of glucose metabolism is much less effective than aerobic glycolysis engaged with oxidative phosphorylation. The latter way of metabolism delivers 36 ATP molecules from 1 molecule of glucose in contrast to anaerobic glycolysis that delivers only 2 ATP molecules [5].

27.3 Reactive Species

These species include ROS (reactive oxygen species) and RNS (reactive nitrogen species), they possess one or more unpaired electrons, and they are particularly reactive and may include ions, atoms, and molecules [4].

During cellular oxygen metabolism, reactive species are normally produced and relate with a significant part of various biological processes such as the stimulation

$$\text{Adenosine-triphosphate} + H_2O \xrightarrow{\uparrow 30.5\ \text{kJ}\cdot\text{mol}^{-1}} \text{adenosine-diphosphate} + \text{Pi}$$

Fig. 27.1 ATP breakdown to ADP with energy release

of glucose transport into cells affecting inter- and intracellular signal transmission [6]. Additionally, in normal responses, they represent a vital component of the innate immune system, thus participating to the defense mechanisms against pathogens [7]. Reactive species may also become involved in abundant pathological processes immune system guided, called "oxidative stress," which is usually harmful for the cells.

27.3.1 Reactive Oxygen Species (ROS)

Normally, in aerobic conditions, the cellular metabolism constitutes itself an important source of ROS. Their origin comes from some cellular systems located not only at organelles such as peroxisomes, membranes of endoplasmic reticulum, and mitochondria but also at plasma membrane and at cytosol. The three major significant reactive oxygen species (ROS) are superoxide anion ($O_2^{\bullet -}$), hydrogen peroxide (H_2O_2), and hydroxyl radical ($OH^{\bullet}$). During I/R the initially produced ROS is $O_2^{\bullet -}$, which is created by the addition of an electron to O_2 [8], either enzymatically, e.g., NADPH oxidase and xanthine oxidoreductase, or non-enzymatically in the mitochondrial respiratory chain.

(a) Enzymatic production of ROS via NAD(P)H oxidase: $2O_2 + NADPH \rightarrow 2O_2^{\bullet -} + NADP^+ + H^+$, where the electron is added in a univalent reduction using as electron donor the NADPH or NADH. This membrane-bound enzyme is located in different cells, e.g., fibroblasts, smooth muscle cells, the endothelial cells, monocytes, polymorphonuclear leukocytes, and macrophages [3]. These cells during phagocytosis liberate a burst of superoxide NAD(P)H oxidase-mediated that leads to bactericidal activity. Although it was considered that only phagocytic cells express this enzyme, a whole group of NAD(P)H oxidases exists, as recent data indicate, that is known as NOX family of NAD(P)H oxidases and mediates diverse biological reactions in various tissues they are expressed [9]. This group of multiprotein complexes comprises the oxidases from Nox-1 to Nox-5 and the Duox-1 and the Duox-2 (dual) oxidases, namely, seven enzymes in all [10]. Mainly the Duox enzymes along with Nox-4 generate hydrogen peroxide, though the other Nox isoenzymes mainly produce superoxide [11, 12]. The most effective substance controlling NAD(P)H oxidase is angiotensin-II [13].

(b) Enzymatic production of ROS via xanthine oxidoreductase (XOR): Probably the most studied ROS-producing enzyme is xanthine oxidoreductase (XOR), which is also an essential source of ROS [14]. The formation of uric acid through hydroxylation of xanthine represents the rate-limiting level of catabolism of purine which is controlled by XOR, a complex molybdo-flavoenzyme. There are two interconvertible forms of XOR enzyme in mammals, the xanthine dehydrogenase (XDH) as prevalent form (about 90%) in normal condition and xanthine oxidase (XO). XDH preferably utilizes NAD^+ as electron acceptor [12], as appears in the following reaction: xanthine + H_2O + NAD^+ $\rightarrow$ uric acid + NADH + H^+. XDH, however, in hypoxic conditions converts to XO which, as the terminal electron acceptor, utilizes O_2, showing therefore its capacity to create ROS [12]. This process has been regarded as the principal mechanism of the oxidative injury. Therefore, in the presence of hypoxanthine and XO, the profuse molecular oxygen reentry produces ROS as follows:

hypoxanthine (or xanthine) + H_2O + O_2 → xanthine (or uric acid) + $O_2^{\bullet-}$ + H_2O_2 [14, 15]. Besides, XO may reduce nitrite to nitric oxide [16], a reaction that provides a mechanistic base for the nitrite utility in the treatment of ischemic conditions. The tissue redox state during ischemia changes from oxidative (increased concentration of NAD^+ in relation to NADH) to reductive (increased concentration of NADH in relation to NAD^+). The accumulation of xanthine intracellularly in the case of altered redox state increases the production of $O_2^{\bullet-}$ from XDH [12, 17], contrary to the prevailing view.

(c) Mitochondrial intracellular release of ROS: It has been shown that in physiological condition, the mitochondrion represents the largest single source of intracellular $O_2^{\bullet-}$ release [18, 19]. Among the four complexes of the electron transport chain in mitochondria, those of NADH ubiquinone oxidoreductase and ubiquinone/cytochrome *c* reductase complexes represent the main sites of a physiologic electron leak that reduces about a 1–3% of oxygen volume consumed by the cell to superoxide; however, the great percentage of oxygen, i.e. about 90%, is converted to water by reduction in the mitochondrial chain. ROS production by the mitochondria increases significantly during I/R [20]. The decreased mitochondrial antioxidant capacity, due to increased endogenous consumption, represents another mechanism that contributes to increased ROS production [20]. In any case, the production of $O_2^{\bullet-}$ in tissues during ischemia seems to involve purine metabolism alterations in ischemic cells [21]. In progressive ischemia, two products of purine catabolism, i.e., hypoxanthine and xanthine, accumulate in tissues. Specifically, it has been recently shown through studies of metabolomic analysis that only three metabolic products, hypoxanthine, xanthine, and succinate, were accumulated in different organ-tissues (heart, liver, kidney, and brain), exposed to I/R [12, 15]. This observation confirms the consideration that purine metabolites represent the substrate of XO-catalyzed ROS generation at reperfusion.

Consequently, net ROS mitochondrial liberation likely mirrors the balance between generation and disposal/clearance [22]. The major part of ROS deliberated in I/R conditions becomes from mitochondria as it has been shown by many studies [18, 19, 22]. This knowledge was accumulated by studies that used inhibitors specifically directed to the various levels of the electron transport chain targeting specific antioxidants to the mitochondria or using CuZnSOD vs. MnSOD for the transgenic overexpression of cytosol vs. mitochondrial-specific isophorms of antioxidant enzymes [19]. Mitochondrial ROS generation was found to become inhibited in studies where agents, which defend the I/R-induced vascular and tissue injury, were given [19, 22].

(d) The system of myeloperoxidase–halide– H_2O_2 release of ROS: The MPO (myeloperoxidase) enzyme is found in the granules of cytoplasm of neutrophils. The conversion of H_2O_2 to hypochlorous (HOCl), a potent oxidant and antimicrobial agent [23], is taking place in the presence of the ubiquitous chloride ion after the reaction: $Cl^- + H_2O_2 + H^+ \rightarrow HOCl + H_2O$.

Superoxide generated by the above principal sources produces hydrogen peroxide through SOD (superoxide dismutases) as follows: $O_2^{\bullet-} + 2H^+ \rightarrow H_2O_2 + O_2$. Hydrogen peroxide ($H_2O_2$) is easily diffused in the plasma membrane. Other ROS are additionally generated from $O_2^{\bullet-}$ and H_2O_2 by Fenton reaction $H_2O_2 + Fe^{2+} \rightarrow OH^\bullet + OH^- + Fe^{3+}$ and/or by Haber-Weiss reaction $O_2^{\bullet-} + H_2O_2 \rightarrow OH^\bullet + OH^- + O_2$.

27.3.2 Reactive Nitrogen Species (RNS)

The term, reactive nitrogen species (RNS), collectively comprises nitric oxide ($NO^\bullet$), nitrogen dioxide ($NO_2^\bullet$), the potent oxidant peroxynitrite ($ONOO^-$), and other nitrogen oxides and metabolites that result when $NO^\bullet$ reacts with $O_2^{\bullet-}$, $H^\bullet O^\bullet$, and $RO^\bullet$. Endogenous synthesis of $NO^\bullet$ is accomplished by several cells and endothelial cells (ECs), in particular. L-Arginine reacting with molecular oxygen in the presence of NOS and NADPH [3, 24, 25] produces L-citrulline, $NO^\bullet$, and other products via a five-electron oxidation reaction as NOS + L-arginine $+2O_2 + 1.5NADPH \rightarrow NO^\bullet + \text{citrulline} + 1.5NADP^+ + 2H_2O$, or, in case of uncoupling states, these enzymes may additionally generate superoxide as NOS (Fe^{++} heme) + $O_2^{\bullet-} \rightarrow$ NOS (Fe^{+++} heme) + $O^{\bullet-}$.

The above reaction is evoked by the enzyme NOS (nitric oxide synthase). There are three isoforms of the enzyme that have been recognized: NOS-1 (nNOS, neuronal), NOS-2 (iNOS, inducible), and NOS-3 (eNOS, endothelial). NOS-1 and NOS-3 provide a more ubiquitous distribution, e.g., they are both found in cardiomyocytes though they are mainly expressed at the brain/neurons and endothelial cell, respectively [26]. They are nominated as constitutive-NOS enzymes and become activated by their connection with the compound Ca^{2+}/calmodulin after the intracellular increase of Ca^{2+}.

Until the intracellular calcium concentration drops, small quantities of $NO^\bullet$ are produced by the activated NOS. This fluctuating release of $NO^\bullet$ allows normal physiological functions to be carried out, like signal transmission and maintenance of a basal vasodilator tone [3, 26, 27]. The vasodilatory effect of $NO^\bullet$ appears to be accompanied by an attenuation of leukocyte-endothelium interaction, of platelet aggregation, of cell adhesion, and of limitation of cell proliferation [28]. Another and last isoform of NOS is the iNOS (inducible or NOS 2). This enzyme is regulated by the inflammatory response and expressed therefore in various tissue cells and particularly in macrophages [29]. The expression of iNOS is regulated through gene transcription and is not influenced by Ca^{2+}/calmodulin levels [25, 28, 29]. For as long as the inflammatory challenge exists, the activation of iNOS continuously produces significant quantities of $NO^{\bullet}$ that suppress and kill pathogens. Protein kinases control the activity of iNOS through phosphorylation/dephosphorylation with the former forming to represent the enzyme of diminished activity.

While signaling molecules exert their action via receptors, $NO^{\bullet}$ in contrast passes through membranes and diffuses out of the cell where it was generated. It reaches the target cells where it exerts its physiological effects, i.e., signal transmission, chemical reactions with proteins, nucleic acids, ROS, and superoxide in particular, or reacts with its molecular targets [30]. The radical to radical chemical reaction of $NO^{\bullet}$ with superoxide is of particular importance in the $NO^{\bullet}$-related toxicity since they form the extremely toxic oxidant (not radical) peroxynitrite quickly. It may attack and distract different biologic molecules and is produced in various pathological and inflammatory processes [3]. In conclusion, all isoforms of NOS are homologous, simply as it has been shown, and they differ in the controls and activities and, particularly, in the amounts of $NO^{\bullet}$ produced.

27.4 Endogenous Antioxidants

27.4.1 Enzymatic

As referred above, an important enzymatic antioxidant of first-line defense is superoxide dismutase (either SOD-Cu/Zn intra−/extracellular or SOD-Mn mitochondrial). It dismutates superoxide to hydrogen peroxide (H_2O_2) and oxygen. Hydrogen peroxide via Fenton reaction results either to hydroxyl radical ($HO^{\bullet}$) or under two other important enzymatic antioxidants, catalase (CAT of peroxysomes) and glutathione peroxidase (GPx of mitochondria), and H_2O_2 is further reduced to water and molecular O_2. The oxidized glutathione, termed glutathione disulfide (GSSG), turns by reduction to its reduced form (GSH), taking up the hydrogen from NADPH which converts to $NADP^+$, by the intermediate of GR (glutathione reductase). Thus, the cysteine thiol group of a new GSH molecule is ready to react with next free radicals. NADPH is regenerated at the oxidative pathway pentose phosphate, ready to reduce a GSSG to GSH again [24, 31].

27.4.2 Non-enzymatic

The non-enzymatic antioxidant substances are in common either proteins or molecules of low molecular weight, and they offer efficient modes of intra- and extracellular defense against both RNS and ROS. They are the following [23, 24]:

1. a. Metal-binding proteins (MBPs) such as albumin (ALB), ferritin (FER), ceruloplasmin (CP), myoglobin (MB), metallothioneins (MTs), transferrin (TF), and lactoferrin (LTF)

Cysteine contained in albumin and particularly in the low molecular weight MTs discloses sulfhydryl (-SH) groups that are capable of scavenging hydroxyl radicals. The abundant -SH groups they possess provide their principal antioxidant activity [32].

Conversely, by binding iron ions (Fe^{2+}) and free copper (Cu^{2+}), ceruloplasmin performs as an inhibitor of reactive species or, alternatively, as an antioxidant chain-breaking [33].

Albumin (ALB) is an antioxidant protein possessing multiple functions, i.e., it reacts with hydroxyl radicals as a true scavenger and may also bind with redox metals (Fe^{2+} and Cu^{2+}) [34, 35]. An effective NO scavenger is myoglobin, another MBP [36]. The protection of cells from toxic metals, e.g., Cu, Zn, and Cd, is coming from metallothionein, a family of cysteine-rich proteins localized to the membrane of the Golgi apparatus; it is also effective against superoxide as scavenger [37].

Additionally, ferritin (FER), lactoferrin (LTF) and transferrin (TF) may be efficient in inhibiting free radicals in the Fenton/Haber-Weiss reaction [38]. In this reaction, in a first step ferric ion reduces to ferrous ion via reaction with superoxide that is neutralized as follows: $O_2^{\bullet-} + Fe^{3+} \rightarrow Fe^{2+} + O_2$. In a second step, ferrous ion (Fe^{2+}) reacts with H_2O_2 to form OH^- anion and harmful $OH^\bullet$ radical, converting ferrous back to ferric ion, as follows: $Fe^{2+} + H_2O_2 \rightarrow Fe^{3+} + OH^- + OH^\bullet$. Therefore, these MBPs are classified as antioxidants, because of their capacity to bind iron [39–41].

b. Glutathione (GSH) is a soluble antioxidant that its synthesis mainly occurs in hepatocytes and its high cellular concentrations are found in the cytoplasm and in cellular organelles such as nucleus and mitochondria [24]. Reduced GSH prevails by far over oxidized (GSSG) normally, representing till 98% of the total glutathione reserves. GSH, as an antioxidant, is capable of reducing ROS throughout both enzymatic and non-enzymatic reactions. This substance has the capability of defending the thiol groups of proteins from being oxidatively destructed, thanks to the thiol groups that possess participating in its molecule [42]. Its reduced levels reflect a good index of oxidative aggression [43].

c. Uric acid (UA) scavenges different types of reactive species [44, 45] as hydroxyl radical, lipid peroxides, and the oxidant substance peroxynitrite [24], despite some adverse effects [44]. Probably, it has also the ability to scavenge carbonate ions and nitrogen dioxide [42]. With copper and iron ions, it may form stable complexes, inhibiting by this way Fenton and the Haber-Weiss reactions which produce free radicals [43].

d. Coenzyme Q10 (CoQ10) generally participates in the transport of electrons in both places, outside mitochondria and in the mitochondrial respiratory chain [24, 46]. CoQ10 participates in redox reaction particularly of dehydrogenases, in non-heme proteins, and in cytochromes, and its activity appears in the lipid profile [47].

e. Other non-enzymatic antioxidants are melatonin (MEL), bilirubin (BIL), polyamines (PAs).

27.5 Ischemia: Cell Suffering-Cell Priming

In the clinical practice, tissue O_2 deprivation may occur secondary to complete or relative blood flow impairment, representing the "no flow" or "low-flow" condition, respectively. "No flow" may concern either individual organs, e.g., intestine (secondary to mesenteric artery embolism), cardiac infarction (coronary artery occlusion), or the whole body (cardiac arrest). "Low flow" may concern individual organs, e.g., angina (partial occlusion of coronary arteries), or the whole body, e.g., generalized tissue hypoperfusion and tissue hypoxia (continuing hemorrhage and hemorrhagic shock). Apparently, in low-flow conditions, the duration of the insult is of significant importance. Finally, the relative or complete ATP reserve exhaustion determines the reversibility or irreversibility of cell damage and parenchymal tissue destruction during evolutionary ischemia.

Continuous hemorrhage is one of the main causes responsible for the progressive impairment of tissue blood flow and, therefore, represents a good paradigm of the pathophysiologic biochemical consequences attributed to the progressively developed tissue hypoxia [48]. In cellular level during continuing hemorrhage, the biochemical disorders run two phases, the cell suffering and the cell priming.

27.5.1 Cell Suffering

As the circulating blood volume decreases, oxygen delivery (DO_2) decreases as well, but oxygen consumption (VO_2) remains constant, for a considerable amount of blood loss, in a formerly healthy patient (solid line - Fig. 27.2) [49]. This DO_2 reduction is balanced both by the increase of oxygen extraction ratio (O_2ER – the ratio of VO_2 over DO_2) with a corresponding fall of mixed venous blood O_2 saturation and by the counterbalanced mechanisms (tachycardia, vasoconstriction, etc.).

Aerobic metabolism therefore is maintained, and VO_2 remains independent of DO_2 with no cell suffering. However, further reduction of the DO_2 to that point level

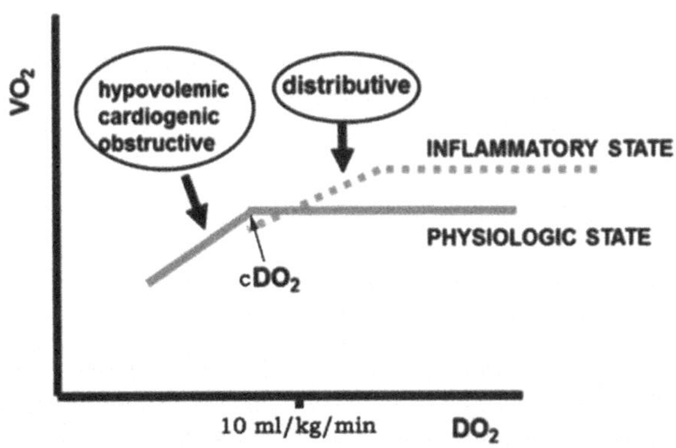

Fig. 27.2 Oxygen consumption to oxygen delivery relationship. (Modified from Ref. [49])

where maximum O_2ER is reached (60–70%) is called critical DO_2 (cDO_2) [50] and coincides with less than 10 mL of O_2/kg of body weight.

Beyond this point, any further decline in DO_2 signifies a blood volume loss of approximately 50% with a substantial reduction in cardiac output and in mixed venous oxygen saturation [48]. Anaerobic metabolism begins to take over from aerobic metabolism because of tissue hypoxia, and there is an abrupt increase in blood lactate concentrations. Cell suffering coincides with this *mild to moderate ischemia*, where ATP is consumed and reliant upon the span of the ischemic insult, and its consumption may exceed its production. Then, its intracellular concentration decreases, the extracellular glucose rapidly declines, and similarly the intracellular glycogen reserves rapidly consume. Besides, ATP reduces proportionally to the magnitude of ischemia, but it is produced in some degree from the fatty acid β-oxidation that still fuels oxidative phosphorylation [5].

27.5.2 Cell Priming

Further drop of DO_2 leads to deep tissue hypoxia and the severely O_2-deprived cell to prime for generation of ROS, upon O_2 reentry during reperfusion. Specifically, cellular priming consists of ATP down-degradation, initially to the level of adenosine [48]. This is because as ischemia deepens and intracellular concentrations of ATP shrink, the cells draw the necessary energy from ADP, from the pyrophosphate bonds in particular, which degrade to AMP and finally to adenosine. The latter substance freely passes cell membranes diminishing noticeably the intracellular nucleotide reserves of adenine, which are the precursors of ATP synthesis [5]. The quantity of adenosine remaining in the cell downgrades to hypoxanthine with concurrent XO accumulation. Moreover, beta-oxidation of fatty acids runs out, since the key enzymes for its proper function are not produced due to the privation of the cofactors NADH, H^+, and FAD^+, which in normal condition are restored via oxidative phosphorylation.

Cell priming during severe ischemia, apart the hypoxanthine, xanthine, and xanthine oxidase intracellular high concentration, includes important other derangements, some of which are the following:

27.5.2.1 Cell Membrane Destruction

Because beta-oxidation has stopped working, the cellular concentrations of amphiphilic compounds fatty acids, acyl-CoA, and acylcarnitine increase progressively and readily dissolve into cell membranes. The functional properties of membrane proteins are affected, and a gap junction conductance reduction appears, which is time-dependent reversible [53].

27.5.2.2 Na^+-/K^+-ATPase-Reduced Activity

The activity of the pumps Na^+-/K^+-ATPase and Ca^{2+}-ATPase decreases, and the same effect occurs for the activation of potassium channels, ATP-dependent. Therefore, the current which corrects potassium reduces, extending the Na^+ channel opening, inhibiting their inactivation [5].

27.5.2.3 Intracellular Acidosis

Cellular ischemia is accompanied by intracellular acidosis which represents one of its principal features [5]. The exhaustion of the buffering capacity of the cell comes rapidly on, because of the increased proton production secondary to metabolic changes. The proper functioning of the cell is directly and indirectly involved by intracellular acidosis by numerous ways: Increasing the production of free radical through the increases of the intracellular Na^+, induced by Na^+/H^+ and Na^+/Ca^{2+} exchangers activation; modifying the affinity of troponin C and various enzymes; changing the tertiary structure of proteins; suspending enzymes; and distracting the function of carriers and pumps of the sarcoplasmic reticulum [51]. In the presence of lactate dehydrogenase, pyruvate converts to lactate, and this reaction consists of the principal source of protons during ischemia. Since lactate's extracellular concentration increases, lactate/proton cotransporter is no more efficacious impeding thus the removal of protons. Moreover, the remaining metabolic activity further supplies acidosis, since a proton is released by the hydrolysis of each molecule of ATP [5].

27.6 Reperfusion: Free Radical Burst

The massive reentry of molecular oxygen in deprived from O_2 cells but rich in hypoxanthine and xanthine oxidase upon reperfusion results in the following paradox: instead of ascending the route to establish resynthesis of ATP, the reaction diverts toward the conversion of hypoxanthine to uric acid and xanthine, releasing the generation of ROS (Fig. 27.3).

The high body concentration of ROS can modify the structure of DNA, of lipids, and of proteins, may activate some transcription factors induced by stress, and may produce both cytokine classes, pro-inflammatory and anti-inflammatory.

ROS may disorder the bilayer arrangement of membranes by its lipid peroxidation and therefore, disable the activity of receptors and enzymes bound on the membranes [31]. Either by adding on a polyunsaturated fatty acid (PUFA), an oxygen radical, or by removing a hydrogen, a chain reaction is begun designated as lipid peroxidation. Since PUFAs are more sensitive than the saturated fatty acids, they are easily affected and undergone oxidative destruction. A rearrangement of the double bonds stabilizes the free electron on the carbon, producing a conjugated diene and next, peroxyl radicals (ROO·) in combination with oxygen. Since this radical is in itself capable to remove an atom of hydrogen from another PUFA, the chain reaction continues producing in each turn additional peroxyl radicals [52].

Alcoxyl (RO·) or hydroxyl (HO·) radicals are produced by complexes of peroxyl radicals with transition metals. The first relatively stable product of lipid peroxidation chain reaction is lipid hydroperoxide (ROOH) [52, 53]. MDA and unsaturated aldehydes, products of lipid peroxidation, are efficient of deactivating cellular proteins by creating protein cross-linkages [54]. 4-Hydroxy-2-nonenal and other hydroxyl-alkenals cause depletion of intracellular GSH, peroxide production, and induce fibronectin production [54, 55, 56, 57].

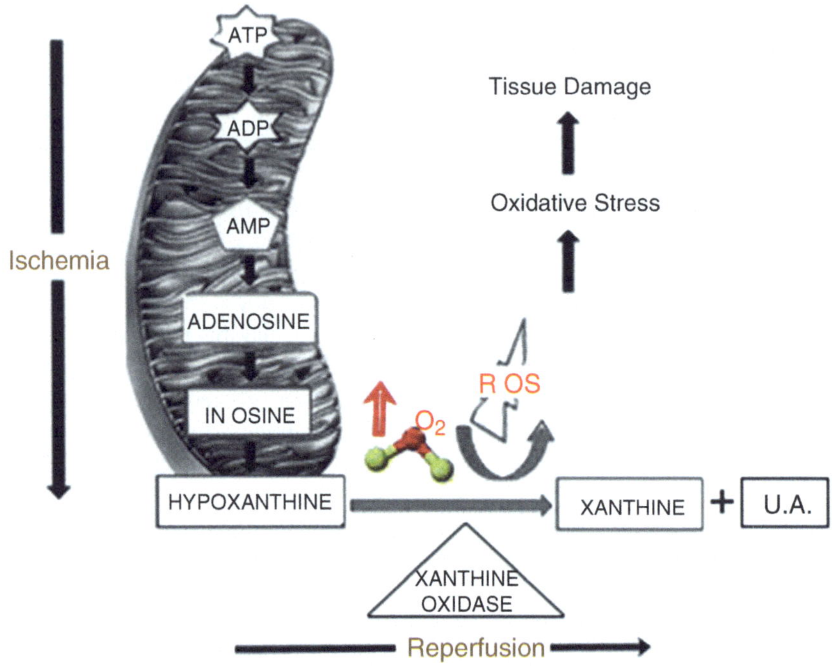

Fig. 27.3 In severe cellular ischemic state, not only ATP synthesis has stopped but ATP degrades initially to adenosine and thereafter to hypoxanthine. XDH has been largely converted to XO. Abrupt and in abundance reintroduction of O_2 at resuscitation results in xanthine and uric acid formation with burst of ROS and tissue damage. (Figure modified from Ref. [1])

Expression of genes engaged in signal transduction may become stimulated by increased ROS [58]. An increased GSH/GSSG ratio significantly protects the cell from oxidative attack which results in cell destruction. When this ratio decreases, the sensitive to intracellular redox transcription factors activate. These factors (NF-kB, AP-1, NFAT and hypoxia-inducible factor 1) are mainly involved in the inflammatory response [31].

The damage that ROS cause to proteins is serious since they may hit and fragment the peptide chain, alter the protein electrical charge, and oxidize selective amino acids, making thus proteins sensitive to proteolysis by appropriate proteases. In the peptide chain, the sulfhydryl groups of methionine and cysteine are more easily oxidized resulting in changes of conformation and of folding, leading finally to protein degradation [59].

Molecular products, markers of oxidative aggression, such as nitrotyrosine, oxidized glutathione (GSSG), H_2O_2, $O_2^{\bullet-}$, and MDA may be measured in samples of plasma or bronchoalveolar lavage fluid by established methods [31, 56, 57].

27.7 Therapeutic Trends of the I/R Injury

On all experimental studies on I/R injury, there are some common factors that need to be highlighted. Contemplating the experimental model, for tissue injury to occur on reperfusion, ischemia should be of sufficient intensity and duration for the following to be fulfilled: first, tissue cell priming; second, exhaustion of antioxidant capacity; and third, the substrate hypoxanthine and the enzyme xanthine oxidase to be accumulated. A forth factor needed, for injury to become ensured, is the presence of sufficient molecular O_2 as an electron acceptor upon reperfusion. Then, ROS burst is elicited in a place where antioxidant defenses are lacking and tissue damage occurs. The knowledge of this sequence has led medical research to focus on the development of potential strategies, aimed at eliminating the effects of reactive oxygen species, which are discussed below. However, studies manipulating the fourth, factor i.e., O_2, are lacking.

27.7.1 Regional Insults: Organ System I/R

Suggested methods [1] include the use of antioxidants to minimize the oxidative stress [60–63], scavengers for the removal of metabolic waste, and preconditioning techniques (ischemic, hypoxic, pharmacologic, and remote ischemic preconditioning) to prepare cells to better respond to the forthcoming stress [64–73].

Despite the proven beneficial effects, all the above strategies share one common disadvantage: they lack effectiveness when they are applied after or during reperfusion/resuscitation which limits their usefulness in the clinical setting [1]. Similarly, antioxidants should be administered ideally before ischemia and reperfusion in order to achieve their maximum effect. In fact, most available evidence regarding their favorable effects derives from studies in which antioxidants were used as pretreatment [61, 63]. Moreover, their use, even in combination with scavengers, does not completely abolish the ensuing injury. The same applies to the use of preconditioning techniques. The rationale of these techniques is to "premedicate" the patient, an option which may not be feasible in all clinical scenarios [1]. Therefore, the application of these strategies in the clinical setting may be limited [70–72, 74, 75]. A recent meta-analysis questioned the efficacy of ischemic preconditioning in the setting of liver surgery [76]. Similarly, remote ischemic preconditioning, a technique that held great promise for its demonstrated favorable effects, did not exert the expected outcomes when tested in clinical trials [77–79]. An important thing that should be considered is that once reperfusion injury has begun, many pathways evolve. Whatever significant intervention, may moderate the injury temporarily or partially, because, even if one or more paths have been eventually blocked, however the others, remain evolving.

Nevertheless, in a victim with regional or generalized ischemia, the only influential factor, among those aforementioned, necessary to induce injury on reperfusion is the way that oxygen is given during the therapeutic management. Considering that the needs for O_2 of the adapted to ischemia tissues are minimal, comparatively excessive high tissue partial pressure of O_2 (P_tO_2) could be created upon tissue reperfusion, leading to ROS burst, even if the victim breathes air atmosphere during resuscitation. In contrast, supplying initially very low and progressively increasing concentrations of oxygen coupled with effective tissue perfusion should permit, theoretically, the gradual restoration of energy resources, attenuating ROS generation and reducing the I/R injury.

We tested the above hypothesis in an experimental model of hemorrhagic shock. The intention was to give a sufficient tissue perfusion during resuscitation and, at the same time, to manipulate the oxygen content in the initial blood perfusate, in order to meet with the adapted – at low cellular energy state – needs during ischemia. It is about gradually increasing the O_2 content of the reperfusate blood from a lower level corresponding to a PaO_2 of 30–35 mmHg initially at reperfusion, to gradually achieve a PaO_2 of 95–100 mmHg at the end of the resuscitation period [1]. It is thought that, resuscitating in this way, gradual restoration of cellular energy resources may occur, elevating the levels of ATP resynthesis (Fig. 27.4). By this

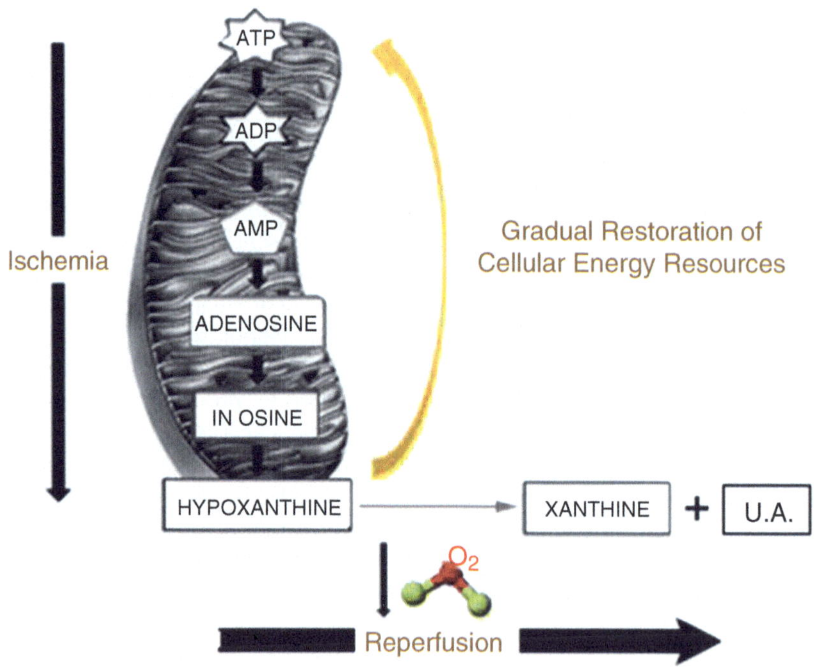

Fig. 27.4 Here is presented a possible mechanism of the favorable effect exerted by hypoxemic resuscitation. Low and progressively increasing O_2 content to the tissues, combined with adequate tissue perfusion, reduces ROS generation. It is suggested that less hypoxanthine converts to xanthine and therefore accumulates intracellularly, eventually resulting to gradual restoration of cellular energy resources. (Figure modified from [1])

approach, the primed cells rather ameliorate turning to suffering cells, in a gradual process, before they finally become normal.

This hypothesis was initially tested in the most sensitive to oxygen devoid organ, i.e., the brain [56, 80]. Prior studies tested the effect of hypoxemic [81] or hyperbaric [82] reperfusion reporting no improvement either of acute brain recovery the former or promotion of brain lipid peroxidation the latter. However, global cerebral ischemia was produced by raising intracranial pressure either to 100 mmHg above arterial pressure [81] or equal to MAP [82] with rapid infusion of artificial cerebrospinal fluid into a lateral ventricle. In this way, mechanical might overshadow ischemic injury of the brain, and besides, the post-resuscitative syndrome was not reliably reproduced since no systemic ischemia occurred, as it happens in cardiac arrest [77].

In male pigs, a different ischemic brain insult of 10 min was produced by ligation of both carotids, systemic shock with a mean arterial pressure of 15 to 35 mmHg, and disconnection from the respirator under paralysis [56]. A hypoxemic or hyperoxemic reperfusion of 60 min was tested thereafter. The hypoxemic animals had significantly better overall neurologic performance than hyperoxemic and was similar to the sham-operated animals. Also, the hypoxemic animals had less blood MDA and hydroxyalkelans compared with the hyperoxemic animals. Similarly, the histopathological changes were significantly smaller, specifically the neuronal degeneration [80]. These results may imply that the progressive reintroduction of O_2 to the ischemic brain tissue ensured by hypoxemic reperfusion attenuates the oxidative injury after a severe global brain insult.

On the same issue [56, 80], there are at least three studies supporting the notion that ventilating with air atmosphere versus 100% O_2 during resuscitation after cardiac arrest may be advantageous with respect to neurological outcome. Mickel, for example [83], showed a threefold decrease in 14-day mortality in gerbils when exposed to an air atmosphere instead of to 100% O_2 after ischemia. Zwemer et al. [84] similarly showed that hyperoxically resuscitated dogs sustained significantly worse neurological deficit at 12 and 24 h than did both antioxidant-pretreated hyperoxically resuscitated and normoxically resuscitated dogs after cardiac arrest. Similarly, Liu et al. [85] tested normoxic versus hyperoxic reperfusion after cardiac arrest showing a better outcome in the normoxically treated animals together with lower levels of oxidized brain lipids. It seems, therefore, that the normal – compared to the high – oxygen mixture ventilation during reperfusion favors diminished ROS formation and lower cerebral lipid peroxidation. These results may lead to the rationale that further diminishment of elemental tissue O_2 concentration may further minimize the postarrest cerebral damage [56].

In the same line of evidence were the results of another study [86] with regional intestinal ischemia of 120 min introduced by superior mesenteric artery clamping, followed by a period of 120 min of hypoxemic or normoxemic reperfusion. At the end of reperfusion, significantly less hypoxemic animals had positive portal Limulus test (detection of endotoxin), less portal IL-1β levels, and higher PaO_2/FiO_2 ratio level. These results dictate that the intestinal mucosa and lungs of hypoxemic

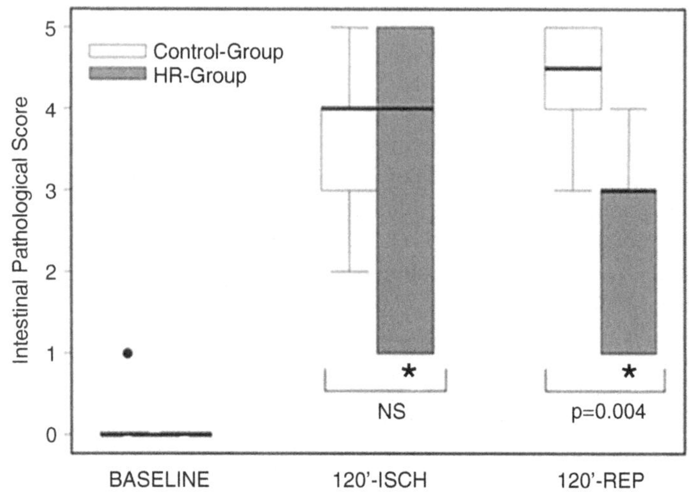

Fig. 27.5 Box plots showing the median intestinal pathologic scores (line), interquartile range (box), and 5th–95th percentiles (whiskers) at the three conditions of the experiment. Circles denote outliers. HR, hypoxemic reperfusion; ISCH, ischemia; REP, reperfusion. Control-Group (n = 16), HR-Group (n = 9). ∗ Difference $p = 0.03$. (Figure modified from Ref. [86])

animals were preserved since the intestinal barrier disruption, which induces mucosal permeability and bacterial translocation, was obviated. These results are further confirmed by the intestinal and pulmonary histopathologic scores. The intestinal scores, shown in Fig. 27.5 [86], at the 120 min of ischemia were high enough in both groups and reveal the severity of injury introduced by ischemia. However, the reversibility of this injury becomes apparent, since the scores differed significantly at 120 min of reperfusion being lower in hypoxemic animals and higher in normoxemic. Similar findings were exhibited in pulmonary histology, highlighting the attenuation of injury in remote organs, as well.

From these initial experimental studies, it seems that severely ischemic cells, when they become well perfused but initially still devoid of oxygen, which is increasing gradually thereafter, divert to a lesser degree the accumulated hypoxanthine toward xanthine, uric acid, and generation of ROS, as it is suggested in Fig. 27.5. Since the regional ischemic experimental models carry almost no mortality because the injury caused is limited, mortality should be sought in a different experimental model.

27.7.2 Generalized Insults: Hemorrhagic Shock-Whole Body I/R

This could be a generalized model of injury introduced by severe hemorrhagic shock followed by resuscitation, to conclude about the importance of hypoxemic resuscitation over the usual standard of care. Hemorrhagic shock/resuscitation may be considered as a global hypoxia/reoxygenation injury or, in other words, as a generalized I/R insult [57, 87]. In hemorrhagic shock, reoxygenation of previously ischemic organ-tissues follows the therapeutic process. On clinical and experimental grounds, controversy exists on the type of fluid selection (crystalloid vs. colloid), on fluid quantity (low volume vs. large volume), and on the resuscitation strategies (prompt vs. delayed, hypotensive vs. normotensive), for the effective restoration of hemodynamic stability [57, 88–93]. However, newer studies are more conclusive on the resuscitation pattern, indicating rather a trend toward small hypertonic volumes of saline, permissive hypotension, or delayed resuscitation; these are the limited patterns of resuscitation from hemorrhagic shock; certainly, resuscitation should have been accompanied or followed by prompt and definitive control of hemorrhage [91, 92, 94]. These strategies seem to tend over a model of progressive reintroduction of oxygen to ischemic tissues [43]. Additionally, delayed shock resuscitation of trauma patients has been associated with a reduction in mortality [89], and small volume (hypertonic saline) resuscitation has been shown to reduce systemic and pulmonary inflammatory response [95].

Studies of hemorrhagic shock resuscitation with a more direct method of ROS measurement, assessment of organ failure, and the incidence of ARDS and of hepatic involvement with mortality assessment were necessary. Incubating whole blood or collected fluids (e.g., bronchoalveolar fluid - BAL fluid) with dichlorohydrofluoresceine diacetate as a probe, geometric mean fluorescence intensity (GMFI) by flow cytometry [57] is measured. This is possible, since the probe is oxidized by ROS in cytoplasm to 2′-7′-dichlorofluorescein, a highly fluorescent compound thereafter ROS is measured, analyzing the GMFI. The GMFI in the blood of rabbits subjected to hemorrhagic shock to mean arterial pressure of 40 mmHg for 60 min [57] was similar at baseline, and at the end of shock, while at resuscitation, it increased in a stepwise pattern reaching fivefold at 120 min of resuscitation in control animals but only 1.5-fold in hypoxemic animals ($p < 0.001$).

This was accompanied by complete hemodynamic restoration of hypoxemic animals at 120 min of reperfusion in contrast to normoxemic animals. The latter presented hemodynamic instability after 60 min of resuscitation, to unresponsive fluids ($p < 0.05$), with increased mortality (4/10 animals) in contrast to hypoxemic (1/10). Malondialdehyde (MDA) serum levels also achieved the same pattern ($p < 0.001$). The ratio of reduced to total glutathione was significantly greater at all time points of resuscitation in hypoxemic animals compared with controls ($p < 0.05$), while at the end of ischemia, it was similar between the groups and comparable to the sham animals.

Hemorrhagic shock and resuscitation promote the development of multiple organ dysfunction with the lungs to be among the most promptly and most commonly affected organs [96, 97]. Neutrophil accumulation increases generation of reactive oxygen species (ROS), and the local release of proinflammatory cytokines contributes to the pulmonary inflammatory process that characterizes lung injury [96, 98]. Among cytokines, interleukin IL-8 has been shown to play a predominant role [96, 99]. Apart from acting as a potent chemoattractant for neutrophils, IL-8 levels correlate with lung function parameters in the setting of ischemia/reperfusion injury caused by lung transplantation [100] and may additionally enhance the oxidative burst produced by neutrophils [101].

Therefore, it was sought to investigate whether hypoxemic resuscitation from hemorrhagic shock moderates the development of lung injury and ARDS. Oxidative parameters of BAL fluid supernatant from animals, similarly subjected to hemorrhagic shock and resuscitation as aforementioned, were measured [96]. In the hypoxemic animals, the GMFI of polymorphonuclear cells and macrophages as well as the MDA was not increased, while GSH was significantly higher at resuscitation in contrast with the controls, in which GMFI values exhibited increases of five- to sevenfold ($p < 0.05$). Interestingly, there were similar findings when BAL fluid from these animals was co-intubated for cell-stimulation assays on monocyte-like cells, the human U937 cell line. In fact, the inflammatory cytokines (IL-1β, TNFα, and IL-6) were increased in both BAL fluid and U937 incubation supernatant, with IL-8 to increase excessively ($p < 0.05$).

Similarly, neutrophil infiltration of the lung interstitium and alveoli, lung parenchyma MPO, and wet-to-dry lung weight ratio differed significantly ($p < 0.05$) being higher in the control group. Lung histopathology is delineated in Fig. 27.6, where the evidence of co-localization of IL-8 and nitrotyrosine (a footprint of RNS effect) in the lung appears. In Fig. 27.6D, there is a normal expression of IL-8, as the light brown staining is shown, which is not associated with nitrotyrosine detection (Fig. 27.6E). Cytokine IL-8 staining was lower in hypoxemic animals compared with normoxemic animals. The more intense IL-8 staining in a slice of normoxemic animal (Fig. 27.6F) is in contrast to the mildly stained slice from a hypoxemic animal (Fig. 27.6H).

A respective difference in nitrotyrosine brown-colored staining between the two groups (Fig. 27.6G vs. I) was also observed, which colocalizes with IL-8 detection, and its intensity corresponds to the degree of IL-8 expression. Findings other than oxidative indices point to the same direction and also favor the hypoxemic group: (a) lower inflammatory state of the lungs, i.e., lower cytokine concentration in the BAL fluid as already mentioned, lower tissue MPO activity, and lower degree of inflammatory infiltration on tissue histology; (b) lower degree of pulmonary edema, evidenced by the lower wet/dry lung weight ratio; and (c) lower degree of protein nitration, immunohistochemically detected.

Continuing in the field of resuscitation from shock and lung injury, early signs of ARDS development were explored [102]. Comparisons of lung histology scores, semiquantitative expression of ICAM, and VCAM in inflammatory cells were studied. Similarly, the number of the same receptors' expression in endothelial cells of rabbits subjected to hemorrhagic shock and normoxemic or hypoxemic resuscitation is presented in Table 27.1. In that study using indicator-dilution

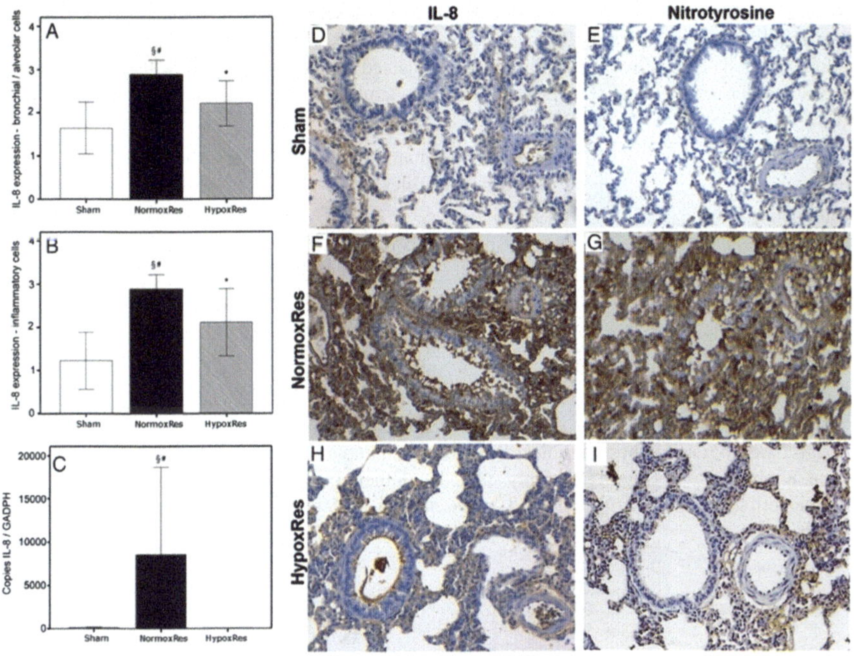

Fig. 27.6 Evidence of colocalization of IL-8 and nitrotyrosine in the lung. Immunohistochemical IL-8 score of (**A**) bronchial or alveolar epithelium, (**B**) inflammatory cells, and (**C**) IL-8 gene expression in the lungs of sham-operated rabbits ($n = 9$), rabbits subjected to normoxemic resuscitation (NormoxRes, $n = 10$), and rabbits subjected to hypoxemic resuscitation (HypoxRes, $n = 10$). Immunohistochemical IL-8 expression and nitrotyrosine detection: (**D, E**) from sham-operated group, (**F, G**) from NormoxRes group, and (**H, I**) from HypoxRes group. $*p < 0.05$ comparing sham to HypoxRes group, $^{\S}p < 0.05$ comparing NormoxRes to HypoxRes group, and $^{\#}p < 0.05$ comparing NormoxRes to sham group. (Figure from Ref. [96])

techniques by the substrate ^{3}H-benzoyl-Phe-Ala-Pro, capillary endothelial angiotensin-converting enzyme activity was additionally measured. The increase of parameters presented in Table 27.1 and the decrease of angiotensin-converting enzyme were observed in the normoxemic animals and were significantly attenuated in the hypoxemic animals.

In a similar experimental protocol of resuscitation from hemorrhagic shock [103], the nitrosative and oxidative stresses on ischemic liver injury were studied. Particular emphasis was given on the hepatic antioxidant capacity that may change in I/R injury. Briefly, on Fig. 27.7, the hepatic tissue levels of MDA and GSH and the percentile ratio reduced to total GSH (R/T) are presented. In panel D, the total antioxidant capacity (TAC) is shown, that appears to be coherent with the other parameters already referred.

Biopsies from normoxemic group demonstrated significant evidence of injury with severe sinusoidal/vascular congestion and marked vacuolization focally

Table 27.1 Comparisons of cumulative lung histology scores, semiquantitative expression of ICAM and VCAM in inflammatory cells, and the number of the same receptors' expression, in endothelial cells of positive vessels in rabbits subjected to hemorrhagic shock that subsequently received either normoxemic (Normox-Res) or hypoxemic (Hypox-Res) resuscitation. (From Ref. 102)

Parameter	Normox-Res Median (25th–75th percentiles)	Hypox-Res Median (25th–75th percentiles)	*p*
Cumulative lung histopathological score	10 (9–11)	7 (4–8.5)	<0.01
VCAM-1 positive vessels[a]	47 (33–56)	24 (21.5–27.5)	<0.05
ICAM-1 positive vessels[a]	88 (65–110)	25 (21–30)	<0.01
VCAM-1 positivity in inflammatory cells[b]	3 (2–3)	1 (1–1)	<0.01
ICAM-1 positivity in inflammatory cells[b]	2 (2–3)	1 (1–1)	<0.01

VCAM-1 vascular cell adhesion molecule-1, *ICAM-1* intercellular adhesion molecule-1
[a]Total number of vessels with positively stained endothelial cells in five random fields at ×200
[b]semiquantitative assessment (score 1–3) of extent of ICAM-1 and VCAM-1 expression in inflammatory cells

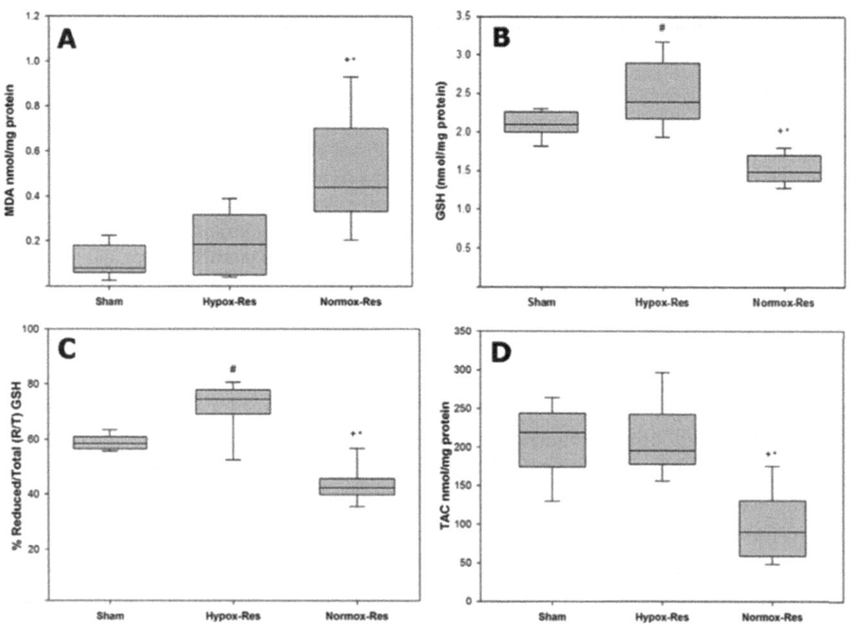

Fig. 27.7 Indices of oxidative stress. Box plots showing the median (lines), interquartile ranges (boxes), and the 5 and 95 percentiles (whiskers) of the three groups at the end of the experiment. $^{+}p < 0.05$ sham vs. Normox-Res; $*p < 0.05$ Normox-Res vs. Hypox-Res. (Figure from Ref. [103])

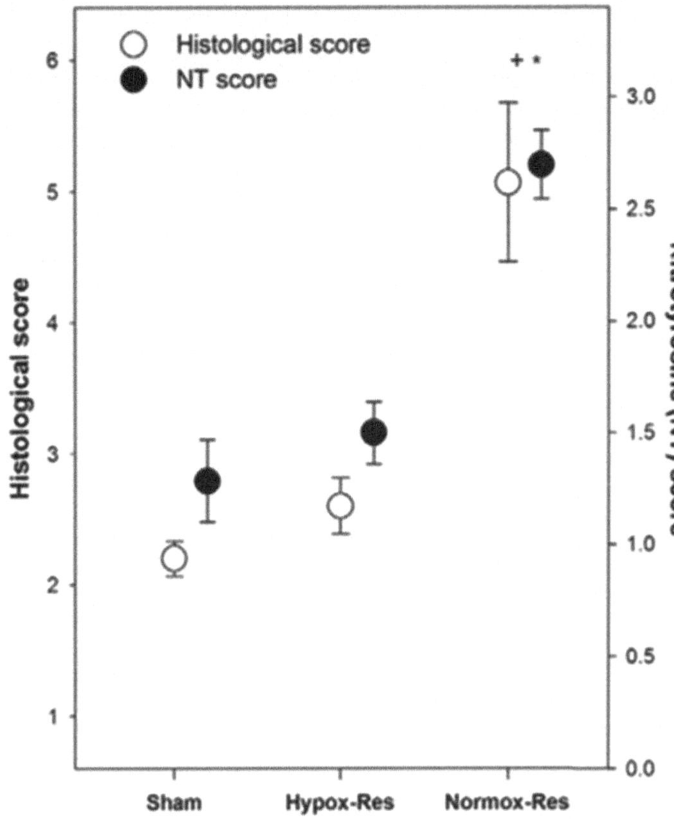

Fig. 27.8 The differences of histologic and NT scores of livers from animals subjected to hemorrhagic shock and resuscitation of normoxemic, hypoxemic, and sham animals are shown. $^{+}p < 0.05$ sham vs. Normox-Res; $*p < 0.05$ Normox-Res vs. Hypox-Res. (Figure from Ref. [103])

associated with minimal hepatocyte necrosis, in contrast with hypoxemic animals. Hepatic mean nitrotyrosine score was significantly increased in normoxemic (2.7 ± 0.15) compared with hypoxemic (1.5 ± 0.14, $p < 0.05$) and sham (1.29 ± 1.18, $p < 0.05$) groups (Fig. 27.8).

27.8 Conclusions

Much knowledge has been accumulated on theoretical and physiological background, for the implication and importance of oxidative and nitrosative stresses on I/R injury. Basic research for many years has been concentrated on the clarification of the pathophysiology involved in the I/R injury, so that the prevention and management become possible. The interesting study of Bickel [89] is one with a clear decrease in mortality. Similarly, many studies assessing the effect of limited resuscitation from hemorrhagic shock present favorable results in outcome. It should not escape our attention the fact that the results of the aforementioned studies represent a clear example of progressive oxygen re-entry to the ischemic tissues. However, many questions still exist on clinical grounds that should be resolved before methods with considerable theoretical interest become eventually considered for clinical evaluation.

References

1. Tasoulis MK, Douzinas EE (2016) Hypoxemic reperfusion of ischemic states: an alternative approach for the attenuation of oxidative stress mediated reperfusion injury. J Biomed Sci 23:7–13
2. Halliwell B (1994) Free radicals, antioxidants, and human disease: curiosity, cause, or consequence? Lancet 344(8924):721–724
3. Dhawan V (2013) Reactive oxygen and nitrogen species: general considerations. In: Studies on respiratory disorders. Springer, New York, pp 27–48
4. Weidinger A, Kozlov AV (2015) Biological activities of reactive oxygen and nitrogen species: oxidative stress versus signal transduction. Biomolecules 5(2):472–484
5. Gourdin M, Dubois P (2013) Impact of Ischemia on Cellular Metabolism. Chapter 1, IntechOpen https://doi.org/10.5772/54509
6. Łuszczewski A, Matyska-Piekarska E et al (2007) Reactive oxygen species physiological and pathological function in the human body. Reumatolog 45(5):284–289
7. Rosen H, Klebanoff SJ, Wang Y et al (2009) Methionine oxidation contributes to bacterial killing by the myeloperoxidase system of neutrophils. Proc Natl Acad Sci 106(44):18686–18691
8. Miller DM, Buettner GR, Aust SD (1990) Transition metals as catalysts of "autoxidation" reactions. Free Radic Biol Med 8:95–108
9. Sumimoto H, Miyano K, Takeya R (2005) Molecular composition and regulation of the Nox family NAD(P)H oxidases. Biochem Biophys Res Commun 338(1):677–686
10. Kahles T, Brandes RP (2013) Which NADPH oxidase isoform is relevant for ischemic stroke? The case for nox 2. Antioxid Redox Signal 18(12):1400–1417
11. Brandes RP, Weissmann N, Schrode K (2010) NADPH oxidases in cardiovascular disease. Free Radic Biol Med 49(5):687–706
12. Granger DN, Kvietys PR (2015) Reperfusion injury and reactive oxygen species: the evolution of a concept. Redox Biol 6:524–551
13. Crosswhite P, Sun Z (2010) Nitric oxide, oxidative stress and inflammation in pulmonary arterial hypertension. J Hypertens 28(2):201–212
14. Nishino T, Okamoto K, Eger BT et al (2008) Mammalian xanthine oxidoreductase – mechanism of transition from xanthine dehydrogenase to xanthine oxidase. FEBS J 275(13):3278–3289
15. Chouchani ET, Pell VR, Gaude E et al (2014) Ischaemic accumulation of succinate controls reperfusion injury through mitochondrial ros. Nature 515:431–435
16. Golwala NH, Hodenette C, Murthy SN et al (2009) Vascular responses to nitrite are mediated by xanthine oxidoreductase and mitochondrial aldehyde dehydrogenase in the rat. Can J Physiol Pharmacol 87:1095–1101
17. Lee MC, Velayutham M, Komatsu T (2014) Measurement and characterization of superoxide generation from xanthine dehydrogenase: a redox-regulated pathway of radical generation in ischemic tissues. Biochemistry 53:6615–6623
18. Lee HL, Chen CL, Yeh ST et al (2012) Biphasic modulation of the mitochondrial electron transport chain in myocardial ischemia and reperfusion. Am J Phys 302:H1410–H1422

19. Perrelli MG, Pagliaro P, Penna C et al (2011) Ischemia/reperfusion injury and cardioprotective mechanisms: role of mitochondria and reactive oxygen species. World J Cardiol 3:186–200
20. Stowe DF, Camara KS (2009) Mitochondrial reactive oxygen species production in excitable cells: modulators of mitochondrial and cell function. Antioxid Redox Signal 11:1373–1414
21. Mc Cord JM (1985) Oxygen-derived free radicals in postischemic tissue injury. N Engl J Med 312:159–163
22. Kalogeris T, Baines CP, Krenz M et al (2012) Cell biology of ischemia/reperfusion injury. Int Rev Cell Mol Biol 298:229–317. Review
23. Babior BM (1999) NADPH oxidase: an update. Blood 93:1464–1476
24. Mironczuk-Chodakowska I, Witkowska AM, Zujko ME (2018) Endogenous non-enzymatic antioxidants in the human body. Adv Med Sci 63:68–78
25. Alderton WK, Cooper CE, Knowles RG (2001) Nitric oxide synthases: structure, function and inhibition. Biochem J 357:593–615
26. Bredt DS (1990) Isolation of nitric oxide synthetase, a calmodulin-requiring enzyme. Proc Natl Acad Sci U S A 87(2):682–685
27. Busse R (1990) Calcium-dependent nitric oxide synthesis in endothelial cytosol is mediated by calmodulin. FEBS J 265(1–2):133–136
28. Radomski MW, Palmer RM, Moncada S (1987) The anti-aggregating properties of vascular endothelium: interactions between prostacyclin and nitric oxide. Br J Pharmacol 92(3):639–646
29. Chen K, Popel AS (2006) Theoretical analysis of biochemical pathways of nitric oxide release from vascular endothelial cells. Free Radic Biol Med 41(4):668–680
30. Ignarro LJ, Kadowitz PJ (1985) The pharmacological and physiological role of cyclic GMP in vascular smooth muscle relaxation. Annu Rev Pharmacol Toxicol 25:171–191
31. Birben E, Sahiner UM, Sackesen C et al (2012) Oxidative stress and antioxidant defense. World Allergy Organ J 5(1):9–19
32. Babula P, Masarik M, Adam V et al (2012) Mammalian metallothioneins: properties and functions. Metallomics 4(8):739–750
33. Hineno A, Kaneko K, Yoshida K et al (2011) Ceruloplasmin protects against rotenone-induced oxidative stress and neurotoxicity. Neurochem Res 36(11):2127–2135
34. Vincent JL (2009) Relevance of albumin in modern critical care medicine. Best Pract Res Clin Anaesthesiol 23(2):183–191
35. Plantier JL, Duretz V, Devos V et al (2016) Comparison of antioxidant properties of different therapeutic albumin preparations. Biologicals 44(4):226–233
36. Kreutzer U, Jue T (2004) Role of myoglobin as a scavenger of cellular NO in myocardium. Am J Physiol Heart Circ Physiol 286(3):985–991
37. Irato P, Santovito G, Piccinni E et al (2001) Oxidative burst and metallothionein as a scavenger in macrophages. Immunol Cell Biol 79(3):251–254
38. Faisal AA, Al-Salih RMH, Assi AN (2015) Study of effect of thyroidectomy on serum oxidant antioxidants status. Int J Curr Microbiol App Sci 4(4):1051–1060
39. Anderson CP, Shen M, Eisenstein RS et al (2012) Mammalian iron metabolism and its control by iron regulatory proteins. Biochim Biophys Acta 1823(9):1468–1483
40. Chauhan A, Chauhan V, Brown WT et al (2004) Oxidative stress in autism: increased lipid peroxidation and reduced serum levels of ceruloplasmin and transferrin the antioxidant proteins. Life Sci 75:2539–2549
41. Talebi R, Ahmadi A, Afraz F et al (2016) Parkinson's disease and lactoferrin: analysis of dependent protein networks. Gene Rep 4:177–183
42. Samuelsson M, Vainikka L, Öllinger K (2011) Glutathione in the blood and cerebrospinal fluid: a study in healthy male volunteers. Neuropeptides 45(4):287–292
43. Douzinas EE, Livaditi O, Xiarchos AG et al (2006) The effect of hypoxemic resuscitation of hemorrhagic shock on hemodynamic stabilization and inflammatory response: a pilot study in a rat experimental model. J Trauma 61(4):918–923

44. Sautin YY, Nakagawa T, Zharikov S et al (2007) Adverse effects of the classic antioxidant uric acid in adipocytes: NADPH oxidase-mediated oxidative/nitrosative stress. Am J Phys Cell Physiol 293(2):C584–C5C5
45. Stinefelt B, Leonard SS, Blemings KP et al (2005) Free radical scavenging, DNA protection, and inhibition of lipid peroxidation mediated by uric acid. Ann Clin Lab Sci 35(1):37–45
46. Bentinger M, Tekle M, Dallner (2010) Coenzyme Q–biosynthesis and functions. Biochem Biophys Res Commun 396(1):74–79
47. Acosta JM, Vazquez Fonseca L, Desbats MA et al (2016) Coenzyme Q biosynthesis in health and disease. Biochim Biophys Acta 8:1079–1085
48. Douzinas E (2009) Progressive hemorrhage: administer oxygen or early resuscitation? Intensive Care Med 35:1664–1666
49. Vincent JL, De Backer D (2014) My paper 20 years later: effects of dobutamine on the VO_2/DO_2 relationship. Intensive Care Med 40:1643–1648
50. Schumacher PT, Cain SM (1987) The concept of a critical DO2. Intensive Care Med 13:223–229
51. Martin C, Riou B, Vallet B (2006) Physiologie humaine appliquée (First Editions, Chapter 17:217–227)
52. Halliwell B, Gutteridge JMC (1984) Oxygen toxicity, oxygen radicals, transition metals and disease. Biochem J 218:1–14
53. Repetto M, Semprine J, Boveris A (2012) Lipid peroxidation: chemical mechanism, biological implications and analytical determination. IntechOpen https://doi.org/10.5772/45943
54. Esterbauer H, Koller E, Slee RG et al (1986) Possible involvement of the lipid-peroxidation product 4-hydroxynonenal in the formation of fluorescent chromolipids. Biochem J 239:405–409
55. Tsukagoshi H, Kawata T, Shimizu Y et al (2002) 4-Hydroxy-2-nonenal enhances fibronectin production by IMR-90 human lung fibroblasts partly via activation of epidermal growth factor receptor-linked extracellular signal-regulated kinase p44/42 pathway. Toxicol Appl Pharmacol 184:127–135
56. Douzinas EE, Andrianakis I, Pitaridis MT et al (2001) The effect of hypoxemic reperfusion on cerebral protection after a severe global ischemic brain insult. Intensive Care Med 27:269–275
57. Douzinas EE, Livaditi O, Andrianakis I et al (2008) The effect of hypoxemic resuscitation from hemorrhagic shock on blood pressure restoration and on oxidative and inflammatory responses. Intensive Care Med 34:1133–1141
58. Poli G, Leonarduzzi G, Biasi F et al (2004) Oxidative stress and cell signalling. Curr Med Chem 11:1163–1182
59. Davies KJ (1987) Protein damage and degradation by oxygen radicals. I. General aspects. J Biol Chem 262:9895–9901
60. Ambler SK, Hodges YK, Jones GM et al (2008) Prolonged administration of a dithiol antioxidant protects against ventricular remodeling due to ischemia-reperfusion in mice. Am J Physiol Heart Circ Physiol 295(3):H1303–H1310
61. Dare AJ, Bolton EA, Pettigrew GJ et al (2015) Protection against renal ischemia-reperfusion injury in vivo by the mitochondria targeted antioxidant MitoQ. Redox Biol 5:163–168
62. Song J, Park J, Oh Y et al (2015) Glutathione suppresses cerebral infarct volume and cell death after ischemic injury: involvement of FOXO3 inactivation and Bcl2 expression. Oxidative Med Cell Longev 2015:426069
63. Zhang Z, Yan J, Taheri S et al (2014) Hypoxia-inducible factor 1 contributes to N-acetylcysteine's protection in stroke. Free Radic Biol Med 68:8–21
64. Chen CL, Zheng H, Xuan Y et al (2015) The cardioprotective effect of hypoxic and ischemic preconditioning in dogs with myocardial ischemia-reperfusion injury using a double-bypass model. Life Sci 141:25–31
65. Ji YY, Wang ZD, Wang SF et al (2015) Ischemic preconditioning ameliorates intestinal injury induced by ischemia-reperfusion in rats. World J Gastroenterol 21(26):8081–8088

66. Lu MJ, Chen YS, Huang HS et al (2014) Hypoxic preconditioning protects rat hearts against ischemia-reperfusion injury via the arachidonate12-lipoxygenase/transient receptor potential vanilloid 1 pathway. Basic Res Cardiol 109(4):414–428
67. Chouker A, Ohta A, Martignoni A et al (2012) In vivo hypoxic preconditioning protects from warm liver ischemia-reperfusion injury through the adenosine A2B receptor. Transplantation 94(9):894–902
68. Qiao S, Mao X, Wang Y et al (2016) Remifentanil preconditioning reduces postischemic myocardial infarction and improves left ventricular performance via activation of the Janus activated kinase-2/signal transducers and activators of Transcription-3 signal pathway and subsequent inhibition of glycogen synthase kinase-3beta in rats. Crit Care Med 44(3):e131–e145
69. Savvanis S, Nastos C, Tasoulis MK et al (2014) Sildenafil attenuates hepatocellular injury after liver ischemia reperfusion in rats: a preliminary study. Oxidative Med Cell Longev 2014:161942
70. Mahfoudh-Boussaid A, Zaouali MA, Hadj-Ayed K et al (2012) Ischemic preconditioning reduces endoplasmic reticulum stress and upregulates hypoxia inducible factor-1alpha in ischemic kidney: the role of nitric oxide. J Biomed Sci 19:7–15
71. Hu X, Yang Z, Yang M et al (2014) Remote ischemic preconditioning mitigates myocardial and neurological dysfunction via K(ATP) channel activation in a rat model of hemorrhagic shock. Shock 42(3):228–233
72. Cai Z, Luo W, Zhan H et al (2013) Hypoxia-inducible factor 1 is required for remote ischemic preconditioning of the heart. Proc Natl Acad Sci U S A 110(43):17462–17467
73. Abu-Amara M, Yang SY, Quaglia A et al (2012) The hepatic soluble guanylyl cyclase-cyclic guanosine monophosphate pathway mediates the protection of remote ischemic preconditioning on the microcirculation in liver ischemia-reperfusion injury. Transplantation 93(9):880–886
74. Frohlich GM, Meier P, White SK et al (2013) Myocardial reperfusion injury: looking beyond primary PCI. Eur Heart J 34(23):1714–1722
75. Abu-Amara M, Gurusamy K, Hori S et al (2010) Systematic review of randomized controlled trials of pharmacological interventions to reduce ischaemia-reperfusion injury in elective liver resection with vascular occlusion. HPB (Oxford) 12(1):4–14
76. O'Neill S, Leuschner S, McNally SJ et al (2013) Meta-analysis of ischaemic preconditioning for liver resections. Br J Surg 100(13):1689–1700
77. Nicholson ML, Pattenden CJ, Barlow AD et al (2015) A double blind randomized clinical trial of remote ischemic conditioning in live donor renal transplantation. Medicine 94(31):e1316
78. Meybohm P, Bein B, Brosteanu O et al (2015) A multicenter trial of remote ischemic preconditioning for heart surgery. N Engl J Med 373(15):1397–1407
79. Hausenloy DJ, Candilio L, Evans R et al (2015) Remote ischemic preconditioning and outcomes of cardiac surgery. N Engl J Med 373(15):1408–1417
80. Douzinas EE, Patsouris E, Kypriades E et al (2001) Hypoxemic reperfusion ameliorates the histopathologic changes in the brain after a severe global cerebral ischemic insult. Intensive Care Med 27:905–910
81. Ulatowski JA, Kirsch JR, Traystman RJ (1994) Hypoxic reperfusion after ischemia in swine does not improve acute brain recovery. Am J Physiol 267 (Heart Circ Physiol 36):H1880-H1887
82. Mink RB, Dutka AJ (1995) Hyperbaric oxygen after global cerebral ischemia in rabbits does not promote brain lipid peroxidation. Crit Care Med 23:1398–1404
83. Mickel HS, Vaishnav YN, Kempski O et al (1987) Breathing 100% oxygen after global brain ischemia in Mongolian gerbils results in increased lipid peroxidation and increased mortality. Stroke 18:426–430
84. Zwemer CF, Whitesall SE, D'Alecy LG (1994) Cardiopulmonary-cerebral resuscitation with 100% oxygen exacerbates neurological dysfunction following nine minutes of normothermic cardiac arrest on dogs. Resuscitation 27:159–170

85. Liu Y, Rosenthal RE, Haywood Y et al (1998) Normoxic ventilation after cardiac arrest reduces oxidation of brain lipids and improves neurological outcome. Stroke 29:1679–1686
86. Douzinas EE, Kollias S, Tiniakos D et al (2004) Hypoxemic reperfusion after 120 mins of intestinal ischemia attenuates the histopathologic and inflammatory response. Crit Care Med 32:2279–2283
87. Li C, Jackson RM (2002) Reactive species mechanisms of cellular hypoxia-reoxygenation injury. Am J Phys Cell Physiol 282:C227–C241
88. Roberts I, Alderson P, Bunn F et al (2004) Colloids versus crystalloids for fluid resuscitation in critically ill patients. Cochrane Database Syst Rev CD000567
89. Bickell WH, Wall MJ Jr, Pepe PE et al (1994) Immediate versus delayed resuscitation for hypotensive patients with penetrating torso injuries. N Engl J Med 331:1105–1109
90. Cavanaugh BP, Meyer LJ (2005) Normalizing physiological variables in acute illness: five reasons for caution. Intensive Care Med 31:1161–1111
91. Roberts I, Evans P, Bunn F et al (2001) Is the normalization of blood pressure in bleeding trauma patients harmful? Lancet 357:385–387
92. Stern SA (2001) Low-volume fluid resuscitation for presumed hemorrhagic shock: helpful or harmful? Curr Opin Crit Care 7:422–430
93. Shafi S, Kauder D (2004) Fluid resuscitation and blood transfusion in patients with polytrauma. Clin Orthop 422:37–42
94. Alan HB (2011) Advances in resuscitation strategies. Int J Surg 9:5–12
95. Gurfinkel V, Poggetti RS, Fontes B et al (2003) Hypertonic saline improves tissue oxygenation and reduces systemic and pulmonary inflammatory response caused by hemorrhagic shock. J Trauma 54:1137–1145
96. Douzinas EE, Betrosian A, Giamarellos-Bourboulis EJ et al (2011) Hypoxemic resuscitation from hemorrhagic shock prevents lung injury and attenuates oxidative response and IL-8 overexpression. Free Radic Biol Med 50:245–253
97. Sauaia A, Moore FA, Moore EE et al (1996) Early risk factors for postinjury multiple organ failure. World J Surg 20:392–400
98. Abraham E, Carmody A, Shenkar R et al (2000) Neutrophils as early immunologic effectors in hemorrhage- or endotoxemia-induced acute lung injury. Am J Phys Lung Cell Mol Phys 279:L1137–L1145
99. Ware LB, Matthay MA (2000) The acute respiratory distress syndrome. N Engl J Med 342:1334–1349
100. De Perrot M, Sekine Y, Fischer S et al (2002) Interleukin-8 release during early reperfusion predicts graft function in human lung transplantation. Am J Respir Crit Care Med 165:211–215
101. Mitchell GB, Albright BN, Caswell JL (2003) Effect of interleukin-8 and granulocyte colony-stimulating factor on priming and activation of bovine neutrophils. Infect Immun 71:1643–1649
102. Douzinas EE, Orfanos SE, Livaditi O et al (2009) Hypoxemic resuscitation prevents pulmonary capillary endothelial dysfunction induced by normoxemic resuscitation from hemorrhagic shock. Crit Care Med 37:869–875
103. Douzinas EE, Livaditi O, Tasoulis MK et al (2012) Nitrosative and oxidative stresses contribute to post-ischemic liver injury following severe hemorrhagic shock: the role of hypoxemic resuscitation. PLoS One 7(3):e32968

Targeting Mitochondria for Therapy of Cardiovascular Disease

28

Biaus Samanta, Satabdi Banerjee, Suman K. Nandy, and Sajal Chakraborti

Abstract

Mitochondria play a crucial role in regulation of rhythmical contraction of myocardium, myocardiocyte physiology, stress response and redox signaling cascades, and overall heart function, principally by meeting the energy demand through oxidative phosphorylation. Mitochondrial dysfunction and subsequent imbalance in ATP supply often leads to diseased condition. Although cardiovascular diseases are attributed to almost one third of annual global death, universally accepted strategies for treatment of myocardial cardiomyopathies are yet to be established. This review summarizes the classical and futuristic therapies for treatment of heart diseases.

Keywords

Cardiovascular disease (CVD) · Mitochondria · ROS · Myocardial cardiomyopathy

All the authors contributed equally to this article.

B. Samanta
Department of Cardiology, Medical College, Kolkata, West Bengal, India

S. Banerjee
Department of Environmental Management, William Carey University, Shillong, Meghalaya, India

S. K. Nandy (✉)
Bioinformatics Infrastructure Facility (BIF), North-Eastern Hill University (NEHU), Tura, Meghalaya, India

S. Chakraborti
Department of Biochemistry and Biophysics, University of Kalyani, Kalyani, West Bengal, India

S. Chakraborti et al. (eds.), *Modulation of Oxidative Stress in Heart Disease*,
https://doi.org/10.1007/978-981-13-8946-7_28

28.1 Introduction

The heart is the major pumping station or in the words of Aristotle is the body's furnace, radiating energy in the form of heat [1]. The human heart pumps ≈10 tons of blood and contracts more than 1 million times per day to ensure the oxygen supply to each cell of the body. To meet the energy demand for this regular rhythmical contraction, the human heart hydrolyses ≈6 kg adenosine triphosphate (ATP), the fuel of the furnace, on a daily basis, and as a matter of fact, myocardium has evolved into a tissue with the highest metabolic rate but almost with no energy reserve [2, 3]. This huge amount of ATP is generated through mitochondrial metabolism, mostly by mitochondrial oxidative phosphorylation (OXPHOS). As a result, cardiomyocytes are characterized with high mitochondrial density, accounting for almost one third of the cell volume [4]. Mitochondrial density might increase according to energy demand through mitochondrial biogenesis, while insufficient supply of ATP often leads to a variety of pathophysiological conditions.

28.2 Cardiac Energy Metabolism

The human heart manages energy supply through three different stages such as (1) catabolism of glucose and fatty acids (FAs) and generation of high-energy equivalents, (2) ATP synthesis, and (3) utilization of ATP for myocardial contraction through coordination with various metabolic pathways. At first step, β-oxidation and glycolysis are mainly engaged in oxidation of FAs and glucose, originated from the consumed food through intestinal breakdown of carbohydrate and lipolysis of fat [2]. All kinds of carbohydrates are converted into pyruvate through glycolysis in cytosol and readily transported into mitochondrial matrix (MM) by specific mitochondrial pyruvate carrier, where pyruvate is irreversibly decaboxylated into acetyl-CoA by multimeric pyruvate dehydrogenase complex (PDHC) [5]. On the other hand, triglycerides form various animal and plant sources are digested into FAs by lipases and released in the bloodstream through the intestinal wall. While the short- and medium-chain FAs can passively diffuse through mitochondrial membranes, the long- and very-long-chain FAs are esterified with coenzyme-A by acyl-CoA synthase at the expense of ATP and transferred into mitochondria through carnitine shuttle, which accept carnitine in exchange of FAs [6, 7]. Fatty acid β-oxidation (FAO) is a spiral process, where the β-carbon of the FA serves as the reaction center, hence the name, and at the end of each spiral, the FA gets shortened by two carbons, releasing one molecule of acetyl-coA along with $FADH_2$ and NADH. The process goes on till the FA is oxidized into two-/three-carbon group [8, 9]. Acetyl-coA, the intermediate metabolite of both carbohydrate and FA metabolism, can readily be directed to tricarboxylic acid (TCA) cycle in MM to generate the reducing equivalents NADH and $FADH_2$. FAO is observed to be the major contributor of ATP production in healthy human heart despite its lower cardiac efficiency, i.e., higher utilization of O_2 for ATP production compared to carbohydrate source. However, the equilibrium may shift in either ways depending upon the food intake, physiological transition, and/or perturbation and cardiac development [8, 10].

The electron donors produced in TCA cycle, the common step in amino acids, glucose, and FA metabolism play important roles for the cardiac mitochondrial OXPHOS. The OXPHOS is tightly coupled to the high-energy-reducing equivalents synthesized in TCA cycle for ATP generation [11]. Different tissues illustrate dissimilar OXPHOS capacity depending upon the concentration of electron transport chain (ETC) complexes and their activity, variation in mitochondrial content, etc. While the cardiac and skeletal muscle portrays the highest capacity as well as utmost sensitivity to OXPHOS defects, the liver and kidney record the least, and the brain tissue comes in between [12].

The third step documents transportation and dispersion of high-energy phosphates from mitochondria to cytosol and their consumption for muscle contraction. Several kinases such as creatine kinase (CK), adenylate kinase (AK), and nucleoside diphosphate kinase (NDPK) are present in between MIM and the outer membrane region, which couple OXPHOS to ATP consumption for cardiac contraction and facilitate the maintenance of high-energy phosphate reservoir in cytosol. MtCK uses mitochondria-generated ATPs to reversibly phosphorylate creatine, synthesized and transported from the kidney and liver to cardiomyocytes, into phosphocreatine (PCr) and ADP. Adenine nucleotide translocator mediates the ATP supply to mtCK and transfer of ADP back to MM. PCr holds the double advantage over ATP, its energy-efficient diffusibility through voltage-dependent anion channels, and the metabolically inert nature supports its accumulation in high concentration at cytosol. Myofibril CK catalyzes the reverse reaction to generate ATP from PCr and helps to retain the normal ATP pool [13]. NDPKs reversibly exchange γ-phosphate between mitochondrial ATP and other nucleoside triphosphate (NTP) to preserve the NTP concentration necessary for protein synthesis, DNA replication, and supply of ATP through reverse reaction [14]. AK catalyzes phosphotransfer within the adenine nucleotides. Recycling of ADP by the kinases helps to regulate OXPHOS. In a healthy heart, the mtCK shuttle serves as an energy buffer and is mostly responsible for the outward energy flux from mitochondria and successive utilization by myofibrillar ATPases to supply the mechanical energy for contraction.

28.3 ROS Formation and Its Effect in Cardiac Health

Superoxide radical ($O_2^{\bullet-}$), hydrogen peroxide (H_2O_2), and hydroxyl radical ($OH^{\bullet}$) are the primary reactive oxygen species (ROS). One of the major sources of ROS is mitochondrial ETC where ROS is produced as a byproduct of normal metabolism under standard physiological conditions through flavoenzyme complexes I, II, and III [15–17]. $O_2^{\bullet-}$ produced at different steps of ETC can be converted to H_2O_2 by the mitochondrial Mn and cytosolic Cu/Zn superoxide dismutases (SOD) ($2O_2^{\bullet-} + 2H^+ \rightarrow H_2O_2 + O_2$). Highly reactive $OH^{\bullet}$ are produced from Fenton reaction involving metal ions like Fe (II) or Cu (I) and H_2O_2($H_2O_2 + Fe^{2+} \rightarrow OH^- + OH^{\bullet} + Fe^{3+}$), which causes localized damage to DNA, proteins, and lipids owing to their exceptionally short lifespan (half-life ≈10^{-9} s) [18]. On the other hand, $O_2^{\bullet-}$ and H_2O_2 can be diffused across mitochondrial membrane to cytosol. In mammals, the rate of mitochondrial superoxide generation is inversely correlated with the maximum lifespan potential of the species [19]. The ROS generation in mitochondria can be increased for several reasons like abnormality in ETC function and under pathophysiological conditions like heart failure (HF) and ageing.

Cardiomyocytes are heavily affected by ROS-mediated carbonylation, nitration, and peroxidation of membrane proteins and phospholipids [20]. Fe-S clusters of enzymes are known to be targeted by $O_2^{\bullet-}$. Many in vivo and in vitro studies depict similar alterations in myocardial ETC complexes and loss of enzymatic activity [21]. Cardiolipin is another primary target of $O_2^{\bullet-}$ damage – mainly through decline in complex I activity [22, 23]. Oxidative stress (OS) induced single- and double-strand breaks, base damage, and modifications in mitochondrial DNA (mtDNA). A modification of mtDNA-encoded subunits of flavoenzyme complexes I, III, and IV affects the electron transfer efficiency resulting in a vicious cycle of ROS generation and mitochondrial dysfunction [24–26].

In addition to the direct deleterious effects, ROS like $O_2^{\bullet-}$ and OH· influence the regulation of multiple intracellular signaling pathways that regulate different cellular and subcellular events such as cardio protection, cell survival, apoptosis, and necrosis. Enhanced ROS generation could play a role in cardiac remodeling through overexpression of proapoptotic proteins and perturbation of mitochondrial signaling cascades by activating proteases like matrix metalloproteins (MMPs). Cardiac damages from ischemia/reperfusion or associated with Friedreich ataxia are also attributed to ROS level [27]. ROS is known to contribute to coronary artery diseases such as atherosclerosis, coronary thrombosis, cardiac hypertrophy, and failure by oxidizing the low-density lipoproteins and proteases like MMPs [28, 29].

28.4 Therapeutic Applications

Mitochondrial dysfunction plays a critical role in various cardiac disorders and their pathophysiology. A proper understanding of the molecular basis of the mitochondrial defects will ensure early diagnosis and better treatment of cardiovascular diseases (CVDs). Till date, there are no known unanimously accepted therapies for cardiomyopathy (MCM); rather, the popular therapeutic approaches only provide some sort of relief at the initial stages of the disease [30]. Presently, diverse combinations of vitamins and metabolic cofactors, such as folic acid, coenzyme Q10 (CoQ10), thiamine (vitamin B1), riboflavin (vitamin B2), ascorbate (vitamin C), tocopherol (vitamin E), succinate, menadione, L-carnitine, etc., are used to ameliorate MCM in cases to some extent. These compounds possibly influence mitochondrial physiology in more than one way either by decreasing mitochondrial ROS generation or enhancing the activity of SOD to scavenge ROS [30–34]. Despite considerable improvement in conventional therapies over the last two decades, CVD still remains as one of the major causes of global morbidity and mortality, representing 31% of all global death [35]. This statistics not only portrays the limitations of current therapies but also points out the urgent need of new strategies in treatment of heart diseases. Mitochondria-specific drug delivery systems, stem cell transplant, and gene therapy may have the answer.

28.4.1 Antioxidant Treatment

Increased levels of oxygen free radicals are one of the instrumental factors in pathogenesis of heart diseases, and mitochondria itself serve as the prime source of ROS. Thus, antioxidants have gained interest among medical scientists as a choice for treatment of heart diseases [27, 36].

Chronic increases in ROS level initiate a vicious cycle of mtDNA damage and decline in mitochondrial function and in turn cause further increase in ROS generation leading to cellular injury. Tsutsui et al. [31] demonstrated that overexpression of endogenous mitochondrial antioxidant protein peroxiredoxin-3 (Prx-3) and mitochondrial transcription factor A (TFAM) can prevent mtDNA damage as well as restore mitochondrial function. Thus, activation of Prx-3 and TFAM gene expression have emerged as prospective therapies for HF patients for controling oxidative stress and mtDNA damage.

Prx-3, vitamin E, and dexrazoxane are the other known antioxidants that are used in addition to other drugs for treatment of different heart diseases [31–34]. CoQ10 or ubiquinone, a potent antioxidant and a common lipophilic component of cellular membranes, depicts structural resemblance with vitamin K and is also be capable of boosting cardiac functions in multiple ways [37]. The recent controlled rosuvastatin multinational study (CORONA) and Q-YMBIO trial in heart failure patients with CoQ_{10} supplementation suggest the effectiveness of CoQ_{10} in controlling adverse cardiovascular events [38, 39]. Treatment with CoQ10 reduces cardiac conduction abnormalities in mitochondrial myopathy [40, 41]. Idebenone, the synthetic analogue of CoQ10, has shown some improvement of different types of cardiac diseases [42].

Vitamin E has been shown to prevent atherosclerotic plaque formation in mouse models. Consumption of foods rich in alpha tocopherol, the most abundant form of vitamin E found naturally, has been observed to be associated with lower risk of coronary heart disease in middle-aged to older population. However supplementation of vitamin E has failed to demonstrate any noticeable advantage in the primary and secondary prevention of CVDs; rather, it might be associated with an increase in total mortality, HF, and hemorrhagic stroke [43].

Dexrazoxane and metallothinonine protect cardiac tissue from oxidative damage by sequestering metal ions from reacting with hydrogen peroxide and superoxide. Efficacy of dexrazoxane has been proven in doxorubicin-induced heart failure both in animal and human trials. Metallothionein also provides additional protection by directly reacting with ROS. However, its cardioprotective potential is yet to be verified [44–46].

Xanthine oxidase inhibitor and poly (ADP ribose) polymerase (PARP) inhibitor prevent $O_2^{\bullet-}$ formations and limit ROS-induced cardiovascular injury. PARP inhibitors improve vascular relaxation and emerged as an exciting prospect for therapeutic intervention in cardiac diseases [47, 48].

Glutathione, a naturally occurring antioxidant, also plays an important role in preserving the reduced state of the other antioxidants, for example, N-acetyl cysteine and vice versa. Selenium-dependent glutathione peroxidase protects cells against the damage induced by products generated by lipid peroxidase [49]. Damy et al. [50] have shown that glutathione deficiency correlates with functional status of patients with cardiac diseases, which indicates that serum glutathione level could be considered as a prospective biomarker in asymptomatic heart patients.

HF patients often suffer from enhanced plasma catecholamine levels, atheroma formation, endothelial dysfunction, and higher ROS generation as a result of sympathetic stimulation. β-blockers and angiotensin-converting enzyme (ACE) inhibitors like captopril and enalapril and ANGII receptor blocker like olmesartan are recommended as effective medication for hypertensive and HF patients. β-blockers act as competitive antagonists to block the receptor sites of catecholamine. However, the antioxidative effect of β-blocker remains arguable due to the contradictory results reported by different groups [51]. ACE inhibitor therapy, on the other hand, increases nitric oxide (NO) bioactivity to modulate ROS generation and activates overexpression of endothelial NO synthase, which subsequently decreases endothelial dysfunction and attenuates myocardial remodeling [52].

28.4.2 Metabolic Modulators

Normally mitochondria meet the high-energy demand of mammalian hearts through fatty acid and glucose metabolism by maintaining a dynamic balance between the myocardial energy substrates such as glucose and fatty acids according to physiological and pathological demand. But, in diabetic heart patients, the glucose utilization is heavily constrained due to insulin resistance, scarcity of glucose transporter content, and impaired pyruvate dehydrogenase (PDH) activity. Thus, fatty acid become almost the sole substrate for ATP synthesis for diabetic patients and that eventually leads to an increase in mitochondrial ROS generation, decrease in cardiac efficiency, and onset of mitochondrial uncoupling by activating uncoupling protein and adenine nucleotide translocator [53].

Modulation of glucose and fatty acid oxidation (FAO) by various agents and protocols offer possible targets for therapeutic interventions. Partial FAO inhibitors mainly act by increasing glucose and pyruvate oxidation, controlling ROS generation, and restoring cardiac functions. For example, etomoxir, a transcriptional modulator, has shown promising results in animal trials by blocking the supply of fatty acids from cytoplasm to mitochondria through inhibition of the enzyme carnitine palmytoyltransferase-1 (CPT1) [54]. Perhexiline is an antianginal drug effective against refractory angina and chronic heart failure and also inhibits CPT1 and favors carbohydrate utilization but rarely used due to its associated hepatotoxicity and neurotoxicity, although recently it gained importance and reintroduced after dose modifications [55]. Ranolazine, another antianginal agent, favors glucose oxidation (GO) over FAO by reducing acetyl-CoA content causing indirect activation of PDH [56]. Several clinical studies have demonstrated protective effects of trimetazidine against myocardial ischemia, angina, and diabetic cardiomyopathy through inhibition of long-chain fatty acid oxidation [56, 57]. Omega-3 fatty acid is used regularly as secondary prevention in coronary heart disease (CHD) and HF patients despite the contradictory reports about its beneficial effects [58–60].

28.4.3 Cardioprotective Agents

Repurposing of drugs has shown cardioprotective ability in various animal trials by modulating a variety of signaling pathways. Calcium channel blockers verapamil and diltiazem serve as anti-ischemic and anti-arrythmogenic agents and act by reducing the calcium overload in myocardial cells and protecting them from ROS-mediated cardiac diseases [61–64].

Volatile anesthetic agents like sevoflurane, isoflurane, and halothane offer cardioprotection (CP) by diminishing myocardial oxygen demand, triggering mitochondrial ATP-sensitive potassium (mitoK_{ATP}) channels, reducing ROS generation, and decreasing mitochondrial and cytosolic calcium overload, thereby shifting mitochondrial bioenergetics [65]. Restoration of myocardial functions by sevoflurane has been confirmed by clinical trials in coronary artery bypass graft surgery patients [66].

MitoK_{ATP} channel opener nicorandil has shown to reduce myocardial cell apoptosis in CHD patients [67]. Sato et al. [68] have demonstrated cardioprotective effect of nicorandil in rabbit ventricular myocytes by selective activation of mito-K_{ATP} channels. Cyclosporin A and sanglifehrin A were shown to elicit CP in reperfusion injury by reducing mitochondrial swelling and regulating mitochondrial permeability transition pore (mPTP) – an early step in mitochondrial apoptosis. Chemical uncouplers like dinitrophenol and carbonyl cyanide m-chlorophenylhydrazone (CCCP) have been portrayed CP in animal model systems [69–71]. Grape seed proanthocyanin extract, known for its strong antioxidant property, has evidenced CP by decreasing ROS generation, apoptotic markers, and infract size [72]. Likewise, glucose-insulin-potassium infusion therapy in initial stages of reperfusion successfully limits infract size and offers CP through cell-survival pathways, mediated by protein kinase B (Akt) and ribosomal S6kinase (p70S6 kinase) [73]. Several growth factors such as insulin-like growth factor (IGF-1), fibroblast growth factor (FGF), and transforming growth factor (TGF) also offer CP against OS via attenuation of apoptosis of myocytes. IGF-1 activates various cardioprotective kinases like phosphatidylinositol-3-kinase and serine- threonine kinases. Somatic gene transfer of growth factors is considered safer in comparison to systemic delivery, as the latter might increase the concentration of growth factors in serum leading to cardiac hypertrophy and HF [74–76].

Coronary revascularization by stenting has shown definite mortality benefit in acute myocardial infarction but provides only symptomatic improvement in coronary artery diseases [77]. Ischemic preconditioning therapy, causing stress by exercise or adenosine and/or cariporide administration, provides mitochondrial CP by limiting infract size and has shown promising results in angina patients [78].

28.4.4 Gene Therapy

Success of gene therapy depends on two things: (1) proper identification of genetic defects that could lead to improved therapies and (2) selection of ideal vectors. Present understanding of molecular pathways and pathogenesis of various cardiac diseases have enabled us to find out specific targets for gene therapy.

28.4.4.1 Targets of Gene Therapy

Dysregulations of several calcium-handling proteins have been shown to link with HF. Sarco/endoplasmic reticulum (S/ER) Ca^{2+}ATPases (SERCA2a) transfer Ca^{2+} ion from the cytosol to the sarcoplasmic reticulum, thereby controlling cardiomyocyte contraction/relaxation cycle. In a failing heart, decrease in SERCA2a and other calcium-binding protein levels as well as dephosphorylation of the regulatory protein phospholamban (PLB/PLN) promotes intracellular Ca^{2+} dysregulation. Overexpression of SERCA2a or PLN ablation has been beneficial for cardiac hemodynamics and also in the prevention of ventricular dilatation in animal models. CUPID (Calcium Up-Regulation by Percutaneous Administration of Gene Therapy in Cardiac Disease) trial with advanced HF patients was launched in 2007. This showed promising results in the first phase of SERCA2a cDNA gene transfer, which elicited significantly ameliorated lesser number of cardiovascular disease events than placebo [79, 80]. But, a similar study using CUPID2 with a large population and long-term follow-up failed to recognize any significant improvement in patients with advanced heart failure [81]. Overexpression of S100 calcium-binding protein A1 (S100A1) by gene transfer also found to increase SERCA2a activity, induce better systolic and diastolic activity, and improve calcium cycling plus mitochondrial ATP synthesis in rat and pig hearts [82].

In failing heart, mitochondrial apoptosis is the most prevalent form of cell death. Presently gene therapy is targeting several signal pathways that attenuate apoptosis and restore homeostasis in myocardium. Pathological heart is characterized by significantly decreased expression of antiapoptotic protein BCL-2 (B-cell lymphoma 2). Overexpression of BCL-2 decreases fibrosis, limits infarct size, attenuates cytochrome C-induced caspase-9-dependent cardiomyocyte apoptosis, and preserves cardiac function [83, 84]. Caspase activation is often characterized by loss of contractility in cardiomyocytes, destruction of sarcomeric organization, and subsequent inception of apoptotic pathway. Intracoronary delivery of p35, a caspase-3 inhibitor, has been shown to prevent the onset of heart failure by decreasing cell autophagy [85]. Mitochondrial uncoupling protein (UCP2) overexpression restores mitochondrial inner membrane potential by reducing ROS formation and calcium overloading [86].

Gene products affecting aerobic metabolism in cardiac tissues have also recently proved to be an effective target of gene therapy and effectively treat HF, ischemia, and hypertrophy. E1α segment of pyruvate dehydrogenase complex (PDHC) plays a regulatory role via reversible phosphorylation. In E1α-deficient patients, the segment has been transduced, and PDHC activity has been restored in part by decreasing anaerobic metabolism and ROS injury [87].

Gene therapy-induced overexpression of critical antioxidant enzymes like superoxide dismutase and hemeoxygenase type 1 has shown to provide CP against myocardial ischemia by reducing infract size, production of ROS, inflammation, and apoptotic cell death. Pre-event delivery of such gene as a preventive measure may be useful for long-term protection against CP [88].

In case of cardiac ischemic injury, trials with HF patients demonstrated the importance of chemokine stromal cell derived factor 1 (SDF-1) and its receptor (CXCR4) in tissue repair by limiting remodeling. STOP-HF trail illustrated overexpression of SDF-1 leads to enhancement in ejection fraction (EF) and decrease in left ventricle (LV) size even in patients with lowest LVEF after 1 year of plasmid SDF-1 administration [89].

28.4.4.2 Selection of Vectors

Vectors of gene therapy can be of two types: viral and nonviral. Viral vectors are suitable in chronic diseases like HF where extensive and transgenic expression is needed. Viral vectors cause more widespread but slow transfection, activation of the immune system, possess higher biosafety risks along with prolonged transgenic expression and higher gene transfer efficiency. Among viral vectors, adenoviral vector, lentivirus, and adeno-associated virus have been shown to provide promising results [90, 91]. Nonviral vectors are suitable for transient expression of certain genes used in certain disease conditions where short-lived expression is sufficient for desired phenotypic effect. Nonviral vectors are easy to produce, have smaller cassette size, induce lower inflammatory response, remain localized, and cause short-lived transgenic expression due to intracellular degradation. Among nonviral vectors, naked plasmid DNA and small interfering RNA (siRNA) have shown promising results in animal studies [92, 93].

Several other strategies were employed to repair and replace defective mitochondrial genes that could be important for treatment of mitochondrial-mediated heart disease (CVD). However, none of them have been examined in vivo. Electroporation of nucleic acid is an effective gene delivery method for nucleus, but fall short in mitochondrial gene therapy [94]. Another promising strategy involves use of peptide nucleic acid (PNA) as selective antisense inhibitor of pathogenic mtDNA. Initial studies have demonstrated decrease in replication of pathogenic allele 8344 responsible for MERRF (myoclonic epilepsy with ragged red fibers) when tested in vitro, but failed to reciprocate the same in cultured cells [95]. After this setback, PNA-oligonucleotide construct in cationic liposome was created for selective mitochondrial targeting, which showed promising results in cultured cells [96]. Transfection effect was further enhanced by PNA-oligonucleotide polyethylenimine (PEI) [97]. Increasing ratio of wild- to mutant-type genome aka "gene shifting" using pharmacological and molecular approaches has gained interest in recent times. When cell culture containing both mutant- and wild-type allele for *Leigh syndrome* is grown in the presence of the mitochondrial ATPase inhibitor oligomycin, a significant increase in wild-type allele is illustrated. DQAsome nanocarrier is another drug delivery system that can transfect cells with high efficiency and selectively target mitochondria [98]. Use of multifunctional multilamellar vesicles (MMV) have been found effective in mtDNA delivery [99]. Cell permeable synthetic antioxidants like metirosine, which have shown effectiveness in mitochondrial inner membrane and restore contractility of cardiomyocyte, might be beneficial for treating OS-related damages [100].

28.4.5 Stem Cell Therapy

In the last two decades, stem cell therapy has come up as a prospective alternative therapy for CVD patients mainly aimed to restore myocardial function through cardiomyocyte regeneration. Once a myocardial cell dies in heart attack, scar tissue replaces it. A dominant myocardial scar leads to poor functional capacity and decreases EF. In a fibrotic environment, aggregation of type 1 collagen decreases expression of growth factors, angiogenic factors leading to further loss of myocardial cells. Thus, reduction of fibrosis directly enhances endogenous myogenesis. Three mechanisms have generally been adopted for regeneration of cardiac muscle. These are (1) controlling fibrosis, (2) facilitating angiogenesis, and (3) improving contractile function [101]. Further, the notion of viewing adult heart as a terminally differentiated organ has changed with the discovery of endogenous cardiac stem cells (CSC), which can support myocardial regeneration in both normal and pathological heart [102]. The current stem cell research is mostly focused on nuclear genes of stem cells. However, mutations and deletions in mtDNA as well as mitochondrial function and dynamics in stem cells and progenitor cells are relatively unknown [103–105].

Many studies have shown lower rate of apoptosis, increase in angiogenesis, and decrease in scar tissue following mesenchymal stem cell (MSC) transplant. Gnecchi et al. [107] showed that rat cardiomyocytes conditioned by hypoxic MSC impart CP by decreasing apoptosis and necrosis [106]. Amado et al. [108] reported that improvement in EF and restoration of normal cardiac function after allogenic MSC implantation in swine. Markel et al. [109] have marked vascular endothelial growth factor (VEGF) as a significant paracrine factor in MSC-mediated CP for post-ischemic myocardial recovery. Adipose derived stem cells secrete angiogenic and antiapoptotic growth factors offer CP by reducing cell apoptosis. This cardioprotective effect was shown to be further augmented by engineered overexpression of Akt-1 in MSC, where growth factors, cytokines, and other paracrine factors influence the survival of existing cardiomyocytes [110]. Willems et al. [111] have demonstrated that hydropyridine small molecules could play a key role in controlling cardiomyocytes differentiation through TGF-β signaling pathway.

Several studies have evidenced that for any meaningful posttransplant cardiac repair, neoangiogenesis is a must requirement. Neomyogenesis occurs by two related mechanisms: stimulation of endogenous cardiac stem cells (c-kit$^+$ CSC and other lineages) and enhancement of myocyte cell cycling. Neoangiogenesis in bone marrow (BM)-derived stem cell transplant is evidenced, where stem cells act as pericytes [112]. A 20-fold increase in the endogenous CSC population in MSC-treated pigs has been observed in comparison to the controls, and the cardiac stem cells (c-kit$^+$) possess much greater capacity for myocyte lineage commitment [113]. Loffredo et al. [114] documented the capacity of BM-derived c-kit$^+$ cells to stimulate the endogenous CSC in post-infract heart resulting increased progenitor activity a phenomenon that MSCs failed to achieve. Treatment with intracoronary injections of MSCs led to an improvement in regional wall thickening in hibernating myocardium BM progenitor cells and resident stem cells [101, 115].

References

1. Miksanek T (2011) The sublime engine: a biography of the human heart. JAMA 305:2580
2. Neubauer S (2007) The failing heart – an engine out of fuel. N Engl J Med Overseas Ed 356:1140–1151
3. Wang Z, Ying Z, Bosy-Westphal A et al (2010) Specific metabolic rates of major organs and tissues across adulthood: evaluation by mechanistic model of resting energy expenditure. Am J Clin Nutr 92:1369–1377
4. Piquereau J, Caffin F, Novotova M et al (2013) Mitochondrial dynamics in the adult cardiomyocytes: which roles for a highly specialized cell? Front Physiol 4:102
5. McCommis KS, Finck BN (2015) Mitochondrial pyruvate transport: a historical perspective and future research directions. Biochem J 466:443–454
6. Hamilton JA, Johnson RA, Corkey B et al (2001) Fatty acid transport. J Mol Neurosci 16:99–108
7. Ramsay RR, Gandour RD, van der Leij FR (2001) Molecular enzymology of carnitine transfer and transport. Biochim Biophys Acta, Proteins Proteomics 1546:21–43
8. Grynberg A, Demaison L (1996) Fatty acid oxidation in the heart. J Cardiovasc Pharmacol 28:11–17
9. Moczulski D, Majak I, Mamczur D (2009) An overview of beta-oxidation disorders. Postepy Hig Med Dosw (Online) 63:266–277
10. Lopaschuk GD, Collins-Nakai RL, Itoi T (1992) Developmental changes in energy substrate use by the heart. Cardiovasc Res 26:1172–1180
11. Hatefi Y (1985) The mitochondrial electron transport and oxidative phosphorylation system. Annu Rev Biochem 54:1015–1069
12. Benard G, Faustin B, Passerieux E et al (2006) Physiological diversity of mitochondrial oxidative phosphorylation. Am J Phys Cell Physiol 291:C1172–C1182
13. Wallimann T, Wyss M, Brdiczka D et al (1992) Intracellular compartmentation, structure and function of creatine kinase isoenzymes in tissues with high and fluctuating energy demands: the 'phosphocreatinecircuit' for cellular energy homeostasis. Biochem J 281:21
14. Lacombe ML, Munier A, Mehus JG et al (2000) The human Nm23/nucleoside diphosphate kinases. J Bioenerg Biomembr 32:247–258
15. Chen Q, Vazquez EJ, Moghaddas S et al (2003) Production of reactive oxygen species by mitochondria: central role of complex III. J Biol Chem 278:36027–36031
16. Herrero A, Barja G (2000) Localization of the site of oxygen radical generation inside the complex I of heart and nonsynaptic brain mammalian mitochondria. J Bioenerg Biomembr 32:609–615
17. McLennan HR, DegliEsposti M (2000) The contribution of mitochondrial respiratory complexes to the production of reactive oxygen species. J Bioenerg Biomembr 32:153–162

18. Pryor WA (1986) Oxy-radicals and related species: their formation, lifetimes, and reactions. Annu Rev Physiol 48:657–667
19. Sohal RS, Svensson I, Sohal BH (1989) Superoxide anion radical production in different animal species. Mech Ageing Dev 49:129–135
20. Stadtman ER, Berlett BS (1998) Reactive oxygen-mediated protein oxidation in aging and disease. Drug Metab Rev 30:225–243
21. Choksi KB, Boylston WH, Rabek JP et al (2004) Oxidatively damaged proteins of heart mitochondrial electron transport complexes. Biochim Biophys Acta 1688:95–101
22. Petrosillo G, Ruggiero FM, Pistolese M et al (2001) Reactive oxygen species generated from the mitochondrial electron transport chain induce cytochrome c dissociation from beef-heart submitochondrial particles via cardiolipin peroxidation. Possible role in the apoptosis. FEBS Lett 509:435–438
23. Paradies G, Petrosillo G, Pistolese M et al (2002) Reactive oxygen species affect mitochondrial electron transport complex I activity through oxidative cardiolipin damage. Gene 286:135–141
24. Shen Z, Wu W, Hazen SL (2000) Activated leukocytes oxidatively damage DNA, RNA, and the nucleotide pool through halide-dependent formation of hydroxyl radical. Biochemistry 39:5474–5482
25. LeDoux SP, Wilson GL (2001) Base excision repair of mitochondrial DNA damage in mammalian cells. Prog Nucleic Acid Res Mol Biol 68:273–284
26. Yakes FM, Van Houten B (1997) Mitochondrial DNA damage is more extensive and persists longer than nuclear DNA damage in human cells following oxidative stress. Proc Natl Acad Sci U S A 94:514–519
27. Cooper JM, Schapira AH (2003) Friedreich's Ataxia: disease mechanisms, antioxidant and Coenzyme Q10 therapy. Biofactors 18:163–171
28. Nakagami H, Liao JK (2004) Statins and myocardial hypertrophy. Coron Artery Dis 15:247–250
29. Witztum JL, Steinberg D (1991) Role of oxidized low density lipoprotein in atherogenesis. J Clin Invest 88:1785–1792
30. DiMauro S, Mancuso M, Naini A (2004) Mitochondrial encephalomyopathies: therapeutic approach. Ann N Y Acad Sci 1011:232–245
31. Tsutsui H, Kinugawa S, Matsushima S (2009) Mitochondrial oxidative stress and dysfunction in myocardial remodelling. Cardiovasc Res 81:449–456
32. Lerman-Sagie T, Rustin P, Lev D et al (2001) Dramatic improvement in mitochondrial cardiomyopathy following treatment with idebenone. J Inherit Metab Dis 24:28–34
33. Sayed-Ahmed MM, Salman TM, Gaballah HE (2001) Propionyl-L-carnitine as protector against adriamycin-induced cardiomyopathy. Pharmacol Res 43:513–520
34. Shite J, Qin F, Mao W (2001) Antioxidant vitamins attenuate oxidative stress and cardiac dysfunction in tachycardia-induced cardiomyopathy. J Am Coll Cardiol 38:1734–1740
35. Roth GA, Johnson C, Abajobir A et al (2017) Global, regional, and national burden of cardiovascular diseases for 10 causes, 1990 to 2015. Am J Cardiol 70(1):1–25. 23715
36. Ferrari R, Guardigli G, Mele D et al (2004) Oxidative stress during myocardial ischemia and heart failure. Curr Pharm Des 10:1699–1711
37. Sharma A, Fonarow GC, Butler J et al (2016) Coenzyme Q10 and heart failure: a state-of-the-art review. Circ Heart Fail 9:e002639
38. McMurray JJ, Dunselman P, Wedel H et al (2010) Coenzyme Q10, rosuvastatin, and clinical outcomes in heart failure: a pre-specified substudy of CORONA (Controlled Rosuvastatin Multinational Study in heart failure). J Am Coll Cardiol 56:1196–1204
39. Mortensen SA, Rosenfeldt F, Kumar A et al (2014) Q-SYMBIO Study Investigators. The effect of coenzyme Q10 on morbidity and mortality in chronic heart failure: results from Q-SYMBIO: a randomized double-blind trial. JACC Heart Fail 2:641–649
40. Shoffner JM, Wallace DC (1994) Oxidative phosphorylation diseases and mitochondrial DNA mutations: diagnosis and treatment. Annu Rev Nutr 14:535–568

41. Ogasahara S, Yorifuji S, Nishikawa Y et al (1985) Improvement of abnormal pyruvate metabolism and cardiac conduction defect with coenzyme Q10 in Kearns-Sayre syndrome. Neurology 35:372–377
42. Geromel V, Darin N, Chretien D et al (2002) Coenzyme Q(10) and idebenonein the therapy of respiratory chain diseases: rationale and comparative benefits. Mol Genet Metab 77:21–30
43. Lonn E, Bosch J, Yusuf S et al (2005) Effects of long-term vitamin E supplementation on cardiovascular events and cancer: a randomized controlled trial. JAMA 293:1338–1347
44. Lipshultz SE, Rifai N, Dalton VM et al (2004) The effect of dexrazoxaneon myocardial injury in doxorubicin-treated children with acute lymphoblastic leukemia. N Engl J Med 351:145–153
45. Kang YJ (1999) The antioxidant function of metallothionein in the heart. Proc Soc Exp Biol Med 222:263–273
46. Ali MM, Frei E, Straub J et al (2002) Induction of metallothionein by zinc protects from daunorubicin toxicity in rats. Toxicology 179:85–93
47. Ungvari Z, Gupte SA, Recchia FA et al (2005) Role of oxidative-nitrosative stress and downstream pathways in various forms of cardiomyopathy and heart failure. Curr Vasc Pharmacol 3:221–229
48. Pacher P, Liaudet L, Mabley JG et al (2006) Beneficial effects of a novel ultrapotent poly(ADP-ribose)polymerase inhibitor in murine models of heart failure. Int J Mol Med 17:369–375
49. Thomas JP, Geiger PG, Girotti AW (1993) Lethal damage to endothelial cells by oxidized low density lipoprotein: role of selenoperoxidases in cytoprotection against lipid hydroperoxide- and iron-mediated reaction. J Lipid Res 34:479–490
50. Damy T, Kirsch M, Khouzami L et al (2009) Glutathione deficiency in cardiac patients is related to the functional status and structural cardiac abnormalities. PLoS One 4:e4871
51. Chin BS, Langford NJ, Nuttall SL et al (2003) Anti-oxidative properties of beta-blockers and angiotensin-converting enzyme inhibitors in congestive heart failure. Eur J Heart Fail 5:171–174
52. Bauersachs J, Widder JD (2008) Endothelial dysfunction in heart failure. Pharmacol Rep 60:119–126
53. Duncan JG (2011) Mitochondrial dysfunction in diabetic cardiomyopathy. Biochim Biophys Acta, Mol Cell Res 1813:1351–1359
54. Zarain-Herzberg A, Rupp H (1999) Transcriptional modulators targeted at fuel metabolism of hypertrophied heart. Am J Cardiol 83:31–37
55. Ashrafian H, Horowitz JD, Frenneaux MP (2007) Perhexiline. Cardiovasc Drug Rev 25:76–97
56. Rupp H, Zarain-Herzberg A, Maisch B (2002) The use of partial fatty acid oxidation inhibitors for metabolic therapy of angina pectoris and heart failure. Herz 27:621–636
57. Fragasso G, PiattiMd PM, Monti L et al (2003) Short- and long-term beneficial effects of trimetazidine in patients with diabetes and ischemic cardiomyopathy. Am Heart J 146:854
58. Chung MK (2004) Vitamins, supplements, herbal medicines, and arrhythmias. Cardiol Rev 12:73–84
59. Tavazzi L, Tognoni G, Franzosi MG et al (2004) Rationale and design of the GISSI heart failure trial: a large trial to assess the effects of n-3 polyunsaturated fatty acids and rosuvastatin in symptomatic congestive heart failure. Eur J Heart Fail 6:635–641
60. Siscovick DS, Barringer TA, Fretts AM et al (2017) Omega-3 polyunsaturated fatty acid (fish oil) supplementation and the prevention of clinical cardiovascular disease: a science advisory from the American Heart Association. Circulation 135:e867–e884
61. Inoue K, Ando S, Itagaki T et al (2003) Intracellular calcium increasing at the beginning of reperfusion assists the early recovery of myocardial contractility after diltiazem cardioplegia. Jpn J Thorac Cardiovasc Surg 51:98–103
62. Bertolet BD (1999) Calcium antagonists in the post-myocardial infarction setting. Drugs Aging 15:461–470
63. Theroux P, Gregoire J, Chin C (1998) Intravenous diltiazem in acute myocardial infarction. Diltiazem as adjunctive therapy to activase (DATA) trial. J Am Coll Cardiol 32:620–628

64. Pizzetti G, Mailhac A, Li Volsi L et al (2001) Beneficial effects of diltiazem during myocardial reperfusion: a randomized trial in acute myocardial infarction. Ital Heart J 2:757–765
65. Stowe DF, Kevin LG (2004) Cardiac preconditioning by volatile anesthetic agents: a defining role for altered mitochondrial bioenergetics. Antioxid Redox Signal 6:439–448
66. Julier K, da Silva R, Garcia C et al (2003) Preconditioning by sevoflurane decreases biochemical markers for myocardial and renal dysfunction in coronary artery bypass graft surgery: a double-blinded, placebo-controlled, multicenter study. Anesthesiology 98:1315–1327
67. Argaud L, Ovize M (2004) How to use the paradigm of ischemic preconditioning to protect the heart? Med Sci (Paris) 20:521–525
68. Sato T, Sasaki N, O'Rourke B et al (2000) Nicorandil, a potent cardioprotective agent, acts by opening mitochondrial ATP-dependent potassium channels. J Am Coll Cardiol 35:514–518
69. Halestrap AP, Clarke SJ, Javadov SA (2004) Mitochondrial permeability transition pore opening during myocardial reperfusion–a target for cardioprotection. Cardiovasc Res 61:372–385
70. Minners J, van den Bos EJ, Yellon DM et al (2000) Dinitrophenol, cyclosporin A, and trimetazidine modulate preconditioning in the isolated rat heart: support for a mitochondrial role in cardioprotection. Cardiovasc Res 47:68–73
71. Ganote CE, Armstrong SC (2003) Effects of CCCP-induced mitochondrial uncoupling and cyclosporin A on cell volume, cell injury and preconditioning protection of isolated rabbit cardiomyocytes. J Mol Cell Cardiol 35:749–759
72. Bagchi D, Sen CK, Ray SD et al (2003) Molecular mechanisms of cardioprotection by a novel grape seed proanthocyanidin extract. Mutat Res 523:87–97
73. Jonassen AK, Sack MN, Mjos OD et al (2001) Myocardial protection by insulin at reperfusion requires early administration and is mediated via Akt and p70s6 kinase cell-survival signaling. Circ Res 89:1191–1198
74. Suzuki YJ (2003) Growth factor signaling for cardioprotection against oxidative stress-induced apoptosis. Antioxid Redox Signal 5:741–749
75. Chao W, Matsui T, Novikov MS et al (2003) Strategic advantages of insulin-like growth factor-I expression for cardioprotection. J Gene Med 5:277–286
76. Matsui T, Li L, Wu JC et al (2002) Phenotypic spectrum caused by transgenic overexpression of activated Akt in the heart. J Biol Chem 277:22896–22901
77. Serruys PW, Morice MC, Kappetein AP et al (2009) Percutaneous coronary intervention versus coronary-artery bypass grafting for severe coronary artery disease. N Engl J Med 360:961–972
78. Xu Z, Jiao Z, Cohen MV et al (2002) Protection from AMP 579 can be added to that from either cariporide or ischemic preconditioning in ischemic rabbit heart. J Cardiovasc Pharmacol 40:510–518
79. Jessup M, Greenberg B, Mancini D et al (2011) Calcium Upregulation by Percutaneous Administration of Gene Therapy in Cardiac Disease (CUPID): a phase 2 trial of intracoronary gene therapy of sarcoplasmic reticulum Ca2+-ATPase in patients with advanced heart failure. Circulation 124:304–313
80. Zsebo K, Yaroshinsky A, Rudy JJ et al (2014) Long-term effects of AAV1/SERCA2a gene transfer in patients with severe heart failure: analysis of recurrent cardiovascular events and mortality. Circ Res 114:101–108
81. Greenberg B, Butler J, Felker GM et al (2016) Calcium upregulation by percutaneous administration of gene therapy in patients with cardiac disease (CUPID 2): a randomised, multinational, double-blind, placebo-controlled, phase 2b trial. Lancet 387:1178–1186
82. Pleger ST, Shan C, Ksienzyk J et al (2011) Cardiac AAV9-S100A1 gene therapy rescues post-ischemic heart failure in a preclinical large animal model. Sci Transl Med 3:92ra64
83. Tanaka M, Nakae S, Terry RD et al (2004) Cardiomyocyte-specific Bcl-2 overexpression attenuates ischemia-reperfusion injury, immune response during acute rejection, and graft coronary artery disease. Blood 104:3789–3796
84. Chatterjee S, Stewart AS, Bish LT et al (2002) Viral gene transfer of the antiapoptotic factor Bcl-2 protects against chronic postischemic heart failure. Circulation 106:1212–1217

85. Laugwitz KL, Moretti A, Weig HJ et al (2001) Blocking caspase-activated apoptosis improves contractility in failing myocardium. Hum Gene Ther 12:2051–2063
86. Teshima Y, Akao M, Jones SP et al (2003) Uncoupling protein-2 overexpression inhibits mitochondrial death pathway in cardiomyocytes. Circ Res 93:192–200
87. Stacpoole PW, Owen R, Flotte TR (2003) The pyruvate dehydrogenase complex as a target for gene therapy. Curr Gene Ther 3:239–245
88. Melo LG, Agrawal R, Zhang L et al (2002) Gene therapy strategy for long-term myocardial protection using adeno-associated virusmediated delivery of hemeoxygenase gene. Circulation 105:602–607
89. Chung ES, Miller L, Patel AN (2015) Changes in ventricular remodelling and clinical status during the year following a single administration of stromal cell-derived factor-1 non-viral gene therapy in chronic ischaemic heart failure patients: the STOP-HF randomized Phase II trial. Eur Heart J 36:2228–2238
90. Pislaru S, Janssens SP, Gersh BJ et al (2002) Defining gene transfer before expecting gene therapy: putting the horse before the cart. Circulation 106:631–636
91. Isner JM, Vale PR, Symes JF et al (2001) Assessment of risks associated with cardiovascular gene therapy in human subjects. Circ Res 89:389–400
92. Baumgartner I, Isner JM (2001) Somatic gene therapy in the cardiovascular system. Annu Rev Physiol 63:427–450
93. Morishita R, Higaki J, Tomita N et al (1998) Application of transcription factor "decoy" strategy as means of gene therapy and study of gene expression in cardiovascular disease. Circ Res 82:1023–1028
94. McGregor A, Temperley R, Chrzanowska-Lightowlers Z et al (2001) Absence of expression from RNA internalised into electroporated mammalian mitochondria. Mol Gen Genomics 265:721–729
95. Chinnery PF, Taylor RW, Diekert K et al (1999) Peptide nucleic acid delivery to human mitochondria. Gene Ther 6:1919–1928
96. Muratovska A, Lightowlers RN, Taylor RW et al (2001) Targeting peptide nucleic acid (PNA) oligomers to mitochondria within cells by conjugation to lipophilic cations: implications for mitochondrial DNA replication, expression and disease. Nucleic Acids Res 29:1852–1863
97. Flierl A, Jackson C, Cottrell B et al (2003) Targeted delivery of DNA to the mitochondrial compartment viaimport sequence-conjugated peptide nucleic acid. Mol Ther 7:550–557
98. D'Souza GG, Rammohan R, Cheng SM et al (2003) DQAsome-mediated delivery of plasmid DNA toward mitochondria in living cells. J Control Release 92:189–197
99. Smith RA, Porteous CM, Gane AM et al (2003) Delivery of bioactive molecules to mitochondria in vivo. Proc Natl Acad Sci U S A 100:5407–5412
100. Zhao K, Zhao GM, Wu D et al (2004) Cell-permeable peptide antioxidant stargeted to inner mitochondrial membrane inhibit mitochondrial swelling, oxidative cell death, and reperfusion injury. J Biol Chem 279:34682–34690
101. Karantalis V, Hare JM (2015) Use of mesenchymal stem cells for therapy of cardiac disease. Circ Res 116:1413–1430
102. Beltrami AP, Barlucchi L, Torella D et al (2003) Adult cardiac stem cells are multipotent and support myocardial regeneration. Cell 114:763–776
103. Maitra A, Arking DE, Shivapurkar N et al (2005) Genomic alterations in cultured human embryonic stem cells. Nat Genet 37:1099–1103
104. Falk MJ, Sondheimer N (2010) Mitochondrial genetic diseases. Curr Opin Pediatr 22:711–716
105. Davis RL, Liang C, Sue CM (2018) Mitochondrial diseases. Handb Clin Neurol 147:125–141
106. Gnecchi M, He H, Liang OD et al (2005) Paracrine action accounts for marked protection of ischemic heart by Akt-modified mesenchymal stem cells. Nat Med 11:367–368
107. Amado LC, Saliaris AP, Schuleri KH et al (2005) Cardiac repair with intramyocardial injection of allogeneic mesenchymal stem cells after myocardial infarction. Proc Natl Acad Sci 102:11474–11479

108. Markel TA, Wang Y, Herrmann JL et al (2008) VEGF is critical for stem cell-mediated cardioprotection and a crucial paracrine factor for defining the age threshold in adult and neonatal stem cell function. Am J Physiol Heart Circ Physiol 295:H2308–H2314
109. Rehman J, Traktuev D, Li J et al (2004) Secretion of angiogenic and antiapoptotic factors by human adipose stromal cells. Circulation 109:1292–1298
110. Noiseux N, Gnecchi M, Lopez-Ilasaca M et al (2006) Mesenchymal stem cells overexpressing Akt dramatically repair infarcted myocardium and improve cardiac function despite infrequent cellular fusion or differentiation. Mol Ther 14:840–850
111. Willems E, Cabral-Teixeira J, Schade D et al (2012) Small molecule-mediated TGFβ Type II receptor degradation promotes cardiomyogenesis in embryonic stem cells. Cell Stem Cell 11:242–252
112. Tse HF, Yiu KH, Lau CP (2007) Bone marrow stem cell therapy for myocardial angiogenesis. Curr Vasc Pharmacol 5:103–112
113. Hatzistergos KE, Quevedo H, Oskouei BN et al (2010) Bone marrow mesenchymal stem cells stimulate cardiac stem cell proliferation and differentiation. Circ Res 107:913–922
114. Loffredo FS, Steinhauser ML, Gannon J et al (2011) Bone marrow-derived cell therapy stimulates endogenous cardiomyocyte progenitors and promotes cardiac repair. Cell Stem Cell 8:389–398
115. Suzuki G, Iyer V, Lee TC et al (2011) Autologous mesenchymal stem cells mobilize ckit+ and cd133+ bone marrow progenitor cells and improve regional function in hibernating myocardium. Circ Res 109:1044–1054

Correction to: Modulation of miRNA in Oxidative Stress-Induced Cardiac Remodeling

Sudhiranjan Gupta

Correction to:
Chapter 3 in: S. Chakraborti et al. (eds.), *Modulation of Oxidative Stress in Heart Disease*, https://doi.org/10.1007/978-981-13-8946-7_3

The figure legends of Figs. 3.2 and 3.3 of this chapter were swapped inadvertently and published with errors. The correct presentation is given here.

The updated version of this chapter can be found at
https://doi.org/10.1007/978-981-13-8946-7_3

S. Chakraborti et al. (eds.), *Modulation of Oxidative Stress in Heart Disease*,
https://doi.org/10.1007/978-981-13-8946-7_29

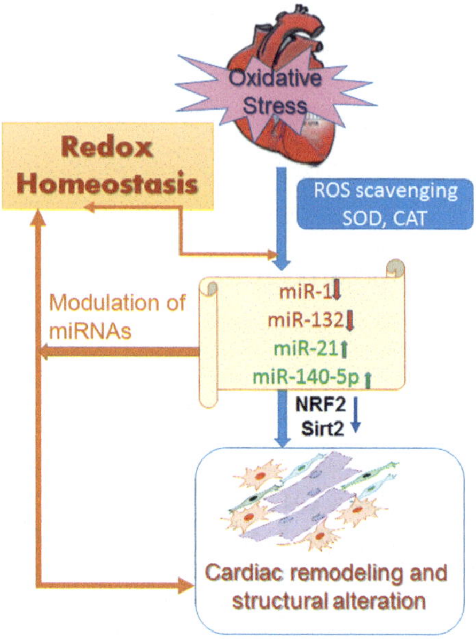

Fig. 3.2 Schematic presentation of ROS-mediated miRNA modulation in cardiac remodeling. ROS can modulate the miRNAs through posttranscriptional regulation of NRF2 and Sirt2 mRNAs. Green arrow indicates upregulatory pattern and red arrow indicated downregulatory pattern

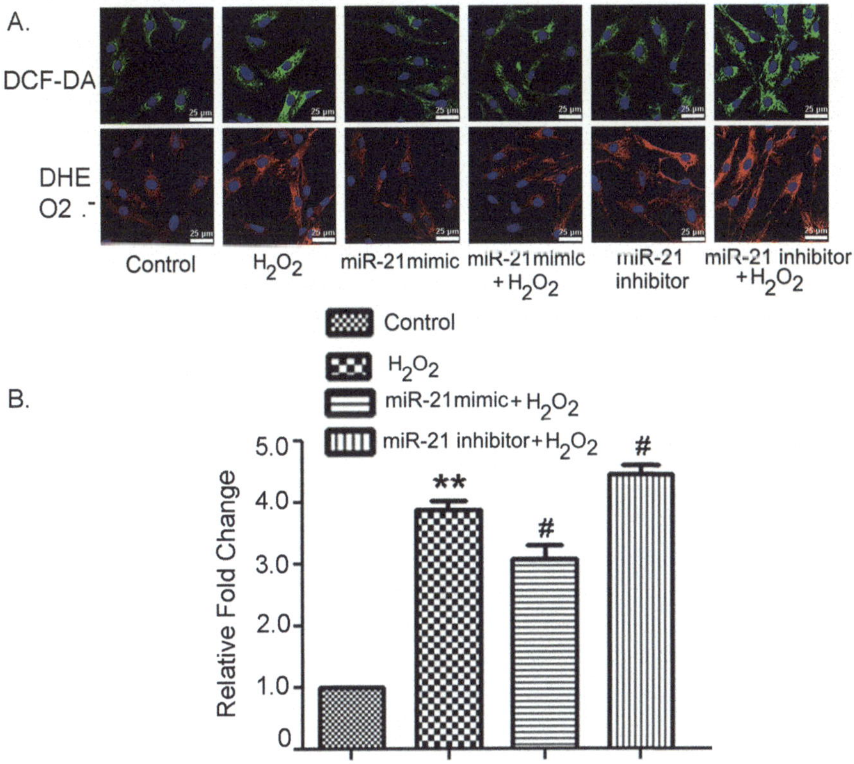

Fig. 3.3 Overexpression of miR-21 attenuates H_2O_2-induced ROS level in neonatal cardiomyocytes. The ROS level was measured in transfected neonatal cardiomyocytes with miR-21 mimetic and inhibitor followed by H_2O_2 treatment for 24 h by confocal microscopy and fluorimetry. (**a**) Representative confocal microscopy images of cardiomyocytes stained with DCFH-DA and DHE, respectively, showing the activity of H_2O_2 and O_2^-. (**b**) Effect of miR-21 mimetic and inhibitor on generation of ROS in cardiomyocytes treated with H_2O_2 by fluorimetry. The data presented are mean ± SE. $**P < 0.01$ vs. control, $^{\#}P < 0.05$ vs. H_2O_2 treatment ($n = 3$). (Adopted from Wei C, Li L, Kim IK, Sun P, Gupta S. Free Radic Res. 2014, Reference 28)

The manufacturer's authorised representative in the EU is Springer Nature Customer Service Centre GmbH, Europaplatz 3, 69115 Heidelberg, Germany. If you have any concerns regarding our products, please contact ProductSafety@springernature.com

Printed and bound by CPI Group (UK) Ltd, Croydon, CR0 4YY

25/09/2026

02222886-0012